Ludwig Hartinger

Handbook of Effluent Treatment and Recycling for the Metal Finishing Industry

2nd Edition

Authorised translation of Handbuch der Abwasser und Recycling Technik – German edition published
*by: **Verlag Carl Hanser**, Munich*

Translation: ***Dr Anselm Kuhn FIMF***

ASM INTERNATIONAL
Materials Park, Ohio 44073

FINISHING PUBLICATIONS LTD
Stevenage, Herts SG1 4BL, UK – Tel: (01438) 745115
Fax: (01438) 364536

Co-published in the USA with ASM International.

This book is an authorised translation of Handbuch der Abwasser und Recycling Technik
*by: **Verlag Carl Hanser**, Munich.*

British Library Cataloguing Publication Data:

A catalogue record for this book is available from the British Library.

ISBN 0-904477-14-2

Typeset by Pantile Publishing, Knebworth, Herts, UK.
Printed and bound by Redwood Books, Trowbridge, Wiltshire.

Preface to the English Edition

This is the authorised English-language translation of the German *"Handbuch der Abwasser und Recyclingtechnik"* (2nd Edition, published by Carl Hanser Verlag, Munich). With one exception, only very minor changes have been made. The chapter dealing with legislative aspects was, in the German edition, framed in terms of German legislation. In this translated edition, it has been wholly re-written to reflect US, Canadian, UK and E.C. regulations. Apart from this section, authorship of the entire work remains, as for the German edition, fully the work of Dr Hartinger himself.

COMMENTS ON THE CONTENTS OF THIS BOOK, ITS LAY-OUT AND USE.

The scope of this book includes all aspects of the treatment of effluents arising from various branches of the metal-working and metal-finishing industries in which aqueous treatment methods are used. In sympathy with modern legislative trends, the emphasis is on means of avoiding or at least minimising the production of effluent. This includes means of increasing the life of process and other baths as well as recycling, wherever possible. The ultimate goal is to reduce to the barest minimum, the amounts of effluent or harmful solid wastes formed and to minimise the extent of treatment necessary.

By far the greatest volume of effluent produced from the various categories of metal-working industries arises from chemical or electrolytic surface finishing operations in which aqueous solutions are used. Within this category should also be included operations connected with the metallising of non-metallic substrates, e.g plastics.
In broad terms, these processes can be grouped as:

a) electrolytic (electrochemical), including:

- electrodeposition of metals

- anodic metal removal processes including electro-deburring,

- electropolishing, etching, machining, turning, grinding, honing as well as electrolytic stripping of previously coated materials (e.g. for re-working).

- surface conversion processes such as anodising of aluminium and other light metals.

b) chemical surface treatment processes including:

- electroless deposition of copper, nickel & other metals, not least in the printed circuit board industry.
- metal removal processes including etching, pickling, chemical deburring, polishing (brightening), chemical machining and chemical stripping.
- surface conversion processes, e.g. chromating, phosphating,
- chemical blacking etc.

Implicit in both of the two categories above, are the various pre- and post-treatment stages associated with the various processes, as well as intermediate rinses etc.

In addition, effluents arise from the pre-treatment and coating of metals with non-metallic substances such as painting and other types of organic finishing, vitreous enamelling as well as coating processes based on immersion in molten metal baths, for example hot-dip galvanising. Finally, one should not overlook the arising of effluents from various metal machining and metal forming operations in terms of cleaning, cooling, lubrication and heat-treatment.

The Handbook is divided into eight major chapters, namely:

- Classification of effluents – their contents and concentrations and origins.

- Effluent treatment methods – an overview of the numerous physical and chemical processes used.

- Techniques of effluent treatment – a description of the most important methods such as batch and continuous processing, post-treatment by final filtration and the use of selective ion-exchange, as well as sludge handling.

- Ion-exchange. Ion exchange resins and their function, recirculation of rinsewater and methods used for this. The use of ion exchange for recycling is covered in Chapter 7, for effluent treatment in Chapter 4.

- Rinse techniques. In terms of legislative compliance, these are virtually mandatory, both for batch treatment and recycling type systems.

- Recycling systems. Options are described based on ion-exchange, precipitation, re-crystallisation, electrolysis, electrodialysis, membrane processes, thermal and other methods, for the recovery of solids and process solutions from effluents or for the life extension of otherwise short-lived process solutions.

– Special processing methods related to specific finishing operations which can give rise to effluents. Included in this category are:

- mechanical treatments
- heat treatments
- etching and other metal removal processes
- hot dip galvanising
- vitreous enamelling
- organic finishing, including pre-treatment
- anodising
- electroplating
- printed circuit board manufacture
- battery manufacture.

Insofar as specific methods are used in the above categories, they are treated in Chapters 2 to 7.

It is the intention in adopting the layout described above, to allow readers not only to further acquaint themselves with the fundamentals of various methods but just as much to learn about specific technologies as used in effluent treatment and recycling. For specific queries, the subject index provides an additional key.

Contents

1 Environmental Legislation for the Metal Finishing and Metal Working Industries*

1.1 GENERAL COMMENTS

There are, perhaps mainly in Germany but certainly also in the USA, proprietors and managers of Metal Finishing and Metal-Working plants whose personal idealism and commitment to the good of the Environment is such that these ideals motivate the effluent treatment and recycling technology they install. Such is the strength of their personal feelings that they may adopt a technology even more effective than that required by existing or imminent legislation and by the same token, such installations might not meet with the approval of an accountant whose perspective was simply that of an annual Balance Sheet. But by and large, management takes the view that their duty is to comply with legislation – and no more.

1.2 THE USA VIEW

Metal finishers in the United States are subject to a variety of environmental and related governmental regulatory programs operating at the federal, state, and local levels. Table 1:1 identifies the most pertinent environmental and related regulatory programs that significantly impact metal finishers in the United States. The purpose of this chapter is to provide some general discussion of these environmental programs, and their basic regulatory framework.

---

* This chapter was contributed by Gary Lindgren Vice President, Environmental Compliance for Heritage Environmental Services, Inc., Indianapolis, Indiana, and replaces that in the original German text, which was written in terms of German legislation.

TABLE 1.1:
ENVIRONMENTAL AND RELATED
REGULATORY PROGRAMS

Air Pollution Control
Water Pollution Control
Solid and Hazardous Waste Management
Hazardous Materials Transportation
Toxic Substances (PCBs)
Community Right-to-Know
Chemical Release Reporting
Spill Prevention and Response
Water Withdrawal and Supply
Chemical Hazard Communication
Employee Health and Safety
Flammable Liquid Storage

Conceptually, these regulatory programs exist to minimize both controlled (routine) and uncontrolled (non-routine or accidental) releases of the various chemicals used in the metal finishing process. Routine air emissions and wastewater discharges are typically controlled through permit(s) issued by the state and/or local environmental regulatory agencies.

These permits typically identify the nature of the discharge allowed under the permit, the maximum amount of discharge per unit of time (expressed on either a concentration or mass basis), monitoring and analytical requirements, and record-keeping and reporting requirements. Oftentimes, some form of pollution control system (e.g., a particulate baghouse, an acid gas scrubber, or a wastewater treatment system) is necessary to reduce the concentration or mass of the regulated constituent(s) or parameter(s) to the levels allowed under such a permit.

Accidental releases are dealt with under the various programs by requirements imposed on the regulated party to take measures to minimize the possibility of an accidental release, to contain and control any accidental releases that might occur, to report the occurrence of accidental releases to the regulatory agencies, and to clean up any contamination resulting form such releases.

Accidental release requirements imposed under the various regulatory programs include both procedural and substantive elements. Procedural elements include requirements to develop spill response plans; to report releases to various regulatory agencies; and to train employees on measures to both prevent and respond to any such releases. Substantive elements include requirements to purchase/install spill prevention equipment and controls, such as tank high level alarms; to obtain spill response supplies such as absorbents, spill booms, and employee personal protective equipment (PPE) to be used in spill response; to install secondary containment around bulk storage tanks and for loading/

unloading areas; and to perform engineering evaluations of potential hazards associated with each process involving hazardous chemicals.

Before discussing specific regulatory programs, it is appropriate to discuss some general considerations that apply to metal finishers in the United States. The elements of environmental compliance are identified, in summary form, on Table 1.2. The most important consideration in managing environmental compliance at a metal finisher is that compliance is expected with respect to all applicable regulatory programs. Efforts and results with respect to one regulatory program are not adequate reasons for lack of effort and results with respect to other regulatory programs.

TABLE 1.2:
ENVIRONMENTAL COMPLIANCE ELEMENTS

Statutes (Federal/State/Local)
Regulations Federal/State/Local)
Permit Conditions
Consent Decrees/Agreed Order Provisions
Common Law Duties
Consensus Codes
Best Management Practices

The second consideration relates to the underlying structure of the programs themselves. Each program consists of a statute or law, and rules or regulations written under the authority provided by the enabling statute or law. These regulations provide the detailed requirements necessary for compliance with the statute's intent. It is assumed that compliance with the regulations is adequate to achieve the statutory goals.

The third consideration is that each program can exist at each level of government (federal, state and local), and that there can be duplicative, overlapping, and sometimes conflicting requirements regarding a single program (e.g., water pollution control or hazardous waste management) at the various levels of government.

The fourth consideration is that the regulations themselves are often confusing to the so-called "reasonable person," and occasionally have "grey areas" and language that requires interpretation to determine the conduct or actions expected by the regulatory agency. This aspect of environmental regulations in the United States has led to much work for lawyers and consultants, especially with the monetary penalties often assessed for violations of such regulations. Lawyers and consultants rely on the history of the program, proposed rule language, the preamble language for the final rule, internal regulatory agency memoranda, and knowledge of how the regulatory agency has applied and enforced the regulatory provision in other situations in assisting the metal finisher with legitimate questions as to how certain regulatory provisions apply to their facility.

The fifth consideration has to do with permit language and permit conditions. As indicated earlier, wastewater discharges and air emissions are often subject to permitting requirements. The regulatory agencies often go beyond the written regulatory requirements in developing permit language and permit conditions.

Many metal finishing permit holders fail to read the "fine print" of permit terms and conditions to consider whether and how they can comply with the requirements hidden in what appears to be boilerplate language. If the problematic permit conditions are not identified and commented on during the draft permit stage, the metal finisher loses the ability to appeal the inclusion of such conditions within the permit, and can be forced to comply with the permit conditions as written.

A sixth consideration is that compliance with regulatory requirements and permit language is not all there is to compliance. There are common law standards of care designed to ensure that neighbouring property owners are not hindered in the use and enjoyment of their property. Common law liabilities can be faced by the metal finisher for problems such as nuisance odours and groundwater contamination, even though the metal finisher is in compliance with all regulatory requirements and permit conditions. Further, insurance companies providing business liability, fire insurance, and workers compensation coverage can impose the recommendations of non-regulatory consensus codes promulgated by such organisations as the National Fire Prevention Association (NFPA) as a condition of insurance coverage, or as requirement for lower premiums. Consensus codes promulgated by such organisations as the National Fire Prevention Association (NFPA) as a condition of insurance coverage, or as a requirement for lower premiums. Consensus codes and other voluntary standards promulgated by such organisations as NFPA, API, UL, ASTM, and NSF are often used as examples of best management practices, and imposed as de-facto regulations by insurance companies, customers, suppliers, fire departments, and other non-regulatory agencies. Regulatory agencies often evaluate consensus codes and other best management practices for use as current enforcement policies, as guidance for potential future regulatory initiatives, and as sources of additional requirements to be imposed as permit conditions.

As mentioned previously, metal finishers in the United States are subject to numerous environmental laws and regulations. Table 1.3 identifies the most significant laws, at the federal level, in this regard. All of the regulatory programs established by these laws or statutes are administered and enforced, at the federal level, by the U.S. Environmental Protection Agency (U.S. EPA), an administrative agency of the federal government. The sole exceptions are the Occupational Safety and Health Act (OSHA), administered at the federal level by U.S. OSHA, and the Hazardous Materials Transportation Act (HMTA) administered at the federal level by U.S. DOT.

Of the environmental laws identified in Table 1.3, the Clean Water Act (CWA), the Clean Air Act (CAA), and the Resource Conservation and Recovery Act (RCRA), as amended, have the most impact on the daily operations and activities of metal finishers. All three laws have provisions for U.S. EPA to delegate authority for administering and enforcing these laws to state regulatory agencies. Further, the Clean Water Act as it relates to industrial discharges to municipal sewer systems, is typically administered at the local (city) level of government.

TABLE 1.3:
SELECTED FEDERAL ENVIRONMENTAL LAWS
AFFECTING METAL FINISHERS

◆ **CLEAN WATER ACT (CWA)** – Permits for wastewater discharges to sewers or streams. Reporting requirements for releases of oil and hazardous substances to the water of the United States.

◆ **CLEAN AIR ACT (CAA)** – Permits for air emissions of criteria pollutants and hazardous air pollutants. Requirements to minimize accidental releases of specified chemicals.

◆ **RESOURCE CONSERVATION AND RECOVERY ACT (RCRA)** – Solid and hazardous waste management and underground storage tanks.

◆ **TOXIC SUBSTANCES CONTROL ACT (TSCA)** – Limitations on the use of PCBs, and the requirements for the storage and disposal of PC wastes.

◆ **HAZARDOUS MATERIALS TRANSPORTATION ACT (HMTA)** – Transportation of hazardous materials and hazardous waste.

◆ **OCCUPATIONAL SAFETY AND HEALTH ACT (OSHA)** – Workplace health and safety.

◆ **SAFE DRINKING WATER ACT (SDWA)** – Drinking water quality and underground injection control.

◆ **COMPREHENSIVE ENVIRONMENTAL RESPONSE, COMPENSATIONS, AND LIABILITY ACT (CERCLA OR "SUPERFUND")** – Release reporting/response and assessment and remediation of uncontrolled hazardous waste sites.

◆ **EMERGENCY PLANNING AND COMMUNITY RIGHT-TO-KNOW ACT OF 1986 (EPCRA or "Title III")** – Chemical inventory reporting and toxic chemical emissions reporting to governmental agencies.

The Clean Water Act, as it relates to metal finishing operations, imposes controls and limitations on the discharge of wastewaters from the various activities associated with metal finishing. These controls and limitations apply regardless of whether the wastewater discharge is to the sanitary sewer system of a municipality, or to the "waters of the United States" (e.g., streams, rivers, ditches, storm sewers). Discharges to municipal sewer systems (termed "indirect discharges") are commonly controlled under the provisions of the pretreatment programs required of states and/or local governments by U.S. EPA. The federal discharge limits applying to the Common Metals Subcategory (>10,000 gpd discharge), by which metal finishers are regulated, are found at Table 1.4. Total Toxic Organics (TTO) analyses can also be required, depending upon the presence of organic chemicals in the wastewater discharge. Municipal sewer ordinances and factors related to the operation of the sewage treatment plant receiving the metal finishing wastewater (termed a "publicly owned treatment works" or POTW) can result in more stringent discharge limits, additional regulated constituents, flow restrictions, or other additional requirements to be imposed on the discharges of metal finishers.

TABLE 1.4:
PRETREATMENT STANDARDS FOR EXISTING SOURCES (PSES) OF METAL FINISHING WASTEWATER DISCHARGES

(Common Metals Subcategory Discharging 10,000 gallons or more per day)
(40 CFR 413)

Constituent	(Max for any 1 day (mg/litre)	Average of Daily Values for 4 Consecutive Monitoring Days Shall not Exceed (mg/litre)
CN, Total	1.9	1.0
Cu	4.5	2.7
Ni	4.1	2.6
Cr	7.0	4.0
Zn	4.2	2.6
Pb	.6	.4
Cd	1.2	.7
Total Metals	10.5	6.8

Discharges directly to the waters of the United States (termed "direct discharges") are controlled under CWA National Pollutant Discharge Elimination System (NPDES) permits, issued at either the federal or state level, depending upon whether the state in which the metal finisher is located has received authorisation to administer the Clean Water Act program. Discharge limitations under NPDES permits are determined on a site-specific basis, and are affected by the size and quality of the receiving water body. Permit terms will, however, be no less stringent for those applying to indirect discharges to POTWs under the CWA pretreatment program.

The Clean Water Act can also impact metal finishers in terms of requirements for so-called stormwater NPDES permits. These are required when channelized stormwater discharges from parking lots and building roofs can contain contaminants from raw material and finished goods storage and use at industrial facilities such as metal finishing operations.

The Clean Air Act impacts metal finishers in terms of required permits or registrations for on-site boilers, and permits and potential air pollution controls for emissions from plating line exhaust ventilation systems. Exhaust from other related activities, such as paint booths and driers can also require permits. Emissions of certain pollutants (either "criteria pollutants" or "hazardous air pollutants") above de minimis thresholds are to be controlled to levels equivalent to those achievable to specific levels of technology. A list of criteria pollutants can be found at Table 1.5. A listing of hazardous air pollutants is found at Table 1.6.

TABLE 1.5:
CRITERIA POLLUTANTS

(40 CFR Part 50)

◆ Sulfur Dioxide	SO_2
◆ Carbon Monoxide	CO
◆ Particulate Matter (less than 10 microns)	PM_{10}
◆ Nitrogen Dioxide	NO_2
◆ Ozone	O_3
◆ Lead	Pb
◆ Particulate Matter	PM

TABLE 1.6:
LIST OF HAZARDOUS AIR POLLUTANTS REGULATED BY THE CLEAN AIR ACT AMENDMENTS OF 1990

CAS* Number	Chemical Name	CAS* Number	Chemical Name
75070	Acetaldehyde	156627	Calcium cyanamide
60355	Acetamide	105602	Caprolactam
75058	Acetonitrile	133062	Captan
98862	Acetophenone	63252	Carbaryl
53963	2-Acetylaminofluorene	75150	Carbon disulfide
107028	Acrolein	56235	Carbon tetrachloride
79061	Acrylamide	463581	Carbonylsulfide
79107	Acrylic acid	120809	Catechol
107131	Acrylonitrile	133904	Chloramben
107051	Allyl chloride	57749	Chlordane
92671	4-Aminobiphenyl	7782505	Chlorine
62533	Aniline	79118	Chloroacetic acid
90040	o-Anisidine	532274	2-Chloroacetophenone
1332214	Asbestos	108907	Chlorobenzene
71432	Benzene (including benzene from gasoline)	510156	Chlorobenzilate
		67663	Chloroform
92875	Benzidine	107302	Chloromethyl methyl ether
98077	Benzotrichloride	126998	Chloroprene
100447	Benzyl chloride	1319773	Cresols/Cresylic acid (isomers and mixture)
92524	Biphenyl		
117817	Bis(2-ethylhexylDphthalate (DEHP)	95487	o-Cresol
542881	Bis(chloromethyl)ether	108394	m-Cresol
75252	Bromoform	106445	p-Cresol
106990	1,3-Butadiene	98828	Cumene

Continued

TABLE 6.6: *(CONTINUED)*
LIST OF HAZARDOUS AIR POLLUTANTS REGULATED BY THE
CLEAN AIR ACT AMENDMENTS OF 1990

CAS* Number	Chemical Name	CAS* Number	Chemical Name
94757	2,4-D, salts and esters	7664393	Hydrogen fluoride (Hydrofluoric acid)
3547044	DDE	123319	Hydroquinone
334883	Diazomethane	78591	Isophorone
132649	Dibenzofurans	58899	Lindane (all isomers)
96128	1,2-Dibromo-3-chloropropane	108316	Maleic anhydride
84742	Dibutylphthalate	67561	Methanol
106467	1,4-Dichlorobenzene(p)	72435	Methoxychlor
91941	3,3-Dichlorobenzidene	74839	Methyl bromide (Bromomethane)
111444	Dichloroethyl ether (Bis(2-chloroethyl) ether)	74873	Methyl chloride (Chloromethane)
542756	1,3-Dichloropropene	71556	Methyl chloroform (1,1,1-Trichloroethane)
62737	Dichlorvos	78933	Methyl ethyl ketone (2-Butanone)
111422	Diethanolamine	60344	Methyl hydrazine
121697	N,N-Diethylaniline (N,N-Dimethylaniline)	74884	Methyl iodide (Iodomethane)
64675	Diethyl sulfate	108101	Methyl isobutyl ketone (Hexone)
119904	3,3-Dimethoxybenzidine	624839	Methyl isocyanate
60117	Dimethyl aminoazobenzene	80626	Methyl methacrylate
119937	3,3-Dimethyl benzidine	1634044	Methyl tert butyl ether
79447	Dimethyl carbamoyl chloride	101144	4,4-Methylene bis(2-chloroaniline)
68122	Dimethyl formamide	75092	Methylene chloride (Dichloromethane)
57147	1,1-Dimethyl hydrazine	101688	Methylene diphenyl diisocyanate (MDI)
131113	Dimethyl phthalate		
77781	Dimethyl sulfate	101779	4,4'-Methylenedianiline
534521	4,6-Dinitro-o-cresol, and salts	91203	Naphthalene
51285	2,4-Dinitrophenol	98953	Nitrobenzene
121142	2,4-Dinitrotoluene	92933	4-Nitrobiphenyl
123911	1,4-Dioxane (1,4-Diethyleneoxide)	100027	4-Nitrophenol
122667	1,2-Diphonylhydrazino	79469	2-Nitropropane
106898	Epichlorohydrin (1-Chloro-2,3-epoxypropane)	684935	N-Nitroso-N-methylurea
106887	1,2-Epoxybutane	62759	N-Nitrosodimethylamine
140885	Ethyl acrylate	59892	N-Nitrosomorpholine
100414	Ethyl benzene	56382	Parathion
51796	Ethyl carbamate (Urethane)	82688	Pentachloronitrobenzene (Quintobenzene)
75003	Ethyl chloride (Chloroethane)		
106934	Ethylene dibromide (Dibromoethane)	87865	Pentachlorophenol
107062	Ethylene dichloride (1,2-Dichloroethane)	108952	Phenol
107211	Ethylene glycol	106503	p-Phenylenediamine
151564	Ethylene imine (Aziridine)	75445	Phosgene
75218	Ethylene oxide	7803512	Phosphine
96457	Ethylene thiourea	7723140	Phosphorus
75343	Ethylidene dichloride (1,1-Dichloroethane)	85449	Phthalic anhydride
50000	Formaldehyde	1336363	Polychlorinated biphenyls (Aroclors)
76448	Heptachlor	1120714	1,3-Propane sultone
118741	Hexachlorobenzene	57578	beta-Propiolactone
87683	Hexachlorobutadiene	123386	Propionaldehyde
77474	Hexachlorocyclopentadiene	114261	Propoxur (Baygon)
67721	Hexachloroethane	78875	Propylene dichloride (1,2-Dichloropropane)
822060	Hexamethylene-1,6-diisocyanate	75569	Propylene oxide
680319	Hexamethylphosphoramide	75558	1,2-Propylenimine (2-Methyl aziridine)
110543	Hexane		
302012	Hydrazine	91225	Quinoline
7647010	Hydrochloric acid	106514	Quinone *Continued*

TABLE 1.6: *(CONTINUED)*
LIST OF HAZARDOUS AIR POLLUTANTS REGULATED BY THE CLEAN AIR ACT AMENDMENTS OF 1990

CAS* Number	Chemical Name	CAS* Number	Chemical Name
100425	Styrene	75354	Vinylidene chloride
96093	Styrene oxide		(1,1-Dichloroethylene)
1746016	2,3,7,8-Tetrachlorodibenso-p-dioxin	1330207	Xylenes (isomers and mixture)
79345	1,1,2,2-Tetrachloroethane	95476	o-Xylenes
127184	Tetrachloroethylene	108383	m-Xylenes
	(Perchloroethylene)	106423	p-Xylenes
7550450	Titanium tetrachloride	0	Antimony Compounds
108883	Toluene	0	Arsenic Compounds
95807	2,4-Toluene diamine		(inorganic including arsine)
584849	2,4-Toluene diisocyanate	0	Cadmium Compounds
95534	o-Toluidine	0	Chromium Compounds
8001352	Toxaphene (chlorinated camphene)	0	Cobalt Compounds
120821	1,2,4-Trichlorobenzene	0	Coke Oven Emissions
79005	1,1,2-Trichloroethane	0	Cyanide Compounds◆
79016	Trichloroethylene	0	Glycol ethers❖
95954	2,4,5-Trichlorophenol	0	Lead Compounds
88062	2,4,6-Trichlorophenol	0	Manganese Compounds
121448	Triethylamine	0	Mercury Compounds
1582098	Trifluralin	0	Fine mineral fibers✛
540841	2,2,4-Trimethylpentane	0	Nickel Compounds
108054	Vinyl acetate	0	Polycylic Organic Matter§
593602	Vinyl bromide	0	Radionuclides (including radon)#
75014	Vinyl chloride	0	Selenium Compounds

Source: Public Law 101-549, Title II.

Note: For all listings containing the word "compounds" and for glycol ethers, the following applies: Unless otherwise specified, these listings are defined as including any unique chemical substance containing the named chemical (that is, antimony, arsenic, and so forth) as part of that chemical's infrastructure.

*   Chemical Abstract Service

◆   X'CN where X = H' or any other group where a formal dissociation may occur. For example, KCN or Ca(CN)$_2$.

❖   Includes mono- and di- ethers of ethylene glycol, diethylene glycol, and triethylene glycol R-(OCH2CH2) -OR' where

        n = 1,2, or 3
        R = alkyl or aryl groups
        R' = R, H, or groups which, when removed, yield glycol ethers with the structure.
        R-(OCH2CH) -OH. Polymers are excluded from the glycol category.

✛   Includes mineral fiber emissions from facilities manufacturing or processing glass, rock, or slag fibers (or other mineral-derived fibers) of an average diameter of 1 micrometer or less.

§   Includes organic compounds with more than one benzene ring and with boiling point greater than or equal to 100°C.

A type of atom, which spontaneously undergoes radioactive decay.

The CAA requirements can impose significant delays in obtaining construction permits for major new sources of air pollution or what are considered major modifications of existing sources. The additional operating permit requirements that will be imposed under Title V of the Clean Air Act Amendments of 1990 will result in significantly greater operational and administrative burdens on metal finishers.

The Resource Conservation and Recovery Act (RCRA) has had a major impact on the operations and waste handling practices of metal finishers. Many of the process wastes and pollution control residuals from metal finishing are considered hazardous wastes and thus are subject to the RCRA regulatory requirements. RCRA permitting requirements do not apply to wastewater treatment tanks regulated under the Clean Water Act, nor do they apply to accumulation of hazardous wastes in containers or tanks for time periods less than 90 days. However, RCRA regulations require metal finishers (and other hazardous waste generators) to comply with a variety of administrative and substantive requirements with regard to hazardous waste management under both normal circumstances and during emergency situations (e.g., releases or fires).

Hazardous waste management regulations under RCRA cause metal finishing wastes to be regulated as hazardous waste in two ways. First, the regulations identify certain common metal finishing wastes as hazardous by virtue of the list of hazardous wastes found at 40 CFR 261.31. In other words, if a metal finishing waste meets the description of any hazardous waste listing found in the regulations, it is a listed hazardous waste. Hazardous waste listings impacting the metal finishing industry are found at Table 1.7. For wastes not captured by the listing descriptions, metal finishers are to either test their wastes or use their knowledge of these wastes in order to determine whether they are characteristically hazardous waste. These characteristics of hazardous waste are described in Table 1.8. The list of constituents and their regulatory limits are found at Table 1.9. If a waste possesses any one (or more) of the characteristics of a hazardous waste, it is a hazardous waste.

As potential generators of hazardous waste, metal finishers are to first determine all the wastes their activities create. Second, metal finishers are to determine, for each waste, whether it is a listed hazardous waste generated in a calendar month, and they must obtain an EPA Identification Number. Further, metal finishers have the option of either structuring their on-site hazardous waste management activities so as to meet certain permit exemptions, or otherwise obtain a RCRA permit (often termed a "Part B permit").

Most metal finishers have opted to structure their activities so as to legally avoid RCRA permit requirements. This typically results in the on-site collection (termed "accumulation") of hazardous wastes in containers or tanks meeting specific regulatory criteria, followed by quarterly (90 day) shipments of hazardous wastes off-site to Part B RCRA permitted treatment, storage, or disposal (TSD) facilities. Transportation to such facilities is to be accompanied by a shipping document called a Uniform Hazardous Waste Manifest (UHWM), which assists the metal finisher in verifying that their hazardous wastes were actually received by the off-site RCRA permitted TSD facility.

TABLE 1.7:
HAZARDOUS WASTE LISTINGS POTENTIALLY
APPLICABLE TO METAL FINISHERS
(40 CFR 261.31)

Industry and EPA Hazardous Waste No.	Hazardous Waste
F006	Wastewater treatment sludges from electroplating operations except from the following processes: (1) Sulfuric acid anodizing of aluminium: (2) tin plating on carbon steel: (3) zinc plating (segregated basis) on carbon steel; (4) aluminum or zinc-aluminum plating on carbon steel; (5) cleaning/stripping associated with tin, zinc and aluminum plating on carbon steel; and (6) chemical etching and milling of aluminum.
F007	Spent cyanide plating bath solutions from electroplating operations.
F008	Plating bath residues form the bottom of plating baths from electroplating operations where cyanides are used in the process.
F009	Spent stripping and cleaning bath solutions from electroplating operations where cyanides are used in the process.
F010	Quenching bath residues from oil baths from metal heat treating operations where cyanides are used in the process.
F011	Spent cyanide solutions from salt bath pot cleaning from metal heat treating operations.
F012	Quenching waste water treatment sludges from metal heat treating operations where cyanides are used in the process.
F019	Wastewater treatment sludges from the chemical conversion coating of aluminum except from zirconium phosphating in aluminum can washing when such phosphating is an exclusive conversion coating process.

The Hazardous Materials Transportation Act (HMTA) governs the interstate and intrastate transportation of hazardous wastes. The HMTA regulatory program establishes requirements for shipping descriptions, shipping papers, proper containers, container marking and hazard labelling, placarding of transportation vehicles and reporting of incidents during transportation. The HMTA regulatory program is administered at the federal level by the U.S. Department of Transportation (U.S. DOT). As indicated earlier, the transportation of hazardous wastes is to be accompanied by a shipping document called a Uniform Hazardous Waste Manifest. These manifest is where the materials are identified and described according to the U.S. DOT regulations. The metal finisher is required to certify, by signature on the manifest, that the shipment is properly classified, described, packaged, marked, and labelled and otherwise in proper condition for transportation in accordance with the applicable regulations.

TABLE 1.8:
40 CFR PART 261, SUBPART C- CHARACTERISTICS OF HAZARDOUS WASTE

IGNITABILITY (I) D001 (40 CFR 261.21)

◆ Liquid, with closed cup flashpoint < 140°F.

◆ Nonliquid, capable of spontaneous and sustained combustion and when ignited, burns so vigorously and persistently as to create a hazard [Ignitable Reactives Subcategory].

◆ DOT ignitable compressed gas.

◆ DOT oxidizer.

CORROSIVITY (C) D002 (40 CFR 261.22)

◆ Aqueous, pH ≤ 2 or pH ≥ 12.5

◆ Liquid, corrodes SAE 1020 steel > 1/4" per year at 130°F [Other Corrosives Subcategory].

REACTIVITY (R) D003 (40 CFR 261.23)

◆ Unstable, reacts violently [Other Reactives Subcategory].

◆ Water-reactive or forms potentially explosive mixtures with water.

◆ Forms toxic gases, vapors, or fumes endangering health when mixed with water [Water Reactives Subcategory].

◆ Cyanide or sulfide containing waste which can generate toxic gases under certain conditions.

  ● EPA Guidance:

  ● Reactive Cyanide (SW846-9010): 250 mg HCN/kg Waste

  ● Reactive Sulfide (SW846-9030): 500 mg H_2S/kg Waste

◆ Detonates or explodes

◆ DOT Explosive (Class A or B)

TOXICITY (E) D004-D043 (40 CFR 261.24)

◆ Toxicity Characteristic Leaching Procedure (TCLP) extract of waste analyzed for specified heavy metals and toxic organics. Zero headspace extraction (ZHE) required for volatile organic constituents. Non-ZHE extraction for metals, semivolatile organics, and pesticides. Extensive Quality Assurance (OA) requirements imposed on analytical laboratories.

◆ TCLP extraction developed to simulate effects of waste mismanagement upon the groundwater in a municipal landfill co-disposal scenario. Regulatory thresholds are derived by multiplying the chronic toxicity level (typically MCLs) by the Dilution Attenuation Factor (DAF). DAF is currently set at 100.

TABLE 1.9:
TOXICITY CHARACTERISTIC CONSTITUENTS
AND REGULATORY THRESHOLDS

EPA HW Number	Constituent	Regulatory Threshold (mg/litre)
D004	Arsenic	5.0
D005	Barium	100.0
D018	Benzene	0.5
D006	Cadmium	1.0
D019	Carbon Tetrachloride	0.5
D020	Chlordane	0.03
D021	Chlorobenzene	100.0
D022	Chloroform	6.0
D007	Chromium	5.0
D023[2]	o-Cresol	200.0
D024[2]	m-Cresol	200.0
D025[2]	p-Cresol	200.0
D026[2]	Cresol	200.0
D016	2,4-D	10.0
D027	1,4-Dichlorobenzene	7.5
D028	1,2-Dichloroethane	0.5
D029	1,1-Dichloroethylene	0.7
D030[1]	2,4-Dinitrotoluene	0.13
D012	Endrin	0.02
D031	Heptachlor (and its hydroxide)	0.008
D032	Hexachlorobenzene	0.13
D033	Hexachlorobutadiene	0.5
D034	Hexachloroethane	3.0
D008	Lead	5.0
D013	Lindane	0.4
D009	Mercury	0.2
D014	Methoxychlor	10.0
D035	Methylethylketone	200.0
D036	Nitrobenzene	2.0
D037	Pentachlorophenol	100.0
D038[1]	Pyridine	5.0
D010	Selenium	1.0
D011	Silver	5.0
D039	Tetrachloroethylene (Perchloroethylene)	0.7
D015	Toxaphene	0.5
D040	Trichloroethylene	0.5
D041	2,4,5-Trichlorophenol	400.0
D042[1]	2,4,6-Trichlorophenol	2.0
D017[1]	2,4,5-TP (Silvex)	1.0
D043[1]	Vinyl Chloride	0.2

1   Quantitation limit is greater than the calculated regulatory level.

2   If o-, m-, and p-Cresol concentrations cannot be differentiated, the total cresol (D026) concentration is used. The regulatory level of total cresol is 200 mg/litre.

The HMTA regulatory program, along with other DOT regulations, has impacted the transportation market-place, resulting in metal finishers, along with other hazardous waste generators, having to pay premium freight rates for the delivery of their hazardous wastes to commercial TSD facilities.

The Comprehensive Environmental Response, Compensation, and Liability Act (CERCLA or "Superfund") has impacted the liabilities of metal finishers for the off-site disposal of their hazardous wastes. CERCLA established a strict, joint and several, and retroactive liability system for responding to releases and controlling the environmental impacts from hazardous waste disposal facilities, including municipal landfills that accepted industrial wastes prior to the promulgation of RCRA regulations in 1980.

Many metal finishers found out about their increased liabilities when U.S. EPA began to identify and investigate abandoned industrial waste facilities and old municipal waste landfills. EPA sent them information requests and administrative orders requiring them, alone or in concert with other "potentially responsible parties" or PRPs, to provide information and even to specify their willingness to co-operate with the U.S. EPA in addressing the environmental issues at the site. Such enormous liabilities often arose out of waste disposal shipments during the 1960s and 1970s that were perfectly legal at the time. As a result, metal finishers have been required to spend significant amounts of money to deal with past waste disposal activities at third party facilities. Those expenditures were not only for investigation and remediation, but also for attorneys and consultants to negotiate their percentage of the costs of the environmental restoration of the third party facility. Typically, liability for "Superfund" sites have been allocated on a volumetric basis. In other words, a metal finisher that sent 1% of the total volume of waste received at the site pays 1% of the total costs of investigation, remediation, and EPA oversight costs.

CERCLA certainly has had a "retroactive" impact, in changing the liabilities for past disposal activities. CERCLA also has an impact on on-going activities, in that it has established an obligation to report any release of "reportable quantities" (RQs) of hazardous substances to the environment. RQ releases must also, of course, be contained and cleaned up after reporting the incident.

The Emergency Planning and Community Right to Know Act (EPCRA) is actually an amendment to the CERCLA statute.

1.3   THE WIDER PERSPECTIVE AND THE FUTURE

Much though almost all in the industry regret the fact, there are no indications that a "Steady State" has been reached in Environmental legislation. In the main, the world scene is dominated by US legislation on the one hand, with European Union (EU) legislation on the other. In the latter case, by far the strictest laws are those prevalent in Germany. With harmonisation the long-term aim in the

E.U, it seems likely that E.U legislation will broadly at least, be modelled on German law. Against this, politicians pay lip service, if nothing more, to a need for "level playing fields" and there is a general expectation that US and EU legislations in these areas will converge. In Britain, a seminal court case (Eastern Counties Leather vs. Cambridge Water Co) was fought through the entire court system to the highest level. The defendants, a tannery company, sold their premises to a water company who sank a borehole. Water from this was so contaminated from chemicals used by the tannery (which had at all times complied with the laws currently in operation) as to be unusable. The UK Courts found (after a reversal and a restitution) for the tannery company. The House of Lords ruled that the tannery, having complied with the laws, could not be subjected to retrospective legislation and could not be penalised for what a reasonable person, at the time in question, would not have foreseen. This clear finding is diametrically opposed to US law as it presently stands, where liability for contaminated land can stretch far back into past years. Whether the EU will adopt the UK or the US position, remains unclear. Issues such as this, and also the whole question of subsidies and grants for environmental improvements (where, for example, the French have been active as one might expect) will doubtless continue to be debated hotly for many years to come. Last but not least, there continues to be a low key "back-lash" against perceived over-severe and pointless legislation. In the USA, a local group of Metal Finishers grouped together and successfully demonstrated that they were not the major source of chromium and other heavy metals in the local rivers and that proposed penalties by the local water authority were needlessly onerous.

We can predict with great confidence not only advance in the technology of effluent treatment and recycling, but also much turbulence in the legislation which surrounds and governs these areas.

NOTES:

2 Introduction and the Composition of Effluents

2.1 INTRODUCTORY COMMENTS – EFFLUENT AND THE ENVIRONMENT

Even today, there are parts of the world where industrial effluents can be seen discharging, totally untreated, into the surrounding environment. Fortunately, there is now a far greater awareness of the hideous damage which such irresponsible practices can cause. The scope of this book does not, in the main, include airborne effluents such as VOC's, which wreak their own particular damage on our immediate atmosphere, not to mention the equally serious upper-atmosphere effects in which the ozone layer is impaired or the greenhouse effect enhanced.

In dealing mainly with aqueous effluents, the book is concerned with avoiding damage to biological effluent treatment plants and, perhaps more importantly, to the damaging distribution of heavy metals and other toxic species, primarily through watercourses, into our entire Eco-system and the food chain. Sadly, it is still true that too many lay people fail to appreciate that *"terra firma"* is not a bottomless sump, capable of accepting unlimited quantities of toxic species.

In the USA, in Germany, in the UK and elsewhere, a horrifying picture has emerged in recent years of whole areas, some industrial, some post-industrial, some now converted to residential use, whose soils are heavily contaminated with toxic species both organic and inorganic. Under recently framed legislation, such sites are now being listed and, in due course, will be remediated – but the cost will be huge. It is to avoid the repetition and extension of this, that effluent treatment practices as described in this book, are now essential. The cost of their implementation is not insignificant – but measured against the human costs (the tragedy of Love Field in the US comes to mind) it is clear that we have no option but to address the problem.

There is one form of effluent treatment which readers will not find described in detail in this book. The disposal of solid wastes or sludges to landfill sites, or their discharge at sea, remains, in some countries, an economically attractive option. Increasingly, however, it is an option no longer available. Because of the growing awareness of the relative ease with which water tables can be contaminated, authorities in all countries are ever more reluctant to tolerate landfill disposal. Authorised sites are fewer and further apart, and the regulations governing the packaging of hazardous wastes are becoming ever stricter. As population densities increase, it becomes progressively less attractive to permit operation of such sites. Equally, it remains open to question whether toxic wastes can indeed be packaged so that the packaging integrity can be guaranteed in perpetuity. Should this not be the case, such landfill sites could truly be described as *"time bombs"*.

Dumping at sea is also increasingly the subject of international agreements. At this time, evidence as to the harmfulness of such practices is ambivalent. Anecdotal evidence suggests that, at least in offshore waters, there are gross consequences in terms of damage to the marine life cycle. In fairness, however, the potential of the wider oceans as a *"sump"* for our industrial wastes, is still an open question. In fairness too, it should be stated that on ocean floors, naturally occurring sediments with significant concentrations of heavy metals can be found. To the dedicated effluent treatment engineer, such comments may be unpalatable. However to conceal the question would be dishonest.

2.2   A PREFACE ON EFFLUENT TREATMENT AND ITS TECHNOLOGY

Before plunging into the many-threaded details of effluent treatment technology, it is as well to summarise the comparatively simple underlying themes.

Seen from afar, all effluent treatment methods can be summarised as one of the following:

i)   Use of a chemical reaction to convert a toxic species to one less toxic (or non-toxic). Oxidation of cyanide would be an example of this.

ii)   Use of a chemical or physical process to convert an effluent:

a)   to a form which can be recycled or otherwise re-used.

b)   to a form in which it can more easily be transported for some form of re-processing or possibly disposal to landfill.

iii)   Use of a chemical and/or physical process to prevent a solution from becoming an effluent or, at worst, deferring the time before this becomes the case (e.g. extending the life of an electroless nickel bath).

At the risk of stating the obvious, there are very few effluents which, after concentration and/or purification, cannot in some way be made use of. As environmental legislation bites ever harder, the re-use or recycling option becomes ever more attractive.

In offering the reader this very detailed treatment of the subject, it is fair that we ask, on their behalf, the following question. *"How mature do we believe the technology to be?"* In answer, the conviction is expressed that the technology is a relatively mature one. By which is meant that there are no indications on the horizon of any revolutionary advances in the technologies available to us. Most of the methods described in the following chapters will doubtless continue to evolve – but if they do so out of present recognition, it will be to the surprise of those in the field.

By contrast, the legislative framework against which we have to consider effluent treatment technologies, remains quite unpredictable. By simple extrapolation of present trends, it could be said that, in comparatively few years, discharge of virtually any effluent might well be totally prohibited in many industrial countries. Against this, there are the beginnings of a ground swell of reaction. It is argued (*in total correctness*) that, in Germany for example, permitted levels of metal ion concentration in effluent discharges are lower than those found in some potable waters or town mains waters feeding the plant in which the effluents arise, or that they are lower than concentrations of the same metals naturally found in bread or rice.

Let us therefore remind ourselves of certain maxims:

1. The cost of effluent treatment (in terms of mass of toxic species removed) increases as ever lower final concentrations are specified, in some cases in exponential fashion.

2. Medical science is slowly discovering that metal ions, though toxic in larger concentrations, are beneficial, in some cases essential, at *"trace"* level. Copper and selenium are just two examples.

3. The pressure for environmental legislation of ever-greater severity is to no small extent derived from lay people who cannot be said to be properly equipped to make reasoned, as opposed to emotive judgements.

As in so many other aspects of human endeavour, it is the personal and political element which is the least predictable as one looks to the future!

The range of different species found in effluents from the metal-working industries is large. Taking, as a basis for this, the Periodic Table, it is seen that of the naturally-occurring elements, apart from the Lanthanides (rare earths), out of the remaining 77 elements, 34 are found in effluents from metal-working industries, of which 23 are most common. 26 elements occur either as cations or compounds, of which 16 are frequently encountered. Eight elements are found only in the anionic state, of which 7 are very commonly occurring.

The following, where present in an effluent, call for treatment measures: acids, alkalis, metals, cyanides, chromates, nitrites, free fluoride, hydrocarbons and halocarbons. This list should be treated as a bare minimum. Other species whose discharge may be subject to legislation include ammonia or its ions, nitrates, phosphates, sulphates, a wide range of organic compounds, any effluent with significant BOD or COD values (Biological or Chemical Oxygen Demand) and effluents at temperatures significantly greater than ambient. Table 2.1 gives more specific information on these and other compounds.

Since specialist treatment of many of the above-mentioned species is necessary, individual process streams are often fed to specific treatment stages. In some countries (Germany for example), legislation specifies that complexed metal ions, cadmium, mercury and chlorocarbons must be separately processed.

Special legislation also applies, in many countries, to the volatile halocarbons and to absorbable organic chloro-compounds (AOX). The best approach, as so often, is to avoid their use in the first place. The discharge of such compounds to atmosphere as VOC's (Volatile Organic Compounds) though falling outside the scope of this book, should also be recognised as illegal in many countries – absorption onto activated charcoal or pyrolysis are the main treatment methods.

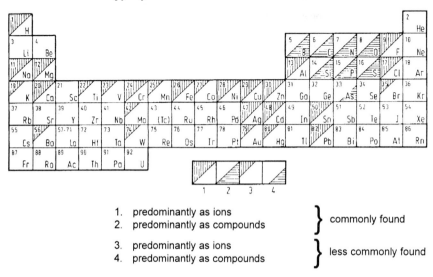

1. predominantly as ions
2. predominantly as compounds   } commonly found

3. predominantly as ions
4. predominantly as compounds   } less commonly found

TABLE 2.1:
TOXIC SPECIES FORMED IN METAL-WORKING INDUSTRIES AND THEIR ORIGINS

Main constituent calling for effluent separation or special treatment	Process solution	Secondary species with possible effluent implications	
		From Solution	Resulting from treatment
Acids	single & mixed pickling baths	F^-, PO_4^{---}, Cr^{6+}	Me^{z+}, scale
	pickling + degreasing baths		Me^{z+}, oil, grease, scale
	pickling of non-ferrous metals (Cu, etc)		Me^{z+} (NO_2^-)
	phosphating baths	PO_4^{---}, Me^{z+}	Me^{z+}
	etch baths	Me^{z+}	Me^{z+}

Table 2.1 Continued

Main constituent calling for effluent separation or special treatment	Process solution	Secondary species with possible effluent implications	
		From Solution	Resulting from treatment
Acids (cont'd)	chemical polishing baths	$(Cr^{6+}$, organics$)$ PO_4^{---},	Me^{z+}
	brightening baths	(F^-, Cr^{6+})	Me^{z+}
	acid baths for metal electrodep'n (Cu, Ni, Zn, Sn, Cd, Pb)	Me^{z+}	
	electroless plating baths, immersion plating	Me^{z+}, CXL reductants	Me^{z+}
	Anodising baths	(org. acids, Cr^{6+})	Me^{z+}
	Anodising & dyeing baths	Me^{z+}, (org. acids)	Me^{z+}
	acid pickling baths	(F^-)	Me^{z+}
Alkalis	Cleaning & degreasing baths	(PO_4^{---}, CN^-, CXL) surfact.	Me^{z+}oil, soil
	Electrolytic descaling baths	CXL (PO_4^{---}, CN^-)	Me^{z+}oil, scale
	Electrodep'n baths (Cu, Zn, Ag)	Me^{z+},CXL (PO_4^{---})	
	pickling & etching baths	(CXL) (CXL, F^-, NO_3^-)	Me^{z+} $Me^{z+}, (NO_2^-)$
Cyanides	Electrolytic cleaning & degreasing baths	(PO_4^{---}, CXL)	oil, grease (emulsion?) soil, scale
	pickling baths	Alkalis	Me^{z+}
	Electrodep'n baths (Cu, Zn, Cd, Ag, Au brass, bronze)	Me^{z+},(CXL)	Me^{z+}
	Quench baths for heat treatment		Alkalis, scale $[Fe(CN)_6]^{3-(4-)}$ (Ba^{++}, NO^-, Me^{z+})
Chromates	Electrodep'n of Cr (hard, decorative)	$H^+, (SiF_6^{--})$	Me^{z+}
	Chromate conv'n baths for Zn,Cd, Al,Mg, Ag)	$H^+, (F^-)$	Me^{z+}
	Etching & pickling baths (also for metallising plastics)	H^+	Me^{z+}

Table 2.1 Continued

Main constituent calling for effluent separation or special treatment	Process solution	Secondary species with possible effluent implications	
		From Solution	Resulting from treatment
Chromates (cont'd)	Brightening & polishing baths	H^+, $(F^-$, $(PO^{---}$, NH_3, NO_2^- (organ. Subst.)	Me^{z+}
	Al anodising baths	H^+	Me^{z+}
Fluorides	Mixed acid etch baths for stainless steels	H^+, (PO_4^{---})	Me^{z+} (NO_2^-)
	Brightening & polishing baths	H^+, (PO_4^{---})	Me^{z+}
	Pickling baths	H^+	Me^{z+}
Nitrites	Metal bronzing & blacking baths	Alkalis (CXL, PO_4^{---})	
	Quench baths for heat treatment		Alkalis, scale $[Fe(CN)_6]^{3-(4-)}$, $(Ba^{++}, CN^-\ Me^{z+})$
Metals	All electroplating & electroless baths	Me^{z+}, $(H^+$, OH^-, CN^-, CXL organ. Subst.)	
	Most other processing baths in this table	$(H^+$, OH^-, CXL, $F^-)$	Me^{z+}
Complexants (organic)	Cleaning & degreasing baths	Alkalis, $(PO_4^{--}$, $CN^-)$ surfacts.	oil, grease, (emulsion?), soil, (Me^{z+})
	Alkaline & neutral descaling & derusting baths	Alkalis, $(PO_4^{--}$, $CN^-)$	ditto (scale)
	Pickling baths	(Alkalis)	Me^{z+}
	Barrel polishing baths	Alkalis (PO_4^{---})	Me^{z+} oil, grease, swarf
	Alkaline electrodep'n baths (Cu, Zn)	Me^{z+}, Alkalis, $(CN^-$, $PO_4^{---})$	
	Electroless dep'n baths (Cu, Ni)	Me^{z+}, reductant (NH_3)	
	Neutral & alkaline metal stripping baths	(Alkalis, CN^-, organics	Me^{z+}
Ammonia/ Ammonium*)	Baths for electroless metal dep'n (Ni)	Me^{z+}, CXL, strong reductants	
	Weakly acid Zn dep'n bath	Me^{z+}, (CXL)	

Table 2.1 Continued

Main constituent calling for effluent separation or special treatment	Process solution	Secondary species with possible effluent implications	
		From Solution	Resulting from treatment
Ammonia/ Ammonium*) cont'd)	Ammoniacal etch baths	Me^{z+}	Me^{z+}
	Fluxing baths	(Me^{z+})	Me^{z+}
Hydrocarbons	Degreasing & cleaning baths	Alkalis, $(PO_4^{--}$, CN^-, CXL),	Soil (Me^{z+})
	Machining coolants	Oil, surfacts (organics)	Soil
COD	Cleaning & degreasing baths	Alkalis, $(PO_4^{--}$, CN^-, CXL), surfacts.	Oils, greases, soil (Me^{z+})
	Electroless dep'n baths (Cu, Ni)	Me^{z+} reductants (NH_3)	
	Machining coolants	Oils, surfacts. (organics)	Soil
	Pickling + degreasing sol'n	H^+, surfacts.	Me^{z+}
	Ferrous pickling baths	H^+, (F^-)	Me^{z+} (oxid'ble)
	Anodising baths with org. acids	H^+, organics	Me^{z+}
	Polishing & brightening baths with organics	H^+, organics (F^-)	Me^{z+}
	Alkaline resist developers & strippers	Alkalis	Me^{z+}
	Paint stripping baths	Alkalis	organics (Me solid)
	Various other sol'ns	Me^{z+}organics (CXL)	
Phosphates	Cleaning & degreasing baths	Alkalis (CXL)	oil, greases (emulsion?)
	Phosphoric acid etch baths	H^+	Me^{z+}
	Phosphating baths	H^+, Me^{z+}	Me^{z+}
Sulphates	Most acid electrolytes	H^+, Me^{z+}, etc.	
	Anodising baths	H^+ (organics)	Me^{z+}
	Sulphuric acid etching & pickling baths	H^+	Me^{z+}

Substances in brackets may occur.
*) Ammonia from cyanide destruction.
**) Ammonia and plant containing effluent adds to COD.   CXL = Complexant.

2.3   EFFLUENT CONCENTRATIONS AND LOADINGS

Effluents can contain toxic species in a very wide concentration range. A conventional classification is:

- diluted effluent (rinsewaters) with ca. 50–500 mg/litre dissolved species

- Semi-concentrated, with 10 to 100 gm/litre dissolved species

- Concentrated, with > 100 gm/litre dissolved species.

During rinsing sequences, progressively more diluted baths will be found but a given maximum permitted concentration is always specified for the final rinse stage. However effluents produced from recycled rinsing systems using ion-exchangers will usually be significantly more concentrated. Insofar as environmental legislation in some countries enforces water saving measures, there are more and more plants where no dilute effluents arise and the only effluents are those which would be classified as concentrated or semi-concentrated. In cases where dilute effluents continue to be released, these may or may not contain harmful species at concentrations calling for treatment prior to discharge. Whatever the position, it remains broadly true that there is a universal trend to discharge of effluent in ever-more concentrated form. Using the definitions above, we may list the following typical sources or types of effluent:

Semi-concentrated:

- Spent or no-longer useable from less-concentrated short-lived process solutions, e.g. hot degreasing baths, chromating solutions, pickling baths, to the extent that these cannot be regenerated or revived.
- The contents of static rinse baths or chemical rinses.
- Rinsewaters from cascaded systems.
- Regenerated concentrates from circulating ion-exchange loops & ion exchangers used in recycling.
- Less concentrated solutions from recycling systems.
- Cleaning water residues and sump waters.
- Solutions from effluent gas scrubbers.

Included in the category as concentrated solutions are:
- Electrolyte solutions no longer useable and those which cannot be cost-effectively recycled.

- Concentrated acid or alkaline etch baths insofar as they cannot be cost-effectively reclaimed.

- Concentrated residue solutions from recycling systems.

There is nothing to be gained by seeking to distinguish between concentrated and semi-concentrated solutions and the examples above should be seen as purely illustrative. Under some legislations, solutions in the above categories are no longer viewed as effluents, but rather considered as (*wet*) solid wastes and thus become subject to regulations governing this type of discharge. It must be admitted that, under certain legislations, a degree of self-contradiction arises. On the one hand, effluent producers are enjoined to employ all reasonable means to save water, recycle or regenerate solutions. On the other hand, as a result of employing such measures, they are *"rewarded"* by having their effluent designated as *"solid waste"* for which much stricter regulations apply. This apart, it must be asked how sensible it is to require, under the law, that concentrated effluents are then designated as "solid wastes" which then have to be transported by road to a centralised treatment plant. Finally, there is the management time required for completion of the paperwork and, as with all freight movement operations, an element of risk involved.

It could be argued that, since concentrated solutions present a potential danger to watercourses and public sewage treatment plants if they are not first treated correctly (*or indeed treated at all!*), permission to dispose to an outside landfill or processing plant should be restricted to those plants unable to treat them in-house for the foreseeable future.

All in all, where legislation is in effect which restricts the total burden of effluent discharge, the end-result is a beneficial one as the total mass of discharged species falls. In the case of effluent first treated in-house, the volume of liquid and the load on public sewerage and treatment facilities is reduced. In the case of discharge of treated effluent, there are benefits in terms of improved water quality and reduced amounts of sedimenting material.

Treatment of concentrated effluents on a batch basis offers, in comparison to processing of large volumes of dilute effluent on a continuous basis, a significant benefit in terms of reduced total waste burden. This is most obvious in the case of species treated by precipitation. Thus, for example, consider 1 cubic metre of solution containing copper at 10 gm/litre. This concentration can be reduced, by precipitation to 0.5 mg/litre, at least in theory. The total effluent load will then be 0.5 × 1000 = 0.5 gm. For 100 cubic metres of solution with 100 mg/litre copper ions, the total copper content is the same. The resulting end solids mass would be 0.5 mg × 100,000 = 50 gm in the second case. The two figures, derived from the same initial mass of copper, differ by ×100.

In practice, there are certain reasons why such results might not be obtained, as discussed in Section 3.7.7.1. In addition, though, the faster kinetics obtaining in the more concentrated solutions found in batch processing as well as absence of dwell time factors (Section 5.2.1) in this mode, all result in higher reaction rates and work to minimise the overall effluent load for processing.

Reducing the mass of toxic species to be handled does not necessarily reduce the mass of neutral salts passing through the plant. It will usually lead to an increased concentration of such species. In considering physical processing methods, it must be borne in mind that, to the extent that even these often use modest amounts of chemicals for conditioning and their effluents too, may then require treatment before discharge. All in all, it must be recognised that as tighter conditions are imposed on effluent discharge, there are consequences in terms of generation of additional solid wastes. Ideally, the goal should be ever-improving effluent quality with decreasing solid waste production. However in order to achieve such a goal, one is forced to resort to ever more expensive equipment in an overall strategy which makes little economic sense. Thus a technology based on low volume discharge and low solids production is a better approach than one based on so-called *"zero-discharge"*. The author is more than apprehensive as regards visions of "hydraulically insulated" plants in which a constant volume of water is recycled between effluent and process liquid status. There is an inevitable danger that, with time, such an unchanged volume of water becomes contaminated, possibly by species unknown. Without the option of continuously replacing a portion of this by fresh water, the danger of progressively more contaminated water leading, in the final resort, to collapse of the process, is a potent one. One method for such total recycling is based on the use of distillation or evaporative water recovery. However this implies a release to atmosphere of some of the vapour-phase, should this be legally allowed. In the case of evaporation of effluent, the vapour phase is far from being of distilled water quality and could well be nothing other than a dilute solution (Section 7.8.2.1.2) in which species *(possibly of unknown composition)* have accumulated. It is therefore suggested that any vision of a zero-discharge effluent operation based on evaporation is, in most cases, a nonsense except where the resulting waste product is both highly water-soluble and also quite harmless..

3 Effluent Treatment Chemical and Physical Methods for Detoxification of Harmful Species or their Elimination

3.1 THEORETICAL BASIS OF CHEMICAL REACTIONS

In effluent treatment technology, certain conditions must pre-exist if chemical reactions are to be employed with success:

- the reaction must take place as fast as possible. This is especially true for continuous type process plant.

- the equilibrium position of the reaction must lie as far as possible in the desired direction.

- no toxic species should be formed from the reaction.

- it is desirable that the course of the reaction is one which can be followed instrumentally.

All of these requirements are linked with the choice of reaction chemicals selected. Fulfillment of the first two conditions is most important.

The two sections which follow deal with reaction kinetics (and rates) and with chemical thermodynamics (equilibrium phenomena) but the treatment is limited to essentials which readers may require in order to appreciate the numerous example reactions which are cited in the following text. Corresponding treatments of the ion-exchange process follow elsewhere in the book.

3.1.1   Reaction Rates [1,2]

In practice, reaction rates depend on:

1. The rate of mass-transport including the rate of mixing of reacting species, one of which will usually be the effluent, the other the treatment chemical(s). Mass-transport itself can be broken down as convective flow, turbulent flow at the boundaries of convected masses together with diffusion. The first two of these are usually connected in effluent treatment with mixing, a discipline forming part of reactor technology.

2. The reaction rate itself. This assumes that mixing has already taken place. The reaction rate is a function of the reaction order, reaction molecularity, temperature and any catalytic effects. These terms are explained below.

"Molecularity" describes the number of molecules, whether of the same or different types which are involved in a single collision in order to bring about the reaction. Thus it is common to refer to mono-, bi-, or tri-molecular reactions. It should be emphasised that "molecularity" is a different concept from "reaction order", indeed only in the case of very simple reactions are the two likely to be identical. Reaction Order can only be used as a general indication as to the relationship over time, of concentration of reactants or products.

Let us now consider a simple bimolecular reaction in which eqivalent amounts of "a" and "b" of the species A and B are transformed into equivalents "c" and "d" of species C and D as shown in Equation 3.1.

$$aA + bB \rightleftharpoons cC + dD \qquad\qquad 3.1$$

The rate of reaction going from left to right can be followed in terms of the decreasing concentration of A (c_A) over a specified time "t". The corresponding reaction rate $r_\rightarrow$ is given by the concentration-time differential – dc_A/dt. This will almost certainly be itself a function of total elapsed time, i.e. the reaction rate is not a constant. The minus sign indicates that concentration decreases with time. Instead of using the term c_A, c_B could be used (this term likewise decreasing with time) or c_C or c_D which, by contrast, increase with time. The last two would, of course, have positive rather than negative signs. Since reaction rate is proportional to the collision number, it is thus related to concentration of reactant species and a constant for a particular reaction, the rate constant $k_\rightarrow$. In terms of the reaction 3.1 going from left to right, we can thus write:

$$r_\rightarrow = -\frac{dc_A}{dt} = -\frac{dc_B}{dt} = \frac{dc_C}{dt} = \frac{dc_D}{dt} = k_\rightarrow \cdot c_A^a \cdot c_B^b \qquad\qquad 3.2$$

as an expression for the rate. Because rate constants are temperature dependant, such an expression will be valid only at a single temperature. Strictly speaking, there are no reactions which proceed in one direction only and once species C and D have formed, they can themselves react to form A and B once more until an equilibrium is reached. An expression for such a back-reaction, derived in the same way as Equation 3.2 is shown by Equation 3.3 below.

$$r_{\leftarrow} = \frac{dc_A}{dt} = k_{\leftarrow} \cdot c_C^c \cdot c_D^d \qquad 3.3$$

After a given time of reaction, so much of C and D are formed that they produce A and B at the same rate that they themselves are formed. $r_{\rightarrow}$ and $r_{\leftarrow}$ are, in this condition, equal to one another. At a given temperature, an equilibrium (which is itself temperature dependant) is reached, and Equation 3.4 holds.

$$\frac{c_C^c \cdot c_D^d}{c_A^a \cdot c_B^b} = \frac{k_{\rightarrow}}{k_{\leftarrow}} = K_c \qquad 3.4$$

K_c is the equilibrium constant for this chemical reaction and, just as are $k_{\rightarrow}$ and $k_{\leftarrow}$, is temperature dependant. Equation 3.4 represents the Mass Action equation for reaction 3.1 and this equation and variants of it, specific to dissociation, solubility product as a stability constant for complexed species or to describe phase distributions, will be encountered in later chapters. The preceding treatment has served to describe the rate-determining kinetic factors in a reaction. For reaction kinetics in general, Reaction Orders have been established which allow the classification of individual reactions in terms of empirically determined reaction rate vs. time behaviour. In terms of the treatment above (Equation 3.2), the following holds. If substance A, at initial concentration c_{A0} reacts in monomolecular fashion to form one or more products, the reaction rate is then dependant on the initial concentration of reactant A and on the reaction specific rate constant k_1. We may then speak of a First Order reaction, characterised by an equation of type:

$$- \frac{dc_A}{dt} = k_1 \cdot c_A \quad \text{or} \quad \frac{dc_A}{c_A} = - k_1 dt \qquad 3.5$$

Integrating and taking the special case $c_A = c_{A0}$, one finds, at time "t" a residual concentration of A as expressed by:

$$c_A = c_{A0} \cdot e^{-k_1 t} \quad \text{or} \quad \ln \frac{c_{A0}}{c_A} = k_1 t \qquad 3.6$$

Thus the entire course of reaction can be described by just two parameters, c_{A0} and k_1. The dimensions of k_1 are s^{-1}. The classical example of a First Order reaction is that of radioactive decay. However there are numerous other, chemical reactions, many of them bimolecular in terms of their collision number, which display similar kinetics. Included in these, for example, are reactions in which changes in concentration of one of the participating species has no effect on the rate, such as water in the acid saponification of an ester. In effluent treatment, hydrolysis reactions behave in this way too since the hydroxyl ion concentration is usually maintained at a constant value. Examples here include hydrolysis of cyanogen chloride to form cyanate or the hydrolysis of metal salts in a neutralisation reaction.

Bimolecular reactions are found not only in systems with a single product but also in those having two reaction products. They are often Second Order reactions and their overall rate is proportional to the product of the concentrations of both reacting species or, in the case of a single reactant, the concentration of that species squared $c_A \cdot c_A = c_A^2$. The expression for rate as a function of elapsed time is then:

$$\frac{dc_A}{dt} = k_2 \cdot c_A^2 \quad \text{bzw.} \quad \frac{dc_A}{c_A^2} = -k_2 dt \qquad 3.7$$

Integrating, and setting $c_A = c_{A0}$ for $t = 0$, an expression involving concentrations results:

$$\frac{1}{c_A} = k_2 t + \frac{1}{c_{A0}} \qquad 3.8$$

In this case, K_2 has dimensions litre. $Mol^{-1}.s^{-1}$. In the case of two different reactants, the expression

$$-\frac{dc_A}{dt} = k_2 \cdot c_A \cdot c_B \qquad 3.9$$

is found. This form of equation involving time is the most commonly applicable term for simple reactions. In the case $c_A = c_B$, Equation 3.9 reverts to the simpler form of Equation 3.8.

Termolecular (trimolecular) reactions occur only rarely. The probability of three reactant molecules colliding simultaneously is extremely small. Reactions which, in terms of their stoicheometry, appear to be termolecular are usually made up of two bimolecular reactions. In its simplest form, a Third Order reaction with a single reactant species is expresssed as:

$$-\frac{dc_A}{dt} = k_3 \cdot c_A^3 \quad \text{or} \quad \frac{dc_A}{c_A^3} = -k_3 dt \qquad 3.10$$

In the case of more than one reactant species, instead of the term c_A^3, one has the product term $c_A \cdot c_B$ of $c_A \cdot c_B \cdot c_C$. Integrating and setting $c_A = c_{A0}$, one again arrives at an expression involving concentrations:

$$\frac{1}{c_A^2} = 2k_3 t + \frac{1}{c_{A0}^2} \qquad 3.11$$

The dimensions of the rate constant k_3 in this cas are litre^2 Mol^{-2} s^{-1}. Fig 3.1 shows concentration vs. time plots for reactions of various orders on a comparable basis. In order to produce such a comparison, the comments on the horizontal axis should be noted. From these plots, the following predictions are possible:

- all reactions other than zero order proceed to completion only at infinite time. The approach of the plots to the "x" axis is asymptotic

- the higher the reaction order, the longer the time of reaction

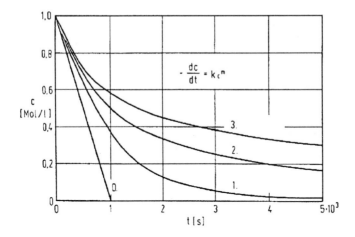

$$-\frac{dc}{dt} = k\,c^m$$

Fig 3.1   Comparison of the rates of reactions of different order (As text).

As has been noted earlier, the concepts of molecularity and reaction order are not synonymous and these terms can differ for any given reaction. Indeed, insofar as many overall reactions consist of discrete individual stages, this is often the case. Other cases to be borne in mind are firstly those reactions which do not proceed to completion because the equilibrium position (nett result of forward and backward reactions) is reached before this and "parallel reactions" in which a single reactant produces several products or where a single product is formed from several reactants.

An example of the first of these is provided by numerous ion-exchange processes when carried out in the batch mode. Only in the continuous column mode, where reaction products are continuously abstracted, is the equilibrium disturbed and in effect never reached so that, as written, the equation proceeds to completion (left to right). In the context of effluent treatment, parallel reactions arise only where catalysis is involved.

A third type of composite reaction is that involving sequential reactions. In this case, one or more of the reaction products itself reacts with one or more of the original reactants or with another product molecule (which can include one of the same type as itself). In such a sequence, one or other of the stages will be slower than the rest and this the slowest is known as the "rate determining step". In other words, the reaction cannot proceed any faster than the rate of the slowest stage. Sequential reactions, including any reverse reactions, need not be of the same order as the initial step. A typical example of a sequential reaction is the oxidation of cyanide with chlorine or sodium hypochlorite.

A special case of sequential reaction is the "chain reaction". In this case, reactive atoms, radicals or ions are formed (propagating species) each of these initiating another round in the sequence, during which a fresh active species is formed and so on. Such reactions continue until the supply of reactants is exhausted or until, for some other reason, the active species are consumed (e.g. by reacting with one another instead of with reactant molecules). An example of such a reaction in modern effluent treatment technology is the oxidation of various substances with hydrogen peroxide or with ozone, while irradiating the reaction with ultra-violet light. In cases where the individual reaction steps within an overall process have not been identified, only the overall reaction can be considered. Even in such cases, provided the initial concentrations of the reacting species and the overall reaction stoicheometry is known, an empirical kinetic equation can be assigned which represents a reaction of the "n"th order ($0 \leq n \leq 3$):

$$- \frac{dc}{dt} = k_n \cdot c^n \qquad\qquad 3.12$$

In such cases, it is not necessary that "n" be an integer.

In order to determine an order of reaction, the concentration of each reacting species must be determined as a function of time. A reaction order can then be assigned by matching this data to the rate equations for various orders. The correct equation is that for which a constant value of "k" is obtained at all values of elapsed reaction time. Alternatively, a graphic procedure can be used. By plotting Equations 3.6, 3.8 or 3.11, a series of straight lines are produced. Their slopes correspond to various reaction orders. Last but not least, there are a number of commercially available computer software programs which accept concentration vs. time data and use this to derive most if not all the information referred to above, e.g. rate constant, order of reaction etc.

In general, reaction rate is influenced by:

- temperature. According to Van't Hoff's rule, reaction rate increases with temperature by $\times 2$ to $\times 4$ for every 10°C increase. (Strictly speaking, the size of this effect depends on the activation energy). Change in temperature will also affect the position of the equilbrium. (see the following section).

- catalysts. In contrast to changes in temperature, these do not affect the equilibrium position. A distinction is usually made between homogeneous and heterogeneous catalysis.

As seen in the following section, the effect of temperature is a major influence in determining the position of an equilibrium. This is quite distinct from its effect on reaction rate, as mentioned above. Such effects, as was made clear in the preceding sections, can only be determined by measurement of rate constants over a range of temperatures. According to the theory of Arrhenius, reactant molecules can only react when they possess a minimum kinetic energy E_A which matches or exceeds the mean activation energy level (E_v) for a given stage of the reaction. Many systems therefore require supply of external energy, e.g. by raising the temperature, before they can proceed at a significant rate. Should the resulting reaction be exothermic, such energy is then supplied by the reaction itself. This applies where the enthalpy of reaction (ΔH) is greater than E_A. Such an energy release is also produced in respect of the activation energy (E_A) for the endothermic back-reaction (see Fig. 3.2) and this in turn promotes the forward reaction. The Van't Hoff Law governing the effect of temperature on activation energy applies to all these effects.

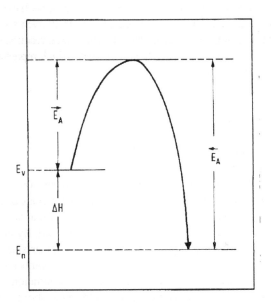

Fig. 3.2 *Schematic representation of the course of a reversible chemical reaction showing the activation energy barrier.*

E_v Average energy of reactants prior to reaction (or after the back-reaction).
E_n Average energy after the forward reaction (prior to back reaction).

The temperature dependance of the rate constant "k" is given by the expression:

$$k = k_0 \cdot e^{-\frac{E_A}{RT}} \qquad 3.13$$

In cases where the order of reaction is known, E_A can then be derived if values of "k" at different temperatures are known. If ln k is plotted vs T^{-1} , this should be a straight line, from the gradient of which an activation energy is derived.

In the case of homogeneous catalysis, the reactants form intermediates with the catalyst, thereby lowering the effective activation energy. The implication of this is that the collision number characteristic of the reaction, increases. The overall reaction proceeding via the catalyst has the same starting and end products as would be formed in absence of catalyst, it simply takes place rather more rapidly. In this way, it can be characterised as a parallel reaction. After the reaction is complete, the catalyst itself remains unchanged and unconsumed, in its pristine state. An example of such a reaction is the oxidation of cyanide by peroxo compounds, in presence of copper ions.

Heterogeneous catalysis is governed by the area and activity of catalyst surface available for reaction, which must be capable of absorbing at least one reactant. This gives rise to local increase of concentration in the near-catalyst surface space and thus favours collision probability. In addition, the heat of adsorption produced can, in many cases, be converted into useful activation energy, thus increasing the number of reactant species with sufficient energy to cross the activation energy barrier and thus react.

As long as the concentrations of reactants remain very small, the surface of the catalyst is only partially occupied by adsorbed species. An equilibrium is set up between species in solution and those adsorbed on the catalyst surface. This is known as an adsorption isotherm (see Section 6.7). In such a case, the rate of reaction is increased as reactant concentration increases, but not necessarily proportionately so. The equation representing such a situation is known as zero order, and is shown in Equation 3.21.

$$- \frac{dc_A}{dt} = k_0 \qquad\qquad 3.14$$

By integrating and setting $c_A = c_{A0}$ for $t = 0$, one obtains the equation:

$$c_A = c_{A0} - k_0 t \qquad\qquad 3.15$$

k_0 here has dimensions litre^{-1}.Mol.sec^{-1}. Graphically, this is a straight line, indicating that the reaction will proceed to completion in a finite time. (see Fig. 3.1). Towards the end of the reaction, the relationship may no longer hold, since concentrations of reactant fall to a low value and the condition, assumed above, that the catalyst surface is fully covered by adsorbed species, may no longer hold. The process then switches into a higher Reaction Order.

A good example of a zero order reaction is the oxidation of cyanide using ozone, given that a sufficiently high ozone concentration is present in the air carrying it. The reaction between the highly active ozone and the cyanide can only take place at the phase boundary. The reaction rate thus depends on the surface area, as long as a reasonably high ozone concentration in the incoming air is maintained at a more or less constant value. (Two-phase systems such as this often involve a reaction between O_3 (gas) and O_3 (dissol.) where the latter species is the reactive one. Electrode reactions (see Sections 3.2.2.5 and 7.6.1) are also often zero order, so long as concentrations in solution are sufficiently high. A further example, in the field of heterogeneous catalysis, is the oxidation of cyanides and other species by air over an activated carbon surface.

3.1.2   Reaction Equilibria [2,3]

In virtually all cases of chemical reactions involved in effluent treatment, it is desirable that the equilibrium lies as far as possible to the right-hand side of thr written equation in order that as little as possible of the toxic starting reactants remain at harmful concentration levels. So strict is the latest effluent legislation that it is not always possible to achieve this in a single reaction. Thus precipitation of noxious species may need to be followed by a "polishing" using selectively operative ion-exchange.

The principles of reaction kinetics, in the previous sections, allow an estimate of the rates of a particular reaction. They permit no insight, however, into reaction equilibria or the prediction of the equilibrium constant itself. Indeed, kinetics alone will not even predict whether a reaction proceeds from left to right or vice-versa, under a given set of conditions. Equilibrium phenomena can only be predicted using chemical thermodynamics, which are based on the relative Free Energies of reactants and products.

Assume that two reactant species (elements, molecules or ions) A and B react with one another at constant pressure, to form products C and D, with no external energy input. Energy (usual in the form of heat) may then be released. This is usually designated as enthalpy of reaction (ΔH). The following examples all assume constant pressure conditions. In most cases this is a valid assumption for effluent treatment reactions which operate under broadly isobaric conditions. Convention in thermodynamics is that a minus sign $- \Delta H$ denotes a release of energy, i.e. the reaction is exothermic. The system thus contains less potential energy at equilibrium than at its initiation. Conversely, a $+$ sign denotes an endothermic reaction, i.e. one requiring a supply of external energy if it is to proceed. Returning to Equation 3.1, we can write this as:

$$aA + bB \rightleftharpoons cC + dD; \qquad - \Delta H \qquad\qquad 3.16$$

for an exothermic reaction. ΔH is here the difference between the free enthalpy of reactants and products. Its dimensions are in kJ. Mol^{-1}. If the free enthalpies of all species in the reaction are known, it is a simple matter to calculate the overall enthalpy change for the reaction and thus whether it is endo- or exo-thermic. Care must be taken that all values used in such a calculation relate to the same conditions of pressure and temperature, as well as correct physical state. The enthalpy of a compound in its solid state can differ substantially from that in the dissolved state. Values for all but the most complex species can readily be found in published compilations, at a standard pressure (1 bar) and temperature (298°K) as one or more defined states. These are known as Standard Free Enthalpies (ΔH^0).

Free energy is the the standard parameter for characterising a chemical reaction in a state of equilibrium. Thermodynamics (as all textbooks on the subject show) allow us to derive the following equation:

$$\ln \frac{a_C^c \cdot a_D^d}{a_A^a \cdot a_B^b} = \ln K_c = - \frac{\Delta H}{RT} \qquad\qquad 3.17$$

It will be noted in this equation that activities (a) rather than concentrations (c) have been used. (See Section 3.7.3)

From Equation 3.17, an equilibrium constant can be calculated and collations of these, as far as they are available, are available (4). Where values have not been published, it is possible to deduce them either by use of the Gibbs-Helmholtz Equation (3.18)

$$\Delta G_T^0 = \Delta H^0 - T \Delta S^0 \qquad\qquad 3.18$$

or by deriving a value of ΔG^0. This can be done by devising an equation in which a compound, whose free energy of reaction is required, is synthesised from its constituent elements. It should be emphasised that, providing the stoicheometry in such a reaction is rigorous, it is quite immaterial whether the reaction can take place in practice or not.

In Equation 3.18 ΔH^0 is the standard molar enthalpy, ΔS^0 the standard molar entropy and T the absolute temperature.

To obtain values of free enthalpies of reaction at temperatures other than 298°K, data is used for the temperature dependance of ΔH^0 and ΔS^0, using the Gibbs-Helmholtz equation. In addition, the temperature dependance of the specific heat Cp between 298°K and the relevant temperature, must be known. This can be derived from the following equation:

$$\Delta G_T^0 = \Delta H_{298}^0 - T\Delta S_{298}^0 - \int_{298}^{T}\!\!\int \Delta G_p d \ln T dt \qquad\qquad 3.19$$

From the foregoing, it can be seen how values for an equilibrium constant can be obtained from thermodynamic data. In general, carrying out the calculations is simpler and quicker than experimental measurements, often under circumscribed conditions. Indeed, use of chemical (as opposed to instrumental) analysis can lead to problems since the analytical sampling may change the concentrations in solution. However the very fact that equilibria in effluent treatment reactions lie so heavily to the r.h.s makes even the use of instrumental analysis methods extremely difficult and often inaccurate.

It can be seen, from Equations 3.17 and 3.18 that, for an exothermic reaction, increase in temperature will push the equilibrium value in the wrong direction, i.e to the left. Increasing temperature reduces the free enthalpy and decreases the equilibrium constant. In effluent treatment reactions, temperature increases are, with a handful of exceptions, only rarely important. Only where it is important to speed up a reaction or in the case of diffusion-controlled chemical processes (e.g. ion-exchange) or where a species can only be formed in an endothermic reaction, is external heat ever employed and then only to the extent there is no significant change in the equilibrium value.

Back in the last century, Le Chatelier formulated his Rule which expresses in words what is otherwise only mathematically definable, namely:

"If a change in external conditions is applied to a system at equilibrium, it will react so as to oppose any such change".

As an example of this, if heat is applied to a reaction in equilibrium, the reaction will proceed in whichever direction (left to right or v.v) it is endothermic.

The application of thermal energy to reactions used in effluent treatment is very rarely of interest, since such reactions have in any case been selected on the basis their equilibrium position lies to the r.h.s, i.e. they are exothermic. Far more effective, as indeed Le Chatelier's Principle indicates, is to increase the concentrations of reacting species, a measure which is specially effective in modifying the equilibrium position of heterogeneous reactions.

3.2   CYANIDE DESTRUCTION

Reactions for the destruction of cyanides have been far more thoroughly studied than most others in the realm of effluent treatment. This can be understood in that cyanides were, until recently, the most toxic of all inorganic species used in industry or found in effluents. It is also true that the reaction mechanisms involved were far more complex than most others in treatment of inorganic substances and required much work for an elucidation.

Today, there are other species, such as heavy metal ions or chlorinated hydrocarbons, which cause as much or even greater concern. This is so because removal of such species causes far greater difficulties than the oxidative destruction of cyanides.

The reactant species is the cyanide anion – this is the toxic species (HCN and cyanogen are excluded from this statement). In the case of alkaline oxidation to cyanate, the overall redox equation is:

$$CN^- + 2OH^- \rightleftharpoons CNO^- + H_2O + 2e^- \qquad\qquad 3.20$$

The redox potential of this is -0.97 w.r.t N.H.E i.e. all oxidants whose standard potential is more positive than this, are potentially capable of oxidising cyanide to cyanate. Such oxidising agents must be able to accept the electrons formed in Eq.uation 3.20 in order to function. The equation shows, albeit in its modified form (Equation 3.23) that hydrogen ions are formed which, if the reaction is to proceed smoothly, must be taken up by the hydroxyl ions found in alkalis. This is one reason why the destruction is always carried out in alkali. The other reason is, of course, to prevent formation of HCN. Once the cyanate ion is formed, the toxicity of which is around a thousand times less, the cyanide destruction process is all but completed.

Using the thermal destruction method, a two-stage process is involved. In the first stage, cyanate is formed via the cyanogen chloride, in a chlorine/chloride containing medium. The cyanate then hydrolyses and is oxidised to carbon dioxide and nitrogen. Each of these reaction proceeds at a different rate and follows its own time-concentration relationship. Because of the complex situation in which both sequential and parallel reactions take place, it is still true today that the total reaction mechanism is not understood in detail.

3.2.1   Cyanide Destruction Using Hypochlorite

Reactions involving hypochlorite (*"bleach solution"*) are almost identical with those using gaseous chlorine injected into solution. The same reactions also apply to electrolytic cells in which chlorine is electrogenerated, although in this latter case, some direct anodic oxidation of cyanide ions at the anode may also occur (see Section 3.2.2.5). In all cases, the primary reaction of chlorine gas and sodium hydroxide is to form the hypochlorite ion:

$$2\ NaOH + Cl_2 \rightleftharpoons NaCl + NaOCl \qquad\qquad 3.21$$

Equivalent amounts of the chloride and hypochlorite ion are formed. It is for the same reason that commercially available hypochlorite solutions also contain NaCl. As freshly manufactured, such solutions contain ca 13% available chlorine. It should be noted that, on storage, other reactions take place, e.g. formation of chlorate and these solutions lose their oxidising power.

The destruction of cyanide using NaOCl proceeds according to the following equations:

I. $2\ CN^- + 2\ OCl^- + 2\ H_2O \rightarrow 2\ ClCN + 4\ OH^-$ 3.22

II. $2\ ClCN\quad + \quad 4\ OH^- \quad \rightarrow 2\ CNO^- + 2\ Cl^- + 2\ H_2O$

$\qquad\qquad\qquad\qquad\qquad\qquad\qquad\qquad\qquad\qquad\qquad\qquad$ 3.23

I. + II. $2\ CN^- \quad + \quad 2\ OCl^- \quad \rightarrow 2\ CNO^- + 2\ Cl^-$ 3.24

These show how the cyanide is oxidised first to cyanogen chloride, then hydrolysed to cyanate. In the following equations, it is seen how the cyanate can either be further oxidised or decomposed by hydrolysis:

III. $2\ CNO^- + 3\ OCl^- + H_2O \rightarrow N_2 + 2\ CO_2 + 3\ Cl^- + 2\ OH^-$ 3.25

$$
\text{IV.}\quad CNO^- + 2\ H_2O \underset{\overset{H^+}{\big\uparrow\big\downarrow}OH^-}{\overset{\displaystyle\xrightarrow{H^+}\ NH_4^+ + HCO_3^-}{\xrightarrow{OH^-}\ NH_3 + CO_3^{--}}} \qquad 3.26
$$

In overall terms, the following set of sequential and parallel reactions can be postulated and the process was studied in detail by Dodge & Zabban [5] and numerous subsequent authors [6–12].

$$
CN^- \xrightarrow{\ Oxidation\ } ClCN \xrightarrow{\ Hydrolysis\ } CNO^-
\begin{cases}
\xrightarrow{Oxidation} N_2,\ CO_2 \\[4pt]
\xrightarrow{Hydrolys} NH_3,\ CO_3^{--} \\
\qquad\quad NH_4,\ HCO_3^-
\end{cases}
\qquad 3.27
$$

In the following sections, the individual reactions are considered in greater detail.

3.2.1.1 Oxidation of Cyanide to Cyanogen Chloride (Reaction I)

This reaction is often described as proceeding "spontaneously". It takes place so rapidly that is never rate-determing in the overall mechanism. The actual rate is determined by mixing factors, i.e. the reactor technology employed. According to the equation, one would expect the reaction to go fastest in acid solution. However so rapid is the process even in alkali, that this effect is not experimentally determinable, still less of any practical importance. The equilibrium position lies hard to the right. Since the pH dependance of the reaction is not readily determinable, it is the pH requirement of the following stage, i.e. pH > 10 which allows the beginning of the cyanide destruction process to begin and to be maintained. This also ensures that no HCN is formed. According to Delphin [13], at a pH of 9.3 (10^{-2} Mol litre^{-1}) around 50% of total cyanide is in the form of undissociated HCN or dissociated CN ions. Above this value, an increasing proportion of cyanide ions are formed.

Some publications suggest that cyanate is formed not only when peroxo compounds are used to oxidise cyanides but also when these are directly oxidised. On further consideration, it is seen that this reaction, since it cannot be rate-determining in alkaline conditions, is of no practical interest. Stumm et al (14) report that cyanogen chloride is also formed during cyanide destruction. This species can readily be confirmed, at least using more concentrated solutions, on account of its lacrimatory properties.

Cyanogen chloride is an extremely toxic gas, with a very high solubility in water of approx 25 litres gas per litre of water at 20°C [8]. Its vapour pressure increases with increase in temperature and for this reason, it is unwise to use solutions at more than 50°C. [14]. According to Dodge & Zabban [5], temperatures greater than 38°C encourage a rapid evaporative loss of the cyanogen chloride. Concentrations in air of as little as 2.5 mg per cubic metre can be detected by their lacrimatory effect while at a concentration of 400 mg per cubic metre, death occurs after a few minutes. Since reactions 3.21 and 3.22 are exothermic, it follows that increases in temperature are undesirable, especially in the case of more concentrated cyanide solutions. At concentrations up to 1 gm/litre, one would not normally expect temperature increases greater than 5°C. For concentrations up to 5 gm/litre, this figure can be up to 20°C [5]. One should also mention certain side reactions which, if the process is operated correctly, will not be observed. Thus, in acidic media, ClCN can polymerise at higher concentrations to form a six-membered ring of cyanuric chloride [8]. In presence of larger excess concentration of HOCl and with passage of time, nitrate may form, according to Weber [8], who explains this in terms of the following reaction:

$$2\ CN^- + 10\ OCl^- \rightarrow 2\ NO^- + 2\ CO_2 + 10\ Cl^- \qquad 3.28$$

It is also possible that nitrate is formed by oxidation of ammonia formed in the reaction (e.g. Equations 3.26 & 3.27). Then too, the formation of nitrogen trichloride, (NCl_3) is possible, but only at pH > 4.5. Between pH 4.5 and 5, $NHCl_2$ is formed, above pH 8.5, NH_2Cl (15). In the case of cyanide destruction carried out at pH 10 or so, as is the normal practice, it is doubtfuil whether any significant quantities of nitrogen-chlorine products are formed, since they would then react with ammonia formed from hydrolysis of cyanate to produce nitrogen.

3.2.1.2   Hydrolysis of Cyanogen Chloride to form Cyanate (Reaction II)

This stage of the overall reaction broadly concludes the destruction process. As Reaction 3.23 indicates, ClCN can react with hydroxyl ions, i.e the reaction is favoured in alkaline medium. In consequence, its rate is strongly dependant on pH. According to Dodge & Zabban [5], the reaction proceeds only slowly below pH 8. As pH increases, so does the rate which, at pH 10, is sufficiently fast to be practically useful. Between pH 10 and 12, no further changes in rate are observed and these observations are reflected in industrial practice. A very thorough study of this reaction was carried out by Stumm et al. [14]. Their findings, expressed in terms of ClCN consumption as seen in Equation 3.23 can be seen in Fig. 3.3. According to this, the reaction requires 45 min. at pH 10.85, 15 min. at pH 12.5, these results not in presence of hypochlorite excess which would normally be present in practice. Lohmann & Henn [12] carried out similar measurements in which they followed the decrease of cyanide concentration with time, this being possible due to the very rapid reaction to form ClCN. Their results, which relate only to the narrow range of conditions of practical importance, are seen in Fig. 3.4. A 2% excess of hypochlorite was maintained during these experiments. Increase in temperature brings significant increases in reaction rate. At only 26°C and pH 10, the rate is as fast as at pH 11.5 and 18°C to 20°C

In contrast to Stumm et al., Eden and co-workers [16] carried out similar studies but with higher concentrations of excess NaOCl (50%.) At pH 8.9, they report reaction times of 8 min. (pH 9.9) and 14 min. (pH 8.9). The kinetics of this reaction were also studied by Price et al. [17] who obtained an equilibrium constant k for the second-order reaction = 6×10.2 Mol^{-1}.min.$^{-1}$.

$$- \frac{dc_{ClCN}}{dt} = k \cdot c_{ClCN} \cdot c_{OH^-} \qquad\qquad 3.29$$

This equation illustrates extremely well, the pH dependance of the reaction. It predicts that an increase of pH by a single unit (a ten-fold increase in OH$^-$ conc'n) will increase the rate of reaction by the same amount – a factor 10. It should be noted that the concentration of neutral salts is increased by the same amount. In practice, since pH – and thus [OH] are held constant, Equation 3.29 can be simplified to one which is pseudo-first order:

thus:

$$k \cdot c_{OH^-} = k', \qquad\qquad - \frac{dc_{ClCN}}{dt} = k' \cdot c_{ClCN} \qquad\qquad 3.30$$

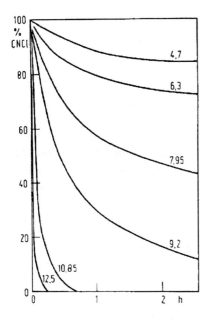

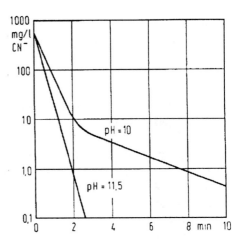

Fig. 3.4   Reaction time for oxidation of cyanide to cyanate as a function of pH, in presence of a 2% excess of NaOCl. (data of Lohmann & Henn [12]).

Fig. 3.3   Hydrolysis rate of CNCl to cyanate as a function of pH (from Stumm et al. [14].)

At pH = 11, k′ has the value 0.6 min.$^{-1}$. If the equation is integrated and a value of 0.1 mg/litre inserted for the ClCN concentration at the end of the run, an expression is obtained for the time required to treat a solution of specified ClCN concentration.

$$t = \frac{2,303}{k'} \cdot \log \frac{c_{ClCN,Start}}{c_{ClCN,End}} \qquad 3.31$$

Typical reaction times for initial ClCN concentrations of 30 to 100 mg/litre are 9.5 to 11.5 min. Bringing together these various findings, a reaction time for cyanide destruction up to formation of cyanate of 1 to 20 min. at pH ≥ 10 can be predicted. To the extent that mixing kinetics are not specified, a degree of imprecision remains. However using ideal mixing conditions, observed times are close to the lower end of the range specified above, in presence of the excess of NaOCl necessary for the operation (a few mg/litre). A number of Statutory bodies specify, erring on the side of safety, a residence time of 40 to 60 min.

3.2.1.3   Oxidation of Cyanide to CO_2 and Nitrogen (Reaction III)

The third stage in the cyanide oxidation occurs, in presence of NaOCl excess, in parallel to Reaction IV. (see below). As seen from Equation 3.25, OH$^-$ ions are formed as a reaction product, i.e. the reaction will be favoured by decreased pH as is found in practice. Stumm *et al.* [14] report the process to react exceedingly slowly in alkaline medium and then only when substantial chlorine excess (OCl⁻) is present. Dodge & Zabban [5] provide more detailed results. Between pH 10 and 12, the reaction is very slow (ca 20 hrs) and takes place quantitatively only in presence of large excess of NaOCl. (c.f Fig. 3.5) At higher initial concentrations −10 to 50 gm/ litre CNO⁻, the reaction proceeds much more rapidly in the first few hours. This can be attributed to an increase in temperature and also to the parallel reaction in which cyanate is hydrolysed. From pH 9 to 10.5, a marked acceleration of the reaction is noted as pH decreases. (c.f Fig 3.6) showing both the data of Dodge & Zabban and the more recent data of Kieszkowski [18] which confirms the earlier studies.) At pH < 8.5, the reaction is so fast that it is complete within an hour, perhaps even 30 minutes.

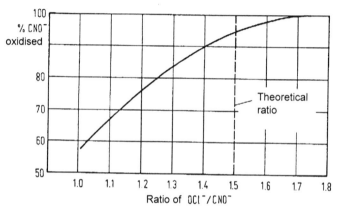

Fig. 3.5   Oxidation of cyanate by NaOCl. Effect of the ratio [OCl⁻]:[CNO⁻]. Initial conc'n: 1 gm/litre CNO⁻; pH = 11 (falling during the reaction to ca. 9.6), reaction time 20 hrs (from Dodge & Zabban [5]).

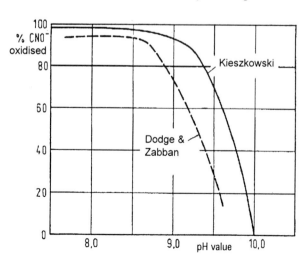

Fig. 3.6.   Effect of pH on the reaction rate of cyanate oxidation by NaOCl. Initial conc'n = 100 mg/litre CNO⁻. Reaction time 1 hr. (data from Dodge & Zabban [5] and Kieszkowski [18]).

Similar findings have been reported by Weiner & Leib [19], i.e. that using stoicheometric amounts of NaOCl with metal cyanide solutions results in an incomplete destruction process. This presumably can be linked to a shortfall of NaOCl, the result of the process shown in Equation 3.25 during the time that the process shown in Equation 3.22 is not completed. An increase in temperature will also result in a pronounced increase in reaction rate.

Price *et al.* [17] also reported that an oxidation to nitrogen and CO_2 was complete in less than 30 min. with consumption of corresponding amounts of NaOCl. They interpret this process in terms of the oxidation of ClCN as shown in the equation below which could also be seen as an overall reaction in which Steps II and III are combined.

$$2\ ClCN + 3\ OCl^- + 2\ OH^- \rightarrow N_2 + 2\ CO_2 + 5\ Cl^- + H_2O \qquad 3.32$$

In theory, oxidation of 1 kg of cyanide (as CN^-) to cyanate would require 13.5 litres of NaOCl (at 13% free chlorine). Since, however, there are other reactions which consume free chlorine, such as the oxidation of 1-valent to 2-valent copper (copper cyanide is a Cu(I) salt), a partial reaction according to equ'n. 3.25 not to mention the oxidation of any organic species present, the actual consumption of NaOCl is far greater than the volume mentioned above and one should reckon on around 26 litres for complete oxidation of cyanide to nitrogen and CO_2.

3.2.1.4 Hydrolysis of Cyanate to Ammonia and Carbon Dioxide (Reaction IV).

In alkaline medium, cyanate is a relatively stable compound. It hydrolyses. as seen in Equation 3.26 increasingly fast as pH decreases, i.e. it is favoured by the same conditions which favour Reaction III. This is confirmed by Dodge & Zabban [5)]who found the lower the pH the faster the hydrolysis rate. An acceptably rapid rate is found at pH = 2 to 3, with the reaction going to completion in 5 to 15 min. Above pH 7, cyanate has been described as a relatively stable species [20]. In concentrated cyanate solution, acidification can result in precipitation of a white substance. This is cyanuric acid.

Resnick *et al.* [20] determined the hydrolysis rate as a function of pH (see Fig. 3.7).
They found the rate governed by an equation:

$$-\frac{dc_{CNO^-}}{dt} = k \cdot c_{CNO^-} \cdot c_{H^+}^n \qquad\qquad 3.33$$

For n = 0.5, k can be considered as being constant with a value of 1.8 at 25°C. These
authors also confirmed that the reverse reaction, cyanide formation from cyanate,
does not take place even under photolysis even though such a reaction would be
thermodynamically favoured with an activation energy of 294 kJ/Mol. It is known
from practical experience that when more concentrated solutions are treated, especially under conditions favouring short reaction times, where increases in temperature are noted, substantial amounts of ammonia are formed. Less welcome, in such
circumstances, are problems associated with formation of ammine complexes which
form with most divalent metals, giving rise to problems in neutralisation.

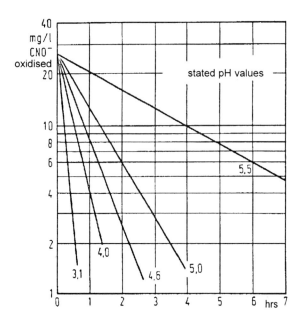

Fig. 3.7   Hydrolysis rate of cyanate as function of pH (from Resnick et al. [20]).

3.2.1.5   Comments on the Overall Reaction

Mechanistic studies by the author on the oxidation of cyanide by hypochlorite have
revealed a number of other aspects affecting the overall reaction rate. If sodium
cyanide solutions are titrated with NaOCl, changes in pH can be followed which
allow insights and conclusions as regards Equations 3.22 to 3.26. If dosing rates
are selected which are, on the one hand, slow enough to allow establishment
of equilibrium but fast enough, on the other, to avoid significant overlapping of

sequential reactions, one obtains a family of of curves for different initial pH values as shown in Fig. 3.8 [21]. At lower pH values, 7 to 9, the pH increases initially due to the alkalinity of the hypochlorite and spontaneous formation of OH⁻ ions in accord with Equation. 3.22. This is followed by an arrest which could be attributed to an uptake of OH⁻ ions (Equation 3.23). Cyanate formation is completed after an equivalent amount of NaOCl is added (at pH = 9). At lower pH values, OCl⁻ consumption is slightly larger, since Reaction 3.23 only proceeds rapidly when pH 10 has been reached and exceeded. At this point in time, only reactions 3.25 and 3.26 can take place. Just before a second equivalent of hypochlorite has been added (Equation 3.25 implies that Equation 3.5 equivalents are required overall), a spontaneous pH decreases takes place in a solution with very slight buffering capacity. This is accompanied by vigorous evolution of gas bubbles (nitrogen and carbon dioxide) as shown in Equation 3.25. As progressively higher initial pH values are used, this stage, up to the consumption of two equivalents of NaOCl moves towards the right. This can be explained in that, as the pH increases, the rate of reaction 3.25 increases at the expense of Reaction 3.26, this resulting in the consumption of somewhat more hypochlorite.

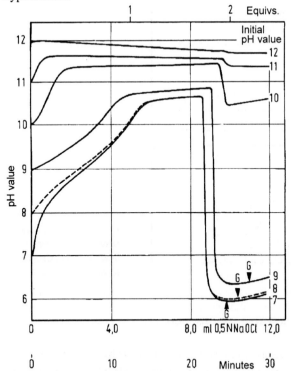

Fig. 3.8 pH changes during the destruction of NaCN using NaOCl for a range of initial pH values. 65 mg CN⁻, dose rate 0.4 ml/min. Chart recorder speed 300 mm/hr. (from Gasen [21].

Both Reactions 3.25 and 3.26 are favoured by acid medium. If the whole exercise is carried out at ten times greater NaOCl dose rate at pH 10, the resulting curve for pH 10 is much sharper than that shown in Fig. 3.8 and gassing is clearly visible after addition of 2.2 equivalents of hypochlorite. If the whole reaction is allowed to take place spontaneously, equilibrium is reached, after the pH maximum is observed, in about 1.8 minutes. As experiments using the same reactor system and same stirring regime but for a simple acid-base neutralisation reaction showed, this time value represents completion of the mixing process. The implication is that the overall cyanide destruction process is an extremely fast one. Redox potential measurements under the same conditions revealed that potential jumps coincided closely with those of pH. These findings summarise, to the author's best knowledge, research on the hypochlorite oxidation of sodium cyanide solutions.

3.2.1.6   Behaviour of Metal Cyanide Complexes

Destruction of metal cyanide complexes (see also Section 3.9.3.1.2) is basically no different from that of the simple alkaline cyanide, at least insofar as the previously cited equations involving ionic species are concerned. These assumed that only the dissociated cyanide (CN^-) was the reactive species. Dissociation of complexed metal cyanides takes place, as is the case for all complexed compounds with unidentate ligands, in stepwise fashion. The equation below shows this in summarised form, going first to the intermediate, a poorly soluble metal cyanide salt and then on to complete dissociation.

$$[Me(CN)_{(z+y)}]^{y-} \rightleftharpoons Me(CN)_z + yCN^- \rightleftharpoons Me^{z+} + (z+y)CN^- \qquad 3.34$$

All metal cyanide complexes are extremely stable and dissociate only very slightly to their constituent species as seen on the right hand side of Equation 3.34. This is represented by the following equilibrium equation:

$$\frac{c_{Me^{z+}} \cdot c_{CN^-}^{(z+y)}}{c_{[Me(CN)_{(z+y)}]^{y-}}} = K_D \qquad 3.35$$

The dissociation constant K_D is the reciprocal of the stability constant for cyano complexes listed in Table 3.1 and is thus extremely small. If this is used to calculate the actual concentration of dissociated cyanide the values in Table 3.1 are found [22]. Low concentrations of a co-reactant of the cyanide also signify, as Section 3.1.1 implies, a low reaction rate. This is specially true for the very stable cyanide complexes of mercury and iron. It is, however, true that longer reaction times are also required for nickel complexes even though this species has a dissociation constant of comparable magnitude to its copper analogue. Destruction of cyanide complexes of zinc, cadmium and copper pose no problems. Dodge & Zabban [5] quote, for oxidation to the cyanate stage, reaction times of 3 minutes for Zn & Cu, and approx. 5 minutes for Cd complexes all at initial concentrations of 100 to 50,000 mg/litre. Similar findings are reported by Stumm et al. [14] for zinc & cadmium cyanide complexes. In the same way, Lohmann & Henn [12] gave values for Cu & Zn cyanide complexes at initial concentrations of 100 to 200 mg/litre CN^- and pH 11.5 of 6 to 8 minutes, in presence of 2% excess chlorine.

TABLE 3.1
CONCENTRATIONS OF FREE CYANIDE IN SOLUTIONS OF VARIOUS
CONCENTRATED METAL-CYANIDE COMPLEXES (AFTER MIRSCH [22])

Metal cyanide complex	Dissociation constant	mg/litre free CN in solution of complexed metal cyanide with total cyanide concentration10,100, 1000,100,000 mg/litre.			
		10 mg/l	100 mg/l	1000 mg/l	100,000 mg/l
$[Hg(CN)_4]^{--}$	$4 \cdot 10^{-42}$	0.00003	0.000045	0.00007	0.00018
$[Ag(CN)_2]^{-}$	$1 \cdot 10^{-21}$	0.0002	0.0004	0.0009	0.004
$[Fe(CN)_6]^{---}$	$1 \cdot 10^{-36}$	0.061	0.085	0.117	0.227
$[Ni(CN)_4]^{--}$	$1 \cdot 10^{-22}$	0.215	0.340	0.54	1.324
$[Cu(CN)_4]^{---}$	$1 \cdot 10^{-22}$	0.215	0.340	0.54	1.324
$[Zn(CN)_4]^{--}$	$1.3 \cdot 10^{-17}$	2.26	3.59	5.68	14.28
$[Cd(CN)_4]^{--}$	$1.4 \cdot 10^{-17}$	2.30	3.64	5.77	14.49

From these relatively rapid reactiuon times, it follows that complexes of copper, zinc and cadmium dissociate very rapidly. If published data from a wide range of authors is correlated with that for the three metals named above, it is seen that the reaction rates are all of the same order of magnitude. The same is not true for the metal-cyanide complexes of silver, iron and nickel. The extremely slow reaction of the Ag complex, long recognised from practical experience, can be linked to its extremely small dissociation constant. As seen in Table 3.1, expressed in terms of free cyanide concentration, it is two orders of magnitude less than the value for iron.

Silver cyanide complexes must always be destroyed and this raises the question why the analogous nickel complexes are so difficult to break up and destroy, while those based on iron are virtually impossible to treat. The Fe(II) complex shown in Table 3.1 only reacts with NaOCl in the sense that Fe(II) is oxidised to Fe(III), this being a slightly less stable entity (see Section 3.2.3). The reason why the iron complex is so stable is that it is not an addition complex, rather it is a penetration type complex. (see Section 3.9.1) So stable are these that they scarcely break down even in concentrated sulphuric acid. Cyanide bound as hexacyanoferrate is designated as "non-chlorine oxidisable cyanide". Because of its very low dissociation constant, it is not toxic. Since the iron cyanide complexes dissociate to form free cyanide to a greater extent than the silver species, the reason for the extreme difficulty in breaking them down must lie in the very low rate of their dissociation. The same applies to nickel complexes. While this does not form a penetration complex like iron, it tends in that direction. (various intermediate types of complex between these two extremes are recognised). While the cyano complex of nickel dissociates relatively easily (the same order of magnitude as the copper complex in terms of the equilibrium position), it does so extremely slowly and oxidation times such as those found for simple cyanide ions are never found and this makes their practical treatment difficult. Dodge & Zabban [5] quote a reaction time of 40 minutes for complexed nickel cyanide in presence of a 20% excess of NaOCl. At higher concentrations, times are shorter – typically 2 to 3 minutes. The process is accompanied by precipitation of black hydrated nickel oxide. Schlegel [21] reports the successful oxidation of a tetracyano nickel complex at unspecified excess chlorine concentration. The time required was, however, 12 hours. Oehme [24] plotted the oxidation of the same complex at various excess NaOCl concentrations. His data are summarised in Table 3.2.

TABLE 3.2

DEPENDANCE OF THE REACTION TIME FOR OXIDATION OF TETRACYANO NICKEL COMPLEX ON EXCESS HYPOCHLORITE CONC'N. (AFTER OEHME *ET AL.* [24]).

Test conditions	Stoicheometric excess of OCl⁻	Reaction time t, (min).
200 mg/litre tetracyanonickelate, pH = 11 (20°C)	1 fold	–
	2 fold	80 . . . 90
	10 fold	19

A better explanation for the problems in destroying the tetracyano nickel complex has been offered by Shashikant [25]. He derived a value of 10^{41} for the stability constant of the Ni complex. This may be compared with values of 10^{17} for the Zn complex, 10^{18} for the Cd complex and 10^{25} for the Cu complex. It will be noted that all these values are higher than those quoted in Table 3.1. Values for the activation energies for hypochlorite oxidation at pH 12 follow the ranking above, i.e. 5.9 kJ/Mol for the Cu complex and 34.7 kJ/Mol for the Ni complex. It should be noted, however, that certain assertions by this author are not supported by observations from several other sources. These include a suggestion that time for oxidation of the Cu complex is little less than that for the Ni complex and that, in terms of time, the three metal cyanide complexes Ag, Zn, Cd behave very similarly. At pH 11, the destruction of the Ni cyanide complex is a function of time and excess NaOCl concentration. In the case of batch treatment, it is simple to establish the reaction parameters. The process is fairly slow (best carried out overnight) and NaOCl should be added such that excess of this species is present, even after the reaction is complete. It appears not to be possible to carry out this process on a continuous basis. In terms of managing effluent streams in-house, any mixing of nickel containing streams with those containing cyanide represents the worst possible outcome and should be avoided at all costs. Oehme has described [26] the use of amperometric measurements during the oxidation of complexed Ni cyanide. These can be used to determine oxidation times, at any given ratio of free hypochlorite to complex and can also show how much hypochlorite has been used. As seen in Equation 3.34, the dissociation of complexed metal cyanides proceeds via the metal cyanide salt. These, when based on heavy metals, are poorly soluble and tend, like their complexed precursors, to dissociate very slowly into their constituent species. This process, usually limited by achievement of equilibrium, is described by Equation 3.36 where L is the solubility product.

$$Me(CN)_z \rightleftharpoons Me^{z+} + z\,CN^- ; \qquad c_{Me^{z+}} \cdot c_{CN^-}^z = L \qquad\qquad 3.36$$

Some values of L for metal cyanides are found in Table 3.3. There is always a danger that, once these compunds have been formed in their poorly-soluble state, they may become mixed with other precipitated species in the sludge formed. In a batch treatment mode, where an equilibrium can be reached as shown in Equation 3.36, the longer dwell times possible can permit complete destruction of the metal cyanide complex. The problem with continuous treatment is that free cyanide may form in the sludges and in the supernatant liquid as well. Weiner & Leiss [19] found exactly this in the case of silver where some 15% of total cyanide was found in the sludges. The author too, has several times found free cyanide in plating sludges, in some cases derived from CuCN [21], see also Section 3.2.4.1. There is a further problem caused by the fact that, during the alkalinisation of the effluent required for cyanide destruction, many of the simple salts causing temporary hardness also precipitate. These act as nuclei on and in which metal cyanide crystallites are incorporated. The cyanide-containing species in this form are poorly accessible to the NaOCl in solution and so render the oxidation slower or less complete.

TABLE 3.3
SOLUBILITY PRODUCT VALUES FOR SOME METAL CYANIDES
(20 TO 25°C) (WEINER & LEISS, [19]

Metal cyanide	$L_{Me(CN)_x}$
HgCN	10^{-39}
CuCN	10^{-19}
AgCN	$10^{-14} \ldots 10^{-16}$
Zn(CN)$_2$	10^{-13}

3.2.1.7   General Problems in Cyanide Destruction Using Hypochlorite

The introduction, a few years ago, of a new parameter AOX (absorbable organic chloro compounds) and the need for analytical equipment to measure these values, resulted in a totally new problem when using NaOCl for effluent treatment. Most effluents also contain organic species, even if only in very low concentrations and these will, in presence of NaOCl, become partially or fully chlorinated. As such, they are included in the overall measured AOX value [27]. Together with AOX type species, halogenated (usually chlorinated) hydrocarbons may also be present. Whereas AOX concentrations of 1 mg/litre are permitted under many legislations, total per-mitted halocarbon concentrations may be as low as 0.1 mg/litre. The most commonly formed AOX compound after hypochlorite treatment is chloroform [27]. Chlorina-tion of organics is a complex process forming, in most cases, a whole range of compounds with varying amounts of chlorine. Some typical examples are shown below:

- chlorination of saturated aliphatic long chains:

$$- - CH_2 - CH_2 - - + Cl_2 \rightarrow - - CHCl - CH_2 - - + HCl \qquad 3.37$$

- chlorination of unsaturated aliphatic long chains:

$$- - CH = CH - - + Cl_2 \rightarrow - - CHCl - CHCl - - \qquad 3.38$$

Even simple additions, e.g. of HCl, can result in the chlorination of unsaturated organic compounds:

$$- - CH = CH - - + HCl \rightarrow - - CH_2 - CHCl - - \qquad 3.39$$

Thus alcohols, aldehydes and carboxylic acids can all be chlorinated as can aromatic compunds, in which case chlor-phenols are produced. The predominance of chloro-form as a product suggests that the chlorination process takes place in conjunction with a bond-breaking step. This continues until a species (chloroform) with only one carbon atom remains and there are no further C-C bonds to be broken. Supporting this hypothesis is the fact that, even when cyanide is totally absent, as is the case in treatment of potable water, chloroform can be detected. The species is also found if pure cyanide is oxidised by hypochlorite, say at pH 11. The concentration of chloro-form increases with time, even after many hours [28]. In view of the fact that cyanogen chloride is spontaneously formed, only this species or cyanate can be considered as the starting product for chloroform. In terms of AOX formation, the hardest hit plants are those which have introduced water-saving measures and, as a result, are processing more concentrated effluents. The use of hypochlorite for cya-nide destruction can lead, in some cases, to AOX or chloroform concentrations of up to several hundred mg/litre. In cases where chloride-containing electrolytes are used, it is possible that free chlorine will be formed as an anodic reaction in the electrolysis cell. Where organic additives are also present, such as brighteners or surfactants, AOX compounds may be formed. Free chlorine can also be liberated from chloride solutions by chemical oxidation and here too, AOX concentrations in excess of 1 mg/litre may be found. At present, no means are known by which formation of AOX type compounds can be prevented, especially where hypochlorite is used to treat cyanide-containing solutions. The use of activated carbon in order to adsorb chlorin-ated organics, simply transforms the problem to one of solid waste disposal. In practice too, effluents contain so many adsorbable species whose total concentration would typically be ten times higher than chlorinated organics alone. In consequence, the lifetime of the activated carbon is very short [29] and this approach is not realistic in most cases.

3.2.2   Oxidation of cyanides with oxygen and peroxo compounds.

The processes treated in this section cover the use of oxygen (gaseous or from a peroxo compound) or other means of electron withdrawal, to form cyanate (see Equation 3.20) without any apparent formation of intermediate. A major advantage of these methods is that little if any increase in the ionic strength of the effluent is produced, apart from any additions to modify the pH. The further oxidation of cyanate to nitrogen and carbon dioxide is not, in all cases, possible. In terms of removal of toxic species, formation of the cyanate is sufficient, especially since hydrolysis of this forms ammoniacal compounds and carbon dioxide, this final stage being shown in Equation 3.26.

3.2.2.1   Cyanide destruction using hydrogen peroxide

Hydrogen peroxide has a tendency to form oxygen in a strongly exothermic reaction, although for this to happen the temperature must first be raised or a suitable catalyst must be present. Under normal conditions, the decomposition rate of hydrogen peroxide is very small (i.e. the compound is metastable) and it can be considered as relatively stable. When contacted with species whose normal electrode potential is more negative than its' own, it functions as an oxidising agent. The normal electrode potential (NHE) of hydrogen peroxide in acid medium is $+1.776$ V. In alkali, it is $+0.878$ V. These potentials correspond to the redox reaction shown below:

$$H_2O_2 + 2\,H^+ + e^- \rightleftharpoons 2\,H_2O \qquad\qquad 3.40$$

Thus, in contact with cyanide, hydrogen peroxide forms cyanate being itself reduced to water.

$$CN^- + H_2O_2 \rightarrow CNO^- + H_2O \qquad\qquad 3.41$$

Further oxidation of cyanate to nitrogen and carbon dioxide does not occur, though, as Equation 3.26 shows, it hydrolyses to form ammoniacal species and carbon dioxide. According to Oehme et al. [24] this reaction proceeds over a wide pH range - from pH 3 to 12, the optimum being pH 4 to 5. Knorre [30], by contrast, suggests pH 9 to 10 is preferable. Since it is essential to avoid formation of hydrogen cyanide, alkaline conditions are in any case required, typically pH = 10. The question of optimum pH is thus purely academic. In addition, lowering the pH reduces the stability of hydrogen peroxide [31]. The rate of Equation 3.41 also depends on various other parameters including concentration of both cyanide and excess peroxide, temperature and presence of catalysts.. At lower cyanide concentrations (> ca 0.5 g/litre), the rate is slow and use of catalyst additions or increased temperature are needed. The catalytic action of copper salts (100 to 200 mg/litre Cu^{++}) has long been known. Their use suffers the drawback that copper ions react with the ammonia, seen as a product in Equation 3.26 to form a tetrammino copper complex, which can be troublesome when neutralisation and precipitation are subsequently used. A more recently adopted catalyst is the patented [32] iodo complex of silver. At even lower cyanide concentrations, a few mg/litre, as for example found in blast furnace gas washings, reaction times of 110 to 180 minutes are typical using substantial excess peroxide but without catalyst [33]. For even smaller cyanide concentrations, reaction rates are so small that several hours may be required for completion of the oxidation. For intermediate concentrations, around 5 gm/litre and those shown in Fig. 3.9, typical reaction times are 30 to 90 minutes. [24, 30, 34]. The reaction can be accelerated by presence of peroxide excess and increase in temperature. Under these conditions, no catalyst addition is necessary. In order to ensure reaction times of 60 minutes or less, the temperature should be raised to 40°C insofar as the exothermic nature of the reaction itself does not achieve this. At higher concentrations (in excess of 10 gm/litre CN^-), higher temperatures of 80°C or even more, are found, but these should be avoided, by moderating the pace of the reaction [31]. In practice, the peroxide is added slowly to the solution by dosing at a rate ensuring the temperature is neither too high nor too low. It is wise to allow a little extra time after formal completion of the reaction. As soon as the oxidation to cyanate is complete, there follows, when a catalyst is present, or copper salts or precipitated heavy metal hydroxides (both of which can act as catalysts), a spontaneous decomposition of excess hydrogen peroxide. This is observed both in terms of a gas evolution and also a sharp rise in temperature. The latter phenomenon is utilised in automatically controlled plants. The increase in temperature is also beneficial in bringing to conclusion the reaction shown in Equation 3.26. In potentiometric measurements, a result dependant only on cyanide content is observed since the excess peroxide has no effect on the potentiometric reading. To measure this, a redox indicator species must be added. In order to follow the progress of the reaction instrumentally, both in terms of individual variables and their effect on the overall reaction, it must be carried out on a batch basis. These requirements must be seen in the light of recent legislation which requires effluent to be low in volume, high in concentration rather than *vice-versa,* as indeed is best for peroxide treatment. Where metal cyanide complexes are present, the dissociated free cyanide shown in Equation 3.34 is oxidised. This poses no problems for complexes of copper, zinc and cadmium, indeed in the first-named case, there is an additional catalytic effect. However the copper salt can act to catalytically dissociate the hydrogen peroxide itself, which is inherently wasteful. Use of the method should therefore be carefully considered when treating effluent from plating plants, for this reason. No such objections exist for use of this technique in heat treatment plants where its environmentally friendly basis has been widely welcomed. In the absence of cyanide complexes of nickel and iron, there remain the same problems as were outlined in Section 3.2.1.6.

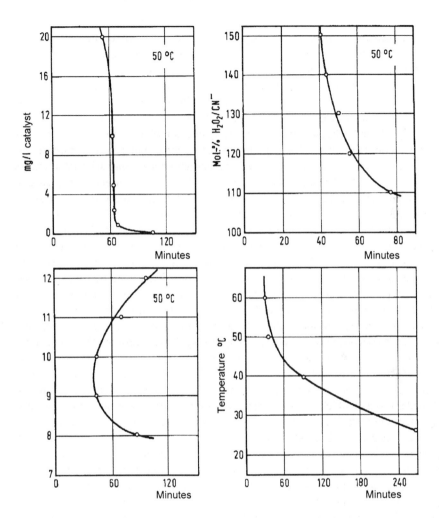

Fig. 3.9 *Dependance of reaction time in the hydrogen peroxide treatment of cyanides on various parameters. Initial concentration: 1.5 gm/litre CN⁻ (after Knorre, [30]).*

Hydrogen peroxide is usually supplied as 35% or 50%. To oxidise 1 kg of cyanide, the theoretical requirement is 2.5 to 3 kg peroxide (35%). In practice, about double this amount is usually needed and for cyanide concentrations less than 1 gm/litre, more still. [34]. Addition of small amounts of alkali silicates (10 ml/litre waterglass) will reduce the amount of peroxide required to 3.5 to 4.5 kg hydrogen peroxide per kg cyanide [35] as long as no heavy metal ions are present.

A useful technique for treating both high and low concentrations of cyanide, avoiding the costs and possible problems associated with catalysts uses a combination of formaldehyde and hydrogen peroxide. This method has been described by Lawes *et al.* [36,37] for treatment of cyanide effluents from zinc and cadmium plating plants. The formaldehyde reacts with cyanide to form formaldehyde-cyanhydrin (glyconitrile) which is then hydrolysed in presence of hydrogen peroxide via the glycollic acid amide to glycollic acid.

$$CN^- + HCHO + H_2O \rightarrow CH_2OH \cdot CN + OH^- \qquad 3.42$$

$$CH_2OH \cdot CN + H_2O \xrightarrow[OH^-]{H_2O_2} CH_2OH \cdot CONH_2 \qquad 3.43$$

$$CH_2OH \cdot CONH_2 + OH^- \xrightarrow{H_2O_2} CH_2OH \cdot COO^- + NH_3 \qquad 3.44$$

During the reaction, zinc and cadmium are precipitated as hydroxides. At the same time, cyanide can react directly with hydrogen peroxide as shown in Equation 3.41 followed by hydrolysis of cyanate as shown in Equation 3.26, Reactions 3.42 to 3.44 are only significant in the case of free cyanide or cyanide complexes with low stability such as zinc or cadmium. They proceed optimally at pH >10<12. Above this value, the peroxide oxidises the formaldehyde. Temperatures above 50ºC are called for and catalyst additions, typically magnesium salts, are required. A further condition is that set ratios of $[H_2O_2]:[CN^-]$ and $[HCHO]:[CN^-]$ should be maintained, typically 1:1 or 1:0.9. Typical reaction rates obtained under these conditions are seen in Fig. 3.10. At low cyanide concentrations (below 100 mg/litre CN^-), hydrogen peroxide should be present in excess – typically up to 100%.

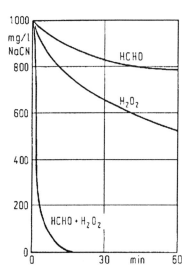

Fig. 3.10 Reaction rates for destruction of cyanide using hydrogen peroxide and formaldehyde or their mixtures. [HCHO]:[CN⁻] = 0.9; [H₂O₂]:[CN] = 1.1. Temperature = 52ºC (after Lawes, Fournier & Mathre [36]).

It should be borne in mind that not only does the latter reaction proceed with some difficulty but also that products with high COD values are formed. This notwithstanding, the method was seriously considered for treatment of cyanide containing effluents from a blast furnace plant.[33] A far simpler means of operating the process is to use it in the immersion dipping mode (see Section 6.3.5). Work emerging from the plating bath, with a film of cyanide electrolyte still adhering to it, is rinsed first in a dilute solution of formaldehyde, then in a dilute hydrogen peroxide bath after which it is rinsed in the normal way. Precipitated metals are removed by recirculation of the solution, using sedimentation (in the formaldehyde) and filtration (in the peroxide) baths [38]. Yet another variant of cyanide destruction using hydrogen peroxide involves simultaneous ultra-violet light irradiation. Photochemical and chain reactions result in formation of OH and O_2H. radicals (see Section 3.11.4). These are extremely strong oxidising agents, capable of breaking up even the most strongly complexed species such as hexacyanoferrate [39]. The ultra-violet light also acts on the hexacyanoferrate, causing it partially to dissociate (see Section 3.2.3). Fig. 3.11 shows the course of oxidation of the hexacyanoferrate (II) formed from the hexacyanoferrate (III).

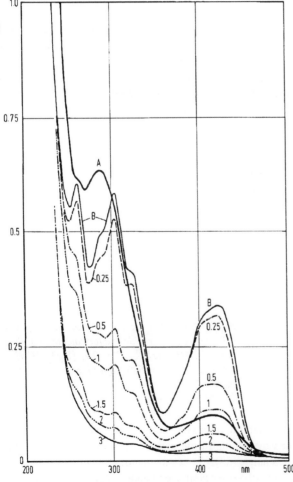

Fig. 3.11   Changes in value of optical extinction during oxidation of cyanide with hydrogen peroxide under   simultaneous irradiation with u.v. light.

A. Initial solution – 100 mg/litre $K_4[Fe(CN)_6]$
B. Initial solution after oxidation with hydrogen peroxide and formation of $K_3[Fe(CN)_6]$. Individual values represent reaction time (hours) using u.v irradiation [39].

3.2.2.2   Destruction of Cyanide Using Mono-peroxy Sulphuric Acid

Mono-peroxy sulphuric acid, also known as peroxy-sulphuric acid or Caro's acid, has found increasing use in recent years for destruction of cyanides. It is used not as the acid, but as an acidic triple salt of composition $2\,KHSO_5 \cdot KHSO_4 \cdot K_2SO_4$ with a $KHSO_5$ content of 45% and it possesses good solubility (250 gm/litre @ 20ºC). [41]. This Caro's salt does not, as might be expected, form an acid sulphate in solution which would then equilibrate with hydrogen peroxide. Rather, it constitutes an acidic salt with the anion HSO_5^-. Above pH 9, this reacts with cyanide, as shown below:

$$CN^- + HSO_5^- + OH^- \rightleftharpoons CNO^- + SO_4^{--} + H_2O. \hspace{2cm} 3.46$$

In favour of this method is a significantly higher reaction rate with cyanide [30]. Use of titanium (IV) chloride to detect hydrogen peroxide gives no result and the system can be followed in terms of the oxidation potential. The reaction rate in alkaline medium is also fast, the destruction process being usually complete within 15 minutes [41]. The process can be further accelerated by use of copper ions as a catalyst, though this is not usually seen as necessary. On account of the acidic character of the salt and the pH change implicit in Equation 3.46, provision should be made for dosing with alkali. If the pH change is accompanied by excess of the Caro's salt, the cyanate will be oxidised to carbon dioxide and nitrogen, although the more rapid hydrolysis leading to formation of ammoniacal compounds is usually favoured.

Use of the Caro's salt is increasingly favoured over use of hydrogen peroxide, not least because it minimises formation of AOX type compounds, at least where batchwise operation is used. In the case of low concentrations of cyanide, excess concentrations of ca 100% are recommended. For 1 gm cyanide, ca 15 gms of Caro's acid are required. This includes a 25% stoicheometric excess [41]. Zinc, cadmium and copper complexes are all readily oxidised though in the case of silver cyanide complexes, large excess concentrations of Caro's salt are needed [11].

Fischer *et al* [33] successfully used Caro's salt to treat a subsidiary effluent as a means of estimating what quantities of hydrogen peroxide and formaldehyde would be required to treat the main effluent line. Being such a strong oxidising agent, Caro's salt will oxidise hydrogen peroxide (to oxygen), chloride (to chlorine) as well as sulphide, sulphate, nitrite and most organic compounds. Since cyanide effluents do not usually contain chlorides, there is no question of the latter species reacting to form AOX type compounds.

A recent observation which should be noted, is that monoperoxy sulphuric acid in the presence of nickel, is catalytically decomposed by the nickel peroxide precipitate, i.e this is a form of autocatalysis. The completion of the cyanide destruction, even after dosing with substantial excess of oxidising agents, cannot be potentiometrically determined by a jump in the positive sense and the destruction reaction itself may remain incomplete [28]. Use of monoperoxy sulphuric acid should thus be preceded by a rigorous analysis for presence of nickel.

3.2.2.3 Cyanide Destruction Using Oxygen

As seen from the equations below, the redox potential of oxygen is $+ 1.229V$ in acid solution, 0.401 in alkali:

$$O_2 + 4\,H^+ + 4e^- \quad 2\,H_2O \qquad\qquad 3.47a$$

$$O_2 + 2\,H_2O + 4e^- \quad 4\,OH^- \qquad\qquad 3.47b$$

In theory therefore, there should be no problems in oxidising cyanide with oxygen or air and this is indeed the case when the othewise slow reaction is catalytically accelerated. In the presence of a solid catalyst on which one or more of the reacting species is adsorbed, a dramatic increase in reaction rate is observed (cf. Section 3.1.1 *"Heterogeneous catalysis"*).

Hartung [41] describes a process developed by Bucksteeg in which the cyanide containing effluent trickles down a vertical column packed with an expanded fine-grained coke-derived carbon. A counter-current of air is forced upwards through the bed. The catalytic cyanide oxidation takes place on the surface of the carbon. This can be represented by combining Equations 3.20 and 3.47b to give 3.48.

$$2\,CN^- + O_2 \;\rightarrow\; 2\,CNO^-. \qquad\qquad 3.48$$

The cyanate formed in the above reaction is then, depending on the pH, oxidised further to ammoniacal compounds and carbon dioxide. There is a limit to the loading of these carbon products − 10 cubic metres of effluent per cubic metre of column filling per 24 hrs would be typical. The method can handle alkaline cyanides as well as cyanide complexes of zinc, cadmium or copper. It cannot, though, be used to process the more stable metal cyanide complexes which are harder to oxidise. In the former case, the metal ions, once the complexing cyanide has been oxidised, precipitate onto the catalyst as hydroxides or as basic salts. In this context, copper salts possess a catalytic activity of their own [42]. These precipitates must, from time to time, depending on the composition of the effluent, be removed with acid. The potential danger of gaseous HCN formation, in the event of residual cyanide being present in the system, should be noted.

In the pH range 5 to 9, reaction 3.48 is virtually unaffected by pH. However at pH 10, the rate drops off sharply. Increase in temperature accelerates the reaction rate, which, for example, is doubled over the range 20 to 80°C. Kuhn & Wilson [43] deduced from this the very low activation energy value of 9.2 kJ/Mol.

An alternative technology has been promoted by Hoeke and Wittbold [44] in which effluent is continuously recirculated through an aerated tank in which activated carbon particles are suspended. Such systems can handle several kilograms of cyanide per tonne of activated carbon in a 3 to 4 hour period. Increasing the mass of activated carbon in use, further increases this figure. The rate controlling step appears to be the dissolution of atmospheric oxygen. This apart, the type of activated carbon used can also affect reaction rates. After an initial period, the reaction rate becomes constant. For fresh carbon it is some 25% higher than when the material has been in service for asome time.

Faul & Kastening [45] examined the catalytic oxidation of cyanide, treating it as simultaneous coupled anodic and cathodic electrochemical processes (see Equations 3.20 and 3.47b). They identified, as specially suitable catalysts, those materials which were stable in alkali and which were effective electrocatalysts for the cathodic reduction of oxygen. These are the same requirements made of fuel cell cathodes and drawing on the findings of fuel cell research, suitable materials for charge transfer reactions are those with high specific surface areas and electrically conductive surfaces. Such criteria are met by activated carbon powder which has been heated at 1100°C in an inert atmosphere. Impregnation with additional catalytically active materials such as platinum, nickel or silver, further increase the activity of the carbon. In practical terms, either packed columns or suspensions are equally useful.

3.2.2.4   Cyanide Destruction Using Ozone

As seen in Equation 3.49, ozone in acid solution has a very positive electrode potential, approx. + 2.07 Volt. Even in alkali, the corresponding figure is +1.24 Volt. These values designate it as one of the most powerful oxidation agents known.

$$O_3 + 2\,H^+ + 2\,e^- \rightleftharpoons O_2 + H_2O \qquad\qquad 3.49$$

The formation reaction of ozone from atomic oxygen is strongly endothermic, and this in part explains its very high oxidation potential and low stability and its very reactive character. In presence of cyanide, ozone reacts to form cyanate, according to the following reaction:

$$CN^- + O_3 \rightarrow CNO^- + O_2 \qquad\qquad 3.50$$

This is followed by hydrolysis of the cyanate as shown in Equation 3.26. In presence of excess ozone, the cyanate is further oxidised, according to Equation 3.51.

$$2\,CNO^- + 3\,O_3 + H_2O \rightarrow 2\,HCO^- + N_2 + O_2 \qquad\qquad 3.51$$

For cyanide concentrations in the range 50 to 100 mg/litre, several authors have confirmed the reaction to be rapid, provided sufficient ozone is introduced, even though several different reactor types were employed. According to Sondak & Dodge [46], typical reaction times using a bubble reactor were 10 minutes. Fabjan [47] reported 10 to 15 min. (packed bed reactor) with Stopka [48] citing 10 to 30 min. (ozone injection with a mixing turbine). According to Sondak & Dodge, the reaction is zero order when cyanide concentration exceeds 4 mg/litre.

$$-\frac{dc_{CN^-}}{dt} = k_0 \qquad\qquad 3.52$$

Only at concentrations less than 4 mg/litre does the reaction become first order. Fabjan & Davies [49] were reluctant to assign the same reaction rate behaviour to complexed and uncomplexed cyanides, although in virtually all cases, the reaction rate plots show hardly any cyanide concentration dependance. The rate does, though, depend linoraly on the rate of ozone supply and/or the ozone content of the air and, of course, on the interfacial contact area of the two phases. Typical ozone concentrations in air were 2 to 3 vol%. When ozone is formed from oxygen, values up to 6 to 8% can be reached. The solubility of the gas in water, as per the Dalton-Henry Laws, increases as its gas-phase concentration grows. Its intrinsic solubility in water is some ten times that of gaseous oxygen. An upper limit to the ozone solubility is imposed by the chemical stability of the gas which decomposes exothermically to form oxygen. It has a half-life in water of around 20 minutes [50]. The theoretical requirement for oxidation of cyanide to cyanate is 1.85 gm ozone/gm cyanide.

The reaction is normally carried out in the pH range 10 to 12. In this range, the effect of pH change on reaction rate has been reported as insignificant [47] or slightly favourable with increasing pH [46]. Bauer [50] recommended an initial pH of 11 to 12 in order that the final pH value be not less than 8. Fabjan [47], on the other hand, suggested that above pH 10, an increased rate of oxygen decomposition would occur. For the oxidation of hexacyanoferrate (II), Stopka [48] found pH 7 to be optimal. No significant temperature effects on reaction rate have been found, although at higher temperatures, the ozone decomposes before it has time to react [46].

Catalytic effects on the ozone oxidation reaction have been studied by Stopka [48] who added metal salts, and by Fabjan & Davies [49] who examined the reactions of metal cyanide complexes. In both cases, copper salts proved the most effective, followed by nickel, whose cyanide complex is normally extremely difficult to oxidise. This is explained by Fabjan & Davies not so much in terms of the stability of the initial complex, but rather in terms of the oxidisability of the central metal atom (Cu to Cu(II), Ni to Ni(III)). In their higher valence states, these do not form stable complexes with cyanide. The very stable cyanide complexes of gold, iron and cobalt are only partially oxidised unless catalyst is added or, most effective of all, ultra-violet light radiation is applied. In these cases, oxidation is complete [49].

In presence of excess ozone, the cyanate - insofar as it has not already hydrolysed - is further oxidised as shown in Equation 3.51. However this second stage oxidation is slow, requiring [46] 5 to 15 times as long as the first stage. This was confirmed by Matsuda et al [51]. The oxidation tends, in alkaline conditions (pH 11) and with excess ozone, to first order behaviour. It is catalysed by tin (II) ions, which accelerate the rate up to six-fold (maximum permitted concentration is 2×10^{-3} Mol). Copper ions, by contrast, though they catalyse reaction 3.50, act to slow down the second stage of the oxidation, possibly because a cyanato-complex is formed. The theoretical ozone requirement for total oxidation of cyanide is 4.6 gm ozone/gm cyanide.

In addition to the reaction products shown in Equations 3.50, 3.51 and 3.26, the presence of ammonium cyanate and urea derived from it, has also been reported. These are, however, oxidised by ozone to nitrogen and carbon dioxide.

All published studies have been based on relatively low cyanide concentrations

(50 to 100 mg/litre), having in mind the problem of ozone production. Because of the high costs of ozone production, there have been proposals for a two-stage treatment process when higher cyanide concentrations (> 500 mg/litre) are involved. These might be based on a first stage using hydrogen peroxide, with ozone used in the second and final stage. [48,49]. It should also be stated that the technological implementation of this otherwise environmentally friendly process are not without their own difficulties.

3.2.2.5 Cyanide Destruction by Anodic Oxidation

Before considering in detail the direct anodic (electrochemical) oxidation of cyanides, as a means for their destruction, some more general comments on the anodic oxidation of chloride ion containing solutions are called for. In normal circumstances, chlorine gas is formed which then dissolves in alkaline solution, to form sodium hypochlorite, as shown in Equation 3.21. For each equivalent of sodium hypochlorite formed, a second equivalent of sodium chloride forms. The hypochlorite then reacts with cyanide, as shown in Section 3.2.1. According to Equation 3.53, the standard electrode potential in alkaline solution is + 0.85 V.

$$Cl^- + 2\,OH^- \rightleftharpoons ClO^- + H_2O + 2\,e^- \qquad\qquad 3.53$$

Leaving aside the very small losses of chlorine implicit in the vaporisation of gaseous chlorine and cyanogen chloride, it will be seen that the bulk of the chlorine is continuously recycled, fresh hypochlorite being formed at each renewed pass through the electrolysis cell. The nett result is that this process requires only electrical energy and no chemicals.The method, analogous to anodic oxidation in absence of chloride ions, is primarily aimed at more concentrated solutions and should only be used when chloride is already present in solution. In no circumstances should chloride be deliberately added, since this would only bring about the additional problems described in Section 3.2.1.7. The method is carried out batchwise or, for larger volumes, by recirculation through the cell. Thus, if one recirculates a solution with 30 gm/litre NaCl through the cell and then the rinse tank, any cyanide in the rinse tank carried over by drag-in is automatically destroyed. Redox potential measurements can be used to monitor and control the process so that between 80 and 200 mg/litre free chlorine (as NaOCl) are available [52]. All in all, this approach is not usually the preferred one and, for low cyanide concentrations, precipitation remains a valuable option.

The reaction rate in the above process increases with NaCl concentration and current density. However in presence of high chlorine concentrations in relation to cyanide present, there is no dependance of reaction rate on cyanide concentration [53,55]. Energy consumption values of between 8 & 40 kWhr/kg cyanide have been quoted [53,54,55]. Hillis [55] compared graphite and platinised titanium anodes, finding the latter can be operated at far higher current densities, so that the mass of cyanide destroyed per unit time per square metre of anode surface, as well as specific energy consumption are all superior.

Other studies on choice of anode materials were carried out by Bustos, Kammel & Steppke [56] who found that surfaces posssessing electrocatalytic properties behaved markedly better than those which did not. They ranked the electrolytic properties of candidate materials in terms their current efficiencies for lowest possible energy consumption. On this basis lithium platinate coated and pyrolytically fired platinised titanium anodes emerged the best. However the long-term performance of these anodes has yet to be fully assessed.

The reaction mechanism for anodic oxidation of cyanide in presence of chloride ions is complex, on account of the numerous reaction pathways (including non-productive side-reactions) which can take place. As the chloride ion concentration decreases, a further set of reactions become important, namely those (see below) operating in chloride-free electrolytes.

Where chloride ions are absent, anodic oxidation is the most environmentally friendly means for cyanide destruction. However it is only fair to make the point that this 2-electron transfer oxidation to form cyanate, is a fairly slow reaction (see Equation 3.20). The theoretical requirement is for 2.06 Amp-hr per kg of cyanide, oxidised to cyanate. At cell voltage values of 2 to 4 volt, this corresponds to 4.1 to 8.2 kWhr.

At the anode, under normal values of overpotential, apart from oxygen evolution, the following reactions can be formulated in presence of cyanide and hydroxyl ions:

Anode reactions with CN^-:

$$2\ CN^- \rightleftharpoons (CN)_2 + 2\ e^- \qquad \qquad 3.54a$$

$$(CN)_2 + 4\ OH^- \rightleftharpoons 2\ CNO^- + 2\ H_2O + 2\ e^- \qquad 3.54b$$

Electron loss leads to formation of the unstable dicyanogen radical which immediately hydrolyses to cyanate [57].

Anode reactions with OH^-:

$$2\ OH^- \rightleftharpoons 2\ OH + 2\ e^- \qquad \qquad 3.55a$$

$$2\ OH \rightarrow H_2O + O \qquad \qquad 3.55b$$

$$CN^- + O \rightarrow CNO^- \qquad \qquad 3.55c$$

Electron loss leads to formation of the strongly oxidising and unstable species OH. which liberates atomic oxygen to oxidise the cyanide.

Summing both reaction schemes, one obtains as overall reaction for anodic oxidation of cyanide in alkali:

$$CN^- + 2\,OH^- \rightleftharpoons CNO^- + H_2O + 2\,e^- \qquad\qquad 3.56$$

If one estimates the standard potentials of reactions 3.54a and 3.54b in alkali by extrapolating from values found in acid solutions and compares these with standard potential values for formation of O' H radicals (+ 2.0V), it is clear that the latter is more positive and the formation of the dicyanogen will be kinetically (or at least thermodynamically) favoured. Formation of dicyanogen was observed by Krusenstern & Mussinger [57] who also studied its hydrolysis as shown in Equation 3.57 to form the oxalate ion:

$$(CN)_2 + 2\,H_2O + 2\,OH^- \rightleftharpoons (COO)_2^{--} + 2\,NH_3 \qquad\qquad 3.57$$

The dicyanogen formation reaction is strongly dependant on the anode material used. It improves progressively as V2A steel is replaced by platinum (platinised Ti) or better still, carbon. However at current densities below 1.5 A/sq. dm, nickel anodes show even higher current efficiencies than the materials mentioned above [57].

In addition to the preceding reactions, there is the hydrolysis of cyanate formed (see Equation 3.26) as well as the anodic oxidation of cyanate:

$$2\,CNO^- + 4\,OH^- \rightarrow N_2 + 2\,CO_2 + 2\,H_2O + 6\,e^- \qquad\qquad 2.58$$

This last reaction has been observed [58,59] but not studied in any detail. Since all solutions contain ammonia after the various reactions have taken place, the cyanide destruction can be deemed to have been completed once the cyanate and the dicyanogen have been hydrolysed. In cases where the cyanide ion is associated with a metal ion, as found in many electroplating baths, the metal will be electrodeposited at the cathode. Such deposits are, when the metal ion concentration is reasonably high, formed at good current efficiencies and are ductile in nature. At lower metal ion concentrations (> ca. 1gm/litre) and higher current densities, the metal ion will deposit as a poorly adherent and powdery deposit. The pH values of these solutions are usually quite high (pH > 11), in consequence of their high cyanide concentra-tions. Easton [58] destroyed concentrated cyanide solutions (45 to 100 gm/litre CN) at temperatures of 93 to 99°C, current loadings of 0.6Amp/litre and anodic current density of 3.8 A/sq. dm. The runs lasted 8 to 17 days, resulting in a final cyanide concentration of less than 1 mg/litre. Normally, an 8 day electrolysis, followed by chemical oxidation, was the preferred approach.

The author and colleagues [39] also established the considerable importance of solution temperature in that an increase from 33 to 50°C approximately doubled the reaction rate. Imposing a larger current across the cell gave added benefits as the Joule heating effect increased the electrolyte temperature.

Jola [59] used current densities of 1 to 4 Amp/sq. dm at 5 to 6 V, until limiting current density was reached. This resulted in constant cyanide destruction rates up to 1 gm/litre. These results imply a zero order reaction. These results would seem to permit cost-effective processing, provided the cation associated with the cyanide was present at sufficient concentration. The energy consumption was calculated at 30 kWh/kg CN.

Kuhn & Biddle [60], in a review, cited numerous other published studies. Since these studies were, in many cases, not comparable, they found that reaction orders were sometimes 0, sometimes 1, and – in cases where a dependance on hydroxyl ion concentration was found – it was 2. In an idealised electrolysis cell, it would be possible to destroy cyanide solutions over a range of concentrations with low energy consumption. Fig. 3.12 shows the relevant data. In this context, the work of Sawyer & Day should also be mentioned (quoted in [60]). They showed that reaction 3.56 is totally irreversible and that it is not possible to form cyanide by cathodic reduction of cyanate.

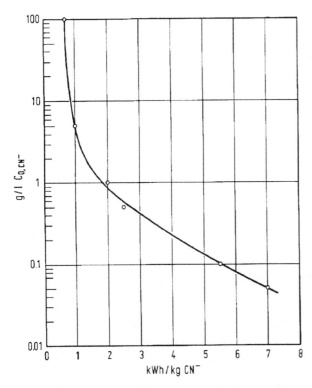

Fig. 3.12. Energy consumption for the destruction of 1 kg of cyanide as a function of initial CN concentration, in an idealised electrolysis cell. Final concentration = 20 mg/litre CN. (after Kuhn & Biddle [60]).

Using an electrolysis cell of the type employed to recover silver from effluent streams, cyanide destruction at the platinised anode surface was reported with a relatively high current efficiency of circa 50% [61].

The author and colleagues [39] have used a wide range of solutions containing

copper or zinc or both metals together, achieving complete cyanide destruction
together with satisfactory metal recovery. Final metal ion and cyanide concentrations
of less than 1 mg/litre were routinely obtained. Nickel, likewise, could be reduced to
a few mg/litre. Iron and chromium, however, which are often introduced into solution from their parent metals, remained in solution, the former as hexacyanoferrate.
Cyanide in this form cannot be anodically oxidised.

Because of the simultaneous removal of metal ions from solution, an option
which is often attractive is to halt the electrolysis at the 100 mg/litre cyanide level
and then complete the cyanide destruction using hydrogen peroxide, itself an environmentally friendly process. Because the metal ions are now absent in solution,
there are no problems of catalytic decomposition of the peroxide.

3.2.3 Cyanide Destruction by Precipitation with Iron (II) Salts

Ferrous ions form, with cyanide, extremely stable complex species, the so-called
hexacyanoferrates or ferrocyanides. (see Sections 3.2.1.6 and 3.9.3.1.2). Depending
as to whether these complexes form from 2 or 3-valent iron, the anionic complex will
be based on a similar valence state.

$$Fe^{++} + 6\ CN^- \rightleftharpoons [Fe(CN)_6]^{----} \qquad\qquad 3.59$$

The potassium salt of this compound, potassium hexacyanoferrate (II) ($K_4[Fe(CN)_6]$,
is also known as potassium ferrocyanide.

$$Fe^{+++} + 6\ CN^- \rightleftharpoons [Fe(CN)_6]^{---} \qquad\qquad 3.60$$

The potassium salt of hexacyanoferrate (III) ($K_3[Fe(CN)_6]$, is reddish in colour.
Instead of forming salts with an alkali metal cation, these complexes can react with
iron ions to form complex salts, in each case with the (II) or (III) valent state. There
are thus four related salts, some of which are relatively insoluble:

a) $Fe^{II}[Fe^{II}(CN)_6]$ Iron (II) hexacyanoferrate (II) – Berlin White
b) $Fe^{III}[Fe^{II}(CN)_6]_3$ Iron (III) hexacyanoferrate (II) – Berlin Blue
c) $Fe^{II}[Fe^{III}(CN)_6]_2$ Iron (II) hexacyanoferrate (III) – Turnbulls Blue
d) $Fe^{III}[Fe^{III}(CN)_6]$ Iron (III) hexacyanoferrate (III)

3

The above compounds a), b), c) are very poorly soluble in water whereas d) dissolves
in water to form a dark brown solution. It is thus not a suitable candidate for cyanide
precipitation. Various equilibria are set up between b) and c) based on electron
transfer reactions. These depend on whether one sets out with (II) or (III) valent Fe or
mixtures of the two valence states. To ensure that the most insoluble species are

formed, excess of Fe-containing species must be present. If this is not the case, the more soluble potassium-iron double salts are formed.

Goetzelmann & Spanier [62] carried out a most thorough study of the precipitation of cyanides as insoluble iron-containing compounds in effluent treatment, reaching the following conclusions:

While the yellow potassium ferrocyanide salt will readily form in the range pH 6.5 to 12, Berlin Blue as an insoluble species only forms in the weakly acid range pH 3 to 5. As pH increases, this compound decomposes into potassium ferrocyanide, at the same time forming ferric hydroxide which is precipitated. This reaction takes place quantitatively at pH 8:

$$Fe_4[Fe(CN)_6]_3 + 12\ OH^- \rightleftharpoons 3\ [Fe(CN)_6]^{----} + 4\ Fe(OH)_3 \qquad 3.61$$

As seen above, a precipitation of Berlin Blue is not involved in this reaction. The pH at which most heavy metals are precipitated from effluents is usually > 8. In the event such sludges come into contact with Berlin Blue either during the sedimentation process or after dumping, the higher alkalinity can result in a re-dissolution of the iron complex.

If Fe(II) salts are used as precipitating agents and neither Fe(III) containing species or oxidising agents are present, then Berlin White can be formed and precipitated, a species insoluble in weakly alkaline media. When precipitated in absence of air, no cyanide (as total CN) is detected below pH 8.5. Even under industrial conditions where Fe(III) containing impurities are present, no cyanide was detected in the filtrate below pH 8. In both these examples, cyanide was only estimated as total CN, i.e. as the non-toxic hexacyanoferrate complex.

Precipitation of Berlin White is a good method for removal of cyanides from concentrated solutions. The precipitate is readily filterable. According to Oehme et al [24], the same method can be used for the cyanide complexes of zinc or cadmium. The Berlin White is precipitated with the hydroxides of the metals in question. In presence of the more stable cyanide complexes of nickel or copper, the precipitate contains in addition to Berlin White, the iron salts of cuprocyanide or nickel cyanide complex as well as copper cyanides.

Using this procedure, the cyanide is separated but not actually detoxified and could, for example, present a hazard in a landfill situation should it come into contact with acid. A possible consequence, in this case, would be introduction of cyanide into the seepage waters from the landfill site. Thermal treatment of such sludges results in decomposition of the cyanide, leaving behind completely harmless residues which can be dumped to landfill without cause for concern.

Comment is also called for here on the decomposition of complexed cyanides using photolytic methods. In practice, the method is relevant only for iron cyanides. The dissociation constant of the hexacyanoferrate ion, normally unmeasurably small (see Table 3.1), is greatly increased in presence of ultra-violet radiation (see Fig. 3.11). According to Weber [63], the extent of photolysis is related to the valence state of the iron in the anionic complex and also the pH. The prediction is for it to increase as pH value decreases and it is larger for the yellow ferrocyanide than for the red ferricyanide. Thus, starting with 1 gm/litre CN in solution, uv irradiation corresponding to 4 hours solar radiation at pH 7 gave a solution with 2 to 3 mg/litre CN. But most important of all, it is the intensity of the radiated light which governs the extent of photolytic decomposition. Bortlisz [64] reports how, starting with 10 mg/litre CN (as potassium hexacyanoferrate (II)) in river water, after 2 days of sunshine some 60% of the cyanide had decomposed at the surface, some 10% being decomposed at a depth of 50 cm. In flowing streams, it is unlikely that any adverse effects will arise from such solar exposure, nor is there any known record of such an event.

3.2.4   Thermal Destruction of Cyanides

Several processes will be considered here, ranging from high temperature hydrolysis of cyanide to its actual combustion. Some of these processes are related to one another in terms of their reaction kinetics. Others are totally unrelated to one another and the only common element is the use of elevated temperatures. In the past, Schindewolf & Bonhoeffer [65] developed a process for the continuous hydrolytic cleavage of cyanide at elevated temperature and pressure, using a tube reactor. At temperatures of 140 to 220°C and pressures up to 100 bar, cyanide hydrolyses relatively rapidly at pH 8 to produce formate and ammonia, as shown in equation 3.62.

$$CN^- + 2\,H_2O \;\rightarrow\; HCOO^- + NH_3 \qquad\qquad 3.62$$

The method can be used at concentrations from 1 mg/litre up to 200gm/litre CN. The reaction is a function of temperature and time with the kinetics being 1st order w.r.t cyanide. The same technology can be used to treat effluents containing cyanides and nitrites such as those typically produced in heat-treatment plants. Here, in a second tube reactor at 150°C, following the first, nitrite oxidises the products of the cyanide hydrolysis reaction according to Equations 3.63 and 3.64. A pH < 6 is required for these reactions and addition of acid may be called for. The end-products are nitrogen, carbon dioxide and water:

$$NH_4^+ + NO_2^- \;\rightarrow\; N_2 + 2\,H_2O \qquad\qquad 3.63$$

$$3\,HCOOH + 2\,NO_2^- + 2\,H^+ \;\rightarrow\; 3\,CO_2 + 4\,H_2O \qquad\qquad 3.64$$

In cases where there is excess nitrite, a further treatment may be necessary as described in Section 3.4 to complete detoxification.

A similar process for cyanide hydrolysis is described in US Pat 4,042,502 [66] operating at 200 to 250°C and pressures of 40 to 140 bar at pH > 7. The reaction is claimed to be complete within 5 minutes.

Tan & Teo [67] studied cyanide hydrolysis in the range 113 to 175°C where reaction rate increased with temperature. However their results do not permit any definite predictions as to the effects of further pH increase as a means of increasing reaction rate. These authors report a lst order reaction for the destruction of free cyanide (k_1 = 180 sec^{-1}) and an activation energy of 94.1 kJ/Mole. Cyano-complexes are more difficult to treat, requiring higher temperatures. The ranking of increased difficulty of destruction follows the magnitude of the stability constant of the complexes; Cu < Zn < Cd < Ag < Ni < Fe, the sole exception being the cuprocyanate. The metals are precipitated as oxides or hydroxides.

One patent application [68] is based on the use of a relatively large continuous flow reactor (20 m^3 for 12m^3/hr) at 175 to 200°C for the destruction of the otherwise very stable nickel cyanide complex. Data in this patent suggests that use of pH >12.5 accelerated the reaction.

For treatment of concentrated cyanide solutions, apart from anodic oxidation or thermal hydrolysis, incineration of HCN is an appropriate process. Jola [69] and Conrad & Jola [70] have reported on this method which is based on catalytic combustion of liberated HCN. Thus concentrated cyanide solutions are treated with fresh mineral acid or with acidic effluents and acidified, depending on the presence of alkali metal cyanides or cyano complexes of the heavy metals, to a pH value of 2 to 6. As already explained in Section 3.2.1.1., at pH 9.3, some 50% of cyanides are in the form of undissociated hydrocyanic acid. At pH 7, the corresponding value is 99%. The HCN containing solution is trickled down a packed bed tower with counter-current flowing air. The air-HCN mixture is passed over a platinum catalyst where it is combusted as shown in Equation 3.65. to form nitrogen and carbon dioxide.

$$4\ HCN + 5\ O_2 \xrightarrow{\ Cat\ } 2\ N_2 + 4\ CO_2 + 2\ H_2O \qquad\qquad 3.65$$

1 cubic metre of cyanide containing effluent requires 100 to 1000 cubic metres of air. The catalyst serves to reduce the activation energy such that the ignition temperature is around 300°C. The thermal energy released by combustion (664.8 kJ/Mol HCN) is sufficient to raise the temperature of the incoming gas to its ignition point. The feed air is carefully metered so that incomplete combustion and formation of carbon monoxide, which would poison the Pt catalyst, is avoided.

Further detoxification of the residual liquor is usually required, in cases where HCN concentration remains above 0.1 mg/litre CN. Jaeckle [71] found that this concentration would be reached at 50°C after 4 hours treatment time. An alternative procedure is to pass such residues, or indeed other dilute effluents, through an ion exchange unit [72] and to feed the resulting concentrate to one of the processes earlier described for such solutions.

In centralised waste treatment plants, a simple cyanide destruction method is to inject the solution into the flame of an oil burner. The destruction of spent oils by their combustion can be used for this purpose. At the typically reached temperature values of 1200 to 1400°C, even the most stable cyanide complexes break down to nitrogen, carbon dioxide and alkali metal salts or metal oxides. However this process is really only suited to use in centralised waste disposal facilities.

Schwarzenbach [72] studied a simple method in which cyanide waste (concentrated solutions or sludges) were treated in a normal tube boiler at combustion temperatures of around 800°C. The water or moisture present formed steam which, as explained above, resulted in hydrolysis. However the process almost certainly also involves straightforward combustion of cyanide and its decomposition products. Ammonia is totally combusted while oxides of nitrogen are not stable at temperatures of 700 to 1000°C, nor could they be determined in the effluent gas which contained, depending on the fuel employed 2.2 to 40 µg/cubic metre of cyanide (measured 1 metre from the stack). Cyanide content of the ash was > 1 mg/kg.

Kieszkowski [73] studied the cyanide sludges formed during cleaning of zinc plating vats. These were heated at 300 to 700°C in flowing air. This led to 50% destruction (at 300°C) and 99.9% destruction (at 700°C) of the cyanides present. In practice, a temperature of 500°C was used when 99.8% of cyanide was destroyed.

Mueller & Witzke [74] destroyed solid cyanide residues from heat treatment plants by heating them, in melt state, to 400°C. For an initial NaCN concentration of 0.9%, decomposition required 30 minutes. In cases where the solid residues also contained nitrates and/or nitrites, these react with cyanide or cyanate at the temperatures used, to oxidise them as shown below:

$$NaNO_3 + NaCN \xrightarrow{\text{ca. } 400°C} N_2 + Na_2CO_3 \qquad 3.66$$

$$3\,NaNO_3 + 5\,NaCNO \xrightarrow{400°C} 4\,N_2 + 4\,Na_2CO_3 + CO_2 \qquad 3.67$$

$$5\,NaNO_2 + 3\,NaCN \xrightarrow{\text{ca. } 400°C} 4\,N_2 + 3\,Na_2CO_3 + Na_2O \qquad 3.68$$

$$NaNO_2 + NaCNO \xrightarrow{400°C} N_2 + Na_2CO_3 \qquad 3.69$$

3.3   DESTRUCTION OF HEXAVALENT CHROMIUM COMPOUNDS

Hexavalent chromium compounds are strong oxidising agents and this alone makes them toxic. The normal detoxification procedure involves their reduction to the trivalent state, after which they can be precipitated as the relatively insoluble chromium hydroxide (see Section 3.7.5.8). Dissolved hexavalent chromium compounds are almost invariably anionic oxo-complexes. Depending on the pH, hexavalent chromium compounds form the yellow chromates (CrO_4) in neutral to alkaline media or the orange-red dichromates (Cr_2O_7) in acid conditions. Higher associated oxides of chromium can also form in acid. The following equation shows the equilibrium between these species. An intermediate which can form, is the hydrogen chromate. However at room temperature, this loses hydrogen to form the dichromate:

$$2\,CrO_4^{--} + 2\,H^+ \rightleftharpoons (2\,HCrO_4^-) \rightleftharpoons Cr_2O_7^{--} + H_2O \qquad 3.70$$

The processes described below start from dichromate or chromate, depending on the pH. In acid, the following redox reaction applies:

$$Cr_2O_7^- + 14\,H^+ + 6\,e^- \rightleftharpoons 2\,Cr^{+++} + 7\,H_2O \qquad\qquad 3.71$$

The strongly positive value of this reaction (+ 1.33V) implies that any reductant whose standard potential in acid media is more negative than this, is capable of reducing the chromate. In practice, virtually all commercially used reducing agents fall within that category. Even hydrogen peroxide, normally seen as an oxidising agent, is capable of reducing chromates in acid media [75]. This apparently environmentally friendly approach has not been used in practice since any excess peroxide would, in the following neutralisation, act as an oxidant and re-form the chromates. To avoid this, it would first need to be removed, e.g. using activated carbon. Redox reactions of this type can also be useful when effluents, after destruction of cyanides, still contain excess oxidising agents. On pooling with chromium-containing solutions after chromate reduction, a weakly alkaline solution results and precipitation can take place. Even dissolved oxygen in alkaline solution can re-oxidise trivalent chromium. This needs to be borne in mind when, after neutralising solutions containing Fe(II), air is blown to oxidise this to Fe(III).

For treatment of chromates in alkali, a general form of redox reaction can be written:

$$CrO_4^- + 4\,H_2O + 3\,e^- \rightleftharpoons Cr(OH)_3 + 5\,OH^- \qquad\qquad 3.72$$

The standard potential for this is − 0.13 V.

3.3.1   Treatment of Chromates with Sulphur Dioxide and Sulphites

Chromate reduction processes based on the use of sulphite and related compounds are widely used. Their great advantage is that the course of the reactions can be sensitively followed by measurement of the redox potential together with the fact that, in the subsequent neutralisation stage, the reducing agent does not give rise to any additional sludges. The following compounds are available to carry out this reaction:

a) *Sulphur dioxide*

   The SO_2 is introduced as a gas into the effluent and readily dissolves in water. At the temperatures usually prevailing (10 to 20°C), its solubility is 56.6 to 39.4 litres/litre, i.e. 150 and 105 gm/litre respectively. Of this, a small proportion reacts with the water to form sulphurous acid, a species in equilibrium with the sulphur dioxide:

$$SO_2 + H_2O \rightleftharpoons H_2SO_3 \qquad\qquad 3.73$$

The sulphurous acid dissociates as shown below to form the acid sulphite ion which is thought to be the actual compound involved in reduction of the chromate.

$$H_2SO_3 \rightleftharpoons H^+ + HSO_3^- \qquad\qquad 3.74$$

b) *Sodium pyrosulphate* $(Na_2S_2O_7)$ or *Sodium Bisulphite* $(NaHSO_3)$

These two compounds are often mistakenly considered as separate reagents. The commercially available salt is the pyrosulphate which, on dissolution in water, hydrolyses to form the bisulphite as shown below:

$$S_2O_5^{--} + H_2O \rightleftharpoons 2\,HSO_3^- \qquad\qquad 3.75$$

c) *Sodium sulphite* (Na_2SO_3)

Sodium sulphite may be used for reduction of chromates in acid conditions. This involves a pH-dependant equilibrium between the sulphite and the bisulphite ions:

$$SO_3^{--} + H^+ \rightleftharpoons HSO_3^- \qquad\qquad 3.76$$

Use of the sodium salt ensures improved utilisation of acid.

Since in all three of the above cases the medium is weak acid, a condition where the bisulphite ion is the preferred species, it is this ion which is normally treated as the actual reductant. Because it is a common factor in those three cases, it allows us to compare the works of different authors who have reported on the use of these reductants.

d) *Sodium Dithionite* $(Na_2S_2O_4)$

This compound is an extremely potent reducing agent which can, for example, reduce copper to its metallic state in acid solutions or react with molecular oxygen. It is used to detoxify chromate solutions when these are alkaline or pH-neutral.

In its hexavalent state, chromium requires 3 electron equivalents for the reduction of each gram-atom of Cr. In addition, the chemically bound oxygen in the chromate ion must also find a new chemical partner. This is expressed in Equations 3.71 (for acid) and 3.72 (for alkali). The reductant must be able to function as an electron donor for the necessary electron equivalents, being itself oxidised as part of the overall reaction. A pre-condition for the reaction to occur spontaneously is thus a sufficiently large difference in Standard Potentials between the two redox half-reactions at the prevailing pH. This condition is met in the case of the couple $Cr_2O_7^{--}/Cr^{+++}$ and HSO_4^-/HSO_3^-. In order to provide the 6 electron equivalents required, 3 equivalents of of bisulphite are necessary:

$$3\,HSO_3^- + 3\,H_2O \rightleftharpoons 3\,HSO_4^- + 6\,H^+ + 6\,e^- \qquad\qquad 3.77$$

Adding the two reactions 3.71 and 3.77, one obtains the overall reaction for reduction of hexavalent chromium with bisulphite in its most general form:

$$Cr_2O_7^{--} + 3\,HSO_3^- + 8\,H^+ \rightleftharpoons 2\,Cr^{+++} + 3\,HSO_4^- + 4\,H_2O \qquad 3.78$$

From this, it is seen that, as a third component of the reaction, an acid is required which is why this process is carried out in acid environment.

Klotter & Mueller [76] who studied the effect of excess bisulphite on the chromate reduction reaction, established that a pH value of not more than 2.5 must be attained if the reaction is to proceed quantitatively (see Fig. 3.13). If the necessary acid is present at the start of the reaction, it takes place rapidly and quantitatively. If, by contrast, the solution is adjusted to pH 2.5, the acid is rapidly consumed, the pH rises and the reaction rate drops away or else the process is prematurely terminated in an incomplete manner. In practice, it is common to maintain the pH at 2.5 by continuous dosing. The dependance of reaction rate on pH has been reported by Grindley [77] and Hill [78] (see Figs. 3.14 and 3.15)

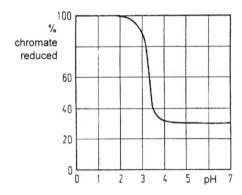

Fig. 3.13   Fraction of reacted chromate as a function of pH, reacted with excess bisulphite, after 2 min. reaction time (after Klotter & Mueller [76]).

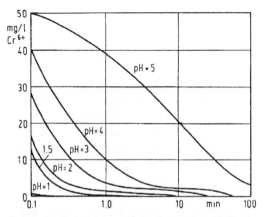

Fig. 3.14   Dependance of reaction time on pH for the reduction of chromates using sulphur dioxide (after Grindley [77]).

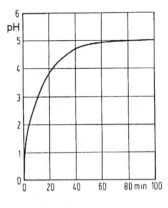

Fig. 3.15   Effect of pH on the reaction rate of chromate reduction with sulphur dioxide (after Hill [78]).

The prolongation of reaction times as chromate concentrations fall to a low value, see in Fig. 3.14 can be attributed to the very modest excess of SO_2 in these studies. In practical effluent treatment, the concentration of the species to be treated must be maintained at a lower value that that of the treatment chemical if an acceptably fast reaction rate is to be maintained throughout the process.

Several authors [76-81] have designated the chromate reduction reaction as "spontaneous" under conditions where sufficient reductant and acid are present. It can be seen from the data of Arnold [81] that reaction rate is proportional to the second power of the hydrogen ion concentration, a finding which can be equally well expressed in terms of pH dependance.

Channon [80] and Hill [78] make the point that, using SO_2 as reactant, this provides not only the reducing action but also a partial source of acid, as shown in Equations 3.73 and 3.74. Looking at the overall reaction here, it is seen (3.79) that – in contrast to Equation 3.78 – the proton requirements are almost fully satisfied.

$$Cr_2O_7^{--} + 3\,SO_2 + 2\,H^+ \rightleftarrows 2\,Cr^{+++} + 3\,SO_4^{--} + H_2O \qquad\qquad 3.79$$

The same can be seen in Fig. 2.16 which shows how, at a given molarity, SO_2 produces lower pH values than $NaHSO_3$.

As to the recommended excess concentrations of reductant, the literature produces divergent suggestions. While Klotter & Mueller [76] used 100% excess, Channon [80] used only the stoicheometrically recommended concentrations. Thanks to the use of automated dosing plants, we know that even very small excess concentrations (a few mg/litre) can cause an excursion of the measured redox potential in a negative direction. This normally signals an excess of reactant and hence the completion of reaction. A small amount of reactant is lost to atmosphere, in consequence of the (very low) vapor pressure of dissolved SO_2 as indicated in Equations 3.73 and 3.74 and this accounts for the slight smell of sulphur dioxide usually detectable in plants using bisulphite to reduce chromates.

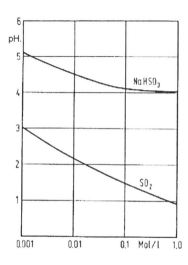

Fig. 3.16   Dependance of pH on concentration of reductant (after Channon [80])

The preceding sections have been concerned with procedures for use in acidic solutions. It is equally possible to reduce chromates in neutral to alkaline media if sufficiently strong reducing agents are used. Sodium dithionite falls into this category and while its reducing action is stronger in acid than in alkali, even in the latter it is strong enough to reduce chromates. In acid, it is itself oxidised to sulphate, yielding 6 electrons in the reaction, in alkali it is oxidised only as far as the sulphite, releasing two electron equivalents. It can also be oxidised to dithionate (S_2O_6) releasing 4 electron equivalents. The redox reaction of dithionite in alkali can be represented as:

$$S_2O_4^{--} + 4\,OH^- \rightleftharpoons 2\,SO_3^{--} + 2\,H_2O + 2\,e^- \qquad 3.80$$

and possesses a standard electrode potential of $-1.12V$. Combining the above equation with 2.72 one obtains the overall equation for reduction of chromate with sodium dithionite in alkaline medium:

$$2\,CrO_4^{--} + 3\,S_2O_4^{--} + 2\,H_2O + 2\,OH^- \rightleftharpoons 2\,Cr(OH)_3 + 6\,SO_3^{--} \qquad 3.81$$

Where it is necessary to treat larger quantities of chromate, care should be taken not to increase too much the COD (Chemical Oxygen Demand) of the solution caused by sulphite formation. It may, in such cases, be preferable to use an alternative process based on addition of Fe(II) described in Section 3.3.2. For the reduction using sodium dithionite in acid, the overall reaction is as shown below:

$$Cr_2O_7^{--} + S_2O_4^{--} + 6\,H^+ \rightleftharpoons 2\,Cr^{+++} + 2\,SO_4^{--} + 3\,H_2O \qquad 3.82$$

3.3.2   Chromate Reduction with Iron (II) Compounds

Iron (II) compounds are well-suited for reduction of chromates in both acid and alkaline conditions

The redox potential of the system Cr_2O_7/Cr^{+++} is $+1.33$ V while, in acid, that for the Fe^{+++}/Fe^{++} pair is $+0.771V$. The oxidation of Fe(II) to (III) releases only one electron equivalent so that larger amounts of reductant are needed than is the case for sulphites.

$$Fe^{++} \rightleftharpoons Fe^{+++} + e^- \qquad 3.83$$

Combining the two redox Equations 3.71 and 3.83, the overall reaction results for reduction of chromate by Fe(II) in acid media.

$$Cr_2O_7^{--} + 6\,Fe^{++} + 14\,H^+ \rightleftharpoons 2\,Cr^{+++} + 6\,Fe^{+++} + 7\,H_2O \qquad 3.84$$

This shows that, for reduction of 2 gramme-atoms of hexavalent chromium one requires not only 6 gm-atoms of iron but also 14 gm-atoms of hydrogen ions. Hence the acid requirement too, is greater than in the case of sulphites. Finally the additional volume of ferrous sludge formed in the subsequent neutralisation should be borne in mind. This particular treatment method can be attractive in plants where spent ferrous pickle liquors are produced which themselves would require treatment. Combining the two processes – chromate reduction and spent pickle liquor treatment – reduces the amount of additional chemicals required.

In acid, this reaction, as is the case with sulphite, only proceeds at an acceptable rate and to completion, if the acid consumed by the reaction is regularly replaced. For this reaction too, pH values of 2.5 or less are called for not only for the reaction itself but also to allow the use of potentiometric methods to follow it, as reported by Bahensky et al [82].

Although Fe(II) salts are unsuitable for chromate reduction in the pH range 2.5 to neutral, they function extremely well in alkaline medium as the Fe(II) hydroxides. Oehme [83] reported that the reaction proceeded best in the range pH 8.5 to 12. Equation 3.72 has a Standard Potential of $-0.13V$ whereas the value for the Fe(II)–Fe(III) pair in alkali, as shown in Equation 3.85, is $-0.59V$.

$$3 \, Fe(OH)_2 + 3 \, OH^- \rightleftharpoons 3 \, Fe(OH)_3 + 3 \, e^- \qquad 3.85$$

Combining Equations 3.72 and 3.85 gives 3.86 which still provides a potential difference to drive the reaction of 0.429V

$$CrO_4^{--} + 3 \, Fe(OH)_2 + 4 \, H_2O \rightleftharpoons Cr(OH)_3 + 3 \, Fe(OH)_3 + 2 \, OH^- \quad 3.86$$

This process offers, as Goetzelmann showed [84], significant advantages in alkaline solutions when smaller quantities of chromium are involved. Typical examples of this might be regenerated liquors from anionic ion exchangers or anodically-operating electrocleaner baths or more variously arising sources of chromium in alkaline effluent streams. This approach avoids the additional costs and environmental disadvantages where acidification is involved, with the resulting formation of large amounts of neutral salts, as is the case where the sulphite route is used. This apart, the Fe(III) hydroxide formed is, in most cases, a useful flocculating agent, especially in cases where the usually difficult-to-filter chromium hydroxide is formed. Any excess of Fe (II) can readily be precipitated as hydroxide at a pH of 9.5 or above.

3.3.3   Cathodic Reduction of Chromates

The cathodic reduction of chromates is shown in Equation 3.71. For each gm-atom of the metal, 3 electron equivalents and 7 gm-atoms of hydrogen are required implying that this reaction can only be carried out in acid medium. Ibl & Frei [85] investigated solutions at pH of 1.5 or below containing 100 and 1000 mg/litre chromate. The process operated satisfactorily down to a concentration of 0.5 mg/litre chromate with mean current efficiencies of 95% It is not clear at what point the reaction became diffusion controlled. For a starting concentration of 100 mg/litre chromate, the limiting current density fell from 0.14 A/sq.dm within 2 hours to 0.014A/sq. dm at which point the chromate concentration mentioned above was reached. The importance of vigorous agitation to maximise mass-transport needs no emphasis and, since the date when this work was completed, many new cell designs have evolved which would allow far higher current efficiencies even at the lowest chromate concentrations, this parameter being related to diffusion limiting current density.

In this work, the hydrogen overvoltage at the graphite cathodes used so large that this reaction did not, under most conditions, compete with the chromate reduction. The reverse is true at the anode, where oxygen evolution overvoltage is so low that this is overwhelmingly the favoured reaction with virtually no re-oxidation of Cr(III) even in an undivided cell. In technical plants, divided cells might be used to minimise this back-reaction. If the conditions of Ibl & Frei's work could be translated to full-scale, one could project a cell with 1.25 sq. m cathode surface being used to treat ca 1 cubic metre of effluent per day to reduce the chromate concentration by two orders of magnitude or so. Energy requirements for this would be around 1.2 kWhr per kg chromic acid.

Heger *et al* [86] used a particulate cathode to study cathodic reduction of chromate (see also Section 7.6.1), i.e one with very large surface area. With this system, they found that the ca 100% cathode current efficiency was virtually unrelated to current density of electrolyte flowrate. At lower current densities and higher chromic acid concentrations, the current efficiency was significantly less than 100%. The authors explained this by suggesting that, at low current densities, the utilisation of the depths of the cathode bed was less than the diameter of the cell containing it, whereby the outermost particles were subjected to a current density in excess of the limiting value for chromate reduction. This conclusion is not one endorsed by many theoretical treatments of such electrode systems. Had they explored larger ranges of parameters such as concentration, current density and electrolyte flowrate, they might well have discovered other features specific to the cell geometry used. These studies did, however, serve to emphasise the very high space-time yields of such three-dimensional electrochemical reactors, in this case ca. 20 gm chromic acid per hour per litre cell volume. Work by Pini [87] using a rotating disk electrode did not result in the high current efficiencies found by Ibl & Frei. In an effort to increase the chromate reduction efficiency, Pini used a sacrificial iron cathode. This, in sulphuric acid, forms the Fe(II) sulphate which will chemically reduce chromate ions as shown in Equation 3.3.2 Use of such an anode also greatly reduces the back-oxidation of trivalent chromium, allowing the use of an undivided cell. Pini went on to suggest the use of scrap iron (in a basket) as anode material.

Schulze [88] reported on work restricted to a reduction of divalent iron. After chromium plating, work was rinsed in a static tank containing Fe(II) and sulphuric acid. This served to reduce the chromate, as shown in Section 3.3.2. Using a built-in electrode system, the Fe(III) formed by reduction of the chromate was itself cathodically re-reduced to Fe(II), at the same time releasing the additional sulphuric acid bound by the Fe(III). sulphuric acid consumed in the chemical reaction (Equation 3.84) must be replaced. Use of iron anodes virtually eliminates losses due to anodic re-oxidation, as described by Pini. There remains, however, a problem of open-circuit dissolution when the cell is not operating. Lead (or alloy) anodes can also be used. These must be protected using a diaphragm, failing which Cr(III) sulphate in solution is re-oxidised to chromate at the anode surface.

The cathodic reduction of chromates [85-86] would seem to be the most environmentally friendly method of treatment, since no additional chemicals are required. There are, though, a number of scale-up difficulties and this is the reason why, as far as is known, the method has yet to be adopted in practice.

3.3.4   Chromate Treatment by Precipitation of Insoluble Chromates

The precipitation of chromate as an insoluble compound is only briefly considered here, since the method regularly features in trade literature. Usually, the precipitated species is the barium chromate. Since this is soluble in acid, the process must be carried out in neutral or weakly acid media. The simple reaction is:

$$CrO_4^{--} + Ba^{++} \rightleftharpoons BaCrO_4 \qquad\qquad 3.87$$

Because of its more attractive price, barium carbonate is the preferred source of barium ions. Formerly, barium sulphide was used but this is no longer acceptable. Because of its ease of solubility, barium chloride is also used as a precipitant. By contrast, barium carbonate can only be used in acid media, on account of its poor solubility (ca. 17 mg/litre). Carbon dioxide is evolved during the dissolution and barium chromate is precipitated during the ensuing neutralisation. The single advantage of barium chromate in this context is its very high rate of sedimentation. Its solubility is ca. 7 mg/litre, i.e. 1.4 mg/litre of the hexavalent chromium, 3.8 mg/litre of barium. The maximum permitted effluent concentrations in Germany lie below these values and for this reason, the use of barium for this purpose is prohibited there.

3.4   TREATMENT OF NITRITES

3.4.1   Nitrite Treatment By Oxidation

By far the best-known and most widely used method for treatment of nitrites is to use the same oxidation agents as were previously suggested for cyanide – sodium hypochlorite or hydrogen peroxide. Air oxidation is also possible, in the presence of a catalyst to accelerate the reaction. The reaction rate is fastest in the pH range 3 to 4, and the following equation represents it:

$$HNO_2 + H_2O \rightleftharpoons NO_3^- + 2\,e^- \qquad\qquad 3.88$$

This generalised redox reaction for nitrite oxidation has a standard potential value of + 0.94V. Sodium hypochlorite in acidic solution has a standard potential of + 1.495V, as represented by Equation 3.89:

$$HClO + H^+ + 2\ e^- \rightleftharpoons Cl^- + H_2O \qquad\qquad 3.89$$

Adding the above two reactions together, we obtain the overall reaction for hypochlorite oxidation of nitrite in acid:

$$HNO_2 + HClO \rightarrow NO_3^- + 2\ H^+ + Cl^- \qquad\qquad 3.90$$

Redox potential measurements are idea as a means for following the course of this reaction. A possible problem, in presence of organic species, is the formation of chlorinated hydrocarbons and the environmental problems with which these are associated (see Section 3.2.1.7)

Using hydrogen peroxide as an oxidant, nitrate is likewise the product formed but in this case without the dangers of forming chlorocarbons. The standard potential of hydrogen peroxide in acid is + 1.776V, its oxidation being shown below:

$$H_2O_2 + 2\ H^+ + 2\ e^- \rightleftharpoons 2\ H_2O \qquad\qquad 3.91$$

Combining Equations 3.88 and 3.91, the overall oxidation process is obtained:

$$HNO_2 + H_2O_2 \rightarrow NO_3^- + H^+ + H_2O \qquad\qquad 3.92$$

Oxidation of 1 kg of nitrite (as NO_2^-) requires, in practice 1.5 to 2 kg of 35% hydrogen peroxide [89].

Because these oxidations are carried out in weak acid media, the danger exists that, in the case of more concentrated nitrite solutions, NOX gases will be liberated. In order to minimise this, Knorre [90] has suggested that the appropriate amount of hydrogen peroxide be first added and only thereafter acidifying the solution. Once the reaction is under way, hydrogen ions are liberated (Equation 3.92) which themselves increase the pH. Where hypochlorite is used, gaseous chlorine is evolved.

If the nitrite oxidation by hydrogen peroxide is initiated at a known pH value, it can then be followed in terms of the pH change, as shown in Equation 3.92, as described by Rodenkirchen [91]. This approach is only applicable to batch reactors and the reaction is controlled using a titration controller (see Section 3.3.3.3). Detection of the process end-point, at least in the case of dilute effluents, is complicated by the buffering capacity of the solution.

Air oxidation of nitrites is also possible when a catalyst is used to speed the reaction. Activated carbon can be used for this. A technology similar to that described for cyanide oxidation, using aerated basins in which activated carbon is dispersed, may be used to oxidise the nitrite to nitrate (see Fig. 3.17) [44].

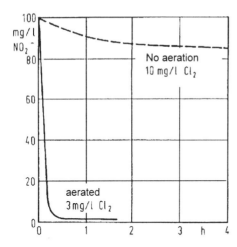

Fig. 3.17   Reaction rate of catalytic nitrite oxidation by air compared with chlorine oxidation at pH 6.5 (after Hoeke & Wittbold [44]).

3.4.2   Treatment of nitrites by reduction

A classic process in organic chemistry is the reaction of acidic amides with nitric acid to form the carboxylic acid derived from the amide and nitrogen. In the same way, the monoamide of sulphonic acid – sulphamic acid (NH_2SO_3H) [92] and urea behave similarly. The two Equations 3.93 and 3.94 should be compared:

$$HNO_2 + NH_2SO_3H \rightarrow N_2 + H_2SO_4 + H_2O \qquad\qquad 3.93$$

$$2\ HNO_2 + O = C\overset{NH_2}{\underset{NH_2}{\diagdown}} \rightarrow 2\ N_2 + CO_2 + 3\ H_2O \qquad\qquad 3.94$$

Both reactions proceed best in the pH range 3 to 4 and are characterised by evolution of gas. In the one case (Equation 3.93), the reaction takes place at ambient temperatures and is essentially complete within 5 to 10 minutes when 1.5 kg of sulphamic acid is added to 1 kg of sodium nitrite. Using urea, temperatures of 60°C are required [93]. Since the progress of these reactions cannot be followed by redox potential measurements, the use of sulphamic acid for nitrite reduction is restricted to batch type operations. If the process parameters are held constant, the end-point in this case is signalled by attainment of a constant pH value.

This treatment method is equally successful when nitroprussates are present as is often the case with effluents from heat treatment plants where hexacyanoferrates and nitrites are found. Nitroprussates are complex compounds in which one or more cyanide ion ligands in the hexacyanoferrates are replaced by nitrite ions. While they are very stable in alkali, the rapidly decompose in weak acid solutions, as are used in oxidative destruction of nitrites. As a result, further easily released cyanide is made available and to treat this, alkali must be added and further oxidant added to treat such solutions. Since nitrates too, can form prussates, this method of cyanide treatment, after acidification, results in further dissociation and free cyanide formation. If the nitrite destruction is carried out using sulphamic acid, no further prussates can form after the second alkali treatment with its coupled cyanide destruction stage [21].

In the case of the destruction of NaCN or NaCNO on the one hand or sodium nitrite on the other, in molten salts, a reduction of the nitrite to nitrogen takes place (Equations 3.66 to 3.69).

A little-known reaction which also results in nitrogen formation is the reaction of nitrite with sulphurous acid in weakly acid solution, according to the equation:

$$2 \ HNO_2 + 3 \ H_2SO_3 \ \rightarrow \ N_2 + 3 \ SO_4^{--} + 6 \ H^+ + H_2O \qquad \qquad 3.94a$$

This reaction has the advantage it can be followed and thus controlled by redox potential measurement. In addition, it allows the simultaneous treatment of both nitrite and chromate.

3.5   OXIDATION OF INORGANIC REDUCING AGENTS

By and large, it is true to say that the effluent from the metal working and metal finishing industries can contain substantial amounts of inorganic reducing agents, many of which can consume oxygen and thus contribute to the COD of such effluents. Typical of such compounds are sulphides, sulphites and Fe(II) compounds and in most cases, they are species involved in some treatment of the metals, e.g. sulphides as precipitating agents, sulphites as used for reduction of chromates and Fe(II) compounds to take up free chlorine following oxidation reactions. Fe(II) salts are also formed from pickling operations of ferrous metals and alloys in non-oxidising acids. In direct discharge of effluents to municipal sewer, these chemicals are included in the overall COD values. In the case of indirect discharge (primary treatment), the are classified as "chemicals with spontaneous oxygen demand".

Sulphides and sulphites can be oxidised by the usual oxidants such as NaOCl or hydrogen peroxide. In the former case, there is always the problem of AOX formation when organic substances are present. Hydrogen peroxide, by contrast, is environmentally friendlier, not making any contribution of its own to the total dissolved salt concentration. The oxidation of sulphides and sulphites is usually carried out in neutral to weakly alkaline media, thus avoiding the liberation of hydrogen sulphide or sulphur dioxide. The simplest means of eliminating sulphides is by addition of iron compounds (see Section 3.10.4). In practice, 1 kg of sulphite (as SO_3) requires 1.5 kg of 35% hydrogen peroxide, for 1 kg of sulphide, 3.5 to 5 kg are required [89].

In many more cases, there is the need to oxidise Fe(II) compounds present in effluent. In addition to the oxidants already mentioned, the use of aeration is widespread and Fe(II) reacts rapidly with dissolved oxygen (from air), the concentration of which is maintained by continuous gas-liquid contact. At an air pressure of

760 Torr, the amount of dissolved oxygen at 10 or 20°C is 11.3 and 9.2 mg/litre
respectively. Freshly precipitated Fe(II) hydroxide which, because of its very low
Fe(III) content, has a greenish tinge, changes colour during air oxidation to dark
green, then almost black until finally, when fully oxidised, it becomes red-brown in
colour. This oxidation is accelerated by blowing in air, being most favoured under
neutral pH conditions. In all cases, the pH value for Fe(III) hydrolysis (3.5) must be
exceeded. The reaction, promoted by the Fe(III) hydrolysis, proceeds as shown
below:

$$2\,Fe^{++} + 5\,H_2O + \frac{1}{2}\,O_2 \rightleftharpoons 2\,Fe(OH)_3 + 4\,H^+ \qquad\qquad 3.95$$

This equation explains the trend to increased acidity, in that the oxidation of each
Fe(II)$^{++}$ ion liberates 2 hydrogen ions and unless these are neutralised by addition of
alkali, the solution can become strongly acid, depending on the iron concentrations,
as seen in Fig. 3.18 [94]. A drop of 3 to 4 pH units is not uncommon. This brings
with it the danger that, where strongly basic hydroxides of other metals are present in
solution, they re-dissolve. It follows that this reaction must be carried out during a
neutralisation operation, not thereafter.

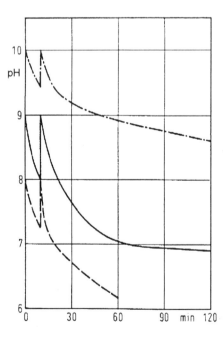

*Fig. 3.18   Drop in pH following the
aeration of iron-containing pickling
effluents at various initial pH values.
400 mg/litre Fe^{++}. After each 10 min.
pH was restored to its initial value.
Aeration was carried out for the first
30 min. in each case [94].*

Because of the modest solubility of oxygen in water, the reaction proceeds slowly,
even when being aerated, a fact which must be borne constantly in mind. Should the
effluent be neutralised without completion of oxidation, a process
typically needing several hours in a settling plant, there is the danger of the oxidation
continuing with the consequent drop in pH and the problems which this causes.

The oxidation is greatly accelerated when typical oxidising agents such as hydrogen peroxide or sodium hypochlorite are used during the neutralisation. The latter has the advantage that a redox reaction occurs even at constant pH, allowing the process to be simply monitored. The overall reaction in the pH range where iron hydrolysis occurs is as follows:

$$2\ Fe^{++} + H_2O_2 + 4\ H_2O \rightarrow 2\ Fe(OH)_3 + 4\ H^+ \qquad\qquad 3.96$$

$$2\ Fe^{++} + OCl^- + 5\ H_2O \rightarrow 2\ Fe(OH)_3 + Cl^- + 4\ H^+ \qquad\qquad 3.97$$

These reactions will also take place in strong acid media. However this results in evolution of gaseous chlorine which effectively precludes such conditions. Fortunately, the presence of Fe ions and Fe(III) hydroxide appears not to catalyse premature decomposition of hydrogen peroxide [95]. Whether hydrogen peroxide or hypochlorite are used as oxidants, a batch-type operation must be used either before or during neutralisation since in these cases too, a strong increase in acidity will otherwise be observed.

3.6   NEUTRALISATION

In the overwhelming number of cases, precipitation is the last chemical process in the overall sequence of effluent treatment. In cases where a previously treated effluent stream or a mixture of several such are at pH neutral or above, such precipitation is a happy coincidence. Even where a process is being operated on a continuous scale with constant drag-out and rinse water transfer, unexpected pH changes can occur, caused by concentrations or depletions in the active species. This effect is most marked around the neutral pH region (6 to 8) as the classical titration curve (Figs. 3.21, 3.22) shows.

Neutralisation of effluents from the metal-working or metal finishing industries is carried out on the basis that these contain dissolved metals. Otherwise expressed, it is not just that such effluents must be treated to achieve an acceptable pH in the broadly neutral range. Rather, a pH must be produced at which the solubility of the precipitated metal oxides or hydroxides is at a minimum. In most cases, such values lie one or more pH units alkali to 7.0. In summary, it is better to refer to "neutralisation-precipitation" rather than simple "neutralisation" in such cases.

3.6.1   pH Values

The hydrogen ion concentration of an aqueous solution is defined in terms of its pH, using the dissociation equation:

$$H_2O \rightleftharpoons H^+ + OH^- \qquad\qquad 3.98$$

Recent researches suggest a more realistic formulation for the hydrated proton is $H_9O_4^+$. However for simplicity, neither this nor the frequently found H_3O^+ terms will be used to replace H^+.

For dissociation of water, the law of Mass Action can be written as:

$$\frac{c_{H^+} \cdot c_{OH^-}}{c_{H_2O}} = K_{H_2O} \qquad\qquad 3.99$$

If c_{H^+} is increased by addition of acid, the c_{OH^-} term decreases proportionately with formation of undissociated water, since the term K_{H_2O} – the dissociation constant of water – remains constant at any given temperature. Since modest additions of salts or alkalis scarcely change the concentration of water – 1 kg of water contains 55.5 gm/mol water – the terms K_{H_2O} and c_{H_2O} can be included in a single constant P, the ionic product of water: This can be expressed as:

$$K_{H_2O} \cdot c_{H_2O} = c_{H^+} \cdot c_{OH^-} = P \qquad\qquad 3.100$$

The magnitude of this can be determined by electrical conductivity measurements. For pure water at 25°C, it is 1×10^{-14}. In pure water, the concentrations of H^+ and OH^- are equal, giving the expressions:

$$c_{H^+} = c_{OH^-}; \quad P = c_{H^+}^2; \quad c_{H^+} = \sqrt{P} = 10^{-7} \qquad\qquad 3.101$$

The latter H^+ ion concentration is that of neutral solutions. For convenience, to avoid having to use large negative exponents, Soerensen devised the concept of pH as a measure of hydrogen ion concentration as the negative of the logarithm (base 10) of the hydrogen ion concentration:

$$pH = -\log c_{H^+} \qquad\qquad pH = -\log a_{H^+} \qquad\qquad 3.102$$

The pH value of pure water at 25°C is 7 (in fact 6.998). At higher ionic strengths, a correction factor "f" is applied to the actual concentrations (see Section 3.7.7.1). This is usually known as the "activity coefficient". Then, in the expressions above and elsewhere, instead of using concentrations "c", one uses "activity", "a" where a=f.c. "a" is then the actual hydrogen ion concentration value as measured with a pH meter. The limits of these parameters are then $f \leq 1.0$ and $a \leq c$. Titrations give the actual "c" values since, by progressively reacting all available H^+ ions to form water, their final concentration is so low that f= 1. However even in strong acids such as HCl, the hydrogen ion concentration is virtually the same as its activity, i.e. f= approx 1. From Equations 3.100 and 3.102 we may write:

0.1   M HCl ... $a_{H^+} \approx 10^{-1}$ ... pH ≈ 1 (in practice 1.08)

0.01   M HCl ... $a_{H^+} \approx 10^{-2}$ ... pH ≈ 2 (in practice 2.02)

0.001 M HCl ... $a_{H^+} \approx 10^{-3}$ ... pH ≈ 3 (in practice 3.0) etc.

Taking into consideration Equations 3.100 and 3.102:

$$\text{H}^+ + \text{OH}^- \underset{\text{Hydrolyse}}{\overset{\text{Neutralisation}}{\rightleftharpoons}} \text{H}_2\text{O} \qquad\qquad 3.103$$

and $\text{pH} + \text{pOH} = \text{pP}$ 3.104

and recognising that the sum of Equation 3.104 must always give 14, one can use pH measurements to calculate a_{OH^-}, a term which is essential in order to calculate the solubility product of a metal hydroxide (see Section 3.7.3). Corresponding to these predictions for strong acids, we can also write similar terms for alkaline solutions:

0.1 M HCl ... $a_{OH^-} \simeq 10^{-1}$ ... $a_{H^+} \simeq 10^{-13}$ ... pH $\cong 13$

0.01 M HCl ... $a_{OH^-} \simeq 10^{-2}$ ... $a_{H^+} \simeq 10^{-12}$ ... pH $\cong 12$

0.001 M HCl ... $a_{OH^-} \simeq 10^{-3}$ ... $a_{H^+} \simeq 10^{-11}$ ... pH $\cong 11$ etc.

These predictions are confirmed by experimental measurements. In this case too, as OH⁻ concentrations increase, there is a slight decrease in activity. Since the ionic activity of H⁺ and OH⁻ ions in pure water at 25°C is 10^{-7}, this being the definition of neutrality, it follows that solutions of pH < 7 are acid, pH >7 are alkaline, being the more acid or alkaline respectively, the greater the divergence of pH from a value of 7.0. As Equation 3.102 indicates, each pH unit corresponds to a ten-fold increase or decrease in hydrogen ion concentration. Using the same definitions as for pH, the term pOH and OH⁻ activity may be derived and used, though these are only occasionally used in practice. (see Fig. 3.19).

Editor's Note:

For further details and experimental methods relating to pH, readers should consult any graduate textbook of Physical Chemistry.

In effluent treatment technology, pH values are invariably measured using glass pH electrodes, except where fluoride ions are present (since these attack glass). (see Section 3.2.3.3).

Both ionic product P and dissociation constant K_{H_2O} are temperature dependant. The degree of dissociation of water increases with increasing temperature. Pure water has a pH of 7.47 at 0°C, while above 25°C, its value is somewhat less than 7.0. This last should be noted where a danger of corrosion exists, for example with boiler feedwater. In effluent treatment too, erroneous pH measurements with up to 0.4 pH units are known. However a distinction must be made between **genuine** effects due to changes in degree of dissociation with temperature on the one hand, and **experimental artefacts** on the other where pH electrodes have not been recalibrated after changes in solution temperature. Thus, when solutions are neutralised on a batch basis, increases in temperature of up to 40°C are common. If the solution feed is slower and cooling down to 15°C occurs, other errors in pH will arise with implications for metal precipitation. The dependance of pH on temperature is seen in Fig. 3.20.

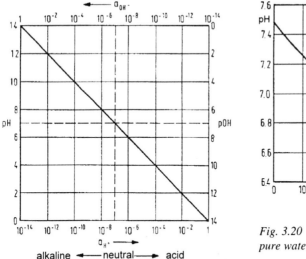

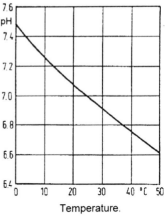

Fig. 3.20   Dependance of pH for
pure water on temperature.

Fig. 3.19   Relation between H⁺ and OH⁻ activities and pH and pOH.

3.6.2   Neutralisation Reaction

Neutralisation is a *"spontaneous"* reaction, i.e. it meeds no energy input such as increased temperature, to take place. The process releases a quantity of heat which is essentially the same for all strong acids being neutralised by a strong base – ca. 57.4 kJ/Mol. This statement is true only if the solutions are sufficiently dilute that near-complete dissociation of acids and bases is achieved. Where this is not the case, the picture is more complex for various reasons which will not be elaborated here.

This constancy of energy release, together with the fact that the neutral salts so formed are themselves essentially completely dissociated, leads to the conclusion that the acid-base neutralisation is, in fact, a simple reaction between hydrogen ions and hydroxyl ions, each of these being, more correctly speaking, hydrated.

$$H^+ + OH^- \rightleftharpoons H_2O \qquad\qquad 3.105$$

In cases where weak acids and/or weak bases take part in the reaction, the enthalpy of reaction is less than 57.4 kJ/Mol in consequence of the energy required for subsequent dissociation of H^+ and OH^- ions. Cases are also known where larger values of reaction enthalpy have been reported. These arise where other strongly exothermic reactions are simultaneously taking place, e.g. when hydrated ions are formed.

The equilibrium position for the reaction of almost completely dissociated strong acids with strong bases is almost totally to the right hand side of the equation, as written. This is best appreciated by consideration of the back reaction – hydrolysis –

having regard to the ionic product of water (see Equation 3.105). Figure 3.21 shows a set of titration curves for strong and weak acids and bases. They model, by and large, the pH changes recorded when a neutralisation operation is carried out in batchwise manner. The logarithmic dependence of pH on hydrogen ion concentration explains the very steeply rising portions of such curves (solid line) in the vicinity of the equivalence (neutral) point. Here, even small additions of strong acid or base respectively, give rise to a large pH change.

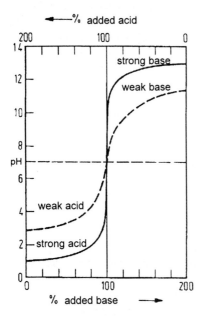

Fig. 3.21 Schematic neutralisation curves

a) strong acid – strong base (-------------);
b) weak acid – strong base (-------------);
c) strong acid – weak base (-------------);
d) weak acid – weak base (-------------)

Study of this plot also suggests that neutralisation of effluent can lead to problems associated with overshooting the desired pH value. In fact, however, there is no difference between the former operation and a laboratory pH titration, provided that certain guidelines are observed, namely:

- Use of a controller with a *"proportional-integral"* mode, i.e. the rate of dosing with acid or alkali as required is not a constant, but rather is inversely proportional to the distance from null point.

- Relating the dosing rate and/or concentration of the neutralising species to the quantity of effluent.

- Matching the mixer capacity to the size of the neutralisation tank. Since the chemical reaction takes place virtually instantaneously, the neutralisation rate is, in practice, governed by the rate at which acid and alkali are contacted with one another.

In respect of rinsewaters, the concentrations of dissolved species are usually very low. Where the neutralising solution is not too concentrated, one can envisage the position as that shown in Fig. 3.21 only with the abscissa stretched out, so that the curves become less steep. This approach, while not altering the overall neutralisation capacity (expressed in Molar terms), leads to a more easily controllable process.

Fig. 3.21 also reveals how weak acids and weak bases behave during neutralisation in that their buffering capacity in the vicinity of the neutralisation point is much greater than is the case for strong acids or alkalis. The term *"buffering"* (for a rigorous treatment see any undergraduate physical chemistry textbook) implies offering a *"resistance"* to pH change as acid or alkali are added. Such behaviour would argue for use of weak bases to neutralise strong acids, ammonium hydroxide comes to mind as a candidate. In practice, though, its use is precluded by various factors (toxicity, cost, its propensity to form complexes). A far more acceptable weak base than caustic soda or calcium hydroxide, is sodium carbonate (soda ash). Being a polyacidic base (i.e. one that is neutralised in two discrete stages), it displays relatively good buffering capacity in and around the neutrality point. sodium carbonate is the salt of a weak acid and strong base, and thus capable of hydrolysing in two stages to form sodium hydroxide. this accounts for its comparatively strong alkaline action.

$$Na_2CO_3 \quad + H_2O \rightleftharpoons NaOH + NaHCO_3 \qquad\qquad 3.106$$

$$Na_2HCO_3 \quad + H_2O \rightleftharpoons NaOH + H_2CO_2 \ (\rightleftharpoons H_2O + CO_2) \qquad 3.107$$

The first of these reactions takes place at pH values of 9.5 to 10.5, the second between pH 6 and 7. The intermediate species, sodium hydrogen carbonate (sodium bicarbonate), at concentrations of 0.1 to 1M, gives a solution of pH 8.5 to 9 as compared with a value of 11 for sodium carbonate. The reactions shown in 3.106 & 3.107 show, going from right to left, a neutralisation process. This takes place, as Fig. 3.22 suggests, in two stages. As a result of the buffering action around the neutralisation point, however, a greater than stoichiometric quantity of neutralising liquid is required. This would not be too significant, were the task simply to adjust the effluent to pH 7 (neutral). In cases where metal ions are also present in the effluent, a much higher pH is required (see Section 3.7) to ensure their most complete removal from solution by precipitation, in some cases in excess of pH 10. This calls for appreciably greater additions of the neutralising solution as Fig. 3.22 shows. An unfortunate consequence of this is an increase in the total concentration of dissolved salts in the effluent. For this reason, soda ash is normally used for neutralising only in those cases where it is advantageous for metal ion precipitation, e.g. where the product (a carbonate or basic carbonate) is more insoluble than the corresponding hydroxide.

Normal practice, then, is to use strong bases such as caustic soda or slurries such as hydrated calcium oxide (slaked lime) for the neutralisation of strong acids. Because of the poor solubility of slaked lime at pH values above 7, more than the stoicheometric amounts are almost invariably required (see Fig. 3.22).

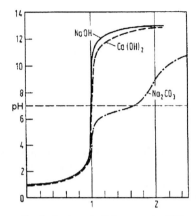

Fig. 3.22 Neutralisation curve for titration of a strong monobasic acid with sodium hydroxide, slaked lime and soda ash.

Equivalent neutralising agent addition

In cases where alkaline effluents have to be neutralised, normal practice is to use strong acids such as sulphuric or hydrochloric. In exceptional cases, carbonic acid (weak acid) is used.

Below are shown a few equations corresponding to the neutralisation of strong acids and alkalis. The resulting neutral salts formed, except where these are precipitated, are effectively completely dissociated.

$$HCl \quad + NaOH \quad \rightleftharpoons \quad NaCl \quad + H_2O \qquad\qquad 3.108$$

$$H_2SO_4 \quad + 2\,NaOH \quad \rightleftharpoons \quad Na_2SO_4 \quad + 2H_2O \qquad\qquad 3.109$$

$$2HNO_3 \quad + Ca(OH)_2 \quad \rightleftharpoons \quad Ca(NO_3)_2 \quad + 2H_2O \qquad\qquad 3.110$$

$$H_2SO_4 \quad + Ca(OH)_2 \quad \rightleftharpoons \quad CaSO_4 \quad + 2H_2O \qquad\qquad 3.111$$

It would be reasonable to expect that polybasic acids such as sulphuric acid (2) or phosphoric acid (3) exhibit characteristics similar to sodium carbonate (a di-acidic base), yielding neutralisation curves with two or three steps corresponding to the sequential neutralisation of successive H ions.

In fact no such steps are observed in the case of sulphuric acid. The first hydrogen ion is already fully dissociated in dilute solutions, as is the 2nd at pH of 2 or above. The implication is that where the two dissociation constants are so close together, no steps are observed in the neutralisation curve which follows the solid line in Fig. 3.21 for the appropriate concentration values.

By contrast, phosphoric acid, if not a weak acid, is not a strong one either. The sequential surrender of its three H ions takes place at more widely separated pH values, as indicated by the three dissociation constants:

$$K_1 = 7.5 \cdot 10^{-3}; \quad K_2 = 1.2 \cdot 10^{-7}; \quad K_3 = 1.8 \cdot 10^{-12}$$

Thus phosphoric acid can form primary, secondary and tertiary salts. In the representation below, the numbers above the equilibrium arrows represent the pH values for the transitions from one form to another, while the numbers below the chemical compounds are the pH values at which the concentrations of these species are at a maximum.

$$H_3PO_4 \underset{4.5}{\overset{2.1}{\rightleftharpoons}} H_2PO_4^- \underset{9.5}{\overset{6.9}{\rightleftharpoons}} HPO_4^{--} \underset{14.5}{\overset{11.7}{\rightleftharpoons}} PO_4^{---} \qquad 3.112$$

What is clearly seen is the spacing between the primary and secondary phosphates, the latter displaying a substantial buffering action at and around pH 6.9. These effects can manifest themselves in the neutralisation of concentrated phosphate ion containing solutions. Examples of this include plants where spent cleaning baths with high phosphate concentrations, phosphoric acid pickling baths or phosphating solutions have to be neutralised.

Other salts possessing a buffering action such as borates or acetates are usually found only in very low concentrations in effluents and their buffering action is thus negligible.

It should thus be recognised that the deionisation of rinsewater using ion-exchange, as described in Section 5.3.2 is, in effect, no more than a neutralisation process. The H^+ and OH^- species liberated in the partial exchange reactions, react together as shown in Equation 3.105 to form undissociated water.

3.6.3   Neutralising Reagents

The overwhelming majority of effluents arising from the metal finishing and metal-working industries are acidic and therefore require alkalis for their neutralisation. Those most widely used have already been mentioned above. In addition, caustic soda liquor is sometimes added to the other alkaline species mentioned. Use of other alkaline compounds such magnesia, dolomite (magnesium carbonate) and calcium carbonate is rare. If there is a trend, it is away from the use of sodium hydroxide to more and more use of slaked lime.

Among the advantages of sodium hydroxide are its very high solubility, the ease with which it can be metered into an effluent using low cost dosing equipment and controllers and simple storage reservoirs. In addition, when used, it does not contribute to the amount of neutralisation sludges formed. Against its use are its relatively high cost and the fact that metal hydroxides formed with its use, often settle less easily and are more difficult to filter than those formed using slaked slime. Where it is used, caustic soda is usually in the form of a concentrated liquor of 40 to 50% strength (at 20°C its solubility is up to 52%). At these strengths, it should be noted, the liquor is extremely viscous and thus ill-suited for dosing. In most cases, before use it is diluted to between 20 and 30%. This results, depending on the degree of dilution, in the release of appreciable heat of hydration.

This heating should not be forgotten, in terms of the temperature resistance of tanks, pipework, pumps and valves used in the plant. As the phase diagram (Fig. 3.23) shows, there exists the possibility of strong caustic liquors crystallising out if cooled below a certain temperature. This is specially true when highly concentrated caustic liquors are stored, and provision for heating in winter should be considered.

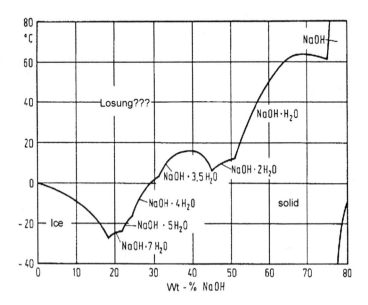

Fig. 3.23   Phase diagram for H₂O-NaOH

Where effluents have to be neutralised, e.g. for cyanide destruction, caustic liquor is almost invariably used.

In cases where basic calcium compounds are chosen for neutralisation, the preferred form is slurried hydrated calcium oxide, so called slaked lime. Its solubility in water at 20°C is only 1.7 gm/litre, this value decreasing as temperature increases. A good grade of lime will have a 95% content of $Ca(OH)_2$, though these tend to be somewhat bulky at around 0.4 kg/litre. The concentrations for slurrying are typically 5 to 15% by weight.

As soon as the slurry is brought into contact with acidic effluent, a rapid dissolution results so that neutralisation immediately follows. Above the neutrality point, there may occur, because of its low solubility, an excess of slaked lime in solution. In any case, pH values > 7 are almost invariably called for when dissolved metal ions are present, as explained above. The use of slaked lime is most cost-effective when large volumes of effluent require treatment, and a ratio of 5.7:1 or more has been suggested [96,97], for the cost of NaOH:CaO on a molar basis.

Nor is the advantage of using lime purely one of cost. Because it is divalent, the calcium ion gives better coagulation during precipitation, and forms, because it is itself poorly soluble, a metal hydroxide precipitate which is more easily sedimentable and filterable. In cases where the effluent contains fluoride ions, or when sulphates or phosphate require removal, the use of lime is virtually obligatory.

In certain cases, for example lead or cadmium, precipitation is favoured by use of soda ash. Its solubility is high, albeit strongly dependent on temperature in the range most commonly found [98]. See also Table 3.4. Because of its higher consumption, as seen in Fig. 3.22 and the resulting greater concentration of dissolved salts, soda is very rarely used alone, for neutralisation, being more often used as an additive to caustic soda. When added to slaked lime slurries, an immediate precipitation of calcium carbonate results.

One should also mention, as additional species used for neutralisation, calcium carbonate (chalk), dolomite (magnesium carbonate) and fired dolomite (magnesium oxide or magnesia). Where these are used (and this is rarely so), it is usually in the form of rods or balls, packed into towers or troughs through which the acid effluent flows, thus constituting a simple form of self-regulation. This does not, however, permit the achieving of the higher than pH 7 values required for metal ion precipitation. More recently, slurries of magnesium hydroxide have been suggested for neutralisation [98a]. It is claimed that this results in metal hydroxide sludges of higher density (more hydroxide, less water) than with conventional reagents. Easier filterability is also claimed. However it is difficult to reach pH values > 9 so that, in many such cases, caustic soda must be added to the magnesium hydroxide.

TABLE 3.4:

DEPENDENCE OF SODIUM CARBONATE SOLUBILITY ON TEMPERATURE (AFTER KOBE & SHEELY [94]).

Temperature (°C)	0	5	10	15	20	25	30
Solubility (gm/100 gm)	6.54	8.2	10.8	14.1	18.1	22.7	28.2

Fig. 3.24 allows one to deduce the amounts of acid or alkali respectively, in their different forms, required for neutralisation.

Insofar as it is necessary to neutralise alkaline effluents, normal practice is to use strong mineral acids such as sulphuric or hydrochloric. On the basis of the same "per kilogram price", sulphuric acid has twice the neutralising capacity of HCl. In spite of this, HCl is frequently used when, for example, there is a danger of exceeding specified sulphate ion concentration discharge limits.

There may be several reasons for not using concentrated sulphuric acid. These include its high viscosity, high concentration and its chemical attack on a wide range of materials. When diluted, a considerable heat of hydration is released. If such a dilution is carried out slowly with intervening periods for cooling down, lead-lined tanks can be used. Fig.3.25 shows typical temperatures reached during such a dilution process [100]. As in the case of caustic soda, care must be taken that such hot solutions do not contact materials which cannot withstand such temperatures. Similar heating effects arise when HCl is diluted. Commonly used dilution ratios are 1:3 for sulphuric and 1:1 for hydrochloric acids.

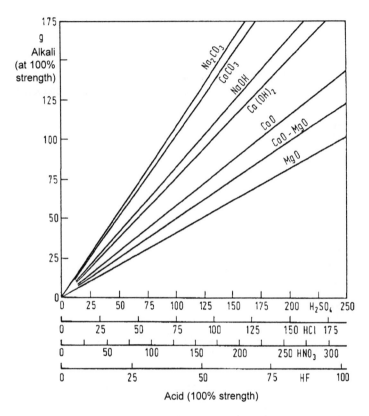

Fig. 3.24 Equivalent amounts of the most commonly used acids and bases (after von Ammon [99])

The advantage of sulphuric acid lies in its lower price. Its disadvantages, as explained above, are problems with heat of dilution and with the added, not always acceptable, sulphate ion concentration. In respect of the latter, the susceptibility of concrete to sulphate attack should not be forgotten.

Hydrochloric acid, on the other hand, must be stored and handled under more or less sealed conditions and the area around storage tanks should be subjected to forced air extraction to minimise corrosion which would otherwise occur.

The neutralisation of spent acid or alkaline solutions calls for a significant outlay for chemicals. Every possible attempt should thus be made to bring about such a neutralisation, as far as possible, by mixing alkaline effluent streams with others which are acidic. A further benefit from so doing is to minimise the total mass of dissolved salts formed.

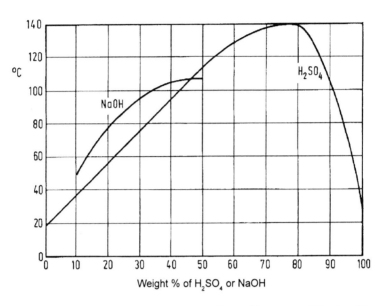

*Fig. 3.25   Temperature increases observed during dilution of sulphuric acid or
during neutralisation with NaOH. (Babcock – Water Handbook [100])*

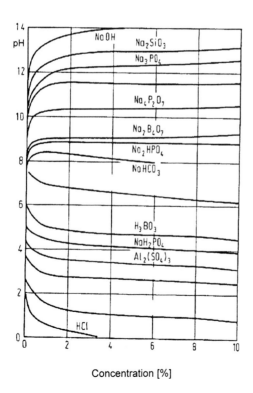

*Fig. 3.36   pH Values of some
alkalis and acids as a function of
their concentrations (after
Schlossberg [101])*

In adopting this policy, however, care must be taken to ensure that metal ions and complexing agents are not allowed to mix as a result since this would call for additional treatment stages. Effluent producers are well-advised to be aware of the acidity and alkalinity of their spent effluent concentrates in order to best plan for their mutual neutralisation, also bearing in mind the volumes in question and the life-expectancies of the various solutions. Fig. 3.26 shows pH vs. concentration for a selection of important chemicals, in the most widely used concentration ranges where they have limited life.

3.7 PRECIPITATION OF METALS WITH ALKALIS (NEUTRALISATION PRECIPITATIONS)

The aim, in the preceding section, was to consider straightforward neutralisation of acids or bases. In the latter case, there is – to the extent that metal ions are present in solution – a simultaneous precipitation of their oxides or hydroxides (and occasionally other metal salts). In the past, such precipitation was considered as an adjunct to the neutralisation reaction. Today, precisely the reverse is true. The effluents are pH adjusted to the point where precipitation of the metal ion (in whatever form) is most efficient, with simultaneous neutralisation of the acid. In cases where the effluent is already alkaline, it may be that the metal ions have already been precipitated, or that they are in some way complexed, especially as hydroxy complexes in cases where amphoteric hydroxides can form. Acid is then added to bring about the optimum pH value for precipitation. Because this optimum pH value can cover a considerable range (depending on the metal in solution), the term *"neutralisation precipitation"* is widely used. In cases where a very high pH is required for this purpose (for example with effluents containing complexed metals), the pH must in due course (after removal of solid precipitates) be adjusted back to a truly neutral; value by acid addition.

3.7.1 Precipitation Reagents and Precipitates

Precipitation of metal ions using alkalis (and in some cases acids) is accomplished with the same reagents as are used for straightforward neutralisations, as described in the previous section, i.e. caustic soda, soda ash (sodium carbonate) or slaked lime. Heavy metals then precipitate as their hydroxides or insoluble basic salts. In some cases, an insoluble salt is formed, e.g. a carbonate, chromate or phosphate. This depends on the anions present in the effluent and their concentration. The formation of basic salts and their stoichiometry is often a function of the pH at which they are precipitated. A condition for successful precipitation is that the metal compounds formed are so insoluble that the residual concentration of dissolved metal ions is small enough to fulfil the legislative requirements.

In the case of trivalent metal ions, there is a tendency for anions to be adsorbed from solution onto the precipitates formed. Divalent metal ions, by contrast, often form basic salts with a range of stoicheometries. Both these processes can occur side by side. It follows that the composition of such precipitates can vary widely.

The behaviour of divalent metals, on precipitation, especially the formation of basic salts has been described in great detail by Feitknecht [102,103]. Using Debye-Scherrer diagrams, he showed that such precipitates consisted of hydroxides, with intercalated salt ions. When caustic soda is used, in quantities less than those for a theoretical complete neutralisation, the product is usually a basic salt with a precise stoichiometric composition. As pH increases, anions are replaced by hydroxyl ions such that the basic salt becomes more and more like the hydroxide. In consequence of the innumerable variables which affect precipitation, a range of structures can result from precipitation. These are manifested in the width and number of X-ray diffraction lines. Because of the incidence of lattice defects, it is possible, especially in the case of divalent metals, for the cations in a lattice to be replaced with those of another metal, but also having the same valence. Because of the importance of this, in ensuring efficient metal removal from solution, the topic is further discussed in Section 3.7.6.

The formation of basic salts during a neutralisation reaction can often be seen during potentiometric titrations of metal salts with NaOH. Precipitation is observed, and the equivalence point recorded before the theoretical amount of alkali has been added. Such observations were reported by Britton [104,105] in the case of all metals studied by him, although he suggested that some further investigations might be desirable.

Schwab & Polydoropoulos [106] failed to detect formation of a basic salt in precipitation of nickel, chromium or iron. In the case of copper, when sulphate ions were present, they observed a pronounced step in the titration curve, suggesting that the basic salt forms when sulphate ions are present in solution. By contrast, as they reported, basic copper chloride forms only after the solution has stood for some time, while basic copper nitrate only forms in presence of high nitrate concentrations. It follows that formation of a basic metal salt is very much dependent on the metal in question and the type of anion and its concentration, in solution. In the case of nickel precipitation, these authors inclined to suggest adsorption of the anions onto the precipitate rather than formation of a basic salt.. In the case of simultaneous precipitation of copper and nickel, Schwab & Brennecke [107], using X-ray methods, found mixed crystals of variable compositions, so confirming the hypothesis of Feitknecht regarding incidence of lattice defects.

Depending on the precipitation conditions, various modifications of a metal hydroxide, its basic salts or mixtures of the two, can form. In some cases, such differences can be readily detected, for example by being more coarsely crystalline or alternatively more amorphous in character. Such differences offer a means by which settling rates or solids content of a sludge can be optimised in an effluent treatment process.

Certain metals, in contact with anions which can be found in effluents, form somewhat insoluble compounds. These include silver salts such as the chloride, phosphate or chromate. Because these are not highly insoluble,, they leave behind them silver ion concentrations in solution of 1.2, 4.9 and 9.1 mg/litre respectively. Only in the case of lead are highly insoluble compounds formed such as the chromate or the phosphate which result in a residual dissolved lead ion concentration of 0.027 and 0.095 mg/litre respectively. That said, it must be pointed out that deliberate addition of chromates or phosphates to form precipitates has no practical significance in effluent treatment technology.

In the context of phosphates, however, the work of Schulte-Schrepping & Deike [108], should be mentioned. They found that strontium and aluminium phosphates could be used as *"collector precipitating agents"* for heavy metals, during neutralisation. The benefits include a broadening of the pH range in which precipitation occurs and a lowered value of the residual metal ion in solution. For initial metal ion concentrations of 10 mg/litre with 15.7 mg/litre phosphate, there is in any case no determinable residual metal ion concentration. In presence of phosphate ions, it is almost invariably the case that highly insoluble basic phosphates are formed. The latter are, as experiences in any phosphating line effluent plant shows, readily sedimented and filtered.

Dyck & Lieser [109] also reported how, when precipitating trivalent iron at 37.2 mg/litre as its hydroxide or phosphate, there was a virtually 100% complete co-precipitation of very small concentrations (0.1 mg/litre) of copper, arsenic or lead which had been added as radio-labelled species. In presence of sodium carbonate, a similar observation was also found for cadmium (see also Section 3.7.5.4).

Only silver gave less impressive results, some 38 to 55% of this element being co-precipitated.

3.7.2   pH Values Required for Precipitation of Metals

In order to convert dissolved metal ions into an insoluble form, a sufficient quantity of anions must be added, with which they can form such insoluble species. Once a sufficient amount of anions have been added, precipitation takes place – the solubility product of the insoluble species has been exceeded. In the case of metal ion removal from effluent, it must be ensured that sufficient amounts of hydroxl ions have been added. In cases where the optimum pH value for precipitation is known, it is simply a matter of monitoring the pH and adding NaOH until the correct value is reached (a procedure which is usually automated). As might be expected, the optimum pH value will depend on the metal being precipitated and depends largely on the basicity of the metal hydroxide formed. In the same way as when acidic solutions are neutralised and water *("hydrogen hydroxide")* is formed, so metal hydroxides form during neutralisation precipitations:

$$H^+ \quad + OH^- \quad \rightleftharpoons HOH \hspace{5cm} 3.113$$

$$Me^{z+} + z\,OH^- \;\rightleftharpoons\; Me(OH)_z \tag{3.114}$$

The more basic is the metal hydroxide, the less is the reaction equilibrium shifted to the right-hand side. To obtain a quantitative precipitation, it is necessary, depending on the basicity of the hydroxide, to ensure a greater or lesser excess of hydroxyl ion, i.e. the optimum pH for precipitation should be higher, the more basic is the metal hydroxide.

The more weakly basic is the metal hydroxide, the sooner (in terms of increasing pH) does it form, using hydroxyl ions formed by dissociation of water, as shown in Equation 3.103, the overall equation for this being:

$$Me^{z+} + z\,H_2O = Me(OH)_z + z\,H^+ \tag{3.115}$$

The two Equations 3.114 and 3.115 describe the same process, with the former more accurately describing the position where pH is above 7, the latter applying to pH values below 7.

If an acid, metal ion containing solution is potentiometrically titrated, a curve such as that idealised in Fig. 3.27 is found. The hatched portion of the curve rises steeply as soon as any acid present has been neutralised. At "A", metal hydroxide begins to precipitate and this continues as further NaOH is added, as represented by the solid portion of the curve, as indicated by Equation 3.114. As soon as the plot begins to curve upwards once more and the point "B" is reached, metal hydroxide precipitation is complete. Further addition of alkali serves only to increase the pH. In theory this should also bring about a further very modest precipitation of metal hydroxide.

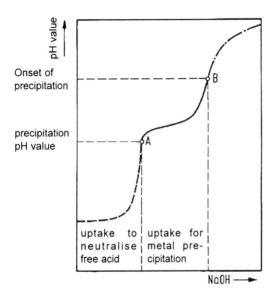

Fig. 3.27 pH Change during a neutralisation precipitation (with metal precipitation occurring).

In practice this is only significant ion exceptional cases. The point *"B"* is known as the *"precipitation pH value"* for a particular metal hydroxide. In order to obtain a quantitative precipitation, this pH value must be reached or better still, somewhat exceeded. By *"quantitative"* is understood both here and in the following sections, that the values shown in Tables A and B below are not exceeded. Depending on the metal, the "precipitation pH" ranges from 3.5 (for trivalent iron) to 10.5 (for cadmium). It should be noted that point *"B"* can only be approximately determined by titration. The author, on the basis of many experimental studies, stresses that a more accurate value can only be obtained by chemical analysis of the metal ion concentration in the filtrates.

Other information can also be derived from a titration plot. For example, there are metal hydroxides which, in presence of excess alkali, form soluble hydroxy-complexes. These include aluminium, tin, chromium and zinc. The formation of such complexes requires additional NaOH (see Section 3.9.3.1.1) and as such can often be seen in the shape of the titration curve except in cases where the complex formation takes place in the most strongly alkaline regions.

In cases where point *"B"* lies to the left-hand side of the expected theoretical equivalence point, the less than expected volume of alkali required can be explained in terms of the formation of a poorly soluble basic salt. In cases where *"B"* lies to the right hand side, it is probably because the alkali is adsorbing onto the precipitate. In cases where chalk (calcium carbonate) is used for neutralisation, there is a shift to the right-hand side because of the low solubility of this species in alkali.

TABLE A – TYPICAL EUROPEAN EFFLUENT DISCHARGE LIMITS

Parameter		Plating	Picking	Anodising	Bronzing	Hot dip galv.	Heat treatment	pcb manufacture	Battery mfr.	Vitreous enamel	Mech. eng'g.	Vibratory finish	Organic coating
		2 hourly mixed or agreed random sampling[1]											
Aluminium	mg/l[2])	3	3	3	–	–	–	–	–	2	3	3	3
Nitrogen as ammonia cmpds	mg/l	100	30	–	30	30	50	50	50	20	30	–	–
COD	mg/l	400	100	100	200	200	400	600	200	100	400	400	300
Iron	mg/l	3	3	–	3	3	–	3	3	3	3	3	3
Fluoride	mg/l	50	20	50	–	50	–	50	–	50	30	–	–
Nitrogen as nitrate	mg/l	–	5	5	5	–	5	–	–	5	5	–	–
Hydrocarbons	mg/l[2])	10	10	10	10	10	10	10	10	10	10	10	10
Phosphorus	mg/l	2	2	2	2	2	2	2	2	2	2	2	2

TABLE B – TECHNICAL ACHIEVABLE DISCHARGE LIMITS

Parameter		Plating	Picking	Anodising	Bronzing	Hot dip galv.	Heat treatment	pcb manufacture	Battery mfr.	Vitreous enamel	Mech. eng'g.	Vibratory finish	Organic coating
		2 hourly mixed or agreed random sampling[1]											
AOX	mg/l[2])	1[10])	1	1	1[10])	1	1[10])	1[10])	1	1	1[10])	1	1
Arsenic	mg/l	0.1	–	–	–	–	–	0.1	0.1	–	–	–	–
Barium	mg/l	–	–	–	–	–	2	–	–	–	–	–	–
Lead	mg/l	0.5	–	–	–	0.5	–	0.5	0.5	0.5	0.5	–	0.5
Cadmium	mg/l	0.2	–	–	–	0.1	–	–	0.2[8])	0.2	0.1	–	0.2
	kg/t[3])	0.3	–	–	–	–	–	–	1.5	–	–	–	–
Free Cl$_2$	mg/l[2])	0.5	0.5	–	0.5	–	0.5	–	–	–	0.5	–	–
Chromium	mg/l	0.5	0.5	0.5	0.5	–	–	0.5	–	0.5	0.5	0.5	0.5
Chromium (VI)	mg/l	0.1	0.1	0.1	0.1	–	–	0.1	–	0.1	0.1	–	0.1
vol. hydrocarbons[6]	mg/l[2])	0.1	0.1	0.1	0.1	0.1	0.1	0.1	0.1	0.1	0.1	0.1	0.1
Cobalt	mg/l	–	–	–	–	–	–	–	–	1	–	–	–
Cyanide (easily released)	mg/l	0.2	–	–	–	–	1	0.2	–	–	0.2	–	–
Fish toxics as dilution factor G$_F$[9]		6	4	2	6	6	6	6	6	4	6	6	6
Copper	mg/l	0.5	0.5	–	–	–	–	0.5	0.5	0.5	0.5	0.5	0.5
Nickel[7]	mg/l	0.5	0.5	–	0.5	–	–	0.5	0.5	0.5	0.5	0.5	0.5
Mercury	mg/l	–	–	–	–	–	–	–	0.05	–	–	–	–
	kg/t[5])	–	–	–	–	–	–	–	0.03	–	–	–	–
Selenium	mg/l	–	–	–	–	–	–	–	–	1	–	–	–
Silver	mg/l	0.1	–	–	–	–	–	0.1	0.1	–	–	–	–
Sulphides	mg/l	1	1	–	1	–	–	1	1	1	–	–	–
Tin	mg/l	2	–	2	–	2	–	2	–	–	–	–	–
Zinc	mg/l	2	2	2	–	2	–	–	2	2	2	2	2

[1]) If no value shown, the species not expected to be present from this source.
[2]) Random samples.
[3]) For batch processing all samples are treated as random.
[4]) For metallising glass, requirements for Cu, Ni and fish toxics only apply. For the latter, dilution factor $G_{2F} = 2$.
[5]) Discharge amounts depend on amounts of Cd or Hg used.
[6]) Sum of trichloroethylene, tetrachloroethylene, 1,1,1.
[7]) For electroless Ni 1 mg/litre.
[8]) For primary battery mfr. 0.1 mg/litre.
[9]) Special local legislation in most cases.
[10]) Levels applied as from 1992.

Finally, titration curves can be used to reveal differences arising from the use of different precipitation reagents. Some metals, when treated with sodium carbonate, form basic carbonates or even, as with cadmium, virtually pure carbonates. This leads to a precipitation pH value more extreme than when NaOH or chalk are used. It has been shown both in practice and in laboratory trials that, after neutralisation precipitations, there is a certain pH recovery, mostly in the first two hours. The following explanations for this have been given:

- initially formed basic salts slowly transform, in the higher pH conditions found at the end of the neutralisation, to hydroxides, consuming available OH$^-$ ions in the process.

- the neutralising alkali is adsorbed at the active surface of the hydroxides.

- the metals are oxidised to a higher valence state, releasing H$^+$ ions as a result (see Section 3.5 – oxidation of Fe(II)).

The transformation of initially formed basic salts into their hydroxides is difficult to follow. A pH retreat from 9 to 8 would correspond, in theory, to ca. 0.17 mg/litre OH$^-$. This amount would be capable of reacting with approx. 0.3 mg/litre of metal ion (assumed the metal is 2-valent, with atomic weight of 60 to 65). In the case of a pH retreat from 8 to 7, the corresponding figure would be only 10% of this.

The pH retreat is specially large when trivalent metal ions are precipitated. In cases where a basic salt is formed, the precipitation pH is reached long before the pH neutral point of 7 and much of the added hydroxyl ions will have been taken up in metal hydroxide formation.

These trivalent metal hydroxides are also among the most bulky and thus offer an exceptionally large surface for adsorption. Much use is made of this in effluent treatment technology and this adsorption of neutralising chemicals on the highly active surface of freshly precipitated metal hydroxides is often put to good use. For further comments see Lutter [110] but also von Kruzenstern et al [111] who reported very large pH retreats after the neutralisation precipitation of chromium hydroxide. This is in agreement with the author's own unpublished findings where adsorption effects on chromic hydroxide were found to be larger than for any other metal investigated. The above theories of adsorption behaviour can be tested using aged precipitates (sludges). These have a reduced surface energy which leads to the desorption of some of the neutralising alkali which is then found in the various types of bound water associated with the precipitate (capillary-bound etc). As a result, the pH of such sludges is often considerably higher than the value at which they were originally precipitated.

3.7.3   Solubility Product

The solubility of a poorly soluble species can be expressed using the law of mass action. The ratio of concentration of ionic species in solution to the concentration of undissolved species gives a (temperature dependant) constant – the Dissociation Constant K_D for the substance in question. Concentrations are expressed in Mol/litre. The dissolved portion of the metal hydroxide dissociates according to Equation 3.116.

$$Me(OH)_z \rightleftharpoons Me^{z+} + z\,OH^-$$ 3.116

The dissociation constant K_D is then defined, after equilibrium has been established, as follows:

$$\frac{c_{Me^{z+}} \cdot c_{OH^-}^z}{c_{Me(OH)_z}} = K_D$$ 3.117

In cases where the solution is saturated, i.e. in contact with undissolved material, the concentration of undissolved $Me(OH)_z$ is constant at any given temperature and is related, in the same manner as was shown for the ionic product of water, with the dissociation constant. This results in a new constant being defined, the solubility product L: Equation 3.118 expresses this:

$$c_{Me^{z+}} \cdot c_{OH^-}^z = K_D \cdot c_{Me(OH)_z} = L_H$$ 3.118

where L_H is the solubility product of the metal hydroxide. The less soluble a substance, the smaller is its solubility product. Strictly speaking, Equation 3.118 is only applicable in ideal, that is to say, very dilute solutions. As the concentration increases, electrostatic inter-ionic forces become significant with the result that the **effective** concentration of species is less than the **true** concentration, as one might determine it analytically. This thereby increases the solubility product and the Mass Action equation ceases to hold. All these problems can be solved by correcting the true concentrations "*c*" of species in solution using *"activity coefficients"*. In an ideal, i.e. very dilute solution, these approximate to 1. As the solution becomes more concentrated, their value falls to perhaps 0.3. In certain exceptional cases, activity coefficient greater than 1 are known. The product of concentration and *"activity coefficient"* is known as *"activity"* and can be considered as a corrected concentration – with which it shares the same dimensions. Equation 3.119 embodies a re-definition of solubility product using these concepts.

$$(f_{Me^{z+}} \cdot c_{Me^{z+}}) \cdot (f_{OH^-} \cdot c_{OH^-})^z = L_H$$ 3.119

Equation 2.120 is an alternative format in which activities are explicitly expressed.

$$a_{Me^{z+}} \cdot a_{OH^-}^z = L_H$$ 3.120

a_{H+} can also be determined by pH measurement since the strict definition of pH is based not on hydrogen ion concentrations, but rather their activities. a_{OH-} can be derived as P/a_{H^+}, as shown in Equation 3.100. Rather than being expressed individually, a mean activity coefficient can be be defined for ion pairs (anion and cation), namely f_m. For a metal hydroxide, $Me(OH)_z$, one then has the expression:

$$f_m = \sqrt[1+z]{f_{Me^{z+}} \cdot f_{OH^-}^z}$$ 3.121

This allows Equation 3.120 to be re-written as:

$$f_m \cdot c_{Me^{z+}} \cdot c_{OH^-}^z = L_H$$ 3.122

Typical values for mean activity coefficients can be found in a number of Physical Chemistry reference publications. Using one of the Equations 3.119, 3.120, 3.122 with such data, the solubility of a species at a given temperature can be ascertained. If then, in a solution already saturated with $Me(OH)_z$, the OH^- concentration is increased by addition of alkali then, because L_H is a constant, there are two possible outcomes:

- because a_{OH^-} increases, $a_{Me^{z+}}$ must decrease, i.e. more $Me(OH)_z$ will precipitate out. In kinetic terms, increased concentration of reactants leads to greater rate of their collision and more rapid formation of precipitant.

- C_{OH^-} increases, as does its activity. In consequence the activity coefficients become smaller.

Both the above actions can take place simultaneously. When effluent concentrations are low, further addition of alkali produces little additional hydroxide precipitation. However a significant change of activity coefficient will occur when species added or already present in solution, do not share a common ion with the substance being precipitated and are thus not involved in any additional precipitation. Such ions do, however, increase the total ionic strength and can affect the precipitating species. The energy required for transfer of ions from the solid (precipitate) to the dissolved state is reduced in this case with the result that the precipitate will more readily re-dissolve. In general, it can be said that:

Increased ionic strength (from non-participating ions) will always increase the solubility of a poorly soluble species. This can be deduced from Equation 3.119. Insofar as the activity coefficients of all ions present in solution are an indicator of the magnitude of electrostatic effects, it is clear that addition of further ionic species, participating or not, will reduce the activity coefficients. If it is desired to maintain a constant value of L_H, then the values of c_{Me2+} and c_{OH^-} must be increased.

On the basis of Debye-Hueckel Scherrer theory, activity coefficients can be predicted from the equation:

$$\log f_{(\pm)} = -A \cdot z_{(\pm)}^2 \cdot \sqrt{J}$$ 3.123

which assumes the ions are infinitely small, with an ionic concentration J. A is a lumped constant which assumes the medium remains the same (in this case, water); $z_{(\pm)}$ is the mean ionic charge and $f_{(\pm)}$ is their mean activity coefficient. J is the total ionic concentration of all non-participating ions present, as defined by Equation 3.124.

$$J = c_1 z_1^2 + c_2 z_2^2 + c_3 z_3^2 \ldots + c_n z_n^2 = \Sigma\, c_i z_i^2 \qquad\qquad 3.124$$

While the foregoing treatment holds for prediction of solubility product in ideal solutions, other factors have to be taken into consideration for more concentrated solutions. See for example the work of Daester & Jola [112]. This shows, as already mentioned, how divalent metals can be precipitated as basic salts. The solubility of such species which incorporate anions on a stoichiometric basis, can be described in terms of the solubility product concept, especially in the initial stages of precipitation. Because there are so many different basic salts, a generalised equation to cover these would be too cumbersome. An idealised solubility product is therefore used, based on the commonly-found composition $MeSO_4 \cdot xMe(OH)_2$. If, for this type of compound, a solubility product for the hydroxide were used, the outcome would be a dependence of solubility on pH. With increasing pH, there would be a trend to high values of solubility product. If, on the other hand, the solubility product is based on the stoicheometry of the basic salt, then – as Feitknecht [102] has shown – an expression with high negative coefficients will result:

$$a_{Me^{z+}}^{(x+1)} \cdot a_{OH^-}^{2x} \cdot a_{SO_4^{2+}} = L_B \qquad\qquad 3.125$$

where L_B is the solubility product of the basic salt.

Where L_B is known, the solubility product for the hydroxide can be determined in the pH range where the basic salt forms. This is done by combining the expressions for L_B and L_H and eliminating the term $a_{(OH^-)}$, as shown in the example below:

$$L_H = \sqrt{\frac{L_B}{a_{Me^{2+}} \cdot a_{SO_4^{2-}}}} \qquad\qquad 3.126$$

In the case of the basic zinc sulphate, $ZnSO_4 \cdot 3Zn(OH)_2$, studies by the above-named author, working in the pH range 5.8 to 6.5 gave values for L_H from $1 \cdot 1^{-18}$ to $4. 10^{-18}$ (mean = 2.5×10^{-18}), with Equation 3.126 giving a value of 1.6×10^{-17}. In terms of the very small absolute values, the differences between measured and predicted values is unimportant. In any case, it should be recognised that solubility product values published by different authors often differ by this order of magnitude. At higher pH values, e.g. 7.5 to 9, in the zinc solution referred to above, there is a predominance of hydroxide and under such conditions, one can reliably use the solubility product of the metal hydroxide without any too great errors arising.

3.7.4   Separation of Precipitates from the Aqueous Phase

In order to remove dissolved metals from effluent, two consecutive steps are required, namely precipitation and a separation of the liquid and solid phases. In some cases, a third step is interposed, namely flocculation.

In cases where precipitation results in formation of very finely divided particles, perhaps even in colloidal form, the effective separation of solid from liquid requires flocculation, which is a physico-chemical process. Once macroscopic particles are formed, whether by precipitation or only after subsequent flocculation of a precipitate, then the phase separation can be brought about by a mechanical action (sedimentation and/or filtration).

This sequence of processes is described below, using a metal hydroxide as model. The same principles apply equally to other insoluble compounds such as metal sulphides or calcium compounds.

3.7.4.1   Precipitation

The rudiments of precipitation have been described in preceding sections. In order to bring about effective precipitation, an excess of the precipitating reagent should be present in solution. Initially, crystal nuclei are formed and these then grow to form larger particles until they are large enough to be filterable or sedimentable. Once this process is under way, conditions are reached in solution which reflect the solubility product equations shown in Section 3.7.3.

Of major importance in formation of the crystallisation nuclei is the so-called "nucleation rate". The greater the excess of precipitating agent, the larger this is. Under constant temperature conditions (which can be assumed for treatment of dilute effluent), the nucleation rate is proportional to the degree of insolubility of the precipitating species and the reactor technology being used.

Onto the crystallisation nuclei, ions of the precipitating species are attracted to form a more or less ordered crystal lattice. The rate of this step, the "crystallisation rate" depends mainly on the rate of diffusion of ions through the diffusion layer. This in turn is a function of the concentration gradient. Between the diffusion step and the final incorporation into the crystal lattice, there is an adsorption stage.

The properties of a precipitated substance are highly dependent on the rate of nucleation and crystal growth. The former process is a chemical one and so more rapid than the diffusion-controlled (i.e. physical) process involved in crystal growth. Where there is a large excess of precipitating agent, the enhanced nucleation rate results in formation of more, though smaller nuclei. More particles thus form, though the growth process comes to a halt when the concentration gradient across the diffusion layer surrounding these particles falls to zero. The outcome is a finely divided precipitate which is not so easy to filter or sediment.

Where the precipitating reagent is not allowed to be present in great excess, fewer nuclei form, but these then grow to form larger particles than under the previously described conditions. Note however that conditions of excess precipitating reagent concentration arise more rapidly in the case where a highly insoluble compound is formed than in those instances where the precipitate is less insoluble. This leads to the rule-of-thumb that the more insoluble a species, the more finely divided is its precipitate. This is the very opposite of what is desirable in effluent treatment, such finely divided precipitates are not only difficult to sediment or filter but also, the dewatering of their sludges is likewise harder. In addition to seeking the coarsest possible particles, it is further highly advantageous if the range of their particle distribution sizes is not too large. It is the proportion of most finely-divided matter which determines the rate of sedimentation and/or filtration and thus their rates and indeed the economics of the phase separation process.

Junghans [113] who has studied these effects in detail, reports how precipitate particle size and its distribution are determined by several factors in the precipitation process. Use of appropriate technologies can beneficially affect this. Thus, to ensure the greatest uniformity of particle size, it is essential to maintain constant precipitation conditions, e.g. by avoiding swings in concentration or pH and by maintaining constant temperature and degree of agitation or stirring. It is clear that continuous processing allows this more easily than batchwise reactors. In the former case, automated pH sensing and dosing allows very tight pH control. The metal ions in the effluent feedwater, even if their concentration is slightly in excess of the instantaneous alkali concentration, are thus rapidly contacted with fresh alkali while avoiding conditions of high alkali excess. The result is a uniform and medium-to-coarse precipitate particle size.

In batch reactors, the concentrations of the precipitating agents are initially in considerable excess, at the end of the process in very small excess. Even during continuous dosing, there can arise locally very high excess alkali concentrations. While precipitates with moderately-sized particles can be obtained, their size distribution spectrum is broad. In practice, this need not cause problems especially if a post-precipitation standing time *(see below)* is arranged and a suitable filter installed to process the contents of the reactor tank.

The formation of the solid precipitate described above does not signal the completion of the process. Finely divided particles have a greater surface energy than those which are coarser. This implies that their re-dissolution rate is also higher. Thus, where fine and coarse particles co-exist in solution, the former, by virtue of their faster re-dissolution, will form ions which then adsorb onto larger particles causing further growth of these. The nett outcome is that, on standing, there is a shift in particle size distribution towards the larger sizes which is most desirable for efficient phase separation.

The precipitate particles will, according to conditions of their formation and their chemical composition, incorporate numerous defect sites. In the case of metal hydroxides and basic metal salts, a high degree of defectiveness is normally found, sometimes there are virtually no lattice structures. This can result in ions migrating to locations in the lattice which, though still defective, are less so than other regions on the surface. The result (and also the driving force for this) is a more ordered and lower surface energy structure.

These two processes, known respectively as external and internal recrystallisation, underlie the so-called *"ageing"* of precipitates. The resulting decrease in surface area and increased ordering of the structure serve to decrease the surface energy of the solid and hence its solubility. Recrystallisation, especially the internal mechanism, often leads to incorporation of foreign ions at defect sites. This process, which has great significance in effluent treatment, is further examined in Section 3.7.6. Recrystallisation processes are sometimes so pronounced that they can be seen with the naked eye, as for example in the case of zinc precipitation.

The following equations are examples of precipitation of a metal as hydroxide or its basic salts.

$$NiSO_4 \quad + 2\ NaOH \quad \rightleftharpoons \quad Ni(OH)_2 \quad + Na_2SO_4 \qquad\qquad 3.127$$

$$2\ FeCl_3 \quad + 3\ Ca(OH)_2 \rightleftharpoons 2\ Fe(OH)_3 \ + 3\ CaCl_2 \qquad\qquad 3.128$$

$$4\ CuSO_4 \quad + 6\ NaOH \quad \rightleftharpoons \quad CuSO_4 \cdot 3\ Cu(OH)_2 + 2\ Na_2SO_4 \quad 3.129$$

$$4\ Cu(NO_3)_2 \ + 3\ Na_2CO_3 + 2\ H_2O$$
$$\rightleftharpoons 2\ CuCO_3 \cdot Cu(OH)_2 + 6\ NaNO_3 + 2\ NaHCO_3 \qquad 3.130$$

pH affects not only the precipitation of the metal from solution but also the physical and chemical properties of the precipitate. The following example demonstrates a transformation of the initially precipitated basic salt into the simple hydroxide, during a neutralisation precipitation:

$$ZnSO_4 \cdot 3\ Zn(OH)_2 + 2\ OH^- \rightleftharpoons 4\ Zn(OH)_2 + SO_4^{--} \qquad\qquad 3.131$$

When the total alkali is added rapidly, the hydroxide is also rapidly formed. Because of the greater rate of nuclei formation, the precipitate is then very finely-divided.

3.7.4.2   Flocculation

The process by which finely-dispersed down to colloidally-sized materials are converted by a physico-chemical method into a form lending itself to sedimentation, is known as *"Flocculation"*. This term embraces two different approaches [114]:

- Coagulation. The electrostatic charges stabilising the finely-divided or colloidal state are neutralised by addition of electrolyte.

- Flocculation. Increase in particle size by binding a large number of the smallest particles together, using long-chain organic molecules.

Finely divided or colloidal particulates are stabilised by their charged state. Like charges repel one another and prevent agglomeration. Coagulation works by neutralising these charges which reside in the diffuse double-layer. This consists of an inner layer adsorbed directly onto the particle surface and an adjacent, less well-defined cluster of charge carriers. The adsorbed layer gives rise to a phase boundary potential, the so-called zeta potential. When this charge is neutralised by adsorption of oppositely charged species, the zeta potential breaks down with formation of smaller or larger-sized flocs.

The effectiveness of the breakdown of the zeta potential by addition of oppositely charged electrolyte is approximately proportional to the exponent of the charge on the counter-ion. This is known as the Schulze-Hardy Rule. In cases where the added electrolyte consists of readily-hydrolysable metal salts such as iron (III) chloride or aluminium sulphate, as is often the case in effluent treatment, it is not only a coagulation but also an adsorption action which takes place. Such a combined action is known as *"adsorption coagulation"*. By adding excess of such metal salts, it often happens that the finer agglomerates are mechanically incorporated into the coarser Fe(III) hydroxide or the aluminium hydroxide and this capture process continues during the sedimentation.

In practice, because it can often be readily seen with the naked eye, coagulation using a hydrolysable metal salt is often known as *"flocculation"*, the metal salts being known as *"flocculating agents"*.

In the precipitation of lead or tin, especially when these are present in low concentrations, also with other metals at the very lowest concentrations, it is usually the case that their precipitates are so fine that sedimentation takes too long to be a useful process while filtration is likewise extremely difficult. In such cases, the use of flocculating agents is mandatory. The precipitates then often adopt the morphology of the flocculating agent (see Section 3.7.5.5.4).

In the treatment of metal ion containing effluents, the term *"flocculation"* is more often used than *"coagulation"*, not least because it is often a question of finely-divided rather than colloidal materials. Flocculation is a process often associated with *"Flocking"* or coagulation but which can be used alone, especially when the reagents used are themselves ionic and thus lead to coagulation.

Most widely used as flocculating agents are organic macromolecules with long-chain structures. They should be water-soluble and capable of adsorbing onto colloidal or finely-divided particles whether by electrostatic or chemical means. It is important that the macromolecule attaches itself to the particle either at one end or along its length, with the other end reaching as far as possible into solution. This can then attach itself in the same way to another particle. By these means, larger agglomerates are formed, linked together by these thread-like molecules. Their larger size promotes faster sedimentation and greater ease of filtration.

Substances which promote flocculation in this way are known as *"flocculating agents"*. In the past, various natural products such as size, alginates, carob seed extract or extracts from aromatic woods were all used. With the exception of one or two chemically treated compounds such as carboxymethylcellulose, none of these are now used. Instead, modern flocculating agents are synthetic macromolecules with much higher efficacy whose properties can, according to the requirements in question, be tailored on demand. They are mainly long chain aliphatics which retain unsaturated bonds after polymerisation with molecular weight in excess of 1 million. The longer the molecule, the better is its flocculating action and the faster the settling rate of the agglomerated species. A limit to this is, however, imposed by the solubility of such species in water. Beyond a given degree of polymerisation, such substances become solid. In speaking of such synthetic flocculating agents, it is customary to distinguish between:

- non-ionic polymers, e.g. polyacrylamide, polyvinyl alcohol, polyethylene oxide

- anionic polymers, e.g. polyacrylate, polymethylacrylate

- cationic polymers, e.g. polyethyleneimine, polymeric compounds with quaternary ammonium groups

- copolymers of ionic and non-ionic polymers.

Starting products, their molecular structures and some of the most important radical groupings for some of these substances are shown in Fig. 3.28. Their molecular weight can exceed 10,000,000. A macromolecule made up of 100,000 monomer units will have the same number of reactive groupings and is ideally, in its extended state, about 25 um long. The properties and use of these flocculating agents have been described in numerous publications. To mention just a few, the names Akyel & Neven [115], Reuter [116, 119], Hanboldt & Vogel [117] and Jola [118] come to mind. In addition, numerous technical publications are produced by the manufacturers, e.g. [120,121].

The flocclation process is, like other steps in the overall metal precipitation sequence, dependant on a number of variables, e.g:

- the type, concentration & surface area of the precipitate to be treated

- the type, concentration and age of the flocculating agent

- the presence of electrolytes and their concentration

- mixing conditions (mixing time, intensity, shear force).

Some of the most important factors arising from these are elaborated below. For metal ion containing effluents such as those typically found in metal finishing, the concentration of flocculants typically required is around 5 to 10 gm/cubic metre. As in all homogeneous reactions, it is important that the flocculant be uniformly distributed through the bulk of solution in the shortest possible time. The polymer molecules are immediately attracted to the precipitate particles where the form the larger flocs.

Monomer Unit	Polymer Structure
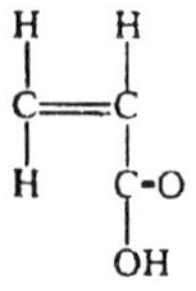 acrylic acid	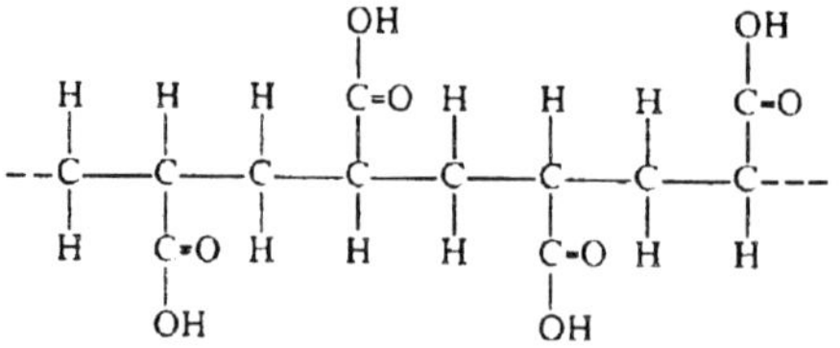 polyacrylic acid
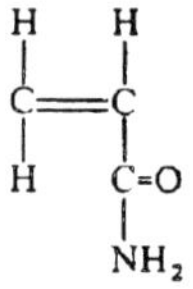 acrylamide	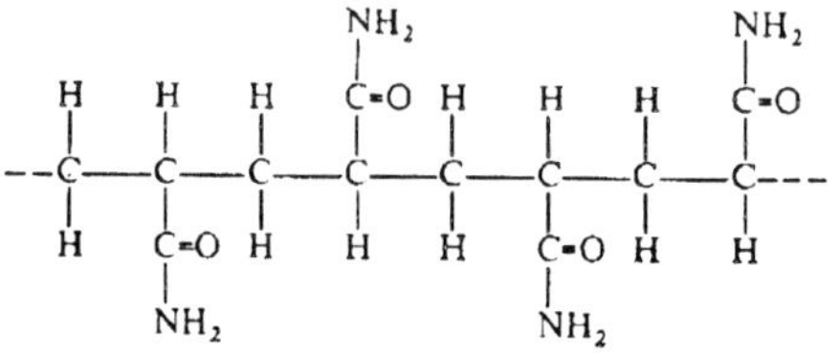 polyacrylamide
acrylic acid + acrylamide + acrylamide	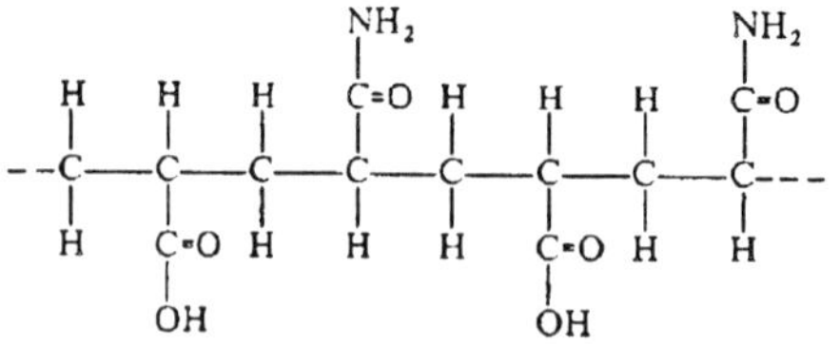 copolymer of acrylic acid

Fig. 3.28 Structure of some synthetic flocculating agents

It should, however, be noted that too strong an agitation action, whether by pumping or raises the risk that the macroflocs are broken down again. Larger particles, by crossing zones of differing liquid velocity, are subjected to shearing forces and thus fractured. This said, the addition of flocculating agents should not be made into a zone where there is laminar flow, since this is not conducive to a uniform distribution of the macromolecules through solution on account of the speed with which they attach themselves to whatever solid they first encounter. Where this happens, their effectiveness is reduced by 20 to 50% [122].

Thus, in order to maximise the benefits from flocculating agents in achieving the optimum particle size, the best results are achieved under conditions of moderate turbulence commensurate with the best contacting time in the reactor. This can, except in very few cases, only be determined by trial and error. Too long a mixing and contacting time will result in break-up of the flocs. The stability of a floc depends on the flocculating agent used. Thus flocs formed with polyacrylamides are much more resistant to mechanical breakdown than those formed using polyethylene oxide [122]. Flocs formed using polyacrylamides can, after breaking up, re-form. When polyethylene oxide is used, any breakdown is irreversible.

To achieve maximum mixing but without mechanical damage, rotary mixers operating at low speed with low circumferential velocities are recommended. Use of mixing channels and vessels with tangential feed has also been found useful [122]. It is also best if the flocculating agent is dosed at more than one point, its solution being made as dilute as possible.

The nature of the forces bonding macromolecules to a colloid or particulate can vary. Where the flocculating agents are ionic, electrostatic forces attract them to oppositely charged sites on the solid surface to form a species analogous to a weakly dissociated salt. The resulting charge neutralisation works to destabilise the colloid where coalescence is prevented by repulsion of like charges on the particles. In the case of non-ionic flocculating agents, bonding can take place to atoms containing unpaired electrons, e.g. nitrogen or oxygen atoms, such bonds being covalent or hydrogen bonds. Then too, chemisorptive bonding is possible for the macromolecules on the surface of the solid species. Nor is it necessary to postulate only one of these bonding processes being in operation at a given time. Where chemical processes are involved in the bonding mechanism, the concentration of macromolecule in solution and the number of active sites it contains are both important in determining the likliehood of a successful collision between macromolecule and suspended solid. In the case of chemisorptive processes, an adsorption isotherm will govern the relationship between concentration of adsorbed species and that in solution.

Where a long-chain macromolecule is attached at each end to a suspended solid particle, a three-dimensional structure begins to form which ultimately results in formation of larger flocs. Such structures can, in turn, take up by adsorption or interstitial inclusions, the finer suspended solids. SEM studies of the flocculation of kaolin suspensions [120] shows the process does not occur by building up of individual molecules but rather from aggregates which are estimated to contain several hundred molecules.

Whether ionic or non-ionic flocculating agents are used, the pH of the effluent is of great importance. In acid solutions, polyacrylates form the less dissociated polyacrylic acids. The molecules tend to lose their linear configuration, which is encouraged by charge repulsion of neighbouring carboxylate groups and tend instead to coil up. Insofar as they are then less effective as flocculating agents, a suspension may remain stable. Most effluent arising from the metal-working industry are, however, neutral or weakly alkaline. The effect of a given concentration of flocculating agent will vary depending on the type used. While formulations based on polyacrylate, when added in small concentrations, give higher sedimentation rates than would be the case for those based on polyacrylamides, a further dosing brings no added benefits. With polyacrylamides, by contrast, the flocculation and sedimentation performance improves as more is added up to the point of so-called "total flocculation".

The greater the mass of suspended material, the more flocculating agent is needed, though not necessarily proportionately so. Thus for a doubled amount of suspended material, several times more flocculating agent is needed to obtain the same sedimentation rate. The same is true when, for a given mass in suspension, its mean particle size is smaller or its surface area in some way increased. [120], the latter being the ultimate explanation in all cases. Fig. 3.29 shows some of the most important parameters.

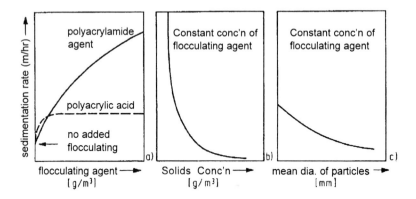

*Fig. 3.29   Effect of some variables on sedimentation rate,
using added flocculating agent [120].*

In cases where very little matter is held in suspension, the flocculating agents cease to be effective since the collision probability of macromolecules and suspended solids is then low. This can be overcome by deliberate addition of solids such as the readily hydrolysable salts of Fe(III) or aluminium.

Because of the broadly linear structure of the macromolecules in solution, solutions containing these can be quite viscous. A stock solution of the flocculating agents will be 0.5 to 1 wt% and this has a shelf-life of several weeks. This is further diluted, to 0.1 – 0.5 wt%, before the actual dosing. The rotation speed of the paddle mixers should not exceed 50 r.p.m and the mixing operation should not last more than one hour. Thus, when this was extended to 5 hours, the efficiency of sedimentation fell by 50%. Very dilute aqueous solutions of macromolecules have a short life and stock solutions should not be diluted to the final concentrations until necessaary.

Under most conditions, metal hydroxides precipitate readily and most of such precipitates readily sediment. A small residue of the most finely-divided particles remain in suspension or settle only very slowly (see Section 3.7.5.8.4). By and large, it is to treat this small residue that flocculating agents are added. The accelerating action of flocculating agents is most markedly visible in the first three minutes following addition but continues, if less markedly so, for up to 10 minutes after addition. This is shown in Fig 3.30 where precipitation with and without addition of flocculating agent is seen. Experience has shown that metal hydroxides respond best to additions of polyacrylamides containing a certain proportion of carboxylate groupings formed by hydrolysis or alternatively, copolymers formed from acrylamide and acrylic acid, i.e. anionic products. Next best after these, is polyacrylamide. Metal hydroxides usually have positively charged species on their surface such as $[Ni(OH)]^+$;

[Cu(OH)]⁺; [Cr(OH)]⁺⁺ or homologues of these [118]. The volume of precipitate formed when flocculating agents are used, is always larger than in cases where it is not used. This can be seen in Fig. 3.30. In the flocculation of metal sulphides, cationic flocculating agents are more effective.

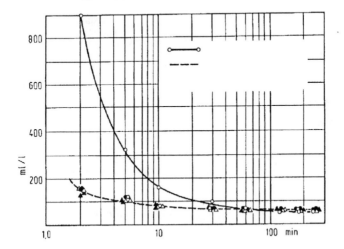

Fig. 3.30 Sedimentation rates of a mixture of several metal hydroxide (100 mg/litre metal) with and without added anionic flocculating agents (10 mg/litre).

3.7.4.3 Sedimentation

Described in this section is the sedimentation, and hence the separation of solids from the aqueous phase, of metal hydroxides or basic salts. Similar comments apply to metal sulphides (Section 3.8.1), certain iron compounds (Section 3.8.2), calcium compounds with fluorine, sulphate or phosphate (Section 3.10) as well as other insoluble species formed during effluent treatment.

Sedimentation is the simplest method for separating a solid from a liquid phase. The process rests on the greater density of the solid which causes it to settle more or less rapidly, to the floor of the container used. If certain assumptions are made – that the particles are spherical, that their diameter does not change and that laminar flow conditions operate, then their rate of sedimentation is given by Stokes' equation:

$$u_s = \frac{g}{18\,\eta}\,(\varrho_F - \varrho_W)\cdot d^2 \cdot F_F \qquad\qquad 3.132$$

where u_s is the settling rate (cms⁻¹); g the gravitational constant (931 cms⁻²); η the dynamic viscosity of the effluent (gcm⁻¹s⁻¹); ϱ_F the density of the solid (gcm⁻³); ϱ_W the density of the liquid (effluent) (gm/cm⁻³); d the diameter of the sedimenting species (cm); and F_F is a constant for a given solid species.

The most important parameters are in practice the difference between the densities of the solid and liquid phases and the particle size. Viscosity (which is temperature dependant) is less important. Also important are the particle morpholgies and the extent to which they interact with one another. In this last, enter effects such as charge attraction or repulsion, hydrodynamic flow patterns especially those which, as concentrations increase, develop around the settling particles and finally, capture processes where smaller particles are taken up by larger, more rapidly settling, ones [124].

So many and varied are these factors, including those discussed in Sections 3.7.4.1 and 3.7.4.2, that calculation of sedimentation rate is not in practice a useful exercise. Additionally, there are kinetic and reactor-based factors such as reduction of floc size by shear action of pump impellers or mixer blades as well as floc size growth from ageing. Normal practice is to measure sedimentation rate empirically under conditions held as nearly as possible constant and using glass measuring cylinders or similar vessels. The following values indicate an order of magnitude of sedimention rates: [125]

sand (particle size 0.1 to 0.3 mm)	25 to 130 m/hr
barytes, finely ground	ca 1 m/hr
precipitated aluminium hydroxide	0.15 to 0.5 m/hr

When sedimentation rates fall to less than 0.03 m/hr, simple sedimentation ceases to be an attractive means for phase separation. At this point, flocculation technology has to be considered.

Experiments to determines sedimentation rates also allow the overall settling behaviour of an insoluble species to be assessed. Thus, the level of sludge in the botton of the vessel is measured at regular intervals and this data is converted into a sedimentation plot. It is customary to distinguish between three types of behaviour [124].

- sedimentation of individual particles, especially at low solids concentrations. These conform most closely to the Stokes Law.

- sedimentation after strong interactions between particles. These become more important as solids concentration increases which in turn affect sedimentation rate. By capture of smaller particles into larger ones, the boundary region between clear water and that containing solids, is readily visible.

- compression. In this regime, sedimentation is very slow. It depends on the increase of pressure due to upper layers of solids on the underlying layers and on the porosity of the sludge, which allows trapped water to escape upwards until finally, channels are formed [126].

The importance, in the overall sedimentation process, of solids interaction and the compression action becomes the more important the bulkier are the settling flocs and such bulkiness is usually found in metal hydroxide precipitates.

Fig. 3.31 shows examples of readily and poorly sedimentable precipitates. These curves cannot, though, be used to indicate the extent to which the supernatant liquid is completely solids-free and thus after what time it would, in practice, be permissible to pipe it to drain without breaching the legal discharge limits. Bearing in mind the now very low concentration limits, there is no alternative to the use of analytical procedures, after first adding acid to dissolve the few remaining solid particles, to decide whether such *"clear"* effluent may be discharged or not. Flocculation and flocculating agents can be used to remove the last of the suspended solids. However a more reliable approach, indeed one that is usually mandatory, is to use filtration to retain these (see Section 3.4.1).

In a few cases, sedimentation is replaced by or enhanced with centrifugation. This has the effect of increasing the gravitational forces acting on the particles by a factor *"z"*. The implications of this on sedimentation rate can be adduced by inspection of Equation 3.132. *"z"* is the *"centrifuge factor"* and is the product of *"d"* the acceleration in the centrifuge and *"g"* the gravitational constant. Use of a centrifuge can increase settling rates several thousand fold.

A quite different phase separation process associated with flocculation is *"Flotation"*. Here, the metal hydroxides onto which other toxic species can be mechanically or adsorptively bound, are exposed to low density substances (air, gases, oil) and so transformed into a mixture lighter than water which thus float to the surface (see Section 3.13.4.4). Flotation, like sedimentation, is governed by Stokes' Law, which has to be modified in that the pseudo-density of the floating substances has to be subtracted from the density of the liquid.

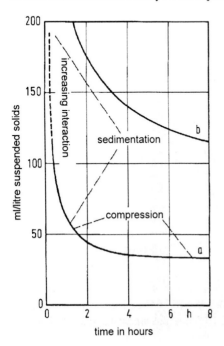

Fig. 3.31   Examples of sedimentation of a precipitate

a) good sedimentation behaviour
b) poor sedimentation behaviour

3.7.5   Neutralisation Precipitation of Specific Metals from Dilute Solution

At the time when the author first began, in 1960, to systematically collect information on the neutralisation precipitation of individual metals from dilute solution (100 mg/litre or less), there was hardly any published work containing solubility curves calculated from solubility products [127,128]. Even these were not of any practical use, since they were not applicable in the lower concentration ranges. At that time, some Water Authorities permitted the use of maximum discharge limits to be calculated using the solubility product. For the much lower limits currently in force, this approach is not useful (see Section 3.7.5.3.3).

Much valuable data was, however available from the work of Feitknecht, Schwab & Naesaenen and co-workers. The primary goal of these studies was in a very different area, e.g. investigations of the morphologies of precipitates, the production of mixed catalysts etc. In 1963, valuable work by Schegel [129] was published which was directly concerned with neutralisation precipitation. The author's own researches [130] in this field were mainly concerned with the following criteria:

- precipitation pH values & pH changes post precipitation.

- composition of precipitates.

- solubility of precipitates as function of pH, calculation of solubility product and comparison with published values.

- sedimentation of precipitates.

These studies were based on use of NaOH, slaked lime and soda as neutralising agents and results were then published [131-133]. Insofar as the scientific results were extrapolated to draw conclusions for practical applications, the subsequent tightening of effluent discharge legislation in many parts of the world now invalidate such conclusions, though not of course the scientific data on which they were based.

3.7.5.1   Copper

3.7.5.1.1   Precipitation pH Value

The potentiometric titration plots in Fig. 3.32 show the change in pH during the neutralisation precipitation of a sulphuric acid solution of copper, using NaOH, soda and slaked lime. The first precipitate is formed, using NaOH or slaked lime, at pH 5.8 to 5.9. In the case of soda, this takes place at pH 5.6. In the case of the first two reagents, copper concentrations of less than 0.5 mg/litre are first found only at pH values of 7.6 or above, as analytically determined.

As explained in Section 3.7.2, if samples precipitated at a given pH value are allowed to stand, this pH then changes with time. This effect is of practical importance since, with such a pH change, the solubility of the precipitate can also change. It is further important to be aware of this phenomenon since the solubility values, obtained some 2 hours after precipitation, should be associated with the pH at which they were measured, not that at which precipitation initially took place.

Fig. 3.33 shows the pH retreat following precipitation of copper at various initial pH values. The most marked retreat is found at initial value of pH 8. This moves, within two hours, to pH 7.5 which is in fact the precipitation pH of copper. Similar findings are obtained using calcium hydroxide but using soda, on account of its buffering capacity, no such changes are reported [134].

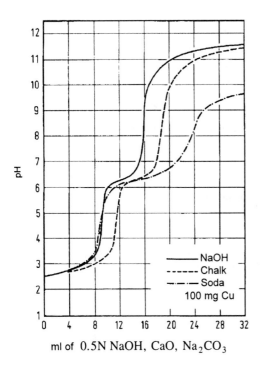

ml of 0.5N NaOH, CaO, Na$_2$CO$_3$

Fig. 3.32   Potentiometric titration of acid copper sulphate solution. (100 mg/litre copper) with various neutralising agents.

3.7.5.1.2 Composition of precipitates

Depending on the anions present in solution, basic salts of copper are formed as titration or neutralisation commences. As the hydroxyl ion concentration in solution increases, the equilibrium moves steadily towards metal hydroxide formation.

The most easily formed species, according to Schwab *et al.* [106], is the basic sulphate, $CuSO_4 \cdot 3Cu(OH)_2$. This forms in the initial stages, so long as sulphate

ions are present. According to Naesaenen [135], its formation is preceded by a
sulphate-richer species, $CuSO_4 \cdot 2Cu(OH)_2$. The author's own studies, are shown in
Fig 3.34. These suggest the formation, in the range pH 6.5 to 7.5, of near-enough
stoicheometric $CuSO_4 \cdot 3Cu(OH)_3$. As pH increases, so the sulphate content falls, as
one might expect. The still relatively sulphate-rich precipitates found at higher pH
values would be less if the precipitates were not immediately separated from the
liquid. The basic salts formed when precipitation takes place are converted, at higher
pH values, as shown below:

$$CuSO_4 \cdot 3\,Cu(OH)_2 + 2\,OH^- \rightleftharpoons 4\,Cu(OH)_2 + SO_4^{--} \qquad\qquad 3.133$$

When chloride ions are present, there forms, according to both Naesaenen [135] and
Schwab [106], a basic chloride $CuCl_2 \cdot 3Cu(OH)_2$ which precipitates, according to
Schwab, only when the mother liquor is contacted for some time with the solid phase
at a lower pH value.

In a solution of nitrate ions, the basic salt $Cu(NO_3)_2 \cdot 3Cu(OH)_2$ is formed,
according to Schwab & Polydoropoulos only when large excess of nitrate ions are
present.

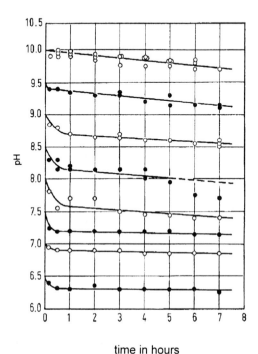

time in hours

*Fig. 3.33 pH change as function
of time of standing and initial pH
after precipitation of copper with
NaOH (100 mg/litre Cu) [134].*

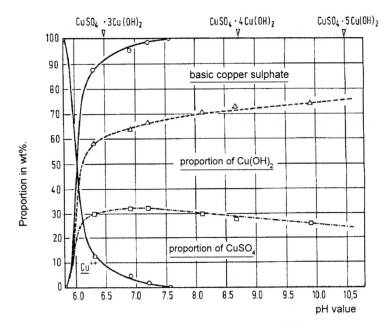

Fig. 3.34   Composition of precipitates formed by by addition of NaOH to acid sulphate copper solutions as a function of pH (1 gm Cu/litre).

Depending on the precipitation conditions and the time of standing, the copper hydroxide can undergo a further transformation, namely loss of water:

$$Cu(OH)_2 \rightleftharpoons CuO + H_2O \qquad\qquad 3.134$$

The blue colour of the copper hydroxide changes, in this process first to brown, then to black. This colour change is the more pronounced, the higher the copper, the concentration and pH values and the longer the time of standing. Schwab and Polydoropoulos noted that such copper hydroxide was stabilised when nickel was co-precipitated as the hydroxide and the frequency with which nickel is found in industrial effluents could well be the explanation why such blackening of copper hydroxides is rarely seen in industrial effluent plants.

The formation of basic metal salts is affected not only by the type of anions present but also by the neutralising chemicals used. When soda is used, a green-blue basic copper carbonate forms (see Equation 3.130). Where other anions are also present, a wide range of basic salts can be formed.

3.7.5.1.3   Solubilities and Solubility Products

The solubilities of copper precipitates formed by addition of NaOH in sulphate solutions and changes of solubility with time, are shown in Fig. 3.35. Combining this information with the pH retreat data shown in Fig. 3.33 for 2 hour standing times using various neutralising agents, one can construct the solubility plots shown in Fig 3.36. [134]

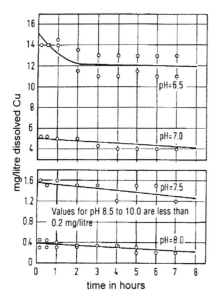

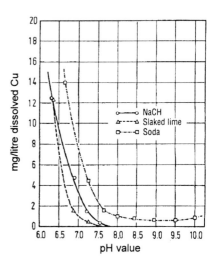

Fig. 3.35   Solubility of precipitates formed by NaOH addition to acid copper sulphate solutions as function of pH and time of standing [134]

Fig. 3.36   As Fig. 3.35 but using also slaked lime or soda. 100 mg/litre Cu, 2 hours standing time.

These results are in good agreement with the findings of Schlegel [129]. The some-what poorer results found by this author can be attributed to the hard water which he used, i.e. its carbonate content. As seen in Fig. 3.36, results using soda as neutralising agent are less satisfactory, barely reaching the 0.5 mg/litre level which is a minimum requirement for effluent discharge.

Solubility product data quoted from older sources are not always reliable and for this reason, Table 3.5 quotes only more recently published values. These are in excellent agreement with values obtained by the author, these being obtained under conditions relevant to effluent treatment.

3.7.5.1.4   Sedimentation

The literature appears to contain no reports of systematic sedimentation studies allowing any form of comparison and the author is forced to rely on his own results [134] most of which were obtained under constant conditions. Fig. 3.37 shows sedimentation curves for copper from acid sulphate using NaOH, slaked lime and soda. The volume of precipitate increases initially with pH until precipitation is complete. Then (for NaOH or slaked lime) after passing through a small maximum, it becomes constant. In the case of soda, no maximum is seen but the supernatant liquor remains cloudy. In this case, a basic copper carbonate is formed, much of it very finely divided. Its solubility too, as Fig. 3.36 shows, is somewhat larger than that found with the other two neutralising agents.

TABLE 3.5:

SOLUBILITY PRODUCT VALUES FOR COPPER HYDROXIDE AND SOME BASIC COPPER SALTS

Sparingly Soluble Compound	Definition of Sol'y Product	Source	Solubility Product (L_H or L_B)
hydroxide	$a_{Cu^{++}} \cdot a_{OH^-}^2 = L_H$	Näsänen [135]	$3.9 \cdot 10^{-19}$
		Feitnecht [136]	$1.6 \cdot 10^{-19}$
		Schwab and Polydoropoulos [103]	10^{-18}
		D'Ans-Lax	$5.6 \cdot 10^{-20}$
		Hartinger [134]	$2 \cdot 10^{-19}$*)
basic sulphate	$a_{Cu^{++}}^4 \cdot a_{OH^-}^6 \cdot a_{SO_4^-} = L_B$	Näsänen [135]	$8.2 \cdot 10^{-63}$**)
		Schwab and Polydoropoulos [106]	$3.7 \cdot 10^{-63}$
		Hartinger [134]	$3.4 \cdot 10^{-63}$*)
basic chloride	$a_{Cu^{++}}^2 \cdot a_{OH^-}^3 \cdot a_{Cl^-} = L_H$	Näsänen [135]	$2.2 \cdot 10^{-34}$
		Schwab and Polydoropoulos [106]	$1.57 \cdot 10^{-34}$
basic nitrate	$a_{Cu^{++}}^2 \cdot a_{OH^-}^3 \cdot a_{NO_3^-} = L_B$	Schwab and Polydoropoulos [106]	$2 \cdot 10^{-30}$

*) Values obtained under effluent treatment conditions

**) Derived from relationship: $a_{Cu^{++}}^2 \cdot a_{OH^-}^3 \cdot a_{SO_4^-}^{0.5} = L_B$.

These copper precipitates are, those formed with soda excepted, very readily sedimentable. Addition of flocculating agent increases floc size but only increases the rate of sedimentation in the first 10 minutes. Flocculating agents based on polyacrylamide and those based on polyacrylates both perform comparably.

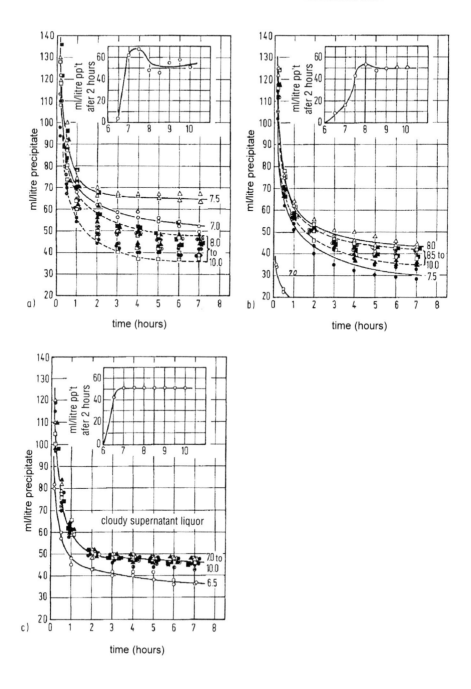

*Fig. 3.37   Sedimentation curves and precipitate volumes for copper precipitated
from acid sulphate solution as function of pH (100 mg/litre Cu) [134].
a) with NaOH   b) slaked lime   c) soda*

3.7.5.2   Zinc

3.7.5.2.1   Precipitation pH Value

Titration curves for zinc sulphate solution neutralised with NaOH, slaked lime and soda are shown in Fig. 3.38. The first sign of cloudiness using the first two of these, appears at pH 7.6. Using soda, because a basic carbonate is formed, turbidity is first seen at pH 7.4. In terms of solubility measurements, values of less than 0.5 mg/litre were not obtained until pH 9.0 had been reached (NaOH or slaked lime) or 8.3 (soda).

Fig. 3.38 shows, as is the case with other metals, that the amount of soda required for completion of precipitation is larger than is the case with NaOH or slaked lime. Clearly visible (again this is true for all strongly basic metals) are the two stages of neutralisation, the acid carbonate and the carbonate itself. The buffering action of the former is also manifest. Such behaviour differs from that shown by copper ions where both stages form in the same pH range [137].

Zinc does resemble copper in that, after attaining the precipitation pH, there is a pH retreat on standing. As Fig. 3.39 using NaOH shows, this is most marked where the initial pH is taken to the precipitation value. Similar results are found using slaked lime while with soda, on account of its buffering action, no significant changes are found.

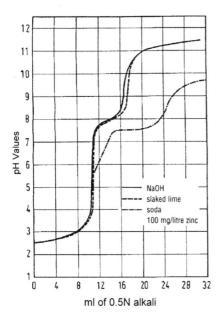

Fig. 3.38   Potentiometric titration of acid zinc sulphate solution (100 mg/litre zinc) with various neutralising agents [137].

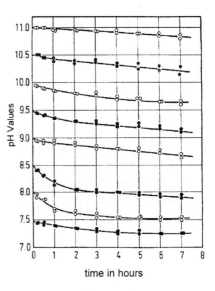

Fig. 3.39   pH change as function of time of standing and initial pH for ppt'n of zinc with NaOH (100 mg/litre Zn) [137]

3.7.5.2.2 Composition of precipitates

Several types of zinc hydroxide and its basic salts are known [138]. Zinc is a divalent
metal with a strong tendency to form basic salts. Feitknecht in particular, studied the
extent to which the precipitate was stoicheometrically constituted on the one hand, or
mixed-crystalline on the other, with a wide range of anion + OH⁻ compositions. He
found that in the presence of monovalent ions ($X = Cl^-$, NO_3^-), compounds such as
$ZnX_2 \cdot Zn(OH)_2$ or $ZnX_2 \cdot 4\,Zn(OH)_2$ were formed, these producing different X-ray
diffraction patterns [138]. Subsequently [139] he reported a basic zinc chloride
$ZnCl_2 \cdot 6\,Zn(OH)_2$.

In the case of sulphate anions, a basic salt $ZnSO_4 \cdot 3\,Zn(OH)_2$ is known [102,138].
As stated earlier, such basic salts form where OH⁻ concentrations are not high. As
[OH⁻] increases, the basic salts tend to transform to the simpler hydroxides, as seen
in Equation 3.131 for zinc sulphate. Feitknecht studied the lattice parameters for both
hydroxide and the basic salts and showed how the transformation to simple hydroxide takes place in the plane of the crystal lattice. It is also possible for hydroxide to
be formed onto the basic salt, both species having similar lattice structures [140].
Kiessig & Reimers [14] showed how formation if the basic zinc sulphate is temperature dependent. At 20°C, the basic salt described above is formed after addition of
1.5 equivalents of NaOH. As this is carried out at higher temperatures, increasing
amounts of NaOH are required until, at above 65°C, the pure hydroxide is formed. A
comment by Geiger & Schneider [142] suggests that the composition of the precipitate is also dependent on the initial zinc concentration and the rate of neutralisation
with no stoicheometrically defined compound being formed. In this context, Fig.
3.40 is relevant, showing the transition from basic salt $ZnSO_4 \cdot 3\,Zn(OH)_2$ to zinc
hydroxide.

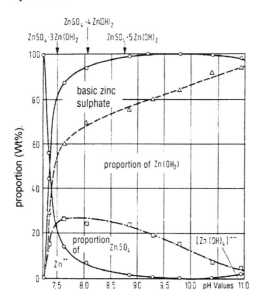

*Fig. 3.40 Composition of
precipitate formed by NaOH
neutralisation of zinc sulphate
solution, as function of pH (1gm/
litre Zn) [137].*

Analyses of precipitates showed, as expected, a decrease in zinc sulphate content as pH increased [137]. This decrease is far more marked than is the case with basic copper sulphate, on account of the lower stability of the basic zinc salt, as shown in Fig. 3.40 where the conditions at higher metal ion (1 gm/litre zinc) were favourable for basic salt formation.

Using soda as a neutralising agent, the author [137] found that precipitates in the pH range 7.0 to 9.5 were more than 90% basic carbonate with a composition $2\,ZnCO_3 \cdot 3Zn(OH)_2$ of the zinc spar type. At lower pH values (up to pH 8), small amounts of basic zinc sulphate were also found.

In addition to its tendency to form basic salts, zinc ions in solution display another pH dependent behaviour which is significant in effluent treatment technology. Zinc can form an *"amphoteric"* hydroxide, i.e. a poorly soluble species which can be dissolved either by acid or by further addition of alkali. The initial reaction is with additional hydroxyl ions to form the soluble tetrahydroxy zincate:

$$Zn^{++} \underset{H^+}{\overset{OH^-}{\rightleftharpoons}} Zn(OH)_2 \underset{H^+}{\overset{OH^-}{\rightleftharpoons}} [Zn(OH)_4]^{--} \qquad\qquad 3.135$$

This complex is sometimes found in effluent when it is strongly alkaline. This then calls for neutralisation using acid as implied by Equation 3.135 going from right to left, until precipitation takes place. The point at which the positively charged zinc ions are in equilibrium with the negatively charged zinc ions is known as the *"isoelectric point"*. It implies a hydrogen ion concentration at which neither anionic nor cationic forms are dissolving. Zinc hydroxide is precipitated as a highly insoluble compound. Similar amphoteric (from the Greek *"both"*) behaviour is displayed by other metals found in effluent and these (as hydroxy-complexes) are treated in Section 3.9.3.1.1.

The readily soluble alkaline zincates form, in presence of calcium ions, the insoluble calcium zincate. The amphoteric character of zinc hydroxide is thus not shown when slaked lime is used for neutralisation:

$$[Zn(OH)_4]^{--} + Ca^{++} \rightleftharpoons Ca[Zn(OH)_4] \qquad\qquad 3.136$$

3.7.5.2.3   Solubility and Solubility Products

Fig. 3.41 shows the solubilities of precipitates formed from zinc neutralised with NaOH, slaked lime or soda as a function of pH and time of standing [137]. No ageing effects are observed in the case of NaOH or soda. The change in solubility using NaOH or at higher pH values is due to the slow equilibrium between hydroxide and hydroxy-zinc complex in favour of the latter. A similar explanation underlies the results found using soda where a transformation at high pH values is slow, this being a heterogeneous reaction. The basic carbonate reacts, at higher pH values to form the more stable hydroxide. The solubilities of the tetrahydroxyzincate formed at high pH as well as the insoluble nature of the corresponding carbonate can both be seen in Fig. 3.42. The data show solubilities 2 hours after precipitation at the pH measured at the same time.

In effluent treatment terms, the increased solubility resulting from formation of the alkali hydroxy-zincate is first seen at pH >10.7. The amphoteric behaviour of the zinc hydroxide is not as marked as is often mistakenly assumed. Its iso-electric point, more clearly shown in semi-logarithmic form in Fig. 3.87 lies between pH 9.6 and 10. The concentration of dissolved zinc at pH values below the iso-electric point is determined by the solubility product, above it, by the dissociation constant K_D of the tetrahydroxy-zincate.

$$\frac{c_{Zn^{++}} \cdot c_{OH^-}^4}{c_{[Zn(OH)_4]^{--}}} = K_D \qquad\qquad 3.137$$

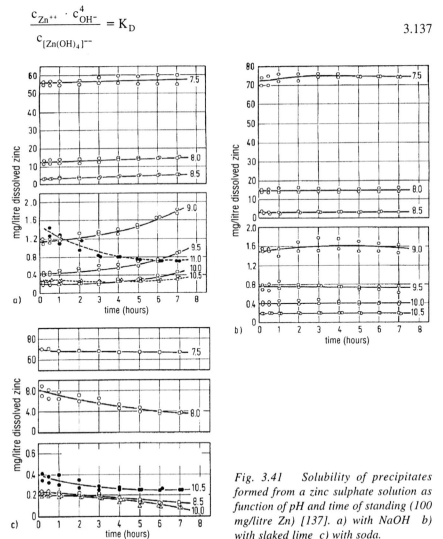

Fig. 3.41   Solubility of precipitates formed from a zinc sulphate solution as function of pH and time of standing (100 mg/litre Zn) [137]. a) with NaOH b) with slaked lime c) with soda.

If the solubility product for zinc hydroxide L_H and the ionic product for water, P, are inserted into Equation 3.137, then, from the concentration of dissolved zinc, a dissociation constant of the tetrahydroxy zincate can be derived.

$$\frac{L_H \cdot P^2}{c_{Zn(gel.)} \cdot c_{H^+}^2} = K_D \qquad\qquad 3.138$$

Using values for the pH range 10.5 to 11, K_D is calculated to be 6×10^{-19}.

Table 3.6 summarises published values for the solubility product of zinc hydroxide and some of its basic salts. These too, are in good agreement with results found by the author.

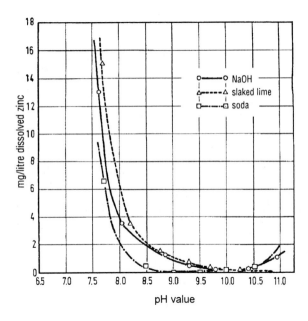

Fig. 3.42   Solubilities of zinc precipitates from sulphate solution as function of pH (100 mg/ litre Zn) after 2 hours standing time [137]

3.7.5.2.4   Sedimentation

Fig 3.43 shows sedimentation curves for zinc neutralised with NaOH or soda at various pH values under otherwise constant conditions [137]. In the case of NaOH, the curves differ significantly in comparison with those for other metals. The volume of precipitate increases up to pH 9, then falling strongly as pH increases further. In addition, at pH values greater than 9, after standing times which grow shorter as the pH increases, there is a compression of the precipitate. This effect is specially pronounced at pH 10.5 to 11.5. The effect can be seen by eye as the bulky and flocculant precipitate is converted to a finer, more crystalline form. This has nothing to do with the high solubility of the zincate, since the solubility of the hydroxide at concentrations typically used is less than 2% even at pH 11.5. Similar behaviour was observed by Feitknecht [143] in the precipitation of zinc from fluoride containing solutions. Using X-ray methods, he found that the denser form was mainly finely dispersed defect-structured zinc oxide. The implication is that this process involves a loss of water, as already described for copper (Equation 3.134). This precipitation, accompanied by the morphological change, may be of limited interest to the effluent treatment technologist. It does, however, suggest a means for more rapid removal of zinc from solution by sedimentation, provided the subsequent correction of pH is carried out.

TABLE 3.6:

SOLUBILITY PRODUCT VALUES FOR ZINC HYDROXIDE AND SOME OF ITS BASIC SALTS

Sparingly Soluble Compound	Definition of Sol'y Product	Source	Solubility Product $(L_H$ or $L_B)$
Hydroxide	$a_{Zn^{++}} \cdot a^2_{OH^-} = L_H$	Feitnecht and	$2.7 \cdot 10^{-17}$
		Häberli [139]	$5.5 \cdot 10^{-16}$ from Cl$^-$ and NO$_3^-$ cont'g sol'ns of different strengths
		Schwab and	
		Polydoropoulos [106]	$4 \cdot 10^{-17}$
		D'Ans-Lax	$1 \cdot 10^{-17})$
		Hartinger [137]	$4 \cdot 10^{-17*})$
Basic sulphate	$a^4_{Zn^{++}} \cdot a^6_{OH^-} \cdot a_{SO_4^-} = L_B$	Feitnecht [102]	$1 \cdot 10^{-57}$
		Hartinger [137]	$2.6 \cdot 10^{-56*})$
Basic chloride	$a^5_{Zn^{++}} \cdot a^8_{OH^-} \cdot a^2_{Cl^-} = L_B$	Feitnecht [102]	$2.8 \cdot 10^{-73}$
	$a^7_{Zn^{++}} \cdot a^{12}_{OH^-} \cdot a^2_{Cl^-} = L_B$	Feitnecht and Häberli [139]	$4.4 \cdot 10^{-105}$
Basic nitrate	$a^5_{Zn^{++}} \cdot a^8_{OH^-} \cdot a^2_{NO_3^-} = L_B$	Feitnecht and Häberli [139]	$6.1 \cdot 10^{-69}$

*) Values obtained under effluent treatment conditions.

Results using slaked lime are similar except that the modified structure forms more slowly. When soda is used for precipitation, very uniform and rapidly sedimentable precipitates are formed.

Polyacrylate flocculating agents showed very little benefits in the first 10 to 30 minutes while addition of polyacrylamides showed no apparent action.

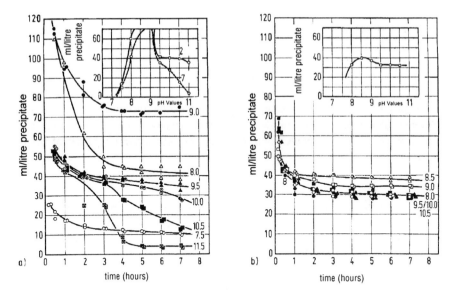

Fig 3.43   Sedimentation curves and precipitate volumes for zinc precipitated from sulphate-containing solution as function of pH. (100 mg/litre zinc) [137].
a) using NaOH   b) using soda

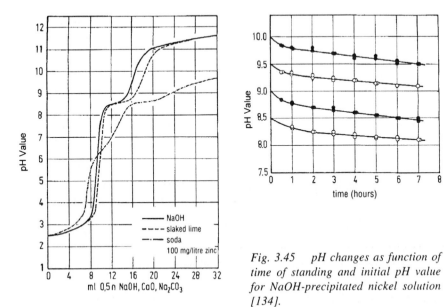

Fig. 3.45   pH changes as function of time of standing and initial pH value for NaOH-precipitated nickel solution [134].

Fig. 3.44   Potentiometric titration of acid nickel sulphate solution (100 mg/litre Ni) using various alkalis [134].

3.7.5.3 Nickel

3.7.5.3.1 Precipitation pH Values

Titration curves for nickel sulphate containing solutions using the three alkaline species shown in Fig. 3.44 shows that, in all cases, precipitation commences at a pH slight less than 8. Precipitation is complete (Ni conc'n < 0.5 mg/litre) at pH > 9.9 as determined by chemical analysis. As with other strongly basic hydroxides, titration using soda solution reveals two stages, the buffering action of the bicarbonate and nickel precipitation being readily distinguishable. pH changes after precipitation are relatively small – around 0.2 to 0.3 pH units after two hours. Slaked lime behaves much as does caustic soda as shown in Fig. 3.45. Where soda is used, pH retreat effects are even smaller. In all cases, however, the pH retreat is most marked at pH 9.0, the point where most of the nickel is precipitated.

3.7.5.3.2 Composition of Precipitates

For nickel in particular, published literature describing the nature of the precipitate is meagre in the extreme. Naesaenen [144], in his determination of nickel and cobalt solubility product values bu NaOH titration, was concerned as to effects from formation of basic salts and, for this reason, carried out the titration in reverse sequence. Schwab & Polydoropoulos [106], titrating nickel solutions from 2.5×10^{-3} to 0.25 Mol/litre found a decrease in the amount of alkali required (pro-rated for the amounts present) of up to 10%. They preferred to explain this in terms of neutral salt adsorption rather than formation of a basic salt.

The author's own studies of precipitates, suggest an onset of precipitation at pH 8.7, the composition of which is $NiSO_4 \cdot 6\ Ni(OH)_2$. At pH 9.75, the compositional ratio is over 10 Mol $Ni(OH)_2$ to 1 of $NiSO_4$ (see Fig. 3.46).

The titration plot shown in Fig. 3.44 corresponds to a precise stoicheometric consumption of alkali in terms of nickel hydroxide formation. When titration occurs more slowly, it appears that the conversion of the initially formed hydroxyl-rich basic nickel salt has been completed by the time the high pH values are attained. At pH values of 9.5 or above, after only a short time of standing, nickel hydroxide is present with some adsorbed sodium sulphate, as was suggested by Schwab.

Though it is difficult to form pure nickel carbonate, a basic carbonate is formed when soda is used for neutralisation, albeit with a very wide compositional spectrum which is best described as $xNiCO_3 \cdot y\ Ni(OH)_2 \cdot z\ H_2O$. The composition shifts, with increasing pH, to ever-greater hydroxide content.

Chemical analyses of precipitates gave, as shown in Fig. 3.47, at pH 8.8 a precipitate of composition $NiCO_3 \cdot 2\ Ni(OH)_2$. The sulphate content of this was so small that, in this case too, adsorption was assumed to be responsible for its presence.

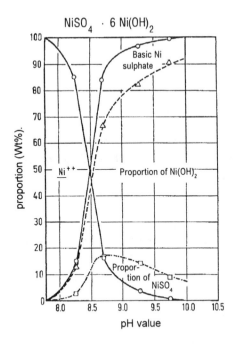

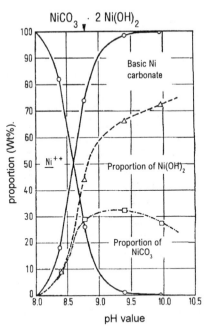

Fig. 3.46   Composition of precipitates formed from nickel sulphate containing solutions using NaOH, as function of pH (100 mg/litre Ni) [134]

Fig. 3.47   Composition of precipitates formed with soda as neutralising agent using NaOH, as function of pH (100 mg/ litre Ni) [134]

3.7.5.3.3   Solubilities & Solubility Products

As the titration curves reproduced above show, the pH needed for precipitation of nickel is high. Fig. 3.48 shows the solubility of precipitates as a function of pH and time of standing for NaOH and soda as neutralising agents. In the case of NaOH, the curves show only a slight increase and, after 3 to 4 hours, a decrease in solubility. The extent of the initial solubilisation of the precipitate is the greater, the closer the pH to the value required for precipitation. The plots for slaked lime show similar trends. It may well be that the attainment of equilibrium requires longer for the strongly basic nickel hydroxide. The decrease in solubility after standing is probably due to the onset of the ageing process, as is the case with other hydroxides. In the case of the basic carbonate, the phenomenon does not appear.

Fig. 3.49 allows a deduction of solubilities after two hours standing time, after correction for pH changes during this time. For efficient nickel precipitation, when no other metals are present, pH values of not less than 10 must be achieved if the residual nickel concentration is to be less than 0.5 mg/litre. The data obtained using NaOH is in good agreement with published work by Schlegel [129].

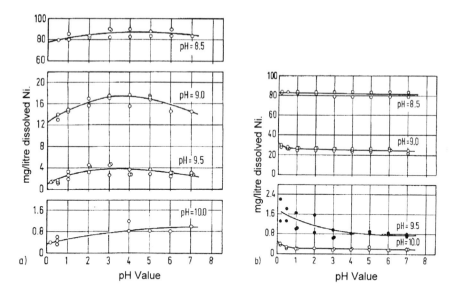

Fig. 3.48   Solubility of precipitates formed from nickel sulphate solutions as
function of time of standing and pH. (100 mg/litre Mi) [134].
a) Using NaOH   b) using soda.

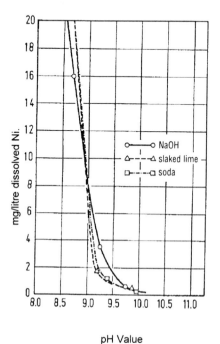

Fig. 3.49   Solubility of precipitates
from nickel sulphate solution using
NaOH, slaked lime or soda as
function of pH (100 mg/litre Ni, data
taken after 2 hours time of standing
[134]).

Solubility product data for nickel hydroxide, drawn from various sources are compared in Table 3.7 with the value obtained under conditions relating to effluent treatment. The latter is an order of magnitude larger than the figure quoted by Naesaenen *et al.* The solubility product for pure nickel carbonate (D'Ans-Lax) is roughly half the magnitude at $1.35 \cdot 10^{-7}$. However the corresponding figure for the basic carbonate formed when soda is used, is much the same as that for the hydroxide itself.

Nickel hydroxide offers a good example of how unreliable are limiting value solubility predictions derived using solubility products, in *"real-life"* effluent situations. Nickel is chosen for this because, of all two-valent metals, it is least likely to form a basic salt when caustic soda is used as a neutralising agent and also because the solubility product values obtained at each pH tend to increase with pH as shown in Fig. 3.50. After determining the actual hydrogen ion concentration from pH measurement, any deviations from solubility product values are dependent only on the analytically determined nickel ion concentrations (not their activities).

TABLE 3.7:
SOLUBILITY PRODUCT VALUES FOR NICKEL HYDROXIDE

Sparingly Soluble Compound	Definition of Sol'y Product	Source	Solubility Product (L_H or L_B)
Hydroxide	$a_{Ni^{++}} \cdot a_{OH^-}^2 = L_H$	Näsänen [144]	$6.2 \cdot 10^{-16}$
		Schwab and Polydoropoulos [106]	$3 \cdot 10^{-16}$
		D'Ans-Lax	$1.6 \cdot 10^{-14}$
		Hartinger [134]	$5.8 \cdot 10^{-15*)}$

*) Values obtained under effluent treatment conditions.

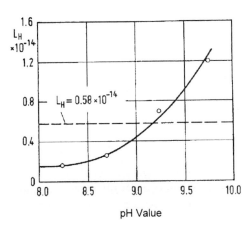

Fig. 3.50 Changes in solubility product value derived from solubility, as function of pH. The dashed line denotes the theoretical prediction [133].

In testing of effluents, it is concentrations not activities which are measured. The discrepancy between these two increases with increasing pH, a result of the increase in total ion concentration of the solution in contact with the precipitate. Thus the calculation of very low solubilities using solubility products will lead to erroneous values, since the total ionic strength, increased as a consequence of the higher pH value, also raises the solubility of the precipitate [133].

3.7.5.3.4   Sedimentation

Sedimentation plots derived using NaOH, slaked lime and soda are shown in Fig. 3.51. Under conditions where quantitative precipitation takes place, there is no significant difference between the sludge volumes in each of the three cases. These volumes are, two hours after formation, of equivalent size and show no dependence on precipitation pH. In contrast to precipitates with NaOH and slaked lime which take longer to reach their final volumes, those formed using soda are rapidly reached. Inspection of the plots shows, however, that the prolonged time of sludge densification is not disadvantageous as compared with the results found using soda. Indeed the reverse can be true in that slow contraction of sludge volumes may be more acceptable. Flocculating agents, at least in the first 30 minutes, significantly accelerate the sedimentation, polyacrylates being most effective.

3.7.5.4   Cadmium

3.7.5.4.1   Precipitation pH Values

The titration curves in Fig. 3.52 show the behaviour of various neutralising agents in precipitation of cadmium salts. It can be seen how, using soda, less extreme values of pH are required than is the case with other reagents [137].

The first signs of solid formation, using NaOH or slaked lime, are observed at pH 9.1 to 9.2. In the case of soda, the equivalent value is pH 7. To bring the concentration below the widely prescribed limit of 0.2 mg/litre, a pH of 10.4 should be attained (in the former case) or pH 8.5 (using soda).

Precipitation of the carbonate-containing precipitate occurs in the buffering range of the bicarbonate. The second step, found between pH 8 and 9, is ascribed to formation of an OH-richer compound.

The most pronounced pH retreats are observed using NaOH or slaked lime, at initial pH values of 9.5 to 10, i.e. in the main precipitation pH range. They are typically up to 0.5 of a pH unit after 2 hrs. The same is seen when soda is used, albeit in the pH range 7.5 to 8. (cf. Fig. 3.53)

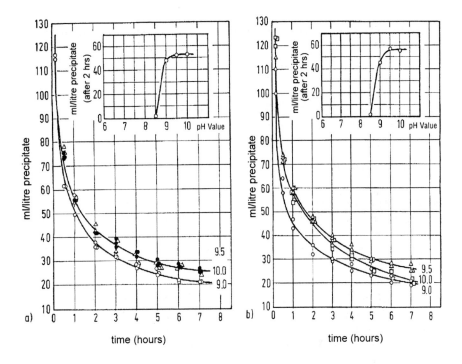

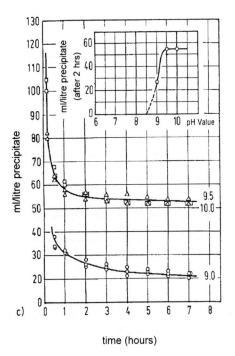

Fig. 3.51 Sedimentation plots and precipitate volumes for nickel precipitation from sulphate solution as function of pH (100 mg/litre Ni) [134].

a) with NaOH
b) with slaked lime
c) with soda.

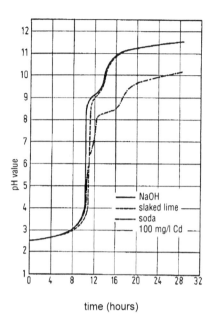

Fig. 3.52   Potentiometric titration plots for acid cadmium sulphate (100 mg/litre Cd) using various reagents [137].

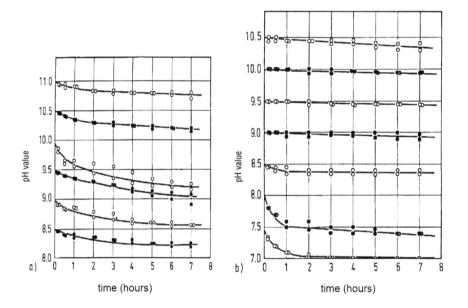

Fig. 3.53   pH changes as function of time of standing an initial pH for cadmium precipitation [137]
a) NaOH   b) soda

3.7.5.4.2   Composition of precipitates

Thanks to the work of Feitknecht [145], we have a good understanding of the composition of cadmium precipitates, especially their basic salts, and the way in which they vary with conditions in which they are formed. He and Gerber were able to distinguish between five variants of basic salts and mixed crystal, starting from chloride solutions alone. These ranged from the salt-rich type ($Cl^-:OH^- = 1$) through to the simple hydroxide. Mixed salts with up to 15% Cl^- are isomorphous with the hydroxide. The composition of precipitates depends above all on prevailing OH^- concentration during precipitation. In some cases, the species formed are labile, undergoing transformations on standing. Chloride-rich species are only formed in the initial stages of precipitation, when precipitation is far from complete and at high initial salt concentrations. In the pH range 8.5 to 9.5, a salt of composition $CdCl_2 \cdot 4$ $Cd(OH)_2$ is formed. As pH increases, the chloride content decreases until the simple hydroxide alone is formed.

Moeller & Rhymer [146] were able to show that no basic nitrates are formed in precipitation from dilute solutions. The same is not true for sulphate and chloride solutions where they report compositions such as $CdSO_4 \cdot 3.5 \ Cd(OH)_2$ or $CdCl_2 \cdot 2Cd(OH)_2$ in line with the findings of Feitknecht [147]. Feitknecht did, however, find that basic nitrates can form such as $Cd(NO_3)_2 \cdot 4 \ Cd(OH)_2$ from concentrated nitrate solutions. The author's own findings in relation to formation of basic sulphates under effluent conditions can be seen in Fig. 3.54. Because cadmium hydroxide is itself a very basic compound, its precipitation commences in a strongly OH^- environment so that the initially formed basic sulphate is rapidly transformed to the true hydroxide. Above pH 11, only this species exists.

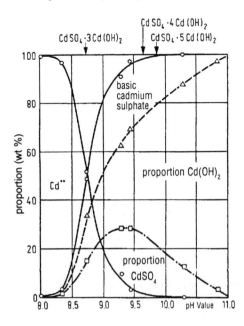

Fig. 3.54   Composition of cadmium salt precipitates from sulphate solutions using NaOH as function of pH (1 gm/litre Cd) [137].

Comparable studies using soda show that, within the limits of error, almost pure carbonate is formed within the pH range 7 to 9. Above pH 9, carbonate content falls away in stepwise fashion in line with the titration curve shown in Fig. 3.52 and the volume ratios seen in Fig. 3.57.

3.7.5.4.3   Solubilities & Solubility Products

Fig. 3.55 shows solubilities as a function of initial pH and time of standing. There was no sign of ageing processes which can affect solubility.

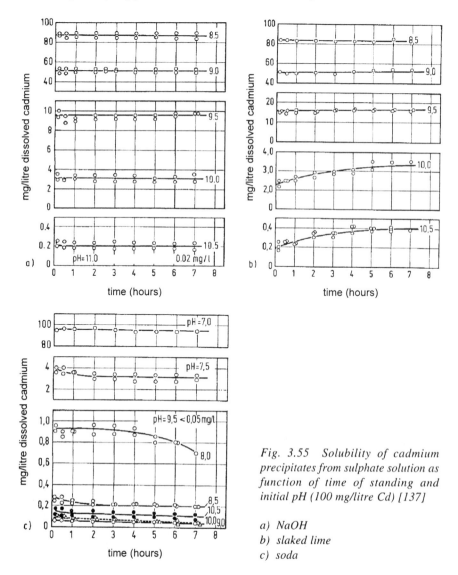

Fig. 3.55 Solubility of cadmium precipitates from sulphate solution as function of time of standing and initial pH (100 mg/litre Cd) [137]

a) NaOH
b) slaked lime
c) soda

Taking the solubilities and pH values found after 2 hours standing, one obtains the plots shown in Fig. 3.56. These show unequivocally how cadmium carbonate or perhaps a low OH⁻ basic carbonate with significantly lower solubility is formed in preference to other basic salts or the hydroxide. This observation has important implications in treatment of cadmium containing effluents. The higher pH values required for precipitation using NaOH or slaked lime are no longer permitted for direct discharge of effluent and thus require greater quantities of neutralising acid. This in turn, creates a danger that metals capable of forming amphoteric hydroxides could, as part of an overall neutralisation, go into solution again. One option here is to use either soda or soda-NaOH mixtures for precipitation. Data for the solubility product of cadmium hydroxide and some of its basic salts are shown in Table 3.8 and these can be compared with values obtained by the author under effluent treatment conditions. For cadmium carbonate, a solubility product value of $2.5 \cdot 10^{-14}$ (D'Ans-Lax) is given – this is comparable with the figure for the hydroxide.

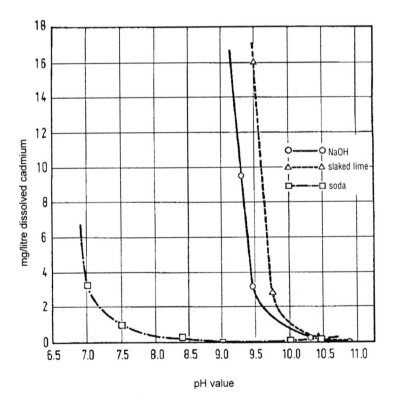

Fig. 3.56   Solubilities of cadmium precipitates from sulphate solutions using NaOH, soda and slaked lime as function of pH (100 mg/litre cadmium, after 2 hours standing time) [137].

TABLE 3.8:

VALUES FOR THE SOLUBILITY PRODUCT OF CADMIUM HYDROXIDE AND SOME BASIC SALTS OF CADMIUM

Sparingly Soluble Compound	Definition of Sol'y Product	Source	Solubility Product $(L_H$ or $L_B)$
Hydroxide	$a_{Cd^{++}} \cdot a_{OH^-}^2 = L_H$	Moeller and Rhymer [146]	$3.2 \cdot 10^{-14}$
(active form)		Feitnecht and Reinmann [148]	$2.2 \cdot 10^{-14}$
(aged)		dto.	$5.9 \cdot 10^{-15}$
Mixed phase at higher			
	pH-values, OH/Cl = 5.7–39	dto.	$34–6.75 \cdot 10^{-15}$
		D'Ans-Lax	$1.2 \cdot 10^{-14}$
		Hartinger [137]	$1.3 \cdot 10^{-14*})$
Basic sulphate	$a_{Cd^{++}}^{4,5} \cdot a_{OH^-}^7 \cdot a_{SO_4^{--}} = L_B$	Feitnecht [147]	$4.4 \cdot 10^{-53}$
		Hartinger	$1.5 \cdot 10^{-55*})$
Basic chloride	$a_{Cd^{++}}^3 \cdot a_{OH^-}^4 \cdot a_{Cl^-}^2 = L_B$	Feitnecht [147]	$1 \cdot 10^{-34}$

*) Values obtained under effluent treatment conditions.

3.7.5.4.4   Sedimentation

Sedimentation plots for cadmium containing species using NaOH or soda as precipitating agents are shown in Fig 3.57 [137].

Using NaOH, the precipitate volumes increase as a function of pH as long as precipitation continues to take place quantitatively. Thereafter, a contraction sets in. Feitknecht, who carried out volume measurements as part of a study of the morphology of the precipitate, obtained similar curves and found that the transformation of the initially-formed species into other morphologies (as implied by the ageing process) was accelerated by increase in OH concentration [145]. By the time pH 11 was reached, the simple hydroxide predominated, having lost some of the water of hydration included in the lattice.

Where soda is used as precipitant, totally different findings have been reported. Up to pH 9, the precipitate is cadmium carbonate. It is finely-dispersed, of crystalline appearance and settles only slowly. Above pH 9, it becomes increasingly flocculated and settles more rapidly. At the same time, the volume of the precipitate increases sharply. This change in behaviour is linked to a progressive transformation of the carbonate into the hydroxide. The disperse precipitates formed up to pH 9 can be readily sedimented by adsorption coagulation, with other readily hydrolysable metals. Where NaOH is used as precipitant, addition of flocculating agents does not appear to significantly improve the sedimentation process.

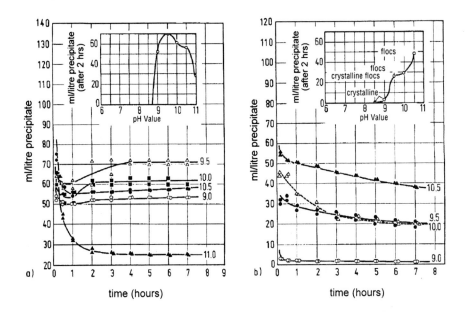

Fig 3.57   Sedimentation plots and precipitate volumes for cadmium from sulphate solutions as function of pH (100 mg/litre Cd) [137]
a) with NaOH   b) with soda.

3.7.5.5   Lead

3.7.5.5.1   Precipitation pH Value

Titration curves obtained with NaOH, slaked lime & soda using nitric acid solutions of lead ions, show a marked resemblance to plots for cadmium. In this case too, the precipitate formed using soda forms at a lower pH than with the other two alkalis. (cf. Fig. 3.58) [149].

Using NaOH or slaked lime, the first visible cloudiness is observed at pH 6.4 to 6.6, with soda the comparable values are 5.1 to 5.2. Notwithstanding the low pH values above, complete precipitation to a level in compliance with commonly required maximum concentrations (0.5 mg/litre) does not occur until pH 10.3. (In the case of soda, this concentration is typically reached at pH of 6.9). The titration curve in this last case shows only a single step since the buffer region and precipitation process overlap one another.

For NaOH, the pH retreat is most marked in the main precipitation range of pH 9 to 9.5. Using soda, a pH increase is recorded, this being most marked at pH 6.5 to 7.5.

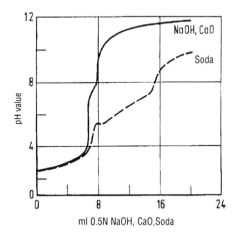

Fig. 3.58 Potentiometric titration of acidified lead nitrate [149].

3.7.5.5.2   Composition of precipitates

Basic lead nitrate, $2\ Pb(NO_3)_2 \cdot Pb(OH)_2$ has the relatively high solubility value of 194 gm/litre. Where precipitation is used to remove lead salts, a hydroxyl-richer compound must be formed. Referring to the titration curve above, 80% of the theoretical requirement of NaOH was used to complete the precipitation. This implies that the precipitate must be either a hydroxyl radical- rich basic nitrate or a mixture of basic nitrate and hydroxide with overall composition $Pb(NO_3)_2 \cdot 5\ Pb(OH)_2$. As is the case for basic salts of other divalent metals, it can be assumed that for lead too, the precipitate composition tends towards that of the simple hydroxide as OH^- concentration in solution increases. There is also a kinetic factor involved in this.

By contrast, the reaction with soda leads to products such as the poorly soluble carbonate of lead (1.5 mg/litre solubility) and the basic carbonate $2\ PbCO_3 \cdot Pb(OH)_2$. Where precipitation is not yet complete, the solid species may consist entirely of carbonate and this, on addition of more soda, is transformed to the basic carbonate.

Basic salts formed with chlorides are also known though their solubility is relatively high. $PbCl_2 \cdot Pb(OH)_2$ – 95 mg/litre; $PbCl_2 \cdot 3\ Pb(OH)_2$ – 56 mg/litre. The most complete precipitation of lead from chloride containing solutions is achieved with hydroxyl ion containing compounds.

Lead differs from most other metals in the range of other sparingly soluble species such as lead chromate which is as insoluble as the carbonate and also its phosphate. The latter species may form as a sludge in effluents containing both lead and phosphate ions. Its solubility – 0.0014 mg/litre – is extremely low. However as a deliberate strategy for lead removal, use of phosphate is precluded both on grounds of cost and also because over-dosing leads to undesirable eutrophication. In the same way, precipitation of the chromate can sometimes occur spontaneously. However the hexavalent chromium should always be treated by reduction before being allowed to contact the lead.

3.7.5.5.3 Solubilities & Solubility Products

Precipitates formed with NaOH and slaked lime are finely dispersed, often partly colloidal. The decrease in solubility due to ageing is thus quite marked if unpredictable. At a pH of 9 to 10.5, the solubility after 2 hours standing is typically 25 to 30% of its initial value. The ageing process involves coagulation of colloidal particles and also the growth of larger particles at the expense of smaller ones. An additional factor is the transformation of initially-formed basic salts of lower OH$^-$ content to those with higher hydroxyl ion compositions.

Solubility values obtained under such conditions are shown in the right-hand curve of Fig. 3.59. The modest gradient as function of pH as well as the high degree of scatter of data points, suggest there is no single value of solubility product to cover the whole range. For the neutral region, a figure of 10^{-17} is quoted. At pH 10, this becomes 10^{-13}. Using soda as precipitant, the same finely-dispersed precipitate is formed but with no colloidal fraction. The solubility curve, for this, is shown on the left hand side of Fig. 3.59.

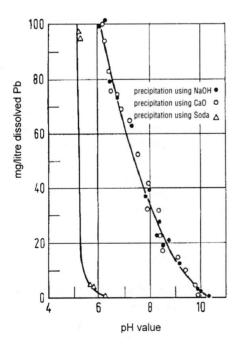

Fig. 3.59 Solubility of precipitates from nitric acid solutions of lead as function of pH (100 mg/litre Pb) after 2 hours standing. [149].

Table 3.9 lists solubility products of sparingly soluble lead species as they occur under effluent treatment conditions. Lead hydroxide behaves amphoterically, albeit only in strongly alkaline solutions. The formation of plumbites and other tetravalent lead species has no practical relevance in effluent treatment.

TABLE 3.9:

SOLUBILITY PRODUCT VALUES FOR LEAD HYDROXIDE & CERTAIN OTHER SPARINGLY SOLUBLE LEAD COMPOUNDS

Sparingly Soluble Compound	Definition of Sol'y Product	Source	Solubility Product (L_H or L_B)
Hydroxide	$a_{Pb^{++}} \cdot a^2_{OH^-} = L_H$ Hartinger [149]	Uhlig $10^{-17} \times 10^{-13*})$	$2.8 \cdot 10^{-16}$
			Carbonate
D'Ans-Lax	$a_{Pb^{++}} \cdot a_{CO_3^{--}} = L$	Uhlig	$1.5 \cdot 10^{-13}$
Sulphate	$a_{Pb^{++}} \cdot a_{SO_4^{--}} = L$	D'Ans-Lax	$1 \cdot 10^{-57}$
Chromate	$a_{Pb^{++}} \cdot a_{CrO_4^{--}} = L$	D'Ans-Lax	$2.6 \cdot 10^{-56*})$

*) Values obtained under effluent treatment conditions.

3.7.5.5.4 Sedimentation

Whichever of the three most commonly used alkalis are employed as precipitating agents, the result is a finely dispersed precipitate which does not settle to any significant extent within practically acceptable times and practically acceptable sedimentation is only accomplished by co-flocculation with another metal hydroxide (adsorption coagulation) such as ferric hydroxide. If, to a solution containing 100 mg/litre lead, increasing amounts of 3-valent iron are added (10, 30, 100 mg/litre) and then neutralised to various final pH values, then the final concentration of lead decreases as pH and iron concentration increase. Satisfactory results are only found, as shown in Table 3.10, at higher pH and iron concentration values. NaOH is, under these conditions, not so well-suited as the other alkalis in common use.

Where effluent is neutralised with soda, after addition of iron, pH values of 7 to 7.5 are adequate under most concentration conditions e.g. 10 to 100 mg/litre iron (III) and 5 to 100 mg/litre Pb, to reduce the final lead concentration below 0.5 mg/litre. The optimum conditions for bringing this about are shown in Fig. 3.60. From the curves can be seen whether, during the sedimentation. Inspection of the curves reveals the extent to which ferric hydroxide or lead-based precipitates dominate the process.

TABLE 3.10:

SOLUBILITY OF LEAD WHEN CO-PRECIPITATED WITH IRON (III) SALTS. ALKALI – NAOH, [PB⁺⁺] = 100 MG/LITRE

mg/litre Fe^{+++}	mg/litre Pb in solution (after 2 hours) at pH:				
	7.0	8.0	8.5	9.0	9.5
10	–	29	15	12	7
30	-	15	9	3.3	1.7
100	2.9	1.1	0.4	0.4	0.4

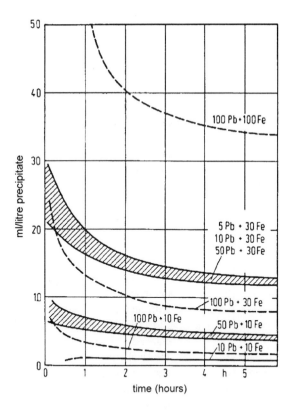

Fig. 3.60   Sedimentation curves for co-precipitation of lead and iron (III) using soda, as function of Pb:Fe ratio (mg/litre) at pH 7 & 8. [149].

In cases where the ratio Fe:Pb is greater than 0.5, the precipitate behaves more or less like iron. In these cases, additional of flocculating agents (e.g. copolymerisate of acrylamide and acrylate) significantly accelerates sedimentation.

Precipitation of lead by its flocculation with iron, using NaOH, involves an adsorption-coagulation process of only partly precipitated, partly colloidal material requiring a relatively large quantity of adsorbent. The case of soda is quite different. No colloidal material is formed in this case, only finely crystalline and highly insoluble lead compounds. In the presence of the voluminous flocs of Fe (III) hydroxide, the interaction is more one of incorporation than adsorption.

Heinzelmann & Ostermann [150] precipitated lead in presence of aluminium salts with comparable success, the aluminium behaving very much like the ferric iron. Initial concerns by these authors that soda might not prove usable, proved unfounded as was later confirmed by Eickershoff [151].

3.7.5.6 Iron

3.7.5.6.1 Precipitation pH

When solutions of iron (III) are titrated with NaOH, slaked lime or soda, the first precipitation is observed at pH 2.8. Initially a light yellow, partly colloidal hydroxide is formed. This becomes darker with increasing pH, finally flocculating as a red-brown solid. Chemical analysis shows precipitation to be complete at pH 3.5 (> 3 mg/litre Fe). The titration curves are shown in Fig. 3.61 [152].

In the case of iron (II), the first sign of precipitation, using slaked lime or NaOH are observed at pH 7. Using NaOH, a basic sulphate forms at pH 4.5, a basic carbonate being formed with soda. Both of these are relatively soluble. Precipitation is only complete above pH 8.9 with NaOH or slaked lime, a slightly lower value of 8.6 being found for soda. (see Fig 3.62)

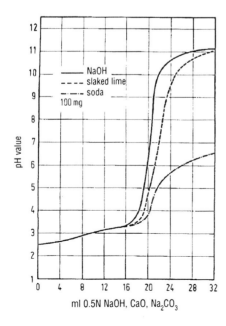

Fig. 3.61 Potentiometric titration curves for acidified iron (III) sulphate solutions (100 mg/litre). [152]

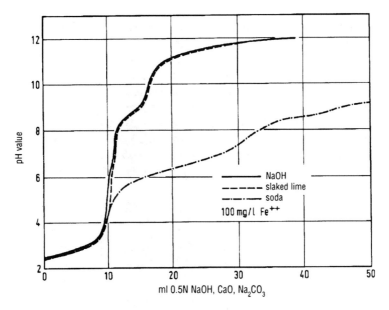

Fig. 3.62   Potentiometric titration curves for
acidified iron (II) sulphate (100 mg/litre Fe).

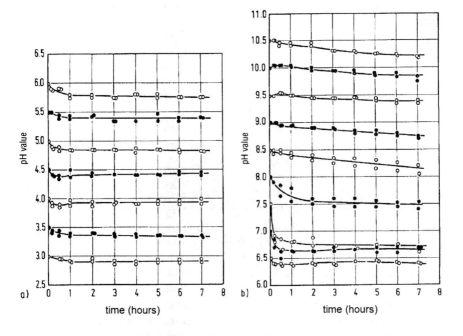

Fig. 3.63   pH change as function of standing time and initial pH
for precipitation of iron (III) using NaOH [152].

The pH retreat following neutralisation precipitation of iron (III) at pH 6.5 is, though precipitation is complete, only minimal. At pH values of 7 to 8, it becomes more pronounced. This is true both for NaOH and for slaked lime. Fig. 3.63 shows this in the case of NaOH. Since it is in the range of neutrality that the smallest changes in hydrogen ion concentration result in the largest pH changes, the phenomenon is most probably related to adsorption of the neutralising agent on the highly active surface of the iron (III) hydroxide. Where soda is used, virtually no pH changes are observed [152] and this is ascribed to the buffering action of the soda.

After precipitation of the iron (II) hydroxide, a marked pH retreat is observed. This is due to the oxidation, by dissolved oxygen of Fe (II) to Fe (III) in the neutral pH region. This oxidation releases 2 hydrogen ions per equivalent of iron oxidised (see Section 3.5).

3.7.5.6.2   Composition of Precipitation Products

On precipitation using any of the three alkalis, compounds of iron (III) precipitate as bulky reddish-brown flocs of the ferric hydroxide. The precipitate contains large amounts of adsorbed and capillary-attracted water and is, because of its highly active surface, ideal for adsorption of other species in solution. Iron (III) compounds are thus frequently employed to promote flocculation of difficult-to-precipitate or colloidal species.

According to Guiter [153], in the initial stages or during incomplete precipitation from sulphate solutions, a basic salt of composition $Fe(OH)SO_4$ forms. The initially colloidal precipitate and the colour changes mentioned previously suggest the existence of another salt composition with higher hydroxyl ion content, having a darker colour. The basic salts of trivalent metals are described in greater detail under the sections on chromium & aluminium.

Chemical analyses of precipitates formed from sulphates, suggest that at pH 3.5, for 17 molecules of $Fe(OH)_3$, there is 1 molecule of $Fe_2(SO_4)_3$. As pH increases, so the sulphate content falls off rapidly so that the precipitate consists of a mixture of the basic sulphate with large excess of the ferric hydroxide, or the basic sulphate is adsorbed on the surface of the hydroxide. Van der Giessen [154] established. using Debye-Scherrer diagrams, that carefully dried iron hydroxide of composition FeOOH had a cubic lattice. The particles have irregular shape and size (ca 20 to 30 Å), as shown by SEM photos. It is thus seen that even those precipitates usually described as being amorphous can, under certain circumstances, assume a crystalline structure.

When soda is used as precipitating agent for iron at concentration of 1 gm/litre Fe^{+++}, a wide range of coloured precipitates can form.. At pH 3, they are dark brown, at pH 4, a yellow-brown, then becoming darker as pH increases, being almost black at and above pH 6. The Fe(III) carbonate is not known. Research suggests that all precipitates contain the same carbonate content – ca 0.7% CO_3. Adsorption is thought to play a major role, although not explaining the colour changes noted above.

Iron (II) hydroxide in its purest state, is white, becoming green to light green with even the slightest amount of oxidation. It is considerably more soluble than the iron (III) hydroxide. On air oxidation, dark green to almost black hydrolysis products form and these, on standing, acquire the reddish-brown colour of iron (III) hydroxide. As shown by Feitknecht & Keller [155], accidentally oxidised (green) iron (II) hydroxide will typically contain 1 to 2.5% trivalent iron. Only above an Fe (III) content of 10% does a brown colour become apparent. Up to this point, X-ray methods fail to reveal any crystal lattice structure. Above the 10% level, the αFeOOH lattice is revealed. The same authors have shown that basic iron (II) salts such as $FeCl_2 \cdot 3 \; Fe(OH)_2$ can be oxidised to form initially the dark green basic Fe (II)-Fe (II) compound. As oxidation proceeds, the lattice structure remains unchanged until, around the 56At% Fe (III) value, γFeOOH structure predominates. Complete oxidation is represented by iron (III) hydroxide which has a structure based on molecular layers with hexagonal symmetry.

Using soda, iron (II) initially forms, in the lower pH range, a basic carbonate which transforms to the hydroxide as pH increases. At extremely high pH, iron (III) hydroxide is relatively soluble and ferrites are formed. Where oxidising conditions simultaneously prevail, ferrates are known to form – in which iron is hexavalent. In the context of effluent treatment, these compounds are of no importance.

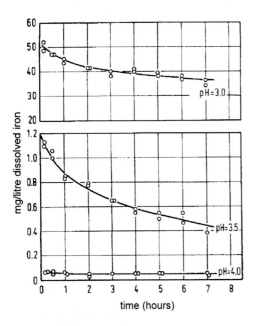

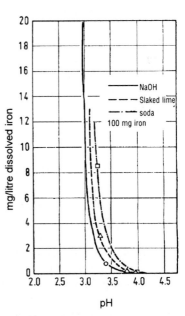

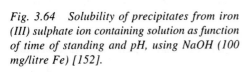

Fig. 3.64 Solubility of precipitates from iron (III) sulphate ion containing solution as function of time of standing and pH, using NaOH (100 mg/litre Fe) [152].

Fig. 3.65 Solubility of precipitates from iron (III) sulphate ion containing solution as function of pH (100 mg/litre Fe 2 hrs standing time)[152].

3.7.5.6.3   Solubilities & Solubility Products

Precipitates of iron (III) show pronounced ageing phenomena. Solubility plots using NaOH, reported as function of initial pH and time of standing show this clearly, as seen in Fig. 3.64. Similar studies with slaked lime or soda confirm the same trends. Curves, mainly in the range of incomplete precipitation, show little difference as between the various alkalis used, as seen in Fig. 3.65. This is linked to the virtual absence of structure in this pH range where precipitates are finely dispersed particles, if not colloidal in form.

Fig 3.66 shows the corresponding curves for iron (II). Using NaOH or slaked lime, the solubility at pH 8.9 is less than 3 mg/litre. Using soda, the same value results at the slightly lower pH of 8.6. See also Fig. 3.66 [21].

Table 3.11 quotes some values for solubility product of iron (II) and (III) hydroxides in comparison with those obtained by the author in effluent treatment studies.

TABLE 3.11:
SOLUBILITY PRODUCT VALUES FOR IRON (II) & (III) HYDROXIDES

Sparingly Soluble Compound	Definition of Sol'y Product	Source	Solubility Product (L_H)
Hydroxide Fe(II)	$a_{Fe^{++}} \cdot a^2_{OH^-} = L_H$	Uhlig	$1.65 \cdot 10^{-15}$
		D'Ans-Lax	$4.8 \cdot 10^{-16}$
		Hartinger [21]	$2 \cdot 10^{-15*})$
Hydroxide Fe(III)	$a_{Fe^{+++}} \cdot a^3_{OH^-} = L_H$	Cooper [156]	
		pH = 3.45	$2 \cdot 10^{-38}$
		pH = 2.32	$6.3 \cdot 10^{-40}$
		Evans and Pryor [157]	$3.16 \cdot 10^{-36}$
		Feitknecht, Michel and Buser [158]	10^{-35}
		D'Ans-Lax	$3.8 \cdot 10^{-38}$
		Hartinger [153]	$8.7 \cdot 10^{-38*})$

*)   Values obtained under effluent treatment conditions.

3.7.5.6.4   Sedimentation

Sedimentation rates using three different alkalis are shown in Fig. 3.67 as function of pH [152]. Once the precipitation pH for iron (III) has been exceeded, using NaOH or slaked lime, a sludge of constant volume results. Using slaked lime, the corresponding volume is smaller and this has been explained in terms of inclusions of undissolved lime particles within the bulky flocs of iron (III) hydroxide which result in a densifying action.

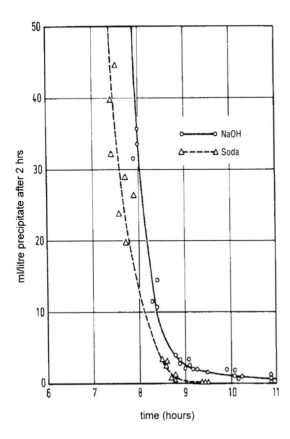

Fig. 3.66   Solubilities of precipitates from sulphate solutions of iron (II) using NaOH or soda as function of pH after 2 hrs time of standing (100 mg/litre Fe) [21].

Using soda, several differences should be noted. Here, even at lower iron concentrations, the colour changes described in Section 3.7.5.6.2 are seen, as function of pH. The supernatant liquors become clear, above pH 6, after some six hours time of standing. Prior to this, they are cloudy, thanks to the presence of very finely dispersed solids which are slow to settle. It would be incorrect to imply that formation of gaseous carbon dioxide exerts any flotative action, since bubbles which might lead to floating froth are only observed at pH values less than 6. Since an iron (III) carbonate appears to be unknown and there is nothing in the literature concerning the formation of basic iron (III) carbonate or its modifications using soda, it is difficult to draw any conclusions as to the above observations.

The degree of compression of the precipitate is quite large in comparison with iron (II) sludges and the voluminous hydroxide flocs display substantial densification on the floor of the containing vessel. Use of flocculating agents is highly advantageous in precipitation of iron (III) hydroxide. Best results are found using polyacrylamides with low content of polyacrylates. Fig. 3.68 shows the use of such compounds. A pH effect is scarcely in evidence. However certain other flocculating agents did, in some cases, show strong pH effects.

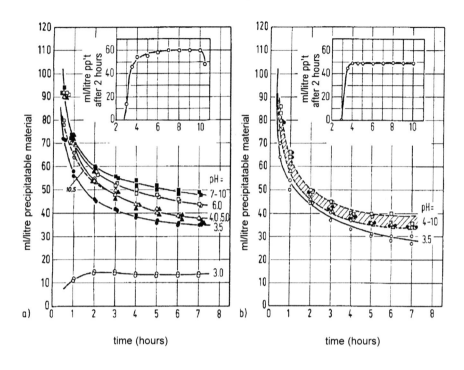

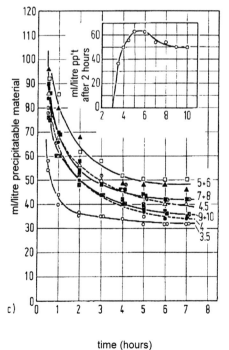

Fig. 3.67   Sedimentation curves &
precipitate volumes for sulphate
solutions of iron (III) as function of
pH (100 mg/litre Fe) [152].

a) with NaOH
b) slaked lime
c) soda

Iron (II) hydroxide can be completely precipitated above pH 8.9. The effect of dissolved air results in its incomplete oxidation to the iron (III) hydroxide, the process being accompanied by a decrease in pH. Thanks to the slow but steady hydrolysis reaction, a complete clarification by sedimentation is not possible. For this reason, it is preferable for a complete oxidation first to take place. However at higher iron concentrations, this is not possible even after long aeration periods, as shown by Morkowski & Thomaschk [159]. More powerful oxidising agents must be used to bring about a complete oxidation.

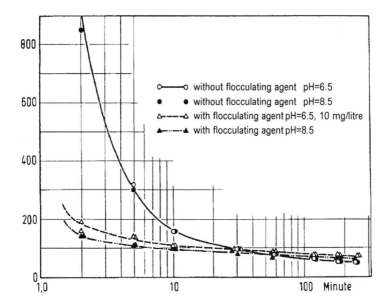

Fig. 3.68   Sedimentation rates for precipitation of iron (III) (100 mg/litre)
with and without added flocculating agents
(polyacrylamide containing some acrylates) at pH 6.5 and 8.5 using NaOH [152].

3.7.5.7   Aluminium

3.7.5.7.1   Precipitation pH values

Fig. 3.69 shows the titration plots obtained when neutralising aluminium solutions with various alkalis. In all cases, the first clouding of solution is seen at pH 4.3. Ap pH 4.8, as analysis of solutions shows, less than 3 mg/litre of metal ions remain in solution. Addition of yet more alkali leads to a further step in the titration plot between pH 9 and 10 (in the case of soda, this is seen hard to the right hand side of the Figure). The additional consumption of hydroxyl ions is due to formation of a soluble hydroxyaluminate. The dissolution of this species sets in around pH 8 but only at pH 9 is it so pronounced that it can be recognised by potentiometric titrations or even by visual observation. Aluminium hydroxide is, like zinc hydroxide, an amphoteric species. Using soda, the pH value for precipitation and for the re-dissolution are so far removed from its buffering pH range that all three steps are clearly visible [152]. pH changes following precipitation with NaOH or slaked lime, are most pronounced around pH 7. This relates, as is the case with iron, not only to the precipitation of the aluminium, but also to the sensitivity of the measurement in this pH range, to adsorption of the precipitating species. This is seen in Fig. 3.70 for aluminium precipitation using NaOH. The change seen at pH 10, notwithstanding a strong buffering action, is due to the uptake of hydroxide ions for formation of the hydroxy-complex, this reaction being most pronounced at this pH.

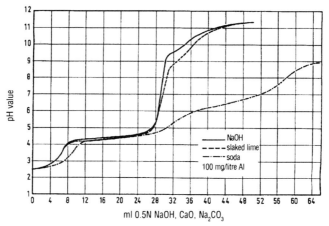

*Fig. 3.69   Potentiometric titration of acidic aluminium sulphate solutions
(100 mg/litre Al) with various alkalis [152].*

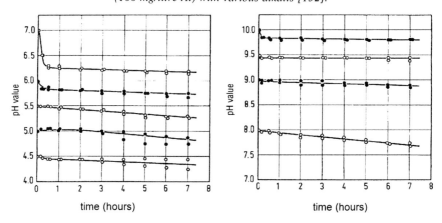

*Fig. 3.70   pH changes as function of time of standing and initial pH value
in precipitation of aluminium with NaOH [152].*

3.7.5.7.2   Composition of precipitates

The precipitates using all three alkalis consist of aluminium hydroxide. At lower pH values, where precipitation is incomplete, basic salts are often formed, with a hydroxide content increasing as pH increases. In aqueous solutions, aluminium ions frequently form a hexa-aquo complex, i.e. six water molecules are oriented as dipoles around the central Al atom, reflecting its hexavalent coordination number. Similar hydration effects are seen in the case of other metal ions (see Section 3.9.1). The process can be represented as follows:

$$[Al(H_2O)_6]^{+++} \rightleftharpoons [Al(H_2O)_5(OH)]^{++} + \rightleftharpoons \qquad 3.139$$
$$[Al(H_2O)_4(OH)_2]^+ + 2\,H^+ \rightleftharpoons [Al(H_2O)_3(OH)_3] + 3\,H^+$$

The condition for the equilibrium to be shifted to the right is for the H^+ equivalent ions formed to be reacted with hydroxide anions as is the case when alkali is added. The species shown on the right hand side as a *triaquo-trihydroxy-complex* is actually no more than the hydroxide. By means of this reaction pathway, a range of basic salts can form, such as $Al(OH)SO_4$ or $Al(OH)_2Cl$ (presence of water molecules here neglected).

The reaction mechanism depicted above, involving aquo-complexes can be equally adapted for formation of any type of basic salt of a metal ion up to the point where the simple hydroxide is formed, provided one knows the co-ordination number of the metal. For amphoteric hydroxides, the transformation set out in Equation 3.139 can also be extended. In the case of aluminium hydroxide, this leads to the di-aquo-tetra-hydroxy aluminate, more informally known as the tetrahydroxyaluminate:

$$[Al(H_2O)_3(OH)_3] \rightleftharpoons [Al(H_2O)_2(OH)_4]^- + H^+ \qquad 3.140$$

Jahr & Pernoll [160] were unable to establish the existence of higher basic hydroxy complexes in spite of using a range of analytical techniques. Jahr & Pletschke [161] suggested that in weakly alkaline conditions, the hydroxyl complex is in equilibrium with a polymeric aquo-hydroxy aluminate. In both polymeric and monomeric forms, this has an octahedral (6 valent) co-ordination. In strong alkaline media, it loses water to form a co-ordination compound with tetrahedral (4-valent) structure. This, in turn, is in equilibrium with its polymeric form. Under extremes of alkalinity, further loss of water occurs with formation of a short-chain polymer in which the units are bonded by mu-type ether bonds rather than the mu-type hydroxy bonds of its precursor (in the equation below, polymers are represented as dimers):

$$[(H_2O)_2(OH)_3Al-(OH)-Al(OH)_3H_2O_2]^- \underset{-OH^-}{\overset{+OH^-}{\rightleftharpoons}} 2\,[Al(H_2O)_2(OH)_4]^-$$
$$\Updownarrow -2\,H_2O$$
$$[(OH)_3Al-(OH)-Al(OH)_3]^- \rightleftharpoons 2\,[Al(OH_4]^-$$
$$\Updownarrow -H_2O$$
$$[(OH)_3Al-O-Al(OH)_3]^{--} \qquad\qquad 3.141$$

The preceding brief survey of the chemistry of amphoteric hydroxides such as those of aluminium should be sufficient to indicate the complexity of reactions which can take place. Aluminium containing effluents arise frequently in alkaline form and in this case, their treatment is as shown in Equation 3.140 but in the backwards sense. Precipitation studies from sulphate solutions using NaOH show that precipitates formed in the pH range 4.5 to 5.0 can contain up to 25% sulphate ions. On a molar basis, however, this would correspond to a ratio of $Al_2SO_4:Al(OH)_3$ = approx. 0.1. The sulphate content falls away rapidly as pH increases, suggesting the existence of a basic salt with substantial quantities of hydroxide. However adsorption phenomena on the strongly active surface of the aluminium hydroxide are an alternative explanation.

3.7.5.7.3   Solubilities and solubility products

Fig. 3.71 shows the solubility of aluminium precipitated using NaOH as a function of pH and time of standing. No significant ageing process is observed, as seen from the negligible change in solubility. Similar data are obtained using slaked lime or soda.

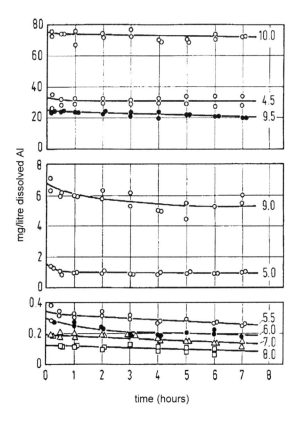

Fig. 3.71 Solubility of precipitates from an aluminium sulphate solution as function of time of standing and initial pH, using NaOH (100 mg/litre Al) [152].

In terms of pH retreat, the solubility curves are also shown in Fig. 3.72 for all three alkaline species but in this case, after two hours standing time. These results show clearly the amphoteric nature of aluminium hydroxide. In spite of this, aluminium can be entirely satisfactorily precipitated in the pH range 5.0 to 8.5 Below pH 5.0, the hydroxide ionises to form aluminium ions while above pH 8.2 aluminates can be detected in solution [152]. At or close to the iso-electric point, both ionic species are in equilibrium with one another and the sparingly soluble hydroxide precipitates. This occurs, as best seen in Fig. 3.91, at pH 6.7. In Table 3.12 are collected some literature values with, for comparison, those obtained under effluent treatment conditions.

The dissociation constant, K_D, for the hydroxyl complex has been calculated, using the aluminium values found at pH 9 to 10.5.

$$\frac{c_{Al^{+++}} \cdot c_{OH^-}^4}{c_{[Al(OH)_4]^-}} = K_D \qquad\qquad 3.142$$

As was earlier shown for the case of zinc, one then obtains:

$$\frac{L_H \cdot P}{c_{Al(gel.)} \cdot c_{H^+}} = K_D = 4 \cdot 10^{-34} \qquad\qquad 3.143$$

In contrast to the hydroxyl complexes formed by zinc or cadmium, aluminium does not form an insoluble calcium salt.

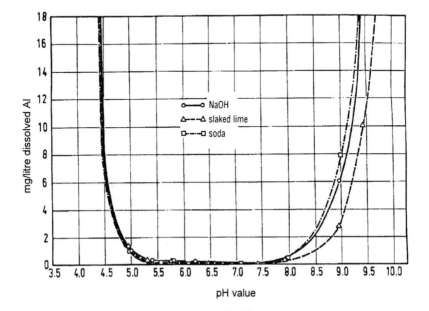

Fig. 3.72   *Solubility of aluminium precipitates from sulphate solution as function of pH (100 mg/litre Al) after 2 hours time of standing [152]*

TABLE 3.12:

SOLUBILITY PRODUCT VALUES FOR ALUMINIUM HYDROXIDE

Sparingly Soluble Compound	Definition of Sol'y Product	Source	Solubility Product (L_H)
Hydroxide	$a_{Al^{+++}} \cdot a_{OH^-}^3 = L_H$	Heyrovsky [162]	$1.06 \cdot 10^{-33}$
		Uhlig	$1.9 \cdot 10^{-33}$
		Kruyt and	
		Troelstra [163]	ca. 10^{-32}
		Lacroix [164]	10^{-34}
		Szabó, Csányi	
		and Kávai [165]	$1.3 \cdot 10^{-33}$
		Hartinger [152]	$2 \cdot 10^{-32}*)$

*) Values obtained under effluent treatment conditions.

3.7.5.7.4   Sedimentation

Aluminium hydroxide tends to settle, as seen in Fig. 3.73, relatively slowly. The precipitate, by comparison with those of other metals at the same concentration in solution (100 mg/litre), is bulky and this is only partly the result of its higher molarity when making comparisons on a weight basis. The plots showing precipitate volume indicate an increase to the point where precipitation is complete, thereafter a decrease at the point where the hydroxy-complex is formed. The plots shown in Fig. 3.73a show what is more or less an inverted solubility curve. Inside the pH range for complete precipitation, the sedimentation properties differ little in the various cases. Results using slaked lime are very similar to those with NaOH [152].

The sedimentation rate of aluminium hydroxide is slow, even in comparison with that of equally voluminous iron hydroxide. The time for sludge densification is long − even after 7 hours it is incomplete. In practical terms, this means that sludge settling plants must be larger than would be the case for comparable amounts of other metals. Use of flocculating agents does bring benefits as seen in Fig. 3.74. Large flocs are immediately formed and as much sedimentation occurs as would take an hour without their use. Here too, polyacrylamides with minimal acrylate content perform best.

3.7.5.8   Chromium

Chromium can only be precipitated in its tri-valent state and where it is present as a hexavalent compound, this must first be reduced. Chromium (VI) salts are all soluble in alkalis.

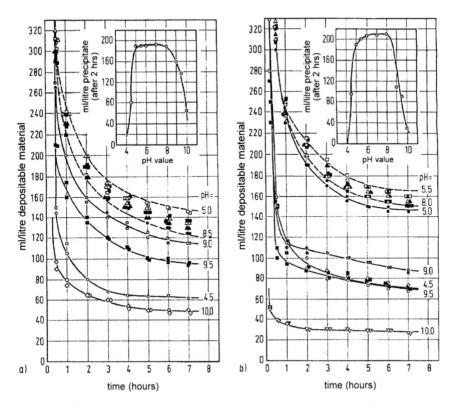

Fig. 3.73 Sedimentation & sludge volumes after precipitation of aluminium (100 mg/litre) from its sulphate solution [152]. a) with NaOH b) with slaked lime.

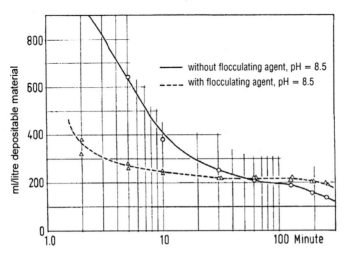

Fig. 3.74 Sedimentation rate of aluminium (100 mg/litre) using NaOH with & without use of flocculating agent (polyacrylamide with some acrylate) at pH 8.5. [152].

Of the tri-valent chromium salts, many exist in one of two isomorphous forms either violet or green. This behaviour is readily explicable in terms of Werner's Co-ordination theory. Like aluminium, Cr (III) ions in aqueous medium have a x-fold co-ordination number, i.e. the metal sits at the centre of an octahedral structure, at each corner being sited, for example, a water molecule (inner sphere co-ordination). The anions of equal and opposite charge are located outside this (outer sphere) shell of hydrated metal ion. This is the violet form. In fact, although it has the lower free energy and is formed from the higher energy isomorph on standing, it is rarely observed in effluent treatment conditions. Where, as is known to be the case for chloride ions, these ions in the outer sphere are exchanged with water molecules in the inner sphere (e.g. by heating), the solution turns green. This transformation is represented by the following equation (see also Section 3.9.1).

$$[Cr(H_2O)_6]Cl_3 \rightleftharpoons [Cr(H_2O)_4Cl_2]Cl \cdot 2\,H_2O \qquad\qquad 3.144$$

3.7.5.8.1   Precipitation pH

Fig. 3.75 shows the titration curves obtained when chromium sulphate is neutralised with various alkalis [166].

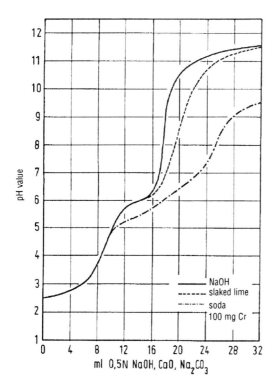

pH value

NaOH
slaked lime
soda
100 mg Cr

ml 0,5N NaOH, CaO, Na$_2$CO$_3$

Fig. 3.75   Potentiometric titration of acidified chromium sulphate solution (100 mg/litre Cr) with various alkalis [166].

As in the case of iron, onset of visible precipitation is preceded by a deepening of colour and then a cloudiness due to the presence of colloidal particles, a sequence indicative of the formation of basic salts. Using NaOH or slaked lime, the first opalescence occurs at pH 5.5, with visible precipitates forming at pH 5.8 to 5.9. Chemical analysis has shown that at pH 6.8 or above, the solubility of chromium is less than 0.5 mg/litre. However this is only the case where pH is held for at least two hours at 6.8 or above. Formation of a hydroxy-complex is not evident from the titration curves since no such compounds are present to any significant extent at pH 10.

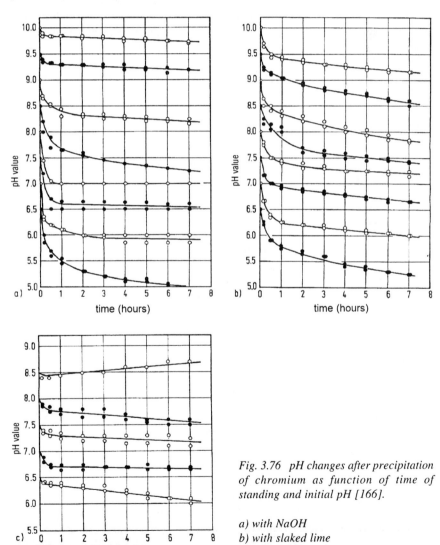

Fig. 3.76 pH changes after precipitation of chromium as function of time of standing and initial pH [166].

a) with NaOH
b) with slaked lime
c) with soda.

Using soda, the onset of opalescence and precipitation begin at slightly lower pH values of 5.2 to 5.4. However even at higher pH values, precipitation is not complete.

The pH changes observed after precipitation of chromium hydroxide are, by virtue of their magnitude and the effect of this on other metals when present, of some significance. Fig. 3.76 shows pH retreat as function of time of standing and initial pH. Differences of behaviour between NaOH and slaked lime at the higher pH values have their origin in the differing chemical reactivities of the precipitates in alkaline solution. The results where soda is used, is due to the buffering capacity of this compound.

Dissolved Cr (III) ions, after neutralisation precipitation, form as finely-dispersed grey-green particles, with a blueish tint. Particle size distribution extends down to the colloidal range. In consequence, the precipitate has an extremely high surface area with a very high surface energy. As such, it is capable of adsorbing large amounts of other species in solution such as neutralising alkali. Such behaviour was confirmed by Krusenshtern & Axmacher [111]. Their findings, albeit at rather high chromium concentrations are shown in Fig. 3.77.

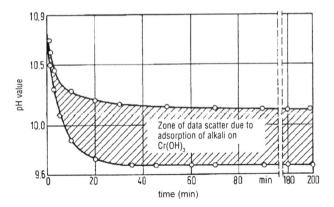

Fig. 3.77 pH changes as function of time of standing after precipitation at pH 10.8 using NaOH or slaked lime (1740 mg/litre Cr) (after Krusenschtern & Axmacher [111])

3.7.5.8.2 Composition of precipitates

The colour changes observed in the initial stages of neutralisation and precipitation are due to partial or complete hydrolysis of the chromium salts and the products formed by this process, in close analogy with the behaviour of iron previously described. The chromium ion, like that of aluminium, as a six-fold co-ordination number and its step-wise hydrolysis to form the hydroxide (triaquo-trihydroxy chromite) is also analogous to aluminium (see Section 3.139). The dissociation constants for this process were determined by Schwab & Polydoropoulos [106]. The series of equilibria which govern the reaction can readily be followed by spectrophotometric methods which show a progressive decrease in the peak heights at shorter wavelengths are matched by increase in peak height at the longer wavelengths as the concentration of cationic species falls away with time. Where the mono- or divalent cations encounter anions in solution, the corresponding basic salts are formed – these are not particularly insoluble. After complete precipitation, the chromium is present as its hydroxide. Chemical analysis of such precipitates is not a fruitful exercise since, on account of the extensive adsorption of whatever species are present in solution, no conclusions can be drawn as to the stoicheometry of basic salts formed.

Chromium hydroxide is amphoteric. Above pH 8.3, its solubility exceeds to 0.5 mg/litre Cr value referred to above. In the formation of the hydroxy-complex, chromium retains its 6-fold co-ordination number to form a tri-valent anion using only hydroxyl radicals. In this, it behaves differently from aluminium. The former reaction can be written as three consecutive processes. These are lumped together in the following equation for formation of the hexahydroxyl chromite ion.

$$[Cr(H_2O)_3(OH)_3] \rightleftharpoons [Cr(OH)_6]^{---} + 3\,H^+ \qquad\qquad 3.145$$

In presence of calcium ions, this hexahydroxyl chromite forms a very insoluble salt, analogous with the tetrahydroxy zincate and this behaviour obscures its amphoteric character when slaked lime is used as a precipitating agent (see also Fig. 3.79

$$2\,[Cr(OH)_6]^{---} + 3\,Ca^{+++} \rightleftharpoons Ca_3[Cr(OH)_6] \qquad\qquad 3.146$$

When chromium ion containing solutions are treated with soda, the relatively soluble carbonate or bicarbonate complexes form. Studies of such precipitates, which are darker than the hydroxides, gave a carbonate content of 7 to 12% CO_3^{--}, increasing as pH increased from 6.5 to 9.0. The supernatant solution remained very cloudy and soda is not a suitable reagent for precipitation of chromium.

3.7.5.8.3   Solubilities and solubility products

Fig. 3.78 shows the solubility of chromium, after precipitation with NaOH or slaked lime as function of time of standing and initial pH. In spite of the very marked pH retreat shown here, no other metals are known to exhibit a change of solubility of comparable magnitude. This ageing process can be ascribed to three types of action. First is a growth of larger particles at the expense of smaller ones, second is adsorption of species in solution and third is the the formation of special complexes (see Section 3.9.3.1.1).

The solubility curves are seen in Fig. 3.79 as function of pH, after 2 hours time of standing [166]. Krusenschtern & Axmacher [111] carried out studies using much lower initial chromium concentrations (10^{-4} N) using NaOH or slaked lime. Their

findings were in broad agreement with those in Figs 3.78 and
3.79. They also showed, as might be expected, that increasing the temperature to
80°C significantly accelerated the ageing process.

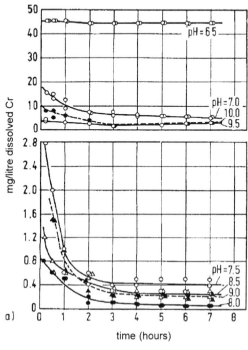

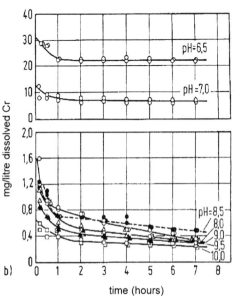

*Fig. 3.78 Solubility of precipitates
from acidified chromium sulphate
solution (100 mg/litre Cr)   as
function of initial pH and time of
standing.*

a) NaOH   b) slaked lime

Table 3.13 lists some values for solubility products from the literature together with comparable values obtained by the author under effluent treatment conditions. The latter suggest a somewhat higher solubility. For the pH range 8.5 to 10.0, the following dissociation constant has been calculated:

$$\frac{c_{Cr^{+++}} \cdot c_{OH^-}^6}{c_{(Cr(OH)^{6]---}}} = K_D \qquad\qquad 3.147$$

When the ionic product of water and solubility product of the hydroxide are inserted into this equation, the dissociation constant can be calculated:

$$\frac{L_H \cdot P^3}{c_{Cr(gel.)} \cdot c_{H^+}^3} = K_D = 5.6 \cdot 10^{-38} \qquad\qquad 3.148$$

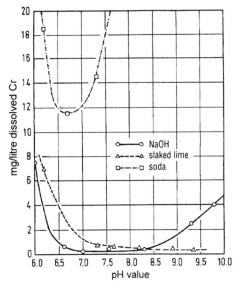

Fig. 3.79 *Solubilities of chromium precipitates from acidified sulphate solution (100 mg/litre Cr) with various reagents as function of pH, after 2 hours time of standing [166]*

TABLE 3.13:
SOME VALUES FOR THE SOLUBILITY PRODUCT OF CHROMIUM HYDROXIDE

Sparingly Soluble Compound	Definition of Sol'y Product	Source	Solubility Product (L_H)
Hydroxide	$a_{Cr^{+++}} \cdot a_{OH^-}^3 = L_H$	Schwab and Polydoropoulos [106]	$3.8 \cdot 10^{-30}$
		Kovalenko [167]	$6.3 \cdot 10^{-31}$
		D'Ans-Lax	$6.7 \cdot 10^{-31}$
		Hartinger [166]	$3 \cdot 10^{-28*)}$

*) Values obtained under effluent treatment conditions.

3.7.5.8.4   Sedimentation

The sedimentation rate for neutralisation of chromium containing solutions using
NaOH or slaked lime, is shown in Fig. 3.80. The two reagents behave very similarly
in regimes where precipitation is complete. No significant further densification of
sediment takes place after 2 hours [166]. The volume of sediment formed using
slaked lime is smaller, making this the more attractive reagent, and Krusenschtern &
Axmacher [111] confirmed this, at higher chromium ion concentrations.

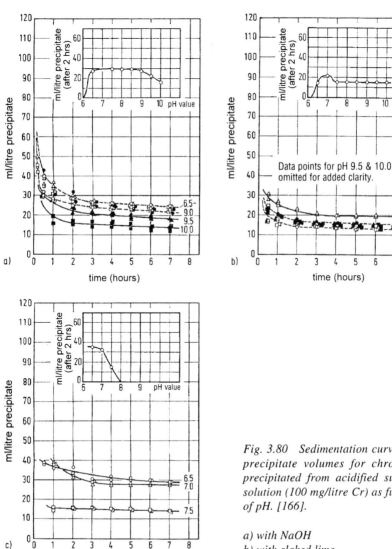

Fig. 3.80   Sedimentation curves and
precipitate volumes for chromium,
precipitated from acidified sulphate
solution (100 mg/litre Cr) as function
of pH. [166].

a) with NaOH
b) with slaked lime
c) with soda.

The sedimentation curves for chromium would compare advantageously with those for other metals, if considered only in terms of what had deposited. However the supernatant solutions remain cloudy and the process is incomplete even after 24 hours. The magnitude of residual Cr above the precipitate is gauged from Fig. 3.81. The total Cr is the sum of unsedimented particles, material in colloidal form and truly dissolved species such as chromium cations, or hydroxy-chromites. Such complex behaviour is, by its nature, irreproducible. The overall picture is nevertheless clear, in showing how – in certain circumstances - chromium is extremely difficult to remove from an effluent. [166].

Use of polyacrylamide flocculating agents greatly improves the flocculation and increases the rate of settling. Even in this case, however, a residual cloudiness persists for far longer than in the case of other metals. Copolymerisates with larger proportions of polyacrylates were much less effective than those with minimal amounts, nor was sodium carboxymethyl cellulose any more effective.

In practice, chromium should always be co-precipitated with other metals where these are present. Even so, care must be taken to maintain conditions for low nucleation rates and conditions of rapid supersaturation, during the neutralisation operation, should always be avoided when chromium ions are present.

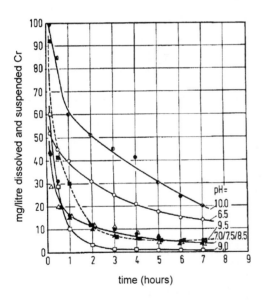

Fig. 3.81   Total chromium content (dissolved & undissolved) in supernatant solutions as function of time of standing and precipitation pH, using NaOH (100 mg/litre Cr). [166].

3.7.5.9   Other Metals

In the following section, comments are offered regarding other metals which arise less commonly in the effluents from metal-working and finishing industries.

Tin, in its 2-valent state, and at initial concentrations of 100 mg/litre, will begin to precipitate in finely-dispersed form at pH 3.8. Above pH 4.3, chemical analysis shows the concentration in solution to be less than 2 mg/litre. Above pH 9.2, the concentration increases as the amphoteric character of the metal manifests itself by formation of hydroxy complexes such as trihydroxy stannite:

$$Sn(OH)_2 + OH^- \rightleftharpoons [Sn(OH)_3]^-$$ 3.148a

This reaction, though not evident by inspection of the titration curve shown in Fig. 3.81, is readily recognisable in the solubility curve seen in Fig. 3.82. In the studies from which these results are drawn, a blanket of inert gas (nitrogen) was used in order to minimise the oxidation of tin (II) to tin (VI) which readily occurs when dissolved oxygen (air) is present in solution. [21]. Since the hydrolysis product of tin (VI) is very much less soluble than those from tin (II), solubilities (and solubility product values derived from these) can be used and compared with those found under effluent treatment conditions, providing the tin is not completely oxidised.

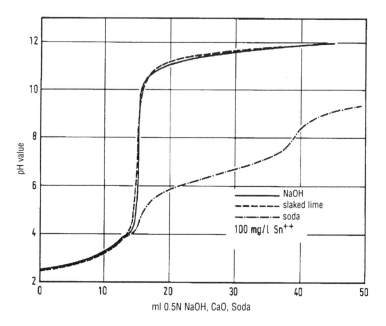

Fig. 3.82 Potentiometric titration of acidified tin (II) sulphate (100 mg/litre Sn) using various alkalis.

When NaOH is used as a precipitating alkali, a pH value of 9.2 should not be exceeded. Use of slaked lime does not bring added benefits by formation of a sparingly soluble calcium hydroxy compound, as is the case for zinc or cadmium (see Fig. 3.83). Although, using slaked lime, the solubility at higher pH values is somewhat less than for NaOH, it is still in excess of 2 mg/litre at pH 9.2. When soda is used, the chemistry is quite different, on account of formation of a relatively soluble carbonate complex. In this case, the 2 mg/litre tin value is exceeded at pH > 7.6. The conclusion is that soda should not be used for metal precipitation where tin (II) compounds are present.

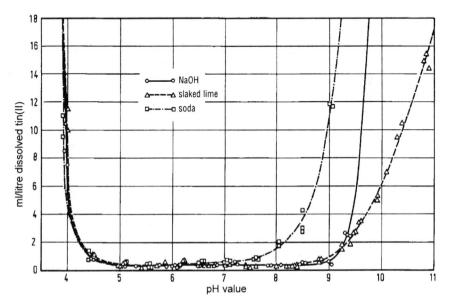

Fig. 3.83   Solubilities of precipitates of tin(II) from sulphate solution as function of pH (100 mg/litre tin) after 2 hours standing time.

Four-valent tin compounds are precipitated, on neutralisation, as the very insoluble species tin oxyhydrate ($SnO_2 \cdot xH_2O$), albeit in finely dispersed, even colloidal form. In practice, this compound will often form when tin (II) is precipitated. In terms of effluent treatment, this imposes certain conditions, as is also the case with lead ions. Thus, where no other metals are present, compounds of iron (III) or aluminium must be added to promote flocculation. Tin (IV) oxide hydrate is, like tin (II) hydroxide, an amphoteric compound, forming, in excess alkali, the soluble hydroxystannate.

The values of solubility product and dissociation constants in Table 3.14 have been calculated from solubilities in the pH range 3.9 to 4.6 (solubility products) and pH 8.5 to 9.6 (dissociation constants). In presence of alkali metal cations, stannites form, as shown in Equation 3.148a. With divalent cations, the corresponding anion is $[Sn(OH)_4]$.

Silver is a metal which, because of its value, should never be found in effluents except in trace quantities. However since it is toxic to fish and the species on which they feed, it must be removed from effluent.

Silver hydroxide, AgOH, is a relatively strong base and is thus not precipitated until pH 9.5. Its formal precipitation pH lies between pH 10 and 11. These precipitates, like those of lead and tin, are finely dispersed and other metals must be present or added, for flocculation to take place. It is better, when possible, to precipitate silver as its chloride but here too, some other species are required for a flocculation action.

The two valent metals cobalt, manganese, magnesium all form relatively strongly basic hydroxides, via a basic salt. According to Britton [104, 105], their precipitation commences at pH values of 6.8, 8.7 and 10.5 respectively. Solubility product values for the hydroxides of these metals as well as some of their sparingly soluble salts are shown in Table 3.14.

TABLE 3.14:

SOLUBILITY PRODUCTS OF SOME SPARINGLY SOLUBLE SPECIES LESS COMMONLY ARISING IN EFFLUENT TREATMENT (FROM D'ANS-LAX)

Metal	Sparingly Soluble Compound	Solubility Product (L_H or L
Tin	$Sn(OH)_2$	$5 \cdot 10^{-26}$
		$6 \cdot 10^{-25}$ [21]
	$Sn(OH)_2$	$1 \cdot 10^{-56}$
Silver	AgOH	$1.24 \cdot 10^{-8}$
	AgCl	$1.61 \cdot 10^{-10}$
	Ag_3PO_4	$1.8 \cdot 10^{-18}$
Cobalt	$Co(OH)_2$	$2 \cdot 10^{-20}$
Manganese	$Mn(OH)_2$	$4 \cdot 10^{-14}$
Magnesium	$Mg(OH)_2$	$5.5 \cdot 10^{-12}$

3.7.5.10 Concluding comments on neutralisation precipitation of individual metals from dilute solution

Concluding the section on precipitation of individual metals, some comments are made on the at times quite divergent recommendations for treatments under extreme conditions. While treatment is often easier when two or more different metal ions are present (see Section 3.7.6), the trends in that case are the same as those mentioned here. Shown in Table 3.15 and Fig. 3.84 are the most important values set out for ease of comparison. This information underpins the requirements set out in Tables A and B (on pages 97 & 98). Table 3.15 summarises the pH values at which precipitation commences as well as the formal precipitation pH's (i.e. the values at which solubility is at a minimum), the pH values at which amphoteric species begin to redissolve as well as the solubility product values determined under effluent treatment conditions and the dissociation constants of the hydroxy complexes.

TABLE 3.15:
SUMMARY OF pH VALUES FOR METAL PRECIPITATION USING NAOH

Metal	pH Value			Sol'y Product of hydroxide**)	Dissoc'n of const. of hydroxy complex**)
	ppt'n onset	complete ppt'n*)	redissol'n		
Tin (II)	3.8	4.3	9.2	$6 \cdot 10^{-25}$	$7 \cdot 10^{-25}$
Copper	5.8	7.6	–	$2 \cdot 10^{-19}$	–
Iron (II)	7.0	8.9	–	$2 \cdot 10^{-15}$	–
Zinc	7.6	9.0	10.8	$4 \cdot 10^{-17}$	$6 \cdot 10^{-19}$
nickel	7.8	9.9	–	$5.8 \cdot 10^{-15}$	–
cadmium	9.1	10.4	–	$1.3 \cdot 10^{-14}$	–
lead	6.5	10.3	–	ca. 10^{-13}	–
iron (III)	2.8	3.5	–	$8.7 \cdot 10^{-38}$	–
aluminium	4.3	4.8	8.7	$2 \cdot 10^{-32}$	$4 \cdot 10^{-34}$
chromium	5.8	6.8	8.3	$3 \cdot 10^{-28}$	$5.6 \cdot 10^{-38}$

*) Values used from Fig. 3.84
**) Values derived under effluent treatment conditions.

Fig. 3.84 affords an overview of the precipitation pH ranges for the most important metals (using NaOH) and the scope for extending these ranges by other means. This graphical presentation provides a simple means for assessing the basicity of the hydroxides. Arranging these in order, a sequence is obtained from the strongly basic hydroxides to those which are used to precipitate the metals. Each hydroxide can, in theory, be used to precipitate the species on its left, being itself dissolved. At constant pH, the solubility of these hydroxides increases in step with their basicity:

$$Me(OH)_4 < Fe(OH)_3 < Sn(OH)_2 < Al(OH)_3 < Cr(OH)_3 < Cu(OH)_2 <$$
$$< Fe(OH)_2 < Zn(OH)_2 < Ni(OH)_2 < Cd(OH)_2 < Pb(OH)_2 < \qquad 3.149$$
$$< Mg(OH)_2 < Ca(OH)_2 < NaOH$$

With very minor exceptions, the most pronounced pH retreats observed after precipitation, follow the same sequence.

From Fig. 3.85, it can be seen that for the divalent metals, the greatest deviations coincide closely with the pH ranges in which most of the metals are precipitated. The magnitude of these deviations also allows us to draw conclusions as to the formation of basic salts of the divalent metals and their heterogeneous transformation to the hydroxide, especially when nickel is used as a reference, bearing in mind that this metal is the least prone to formation of a basic salt. In the case of the trivalent metals,

deviations tend to occur mostly around the pH neutral point. It is here that the
smallest change in hydrogen ion concentration effects the largest change in pH and
thus influences the adsorption of neutralising reagents. Proof of such adsorption can
be found by making pH measurements of the water entrained by an aged precipitate.
It is found that this pH is significantly higher than the value at which the species was
precipitated. The origin of this phenomenon lies in the desorption of the adsorbed
alkali as, during the ageing, the surface area of the precipitate decreases.

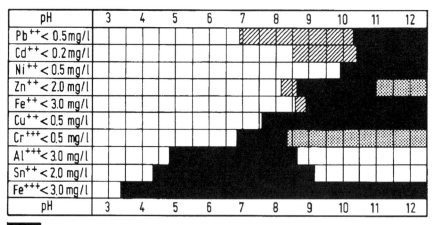

Fig. 3.84 Precipitation pH values for the most important metals,
showing their typically permitted effluent concentrations.

If the concentration vs. pH diagram is re-drawn, using a logarithmic scale for the
metal ion concentration axis and when the theoretically calculated solubility values
derived from solubility product values are inserted, the deviations can be clearly
seen, especially at the lowest concentrations (see Figs. 3.86 to 3.92). These are based
on solubility product values derived under practical conditions. As the precipitation
pH value increases, the solubility values do not invariably coincide with values
expected from theory (for reasons explained in Section 3.73). The solubility products
are derived from values in the mean concentration range and this results in the
intersection points shown. The magnitude of the deviations are an indication of the
errors which result when solubility product data alone is used. These errors are most
pronounced in the case of very low concentrations of metals having an amphoteric
hydroxide.

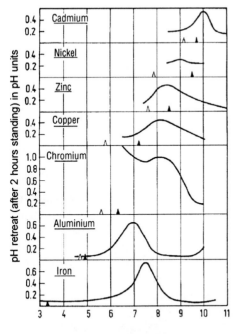

Fig. 3.85 Position and magnitude of pH retreat following precipitation of the most important metals with NaOH (100 mg/litre metal ion) [132].

Δ First precipitation

▲ Solubility of less than 2 mg/litre reached

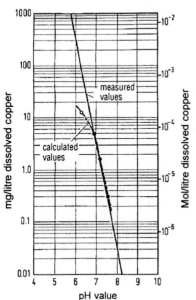

Fig. 3.86 Solubilities after NaOH precipitation of copper ions, from actual measurement and by calculation using solubility product data. [131]

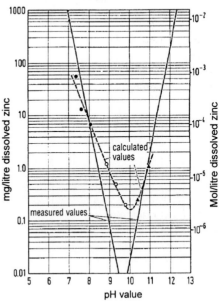

Fig. 3.87 Measured and calculated values (from solubility product & dissociation constants of zincate) for solubility values of zinc after precipitation with NaOH. Legends etc as prev.[131]

Increasingly stringent legislative limits are now very close to the minimal achievable concentration values for the amphoteric hydroxides, leaving little room for error when operating effluent plants in such cases. In respect of other metals, the precipitation pH values have somewhat increased. It is instructive to compare the data in Fig. 3.84 with older regulations, where discharge concentrations of 1 to 3 mg/litre for individual metals were required [168, 169].

The volume of precipitates formed can be calculated using Table 3.16, for effluent metal concentrations of 100 mg/litre of each metal. This also highlights differences when alkalis other than NaOH are used, reflecting, in the main, differences in the composition of the precipitates formed, as is the case for zinc, cadmium, lead or chromium.

Any relationship between the volume of solids formed and the molarity of the initial solution, is evident only in the case of metals much heavier or much lighter than the median value, for example lead or aluminium. In the former case, the precipitates would be even denser, were it not necessary to use other metals to promote flocculation. Even the iron, so commonly used in this case, sediments with lower volume than would be the case were it the sole metal present and this is ascribed to a greater compressibility of the sediment due to the greater weight of the lead-containing flocs.

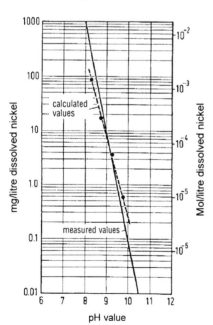

Fig. 3.88   Solubility of nickel after precipitation with NaOH, showing measured values and those calculated using solubility product data. Legends as prev. [131]

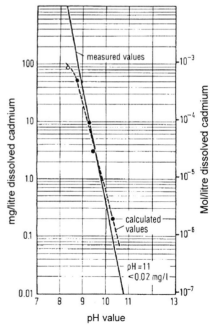

Fig. 3.89   Solubility of cadmium after precipitation with NaOH, showing measured values and those calculated using solubility product data. Legends as prev. [131]

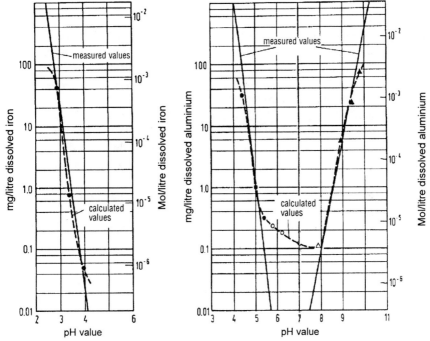

Fig. 3.89 Solubility of iron (III) after precipitation with NaOH, showing measured values and those calculated using solubility product data. Legends as prev. [131]

Fig. 3.91   Measured and calculated (from solubility product and dissociation constant data) value for solubility of aluminium, after precipitation with NaOH [94].

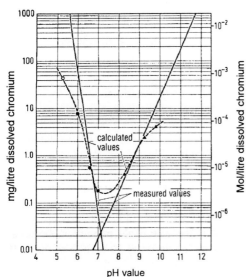

Fig. 3.92   Solubilities of chromium after precipitation by NaOH, from experimental measurements and as derived from solubility products or dissociation constants for the chromites. [131].

TABLE 3.16:

COMPARISON OF PRECIPITATE VOLUMES FOR VARIOUS METAL HYDROXIDES AT 2 HOURS STANDING AFTER PRECIPITATION.
(Concentrations 100 mg/litre of Metal)

Metal	Precipit'n pH value	ppt volume after 2 hrs in ml/litre with NaOH	with soda	Comments
Copper	8.5	50	50	
Zinc	8.5	75	40	
Nickel	9.5	55	55	
Cadmium	10.0	60	30	
Lead	8.0	40*)	10**)	*) with 100 mg/litre Fe^{+++}
				**) with 10 to 30 mg/litre Fe^{+++}
Iron	8.5	60	50	
Aluminium	8.0	190	210	
Chromium	8.5	30*)	0**)	*) not quite complete
				**) colloidal & truly dissolved

*)   Solubility values det'd under effluent conditions are used where available.

**) Values for the most insoluble species are used.

The relatively small volume obtained when chromium is precipitated using NaOH is not due to incomplete precipitation but rather to formation of finely-dispersed particles which lend themselves to denser packing of the precipitate.

3.7.6   Combined precipitation of several metals from dilute solution

In concluding the treatment of precipitation of individual metals (see Fig. 3.84), it might be predicted that precipitation of more than one metal was more difficult, maybe even impossible. Thus, using Fig 3.84, copper & zinc can be precipitated within the acceptable pH limit of 9.5 using NaOH, but copper, zinc, cadmium & nickel together cannot. If soda is used, or an NaOH-soda mixture, then cadmium can also be treated. To remove the nickel, the pH must be raised to 9.8. If chromium were also present, an interfering effect due to formation of it hydroxy-complex would be evident.

Practical experience, however, indicates that pH values of 9.8 or above are not in fact required to precipitate nickel when it is the sole metal present or the main metallic constituent. Thus, for example, where it constitutes 10 to 20% of the total metals present in an effluent, a pH of 8.5 to 9.5 will often suffice to precipitate it down to a level of 0.5 mg/litre or less in solution. As will be seen in the following treatment, there can be more than one explanation for this. Feitknecht [103,140] used X-ray diffraction to study the fine structure of the basic salts of divalent metals. They consist of a hexagonally structured hydroxide lattice with the metal salts housed at interstitial locations. Because a wide range of basic salts share a common stoicheometric composition and since their metallic radii differ little from one another, one can expect them to behave as isomorphs. These compounds may be found only in layers of the metal salts or also with layers of the hydroxide. Feitknecht found that, in presence of nickel or cobalt salts, the basic zinc salts formed a mixed crystal in which there was exchange of metal atoms at the metal sites in the lattice without a modification of the structure of the basic salt. When basic zinc salts and nickel salts were co-precipitated, a mixed crystal was found which appears to exhibit isomorphism in the hydroxide planes of the lattice.

X-ray diffraction studies of such compounds reveal a range of structures and crystal planes, whose fine structure depends mainly on the conditions of precipitation. The higher the density of defects, the higher is the enthalpy of a system and thus, the greater the probability of transformation to a more ordered structure built up from the same ionic components. It seems entirely possible that, where the level of defects is very high, a transformation to a system of lower enthalpy might be driven by steric factors, perhaps incorporating foreign ions in the process.

Further studies were carried out by Schwab & Polydoropoulos [106] addressing the question as to whether, when metals were co-precipitated as their hydroxides, the result was formation of mixed crystals, simple mixtures or new compounds. They found, during titrations of zinc and chromium containing solutions, a plateau at pH 5, corresponding to exactly twice the quantity of hydroxide required to form the chromium hydroxide alone. The precipitate so formed had the composition $2Cr(OH)_3 \cdot 3 Zn(OH)_2$. Any excess zinc then only precipitated when the titration was extended to higher pH values. Similar behaviour was seen in solutions containing nickel and chromium. At pH 5.7, a precipitate is formed of composition $2Cr(OH)_3 \cdot Ni(OH)_2$. The structure, as in the case of the basic salts studied by Feitknecht, is a mixed crystal. Mixtures of copper and chromium too, behave in the same way although a titration does not provide the same information, since the steps for formation of the basic copper salt and the chromium hydroxide more or less coincide. This species can also be described, in some cases after loss of water, as metal chromites of general formula $Me_3(CrO_3)_2 \cdot$ aq.

In other studies, the above-mentioned authors co-precipitated nickel and copper from a solution of their chlorides or nitrates, identifying the precipitates as copper-rich mixed crystals. Nickel ions can be incorporated in the lattice of the basic chloride or nitrate, but not in the basic sulphate. In the latter case, heterogeneous mixtures are found. However nickel is incorporated into the lattice of copper hydroxide when the Ni:Cu ratio is high.

In the case of solutions containing copper and iron (III), heterogeneous mixtures of iron and copper hydroxides are formed. In this case, each metal in turn forms its hydroxide, once a sufficiently high pH is reached. Such behaviour can be used to good advantage in practice, for recovery of valuable metals. Thus nickel-rich

precipitates can be recovered from solutions containing trivalent metals such as
Cr(III) or Fe(III). [170].

The author has studied the solubilities of various metals under conditions of co-
precipitation, under effluent treatment conditions. [21, 131]. In these experiments,
one of the metals – copper, zinc, nickel, cadmium, iron (III), aluminium, chromium
($Me_{(1)}$) was co-precipitated with another metal in this list ($Me_{(2)}$). The condition was
set that $[Me_{(1)}]:[Me_{(2)}]$ was 0:1; 0.5:1; 1:1; and 2:1. Total metal ion concentration
$[Me_{(1)}] + [Me_{(2)}]$ was 100 mg/litre in all cases. Neutralisation was carried out without
regard to the precipitation of the individual metals, to a value of pH 8.5. Figs. 3.93
and 3.94 show typical results.

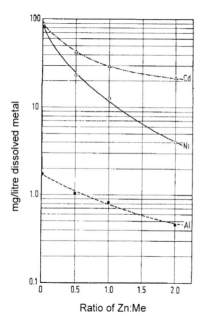

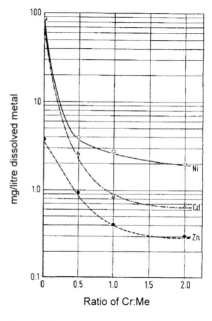

Fig. 3.93 Solubilities of the metals
cadmium, nickel and aluminium after
their common neutralisation precipitation
with zinc using NaOH to pH 8.5.
Concentration of Zn + second metal =
100 mg/litre. [131].

Fig. 3.94   Solubilities of the metals
cadmium, nickel and zinc after each in
turn was co-precipitated with chromium
using NaOH to pH 8.5. Concentration
of Zn + second metal = 100 mg/litre.
[131].

According to Fig. 3.93, it is seen that the solubilities of aluminium, nickel and
cadmium decrease when they are co-precipitated with zinc. Although the solubilities
of the more strongly basic hydroxides of nickel or cadmium not not fall to the
frequently found legal minimum concentrations of 0.5 mg/litre, there is no question
that their solubilities decrease as the quantity of zinc present increases. At a pH of
8.5 and the same concentrations, nickel would only slightly precipitate, cadmium not
at all.

When the metals cadmium, nickel and zinc are precipitated in presence of iron (III), the enhancement is greater than is the case with divalent metals. Given that mixtures containing iron (III) only precipitate to form heterogeneous mixtures, it must be assumed that the enhancement effect of this metal is due to strong adsorption effects.

As seen in Fig. 3.94, the solubility decrease of divalent metals in presence of chromium (III) cannot be ascribed solely to formation of the chromite-type mixed crystals reported by Schwab. This effect might, however, explain why cadmium, in its behaviour, does not follow the usual sequence where metals behave according to the basicity of their hydroxides, when cadmium would be expected to come after nickel.

Results for precipitation with various concentrations of other metals are very similar *.

It is also worth noting that the solubility of nickel in presence of aluminium at a ratio Al:Ni = 0.5 lies below 1 mg/litre. The solubility by co-precipitation of metals which normally require a pH of 8.5 or above, worsens when the total metal ion concentration increases and as the hydrolysis products lose their very bulky form with its highly active surface. However the benefits of co-precipitation very seldom vanish entirely. Schlegel [171] showed how, even with concentrated solutions, nickel is more completely precipitated from pickling baths containing copper and nickel, than from nickel solutions alone.

Other studies by Schulte-Schrepping & Deike [108] reported that heavy divalent metals in presence of phosphate and strontium ions precipitated as mixed crystals with only approx. 0.1 mg/litre remaining in solution. Similar results were found for chromium, albeit in presence of larger quantities of phosphate and aluminium. Dyck & Lieser [109] describe how very low concentrations of copper, zinc, cadmium, lead or silver precipitated to leave less than 0.1 mg/litre in solution, when iron (III) hydroxide or iron (III) phosphate were also formed. Using isotopic radio-labelling methods, it was found that over 90% of Cu, As, Pb were removed with the iron (III) hydroxide, similar values being obtained for zinc and cadmium when carbonate was formed. Formation of the iron phosphate was equally effective for removal of Zn & Cd, Cu or Pb when carbonates were present.

Other studies, for example those of the author *et al.* [28], concerned themselves with co-precipitation of easily (e.g. copper) and more difficultly precipitatable (e.g. nickel) divalent metals. It was found that the lowest final concentrations of the more difficult to precipitate metal resulted when small but roughly equal concentrations of both metals were initially present (see Fig. 3.95). High copper concentrations (1 gm/litre) are effective in removing nickel (at 100 mg/litre) to less than 0.5 mg/litre at pH 9.5. In the reverse sense, when nickel is present at 1 gm/litre, the presence of copper at 500 mg/litre brings little benefit. To reduce nickel levels to 0.5 mg/litre, pH values

---

* It should be noted that because of various factors affecting formation of mixed crystals as well as adsorption phenomena, reproducibility of this type of experiment is sometimes poor.

above 10.5 are necessary whereas for copper, up to an initial concentration of 500 mg/litre, requires only a pH of 8.3 for its removal below the 0.5 mg/litre mark (see Fig. 3.96).

For the purposes of effluent treatment, when two or more metals are co-precipitated, the most favourable conditions can be summarised as follows:

1. Formation of mixed crystals, especially when several divalent metals are jointly present.

2. Formation of compounds [106], especially between a divalent and a trivalent metal (e.g. chromites)

3. Adsorption of metals otherwise requiring a high pH on the active surfaces of metals precipitated at a lower pH, in particular adsorption of divalent metal ions on the hydroxides of a trivalent metal.

In that the adsorption of neutralising alkalis on hydroxide precipitates has been shown, it seems reasonable to assume that metal ions too, adsorb. However for systems where only heterogeneous mixtures are observed, another explanation is called for. Nor should it be assumed that the three phenomena listed above can only operate singly. There seems no reason why two or more might not jointly take place, depending on the conditions of precipitation and the metals and alkalis used.

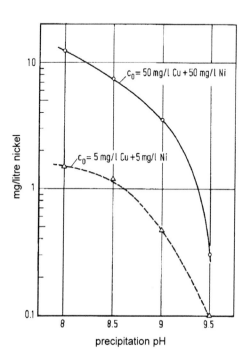

Fig. 3.95   Removal of nickel using NaOH in the presence of equal amounts of copper as function of pH. [28].

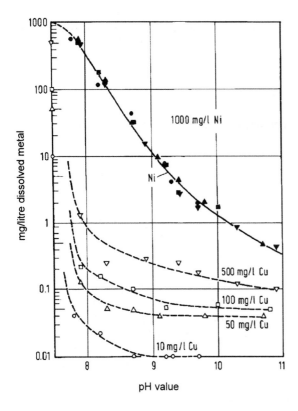

Fig. 3.96   Solubility of nickel and copper in their co-precipitation with NaOH, as function of pH [28].

Just as solubility values benefit by a co-precipitation of metals, so are the processes of flocculation and sedimentation improved. Thus chromium hydroxide, in presence of other metals, will settle without leaving any cloudiness in the supernatant liquor. Precipitation of lead is only possible when other readily flocculating metals are present, as discussed in Section 3.7.5.5.4. The use of flocculating agents to accelerate sedimentation of mixed metal precipitates is more successful than, for example, is the case with a single divalent metal species.

3.7.7   Neutralisation Precipitation of Metals from Concentrated Solutions

Both from theory and practical experience, it is recognised that many chemical reactions, not least precipitations, differ according to whether they are carried out in dilute or concentrated solution. For effluent treatment, redox reactions are most widely used which, because of the thermodynamics involved, tend to go to completion and are virtually irreversible.

In the case of precipitation reactions, by contrast, there is an equilibrium between the reactant, in solution, and the sparingly soluble product. This equilibrium can be affected by other species in solution. Such effects can so alter the precipitation process that it ceases to serve the purpose for which it was intended. Such effects, at higher concentrations, may involve adsorption behaviour and the sedimentation properties of deposits. Where possible, such adverse effects should be avoided in the operation of effluent treatment processes.

3.7.7.1   Effect of Total Ionic Concentration

As the concentration of a metal ion in solution increases, so can it be said that its final concentration in solution after precipitation of its hydroxide or basic salt, will likewise increase. Simply expressed, if a metal ion present at initial concentration of 100 mg/litre can be removed by precipitation at pH 9.0 down to 0.5 mg/litre, the same is no longer true when its initial concentration is 1 gm/litre. The reasons for this are well understood, and are summarised in Section. 3.7.3. The total ionic strength, on which the activity coefficients depend, increases thanks to formation of neutral salts which then dissociate, during the neutralisation precipitation reaction. Neutral salts already present or otherwise added for any reason, have the same effect. The outcome is an increased solubility of the precipitate, resulting from electrostatic forces from solution. One method for minimising this, is is to add to the effluent an excess of those ions which form the solid phase, in most cases, hydroxyl ions. In practice this means increasing the pH value. Using the example quoted above where metal ions were present at 1 gm/litre with neutral salts present, the pH can be increased to 9.5 or 10 and by this means the residual dissolved concentration can usually be brought below 0.5 mg/litre. In cases where the pH selected for a neutralisation precipitation is some way above the precipitation pH for a particular metal, as is often the case with dilute solutions, it is most unlikely that the 0.5 mg/litre concentration will be exceeded.. Conversely, if a neutralisation precipitation is carried out to the pH which, for a dilute solution, results in a 0.5 mg/litre residual concentration then, in the case of more concentrated liquors, it is most likely that this limit will not be met, with higher residual concentrations resulting. This can be illustrated with reference to copper, zinc and nickel.

3.7.7.1.1   Copper

When copper is precipitated, e.g. at pH 8.5, as the copper concentration in solution increases, so does the solubility of the precipitate, as shown in Fig. 3.97. In this case, the increased concentration in the supernatant liquor is modest, corresponding to about 100% over 2.5 decades of concentration increase. In the case of initial concentrations of 1 gm/litre, a final concentration in excess of 1 mg/litre is probable.

If an initial copper concentration of 100 mg/litre is used, and increasing amounts of neutral salt are added, similar results are observed and Fig. 3.98 again shows a 100% increase in solubility as added NaCl concentrations pass through 2.5 decades. In both these cases neutralisation was carried out to pH 8.5, i.e. 1 pH unit greater than the value required to precipitate copper to a residual concentration of 1 mg/litre in dilute solution. NaCl was used as the neutral salt since this species is used for regeneration of ionic exchange units and is typically present in the circuit at concentrations of around 5 gm/litre [172].

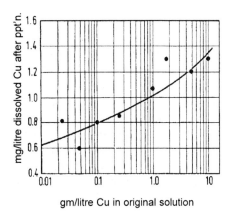

gm/litre Cu in original solution

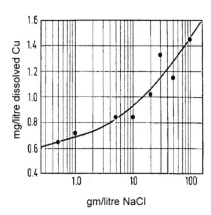

gm/litre NaCl

Fig. 3.98   Relationship between solubility of Cu precipitate and concentration of neutral salt (NaCl) using NaOH as alkali. 100 mg/litre Cu, pH 8.5, 2 hours time of standing. [172]

Fig. 3.97   Relationship between solubility of the copper precipitate and the initial concentration in solution using NaOH to pH of 8.5. 2 hours time of standing [172].

3.7.7.1.2   Zinc

When experiments are carried out using zinc solutions but similar to those described above for copper, similar results are obtained, as shown in Figs. 3.99 and 3.100.

In contrast to copper, however, the curves are seen to be much steeper. Using the same initial concentrations, they are 1000% and 500% steeper respectively. This can be traced back to the precipitation pH used. In the case of dilute zinc solutions, residual concentrations down to 3 mg/litre can be obtained and under these conditions, the benefits of adding neutral salts are immediately visible. In this case, there is no significant excess of hydroxyl ion which might shift the equilibrium in favour of the solid product. In order to make a strict comparison with the copper case, the pH would have to be significantly increased [172].

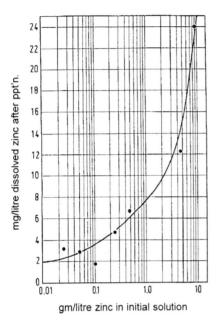

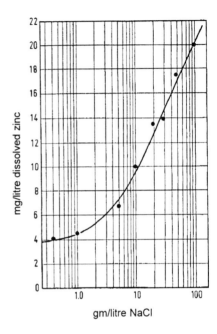

Fig. 3.99   *Relationship between zinc concentration in solution after precipitation and the initial zinc concentration, using NaOH to pH 8.5. 2 hours standing time. [172].*

Fig. 3.100   *Relationship between zinc concentration after precipitation and concentration of neutral salt (NaCl) using NaOH to pH 8.5 and 2 hours standing time. [172].*

3.7.7.1.3   Nickel

Studies of the effect of increasing nickel concentration on precipitation behaviour could only be carried out above pH 9.5, i.e. the precipitation pH for this metal in dilute solution, reflecting the more basic character of the nickel hydroxide. The results are shown in Fig. 3.101 and the same steep increase as was the case for zinc, is seen for the same reasons.

Studies with nickel illustrate an effect which is only slightly perceptible in the case of copper, being somewhat more pronounced with zinc. As time of standing increases, so more and more of the precipitate re-dissolves, this effect being the larger the higher the initial metal ion concentration (Fig. 3.102). This behaviour suggests a process whereby, after onset of supersaturation, a more or less complete precipitation of the nickel hydroxide occurs. Thereafter, electrostatic effects due to solution forces come into play and, to an extent depending on the neutral salt concentration, the precipitate partly redissolves. Associated with this, are pH retreats which, as with the case of copper and zinc, play a part in the re-dissolution. Evidence for these two processes and their association with one another exists [172].

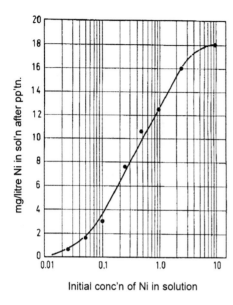

Fig. 3.101   Relationship between solubility of nickel precipitate and its initial concentration in solution, after pp'tn with NaOH to pH 9.5 and 2 hours time of standing. [172].

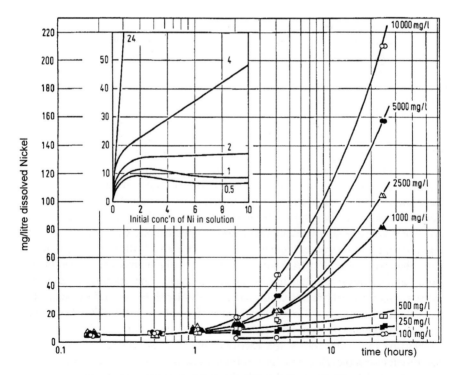

Fig. 3.102 Relationship between solubility of the nickel precipitate and initial Ni concentration and time of standing, using NaOH to pH 9.5 [172].

Schlegel [171] carried out various studies, prompted by a failure to remove sufficient quantities of nickel from ion-exchange regenerating liquors. He diluted these but found, even after increasing the pH, no improvement. The problem was evidently the high concentration of neutral salts in these liquors. Solutions with 100 mg/litre Ni to which increasing amounts of sodium sulphate were added, again show a very steep rise in the nickel solubility curve. Above 20 gm/litre sodium sulphate, a near complete dissolution of the nickel precipitate is found and in this case too, it is necessary for the precipitation to be carried out at higher pH values.

3.7.7.2   Effect of pH

The increase in solubility of metal hydroxides caused by electrostatic effects can be partly, sometimes wholly, offset by increasing the pH. However two points should be noted:

- there is usually an upper pH value which should not be exceeded.

- where amphoteric hydroxides are present, these must not be converted to their more soluble hydroxy-complex derivatives. This would result in exceeding the statutory limits or require the use of slaked lime.

It is true that the very fact of adding neutralising alkalis itself results in increased ionic strength, but this effect is not so marked as to result in increased residual metal ions in solution. As the initial metal ion or neutral salt concentrations increase, so the solubility curves (concentration/pH) shift to the right, to higher pH values.

If pH changes are related to initial metal ion concentration, a distinct minimum will be found, within a given range of concentrations and this has been observed, for example, in the case of zinc, nickel and copper. This minimum is specially marked in the case of zinc, being located in the region 1 to 2 gm/litre zinc (Fig. 3.103). Comparing this with the graphs in the next Section which show the effect on precipitate volume, of solution concentration and the phenomenon becomes clearer (see Fig. 3.107). At lower concentrations, the precipitated zinc hydroxide is voluminous and of high surface activity. At higher concentrations, the solid formed is denser and more tightly packed and thus has lower surface energy. As the concentration increases higher still, there is increasing deviation in the pH plot, though not necessarily because of the large volume of hydroxides formed. At these pH's, though the mass of precipitates formed are larger, their volume is lower, reflecting a denser packing. As Fig. 3.103 shows, the activity of precipitates from 25 to 100 mg/litre zinc solutions is larger than that formed from solutions with 10 gm/litre zinc.

In the case of copper and nickel too, the observed minima in the pH plots occur at those values where inflections are seen in the corresponding plots for precipitate volumes (cf. Figs 3.106 and 3.108).

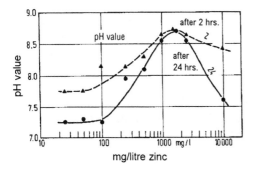

Fig. 3.103 pH changes as function of initial solution concentration in precipitation of zinc with NaOH to pH 8.5 after 2 and 24 hours of standing. [94].

3.7.7.3 Sedimentation and precipitate volumes.

The dependence of sedimentation rate on metal ion concentrations can be seen in Figs. 3.104 and 3.105. for precipitation of copper and iron hydroxides. As is the case for other metals, the time for compression of the precipitate takes longer, the higher the initial metal ion concentration in solution. At high concentrations, the process is often incomplete even after several days. Fig. 3.105 relates to the condition where some turbidity remains in the supernatant liquor. In those cases where iron alone is precipitated, it is advisable to add flocculating agents at the concentrations in question.

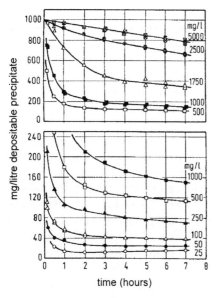

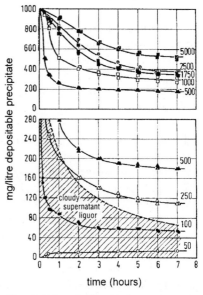

Fig. 3.104 Sedimentation curves as function of initial metal ion concentration,, for precipitation of copper with NaOH to pH 8.5 [134].

Fig. 3.105 Sedimentation curves as function of initial metal ion concentration,, for precipitation of iron (III) with NaOH to pH 8.5 [152].

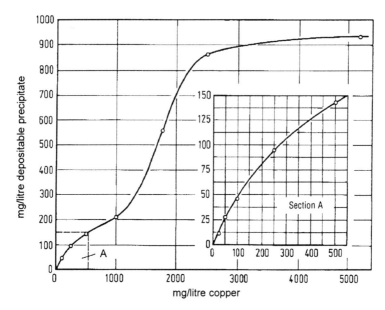

*Fig. 3.106   Precipitate volumes as function of initial metal ion concentration for
precipitation of copper with NaOH to pH 8.5, 2 hrs time of standing [134].*

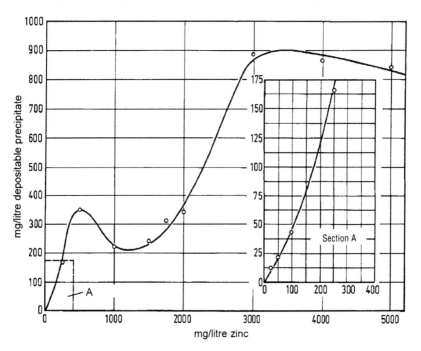

*Fig. 3.106   Precipitate volumes as function of initial metal ion concentration for
precipitation of zinc with NaOH to pH 8.5, 2 hrs time of standing [137].*

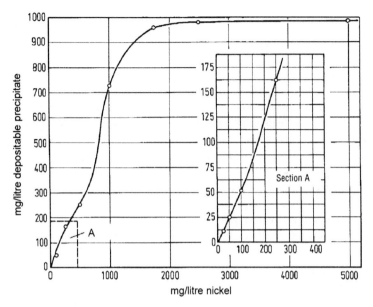

Fig. 3.108   Precipitate volumes as function of initial metal ion concentration for precipitation of nickel with NaOH to pH 9.5, 2 hrs time of standing [134].

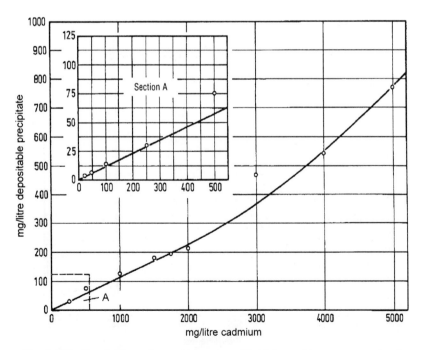

Fig. 3.109   Precipitate volumes as function of initial metal ion concentration for precipitation of cadmium with NaOH to pH 10.5, 2 hrs time of standing [152].

More revealing, in many cases, is an inspection of groups of plots in which the volume of sludge formed at a set time is plotted versus the metal ion concentration (see Figs 3.106 to 3.112). What one expects, and indeed finds is that a relationship between sludge volume and metal ion concentration exists only at low concentrations, say up to 250 mg/litre. At concentrations above this value, the curves flatten out, in the case of copper or zinc, in the latter case passing through a minimum [134, 137, 152, 166]. At higher concentrations, sludges tend to be denser and this is ascribed to release of capillary and adsorbed water and, to some extent, release of chemically bound water. The latter effect can be seen in the case of copper. Precipitates here darken on standing (see also Sections 3.7.5.1.2 and 3.7.5.2.4). Also supporting this hypothesis is the decrease in activity, referred to previously. Similar behaviour in the case of cadmium, iron, aluminium or chromium was either not observed or at best ambiguously so.

As concentration increases further, the volumes of the denser sludge become more voluminous until the curve becomes asymptotic to the metal ion concentration axis. cf the plots for precipitate volumes for the metals cadmium, iron, aluminium and chromium shown in Figs. 3.109 to 3.112.

At very high metal ion concentrations, as found for example in spent pickle liquors or metal plating solutions, a renewed densification of the precipitate is found, since this cannot exceed the total volume of solution. In certain cases, sulphuric acid pickling baths can be directly converted to a near-solid sludge which can be disposed of by tipping. This is brought about by stirring into solution powdered lime. The formation of solid calcium sulphate and loss of water by evaporation, as a result of the heat of neutralisation, make this a practical option [173]

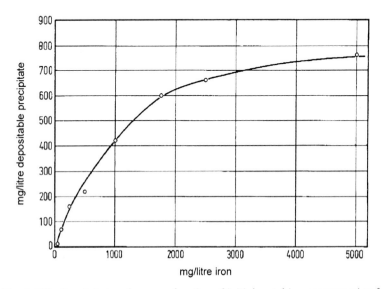

Fig. 3.110   Precipitate volumes as function of initial metal ion concentration for precipitation of iron (III) with NaOH to pH 8.5, 2 hrs time of standing [152].

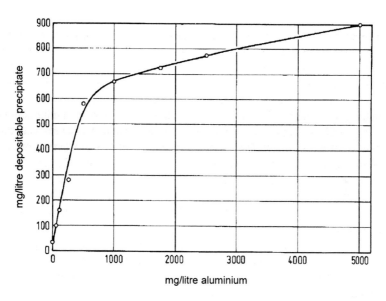

Fig. 3.111 Precipitate volumes as function of initial metal ion concentration for precipitation of aluminium with NaOH to pH 8.5, 2 hrs time of standing [152].

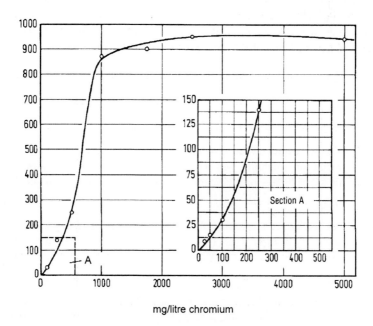

Fig. 3.112 Precipitate volumes as function of initial metal ion concentration for precipitation of chromium with NaOH to pH 8.5, 2 hrs time of standing [166].

The choice of neutralisation alkali can affect both sedimentation rate and precipitate volume. Although Lutter [110], working at initial concentration of 260 mg/litre Cr found no significant differences in sedimentation rates using NaOH or slaked lime, Krusenstern & Axmacher [111], working at 300 mg/litre Cr found that slaked lime worked more effectively in this context. In the first case, the solubility product of calcium sulphate (formed from the chromium sulphate) was not reached, in the latter case it was considerably exceeded. The calcium sulphate adds weight to the chromium hydroxide precipitate, resulting in more rapid precipitation and lower volumes, albeit with higher solids content.

3.7.8   Guidelines for precipitation

From Sections 3.7.1, 3.7.3. 3.7.6 and 3.7.7, valuable conclusions can be drawn for neutralisation precipitation and in what follows, these are drawn up as a set of guidelines. Bearing in mind the importance of complying with effluent discharge limits and legislation, it is suggested that whoever is in charge of effluent treatment should familiarise themselves with these rules, as should those responsible for operating the effluent treatment plant. By these means, operating failures involving precipitation will be held to a minimum.

Rule 1: Selecting the correct precipitation pH. – For dilute solutions containing only a single metal or with only very small amounts of a second metal, neutralisation should be carried out at or above the precipitation pH for that metal (see Sections 3.7.5.10 and preceding sections).

Rule 2: Selecting the best neutralising alkali – Using slaked lime, certain of the amphoteric metals can also be precipitated at higher pH values. Using soda, some divalent metals can be precipitated at lower pH values (see Section 3.7.5.10 & preceding sections).

Rule 3: Where several metal ions are present in solution, a decrease in residual metal ion solubility will result. Presence of a more easily precipitatable metal facilitates precipitation of a more difficult to precipitate metal ion. (see Section 3.7.6).

Rule 4: As the concentration of neutral salts increases, so does the residual solubility of metal ions. This effect can be largely offset by increasing the precipitation pH. (see Section 3.7.7).

Rule 5: The pH retreat following precipitation should always be taken into account. If not, there is a danger that the more strongly basic metal hydroxides re-dissolve. To compensate for the pH retreat, a greater than predicted precipitation pH should be employed. The magnitude of this is best determined empirically (see Section 3.7.5.10 and preceding). In presence of divalent iron, the solution must be oxidised during precipitation. (see Section 3.5).

By heeding these rules, it should be possible to avoid exceeding the prescribed maximum discharge metal ion concentrations. This can, however take place as a result of other factors:

- metals are present in complexed form. In this case other measures are called for (see Section 3.9.5).

- metals are present as sparingly soluble compounds, but in a form making them difficult to separate from the aqueous phase. Legislation in many countries requires, in such cases, that a final filtration is carried out (see Section 3.4.1).

Finally, it should be noted that, in most cases, the Authorities require only a total concentration value for each metal present in the effluent, i.e. the sum of its cationic, complexed and solid forms. However where the permitted concentrations are exceeded, it is for the effluent producer to determine how (in which of the three states listed above) and why, the total limit value is exceeded. Without this insight, it will prove difficult to remedy the situation.

3.8   PRECIPITATION OF METALS USING NON-ALKALINE SUBSTANCES

3.8.1   Precipitation of metals as sulphides

3.8.1.1   Precipitation using sodium sulphide

While there is a huge body of literature dealing with precipitation of metals as their hydroxides, very little has been published relating to the analogous use of sulphides in effluent treatment. In consequence, one is forced to rely on a limited number of papers, above all those produced in the context of analytical chemistry.

Sulphur, as a sulphide, which can be used to precipitate a range of metal ions, forms a weakly dibasic acid in presence of hydrogen ions – hydrogen sulphide. The behaviour of this is summarised in the following equation, the effects of pH on its dissociation being shown in Fig. 3.113.

$$H_2S \rightleftharpoons HS^- + H^+ \rightleftharpoons S^{--} + 2H^+ \qquad\qquad 3.150$$

The equilibrium constants have the following values:

$$\frac{c_{H^+}\ c_{HS^-}}{c_{H_2S}} = K_1 = 9.5 \cdot 10^{-8} \qquad\qquad 3.151$$

$$\frac{c_{H^+} \cdot c_{S^{--}}}{c_{HS^-}} = K_2 = 1 \cdot 10^{-14} \qquad\qquad 3.152$$

$$\frac{c_{H^+} \cdot c_{S^{--}}}{c_{H_2S}} = K_1 \cdot K_2 = K = 9.5 \cdot 10^{-22} \qquad\qquad 3.153$$

With a wide range of metals, it is possible to form their sulphides by a reaction in solution, these being considerably less soluble than their analogous hydroxides. Some of these can be precipitated in acid solutions (the hydrogen sulphide grouping), the others in neutral to alkaline media. Their behaviour in this way is governed by the relevant solubility product. Sulphide precipitation responds to pH as shown by the following equations:

$$Me^{++} + H_2S \rightleftharpoons MeS + 2\ H^+ \qquad\qquad 3.154$$

$$Me^{++} + HS^- \rightleftharpoons MeS + H^+ \qquad\qquad 3.155$$

$$Me^{++} + S^{--} \rightleftharpoons MeS \qquad\qquad 3.156$$

For acid at 1N concentration (pH = ca. 0) and using a value of 3.8 mg/litre for the solubility of hydrogen sulphide at 20°C (= approx. 0.09 Mol/litre), the sulphide ion concentration will, from Equation 3.153, be around 10^{-22}. The solubility product of copper sulphide is around 8×10^{-45}. From these, a copper ion concentration in solution can be predicted:

$$c_{Cu^{++}} = \frac{8 \cdot 10^{-45}}{10^{-22}} = 8 \cdot 10^{-23} \qquad\qquad 3.157$$

This corresponds to 6.4×10^{-19} mg/litre copper ions, an unimaginably low value and, it must be said, one which is not even approached in practice, for various reasons. Care should be taken, as previously pointed out in Section 3.7.5.10 in respect of hydroxides, not to rely on solubility values derived from solubility products. Even so, it is clear that using sulphides allows prescribed maximum permitted metal ion concentrations, such as 0.5 mg/litre for copper, to be achieved with ease. The same calculation would give a value of 0.2 mg/litre for lead with a figure in excess of 1 mg/litre for cadmium. Although both these metals belong to the hydrogen sulphide group, they should be precipitated using sulphides under weakly acid conditions.

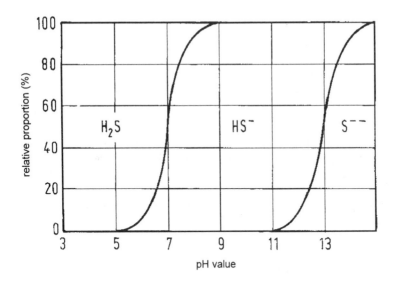

Fig. 3.113   Dependence of hydrogen sulphide acid dissociation on pH.

For metals in the ammonium sulphide classification, precipitation is carried out in neutral or alkaline conditions, since sulphide ion concentration increases with pH with metal ion concentration being correspondingly reduced. The constant derived using Equation 3.157 for individual metals should, however, be smaller than 10^{-10}. This is not difficult, since the reaction is in any case best carried out in neutral to alkaline conditions in order to minimise emissions of hydrogen sulphide gas. Sodium sulphide is the most commonly used precipitant. Table 3.17 presents solubility product data for the metal most widely used in metal-working and finishing industries, Data for the corresponding hydroxides is shown for comparison. The trivalent metals such as aluminium or chromium do not form stable sulphides, hydrolysing rapidly to form the more stable, less soluble hydroxide. Iron (III) salts are initially reduced by sulphide to the 2-valent state, Fe (II) then being precipitated as the sulphide (see also Equations 3.158 to 3.160).   The less soluble is the species precipitated, the more likely are precipitation conditions likely to give rise to high nucleation rates and thus finely dispersed precipitates (see Section 3.7.4.1). This should be noted, especially in relation to precipitation of the most insoluble metal sulphides.   Saracco *et al.* [174] precipitated copper from non-complexing solutions and experienced problems with coagulation. Use of flocculating agents brought little benefit. Addition of NaCl, however, was effective. Since it is known that coagulation of metal sulphides is promoted by cationic flocculating agents due to a collapse of the zeta potential, it can be surmised that the findings of Saracco *et al.* were due to the action of the sodium cation. Better results were obtained using calcium containing compounds or other species which dissociated to form multi-valent cations.

TABLE 3.17:

COMPARISON BETWEEN SOLUBILITY PRODUCTS OF HYDROXIDES AND SULPHIDES OF METALS

Metal	Hydroxide		Sulphide	
	Formula	Sol'y Product	Formula	Product
Aluminum	$Al(OH)_3$*)	$2 \cdot 10^{-32}$	--	--
Lead	$Pb(OH)_2$*)	10^{-7} or 10^{-13}	Pbs	$3 \cdot 10^{-28}$
Cadmium	$Cd(OH)_2$*)	$1.3 \cdot 10^{-14}$	CdS	$5.1 \cdot 10^{-29}$
Chromium	$Cr(OH)_3$*)	$3 \cdot 10^{-28}$	--	--
Iron (II)	$Fe(OH)_3$*)	$2 \cdot 10^{-15}$	FeS	$3.7 \cdot 10^{-19}$
Iron (III)	$Fe(OH)_2$*)	$8.7 \cdot 10^{-38}$	--	--
Copper	$Cu(OH)_2$*)	$2 \cdot 10^{-19}$	CuS	$8 \cdot 10^{-45}$
Nickel	$Ni(OH)_2$*)	$5.8 \cdot 10^{-15}$	NiS**)	$1 \cdot 10^{-26}$
Silver	AgOH	$1.24 \cdot 10^{-8}$	Ag_2S	$1.6 \cdot 10^{-49}$
Zinc	$Zn(OH)_2$*)	$4 \cdot 10^{-17}$	ZnS**)	$6.9 \cdot 10^{-26}$
Tin (II)	$Sn(OH)_2$*)	$6 \cdot 10^{-25}$	SnS	ca. 10^{-20}
Tin (IV)	$Sn(OH)_4$*)	$1 \cdot 10^{-56}$	--	--

*) Solubility product values are those determined under effluent treatment conditions, where available.

**) Values for the more insoluble forms have been used

The author and colleagues [28] recommended iron (III) compounds for flocculation. Not only do the Fe(III) cations as also the positively charged basic salt ions neutralise the negative charge on finely dispersed precipitates, they also (in slight excess) act as flocculating agents. In consequence, readily sedimentable and easily filterable materials are formed.

The work of Saracco also showed that the lowest residual metal ion concentrations were formed in neutral solutions and that low excess sodium sulphide concentrations (1:3) were better than larger excesses (1:5). Various large additions of NaCl from 5 to 20 gm/litre appeared not to affect the solubility of the precipitate. Probably, bearing in mind the very low solubilities in question, any effects lay outside the limits of experimental detection. One would certainly expect such effects to be present. Residual concentrations of copper from 0.1 to 0.2 mg/litre were found. These values, as the following account of the author's researches in presence of complexing agents showed, could be further reduced by use of flocculating agents to remove finely-dispersed or colloidal matter. In the case of the hydroxides with their larger solubility, the effects of change in total ionic strength were very obvious. In the case of less sparingly soluble compounds, e.g. cadmium sulphate (see Section

3.10.2), the effects of total ionic strength are even more evident. This leads to the Rule:

"The effects of dissociating species on the solubility of precipitates increases with the solubility of such precipitates in pure water."

Studies by the author using very low metal ion concentrations (5 mg/litre) in the pH range 7 to 10 with sodium sulphide precipitation in the metal:sulphide ratio 1:2, showed residual metal ion concentrations less than 0.1 mg/litre for copper or zinc in the pH range 7 to 9. For nickel, a similar value was found only at pH 9 to 10. In these studies, sulphide precipitation was followed by 10 mg/litre of Fe (III) with an anionic flocculating agent. The aim was to determine how effective such methods might be. In this process, iron (III) operates as an oxidising agent for excess sulphide which, in alkaline conditions, exists mainly as elementary sulphur in colloidal form. It also functions as a flocculating agent for finely dispersed metal sulphides and colloidal sulphur:

$$S^{--} + 2Fe^{+++} \rightleftharpoons S^{\circ} + 2Fe^{++} \qquad\qquad 3.158$$

$$2S^{--} + 2Fe^{+++} \rightleftharpoons 2FeS \qquad\qquad 3.159$$

$$\overline{\phantom{2S^{--} + 2Fe^{+++}}}$$

$$2S^{--} + 2Fe^{+++} \rightleftharpoons 2FeS \qquad\qquad 3.160$$

This approach to removing excess sulphide operates without problems for metal ion-containing effluents. It has to be said, though, that the method is only suited to batchwise operations since the small and poorly reproducible redox potential is interfered with by the iron hydroxide precipitation, whereas sulphide precipitation at constant pH is readily monitored. In this case too, however, use of a titration controller to follow the reaction (see Section 3.3.3.3) is strongly advised although such device only operate in conjunction with batchwise processing.

There have been numerous studies aimed at carrying out sulphide precipitation with subsequent elimination of excess sulphide on a continuous basis. Fischer [175] proposed a form of effluent treatment based on controlled sulphide treatment at constant pH followed by an adsorption filter. The adsorption bed is of a type that will bind both residual metal ions and sulphide and well as metal sulphides. Used under low hydraulic head, such bed materials are capable of absorbing up to 0.5kg metal sulphide per kg bed material. This technology is mainly designed to deal with effluent containing complexants but is equally suited to maintaining the very low metal ion concentrations demanded in latest legislation.

Another method using sulphide precipitation but avoiding excess sulphide ions, was described by Rudzki [176]. Freshly precipitated iron sulphide is added in a manner such that a modest excess of iron is present. Iron sulphide is itself sparingly soluble, albeit not to the same extent as the sulphides of the toxic heavy metals. Precipitation then takes place as follows:

$$Me^{++} + FeS \rightleftharpoons MeS + Fe^{++} \qquad\qquad 3.161$$

The process cannot easily be followed instrumentally and is thus restricted to batch operation. For subsequent removal of Fe(II), a precipitation with alkali to pH 9 or above is required.

As the result of tighter effluent discharge limits, sulphide precipitation, hitherto used only for effluents containing complexants, is now used also for effluents free from complexants [177,178]. However – since sulphide is itself a Regulated Substance – use of this method brings its own drawbacks. Two options exist:

- total metal precipitation

- post-treatment using a second process.

What should never be done (though it often has been in the past) is to follow the hydroxide precipitation with addition of sulphide. This achieves nothing, since metals precipitated by sulphides are in every case those removed using hydroxides. Not only is nothing gained by this, but also the sulphides do not remove metals such as zinc where hydroxide treatment may not always serve. One exception to this is the case of complexed metals not too strongly bound.

3.8.1.2   Precipitation using organosulphides

In cases where residual metal ion concentrations even lower than those achieved using inorganic sulphides are required, there is only one realistic option. This is to form, by precipitation, species even less soluble than the heavy metal sulphides. A range of organic sulphides exists, with which this can be done. Because these chemicals are very much more expensive than, for example, sodium sulphide, they should not be used for wholesale precipitation of heavy metals from effluent, but should be resorted to only in special cases, e.g. when complexing agents are present. Further comments on these reagents is found in Section 3.9.5.1.2.

The best known organosulphides, apart from mercaptobenzothiazole, are shown in Table 3.18. With the exception of trimercaptotriazine, all have a structure derived from the thiocyanates. Even using the simplest organosulphides, such as thiocarbonic acids, thioalcohols or thioethers, better results are obtained than using sodium sulphide. Leaving aside cellulose xanthogenate and TMT (trismercaptotriazine), not included in the study, the author was able to obtain final copper concentrations as low as 10 μg/litre, based on inlet values of 5 mg/litre Cu and using low concentrations of Fe(III) compounds as flocculating agents, in the pH range 7 to 10. Similar results were found using DMDTC (Table 3.18) adsorbed on bentonite and using mercaptobezothiazole (in this case at pH 7 to 9).

TABLE 3.18:

ORGANOSULPHIDES USED FOR PRECIPITATION OF METALS

Designation	Formula	Comments
Thiocyanate	$S=C\begin{smallmatrix}\diagup NH_2\\[2pt]\diagdown NH_2\end{smallmatrix}$	white crystals
Trimercaptotriazine (TMT)	NaS, N, SNa ring with N, N and S (triazine)	yellow liquid; min. 15% active contents
Cellulose xanthogenate (CX)	$S=C\begin{smallmatrix}\diagup O\text{-Cellulose}\\[2pt]\diagdown SNa\end{smallmatrix}$ (Na)	organic crystals, only stable at low temps.
Starch xanthogenate (SX)	$S=C\begin{smallmatrix}\diagup O\text{-Starch}\\[2pt]\diagdown SNa\end{smallmatrix}$	yellow powder, only stable at low temps.
Mercaptobenzothiazole (MBT)	$HS-C$ benzothiazole ring (N, S)	light yellow poorly soluble powder only sol. above pH 10
Dimethyldithiocarbamate (DMDTC)	$S=C\begin{smallmatrix}\diagup N(CH_3)_2\\[2pt]\diagdown SNa\end{smallmatrix}$	light yellow viscous liq. (40% DMDTC) (Trade Name "Plexon")
Diethyldithiocarbamate (DEDTC)	$S=C\begin{smallmatrix}\diagup N(C_2H5)_2\\[2pt]\diagdown SNa\end{smallmatrix}$	yellowish crystals known as heavy metal reagent.

For nickel, equally good results were found, albeit only using DMDTC or DEDTC as sorption species. The same was true for cadmium (DMDTC only). In the case of zinc or lead, use of DMDTC brings residual metal ion concentrations down only to 100 µm/litre, i.e. not significantly better than using sodium sulphide. In all cases, precipitates formed in presence of iron as a flocculating agent settle well and are easily filterable.

3.8.2 Precipitation of arsenic and other metals as iron compounds

Arsenic occurs extremely rarely in effluents nowadays and then only in connection with processes one might describe as "exotic". However since information for its removal from effluent is available and conditions for its precipitation are very similar to those used to remove certain transition metals, some details will be given.

The metals molybdenum, vanadium & tungsten are very rarely found as components of process baths. However they do occasionally appear as impurities in effluents or rinsewaters, their origins are not always known. They are found after pickling of stainless steels, some of these containing Mo, W or V as alloying constituents. To what extent they are undesirable in effluent is little known. The same is true of conditions for their precipitation. All are most stable in their highest valence state (5 or 6) and exist, in aqueous media in their anionic form (vanadate, molybdate, tungstate). Thus they cannot be precipitated as hydroxides with other metals. Sometimes they are adventitiously co-precipitated with other metals, e.g. from pickling liquors in stainless steel working plants. This happens in most cases when they are present in strong acid solutions with excess of iron which is then dosed with slaked lime to a high pH value.

The author and colleagues [28] sought to determine optimum conditions for precipitation of Mo, W, V. These resemble closely conditions laid down in a patent [179] for treating effluent from steelworks. Iron compounds, slaked lime, chromium (III) compounds already present or formed by reduction will also bring down the three above-named metals. However the resulting residual concentrations of these metals were high (10 mg/litre Mo, 400 mg/litre W, less than 5 to 275 mg/litre V).

3.8.2.1   Arsenic

Arsenic forms two sparingly soluble compounds of potential use in effluent treatment. These are the sulphide (As_2S_3) or iron arsenate ($FeAsO_4$), skorodite. In effluent treatment, residual concentrations less than 0.1 mg/litre As are called for.

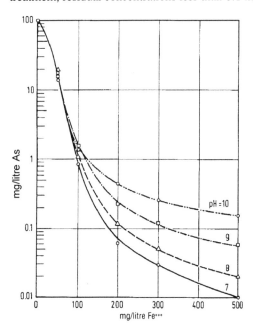

Fig. 3.114 Solution concentration of arsenic after precipitation as the Fe(III) arsenate as function of pH and iron concentration. [28].

In the case of the sulphide, readily sedimentable precipitates are obtained in acid solutions (hydrogen sulphide will be evolved – this is an extremely toxic gas). After neutralisation to pH 7, the residual As concentration is around 0.2 mg/litre.

Should the precipitation take place at higher pH values, the precipitate is finely divided and settles slowly and even after use of iron (III) salts as flocculating agents, a significantly higher residual As concentration is found.

Precipitation of As as the iron arsenate, by contrast, gave much better results. The arsenic must first be oxidised to its 5-valent state, should this not already be the case. The solution must invariably be acidified. Then 3, preferably 4 Moles of iron (III) must be added per Mole As (i.e. 4 times the stoicheometric ratio) and slaked lime added to neutralise the solution to pH 7 (not higher!). The residual solubility of arsenic (as skorodite) is well below the 0.1 mg/litre level as seen in Fig 3.114 which also shows the effect of pH. All reactions take place readily and, in presence of excess Fe(III), the precipitate is easily filterable and settles rapidly [28].

These findings are embodied in a patent [180] which claims residual concentrations of 0.01 to 0.05 mg/litre As when at least 3 times the stoicheometric value of Fe(III) are used at pH 7.

3.8.2.2   Molybdenum

A number of highly insoluble molybdenum-containing minerals occur naturally and these can be used to guide one to potentially useful effluent treatment approaches. These minerals include molybdenite (MoS_2), Powellite ($CaMoO_4$), ferrimolybdite ($Fe_2(MoO_4)_3 \cdot 8H_2O$). Although the hydroxide has been reported, very little is known about about Mo in its ionic form. The element is most stable in its highest (hexavalent) state as the molybdate (MoO_4^{--}). It exists in monomeric form only in strong alkaline solutions. Otherwise, depending on pH, it forms polymolydates such as $HMo_6O_{21}^{5-}$; $Mo_7O_{24}^{6-}$; $Mo_8O_{26}^{4-}$; $HMo_{12}O_{41}^{3-}$; $H_7Mo_{24}O_{78}^{5-}$ etc. [180]. This behaviour (very similar to that exhibited by vanadium and tungsten) is quoted here because it plays an important part in the precipitation of all three metals, a process which begins with acidification.

Use of sulphides to precipitate molybdenum from solution has many times been shown to be of little value. Even under the most favourable pH conditions, residual concentrations of 10 to 20 mg/litre Mo were found. Precipitation as barium molybdate gave a value of 9 mg/litre. The best result was obtained by forming lead molybdate, where the value was 0.5 mg/litre Mo. For obvious reasons, deliberate addition of lead compounds is not a practicable option. However where these ions co-exist with Mo in solution, their effect is beneficial.

Use of slaked lime for neutralising after a prior acidification shows a strong pH dependence with minimum residual solubility values at pH 2.5 to 3. Even so, the actual value is very high - 80 mg/litre Mo. If iron (III) salts are added to the acid solution, somewhat better results are obtained, minimum residual solubility now occurring at pH 4. (Fig. 2.115). Doubling of the iron (III) concentration to twice the stoicheometric value brings slight improvement. Under these conditions, a residual solubility of ca 0.3 mg/litre is found. The precipitate is probably a calcium iron (III) molybdate. Though the precipitate settles well, the drawback of the technique is that the pH value of 4 must be pretty tightly adhered to. In addition, after removal of solids, a further neutralisation is required [28].

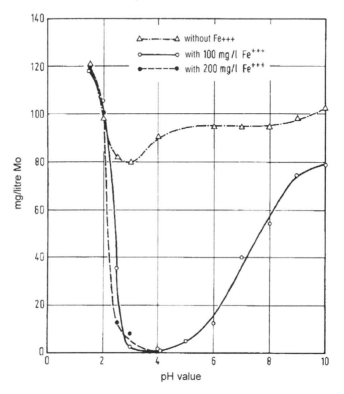

Fig 3.115   Solubility of molybdenum after precipitation with iron (III) compounds as function of pH [28].

3.8.2.3   Vanadium

Vanadium, like Mo, forms its most stable compounds in the highest oxidation state (5-valent), i.e. VO_3^-, the metavanadate. The calcium and iron (III) vanadate salts are very slightly soluble. Alkali metal vanadates, on acidifying, form polyvanadates ($H_2V_{10}O_{28}^{4-}$) and these are readily soluble in strong acid. [182].

Successful precipitation of vanadium using iron salts is strongly dependent on the pH prior to addition of the iron containing solution, presumably reflecting the extent to which the polymeric molybdates are formed. A pH of 1.5 was found best. After neutralisation with slaked lime, very tightly defined minimum solubility regions are found. Using iron (II) as precipitant, this is at pH 4, using iron (III), pH 7. The more iron is added, the better the performance and in this respect, as Fig. 3.116 shows, Fe(II) salts gave the better result notwithstanding the need to operate at pH 4. After their addition, a colour change from green to blue is observed, presumably reflecting a reduction of V(V) to V(IV). In order not to add too much iron, a ratio V:Fe of 1:3 should be set. This corresponds to a stoicheometric excess of ca 2.7:1. When this is followed by neutralisation to pH 7, where, using iron (II), the lowest residual vanadium solubility results, typical values are 0.05 mg/litre V. [28].

Vanadium can also be satisfactorily precipitated as its sulphide (V_2S_5) using sodium sulphide. The solid is separated after removal of excess sulphide and flocculation using iron (III) salts. When precipitation and flocculation are carried out at pH 7, the residual vanadium solubility is less than 0.5 mg/litre.

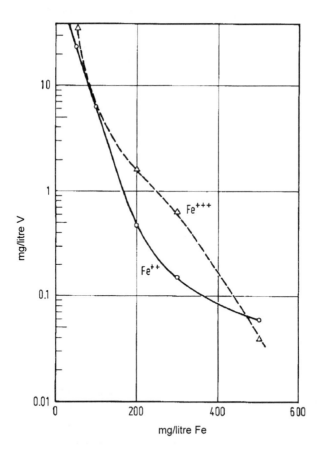

Fig. 3.116   Solubility of vanadium after precipitation using iron (II) or iron (III) as function of amount of added iron. Initial pH 1.5, pH after precipitation adjusted to 4. [28].

3.8.2.4   Tungsten

In the case of tungsten too, its mineralogy serves as a guide to effluent treatments, the best-known insoluble minerals being Scheelite $(CaWO_4)$ and wolframite $(FeMn)WO_4$. The former compound has a solubility of 17.3 mg/litre W. As before, the most stable species are those in their highest (6) valence state, the tungstates. These too, polymerise in acid to form complex polytungstates [183].

The residual solubility of tungsten after precipitation with iron (III) salts decreases as the pH is reduced (see Fig. 3.117) and increases as the concentration of added iron increases. A compromise is to work at pH 6, using a 6.5 times stoicheometric excess of iron which, reflecting the high atomic weight of tungsten, is 200 mg/litre Fe per 100 mg/litre W. The residual solubility of tungsten is then 0.8 mg/litre. The precipitation reaction occurs spontaneously with the precipitate being readily sedimentable and filterable. No effect of the initial pH value on precipitation has been found. Neutralisation can be carried out equally well with slaked lime or NaOH. Iron (II) compounds should not be used for precipitation since they are found less effective than Fe(III) in removal of the tungsten.

As before, tungsten can also be precipitated as its sulphide although, at pH 7, it leaves a comparatively high residual concentration of not less than 30 mg/litre W.

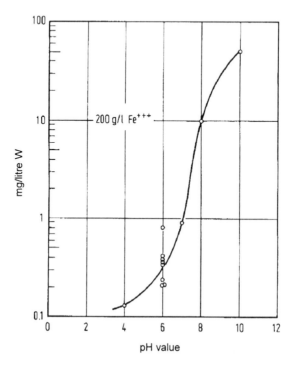

Fig. 3.117   Solubility of tungsten after precipitation with iron (III) as function of pH.

3.8.2.5   Precipitation of Molybdenum-Tungsten-Vanadium Mixtures

The optimum conditions for precipitation of each of the three metals above, differ significantly from one another. However since they often occur together in effluents, it must be seen whether some satisfactory compromise exists for their joint precipitation. In a trial run, a mixture of the three metals in solution was prepared, their total concentration being 100 mg/litre (metal) with 33.3 mg of each metal present. This was acidified to pH 1.5 and 300 mg/litre iron (III) salts added, after which pH was adjusted with slaked lime to pH 6, 7 and 8 in three separate runs. In this way it was sought to establish whether isomorphous effects operated as is the case with hydroxides and their basic salts.

The first conclusion was that by increasing the concentration of iron (III), a significant improvement in removal of molybdenum resulted. Under the conditions described above, at pH 6 with 500 mg/litre Fe(III), a residual Mo value less than 1 mg/litre was found. Doubling the Fe(III) to 1 gm/litre resulted in a tenfold reduction of the Mo to 0.1 mg/litre or less. Even at pH 7, with 1 gm/litre Fe(III), the Mo value was less than 0.5 mg/litre. As for the other two metals, W and V, their residual concentrations were in all cases less than 0.5 mg/litre, in the case of vanadium, very much lower than this.

The experiments showed that it was possible to work at pH values up to 7, but not beyond, on account of the Mo, where large amounts of added iron salt are necessary. In view of the fact that pH 4 is the preferred value for vanadium, pH 6 for tungsten when these occur alone, there is a strong suggestion that isomorphous effects must be operating, even as regards the polymeric Fe(III) metallates.

3.9   METAL COMPLEXES AND THEIR TREATMENT

Virtually everything that has been written up to this point in terms of heavy metal ion precipitation, applies only when these metals are uncomplexed and largely does not hold when they are. The following sections consider procedures in this latter case.

3.9.1   The Chemistry of Metal Complexes (see refs. [184-187])

A metal complex is a compound based on a central metal atom around which are positioned atomic or molecular ions or neutral molecules, such as water. In the latter case, the number of such neutral molecules is unrelated to the valency of the central metal atom. Such complexes may be electrically charged or they can be neutral. In terms of their chemical reactivity, they behave very differently from the central metal in its uncomplexed state, undergoing some of the same reactions as uncomplexed metal but far from all.

Many of the most commonly found anions such as sulphate, fluoborate, phosphate, perchlorate etc are themselves complexes of the parent atom with oxygen and no longer display the typical chemistry of the parent atom.

In the context of effluent treatment, one is almost invariably concerned with metal complexes. In this form, the parent metal is either much harder to precipitate as the hydroxide or, in some cases, this is impossible. Metal complexes consist of the following components:

- an electropositive central metal ion

- ligands. These are ions or molecules arrayed around the central   metal ion in a particular configuration and it is these species which are the actual complexants. The space between these and   the central metal ion is known as the first or inner sphere.

These assemblies can be positively or negatively charged or they may be neutral. In the case of charged complexes, their nett charge is determined by the number of co-ordinated ions arranged in the second or outer sphere and their charge. The nett charge on the complex is the algebraic sum of the charge on the central metal ion and the sum of the charges on the co-ordinating ions. Such complexes are usually strongly dissociated into the charged complex itself and as many counter ions as are needed to assure charge neutrality in solution. These counter ions behave in their usual manner and only in very few cases can they be exchanged with the inner sphere ligands when they lose their normal reaction characteristics. This exchange is often associated with strong colour changes, as in the case of cobalt or chromium complexes (see Section 3.7.5.8).

The number of ligands is variable, depending on the size of the central metal ion, its charge density and stereochemical factors and the existence of repulsive charges. The smaller the central metal ion, the greater its charge density and thus, as a general rule, its propensity to form complexes. This is the reason why so many heavy metals readily form complexes. However because of like charges, repulsive forces operate in these cases, repelling the ligand molecules. This, together with steric factors, results in a range of symmetrical configurations for metal complexes. Most commonly found is the tetrahedral configuration (4-fold co-ordination) and also the octahedral (6-fold co-ordination) structure. More rarely, the metal lies at the centre of a square (4-fold co-ordination) or a cube (8-fold co-ordination). Metal complexes of importance here are almost exclusively tetrahedral (e.g. tetrahydroxyzincate, tetrammino copper, complexes of nitrilotriacetic acid) or octahedral (e.g. hexahydroxychromite, hexacyanoferrate and EDTA). Fig. 3.118 shows the most important co-ordination structures. The screening effect exerted by the ligands on the central metal ion is easily seen.

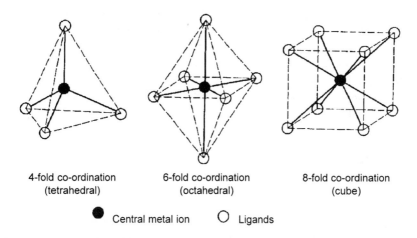

4-fold co-ordination   6-fold co-ordination   8-fold co-ordination
(tetrahedral)   (octahedral)   (cube)

● Central metal ion   ○ Ligands

Fig 3.118 Idealised structures of the most important co-ordination complexes.

The chemical properties, notably stability, of such complexes depend on a number of factors, most notably the type of ligand-metal ion bonding and their dentate characteristics. By the latter, is meant the number of bonds joining a ligand molecule to the central metal ion. As previously mentioned, these complexes dissociate strongly in solution to form the complex anion and the counter-ion (cation). However the anionic complex can itself dissociate (secondary dissociation) into metal ions and ligands. Alternatively, the metal ion can exchange one set of ligands for another (most notably those derived from the solvent, e.g. water). The more stable the complex, the less this will take place and the extent of this process is formally expressed, using the law of Mass Action, in terms of the "stability constant" ("K") for a complex and its magnitude is an index of its stability. Its reciprocal, 1/K is the dissociation constant. In the following example, a divalent metal ion complex with co-ordination number 4 is shown where, for simplicity's sake, the ligands are uncharged molecules, as is the case with aquo- or ammine- complexes. "L" designates the ligand:

$$[MeL_4]^{++} \rightleftharpoons Me^{++} + 4L \ \ \text{(Dissociation)} \hspace{2cm} 3.162$$

$$\frac{c_{[MeL_4]^{++}}}{c_{Me^{++}} \cdot c_L^4} = K \ \ \text{(Stability constant)} \hspace{2cm} 3.163$$

In fact, this dissociation as well as the complex formation takes place stepwise as a function of the complexant concentration (usually determined by the pH) both for unidentate and polydentate ligands.

$$[MeL_4]^{++} \rightleftharpoons [MeL_3]^{++} + L; \qquad \frac{c_{[MeL_4]^{++}}}{c_{[MeL_3]^{++}} \cdot c_L} = K_4 \qquad \text{3.164a}$$

$$[MeL_3]^{++} \rightleftharpoons [MeL_2]^{++} + L; \qquad \frac{c_{[MeL_3]^{++}}}{c_{[MeL_2]^{++}} \cdot c_L} = K_3 \qquad \text{3.164b}$$

$$[MeL_2]^{++} \rightleftharpoons [MeL]^{++} + L; \qquad \frac{c_{[MeL_2]^{++}}}{c_{[MeL]^{++}} \cdot c_L} = K_2 \qquad \text{3.164c}$$

$$[MeL]^{++} \rightleftharpoons Me^{++} + L; \qquad \frac{c_{[Me_L]^{++}}}{c_{Me^{++}} \cdot c_L} = K_1 \qquad \text{3.164d}$$

Reading from top to bottom, the sequence of equations represents the dissociation process. Read from bottom upwards, they show complex formation. The sum of the individual stability constants yields to total stability constant "K" for a fully-complexed complex, i.e. as much ligand is available as the metal ion can co-ordinate.

$$K_1 + K_2 + K_3 + K_4 = K \qquad \text{3.165}$$

Bidentate ligands or those with higher dentate values are bonded to the central metal ion in fewer and fewer steps as their dentate number increases and in the limit, the process is continuous, i.e. stepless. If the ligand is bidentate or greater, it is customary to refer to the complex as a chelate complex or simply as "chelated". Another possibility is that multidentate ligands are co-ordinated not to one but rather to two or more metal ions, thus forming "polynuclear" complexes. In the latter case, ligand bonding is generalised. A typical examples is the olation of chromium hydroxide complexes to form larger molecular aggregates or the formation of inorganic polyacids.

The stability of complexes depends most of all on the type of bonding (covalent, ionic, co-ordinate, donor-acceptor etc). In certain cases, molecules which are electrically neutral overall, become polarised, i.e. a separation of charges is induced and such molecules are then known as dipoles. Typical ligands in which this can take place include water or ammonia, as shown in Fig 3.119.

Other ligands capable of behaving in the same way include alcohols and other electrically neutral molecules. In cases where these are found in the solvent, the complex is known as a "solvated" or "solvation" complex. Complexes formed from water or ammonia are known as aquo- and ammine- complexes respectively. Their stability increases with the dipole moment of the ligand and with the reduction in the diameter of the central metal ion. The overall charge on the complex is the same as that on the central metal ion, since the ligands themselves are electrically neutral in this case.

In some cases, exchange takes place between neutral ligand molecules and anions in the outer sphere. The charge (number of charges and polarity) of the complex ion is then determined by the algebraic sum of charge on the central metal ion and that on the inner sphere ligand ions. Strictly speaking, it can be said that ligand exchange occurs for unidentate ligand complexes in aqueous environment, since the metal ions exist as hydrated ions, i.e. aquo-complexes (see also Section 3.144). Thus in the formation of ammine complexes, associated water molecules are exchanged stepwise with ammonia molecules. Equation 3.166 shows the initial exchange which then continues until the hexammine complex is formed.

$$[Ni(H_2O)_6]^{++} + NH_3 \rightleftharpoons [Ni(H_2O)_5 (NH_3)]^{++} + H_2O \qquad\qquad 3.166$$

While in some complexes, no electron exchange takes place between ligand and central metal ion, in others, bonding occurs via a free electron pair in the outer shell of the ligand and the outer electron shell of the central metal ion. Such complexes are among the most stable, being highly resistant to breakdown by acid.

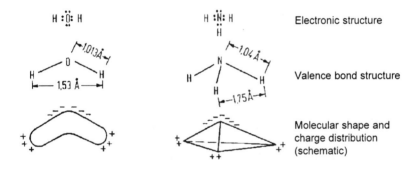

Fig. 3.119   Schematic representation of dipoles in water or ammonia.

A simple inspection of the formula of a complex does not allow any conclusions to be drawn as to the nature of the ligand bonding although, knowing how many electrons are required to fill the outer electron shell of the metal ion and whether the ligands have electron pairs to share, can give a guide to this.

Of great significance in surface treatment processes and effluents from them are chelate complexes. They are characterised by being formed from bidentate or polydentate ligands, all bonded to the same central metal ion. For a chelate to be formed, bonds must be able to form between the ligands so that a cyclic structure can exist. The most stable chelates are those incorporating a 5-membered ring. Because, in such cases, the central metal ion is enclosed in two or more directions, ("in claws"), such complexes are known as chelates. Chelating ligands commonly found in metal-finishing or working include carboxyl groups as well as various atoms with free electron pairs such as nitrogen, oxygen and sulphur.

The great stability of chelate complexes is attributable not to the ligand-metal bond strength but rather to the formation of stable rings. In thermodynamic terms this can be seen in the greater entropy increase, when chelates are formed, than is the case for similar complexes with unidentate ligands. If, for example, one starts with the hexa-aquo complex of nickel and reacts this with ammonia to form the hexa-ammine complex, then 7 free molecules (1 hexa-aquo and 6 ammonia molecules) are converted to the same number (1 hexammine and 6 water molecules). In the case where a chelate is formed from the hexa-dentate tetrakisethylenediamine, from two molecules, 7 are formed by release of the water molecule ligands. This corresponds to a substantial increase in the degree of freedom by the release of these molecules which corresponds in turn to an increase in entropy. If the relationship for free Energy (ΔG) in Equations 3.17 and 3.18 is linked, it is seen that an increase in Free Entropy Difference (ΔS) results in an increase in the magnitude of the stability constant K_c, i.e. a more stable complex.

Fig 3.120 shows two complexes and their stability constants. The difference between these two – log ΔK = 11.1 is known as the "chelating effect".

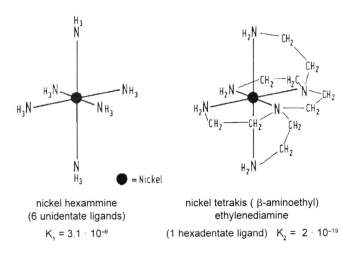

nickel hexammine
(6 unidentate ligands)

$K_1 = 3.1 \cdot 10^{-8}$

nickel tetrakis (β-aminoethyl)
ethylenediamine

(1 hexadentate ligand) $K_2 = 2 \cdot 10^{-19}$

*Fig 3.120   Illustration of the chelating effect for two complexes
both with 6-fold co-ordination.*

3.9.2   Complexants in the metal-working and metal finishing industries

Among the reasons why complexants are used for surface treatment of metals, we can include:

- dissolution of metals, their oxides and other corrosion products in neutral or alkaline media without attack on the substrate.

- retaining in solution, metals which would otherwise precipitate as a result of hydrolysis.

- maintaining, for specific processes, metal ions at low but constant concentration in solution.

- to bind to hardening agents.

In addition, some complexants (e.g. condensed phosphates) exhibit dispersant properties and are thus used in aqueous solution for metal cleaning.

In order to dissolve metals and hold them in solution, complexants are of special importance as a component of metal stripper solutions. As seen in Section 3.9.1, depending on the stability constant of the particular complex, metal ion concentrations can be held extremely low.

TABLE 3.19:
COMPLEXING AGENTS MOST WIDELY USED IN THE
METAL-WORKING & FINISHING INDUSTRIES

Complexant	Formula
Hydroxyl ion	OH^-
Cyanide ion	CN^-
Ammonia	NH_3
Pyrophosphate ion	$P_2O_7^{4-}$
Tripolyphosphate ion	$P_3O_{10}^{5-}$
Polyphosphate ion	$P_nO_{3n+1}^{(n+2)-}$
Triethanolamine (TEA)	$HOCH_2 \cdot CH_2 - N \begin{smallmatrix} CH_2 \cdot CH_2OH \\ CH_2 \cdot CH_2OH \end{smallmatrix}$
Nitrilotriacetic acid (NTA)	$HOOC \cdot CH_2 - N \begin{smallmatrix} CH_2 \cdot COOH \\ CH_2 \cdot COOH \end{smallmatrix}$
Ethylenediamine	$N_2H \cdot CH_2 \cdot CH_2 \cdot NH_2$
Ethylenediamine-tetracetic acid (EDTA)	$\begin{smallmatrix} HOOC \cdot CH_2 \\ HOOC \cdot CH_2 \end{smallmatrix} N \cdot CH_2 \cdot CH_2 \cdot N \begin{smallmatrix} CH_2 \cdot COOH \\ CH_2 \cdot COOH \end{smallmatrix}$
NNN'N'-tetrakis- (2-hydroxy-propyl) ethylenediamine (Quadrol)	$\begin{smallmatrix} CH_3 \cdot CHOH \cdot CH_2 \\ CH_3 \cdot CHOH \cdot CH_2 \end{smallmatrix} N \cdot CH_2 \cdot CH_2 \cdot N \begin{smallmatrix} CH_2 \cdot CHOH \cdot CH_3 \\ CH_2 \cdot CHOH \cdot CH_3 \end{smallmatrix}$
Tartaric acid	$HOOC \cdot CHOH \cdot CHOH \cdot COOH$
Citric acid	$HOOC \cdot CH_2 \cdot \underset{\underset{COOH}{\overset{OH}{\mid}}}{C} \cdot CH_2 \cdot COOH$
Gluconic acid	$HOOC \cdot \underset{OH}{\overset{H}{C}} \cdot \underset{H}{\overset{OH}{C}} \cdot \underset{OH}{\overset{H}{C}} \cdot \underset{OH}{\overset{H}{C}} \cdot CH_2OH$

Use is made of this to impart certain properties to various process baths. In the case of plating baths, it can be used to increase the polarisation thereby improving throwing power. Here it is mainly the inorganic complexants such as cyanide or aquo-complexes which are important. Complexants are used in electroless plating baths to avoid too rapid or even spontaneous metal deposition. Electroless nickel and copper baths (the latter as used in the printed circuit board industry) are examples of this. In the following treatment, the complexants listed in Table 3.19 and the metal complexes they form are discussed (see also Schlegel [188] and Hartinger [189,190]. Not considered here are:

- metal complexes such as anions which only exist in that form or for which specific effluent treatments exist, e.g. for chromate (Section 3.3), vanadate (Section 3.8.2.2), molybdate (3.8.2.3) or tungstate (3.8.2.4).

- metal complexes which only form at very high salt and/or acid concentrations, e.g. fluoro-, chloro- or sulphato- complexes etc.

- metal complexes which are not usually found in effluents from the metal working or finishing industries, e.g. dyes (though see Section 7.7.1) or indicators which are unlikely ever to be found in significant quantities.

3.9.3   Metal complexes

3.9.3.1   Metals complexed with inorganic substances

3.9.3.1.1   Hydroxy complexes

Many metal hydroxides form hydroxy complexes, in presence of excess alkali. The alkali metal salts of these are readily soluble. In some cases, these are found in metal finishing process solutions, e.g. zincates or stannates. They can also form during surface treatment operations or during neutralisation of effluents when these solutions are made more alkaline, e.g. zincates, aluminates, chromites, stannites. The hydroxy complexes are anionic, their charge depending on that of the central metal ion and charges on the associated ligands. Thus:

$$Me(OH)_x + y\,OH^- \rightleftharpoons [Me(OH)_{(x+y)}]^{y-} \qquad\qquad 3.167$$

Hydroxides which can form hydroxy complexes in this way are known as "amphoteric". Their stability constants can be derived from the following equation:

$$\frac{c_{[Me(OH)_{(x+y)}]^{y-}}}{c_{Me^{x+}} \cdot c_{OH^-}^{(x+y)}} = K \qquad\qquad 3.168$$

Those hydroxy complexes most often encountered in effluents are listed in Table 3.20 and the extent to which their formation depends on pH is seen in Fig. 3.121. The corresponding complexes of 2- and 4-valent lead and tin are not considered here since they are not found under typical effluent conditions.

TABLE 3.20:
SOME HYDROXY COMPLEXES AND
THEIR STABILITY CONSTANTS

Metal ion	Hydroxy complex	Stability constant	Complex forms insol. Ca salt
Al^{+++}	$[Al(OH)_4]^-$	$2.5 \cdot 10^{33}$ [152]	no
Cr^{+++}	$[Cr(OH)_6]^{---}$	$1.8 \cdot 10^{37}$ [166]	yes
Zn^{++}	$[Zn(OH)_4]^{--}$	$1.7 \cdot 10^{18}$ [137]	yes
Sn^{++}	$[Sn(OH)_3]^-$	$1.7 \cdot 10^{25}$ [21]	no

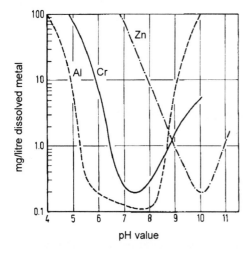

Fig. 3.121   Solubility of some amphoteric metal hydroxides vs. pH

Hydroxy complexes of chromium and zinc form insoluble calcium salts when slaked lime is added so that when this is used to neutralise effluents containing these metals, their amphoteric behaviour is not apparent. However hydroxyaluminate or stannite do not form such insoluble calcium salts.

Lying between the aquo complexes which are always present in aqueous solution and the hydroxy complexes, a series of pH-dependant equilibria exist which result in the following stepwise series of transformations:

$$[Me(H_2O)_x]^{y+} \rightleftharpoons [Me(H_2O)_{(x-1)}(OH)]^{y-1)+} + H^+ \dots \text{usw.} \qquad 3.169$$

If the hydrogen ions released are then neutralised with alkali, i.e. the pH is raised, the equilibrium is shifted to the right to give complexes progressively richer in hydroxyl ligands and finally the hydroxide. Further steps in this progression form the hydroxy complex, as shown below using chromium as example:

3.170

$$[Cr(H_2O_6)]^{+++} \rightleftharpoons [Cr(H_2O)_3(OH)_3] + 3H^+ \rightleftharpoons [Cr(OH)_6]^{---} + 6\,H^+$$

This expression sheds light not only on the behaviour of amphoteric hydroxides when these are acidified, but also on the reaction of acid on dissolved salts of the higher valence metals (see also Section 3.7.5.7.2).

In this context, the phenomenon known as "olation" should be mentioned. Hydroxy complexes can condense with aquo-complexes present in solution, loss of water molecules being the result. The metal containing nuclei are then linked via so-called μ-hydroxy groupings. The result is a multi-nuclear complex which, if the process continues (total olation) ends as a poorly soluble, highly condensed system. Metal hydroxides in a totally olated state can be envisaged as shown below:

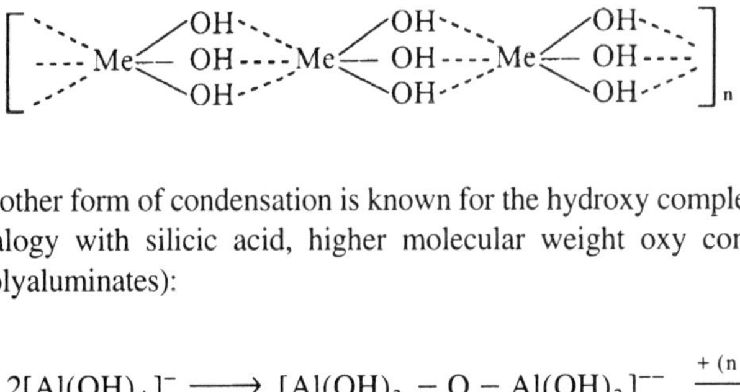

3.171

Another form of condensation is known for the hydroxy complexes of aluminium. In analogy with silicic acid, higher molecular weight oxy compounds are formed (polyaluminates):

$$2[Al(OH)_4]^- \xrightarrow[-H_2O]{} [Al(OH)_3 - O - Al(OH)_3]^{--} \xrightarrow[-(n-2)H_2O]{+(n-2)[Al(OH)_4]^-} \quad 3.172$$

$$[Al_nO_{3n+1}H_{2n+2}]^{n-}$$

By and large, hydroxy complexes present no great problems in effluent treatment since, if the iso-electric point is reached during neutralisation precipitation (see Equation 3.167 reading from right to left), they precipitate as hydroxides.

3.9.3.1.2 Cyanide complexes

Cyanide complexes are an atypical example of complexes found in metal finishing. They are also extremely stable. Until the early 1960's, electroplating of copper, zinc, cadmium, brass, bronze or silver without resorting to cyanide-based solutions was unthinkable, apart from coating of wires and tubes in acid baths and certain uses of acid copper.

Even today when acid plating baths are widely used, electroplating of bright copper and zinc are still largely based on cyanide solutions. Then too, alkali metal cyanides are still very widely used, albeit with slowly decreasing trends, for degreasing, descaling, pickling and stripping baths. And though iron or nickel cyanides are not usually found as components in processing baths, they do occur in effluent streams, especially those from regeneration of ion exchange beds since, in these cases, there is no separation of rinsewaters and thus iron present in mains water can combine with nickel and cyanide to form complexes of the highest stability. By using separate circulation loops, this can be avoided.

When cyanide and metal ions come together, sparingly soluble metal cyanides form initially. Only then are complexes formed where excess cyanide is present, as shown in Equation 3.173.

$$Me(CN)_x + y\,CN^- \rightleftharpoons [Me(CN)_{(x+y)}]^{y-} \qquad\qquad 3.173$$

The most widely found cyanide complexes associated with metal treatments are listed in Table 3.21 Their stability constants can be calculated as follows:

$$\frac{c_{(Me(CN)_{(x+y)}]y^-}}{c_{Me^{x+}} \cdot c_{CN^-}^{(x+y)}} = K \qquad\qquad 3.174$$

Cyanide complexes of copper, zinc or cadmium can readily be destroyed by oxidation. By contrast nickel cyanide complexes are very stable and difficult to destroy and oxidation is only successful over prolonged periods under controlled conditions. Inspection of the stability constants of nickel or copper complexes sheds no light on why the tetracyanonickel complex is so very stable. Doudoroff [191] estimated the order of magnitude of the stability constant of the latter on the basis of its toxicity to fish life at around 10^{30}. It is possible that kinetic rather than thermodynamic factors explain why the complex oxidises so slowly.

The hexacyanoferrate complexes are likewise so difficult to oxidise in the normal way, their degree of dissociation is minute as shown by the stability constant and possibly kinetically hindered as well. These complexes will even survive short exposure in hot concentrated sulphuric acid. Their tiny degree of dissociation is also shown by experiments using fish, which can survive up to 1 hour in a hexacyanoferrate solution with an equivalent cyanide concentration of 500 mg/litre [14]. The hexacyanoferrate (II) complex listed in Table 3.21 is somewhat more stable (or dissociates more slowly) than its hexacyanoferrate (III) analogue.

TABLE 3.21:

SOME CYANIDE COMPLEXES AND THEIR STABILITY CONSTANTS (AFTER MILNE [192]. SEE ALSO KALANI [193])

Metal ion	Cyano complex	Stability constant
Ag^+	$[Ag(CN)_2]^{--}$	$1 \cdot 10^{21}$
Cu^+	$[Cu(CN)_4]^{---}$	10^{22}
Zn^{++}	$[Zn(CN)_4]^{--}$	$7 \cdot 10^{16}$
Cd^{++}	$[Cd(CN)_4]^{--}$	$7.7 \cdot 10^{16}$
Ni^{++}	$[Ni(CN)_4]^{--}$	$1 \cdot 10^{22}$
Hg^{++}	$[Hg(CN)_4]^{--}$	$2.5 \cdot 10^{41}$
Fe^{++}	$[Fe(CN)_6]^{----}$	10^{36}

3.9.3.1.3   Ammine complexes

The origin of ammine complexes found in effluent from metal finishing is often found in ammonia which is formed when cyanides are destroyed (Section 3.2.1.4). It is formed mainly in the treatment of concentrated solutions or regenerated liquors, and this can lead to complex formation with various metals. Ammonia also arises in effluent from the use of process solutions containing ammonium salts. These include alkaline etch baths widely used in the printed circuit board industry, weak acid bright zinc plating baths (of the type based on ammonium salts). To a lesser extent, they are found in brass plating baths and in some metal colouring baths used for non-ferrous metals.

Studies related to the use of acid bright zinc baths containing ammonium salts [194, 195] showed that problems did not arise with residual concentrations of 3 mg/litre zinc and 1 mg/litre copper if the ammonia concentration was less than 200 mg/litre even over prolonged reaction times. However under new legislation in Germany and elsewhere, tighter limits on these metals have been set which, at such ammonia concentrations, would not be met. Since the 1970's, there has been a trend to using acid bright zinc baths containing little or no ammonium ions and which are otherwise free from complexants. Such baths can be used to replace ammonium ion containing bright zinc solutions, where these are still in use. Ammonium salts are often the origin of ammine complex formation though they are not themselves complexants. For this, ammonia itself, rather than the cation, is needed and the equilibrium shown below governs this:

$$NH_4^+ + OH^- \rightleftharpoons NH_3 + H_2O \qquad\qquad 3.175$$

A 1M solution of ammonium chloride (pH 4.9) will contain, as implied by Equation 3.176 below, approx 10^{-5} Mol/litre ammonia. From this equation as from Figs 3.122

and 3.123, it can be seen that interfering effects due to complex formation are unlikely below pH 8.

$$\frac{c_{NH_4^+} \cdot c_{OH^-}}{c_{NH_3}} = K \cdot c_{H_2O} = K' = 1.75 \cdot 10^{-5} \qquad 3.176$$

Metals which can be complexed by ammonia are associated, as seen in Table 3.22 with 2 to 6 molecules of the latter ligand. Stability constants for these are also shown in Table 3.22.

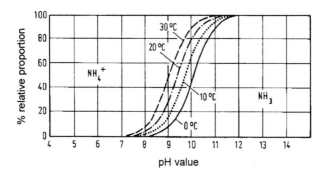

Fig. 3.122 Dependence of the ammonia-ammonium equilibrium on pH.

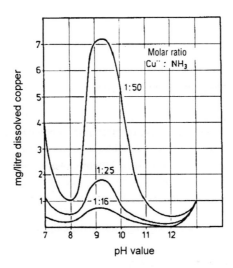

Fig. 3.123 Concentration of dissolved copper as function of pH and the Cu:NH₃ ratio. Initial concentration 10^{-3} Mol/litre Cu, after 24 hrs time of standing [189].

TABLE 3.22:

**SOME AMMINE COMPLEXES AND THEIR STABILITY
CONSTANTS (AFTER SCHWARZENBACH [185])**

Metal ion	Ammine complex	Stability constant
Ag^+	$[Ag(NH_3)_2]^+$	$1.07 \cdot 10^7$
Cu^{++}	$[Cu(NH_3)_4]^{++}$	$7.71 \cdot 10^{12}$
Zn^{++}	$[Zn(NH_3)_4]^{++}$	$1.15 \cdot 10^9$
Cd^{++}	$[Cd(NH_3)_4]^{++}$	$8.32 \cdot 10^6$
Ni^{++}	$[Ni(NH_3)_6]^{++}$	$3.09 \cdot 10^8$
Cr^{+++}	$[Cr(NH_3)_6]^{+++}$	

Fig. 3.123 shows how the formation of tetrammino copper depends on pH at different Cu:NH$_3$ ratios. At pH 7, the residual copper concentration decreases as a result of hydroxide precipitation. At pH 8, a minimum is seen which is followed by a steep rise indicative of complex formation, occurring between pH 8 and 11. At still higher pH values, the residual copper concentration decreases strongly because the hydroxyl ion concentration is now so large that copper hydroxides can form once more or, put another way, the cupric ion concentration is so large that the solubility product of the hydroxide is exceeded. When large excesses of ammonia are present, solubility again increases but this is due to an increased solubility in concentrated solutions which has been earlier discussed.

3.9.3.1.4   Complexes with condensed phosphates

Phosphates find several applications in metal finishing, namely:

- surface pre-treatment in alkaline solutions (cleaning and degreasing)

- phosphating

- electroplating with baths based on pyrophosphate

- chemical brightening and electropolishing in phosphoric acid containing baths.

We are here interested only in condensed phosphates which can form complexes. They are used primarily for cleaning and degreasing of metals since they not only act as dispersants and act as buffering species but also are able to form very stable complexes so keeping otherwise undesirable species in solution. Where effluent containing polyphosphates comes into contact with metal ions, a large range of metal complexes can form, indeed the process solutions listed above, contain pyrophosphate precisely because of its complexing properties.

Orthophosphate is itself a complex, formed by electron donation of four oxygen atoms to a central phosphorus atom. The three electron negative charge required for the fourfold shared oxygen atoms are derived from the cations thus assuring overall charge neutrality. The orthophosphate ion has four-fold co-ordination and thus a tetrahedral structure. It can be written as:

$$
\begin{bmatrix} \ddot{\,}\ddot{O}\ddot{\,} \\ \ddot{O}:\!P\!:\ddot{O} \\ \ddot{\,}\ddot{O}\ddot{\,} \end{bmatrix}^{3-}
\quad \text{or more simply} \quad
\begin{bmatrix} O \\ OPO \\ O \end{bmatrix}^{3-}
$$

In a condensation process, orthophosphates form a chain structured polyphosphate. We are here considering a polynuclear compound in which the individual phosphorus atoms are bound by μ-oxo bridging. These μ-oxo bridges then form a common edge of two tetrahedra.

$$
\begin{bmatrix} O\ O \\ OPOPO \\ O\ O \end{bmatrix}^{4-} ; \quad
\begin{bmatrix} O\ O\ O \\ OPOPOPO \\ O\ O\ O \end{bmatrix}^{5-} ; \quad
\begin{bmatrix} O \quad O \\ OPO \ldots PO \\ O \quad O \end{bmatrix}^{(n+2)-}
$$

(4 Na) -   (5 Na) -   Polyphosphate
Pyrophosphate   Tripolyphosphate

Other polyphosphates such as metaphosphate (cyclic), isometaphosphate (cyclic and chain forms) and various cross-linked products have little importance in the metal finishing industry.

Complex formation with metals stems from donation of free electron pairs from oxygen atoms which, according to the literature, results mainly in complexes with 1:1 or 1:2 compositions [196]. Other stoicheometries have been reported. Table 3.23 lists some well-known formulae, Table 3.24 some stability constants.

Condensed phosphates may dissociate again to form orthophosphoric acid. Increase in pH and temperature favours this. To a smaller extent, acidification at ambient temperature can achieve the same effect. Condensed phosphates, as their monomeric analogues, form insoluble calcium compounds and this is the basis for treating effluents containing polyphosphates.

TABLE 3.23:
FORMULAE FOR SOME METAL COMPLEXES WITH CONDENSED PHOSPHATES (FROM BJERRUM, SCHWARZENBACH, SILLEN [196])

Metal ion	Pyrophosphate	Tripolyphosphate	Polyphosphate*)
Ag^+			$[AgL_2]^{(n+1)-}$
Pb^{++}	$[PbL_2]^{6-}$		$[PbL_2]^{n-}$
	$[PbL]^{2-}$	$[PbL]^{3-}$	
Cd^{++}	$[CdL_2]^{6-}$		
	$[CdL]^{2-}$	$[CdL]^{3-}$	
Ni^{++}	$[NiL_2]^{6-}$		
	$[NiL]^{2-}$	$[NiL]^{3-}$	
	$[Ni_3L_2]^{2-}$	$[Ni_3L_2]^{4-}$	
Zn^{++}	$[ZnL_2]^{6-}$		$[ZnL_2]^{n-}$
	$[ZnL]^{2-}$	$[ZnL]^{3-}$	
Cu^{++}	$[CuL_2]^{6-}$		$[CuL_2]^{n-}$
	$[CuL]^{2-}$	$[CuL]^{3-}$	
	$[Cu_2L_3]^{8-}$	$[Cu_3L_3]^{5-}$	
Fe^{+++}	$[FeL_2]^{5-}$		$[FeL_2]^{(n-1)-}$
	$[FeL]^-$		
Al^{+++}	$[AlL_2]^{5-}$		
	$[AlL]^-$		

*) L = ligand with, on average, 5 phosphorus atoms.

TABLE 3.24:
METAL COMPLEXES WITH PYROPHOSPHATE AND POLYPHOSPHATE AND THEIR STABILITY CONSTANTS (FROM BJERRUM, SCHWARZENBACH & SILLEN [196])

Metal ion	Pyrophosphate		Polyphosphate*)	
	Complex	Stability const.	Complex	Stability const.
Pb^{++}	$[PbL_2]^{6-}$	$2.1 \cdot 10^5$	$[PbL_2]^{n-}$	$3.2 \cdot 10^5$
Cd^{++}	$[CdL_2]^{6-}$	$1.5 \cdot 10^4$		
Zn^{++}	$[ZnL_2]^{6-}$	$3.2 \cdot 10^6$	$[ZnL_2]^{n-}$	10^6
		$1.7 \cdot 10^7$		
		ca. 10^7		
Cu^{++}	$[CuL_2]^{6-}$	$3.2 \cdot 10^9$	$[CuL_2]^{n-}$	$3.2 \cdot 10^5$
		$7.4 \cdot 10^{11}$		
		$4.5 \cdot 10^{12}$		
		10^{13}		
		$7.8 \cdot 10^{10}$		
	$[CuOHL]^{3-}$	$5 \cdot 10^{15}$		
Fe^{+++}	$[FeL_2]^{5-}$	$3.5 \cdot 10^5$	$[FeL_2]^{(n-1)-}$	$3.1 \cdot 10^6$
	$[Fe(OH)_2L]^{3-}$	10^{31}		
Ag^+			$[AgL_2]^{(n+1)-}$	$3.2 \cdot 10^5$

*) L = ligand with on average 5 phosphorus atoms

3.9.3.2 Metal Complexes with Amines

3.9.3.2.1 Complexes with triethanolamine (TEA)

The main use of triethanolamine is as a component in electrolytic and chemical metal stripping baths. It is also found as an additive in corrosion inhibiting formulations [197]. TEA could be predicted to be a complexing agent both because of its tertiary amino nitrogen atom as well as its three alcoholic hydroxy groupings. It can form multi-dentate chelates, up to four in number with stable 5-membered rings.

The literature is somewhat contradictory as regards the stoicheometry of TEA complexes. A literature review by Stammler [198] suggests compositions such as:

$$[AgL]^+, [Cd_2Cl_2]^{++}, [Zn_2L_2Cl_2]^{++}, [Zn_3L(SO_4)(OH)(H_2O)_4]^{+++}, [CuL]^{++},$$

$$[CuL(OH)]^+, [CuL(OH)_2 \text{ and } [CuL(OH)_3]^-$$

With tetravalent tin, an organometallic compound forms having 4 molecules of TEA to 1 of tin. However it is open to question whether the initial concentrations and TEA:Sn ratio, as well as pH have been taken into account in formulating such rules. Stammler himself was far from certain as to how categoric one could be as to the stoicheometries shown above. He did conclude that the reaction of TEA was dependent on the nature of the metal present.

A compilation by Bjerrum, Schwarzenbach & Sillen [196] lists a wide range of metal-organic complexes and their stability constants. Table 3.25 lists values from this source. Bearing in mind that stability increases with Ligand number (dentation) and that these are 4-dentate, their relatively low stability is explained. Only divalent mercury forms a TEA complex with a much higher stability constant of $1.2 \cdot 10^{13}$. The author [21] has reported the stoicheometry and stability of the most important (from an effluent standpoint) copper and nickel TEA complexes at defined metal:TEA ratios and pH values, using potentiometric and spectrophotometric methods. It was found that, with both metals, significant complexation only set in above pH 6. Between pH 9 and 12, complexes with a composition $[Cu_2(TEA)_3]^{4+}$ (binuclear) and $[Cu(TEA)_2]^{++}$ and $[Ni(TEA)_2]^{++}$ were established, broadly in line with the predictions of Stammler. At higher pH values, these complexes were very stable, supporting an argument for their use to remove metals from solution by precipitation and equally their use in chemical metal stripping baths in conjunction with aromatic nitro compounds as depolarisers. As with most other amines, TEA forms an extremely stable complex with iron, more stable than those with copper. The consequence of this is that use of iron as a flocculating agent is not always helpful when TEA is present. The stability constant values shown in Table 3.25 are those obtaining at pH 7 and should not be extrapolated to more alkaline values.

TABLE 3.25:

METAL COMPLEXES WITH TEA AND THEIR STABILITY CONSTANTS (AFTER [196])

Metal ion	Complex	Stability Constant
Ag^+	$[AgL_2]^+$	$4.4 \cdot 10^3$
Zn^{++}	$[ZnL]^{++}$	10^2
Cu^{++}	$[CuL]^{++}$	$1.9 \cdot 10^4$

L = ligand (1 molecule of TEA)

3.9.3.2.2   Complexes with ethylenediamine (EDA)

Ethylenediamine is a well-known component in chemical stripping baths as well as in electroplating and electroless deposition solutions. It is the simplest of the diamines capable of forming a 5-membered ring as a bidentate chelating compound.

Depending on whether 2 or 3 molecules attach to a metal ion, a tetrahedral or octahedral complex forms. Copper is an example of the former, nickel the latter [21]. See Fig. 3.124 as an illustration of this.

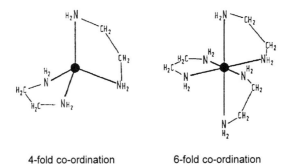

4-fold co-ordination   6-fold co-ordination

Fig. 3.124   Metal chelate complexes of ethylenediamine (schematic).

These complexes are, as is the case for ammonia, positively charged with respect to the central metal ion. Their stability constant can be calculated as follows:

$$\frac{c_{[MeL_{2(3)}]^{x+}}}{c_{Me^{x+}} \cdot c_L^{2(3)}} = K \qquad\qquad 3.177$$

Table 3.26 lists the stability constants for some EDA-based complexes. The listed values are based on those ([196]) obtained at temperatures up to 25°C and up to

0.15M (with the exception of Fe^{++} which was 1M at 30°C). Increasing temperature generally results in lowering of stability constant, as is the case with almost all such complexes. No dependence was found for stability constant on the neutral salt concentration.

With the exception of silver, the chelating effect is more pronounced than for ammine complexes. In the case of nickel, the value is comparable with that for the tetrakis-(β-aminoethyl)-ethylenediamine.

From work by the author, it appears that complex formation becomes significant at pH 3.5 (copper) or 4.5 (nickel). At pH 6 (copper) or 7 (nickel) the final stable complex has formed. The hexadentate nickel complex forms, starting from the hexa-aquo nickel ion, in three stages as pH increases. Walz and Raub [199] found several complexes to co-exist in a nickel-palladium electrolyte operating at pH 3 to 6, including $[Ni(H_2O)_2(EDA)_2]^{++}$ and $[Ni(H_2O)_4(EDA)]^{++}$ as well as the hexa-aquo complexed Ni ion.

TABLE 3.26:
METAL COMPLEXES WITH ETHYLENEDIAMINE AND THEIR STABILITY CONSTANTS [196].

Metal ion	Complex	Stability constant
Ag^+	$[AgL]^+$	$5 \cdot 10^7$
Cd^{++}	$[CdL_2]^{++}$	$1.5 \cdot 10^{12}$
Ni^{++}	$[NiL_2]^{++}$	$2.4 \cdot 10^{19}$
Fe^{++}	$[FeL_2]^{++}$	$3.3 \cdot 10^9$
Zn^{++}	$[ZnL_2]^{++}$	$3 \cdot 10^{14}$
Cu^{++}	$[CuL]^{++}$	$8.9 \cdot 10^{20}$

L= Ligand (1 molecule ethylene diamine)

3.9.3.2.3 Complexes with NNN'N'-tetrakis-(2-hydroxypropyl)-ethylenediamine *(Quadrol)*

NNN'N'-tetrakis-(2-hydroxypropyl)-ethylenediamine is better known by its commercial name "Quadrol". Until recently, the compound was hardly used in the metal finishing and working industries apart from in p.c.b manufacture. In the latter case, it was used as an alternative to EDTA as a component of electroless copper baths.

In its properties, Quadrol behaves much like EDTA, reflecting their very similar structures. In place of the four acetate groups in EDTA, Quadrol has 4 isopropyl groupings (see Table 3.19), i.e. exactly the same number of groupings capable of electron sharing. Thanks to the two nitrogen atoms and 4 oxygen atoms with their free electron pairs, a 5-membered cyclic ring complex can form [200]. As an amine, one would expect such a metal-organic complex to be significantly more stable than the analogous aminoacetic (EDTA) one and this can be seen by comparing Table 3.27 [20] with Table 3.29. From these, it is seen that the selectivity of Quadrol, expressed as the relatively large difference between the stability constants, is greater than that of EDTA. The stability constants in Table 3.27 relate to potentiometrically determined 1:1 complexes [200, 201].

These stoicheometries were confirmed by the author on the basis of absorption spectra. Significant complexation takes place occurs between pH 2.5 and 3 (copper) or pH 3 (nickel). The hexadentate 1:1 copper complex is formed between pH 7 and 12, that for nickel between pH 5 and 9. In the case of copper above pH 10 and nickel, above pH 9, changes in the uv-visible absorption spectrum are seen. The extinction coefficients become larger and, in the case of nickel, the absorption bands shift to longer wavelengths and change their shape. Possibly one or more oxygen atoms are lost from the isopropyl groupings, to be replaced by hydroxyl groups [21].

The equation used to determine the stability constant of the 1:1 complexes formed is analogous to that for NTA (Equation 3.178) while the structure of the stable hexadentate complex at higher pH values resembles that with EDTA (Fig. 3.126, right).

TABLE 3.27:

STABILITY CONSTANTS OF METAL COMPLEXES WITH QUADROL

Metal ion	Stability constant [201]
Ag^+	$1.2 \cdot 10^8$
Pb^{++}	$3.1 \cdot 10^7$
Cd^{++}	$4.2 \cdot 10^7$
Ni^{++}	$7.1 \cdot 10^6$
Zn^{++}	$2.2 \cdot 10^5$
Cu^{++}	$2.6 \cdot 10^9$

3.9.3.3   Metal complexes with aminoacetic acids

3.9.3.3.1   Complexes with nitriloacetic acid (NTA)

The trisodium salt of NTA is used in alkaline cleaning baths as a replacement for polyphosphate and has occasionally been used in chemical metal stripping baths.

The tertiary nitrogen with its 3 carboxyl groupings in NTA enabling the formation of 5-membered cyclic complexes makes NTA a most useful complexant. NTA behaves as a 4-dentate ligand, i.e. all reactive groups participate in complexation. NTA normally forms 1:1 complexes with a metal. Several structures are thus possible. Thus, at higher pH values, 4-dentate binding and tetrahedral structures (see Fig. 3.125 centre) are formed, while at lower pH values when one or more of the carboxyl groups is protonated, a tridentate structure also tetrahedral, can form (Fig. 3.125, left). [202, 203]. Metals with 6-fold co-ordinating capability can also form octahedral complexes with water molecules occupying the two residual positions (see Fig. 3.125, right). Where larger excesses of NTA are present, both copper and nickel can form 1:2 complexes above pH 12. [21].

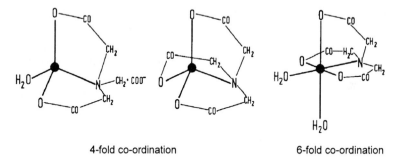

4-fold co-ordination   6-fold co-ordination

Fig. 3.125   Metal chelate complexes based on NTA (schematic)

Values for the stability constants of 2- or 3-valent metal complexes can be calculated, using a simplified approximation for the ionic charge, on the basis of Equation 3.178.

$$\frac{c_{[MeL]}}{c_{Me} \cdot c_L} = K \qquad\qquad 3.178$$

Values for stability constants, all of which are ultimately derived from the compilation of Schwarzenbach *et al,* are listed in Table 3.28. The values are those of the complexes in their most stable state, based on fully deprotonated complexants at high pH values.

TABLE 3.28:

**STABILITY CONSTANTS OF METAL
COMPLEXES FORMED WITH NTA**

Metal ion	Stability constant
Pb^{++}	$2.5 \cdot 10^{11}$ [202, 204]
Cd^{++}	$3.5 \cdot 10^{9}$ [196]
Ni^{++}	$3.2 \cdot 10^{11}$ [202, 204]
Fe^{++}	$6.4 \cdot 10^{8}$ [202, 204]
Zn^{++}	$5 \cdot 10^{10}$ [202, 204]
Cu^{++}	$1 \cdot 10^{13}$ [202, 204]
$Fe^{+++}[FeL]$	$8 \cdot 10^{15}$ [202, 204]
$Fe^{+++}[FeL_2]$	$2.1 \cdot 10^{24}$ [196]

L = Ligand (1 molecule of NTA)

Spectrophotometric studies by the author [21] showed significant complexation to have occurred at pH 0.5 for copper, pH 1.5 for nickel. In the case of both metals, as pH increases, two complexes form which, within their range of stability, have very similar absorption spectra. They are clearly visible in the pH range 2 to 7 and 10 to 12 (copper), and pH 2 to 5 and 8 to 12 (nickel). See also Fig. 3.125.

After EDTA, NTA is one of the strongest complexants used in the metal finishing and working industries. For this reason, the question of its biodegradability was first raised several decades ago. Gudernatsch [205] reported in 1970 that this was possible using a modified form of activated sludge. However when the NTA is complexed with a metal ion, this is either extremely difficult or impossible. More recent studies have considered the effect of NTA as a component of washing powders or cleaning liquids on the aquatic environment. Bernhardt [206] suggested an upper limit be set on such additions, this being reviewed in due course on the basis of experience. Because it is so resistant to biodegradation and because the adaptation process of the micro-organisms in sludge takes a long time, it is possible that excess free NTA might complex metal ions present from other sources, in an effluent, even at NTA concentrations less than 0.1 mg/litre (based on free acid). One is left with concerns regarding the breakdown of the very stable complexes of nickel and mercury. Against this, there would seem to be no serious problems in relation to NTA itself or its complexes with Al, Fe, Cr or Pb.

3.9.3.3.2   Complexes with ethylenediamine tetraacetic acid (EDTA)

EDTA has long been the most widely used complexant in the metal finishing and metal working industry although environmental aspects increasingly weigh against it. Thus it can be found in the following process solutions:

- alkaline cleaning baths. For binding of metals and hardening agents. Less commonly to replace phosphates.

- electrolytic derusting and descaling baths (with polarity reversal).

- electroplating baths. For their beneficial action on the metal deposition process and anode reactions. At times to replace other complexants (e.g. cyanides), especially in alkaline baths.

- electroless plating baths, e.g. copper (for p.c.b manufacture) nickel (coating both metals and plastics). Its main function is to ensure that free metal ion concentrations are held small and constant.

- metal stripping baths., both alkaline and neutral. EDTA is used in combination with depolarising agents such as aromatic nitro compounds.

In Germany, the use of EDTA was prohibited in all of the above, except for electroless copper baths, by legislation published in 1990. Similar prohibitions in other countries are to be expected.

EDTA and its alkali metal salts contain four reactive carboxyl groups and two tertiary nitrogen atoms. All chelate compounds involving both these groupings, form 5-membered rings. Since EDTA can form complexes up to the hexa-dentate type, it is one of the strongest chelating agents. The type of metal ion binding is sometimes a function of the degree of protonation (pH) so that either tetrahedral or octahedral structures can form (see Fig. 3.126). Because of the very great stability of these complexes, it is probably safe to assume that, at high pH values, they are based on hexa-dentate binding with 5-membered rings. All metal-EDTA complexes are, irrespective of pH, 1:1 complexes. The equilibrium equation for the most stable form of complexation is given by Equation: 3.179:

$$Me^{x+} + L^{4-} \rightleftharpoons [MeL]^{(4-x)-}$$

3.179

This equation holds, as for all complexing carboxylic acids, but only above a certain pH at which the carboxyl groups are deprotonated – i.e. at higher pH values. Equation 3.180 gives the stability constant of an EDTA complex:

$$\frac{c_{[MeL]^{(4-x)-}}}{c_{Me^{x+}} \cdot c_L^{4-}} = K$$

3.180

Table 3.29 lists stability constants for the most important metal complexes. If, to a solution, an EDTA complex of lower stability is added and a metal ion with no free EDTA which is capable of forming a more stable EDTA complex, then the first metal is de-complexed, the second metal displacing it in the EDTA complex. Equation 3.181 shows an example of this where complexed zinc is replaced by copper.

$$[ZnL]^{(4-x)-} + Cu^{++} \rightleftharpoons [CuL]^{(4-x)-} + Zn^{++}$$

3.181

Many other such displacements can occur. The reverse action can also occur when, to a solution containing a metal-EDTA complex, a second metal forming a weaker EDTA complex is added. If this second metal ion is present in large excess, it can displace the first. Thus copper can be precipitated from EDTA-containing solutions as the hydroxide, if large amounts of calcium hydroxide are added. Additional dosing with iron (II) salts further enhances such a precipitation.

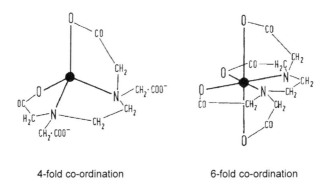

4-fold co-ordination   6-fold co-ordination

Fig. 3.126   Metal chelate complexes of EDTA (schematic)

TABLE 3.29:

STABILITY CONSTANTS OF METAL-EDTA COMPLEXES

Metal ion	Stability constant
Ag^+	$2.1 \cdot 10^7$ [196]
Pb^{++}	$1 \cdot 10^{18}$ [202, 204]
Cd^{++}	$3 \cdot 10^{16}$ [196]
Ni^{++}	$4 \cdot 10^{18}$ [202, 204]
Fe^{++}	$2 \cdot 10^{14}$ [202, 204]
Zn^{++}	$3.1 \cdot 10^{16}$ [202, 204]
Cu^{++}	$6.3 \cdot 10^{18}$ [202, 204]
Cr^{+++}	$1 \cdot 10^{24}$ [203]
Al^{+++}	$1.3 \cdot 10^{16}$ [202, 204]
Fe^{+++}	$1.3 \cdot 10^{25}$ [202, 204]

Metal-EDTA complexes are usually stable over a wide pH range (see Table 3.30).
Spectrophotometric studies [21] showed that significant complexation exists at pH
0.5 both for copper and nickel. A partly-protonated complex of copper or nickel is
stable at pH 1 to 2. One can be very certain that this is a complex of the type shown
in Fig. 3.126 (left) (2 protonated carboxyl groups). The ultra-violet adsorption spectrum of this copper complex in the pH range mentioned above, is identical with that
of the copper complex of ethylenediamine diacetic acid. Above pH 5 (pH 4 in the
case of Ni), the form of the hexadentate complex is as shown in Fig. 3.126 (right) the
stability of which increases further with increasing pH, this reaching a maximum
limiting value at pH 10, as reported in Table 3.29.

It should also be noted that the adsorption spectra of all copper and also nickel complexes are so characteristic that they can be used to identify this type of complex when it occurs in solution.

The very great stability of EDTA complexes has as a consequence that, where EDTA itself is found in effluent, it will complex and thus take into solution any heavy metals present in effluent sludges – indeed it does so more rapidly and completely than almost any other comparable species. To this extent it is unfortunate that, unlike NTA, EDTA is biodegradable and, once this has occurred, the metals remain in solution in the effluent.

TABLE 3.30:

STABILITY RANGES AND COLOURS OF EDTA METAL COMPLEXES (FROM THE BASF TECHNICAL LEAFLET ON "TRILON B PRODUCTS" [204])

Metal ion	pH range in which EDTA complex is stable	Colour of complex
Pb^{++}	2 … 13,5	colourless
Cd^{++}	3.5 … 13	colourless
Ni^{++}	1.5 … 13	blue
Fe^{++}	1 … 12,5	colourless
Zn^{++}	4 … 13	colourless
Cu^{1+}	1.5 … 11,5	blue
Cr^{+++}	1.5 … 5	violet
Al^{+++}	2.5 … 13,5	colourless
Fe^{+++}	1 … 5,5	yellow

3.9.2.4   Metal complexes with carboxylic acids

3.9.3.4.1   Complexes with tartaric acid

Tartaric acid and its salts are used as complexants in copper and nickel electroless plating baths as also, to a lesser extent, in cyanide copper plating baths. They are also used for pickling of copper and its alloys, typically after deposition of these metals. Tartaric acid is also occasionally used as an additive in the electrolytic colouring of anodised aluminium.

Tartaric acid has 4 reactive groupings (2 carboxylic, 2 hydroxyl) which can take part in complexing of metals. Spectrophotometric studies by the author has shown copper to exists as a 1:1 complex, though a 2:1 complex is also known. In the case of nickel, there is even a 4:1 complex. [21]. However the literature [196] also reports 1:2 complexes and others with different stoicheometries.

For chelating to occur, conditions are most favourable when 5-membered rings can form with the metal ion (see Section 3.9.1) and this is the case with tartrates when both hydroxyl radicals or one hydroxyl and the neighbouring carboxyl radicals participate in the binding. The case where both carboxyl groups bind to the same metal ion, to form a much less stable 7-membered ring complex, is also possible. Schoenberg [207] has shown that complexes forming in the first of the two modes described above, can indeed form with copper, whereas those where complexation is in theory possible via 6- or 7-membered rings, do not in fact bind to copper in this way.

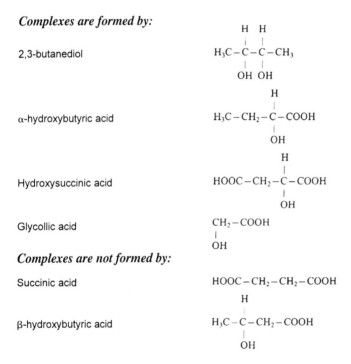

Complexes are formed by:

2,3-butanediol

α-hydroxybutyric acid

Hydroxysuccinic acid

Glycollic acid

Complexes are not formed by:

Succinic acid

β-hydroxybutyric acid

In a typical 4-fold co-ordination of copper, a 1:1 complex is only possible where two co-ordination sites are occupied by water molecules or, at higher pH values, by OH^- radicals. More difficult to envisage, is the structure of metal-rich complexes such as Ni_4L. Only a multi-nuclear structure can readily explain this, in which four nickel ions are bound to a single molecule of tartaric acid via its four complex forming radicals, with three co-ordination sites being occupied by water molecules. Spectrophotometry reveals a significant degree of complexation above pH 3, both for copper and nickel. In the presence of excess complexant, depending on pH, stable complexes can be found, in the case of copper from pH 7 to 10 and pH 11 to 12. In the case of nickel, the analogous pH ranges are 7 to 9 and 10 to 12. It is possible that, in these two cases, the complexes in the lower and upper pH ranges are the diaquo-tartrate and the dihydroxytartrate respectively [21].

Table 3.31 lists stability constants taken from two published compilations [196, 208].

TABLE 3.31:

METAL TARTRATE COMPLEXES AND THEIR STABILITY CONSTANTS

Metal ion	Complex	Stability constant
Zn^{++}	ZnL	$4.8 \cdot 10^2$ [196]
Ni^{++}	NiL	$2.6 \cdot 10^5$ [208]
Cd^{++}	CdL	$1.6 \cdot 10^2$ [208]
Fe^{+++}	FeL	$7.2 \cdot 10^{11}$ [208]
Cu^{++}	Cu_2L_2	$1.6 \cdot 10^8$ [208]
	$Cu(OH)_2L_2$	$7.1 \cdot 10^9$ [208]

L= Ligand (1 molecule tartaric acid)

3.9.3.4.2 Complexes with citric acid

The use-pattern of citric acid is not dissimilar from that of tartaric acid. It is widely used in electroless nickel baths and in pickling formulations. Citric acid contains 3 carboxylic acid and one hydroxyl groupings all of which can participate in complexation. Spectrophotometry reveals a 1:1 complex for both copper and nickel, with an additional 1:4 complex for nickel. [21].

The benefits of 5-membered chelating rings can only be achieved from a bidentate ligand such as citric acid via the carboxyl or hydroxyl groupings on the central carbon atom or by using the hydroxyl group with a terminal carboxyl group. Other-wise unoccupied co-ordination sites are filled with water molecules or, at higher pH values, by OH⁻ ions.

Significant complexation is evident for copper above pH 2.5 and for nickel above pH 3. With excess citric acid, copper forms a stable complex in the pH range 5 to 12 with no change in the observed u.v-visible absorption spectrum. Nickel, by contrast forms three distinct complexes in the pH ranges 3 to 4; 5 to 9 and 11 to 12, as shown by absorption spectra. [21].

Table 3.32 lists some stability constants for metals complexed with citric acid, drawn from published compilations [196, 208].

TABLE 3.32:
CITRIC ACID – METAL COMPLEXES AND
THEIR STABILITY CONSTANTS

Metal ion	Complex	Stability constant
Cu^{++}	CuL	10^{18} [196]
	Cu_2L	$1.25 \cdot 10^8$ [208]
	Cu_2L_2	$1.5 \cdot 10^{13}$ [208]
Zn^{++}	ZnL	$5 \cdot 10^7$ [208]
Pb^{++}	PbL	$4 \cdot 10^{12}$ [208]
Fe^{+++}	FeL	10^{25} [196]

L = Ligand (1 molecule citric acid)

3.9.3.4.3   Complexes with gluconic acid

Sodium gluconate is used in alkaline cleaning baths and in alkaline pickling or brightening baths for pre-treatment of aluminium. In this way, it can bind hardness salts in the water and also minimise build-up of scale on the heating elements, where these are present. It also has useful properties as a dispersant, e.g. for pigments or oxides. Use of gluconates in degreasing baths allows the use of lower operating temperatures and shorter immersion times [209, 210].

Gluconic acid has one carboxyl group and five hydroxyl groupings available for complex formation. At least in theory, bearing in mind their optical activities, it would be possible for two 5-membered ring types to exist using the two ends of the molecule and spectrophotometric studies have confirmed this for copper. In the presence of excess gluconic acid, this appears to convert to a 1:1 complex. In the case of nickel, the 1:1 complex predominates. [21]. These complexes all require additional water molecules to occupy otherwise available sites. At less than stoicheometric ratios of the complexant, metal hydroxides precipitate or other sparingly soluble species as reported by Haarst et al; by Ott and by Jansen and Tervoort (see Section 3.9.5.1.1) who described these in detail.

Significant complexation of nickel and copper commences at pH 3, a stable situation being reached at pH 7 for copper, above pH 8 for nickel [21]. Table 3.33 summarises stability constants drawn from published compilations [196, 208].

TABLE 3.33:
METAL COMPLEXES WITH GLUCONIC ACID AND
THEIR STABILITY CONSTANTS

Metal ion	Complex	Stability constant
Zn^{++}	ZnL	$0.5 \cdot 10^2$ [196]
Cd^{++}	CdL	$1.2 \cdot 10^2$ [208]
Cu^{++}	CuL_2	$2.7 \cdot 10^{19}$ [196]
Ni^{++}	$Ni(OH)_4L$	$2.5 \cdot 10^{29}$ [208]

L = Ligand (1 molecule gluconic acid)

3.9.3.5 Metal complexes formed with other organic species

In this section, various other, less commonly used, organic complexants used in the metal finishing are discussed, to the extent that data are available. These complexants can all be classified in terms of their reactive groupings – amines, aminoacetic acids, aminocarboxylic acids, oxycarboxylic acids and other carboxylic acids. To the extent they are known, the stability constants of the copper complexes of these are listed so that their complexing powers, for this metal at least, can be compared with other species treated previously. Table 3.34 gathers these together.

It is known from practical experience, even where this is not apparent in Tables 3.25 to 3.29, that amines, TEA or Quadrol as well as the very strong complexants NTA and EDTA all form extremely stable complexes with iron (III). Included in this category and shown in Table 3.34 are the amino-acids, DTPA, especially HEDTA as also compounds not treated here such as NN-di(hydroxyethyl)-glycine (DHEG) which are sometimes used in descaling and derusting baths. A few additional comments are in order on the complexation of copper or nickel with the species shown in Table 3.34, as found in studies by the author [21].

TABLE 3.34:
COPPER COMPLEXES FORMED WITH LESS WIDELY USED ORGANICS AND THEIR STABILITY CONSTANTS

Complexant	Formula	Copper complex	
		Comp'n	Stability constant
Oxalic acid	$COOH \cdot COOH$	CuL_2	$3.5 \cdot 10^9$ [208]
Acetic acid	$CH_3 \cdot COOH$	CuL_2	$2.1 \cdot 10^3$ [208]
Glycollic acid	$CH_2OH \cdot COOH$	CuL_2	$5 \cdot 10^3$ [208]
Lactic acid	$CH_3 \cdot CHOH \cdot COOH$	CuL_2	$7 \cdot 10^2$ [196]
Malonic acid	$COOH \cdot CH_2 \cdot COOH$	CuL_2	$1.4 \cdot 10^8$ [196]
Salicylic acid	⟨benzene ring⟩–COOH, –OH	CuL_2	$8 \cdot 10^{16}$ [196]
Aminoacetic acid	$H_2H \cdot CH_2 \cdot COOH$	CuL_2	$1.6 \cdot 10^{15}$ [208]
Imidodiacetic acid	$NH \langle CH_2 \cdot COOH / CH_2 \cdot COOH$	CuL_2	$4.8 \cdot 10^{16}$ [208]
EDDA [1]	$Ac \rangle H \cdot CH_2 \cdot CH_2 \cdot N \langle H / Ac$; H	CuL_2	–
1,2 diaminocyclo-hexane tetraacetic acid	$N \langle CH_2 \cdot COOH / CH_2 \cdot COOH$, $N \langle CH_2 \cdot COOH / CH_2 \cdot COOH$	CuL	$2 \cdot 10^{21}$ [196]
DTPA [2]	$Ac \rangle N \cdot CH_2 \cdot CH_2 \cdot N \cdot CH_2 \cdot CH_2 \cdot N \langle Ac / Ac$; Ac	CuL	$3.2 \cdot 10^{21}$ [196]
HEDTA [3]	$HOCH_2 \cdot CH_2 \rangle N \cdot CH_2 \cdot CH_2 \cdot N \langle Ac / Ac$; Ac	CuL	$2.5 \cdot 10^{17}$ [196]
DETA [4]	$NH \langle CH_2 \cdot CH_2 \cdot NH_2 / CH_2 \cdot CH_2 \cdot NH_2$	CuL	–
TEPA [5]	$NH \langle CH_2 \cdot CH_2 \cdot NH \cdot CH_2 \cdot CH_2 \cdot NH_2 / CH_2 \cdot CH_2 \cdot NH \cdot CH_2 \cdot CH_2 \cdot NH_2$	CuL	

Ac- = acetate group (–CH$_2$ · COOH)
[1] ethylenediaminediacetic acid
[2] diethylenetriaminepentaacetic acid
[3] N-hydroxyethyl-ethylenediamine triacetic acid
[4] diethylene triamine
[5] tetraethylene pentamine

Aminoacetic acid (glycocollic acid) complexes with copper above pH 2.5, with nickel above 5, as observed spectrophotometrically. In the case of copper, this is a 1:2 complex, stable in the pH range 7 to 11. In the case of nickel, it is a 1:3 complex, stable between pH 8 and 12. These stoicheometries were further confirmed by potentiometric studies which also revealed, in the case of nickel, its preferred 6-fold co-ordination.

Imidoacetic acid forms complexes with copper (above pH 1.5) and with nickel (above pH 2) which can be spectrophotometrically observed. In both cases, a stable complex exists in the pH range 3 to 5 (1:1.5 ?). Above pH 7, this forms a stable 1:2 form.

Ethylenediaminediacetic acid (EDDA) likewise forms complexes with copper (above pH 1.5) and nickel (above pH 2). The Cu complex is stable from pH 3 to 5, its absorption spectrum being virtually identical with the 4-dentate EDTA complex (at pH values up to 2). At pH 8 to 12, a stable, possibly 6-dentate hydroxy complex forms. Both this and the Ni complex stable in the pH range 7 to 12 are 1;2 type complexes.

Diethylene triamine (DETA) appears to form a recognisable complex with copper above pH 3, with nickel above pH 5. The former is a 1:1 complex, stable in the range pH 5 to 12, the latter is a 1:2 complex, stable from pH 7 to 12.

Tetraethylene pentamine (TEPA) complexes copper above pH 2.5, nickel above pH 4. Both are 1:1 complexes, stable above pH 4 (copper) or pH 6 (nickel).

3.9.4   Effect of complexing agents on effluent treatment

It has already been pointed out that metals, when complexed, behave quite different-ly from their uncomplexed states and many chemical reactions, e.g. alkaline precipi-tation, no longer take place. Thus, when complexing agents are present, the pre-scribed effluent treatment methods may be very different for a particular metal, from the case where they are absent.

The introduction of complexing agents into a surface treatment process undoubt-edly can bring many benefits. At the same time, the user should always be sure that such benefits are not outweighed by drawbacks in subsequent effluent treatment. In the event it is decided to use complexing agents, the choice of these should be carefully made. Thus use of EDTA containing solutions which might come into contact with metals (except for copper) is increasingly prohibited, except in the case of certain authorised effluent treatment processes. Many complexing agents forming very stable (so-called "hard") complexes with metals can be replaced with others forming less stable complexes. However thought should be given to carrying out such substitution. The use of NTA to replace phosphates in alkaline cleaning baths is, environmentally speaking, pointless, if the NTA comes into contact with metal ions before it is bio-degraded.

As a general principle, it is advisable that effluents containing complexing agents should be kept separate in the same way as this advice was given for cyanide or chromate-containing effluents. This apart, the volume of such effluent should be kept to a minimum thus allowing the complicated treatment necessary to be carried out in a batch reactor. This calls for much the same measures as are described in Section 6 under "Recycling".

Various approaches are currently available for treatment of effluents containing complexants. Almost all have the drawback that, after removal of the metal, the complexant remains intact. While these are not particularly toxic, the problem remains that they will react with most metal ions with which they come into contact. This could happen, for example, when they contact activated sludges in public utility sewage works or sedimented species in watercourses. Thus EDTA or NTA, even at low concentrations, will readily re-dissolve precipitates of copper hydroxide even after several days of ageing, in stoicheometric manner. [21]. The complexing agents themselves can only be removed by precipitation processes or, when they are organic, by oxidation reactions. At least some of these methods have only been used on a laboratory scale.

As pointed out earlier, breach of the prescribed concentration levels for metal ions in effluent discharges (see Section 3.7.8) can result from incorrectly implemented precipitation procedures or by inadequate separation of heavy metal ion containing species. There is a third possible cause for breach of the legislation, namely complexation which prevents precipitation of such heavy metal ions. The simple analytical methods prescribed by some Water Authorities does not allow users to distinguish which of the above three causes is responsible for over-shooting the concentration limits. Defective precipitation or insufficient dwell times can often be pin-pointed by using a colorimeter. Where complexed species are the cause, it is harder to recognise this fact and a laboratory method or use of AAS (atomic adsorption) may be needed, In most cases, it is not possible to remedy the situation unless there is an understanding of what caused it in the first place.

A simple method of estimating a metal complex is to measure the complexing value of an effluent [94]. To 1 litre (or 100 ml) of effluent are added 100 (or 10) mg/litre copper as the sulphate, respectively. pH is adjusted to 8.5. The solution is allowed to equilibrate by standing for two hours at room temperature and is then filtered. If the copper content in the filtrate exceeds 1 mg/litre (for concentrated effluents 2 mg/litre), then it can be said with certainty that complexants are present. Determination of copper must be carried out by AAS or colorimetrically after treating the sample with fuming sulphuric acid. The residual copper concentration in mg/litre is a direct index of the complexing power of the effluent. A value above 100 mg/litre indicates presence of complexants. If the sample is acidified before addition of the copper salt, any complexing action due to polyphosphates is eliminated. If, instead of adding copper sulphate, an iron (III) salt is added, the complexing effects of polyphosphates can be determined even in presence of organic amines (ethylene diamine excepted). Instead of using copper sulphate, a nickel salt can be used and the pH adjusted to 8. Excess of dimethyldithiocarbamate is then added. Nickel concentrations in excess of 1 mg/litre indicate with some confidence the presence of NTA or EDTA.

Such variants of the determination of complexing values coupled with a knowledge of the types of complexant likely to be used in a given process, allow conclusions to be drawn without use of sophisticated or expensive analytical instruments. Use of polarographic methods and, at higher concentrations, spectrophotometric techniques, allows some complexants to be specifically determined, while the copper sulphate approach described above, will usually give a semi-quantitative figure, once its behaviour for a specific situation has been characterised. The method can also be used to determine complexing agents in process solutions which can be important if no information regarding these is available to users.

In general, it can be said that the use of the copper sulphate method allows a quantitative assessment of "hard" complexing agents (NTA or EDTA) or (given some experience) a qualitative assessment for the less powerful complexing agents. If the test gives a positive reaction, then some action is called for by management.

3.9.5   Option for effluent treatment

The following sections apply only to treatment of effluent containing chelated metals with organic complexants and also, to some extent, metal amine complexes. In respect of metal complexes with inorganic compounds, see Sections 3.2.1.6 and 3.9.3.1.2 for cyano complexes, Sections 3.9.3.1.1; 3.7.5.2.2; 3.7.5.8.2 and 3.7.5.9 for hydroxyl complexes and Section 3.9.3.1.4 for phosphate complexes.

Basically, the following processes can be considered for treatment of effluents containing complexed metals:

- metal precipitation using special methods

- metal removal by its reduction to the elemental state

- oxidation of the complexant

- precipitation of the complexant

- thermal treatment, removing the water in the effluent and storing the resulting solid salt. (this last approach is included for completeness only – it is not used in practice).

3.9.5.1   Precipitation of metals

Precipitation of a sparingly soluble metal compound is possible when its solubility product, in presence of the complexant, is exceeded. This would be the case when the metal ion concentration resulting from dissociation of the sparingly soluble metal compound was much smaller than the metal ion concentration corresponding to dissociation of the complex. There is, in effect, a competitive situation between the solubility product L of the sparingly soluble metal compound and the stability constant of the complexed metal:

Solubility product: $\quad c_{Me^{z+}} \cdot c_{A^-}^z = L$ $\hspace{3cm}$ 3.182

A^- is the anion forming the sparingly soluble compound with metal ion Me^{z+} (for simplicity's sake, a monovalent ion is assumed).

Stability constant: $\quad \dfrac{c_{[MeL_x]^{(y-z)-}}}{c_{Me^{z+}} \cdot c_{L^{y-}}^x} = K$ $\hspace{3cm}$ 3.183

Assuming the equilibrium metal ion concentrations resulting from these two equations is equal, we then have:

$$\frac{c_{[MeL_x]^{(y-z)}} \cdot c_{A^-}^z}{c_{L^{y-}}^x} = K \cdot L = K' \hspace{3cm} 3.184$$

If K' becomes very large, the metal complex stays in solution. If it is very small, precipitation occurs. In predictive terms, K' should have a value of 10^{-5} to 10^5. Sparingly soluble compounds are mainly:

- metal hydroxides
- metal sulphides
- metal compounds with organosulphides.

3.9.5.1.1 Precipitation of hydroxides

Where weak or relatively weak metal complexes are involved, e.g. ammoniacal, certain of the amine or pyrophosphate complexes, it is possible to raise the pH to a value where hydroxide precipitation occurs. This is sometimes referred to as "over-alkalinisation" in which the pH is raised to 11 or higher, to those regions where pH measurement becomes difficult. A typical example is the precipitation of copper hydroxide from a copper tetrammine containing solution using NaOH as shown in Fig. 3.123. In similar manner, nickel hydroxide can be precipitated from solutions containing nickel hexammine complexes.

The feasibility of hydroxide precipitation from complex-containing effluents can be established using Equations 3.182 to 3.184 where A^- denotes the OH^- ion. This is perhaps better shown graphically by plotting the pMe value (negative logarithm of the metal ion concentration) derived from Equations 3.182 and 3.183, versus pH.

Fig. 3.127 illustrates this using copper hydroxide, a weak copper complex and the very stable ("hard") EDTA complex of copper.

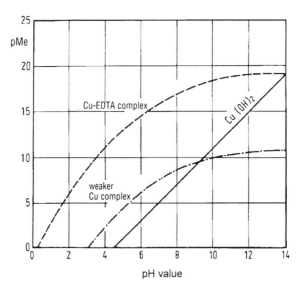

Fig. 3.127   pMe value (see text) of copper hydroxide, a weak copper complex and the strong copper EDTA complex vs. pH. [21].

Hydroxide precipitation is only possible where the curves for the hydroxide and the metal complex intersect [211]. In the case of the EDTA complex, this does not happen as is the case for a weaker copper complex, e.g. the Cu-tetrammine complex. In the case of amphoteric metals such as zinc or chromium, the calcium ion concentration is also important, since if an alkali metallate is formed, the metals will redissolve. If a high enough pH cannot be obtained using slaked lime (because of its limited stability), then additions of NaOH will be necessary. In some cases, it is best to add the requisite amounts of calcium ion in the form of its chloride.

During over-alkalinisation, calcium ions may exert other effects leading to metal hydroxide precipitation or enhancing it:

- formation of sparingly soluble calcium salts with the complexants.

- decomplexing of metals to form, instead, calcium complexes, when large excesses of calcium are present.

- reducing the stability of the complex as a result of increased total ionic strength. A similar effect is found with large excesses of NaOH. However the doubly charged calcium cation is more effective.

The first of these effects operates in the case of polyphosphates which are precipitated in weakly alkaline conditions as sparingly soluble calcium compounds. At high pH values, to the extent they become monomeric, they precipitate as extremely insoluble calcium hydroxyapatites (see Section 3.10.3). The complexant forms part of the equilibrium process and hydroxide precipitation can take place unhindered. Fig. 3.128 shows the precipitation of copper hydroxide from a pyrophosphate containing solution made highly alkaline with NaOH. Also shown is the beneficial effect of added slaked lime [94]. Using slaked lime in this way also allows zinc hydroxide to be precipitated from such solutions.

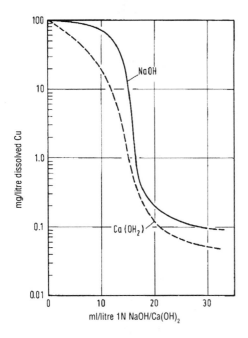

Fig 3.128   Dependence of the residual solubility of copper on the degree of alkalinisation, using NaOH with and without added slaked lime. 100 mg/litre Cu. [189]

Treatment in this way is further assisted by the fact that the metals partly form sparingly soluble salts such as phosphates or basic phosphates. Effluent containing phosphates and heavy metals, after the latter have been precipitated by neutralisation, is also virtually phosphate free, provided the initial phosphate concentration does not exceed the equivalent amount of metals present (see Section 3.10.3). This is confirmed by Wing *et al.* [212] who precipitated copper from pyrophosphate containing solution. They report residual copper concentrations of 0.02 mg/litre and phosphate under 1 mg/litre. According to their study, the optimum excess calcium concentration was 30 mg/litre. A decomplexation of metals to form instead the calcium complex was confirmed even for the strong complexing agents of the aminoacetic type, such as NTA or EDTA. Wing *et al.* also showed [213] that copper could be precipitated from NTA and EDTA containing solutions when slaked lime was added at pH > 11.7. The greater the calcium concentration, the lower the residual copper

value. At $(Ca^{++}):(Cu^{++})$ ratios above 2.5 and initial concentrations in the range 5 to
100 mg/litre copper, residual Cu concentrations less than 0.2 mg/litre were obtained.
(see Fig. 3.129). At pH , 10.5 and initial Cu concentration of 50 mg/litre, no precipitation takes place. At pH 12.5 or above, residual copper concentrations increase
again to 0.5 mg/litre, also increasing with initial copper concentration values. The
same authors report similar results with copper + NTA containing solutions. Less
satisfactory results were found (in that order) with HEDTA, aminodiacetate, tartrate,
citrate and gluconate containing solutions with copper. In the case of solutions
containing TEA or Quadrol, no precipitation was observed. These results suggest
that precipitation is only possible when the nitrogen atom in the complexant is fully
bonded to carboxyl groups. As these are replaced by hydroxyl groups (the same is
true for complexants not containing nitrogen atoms), the complex becomes harder to
precipitate to the point where fully hydroxylated complexants cannot be precipitated.

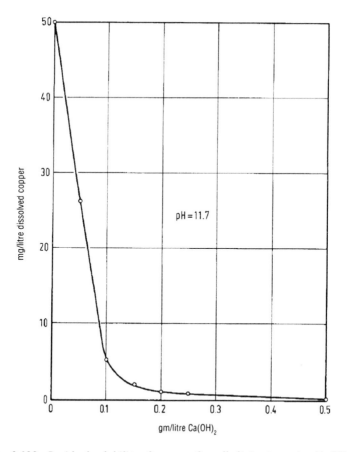

*Fig. 3.129   Residual solubility of copper after alkalinisation using NaOH to
pH 11.7 in presence of EDTA, with added slaked lime. Initial Cu concentration
50 mg/litre. (after Wing et al. [213])*

The author can confirm these results from his own experience where Cu-EDTA or Cu-NTA containing solutions are concerned. As long as no significant excess of complexant was present, residual copper concentrations below 1 mg/litre were found [211]. Similar results were found for the nickel-NTA complexes though not for the more stable Ni-EDTA complex. These findings are in agreement with the work of Wagner [214] who ascertained that the extent to which NTA or EDTA re-dissolved heavy metal compounds present in activated sludges was reduced in the presence of calcium ions.

The partial [213] to satisfactory [94] precipitation of copper from tartrate or citrate-containing solutions by means of slaked lime additions with excess alkalinisation can also be explained in terms of decomplexation of the copper to form instead a calcium complex.

A variant or even improved form of the above precipitation of copper from solutions containing complexants, is to to add iron (II) salts as well as slaked lime. This method is also described by Wing *et al.* [215] after practical experiences in Germany. Wing explains the success of the method in terms of reduction of the Cu(II) ions by the iron (II) so that the copper is precipitated in its monovalent state. The process can be carried out without initial acidification in neutral to acid solutions with equal success. For a reduction to take place in alkali, as shown in Equation 3.185, one would expect formation of a light yellow copper (I) hydrated oxide with a reddish-brown iron (III) hydroxide rather than the mixed colour of blue Cu(II) hydroxide and the grey-green iron (II) hydroxide, as is indeed the case.

$$2\ Cu(OH)_2 + 2\ Fe(OH)_2 \rightleftharpoons Cu_2O \cdot H_2O + 3\ Fe(OH)_3 \qquad 3.185$$

Using the iron-lime approach above, good results are found with copper complexes of ammonia, NTA or EDTA even when these complexants are present in excess. In the case of amines such as TEA or Quadrol, residual copper concentrations are in excess of 1 mg/litre and no excess of complexants must be present. The method can also be used for nickel complexes with ammonia or NTA though not for Ni-EDTA complexes [21]. The author suggests that the benefits of Fe (II) addition in hydroxide precipitation could well be due to the incorporation of residual amounts of the isomorphous copper ion within the severely strained lattice fragments of the iron (II) hydroxide [211]. (see Section 3.7.6). The fact that iron (III) compounds have no effect on precipitation does not speak against either of the two hypotheses. Addition of iron (II) salts to enhance copper hydroxide precipitation enables residual copper concentrations to be brought below 0.5 mg/litre, depending on the initial copper concentrations. Against this, the nuisance values of the resulting iron in the effluent must be balanced and this is specially important in the case of complexants forming a stable complex with iron. These include Quadrol, TEA and gluconate, i,e complexants containing hydroxyl groupings as well as the strongly complexing aminoacetic type species such as NTA and especially EDTA. Final iron concentrations depend partly on the concentration of complexant but also on the pH at the end of the operation [215].

The iron-lime process has significant advantages over the use of slaked lime and extreme alkalinisation. According to Wing *et al.* for initial copper concentrations of 10 mg/litre and additions of ferrous sulphate of 400 mg/litre (based on the heptahydrate, i.e. 80 mg/litre of Fe) or initial Cu concentrations of 50 mg/litre with 1 gm/litre iron sulphate added, a residual copper ion value of less than 0.5 mg/litre results if the final pH is brought to 10.7. This is an operation which may call for close supervision. The process is well-suited as first stage in a batch operation to reduce the consumption, in subsequent stages, of expensive organosulphides. In some cases, such further treatment is unnecessary and this process alone suffices.

Wing & Rayford [216] suggest the use in this process of spent HCl or sulphuric acid iron pickling wastes.

Precipitation of metal hydroxides from effluent containing gluconate complexants is a special case. Haarst *et al.* [217] report that, in such cases using slaked lime and a final pH of 9, the higher the metal:gluconic acid ratio, the better the results. If the metal content is too low, addition of aluminium salts (themselves posing no great problem of toxicity) is suggested to improve the ratio. The author has also studied this, albeit in the inverse sense where alkaline process solutions containing gluconates are used in pre-treatment of aluminium for anodising can have a complexing action on copper ions. Below 100 mg/litre gluconate, when aluminium is present, copper ion concentrations will not normally exceed 0.5 mg/litre. Only for gluconate concentrations above 200 mg/litre would one expect aluminium levels to exceed 3 mg/litre [94].

Ott [218] carried out very precise studies on the effect of gluconates on metal precipitation using alkalis. At pH 9 to 10, in presence of iron ions, other metal ions present could be effectively removed as with hydroxide precipitation. The ratio of iron:gluconic acid is important here – it should be at least 3. Addition of calcium ions, e.g. as slaked lime, allows this to be reduced to 1 or 2:1. It is also important to note that this also reduces the gluconate concentrations, in some cases to only 5% of the initial value. This indicates that, without further characterising it, the precipitate must contain both metal hydroxides and gluconate. The results of Ott were further confirmed by studies of Jansen & Tervoort [209]. They showed that for a metal:gluconic acid ratio of 5 (on a molar basis) over a range of gluconate concentrations from 0.1 to 10 Molar, the metals zinc, nickel, copper could all be precipitated using NaOH to pH 9 or 9.5 when iron was also present. Precipitation of metals and removal of gluconic acid was found to improve as the listed order of metals above with residual metal ion concentrations being less than 0.5 mg in most cases. Bearing in mind that effluent from metal finishing plants using degreasing baths including gluconic acid will usually have metal:gluconic acid ratios around 40:1, one would not expect the gluconic acid to have any interfering effects. However in some metal recycling operations, such excess of metal may not be found. In these cases, the metal shortfall must be made good by adding iron (III) or aluminium salts or by raising the concentration of calcium salts.

Similar studies of metal precipitation using commercial alkaline cleaning or derusting baths were published by Srinivasan [219] who noted the benefits of using slaked lime or addition of calcium, iron (III) or aluminium ion containing solutions.

3.9.5.1.2 Precipitation using sulphides

The use of sulphides to precipitate metals from effluents containing complexants is very recent. Until the adoption (e.g. in Germany) of the new low 0.5 mg/litre metal ion limit, conventional hydroxide precipitation was usually adequate. The use of sulphides in normal circumstances is covered in Section 3.8.1.

When sulphide anions are used in place of hydroxide ions for precipitation, the term: c_A^{z-} in Equations 3.182 and 3.184 should be replaced by c_S. K' in Equation 3.184 is then several orders of magnitude smaller and metals even in very stable complexes can be precipitated. A comparison of the solubility products for a range of metal sulphide and hydroxides is shown in Table 3.17.

Precipitation of metals using sodium sulphide is discussed in Section 3.8.1.1. To the extent metals are part of the hydrogen sulphide grouping, in this case copper and lead being the most important, they are more easily precipitated as the complexants are protonated at low pH values. In its protonated state, complexants are unable to bind to metals since the active grouping involved in complexing are deactivated when protonated, the degree of dissociation of this hydrogen being very small. The following equations show examples based on the complexants cyanide, ammonia and EDTA.

$$CN^- + H^+ \rightleftharpoons HCN \qquad\qquad 3.186$$

$$NH_3 + H^+ \rightleftharpoons NH_4^+ \qquad\qquad 3.187$$

$$(EDTA)^{4-} + 4\,H^+ \rightleftharpoons H_4(EDTA) \qquad\qquad 3.188$$

Thus copper can be precipitated from acid solutions, whether they contain complexants or not. There are, in such cases, some added difficulties and extra cost implications arising from the fact that the solids separation takes place in acid environment. In neutral or alkaline media, the equilibria shown in Equations 3.186 to 3.188 are shifted to the left, thus somewhat restricting the scope for quantitative precipitation. Since nickel, zinc and iron can only be precipitated by sodium sulphide in weak alkaline solutions, stable metal complexes with NTA or EDTA are difficult to treat. It is doubtful whether copper can be quantitatively precipitated and in the case of other metals, precipitation may simply not be possible.

From earlier studies by the author [94], it can be confirmed that, under appropriate conditions, sodium sulphide precipitation can be used satisfactorily, to treat complexes with pyrophosphate, tripolyphosphate, TEA, EDA and gluconic acid. In the case of higher polyphosphates, NTA and EDTA, copper precipitation is doubtful in the former cases, not possible in the latter.

When precipitating agents are used which lead to lower values of K' in line with Equation 3.184, better results are obtained than those with sodium sulphide. These are mainly the organosulphides, the most important of which are listed in Table 3.18.

It might be expected that, as the molecular weight of the precipitate increases, so does the residual metal solubility decrease. Studies by the author [21] do not confirm this. Compounds such as TMT, cellulose and starch xanthogenate are those predicted to give best results. Precipitation of copper from complexant containing solution with thiocarbonic acids, thioalcohols, thioethers and thioesters does not give significantly better results than where sodium sulphide is used, at least to the extent that such precipitations must be carried out in acid medium [220]. Metals in solutions containing weak to medium strength metal complexes can be precipitated using TMT (trimercaptotriazine) at pH 7 to 10. (see Section 3.8.1.2). The normal dose for this is not less than 490 ml TMT per equivalent metal ion. In the case of more strongly bound complexes such as those with EDTA, these can be decomplexed at pH 3 to 4 by first adding iron (III) salts. The previously complexed metals can then be precipitated with TMT and filtered. The Fe(III) EDTA complex formed is then oxidised in two stages using hydrogen peroxide, the ferric hydroxide is precipitated and filtered [221, 222, 223]. By such means copper can be precipitated (directly) from the tetrammine complex or from the EDTA complex (after decomplexation). Silver can be removed from photographic fixing baths (directly or after decomplexing), nickel can be removed from polyamine complexes (after aluminium salt addition) and lead can be removed from cationic electrophoretic painting baths down to residual concentrations of less than 0.5 mg/litre.

The use of xanthates (insoluble starch xanthate and cellulose xanthate) as organosulphide precipitants (see Table 3.18) has also been studied. The results were at best mediocre, at worst bad with starch xanthogenate better than cellulose xanthogenate. It might also be noted that neither of these solutions are very stable and require to be stored at low temperatures. Starch xanthogenate can be used in one interesting way where it is allowed to flow through a filter where, being insoluble, it forms a pre-coat and as such can be used for post-treatment or polishing of effluent [224]. This particular precipitating agent performs best in the weakly acid pH range from pH 3 to 5, depending on the complexant in question.

By and large, the dithiocarbamates have proved the best of the organosulphides for precipitation of metals. Within this category, primacy goes to dimethydithiocarbamate (DMDTC) which performs better than similar homologues with larger alkyl groupings [225]. Using DMDTC (see Table 3.18), copper can be precipitated from any of its complexes as used in metal finishing and surface treatment in neutral to weakly alkaline environments to give residual copper values of less than 0.5 mg/litre. By various modifications of the precipitation process, significantly lower values can be reached. If the initial copper concentrations are low, e.g. a few mg/litre (such concentrations being typical after hydroxide or sulphide precipitation), a second stage precipitation using DMDTC and iron (III) flocculation can be carried out taking the copper concentration down to the order of 10 ug/litre. Similar results are found for nickel, zinc and cadmium with the proviso that aminopolyacetic acid complexants (NTA, EDTA etc.) are not present [21,211]. Copper, although a metal very which complexes very readily, is something of a special case in that its complexes are often easier to handle with correspondingly higher yields, than those of other metals. As precipitated with DMDTC, it is divalent, requiring two equivalents of carbamate to form the sparingly soluble organosulphide:

$$2\,S = C{\overset{\displaystyle N(CH_3)_2}{\underset{\displaystyle SNa}{<}}} + Cu^{++} \rightleftharpoons S = C{\overset{\displaystyle N(CH_3)_2}{\underset{\displaystyle S---Cu---S}{<}}}{\overset{(CH_3)_2N}{>}}C = S \qquad 3.189$$

Fig 3.130 shows how copper can be precipitated even from concentrated solutions and in presence of large excess of EDTA. Precipitation of 1 gm copper requires approx. 8 to 10 gm of the commercially available concentrate [226, 227].

Precipitation of metals other than copper can also be carried out in presence of aminopolyacetic acid, though this requires destabilising of the complex by partial protonation of the complexant, e.g. by adjusting to pH 3. By such means, even metals in the ammonium sulphide grouping can be precipitated as sparingly soluble carbamates in line with Equation 3.189. For this, however, several times the normally required quantity of precipitant are required since it is not stable in acid medium and breaks down, forming toxic carbon disulphide. The concept is thus not a very practicable one (see Equation 3.190).

$$S = C{\overset{\displaystyle N(CH_3)_2}{\underset{\displaystyle SNa}{<}}} + H^+ \rightleftharpoons CS_2 + NH(CH_3)_2 + Na^+ \qquad 3.190$$

The extremely low solubility of precipitates formed with DMDTC can perhaps be explained in terms of formation of larger complex conglomerates in which the central metal ion is additionally bonded to other nitrogen or sulphur atoms in adjacent $Me(DMDTC)_2$ molecules through their unpaired electrons.

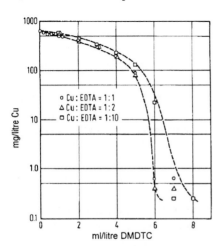

Fig. 3.130 Precipitation of copper with DMDTC from solution containing EDTA. Solubility is plotted for various EDTA concentrations. Precipitation pH = 8. [211].

Wing & Rayford [228] studied the precipitation of metals from solutions containing complexants, using DMDTC. Their paper includes a listing of solubility products for the relevant compounds, drawn from Still [228]. These are in fact larger than those of the simple metal sulphides as implied by the respective solubility products. For copper sulphide, this lies in the range $8. \cdot 10^{-45}$ at $5 \cdot 10^{-18}$ mg/litre, the corresponding figures for DMDTC being $1.58 \cdot 10^{-28}$ at $3.4 \cdot 10^{-5}$ mg/litre copper.

TABLE 3.35:

SOLUBILITY PRODUCTS OF ME(DMDTC)$_2$ PRECIPITATES (AFTER STILL [228].)

Metal	Solubility Product
Silver (Ag^+)	$2.51 \cdot 10^{-21}$
Copper (Cu^{++})	$1.58 \cdot 10^{-28}$
Lead (Pb^{++})	$3.98 \cdot 10^{-24}$
Cadmium (Cd^{++})	$6.31 \cdot 10^{-21}$
Zinc (Zn^{++})	$5.01 \cdot 10^{-18}$

In terms of experimentally observed results, this cannot be correct. The values in Table 3.35 should thus only be used for comparison with one another, in that all were determined under identical conditions. Wherever sulphides of any type are used as precipitating agents, it should be remembered that these are toxic and excess after reaction should be reacted, for example with iron salts as described in Section 3.8.1.1.

3.9.5.2   Reduction of complexed metals

In cases where the central metal ion in a complex can be chemically reduced to its metallic state, this provides a means for its removal from solution. The reduction mechanism involves the metal in its ionic, uncomplexed state as determined by the stability constant. The following reduction mechanisms are important:

- ion exchange (cementation) with a less noble metal
- chemical reduction using a strong reducing agent
- cathodic (electrolytic) reduction.

All of these involve electron transfer. In the first two cases, the process involves a redox system in which the electron donor should be as negative as possible (see Table 7.4). In the last case, electron donation occurs at the cathode.

Whether cementation can be used, can be determined by inspection of the electrode potential series (e.m.f. series) as discussed in Section 7.6.1. Thus zinc can be used to displace all metals up to and including cadmium. The smaller the gap between the two metals (the one displacing the other), the more difficult is such a displacement reaction and satisfactory performance is really only achieved with copper or metals more noble than copper. Cementation (or displacement) occurs when Me_1, the more noble metal (electron acceptor) reacts with Me_2, the less noble electron donor metal, according to the equation:

$$Me_I^{++} + Me_{II}^0 = Me_I^0 + Me_{II}^{++}$$ 3.190a

From work by Abel [229], it is known that copper ions flowing through a reactor with an aluminium mesh can be displaced down to a level of 1 mg/litre. Much the same will happen in acid solutions (pH <2) when ammoniacal compounds or tartrate is present. Since complexants are protonated under these conditions, the reaction corresponds to a straightforward metal salt dissolution.

Chemical reduction is also not without its uses as a technique, not least in cases where electroless deposition from alkaline solutions is used. In practice, this involves mainly copper or nickel. In such cases, one often uses the same reaction employed for the electroless deposition. The approach is mainly used to treat baths which have become unusable or for concentrated rinsewaters. However in order to use this approach, the correct pH and reductant concentrations have to be brought about. In this approach, electron donation is from the reducing agent. As a rule of thumb, this method will work when using a redox system whose Normal Potential is more negative than that of the corresponding Me_0/Me^{z+} reaction (see Tables 7.3 and 7.4).

To treat electroless copper baths, such as are typically used in the printed circuit board industry, an initial addition of 4 gm/litre caustic soda is made, followed by 5 ml formaldehyde per litre (37% v/v). [230]. Suitable precautions must be observed in this operation. The solution should now be contacted with a solid have the greatest possible surface area. Granulated activated carbon may be used for this. It should be activated with a precious metal salt (e.g. palladium) in the same way as items for electroless plating. Less expensive, is the use of silver nitrate. The process can be operated batchwise though it is better to use a recirculatory system in which the solution is pumped through or over a bed of activated carbon. [211]. Copper is formed with simultaneous hydrogen evolution [231] as below:

$$Cu^{++} + 2\ HCHO + 4\ OH^- \rightleftharpoons Cu^0 + 2\ HCOO + H_2 + 2\ H_2O$$ 3.191

The reaction rate can be gauged from Fig 8.38a. The activated carbon granules can be re-used several times and copper deposition continues even when these are thickly coated with copper. As Fig. 8.38a shows, final concentrations of 0.5 mg/litre are easily obtained [211]. The copper metallised activated carbon can, in due course, be shipped to a refinery for pyrolytic processing and metal recovery although, in economic terms, the latter is of no significance. The method does, however, avoid formation of hydroxide sludges and it is possible to recover EDTA from solution, once the copper has been removed.

The method can be improved by use of high surface area stainless steel mesh in place of activated carbon [232]. If then the stainless steel mesh is made the anode in a cell, the copper dissolves and deposits onto a cathode, whence it can be removed as saleable scrap.

Where the solution contains EDTA as a complexant, this can be recovered as described in Section 8.9.4.1. Care should be taken to avoid excessive COD levels following the use of formaldehyde.

Nickel can be removed from spent electroless nickel baths or concentrated rinsewaters in a similar manner, though the process does not take place so readily. pH values are, depending on the species used in the bath make-up, anywhere from weakly acid to strongly alkaline. After setting the correct pH and addition of reducing agent, most of the nickel can be deposited on scrap metal components or on activated non-metallic surfaces. Most commonly used reducing agents are sodium hypophosphite or various boranes (sodium boranate, dimethyl- or diethylamineborane), in some cases hydrazine. All have a Normal Potential, in alkaline solutions, less than - 1 volt. Depending on the reductant used, equations such as:

$$Ni^{++} + H_2PO_2^- + H_2O \rightleftharpoons Ni^0 + H_2PO_3^- + 2\,H^+ \qquad\qquad 3.192$$

$$3\,Ni^{++} + (CH_3)_2NH \cdot BH_3 + 5\,OH^- \rightleftharpoons 3\,Ni^0 + H_3BO_3 + \qquad 3.193$$
$$(CH_3)_2 + 2\,H_2O$$

take place. Hydrazine is used as a reducing agent in chemical rinsing. After the work is dipped in a static rinse tank, it is rinsed with a weakly alkaline hydrazine solution which gives rise to a reduction to form Cu(I) oxide which falls out in powder form [234]. However the known carcinogenic properties of hydrazine have led to the gradual abandoning of this process.

Another strong reducing agent is dithionite ($Na_2S_2O_4$). Depending on the pH, this acts as a 2 electron (alkaline) to 4 electron (acid) acceptor (Equations 3.80 to 3.82). This involves an oxidation to sulphite (SO_3), dithionate (S_2O_6) and finally sulphate (SO_4). The reduction of copper with dithionite is described in a patent [235] where alkaline conditions and a temperature of 60°C is prescribed, the final concentration being less than 1 mg/litre. An excess dithionite concentration of 50 to 500 mg/litre is required and initial copper concentration should not exceed 5 gm/litre. The solid formed consists of metallic copper, the Cu(I) oxide and the sulphide.

These methods seem not to work for other metals. Thus the reduction of nickel salts using sodium boranate in weakly acid solution at elevated temperature gave an unsatisfactory result. The author precipitated copper using dithionite in excess, at room temperatures and was able to obtain final copper concentrations less than 1 mg/litre even when substantial excesses of complexant were present in solution, for all commonly used types [94].

In the presence of a noble metal catalyst such as platinum or palladium, copper and indeed the precious metals themselves can be reduced using gaseous hydrogen at near-atmospheric pressures. Porous carbon sheets with catalysed activated carbon on one side, are supplied with hydrogen gas from the rear. The copper then precipitates on a contacting metal, e.g. copper turnings. Arrays of such carbon sheets can be stacked in a vat with an overflow and the copper-containing solutions is recirculated through these. Previous studies by Faul & Kastening [236] along these lines, were successful, allowing final copper concentrations of 1 mg/litre to be obtained. This required a reactor of 1 cubic metre capacity, supplied with 2 cubic metres gaseous hydrogen per hour. With no catalyst present, the hydrogen reduction requires elevated temperatures and pressures.

Equally environmentally friendly, is the cathodic reduction of metal ions and this can be used for treatment of complexed copper ions (Section 7.6.2.2.2.).

3.9.5.3 Oxidation of complexants

A classical example of an oxidation process is the destruction of cyanide. Many experiments have been carried out on the chemical oxidation of organic complexants using the best-known oxidising agents. However under most conditions, none of these afforded a satisfactory outcome. Only ethylenediamine is susceptible to relatively complete oxidation by hypochlorite in alkaline conditions, with citric acid being moderately well oxidised [94]. At one plant, a semi-concentrated solution of nickel ethylenediamine from the rack plating operation has been jointly destroyed with cyanide concentrates by use of hypochlorite [237].

In such cases, the oxidative destruction of the complexant is the most important aim in order that it cannot, at a subsequent stage, again complex other metal ions, e.g. in sludges or in municipal effluent plants. Metals originally complexed, once the complexant is eliminated, can be removed by traditional methods such as precipitation.

Some effective oxidation processes are known but these do not appear to have been adopted on an industrial scale. They include:

- anodic oxidations
- oxidations using ozone
- oxidation using hydrogen peroxide with simultaneous u.v. irradiation.

The first of these, anodic oxidation, has mainly been used to treat EDTA and its metal complexes. Johnson *et al.* [238] studied the anodic oxidation of EDTA at concentrations of $3.4 \cdot 10^{-5}$ to $3.4 \cdot 10^{-3}$ M in the pH range 0.35 to 3.8. Using paper electrophoresis, they were able to detect a large number of decomposition products. The breakdown proceeds stepwise via a splitting-off of the acetate group leading to formation of carbonic acid and formaldehyde. The following intermediates were also detected – ethylenediamine di- and mono- acetic acids, ethylene diamine & glycine.

Searching for a process capable of addressing the last loophole in the range of existing effluent treatment technologies, namely a method for treating metal complexes other than copper with aminopolyacetic acids, the author *et al.* [39] studied anodic oxidation. Whereas copper can be readily reduced using dimethyldithiocarbamate from effluent containing aminopolyacetates (Section 3.9.5.1.2), the same cannot be satisfactorily achieved for nickel, zinc or cadmium containing solutions.

Anodic oxidation is pH-dependent. If one sets out in the pH range 2.5 to 10, after a short electrolysis time, a value of pH 8 is reached, where oxidation rates are slow. The optimum pH value, as many other authors have confirmed, is in the acid range, especially the strongly acid region. Here, any need to pH adjustment during the electrolysis is eliminated. This apart, it was also shown that the decomposition rate increases with current density but only to a value of 1.5 A/sq. dm (4 to 5 volt). It is also favoured by increase in temperature during the first 15 hours of the process. Platinised titanium anodes performed much better than graphite. The decomposition of EDTA was followed using the complex binding method outlined in Section 3.9.4 and also by potentiometric titration of samples with NaOH. Since some of the intermediates formed during anodic oxidation themselves have complexing properties, a complete effluent treatment calls for metal precipitation by precipitation of the hydroxides or by adding sulphides or dimethyldithiocarbamate (DMDTC). Fig. 3.131 shows the prominent fall in complexation of the nickel by EDTA during the first 12 hours, in terms of its precipitatability by sodium sulphide or DMDTC. No cathodic deposition of nickel takes place at these pH values – a pH of around 2.5 in an unbuffered system is required for this. [246].

Since the anodic oxidation of Ni-EDTA complexes in acid media leads to formation of ethylene diamine as an intermediate, among other products, experiments were carried out with Ni-ethylene diamine complexes, to determine whether these could be anodically oxidised. It was found that while this reaction was extremely slow in acid media, it proceeded at an acceptable rate between pH 7, optimally at pH 10 (as found by ammonia smell and analytical procedures in Section 3.9.4.). At the same time, deposition of a matt though ductile nickel was found, accompanied by precipitation of nickel hydrolysis products [39]. The conclusion is that anodic oxidation of Ni-EDTA complexes calls for a two-stage electrolysis, initially in strongly acid solution, later at pH 10. Following this, residual nickel in solution could easily be reduced to a level of 0.5 mg/litre or less, either by excess alkalinisation with slaked lime or use of sodium sulphide. At the same time, the COD value fell sharply from 7230 to 210 mg/litre. Further studies linked this observation with anodic oxidation in alkaline media.

After these experiments with ethylene diamine, it was found that other amines such as TEA or Quadrol were also best decomposed at pH 10. As was the case with EDTA complexes, cathodic deposition of metallic nickel was accompanied by precipitation of the metal hydroxide and a substantial decrease in COD was observed. In the case of the TEA solutions, residual nickel could be satisfactorily precipitated with sodium sulphide and also quite well removed by use of excess alkali. In the former case, excess sulphide was treated with iron salts which precipitated as an iron complex. In the case of Quadrol, good results were only found by precipitation with DMDTC and this is true even without electrolytic treatment [39]. In the case of oxy-acids such as tartaric or gluconic acids too, anodic oxidation was most successful in the alkaline region, after which the nickel could be precipitated with sodium sulphide. In the case of gluconate, addition of iron salts to take up excess sulphide and form an iron complex is suggested. Where nickel concentrations are low, use of DMDTC without preceding electrolysis is advised. In these cases too, anodic oxidation results in significant reduction of COD value, provided it is carried out in alkali. In acid, very little change in COD is seen. [39]. One patent [239] suggests the following anodic oxidation of EDTA-copper containing effluents, namely 24 hours at pH 10 to 11, the same at pH 6 to 7.5 and finally at pH 0.5 to 3. Current densities of 7 to 20 A/sq. dm were used. These high currents caused the cell temperature to rise to 65 –75°C. Formation of oxides of nitrogen are suggested in this patent to be due to oxidative destruction of the amino groupings. All the authors mentioned above confirm that electrodeposition of the copper in the low pH solutions resulted in concentrations less than 1 mg/litre. Published literature suggests that EDTA destruction takes place in alkaline media with 1 Amp-hr required to destroy 1 gm of EDTA. Platinised titanium is the preferred anode material.

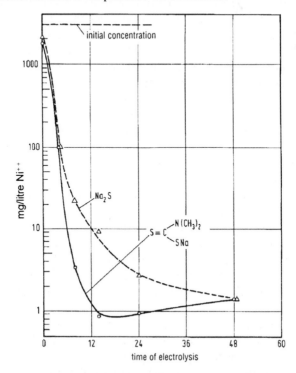

Fig. 3.131 Precipitation of nickel from EDTA containing solutions with sodium sulphide or dimethyldithiocarbamate after various times of anodic oxidation. [39].

Mueller *et al.* [240] carried out further studies on the anodic oxidation of copper-EDTA complex solutions. Using a divided cell, they found a preponderance of oxidation products such as formaldehyde or formate in the anolyte. In the catholyte, glycollic acid, acetate, monomethylamine and ethylene diamine were determined. From these results, they concluded that the overall process also involves cathodic reactions. The rate of destruction increased, in agreement with other authors as well as [239], as pH decreased. The anodes used were graphite, with different types of this material affecting the rate of the process. Packed bed cells performed better than those with platinum electrodes.

Lauksminas & Lauberas [241] studied the anodic oxidation of ethylene diamine and triethanolamine on graphite anodes in an NaCl containing electrolyte. The TEA is broken down, via diethanolamine and monoethanolamine all the way to nitrogen and carbon dioxide. The anodically-evolved chlorine clearly has an important role in these reactions.

The anodic oxidation of EDTA, with its numerous intermediate reactions, often of low efficiency, requires many hours reaction time and consumes much electrical energy. By contrast, oxidation with ozone appears to be a rapid process.

Studies by Fabjan & Marschall [242] using ozone-oxygen mixtures passing over sintered glass "candles" in an EDTA-containing solution showed complete oxidation to form nitrates, carbon dioxide and water. During the reaction, there is a pH shift to lower values, resulting from formation of carbonic and nitric acids.. Runs carried out at pH 5.5 and 10 showed little difference in rate and nature of the end-products (see Fig. 3.132). Using a throughput of 10.7 litres/hour of oxygen having 2.5 to 3 Vol% ozone per litre of solution, destruction was complete after 2.5 hours for an initial EDTA concentration of 3.7 gm/litre. The presence of copper ions in solution accelerated the reaction. However poly-oxy compounds such as glucose or gluconic acid react only slowly under these conditions. Little appears known about the reaction kinetics of these species.

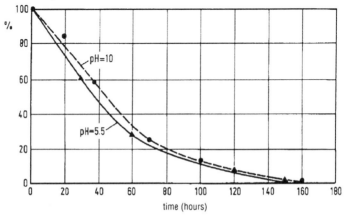

*Fig. 3.132   Oxidative destruction of EDTA with ozone as function of reaction time.
Initial EDTA concentration 3.72 gm/litre. 8 litres/hr oxygen with 2.5 Vol % ozone.
(after Fabjan & Marschall [242]).*

Shambaugh & Melnyk [243] carried out similar studies with a view to establishing the rate constants. Their model compounds were EDTA complexes of manganese, cadmium, nickel and lead at pH 9. The rate constants were of the order of $1.4 \cdot 10^2$ to $3 \cdot 10^2$ litre. Mol^{-1} sec^{-1} for Mn, Cd, Pb complexes, but only 30 for the harder to oxidise nickel complex. They suggest that the reaction rate is limited by mass-transport to the gas-solution interface. At first sight, this is surprising, for if this were the rate-determining step, the greater stability of the Ni-EDTA complex would not significantly affect the kinetics.

Gilbert & Hoffman-Glewe [243a] oxidised NTA with ozone in a batch reactor at pH 3 and 7. In both cases, the first-formed products were iminodiacetic acid and glyoxalic acid. At pH 3, these were further reacted to form oxalic acid, glycine, ammonia and carbon dioxide. At neutral pH values, oxalic and formic acids, ammonia, nitrates and carbon dioxide are the final products. The reaction is faster in weakly acid than in neutral solutions with higher ozone consumption in the latter case. The authors explain these differences in that the higher OH^- concentrations at pH 7 react with ozone under the action of u.v. light to form hydroxyl radicals. Their oxidation potential is greater than that of oxygen by ca 0.8 Volt both in acid and alkaline solution (see Table 7.5). In the oxidation of the less stable cadmium and calcium NTA complexes, the reaction rates were significantly faster.

The destruction of an iron complex as found in photo-processing solutions with ozone has been reported by Bober & Dagon [244].

The use of hydrogen peroxide in presence of ultra-violet light is another means for destroying stable metal-organic complexes, as demonstrated by Ott [245]. Tests carried out on a pilot plant scale showed that most metal complexes are amenable to treatment by these means. At a constant excess hydrogen peroxide concentration the reaction rate is a function of the complexant, the metal complexed, pH and of course the complexant concentration and level of u.v irradiation. Fig. 3.133 shows reaction rates for oxidation of the NTA complexes of Cu, Zn, Ni and Cd in terms of hydroxide precipitation of metals remaining in solution. The optimum pH range was 4 to 6. This is because it is important to avoid hydrolysis of metals after their decomplexation since the resulting turbidity impedes the action of the u.v light. According to these authors, the EDTA or NTA complexes of copper or nickel react more readily when other metal complexes are also present or when Fe(III) salts are added.

The author and colleagues [247] carried out similar research with NTA and EDTA and their metal complexes. In the case of the free aminopolyacids, the oxidation rate was greatest at pH 2 with that for the metal complexes at pH 4 to 5. These rates fall off as a result of metal hydrolysis which largely bars passage of the irradiation. Thanks to light scattering, however, some radiation is passed and continues to act. Using the "binding" analysis method of 3.9.4, reaction rate plots were constructed – they were found to be similar to those of Ott, i.e. the requisite low levels of residual metal ions were not, in most cases, reached after hydroxide precipitation. Use of sulphide proved more successful.

If, using more advanced analytical methods, the concentration of EDTA and its breakdown products is followed, it can be seen that many of the latter themselves

have complexing action. Fig. 3.134 shows that EDTA itself is rapidly reacted, the
same is true for the intermediates, allowing for the reduced peroxide concentrations.
Oxidation of EDTA and its metal complexes led to the following intermediates (as
identified with ion chromatography or polarography) – ethylenediaminodiacetic acid,
NTA, iminodiacetic acid, ethylenediamine, aminoacetic acid, oxyacetic acid, formic
acid, formaldehyde, ammonia, nitrite (ultimate breakdown products are carbon
dioxide, nitrate and water).

It seems likely that using this oxidation method, any bond between reactive
groupings in the molecule may be broken. To this variety must be added the oxidation products of the intermediates themselves – the result is that a very large number
of intermediate products can be identified. Thanks to further oxidation, some of these
will pass beyond the limit of detection. Ott reported an initial decrease in the degree
of complexation when triethanolamine complexes were oxidised. This then increased
once more. Based on the author's own experience, this seems most probably to have
been due to a partial oxidation of triethanolamine to NTA. Divalent metals form a
1:2 complex with the former compound, a 1:1 complex with NTA (see Section
3.9.3.2.1).

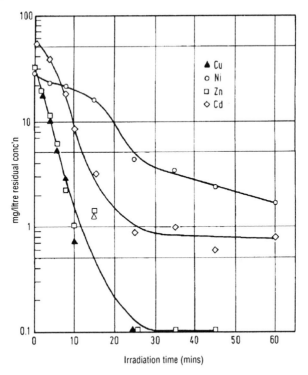

*Fig. 3.133 Destruction of metal-EDTA complexes (defined in terms of metal not
precipitatable at pH 10) by hydrogen peroxide with simultaneous u.v irradiation, as
function of irradiation time. Initial concentrations 0.5 mM metal; 2 mM NTA, 0.1M
hydrogen peroxide, pH 4 (after Ott [245]).*

Studies by the author and colleagues afford insights into the reaction mechanisms. Since neither hydrogen peroxide nor u.v. irradiation alone result in any significant oxidation, either the peroxide or the metal complex must require activation and this requires that light be absorbed. In the case of hydrogen peroxide, this is possible in the short wavelength u.v. (see Section 3.11.4) and the resulting formation of OH radicals is well-known. These are extremely unstable and are very strong oxidising agents with a potential in acid media of + 2.8 V, far greater than that of ozone (+2.07 V). As for the metal-organic complexes, these behave in various ways as regards light absorption in the u.v.

The Cu-EDTA complex absorbs at short wavelengths, Ni-EDTA at longer u.v wavelengths while Cd-EDTA and free EDTA do not absorb either in the u.v or visible regions. In spite of this, all can be readily oxidised by this approach, most easily the cadmium-EDTA complex and the free EDTA itself. It thus appears that absorption and activation of the complex is of no importance.

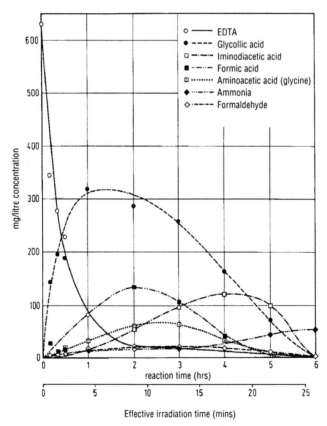

Fig. 3.134 Oxidative destruction of the copper-EDTA complex by hydrogen peroxide with simultaneous u.v irradiation at pH 4. Also shown is the formation of intermediates and their oxidation [247].

The reaction rate depends largely on the intensity of the irradiated light. Treating the radiated energy as quantised allows it to be considered in the same way as one would a chemical reactant (Section 3.11.4). The reaction time is thus determined by the duration of the irradiation of the reactant species and is proportional to their concentration, the light intensity, its wavelength and the construction of the reactor.

Oxidation of the nickel-NTA complex is likewise possible as are complexes of amines. triethanolamine, ethylenediamine and Quadrol with this metal.

In cases where the nickel concentration is not too great, the metal can be directly precipitated from amine complexes using DMDTC.

The three oxidation processes – anodic oxidation, and use of ozone or hydrogen peroxide with u.v irradiation – have all been fairly fully described even though they are, to an extent, still being developed. However little has been published in the metal finishing journals which is surprising since these appear to be the only technologies for addressing the treatment of aminopolyacids and their metal complexes: It is possible that all three processes are based on the same oxidant – the OH radical. While this is known to be involved in the case of ozone reacting with water and when hydrogen peroxide is u.v. irradiated, its involvement could also be assumed in the case of anodic reactions (see Equation 3.55a).

3.9.5.4   Binding of metal complexes or complexing agents

Considered here is the binding of metal-organic complexes or complexing agents by a chemical reaction. Because of their structure, one would not expect adsorptive, i.e. physical, binding to occur.

Strongly acid cation exchangers are capable of binding cationic complexes. However this is only successful in the absence of other cations or at very low concentrations. When such strongly acid cation exchangers are used in their protonated form (e.g. operating in a recirculatory rinse loop), a solution reaction can take place in the reaction (exchange) zone with the resin behaving as an acid. This can result in splitting of weak or moderately strong complexes into metal ions and the complexing ion or molecule, with anionic complexes becoming protonated (see Equations 3.186 to 3.188). While metal ions or complexants which are protonated to form cations, such as ammonia or amines, become absorbed, other complexants pass to the anion exchanger column and, to the extent they are themselves anionic, are absorbed there. Where complexes are present in excess, in effluents, they can react with metal ions absorbed on the cation exchanger to form metal complexes once more. In due course these are absorbed on the anion exchange column where, in the case of anionic species, they are bound. Anion exchange resins have a strong affinity for anionic complexants, especially those using strongly basic resins. The regenerate liquors from ion exchange columns on which metal complexes or complexants are absorbed, must be piped to some form of complex treatment process.

Yet another means for removing complexed metals from effluent is to bind these to complexing type cationic exchange resins. This approach only works complexing function on the resin forms a more strongly bound complex than the original complexant in the effluent., i.e. a decomplexation has to take place. The reaction between a complexing type ion exchange resin in its sodium salt form (RC-Na) and the complex of a divalent metal ($[Me^{II}L_x]$ is used and for simplicity, no charge is shown), is expressed as below:

$$2\,RK - Na + [Me^{II}L_x] \rightleftharpoons (RK)-)_2\,Me^{II} + 2\,NaL_{x/2} \qquad\qquad 3.194$$

The weakly acid cationic exchange resins used for this, usually have iminodiacetate groupings and are usually operated in their sodium form. They are then exposed to a neutral or weakly alkaline effluent. Those metal complexes which are significantly less stable than the metallo-iminodiacetate complexes are converted to the latter form by the resin. In addition, in its normal ion exchange mode, the resin can also bind cationic metal complexes or cationic complexants. Table 3.36 lists the affinities of some weakly acid cationic exchange resins with iminodiacetate groupings to a range of metals, giving also some stability constants for metal-iminodiacetate groupings. The effect of complexing groups on absorption is quite unmistakable (see Section 5.2.3.3.2). Shown below is a typical structure of a benzyl-iminodiacetate grouping of a weakly acidic selective cation exchange resin.

This type of resin is widely used for absorbing residual amounts of metal ions in effluent after treatment and are accepted as selective or final ion exchangers) (see Section 4.4.2).

Following researches by the author with nickel or copper complexes at pH up to 7 and with complexant concentration at twice the level of the metal ions, it was found that the following complexing agents can be bound by ion exchange and/or decomplexing in a manner ranging from very satisfactorily to acceptable: tartaric, gluconic acids, ammonia, ethylene diamine, triethanolamine and Quadrol. While the regenerate liquors from the ion exchange unit were free from traces of carboxylic acids, they did contain varying amounts of ammonia or the amine containing complexants. Thus it was found that EDA was quite strongly complexed, TEA and Quadrol more weakly so and ammonia very weakly, whether these were present in their free form or as metal complexes. In such cases, it may be necessary to regenerate the ion exchange resin using a complexant.

Effluents containing complexes with citric acid, NTA or EDTA are anionic and these cannot, because of their high stability, be absorbed by cationic resins with iminodiacetate groupings.

Regeneration of selective cation exchange resins is usually carried out with mineral acids (HCl, sulphuric). However a pre-conditioning with NaOH is often advisable. Anionic exchange resins with comparable selective properties do not appear to have been developed.

Concentrated solutions of nickel-ethylene diamine complexes can be treated with formaldehyde (itself a hazardous species) as shown by Frick [249]. A water-soluble condensation product analogous with hexamethylene tetramine is formed. In the alkaline conditions used, Cannizzaro and aldol-type condensation reactions also take place and an excess of formaldehyde is required to deal with these. At the same time, nickel hydroxide is precipitated leaving residual Ni concentrations substantially less than 0.5 mg/litre. Excess formaldehyde is then oxidised to formate, using hydrogen peroxide. It should be noted that this results in a high COD.

TABLE 3.36:

COMPARISON OF STABILITY CONSTANTS OF METAL COMPLEXES AND IMINODIACETIC ACID [196] WITH THE AFFINITY SERIES OF A CATION EXCHANGE RESIN CONTAINING IMINODIACETATE GROUPINGS [248].

Affinity series of some metal ions to a weakly acid cation exchange resin containing iminodiacetate groupings	$Cu^{++} > Ni^{++} > Zn^{++} > Cd^{++} > Fe^{++} > Ca^{++}$
stability constants of metal-iminodiacetic acid complexes	$1.6 \cdot 10^{16} \quad 4.1 \cdot 10^{14} \quad 1.5 \cdot 10^{12} \quad 3.4 \cdot 10^{9} \qquad 2.6 \cdot 10^{3}$

Tartrates, in concentrated solutions, can largely be eliminated from effluents by their recovery. Kumar *et al.* [250] describe how saturated copper sulphate solution is added, with stirring to concentrated tartrate solutions (75 to 100 gm/litre) at pH 3 to 5. Temperatures should not exceed 30°C. After the equivalent amount of copper salt has been added, some 90% of the tartrates precipitate and these can be recycled to appropriately formulated electroless copper plating baths.

Precipitation of gluconate-containing solutions has been discussed in Section 3.9.5.1.1.

3.10   PRECIPITATION OF ANIONIC HAZARDOUS SPECIES

Many anionic substances will form more or less insoluble compounds with calcium ions and use can be made of this, whether in direct or indirect discharge effluent systems and also for compounds which are not hazardous. In consequence of the growing use of slaked lime for neutralisation precipitations, many of the compounds discussed below are co-precipitated to an extent determined by their solubility. Care should always be taken that sufficient calcium ions are available for precipitation. Where slaked lime is the form in which calcium ions are added and when pH is monitored, precipitation will only occur when at least the equivalent amount of acid is also present. Should this not be so, additional calcium ions must be added in a more soluble form, e.g. as calcium chloride. Since it is difficult to automatically dispense neutral salts such as calcium chloride because no simple instrumental process control methods are available, this method is usually carried out on a batch basis.

The most common anionic species found in effluent which form sparingly soluble calcium salts include fluorides, sulphates and phosphates. Readers should also refer back to sections dealing with formation of sparingly soluble calcium salts of anionic metal complexes such as hydroxy complexes.

Other important precipitation processes are discussed below. Sulphides, though they do not form sparingly soluble calcium compounds, can also be removed by precipitation. For comments of these and their separation from the liquid phase, see Section 3.7.4.

3.10.1   Precipitation of fluorides

Fluorides can be present in effluents as the free anion or as a ligand in a complex such as hexafluoroaluminate $[AlF_6]^{---}$, tetrafluoborate $[BF_4]^-$ or hexafluosilicate $[SiF_6]^{---}$. Other fluoride metal complexes exist but are of little significance. Most important, for effluent treatment, is the toxic and precipitatable fluoride ion which, in contrast to complexes, forms a highly insoluble calcium salt.

Fluoride is found in effluents from pickling of stainless steels or alloys containing silicon where mixed acids containing HF are used. It is also used in chemical brightening of aluminium. To a lesser extent, it arises from special etch baths. Otherwise it is found only in the form of tetrafluoborate or hexafluosilicic acid complexes which, being stable anionic complexes, contain almost no free fluoride ion.

Fluoride ions are precipitated with calcium salts above pH 7 as the sparingly soluble fluoride:

$$2 \ F^- + Ca^{++} \rightleftharpoons CaF_2 \qquad\qquad 3.195$$

Fig. 3.135 shows such a precipitation as function of slaked lime additions and the resulting pH changes. At pH 8, a fluoride concentration of ca 60 mg/litre is still found. Only at pH 11.5 is the amount of calcium sufficient to reduce the fluoride ion concentration to an acceptably low value as implied by the solubility product of calcium fluoride ($3.4 \cdot 10^{-11}$ M at 18°C [28]). This corresponds to a solubility of 15 mg/litre CaF_2 or ca 7.3 mg/litre fluoride ion. Fraust [251] suggested pH 11.2 as a value for most complete fluoride precipitation, where possible from a concentrated solution. Czukrass & Hompasz [252] maintain that only after addition of a four-fold stoichiometric excess of lime can residual fluoride concentrations be brought down to 30 to 55 mg/litre.

When ammonia is present, fluoride precipitation becomes more difficult [251] and this is also true in presence of hexafluoroaluminates which dissociate to some extent. In neither case can fluoride concentrations as low as those found with free cyanide, be obtained. Calcium fluoride precipitates do not sediment well unless other metal ions (e.g. as hydroxides) or iron (III) salts are also present. Where lead or magnesium ions are present in effluent, these react with fluoride to form insoluble compounds.

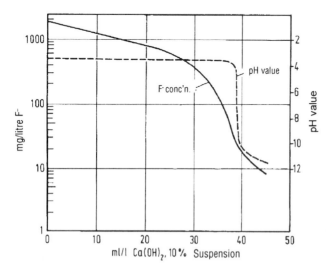

Fig. 3.135 Fluoride concentrations and pH values during neutralisation of HF with slaked lime [28].

3.10.2   Precipitation of sulphates

Sulphate is not only the least harmful of species found in effluent but also one of the most commonly found. It does have one unpleasant aspect – it can attack concrete. According to DIN 4030 [253] water containing 200 to 600 mg/litre sulphate is designated as weakly corrosive, 600 to 3000 mg/litre strongly corrosive and above this as very strongly corrosive. For these reasons, effluent discharges into public sewerage systems are usually restricted to levels of 400 to 600 mg/litre sulphate. Where discharge is not to a public system, there are, in many cases, no set limits. The nature of the attack, which results in formation of the mineral ettringite, is shown below:

$$(3 \text{ CaO} \cdot \text{Al}_2\text{O}_3) \cdot 12 \text{ H}_2\text{O} + 3 \text{ CaSO}_4 \cdot 2 \text{ H}_2\text{O} + 13 \text{ H}_2\text{O} \rightleftharpoons$$

$$(3 \text{ CaO} \cdot \text{Al}_2\text{O}_3 \cdot 3 \text{ CaSO}_4) \cdot 31 \text{ H}_2\text{O}$$

3.196

The high degree of water uptake by the concrete leads to cracking and ultimately to breakdown. Further comment is made below on ettringite.

Sulphate forms sparingly soluble salts with calcium, strontium and barium, the solubility of these falling strongly in the listed order. Thus calcium sulphate has a solubility product of $6.1 \cdot 10^{-5}$, corresponding to a solubility of 1990 mg/litre of the salt, ca 1400 mg/litre on sulphate basis. For barium, the corresponding figures are $1.08 \cdot 10^{-10}$, i.e. 2.3 mg/litre of the salt, 0.95 mg/litre on sulphate basis. However barium is not used in practice, both because of its cost and equally because of its toxicity and calcium is, almost without exception, the preferred material. It is true that toxic metals are precipitated in practice (as described earlier) using sulphides which are themselves toxic. However in this case, excess sulphide is readily removed using iron salts. In the case of sulphates, it makes no sense, environmentally speaking, to use a toxic species to remove a harmless one. Precipitation of sulphate using calcium salts follows the equation:

$$SO_4^{--} + Ca^{++} \rightleftharpoons CaSO_4 \qquad\qquad 3.197$$

Goetzelmann & Hartinger [254] established that, in the case of dilute solutions (up to 4800 mg/litre sulphate), this precipitation can be very slow when slaked lime us used (see Fig. 3.136) and equilibrium is only reached when more concentrated sulphate solutions are used, as seen in Fig. 3.137

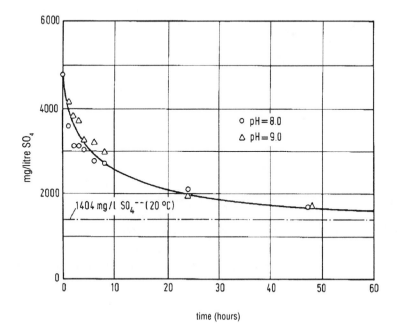

Fig. 3.136 Solubility of calcium sulphate (as sulphate anion) after slaked lime precipitation as function of reaction time. Initial conc'n 4800 mg/litre sulphate, precipitation pH values 8.0 and 9.0 [254].

Batch processing thus has its attractions for sulphate precipitation in part because it is
more suited to use of concentrated solutions which thus save water and partly
because the reaction can be left until precipitation is more or less complete. Morkowski
et al. [255] using concentrated sulphuric acid solutions (up to 20%) were still able to
end up with less than 1800 mg/litre sulphate.

What can cause problems, is when the sulphate to be precipitated from solution is
accompanied by neutral salts other than those based on calcium or sulphate ions. One
then encounters the problems associated with high total ionic strengths discussed in
Section 3.7.7.1. These effects are the more pronounced the less insoluble is the
precipitate, as can be clearly seen in Fig. 3.138. A neutral salt concentration of 0.5
equivalents/litre corresponds to 4000 mg sulphate per litre.

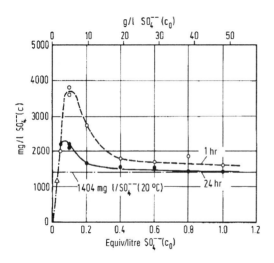

Fig. 3.137   Solubility of calcium   sulphate   (as sulphate) after slaked lime precipitation, as function of initial sulphate concentration. Precipitation pH = 8.0 [254].

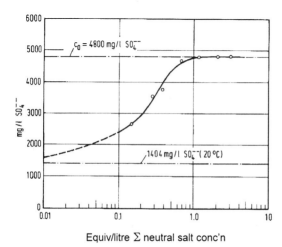

Fig 3.138   Solubility of calcium   sulphate   (as sulphate) after precipitation with calcium chloride (110% theoretical) as function of neutral salt concentration (NaCl). Initial conc'n 4800 mg/litre   sulphate, precipitation pH = 8.0, time of standing – 24 hrs [254]

It follows that use of slaked lime or calcium salts to remove sulphate ions is inadequate in those cases where attack on concrete is possible. Practical experience has shown that in larger towns or indeed anywhere where there is a good flow of water at the outflow point, it is quite rare to observe concrete attack. Thus it may be possible to avoid the capital expense and operating costs entailed if sulphate removal measures are put in place. Overall, various options are available, from use of slaked lime to precipitate sulphate from as concentrated as possible a sulphate solution – at the same time saving water – to replacement of sulphuric acid where used to acidify and treat chromate or nitrite containing effluents, using hydrochloric acid in its place. HCl can likewise be used for general neutralisation of effluent.

A simple but effective means of reducing sulphate discharge was described by Lancy [256]. A static rinse tank is used immediately after a process solution bath holding the sulphate containing solution. Where the latter operates at above-ambient temperatures, the contents of the static rinse can be partly recycled to the process tank. The remainder lends itself well, as a fairly concentrated solution, to treatment as described above. After immersion in this static rinse tank, the work is rinsed in a chemical rinse tank containing calcium hydroxide. Precipitated calcium sulphate is removed using a by-pass loop. A final rinse tank then has to cope with only a small fraction of the initial dragged-out sulphate concentration. In many cases, no additional rinsing is required after the chemical rinse. This is mainly the case with fairly stable metal surfaces, i.e. after sulphuric acid pickling.

Sulphate discharge has other possible legal implications. If a producer of sulphate containing effluents were to use, in their effluent treatment system, materials of construction which might be attacked, they could be liable to pressure from the local authority. Karnowski [257] has described various ways of avoiding damage to sewer pipes by sulphate such as use of coatings or linings. Replacement of concrete by earthenware or plastic piping is another alternative. In cases where a near-complete removal of sulphate is required, a useful approach has been described by Christoe [258] as part of an effluent treatment process. Sulphate ions are precipitated as the ettringite, which is less soluble than calcium sulphate and which was shown in Equation 3.196 as the product formed when concrete is attacked by sulphate. To the sulphate containing effluent is added a soluble aluminium salt (chloride or nitrate) with calcium ions either as a salt or as slaked lime. At pH values above 9.5, ettringite is formed ($3\ CaO \cdot Al_2O_3 \cdot 3\ CaSO_4 \cdot 31\ H_2O$). This reaction requires requires that calcium be in excess by 2M and aluminium by 0.66M per Mol. sulphate. The outcome is sulphate concentrations of less than 400 mg/litre.

Keller et al. [259] describe a four-stage process for sulphate removal. The sulphate containing effluent is neutralised with slaked lime (Stage 1.), crystallised and sedimented in a second reactor (Stage 2.) Precipitation is improved by return of the sludge to the neutralisation stage. Stage 3. is a second precipitation using calcium aluminate at pH 12. The aluminate is added as a suspension and more sulphate is precipitated as calcium aluminomonosulphate ($3\ CaO \cdot Al_2O_3 \cdot CaSO_4 \cdot 2\ H_2O$). In the 4th Stage, this is allowed to settle for 2 to 4 hours. It has a solubility corresponding to ca 5 mg/litre sulphate. It is recommended that this final liquor be mixed with the effluent from Stage 2. so that the overall sulphate concentration in the final effluent is around 400 mg/litre.

Finally, there is a patent [260] based on the above mentioned method but claiming an improved sedimentation method which allows total treatment times of less than five hours. This method too, uses an initial precipitation of calcium sulphate (preferably at 40 to 45°C) as above, followed by precipitation of ettringite using calcium aluminate. The main points of interest are use of an optimum pH of 11.2 to 11.8 which is maintained at these values during the reaction. Without such corrections, the pH will fall and no benefits in reduced sulphate ion concentration are found. Calcium aluminate is added in the same way as the slaked lime, the reaction being completed at 25 to 35°C since a deep minimum in solubility occurs at 30°C. X-ray diffraction studies confirm the precipitate as ettringite. In the presence of 50 mg/litre CaO, this has a solubility of 250 mg/litre, i.e. 5.5 mg/litre as sulphate. The calcium aluminomonosulphate reported by Keller et al. is converted, in sulphate containing waters, to ettringite. Studies with industrial effluents gave residual concentrations between 50 and 450 mg/litre sulphate ion.

3.10.3   Precipitation of phosphate

Phosphates are non-toxic, indeed like nitrogen compounds, they are essential nutrients for micro-organisms and all forms of plant life [261]. Where their concentrations are allowed to become excessive, for example in static or slow-flowing waterways such as lakes or reservoirs, such excess nutrient capacity can have catastrophic effects. A rapid growth of water plants will take place, especially algae. Only 1 gm of phosphate is required for 3 kg of algae. These 3 kg will, in turn, deplete typically 45 cubic metres of water of their oxygen content. This completely destroys the ecological balance in the water (a process known as "eutrophication"). The decrease in dissolved oxygen destroys both animal and plant life and decaying vegetation falls to the floor of the lake etc. At these greater depths of water, only anaerobic processes take place [262]. Associated with certain types of algal growth are toxins directly and strongly poisonous to Man and other mammalian species.

Of the total phosphates discharged, it can be broadly stated that one third derive from agriculture (fertilisers, animal products), one third from domestic faecal matter and one third from domestic washing machines or from industrial effluents [261, 262, 263]. In order to mitigate the latter category, some governments have passed legislation limiting the phosphate content of detergents. In some cases, permitted maximum concentrations are linked to local water hardness levels.

The origin of phosphates from industrial effluents stems from their use in cleaning and degreasing baths, in phosphoric acid pickling, from phosphating operations, in chemical and electrochemical brightening and polishing processes as well as process baths based on pyrophosphate. Of the above, it is the cleaning and degreasing baths as well as heavy-industry phosphating baths which are most significant. Phosphates are precipitated in industrial effluent plants using methods long-established in municipal effluent treatment, though not all such municipal plants incorporate these.

Using slaked lime as a neutralising agent, pH values similar to those for metal hydroxide precipitation are used with a sparingly soluble calcium hydrogen phosphate ($CaHPO_4$) being formed according to the simple equation:

$$H_3PO_4 + Ca(OH)_2 \rightleftharpoons CaHPO_4 + 2\ H_2O \qquad\qquad 3.198$$

This has a solubility product of $5 \cdot 10^{-6}$, corresponding to a solubility of ca. 100 mg/litre or 70 mg/litre as the phosphate anion. If the pH is increased using slaked lime to 9.5, the even less soluble calcium hydroxyphosphate (hydroxyapatite) is formed:

$$3\ H_3PO_4 + 5\ Ca(OH)_2 \rightleftharpoons Ca_5(OH)\ (PO_4)_3 + 9\ H_2O \qquad\qquad 3.199$$

Above pH 10, this has a solubility, according to Poepel [264] of less than 3 mg/litre of phosphate.

An equally effective means of phosphate removal without raising the pH as above, is by addition of a salt of a trivalent metal. Iron (III) or aluminium salts, not being toxic, are ideal candidates for this. Since iron salts hydrolyse at pH 3.5, aluminium salts at pH 5 (Sections 3.7.5.6.1 and 3.7.5.7.1), the precipitate is not a true metal phosphate but rather a basic phosphate. For cost reasons, municipal effluent plants seek to precipitate phosphates at the lowest possible pH value. At pH 7, the most phosphate-rich species precipitated by aluminium salts is 2 $AlPO_4\cdot Al(OH)_3$ [262]. As pH increases, precipitates such as $x.MePO_4\cdot y\ Me(OH)_3)$ are formed with increasing y:x ratios. In practice, metal:phosphate ratios > 1 are not always required. This is because the active flocs of basic phosphate precipitate can themselves remove additional phosphate ions by adsorption [262]. Solubility values of sparingly soluble phosphates are shown in Fig. 3.139. As with other divalent metals, iron (II) salts form sparingly soluble phosphates with a minimum solubility around pH 8, in contrast to the value of pH 4.5 which is the case with iron (III) salts [265]. However the basic iron (II) phosphates sediment less readily [262, 265]. Various oxidation reactions which can take place in the activated sludge stage of an effluent plant can effect oxidation to iron (III), resulting in formation of iron (III) hydroxide.

Polyphosphates pose no problems in municipal effluent plants since, both in use and during and after discharge, they are largely hydrolysed to orthophosphates [261-3, 266]. However in some processes such as those mentioned above, polyphosphates can cause problems due to complex formation. They must therefore be hydrolysed by acidification prior to starting the metal precipitation (see Section 3.9.3.1.4).

As regards orthophosphates, these cause no great problems since virtually all metals used in the metal-working and finishing industries form sparingly soluble basic phosphates on neutralisation. The treated effluent thus contains virtually no

phosphates, especially where slaked lime is used for neutralising. A result of this is
that sludges formed where phosphates are present are rich in these species. In cases
where the effluent contains insufficient metal ions, e.g. in plants which operate
nothing more than a cleaning and degreasing process, additional metals in the form
of iron (III) or aluminium salts must be added and/or slaked lime must be used for
neutralising.

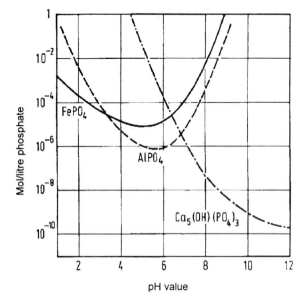

Fig 3.139 Solubility of
various phosphates as
function of pH (after
Leckie & Stumm).

3.10.4   Removal of sulphides

Recent legislation in several countries now includes sulphides as toxic species and
discharge limits around the 1 mg/litre value are or are expected to be implemented.
Sulphides are hardly ever found in the metal-finishing or working industries with the
exception of some compounds used for colouring non-ferrous metals. Increasingly
however, sulphides (both sodium and organo sulphides) are used for precipitating
metals as a second stage in effluent treatment after alkaline precipitation. This requires an excess of sulphide to be added to push the equilibrium in the direction of
sulphide formation. After metal precipitation is complete, this excess sulphide must
be removed before further treatment. Two methods are available.

● precipitation using iron salts

● oxidation.

The classic precipitation method for sulphide removal uses iron (II) salts are shown in Equation 3.159. The solubility product of iron is $3.7 \cdot 10^{-19}$. It follows that inorganic and organic sulphides can be removed almost without trace. Commonly, iron (III) salts are used when the following three functions are carried out:

- oxidation of sulphide in weakly alkaline medium to elemental sulphur which forms as finely dispersed or colloidal material. 2 Mole of iron (III) per Mole sulphide are required (see Equation 3.158).

- precipitation of 1 Mol S by 1 Mol of Fe^{++} formed by the reduction above.

- flocculation of colloidal and finely dispersed solids by excess of iron (III) following it hydrolysis.

The first two stages are shown in Equations 3.158 to 3.160. A special case is the precipitation of metal sulphides from solutions containing complexants (see Section 3.9.5.1.2)

The second means for sulphide removal is by its oxidation. According to Schwarzer [267], this can be carried using hydrogen peroxide at pH of 8.5 or greater to form elemental sulphur. Below pH 8.5, sulphuric acid is formed – the two processes call for very different amounts of hydrogen peroxide:

$$Na_2S + H_2O \xrightarrow{\text{pH} > 8.5} S^{\circ} + 2\ NaOH \qquad\qquad 3.200$$

$$H_2S + 4\ H_2O_2 \xrightarrow{\text{pH} < 8.5} H_2SO_4 + 4\ H_2O \qquad\qquad 3.201$$

It should the noted that the first of these two reactions can only be used after precipitated metal sulphides have been separated. If this is not done, the sulphides will be re-oxidised and the metal content increased.

Alkaline hydrogen peroxide solutions can also be used for wet scrubbing of effluent gases containing hydrogen sulphide [268]. Sodium sulphide is formed which can be oxidised as shown in Equation 3.200. The spent scrubbing water must be recycled to the effluent treatment plant since it may still contain traces of sulphide. Subsequent treatment of the colloidal sulphur is based on use of iron (III) or aluminium salts for flocculation, as described earlier.

Oxidation of sulphide with compounds which release gaseous chlorine or are capable of oxidising chlorides to chlorine is not advised, as explained in Section 3.2.1.7. Under anaerobic conditions it should be noted that sulphides can be formed from sulphates.

3.11   REMOVAL OF NON-IONIC TOXIC SPECIES

3.11.1   Removal of free chlorine

Free or active chlorine can only arise by its use as a strong oxidising agent for destruction of cyanides or nitrites or for oxidation of other oxidisable species such as formaldehyde which can occur in industrial effluents. Gaseous chlorine is hardly ever used today, the most common source of available chlorine being sodium hypochlorite.

The origin of free chlorine in effluent is the result of an excess dosing which is required not only for the reactions above but also to enable the process to be controlled by redox potential measurements. This is specially true of the very slow oxidation of effluents containing nickel-cyanide complexes. In such cases, excess chlorine can be readily taken up by addition of iron (II) compounds. However this is only possible using batch-processing. In most cases, the level of excess chlorine is only a few mg/litre, so low that – by the time the effluent is discharged – it has either been oxidised by air or reacted with miscellaneous compounds in the effluent such as organic bath additives and its concentration is below the permitted limit.

Chlorine can be reduced by other means. In principle, any compound can be used whose Normal Potential in alkali is less than + 0.9V. Virtually all known reducing agents fall within this category (see Table 7.4). Sulphur dioxide, sodium bisulphite or sodium sulphite solutions can all be used, the end product being sulphate. Note, however that the first two of these will cause the solution pH to fall with the result that gaseous chlorine may be evolved.

The best means of chlorine removal is to use hydrogen peroxide and this can be used in strongly alkaline media for even greater concentrations of chlorine or hypochlorite ion. The reaction, shown below, occurs rapidly and more or less stoicheometrically;

$$OCl^- + H_2O_2 = Cl^- + H_2O + O_2 \hspace{3cm} 3.202$$

The advantage of this method is that it does not increase the TDS (total dissolved salt) value of the effluent. The reaction can readily be monitored and controlled by the marked negative change of redox potential as free chlorine is consumed [270]. The reaction illustrates the point that not only the best-known reducing agents but also oxidising agents (providing their redox potential is cathodic to that of hypochlorite) can be used to reduce this species.

It should also be mentioned that chlorine can be adsorbed on activated carbon although this has no application in the context of this book.

Further, it should be noted that chlorine will chlorinate most organic species present in effluent to form AOX (adsorbable organic halo-compounds) (see Section 3.2.1.7). Formation of such chlorinated compounds is generally deemed undesirable.

3.11.2   Removal of ammonia

The sources of ammonia or ammoniacal compounds in the metal-working and finishing industries have earlier been discussed (Section 3.9.3.1.3). While ammonium salts in weakly-acid bright zinc plating baths can be replaced with other buffer compounds, such as boric acid, and the ammoniacal etches used in the pcb industry are increasingly recovered and recycled, major sources of ammonia contamination remain. These include ammonia derived from destruction of cyanides (Section 3.2.1.4). There is no indication that cyanide electrolytes will be totally banished in the immediate future. Nevertheless, the thought remains that, were this to be the case, the problems of AOX compounds would at the same time be mitigated.

Nitrogen, in the form of ammonia is a major nutrient, as mentioned in Section 3.10.3, for plant and algal growth. As such, it can lead to eutrophication in natural waters such as lakes. This apart, ammonia is toxic to fish. To date, apart from some efforts to reduce or eliminate the use of ammonia, effluent treatment to actually remove ammonia are very rare in metal finishing plants. The relatively modest amounts found in metal finishing effluents will often be aerobically oxidised to nitrates in municipal sewage plants. In some modern sewage plants, this may be followed by an anaerobic process leading to formation of nitrogen. In cases where ammonia concentrations exceed the permitted effluent concentrations, it must be removed using a technology well-established in a range of heavy industries. Air is blown through an alkaline solution and this strips out the ammonia which is either discharged direct to air or thermally or catalytically oxidised.

Stripping requires conversion of ammonium ions to ammonia and this involves the equilibrium shown in Equations 3.175 and 3.176 as shown in Fig. 3.122. Whereas ammonium salts are highly water-soluble, ammonia – though itself very soluble – has a high vapour pressure. Dissolved ammonia is thus in equilibrium with gas-phase ammonia adjacent to it:

$$NH_3 \text{ (dissolved)} \rightleftharpoons NH_3 \text{ (gaseous)} \qquad\qquad 3.203$$

If this equilibrium is disturbed, e.g. by removing ammonia from the gas phase, it attempts to re-establish itself. To facilitate mass-transport problems, industrial plants are designed for the greatest possible gas-liquid contact surface. This is usually achieved in one of two ways. Either air, as finely divided as possible, is bubbled through the ammonia solution or the solution is trickled down a packed-bed against a counter-current of air. The efficiency of the overall process rests on two equilibria, the one is chemical (Equation 3.175), the other physical (3.203). As Bischoff [271] showed, using a bubble reactor as an example, once ammonia has been formed, the effect of pH is not large in contrast to those factors which bear on mass-transport, e.g. air throughput rate, bubble diameter and temperature (which increases ammonia vapour pressure). The height of the reactor column and the rate of air supply set limits in that, as air flow increases, bubble diameters will increase and, in the limit, air pockets will form. Fig. 3.140 shows how the rate of ammonia removal increases with airflow rate for initial ammonia concentrations of 1 gm/litre. Similar results are found using the countercurrent air flow packed bed reactor in terms of air flow and temperature effects. However an advantage in this case is that there is no restriction on column height. As this increases, so does the contact area as determined by the packing. The shape and size of packing should be selected to allow good upward air flow but without the danger that liquid droplets are blown upwards with the air.. An airflow rate of 3 metres/sec, similar to that used in evaporator plants, is usually specified.

In making the solution alkaline, it should be noted that this requires addition of NaOH as shown in Equation 3.175. 2.22 kg NaOH are needed per 1 kg ammonia. Account must also be taken of consumption of NaOH by reaction with carbon dioxide in air to form carbonates. Air contains 0.04 wt% carbon dioxide, i.e. 0.33 gm/cubic metre. This an additional 0.6 gm NaOH are required for every cubic metre of air used.

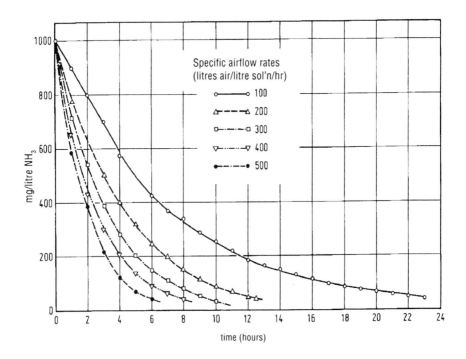

Fig. 3.140 Ammonia concentration as function of stripping time and air volume throughput in a bubble reactor. Initial conc'n 1 gm/litre ammonia. Bubble diameter ca 1 mm, 5 gm/litre NaOH, ambient temperature [271].

In that ammonia stripping is usually carried out on effluent solutions, care must be taken where these also contain metal ions, possibly as amine complexes or surfactants. Effluent from a cyanide destruction operation should be stripped while it is still alkaline and before neutralisation. Much of the metal in this case is present as the hydroxide. In the case of metal-amine complexes, there is a tendency for these solids to clog the packing in packed bed towers so that use of bubble reactors may be preferable in such cases. Where surfactants are present, in some cases even at the very lowest concentrations, foam formation is a possible problem. In extreme cases, this precludes the use of air stripping altogether.

The air emerging from the stripper, carrying the ammonia, may not, under most legislations, be simply vented. Instead, it is wet-scrubbed using a dilute mineral acid solution, usually sulphuric acid. The ammonium salt formed there can usually be reclaimed at a modest value. If the air used for stripping is recirculated, a saving in NaOH results, in that no fresh carbon dioxide enters the system [272].

A related technology is steam-stripping, which allows an aqueous ammonia solution to be recovered [273]. Vacuum stripping [274] is also sometimes used.

Ammonia can also be chemically oxidised to nitrogen using hypochlorite, as shown below:

$$2\ NH_4Cl + 3\ NaOCl \rightleftharpoons N_2 + 3\ NaCl + 2\ HCl + 3\ H_2O \qquad 3.204$$

This involves adding hypochlorite in equivalent amounts to the ammonia and maintaining the pH at 8 to 10. At lower values, up to 6-fold excess of hypochlorite is required [275].

In special cases, a precipitation is possible, using a method known in analytical chemistry where magnesium ammonium phosphate ($MgNH_4PO_4$) is formed. [28]. As shown in Fig. 3.141, this works out best with stoichiometric excesses of magnesium and phosphate where, at room temperature and pH 9, some 30% of ammonia is present at equilibrium.

Using 1.5 times the stoichiometric excess of precipitating agents, there are no problems in reaching nitrogen concentrations less than 20 mg/litre. However larger excesses are required where precipitation pH values other than 9 are used. This is specially important where, in addition to ammoniacal nitrogen, there is also phosphate in the effluent. Magnesium hydroxide (see Section 3.6.3) is specially useful for neutralisation and precipitation in these cases. Where metal ions are also present, it should be noted that these can also form sparingly soluble basic phosphates with phosphate ions in solution thereby hindering precipitation of magnesium ammonium phosphates. In special cases or where the total effluent volume is not large, such precipitation offers a means for removal of ammonia.

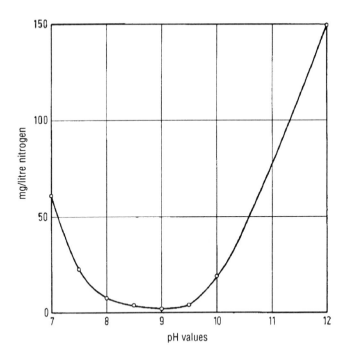

Fig. 3.141. Precipitation of ammonium ions as magnesium ammonium sulphate by addition of twice the stoichiometric quantities of magnesium and phosphate, as function of pH. Initial concentration 200 mg/litre (as nitrogen). [28].

3.11.3 Removal of chlorinated hydrocarbons

Chlorinated (strictly speaking, halogenated) hydrocarbons are classified, in some legislations as:

- volatile halogenated hydrocarbons. These include:
  - dichloromethane
  - 1,1 trichloroethane
  - trichloroethylene *("tri" or "trike")*
  - tetrachloroethylene *("per" or "perchloroethylene").*

For legislative purposes, the total concentration of these is summed, expressed in mg/litre Cl.

- Absorbable organic halocarbons (AOX). This classification includes all halogenated organics which are adsorbed on activated carbon, a property which lends itself both to analysis and identification and to a means of concentrating and thus removing compounds under this heading. It should be emphasised that AOX type halocarbons include the volatile compounds listed above. A few other compounds used in metal-working and finishing should also be mentioned. These include partially fluorinated chlorocarbons such as 1,1,2-trichloro-1,2,2 trifluoroethane (R 113) and trichlorofluoromethane (R11). In addition a range of AOX organic compounds such as chloroform can form when water or effluent are treated with chlorine or chlorine-releasing compounds (Section 3.2.1.7). Similarly, such AOX compounds can form by addition reactions of unsaturated organics in presence of HCl. Then too, AOX compounds are found in cooling oils, while technical grade HCl and technical grade chlorides such as Fe(III) chloride must be included.

Volatile halocarbons and AOX compounds such as those mentioned above are specifically mentioned in German and other environmental legislation. If inadequate effluent treatment and air-scrubbing measures obtain, these compounds can percolate into ground-waters, and it was only quite recently recognised that concrete is permeable to such volatile hydrocarbons. Thus where spillages of volatile organics onto concrete floors with no underlying membrane or cellar occurs, these organics will undoubtedly seep into the soil and so the ground water. The same is true where concrete drain pipes are used or other unsealed pipework. Again, environmental legislation in several European countries prescribes measures to avoid this happening.

Because of their high volatility and in many cases, their low boiling point, volatile chlorocarbons will readily pass into air extraction or stripping systems, then passing into the air. Finally, on account of their high density, they will be washed out of the atmosphere by rainfall and pass, in a finely-divided state, into the soil (see Table 3.37). Effluent treatments involving stripping should thus not use direct untreated discharge to air.

Volatile chlorocarbons are mainly used, in the metal-working industries, in conjunction with mechanical processes for degreasing, application of thin grease layers, stain-free drying as well as in the p.c.b industry for developing and stripping of solvent-based photoresists. Although water-based replacement processes are now available for most of these operations, it remains the case that volatile chlorocarbon solvents offer a range of attractive features, viz:

- unsurpassed solubility for oils and greases
- no explosion risk
- low heating energy requirement
- rapid drying
- degreasing action in vapour as well as liquid phase
- ease of recovery and recycling.

Because of their environmentally unfavourable properties and the high costs of containing and/or recycling or disposing of them, many companies have decided to move to aqueous cleaning technology and in the p.c.b industry there has been widespread move to aqueous alkaline photoresist and etch-resist technology based on the much simpler alkaline developer and stripping processes.

TABLE 3.37:
PROPERTIES OF HALOCARBONS COMMONLY ENCOUNTERED
IN TREATMENT AND RECOVERY OF EFFLUENT AND
EXHAUST AIR [278-280]

Property	tri-chloro-ethy-lene	tetra-chloro-thene	1,1,1-tri-chloro-ethane	Dichloro-methane	R11[1])	R113[2])
sol'y in water (gm/litre)	1.1	0.2	1.3	ca. 20	1.3	1.7
partition coeff't	2.7	1.2	0.7	8.1	–	--
vapour pressure (mbar, 20°C)	77	19	133	473	890	363
saturation conc'n in air (gm/m3)	480	150	725	1650	–	2750
boiling pt. °C.	87	121	74	40	23,8	47,6
Enthalpy of vaporisation [kJ/kg]	239.9	211.0	237.4	329.5	–	146.8
b. pt of azeotrope with water	74	87	65	39	23.2	42.2
Wt% chlorocarbon in azeotrope	91	77	94	98	–	99
	120	150	160	120	–	300
solvent content of oil residue (wt %)	35	40	10	3	–	0
Mol. wt. (gm/ml)	131.4	165.9	133.4	84.9	137.4	187.4
	1.47	1.62	1.33	1.32	1.50	1.58

[1]) trichlorofluoromethane   [2]) 1,1,2-trichloro-1,2,2-trifluoroethane

Effluent contaminated with volatile hydrocarbons can condense out on cold-spots in chilling plants or can dissolve in layers of moisture adhering to parts being cleaned. In the case of smaller plants where regeneration of activated carbon for gaseous emissions is not practicable, there is the option of collecting the small volumes of chlorocarbon contaminated aqueous solutions and periodically disposing of them to centralised toxic waste treatment plants [276].

Larger volumes of effluent arise as condensates where air emissions are cleaned using activated carbon or adsorption resins with these being regenerated by steam. In these cases, the twin problems of effluent treatment and air scrubbing are linked together.

Two options exist for removal of volatile chlorocarbons from effluent – stripping and adsorption. Used alone, the former inevitably leads to contamination of the atmosphere, the latter method requires heavy use of activated carbon. The best approach is to combine the two processes.

Effluent contaminated with volatile chlorocarbons (water condensate and the condensate from regeneration of activated carbon) is initially treated by allowing the two phases to separate under gravity. This phase separation process requires a dwell time of not less than 12 hours in a quiescent tank. The heavier chlorocarbons thus separated are returned to the working baths (degreasing etc), the aqueous phase is stripped and re-used for steam generation in regeneration of the activated carbon.

Stripping of the above-mentioned effluents (condensates, regeneration liquors) is carried out with air. The process is favoured by maximising the interfacial contact area and using large volumes of air at the highest possible temperatures and sub-atmospheric pressures. Counter-current flow packed bed or bubble type reactors can be used. In the latter case, these may be cascaded with air emerging from one unit to re-enter the next [277]. Special reactors such as climbing film types can also be used. In principle, steam stripping may also be used although this can cause problems because of the added water in the emerging gas in cases where an associated absorption stage follows. It can be that another source of chlorocarbon contaminated condensate water is set up if the steam is not released to atmosphere.

Stripping is normally continued until the chlorocarbon concentration falls below 1 mg/litre. The stripping air is passed over an activated carbon bed to remove the chlorocarbon, before being vented to atmosphere. The treated effluent is passed through an activated carbon filter (see below).

The chlorocarbon contaminated air is purified by adsorption onto the activated carbon or resin. The uptake capacity of these is a factor of chlorocarbon concentration, air flow-rate and air temperature while their uptake rate is a function of the surface area of the gas-absorber interface. Increasing water content reduces their capacity and it is thus desirable, if waste heat is available, to warm the air and thus reduce its relative humidity, if possible below the 50% mark [278]. If there is no alternative but to work with moist air, then a first-stage activated carbon column may be used to reduce the moisture content. The capacity of the absorbing species increases with the molecular weight and boiling point of the organic species. In the case of "trike" (trichloroethylene) and where favourable conditions allow a 99% efficient use of the absorber, a figure of 0.15 kg organic/kg absorber can be reached. For perchloroethylene, using activated carbon, the figure would be 0.3 kg/kg [279]. The organic-loaded activated carbon can be regenerated using superheated steam. This process requires 3.5 to 5 kg steam/kg of "per" and 50 to 60 litres of cooling water unless there is already a recirculating heating/cooling loop in place. In the case of activated carbon charged with dichloromethane, a typical condensate would contain 20 gm/litre of this species [279].

The resins best suited for this application are macroporous copolymers of styrene and divinylbenzene. Using 15 kg of steam per cubic metre resin, they can be regenerated leaving only 10 to 15% residual adsorbed material. A typical resin of this type can absorb 85 gm dichloromethane/litre [280] although higher values have been cited in the literature.

Condensate consisting of water with some second-phase volatile chlorocarbon is fed to the settling tank for the two phases to separate. The aqueous phase passes to the steam-raising boiler, the organic phase is distilled, then returned to the degreasing or other bath.

Effluent which has been stripped passes as a separate stream to an activated carbon bed where the last traces of organic are removed. Not all grades of carbon are suited to this operation. They should have a graded pore structure with micropores (up to 20 Å), meso-pores (20 to 500 Å) and macropores >500 Å). This type of material will typically have a surface area of 500 to 1400 square metre/gm, to which the micropores make the greatest contribution. The absorption capacity of these carbons depends on the water solubility, volatility, polarity and molecular weight of the organic. The lower their solubility in water, volatility and polarity and the greater their molecular weight, the better they absorb (tetrachloroethylene > trichloroethylene > 1,1,1 trichloroethane > trichloromethane) [281].

Since the volume of effluent with volatile chlorocarbon content in the range of a few mg to 1 mg/litre after stripping is quite small, the amounts of chlorocarbon recovered from the activated carbon absorption are likewise not large. For this reason it is best to configure the activated carbon beds as a series of small modules rather than a single large one. As the first and second such beds become saturated, they are removed for regeneration, being replaced by modules from further downstream whose absorption capacity has been only slightly utilised [282]. In rough terms (see above), it can be stated that the activated carbon absorbs approx. 10% of its own weight and in a properly operated absorption chain, residual chlorocarbon in the emerging effluent will be less than 0.1 mg/litre.

Because of their relatively low diffusion rates, organic species in water should be allowed a contact time with activated carbon of 15 to 60 minutes with a flow-rate of 5 to 15 metres/hr. For larger absorber plants (above 10 cubic metres of activated carbon), external thermal regeneration is an option. In this case, using temperatures of 700 to 900 ºC, the absorbed species are cracked [281].

Absorbing resins can also be used to retain chlorocarbons and can handle both acid and alkaline streams [283]. However dissolved iron in the feed streams can cause irreversible damage and for this reason, a pH of 1.5 is often recommended [280].

The purification and recovery of volatile chlorocarbons from degreasing baths is traditionally carried out by distillation, using either indirect heating or preferably steam, since over-heating must be avoided. As the boiling point increases because of the increased proportion of oily products on the bath, residual chlorocarbons are recovered as the lower boiling azeotropic mixture without reaching the critical temperature at which degradation sets in. Dichloromethane and 1,1,1 trichloroethane can hydrolyse and in consequence, cannot usually be recovered by the method described above [285]. It can be seen from Table 3.37 to what extent the volatile chlorocarbons, in terms of total hydrocarbon content, can be recovered by distillation. Using azeotropic steam distillation, trichloroethylene and tetrachloroethylene can be removed down to the 3 or 4% level. The residues after distillation are then disposed of as toxic waste material.

Yet another approach to treatment of effluent contaminated with volatile chlorocarbons is aqueous phase oxidation with hydrogen peroxide using u.v radiation. Thus, for example, trichloroethylene will decompose as shown below:

$$C_2HCl_3 + 2\ H_2O_2 \xrightarrow{\ hv\ } 2\ CO_2 + 3\ HCl + 2\ H_2O \qquad\qquad 3.205$$

The HCl so formed is neutralised and using this technique, "tri" and "per" can be rapidly and irreversibly destroyed to levels of less than 10 microgram/litre. This method is specially recommended for treated of contaminated ground-water [39]. In the case of saturated organic compounds such as 1,1,1-trichloroethane, the decomposition process (during which organic radicals are formed) takes considerably longer. The same technology using hydrogen peroxide and u.v light has also been used for destruction of ferrocyanides (Section 3.2.2.1) and strongly bound metal complexes formed with organic chelating agents. (Section 3.9.5.3). Further details are given in the following section.

In trials using electrolytic cells for treatment of tetrachloroethylene in effluent, Becherer [286] established that this resulted in chloride ion formation and in particular, this took place in the catholyte of a divided cell. Using longer electrolysis times resulted in total dechlorination and the conclusion was drawn that the process was a cathodic one in which chlorine was replaced by hydrogen.

Both the last two processes allow volatile chlorocarbons to be destroyed, in contrast to those methods using activated carbon where, after regeneration, the compounds re-appear. In modern plants using volatile hydrocarbons, the trend is to construct them as a more or less sealed unit so that phase separation, stripping, adsorption on activated carbon and its regeneration have least possible contact with the external environment. Within such sealed units, there are similarly closed water and air recirculation loops, all of which serve to minimise solvent loss to atmosphere and the main "product" is the residue remaining after distillation, which is sent out to be incinerated or otherwise treated at a specialist centre [287]. To prevent contamination of the sub-soil, plants are constructed on double-skinned safety containment pans with alarm systems to notify leakages [287a].

3.11.4   Eliminating Processes & Sources of C.O.D Substances

The term COD (Chemical Oxygen Demand) embraces virtually all substances present in effluents which can be oxidised. It should be noted that this includes a number of inorganic compounds [287], namely:

- strong reducing agents such as alkali metal hydrogen sulphites, dithionites, dithionates, thiosulphates, hypophosphites, phosphites, boranes, hydrazine etc. Such compounds are recognised, in some legislations, as "spontaneously oxygen demanding" reflecting the rapidity with which they react with elemental oxygen and oxidising species.

- anionic and non-ionic compounds which, though not themselves reducing agents in a practical sense, can still be oxidised. These include cyanides, nitrites, sulphides, chlorides, hydrogen peroxide etc.

- dissolved metal ions or compounds in their lower valence states, e.g. Cu(I), Fe(II), Sn(II), Pb(II) etc.

The organic compounds discussed here are oxidised, in a COD determination, to form radical ions or molecules derived from their composition, such as CO_2, H_2O, SO_4^{--}, NO_3^{-} (depending on the type of nitrogen compound), PO_4^{--} etc. It should be appreciated that the conditions used for COD determinations (hot concentrated chromic acid in presence of a catalyst), are not feasible as a basis for effluent treatment on a technical scale. Furthermore, effluents produced by the metal finishing and metal-working industries contain predominantly inorganic species, with the notable exceptions of the organic coating and printed circuit board industries. This organic compounds are of only secondary importance. In this category, the following classes are best known:

- surfactants. These are added at low concentrations to virtually all electrolytes. They are present at higher concentrations as emulsifying agents in alkaline cleaning baths and pickling + degreasing solutions.

- brighteners and levelling agents in plating baths.

- pickling inhibitors.

- complexing agents in electroless plating baths and stripping solutions.

- formaldehyde used as reducing agent, mainly for electroless copper in the p.c.b industry.

- water soluble resists (replacing solvent based types) in the p.c.b industry.

- paint components and solvents (from organic coating lines).

- halocarbons.

- cooling oils used in metal cutting, grinding, drilling etc.

A major portion of organic substances in effluents arises in the form of oils and greases from alkaline degreasing baths. In some plants, this is virtually the sole source of organics in effluent, e.g. in metal-working plants.

Formaldehyde, e.g. as used for electroless copper deposition in p.c.b plants, can be rendered harmless by oxidation. Water-soluble resist residues, from the same industry, are mainly precipitated (for further details of both, see Section 8.9.3).

For removal of paint components and non-chlorinated solvents from painting operations, see Section 8.6.2.2.

Removal of oils and greases, including those in emulsified form from alkaline cleaner baths or cooling oils is treated below and also in Section 8.1.2.

In respect of other organic species which are usually found only in trace amounts, no suitable processes have been identified. In cases where the effluent is passed into the municipal sewage system, such compounds will usually be biodegraded in the main treatment plant. In other cases, measures must be taken to control the COD value and these are usually based on either oxidation or absorption processes or a combination of the two. In some cases, a final in-house biological treatment plant is used.

Oxidation processes have been described in several sections of this volume. Most of these are equally well-suited for use with organic compounds although the varying difficulty of oxidation of such compounds must be borne in mind. In the case of gaseous chlorine or compounds which release chlorine, it should be borne in mind that these can form chlorocarbons, especially chloroform. Where the organic compound is not too difficult to oxidise, hydrogen peroxide, peroxysulphuric acid, oxygen or air using activated carbon as a catalyst and of course ozone can all be used. Comments in Section 3.2.2.1 to 3.2.2.4 should be noted (cyanide destruction). For redox potential values of these and other compounds, see Table 7.4 and 7.5.

Mention has earlier been made of the use of hydrogen peroxide under the action of u.v light, this combination can oxidise the most stable of organic molecules such as ferrocyanides (Section 3.2.2.1), EDTA complexes of metals other than copper (Section 3.9.5.3) and chlorocarbons (3.11.3). The mechanism of this process is discussed below.

The oxidation of stable organic or metal-organic species mentioned above with hydrogen peroxide is not, in general, possible without the enhancing action of u.v. and it can fairly be deduced from this that the action of the u.v is to excite one or more of the reactants, so allowing the reaction to proceed. Such a process must involve the absorption of the radiated energy at a given wavelength or over a broader wavelength band. In the case of coloured compounds, visible light can act in the same way but in other cases, only radiation in the u.v is absorbed. Electromagnetic wave theory provides the following equation:

$$c = \nu \cdot \lambda \qquad\qquad 3.206$$

c = velocity of light $(3 \cdot 10^{10}$ cm/sec); ν = frequency (sec^{-1}); λ = wavelength (cm).

This does not, however, provide a quantitative guide to prediction of reaction rates. The Planck-Einstein quantum theory states that:

$$E = h \cdot \nu \qquad\qquad 3.207$$

where E is the Einstein energy of a radiated quantum or photon (J); h is the Planck efficiency constant which implies the minimum energy required for a molecule to react (h = 6.6 · 10^{-34} Js); ν is the radiation frequency.

From this, it follows that the extent of reaction of a compound depends on the frequency of the radiation (or the inverse of the wavelength). Photons can be seen, in terms of Eq. 3.207, as reactants and to react 1 Mole, 6 · 10^{23} photons are required. (This quantity is sometimes known as the Lohschmidt constant). This number of photons can then be designated as the photochemical equivalent.

Except in cases where a chain reaction is triggered by the radiation, such photochemical reactions will have a maximum reaction yield of 1 though usually a smaller value obtains. Considering the radiation to be quantised, it is customary to speak in terms of quantum efficiency ϱ and this is defined as:

$$\varrho = \frac{\text{No. of radiation absorbing molecules reacted}}{\text{Radiation quanta absorbed}} \qquad 3.208$$

As implied by Equations 3.206 and 3.207, the excitation of a molecule depends on the wavelength, with the condition that the molecule must be capable of absorbing the radiation. Information on this is provided by the absorption spectrum of the molecule and the most effective strategy is based on use of the shortest wavelength (i.e. with most energy) which is shown by such a spectrum to be capable of being readily absorbed (a trade-off between better absorption as shown by higher peaks, and shorter wavelength is sometimes called for). The amount of absorption is given by the Beer-Lambert Law:

$$E = \log \frac{I^0}{I} \cdot \varepsilon \cdot c \cdot d \qquad 3.209$$

where E is the Extinction coefficient, I^0 the intensity of the incident light, I the intensity of light emerging from the solution, of path-length d (cm). c is concentration of the absorbing species (Mol/litre) and ϵ is the molar extinction coefficient.

Suitable light sources are those which emit in the u.v-C range (200 to 280 nm). Hydrogen peroxide absorbs in this band and the author and co-workers showed [247] how the oxidation of metal-EDTA complexes proceeds via excitation of the peroxide and not the complex (see Section 3.9.5.3). In this process, OH radicals (hydroxyl radicals) are formed which in turn react with hydrogen peroxide to form O_2H radicals (hydroperoxide radicals). Listed below are some of the processes which can occur or are known to do so when hydrogen peroxide is irradiated [288, 289]. The highly unstable radicals with their oxygen excess react as very strong oxidising agents which virtually all oxidisable species present.

$$H_2O_2 \xrightarrow{h\nu} 2\,OH^{\cdot} \qquad\qquad 3.210$$

$$H_2O_2 + OH^{\cdot} \longrightarrow H_2O + O_2H^{\cdot} \qquad\qquad 3.211$$

$$H_2O_2 + OH^{\cdot} \longrightarrow H_2O + OH^{\cdot} + O_2 \qquad\qquad 3.212$$

$$OH + OH^{\cdot} \longrightarrow H_2O_2 \qquad\qquad 3.213$$

$$O_2H + O_2H^{\cdot} \longrightarrow H_2O_2 + O_2 \qquad\qquad 3.214$$

$$O_2H + OH^{\cdot} \longrightarrow H_2O + O_2 \qquad\qquad 3.215$$

However it is only reaction 3.215 which leads to a dissociative process whereas for Equations 3.211 to 3.214, the product is yet another unstable oxygen-containing radical or re-formation of hydrogen peroxide which re-enters the overall sequence.

By way of illustration, Fig. 3.142 shows the oxidation of pentachlorophenol after Bandemer [290]. He carried out studies with a range of chloro-organic pesticides, reporting a wide range of oxidation rates and an absence of any single reaction order.

In the construction of technical plants using these principles, important factors are the proportion of radiant energy in the u.v – C band (W), the radiation flux density (W/sq. cm), the optical path length through the solution and the means by which hydrogen peroxide is dosed in. Fig. 3.143 shows a photograph of a commercially available unit. The literature suggests that ozone would behave in much the same way as hydrogen peroxide as the following equations show:

$$O_3 \xrightarrow{h\nu} O_2 + O \qquad\qquad 3.216$$

$$O + H_2O \longrightarrow 2\,OH^{\cdot} \qquad\qquad 3.217$$

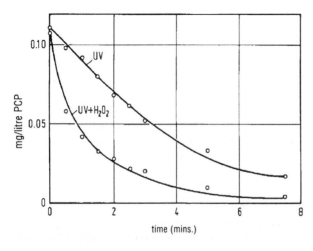

Fig 3.142 Destruction of pentachlorophenol (PCP) as a function of irradiation time with and without added hydrogen peroxide (1 gm/litre) [290].

Fig 3.143. Commercial unit for oxidation using u.v irradiation from 5 high-pressure mercury lamps (not shown) with added hydrogen peroxide. Courtesy GOEMA Dr Goetzelmann GmbH, Vaihingen/Enz, Germany).

The strongly oxidative action of anodic electrolysis can also be attributed to formation of OH˙ radicals (see Equation 3.55a) although other mechanisms have been shown to operate such as those in which PbO_2 is formed and then reacts chemically, thereby acting as a mediator. In many cases [247], anodic oxidation can be as powerful as the u.v light + peroxide combination. Fig.3.144 shows the anodic oxidation of aliphatic alcohols.

OH˙ radicals are also formed in Fenton's reaction where organic compounds react with hydrogen peroxide and iron (II) compounds [267] to form a radical intermediate:

$$H_2O_2 + Fe^{++} \rightarrow \dot{O}H + OH^- + Fe^{+++}$$ 3.218

Little more is known of this reaction for destruction of COD type compounds. In the case of some simple compounds, COD type compounds can be broken down using air over a column of activated charcoal as described in Sections 3.2.2.3 and 3.4.1. However it is not suited to post-treatment on account of the low handling capacity (column processes) or for handling liquors from continuous treatment plants (with powdered carbon). The same is true for so-called "wet oxidation" or "wet combustion" based on the use of air or oxygen in aqueous phase at 150 to 370°C at pressures from 10 to 220 bar [291, 292].

Adsorption type processes are appropriate for the removal of most, if not all types of organic species from effluent. However their use for the removal of non-toxic COD species has no value in cases where there would in any case be a post-treatment in the municipal treatment plant. Moreover, especially in small to medium sized plants, the regeneration of adsorbents is a further problem. Adsorption processes come into their own where there is a need to treat toxic species such as volatile halocarbons, AOX and other specifically toxic substances. Effluent containing such species is best segregated at an early stage in order that the adsorbent is not needlessly loaded with non-toxic compounds. Adsorption of chlorinated hydrocarbons was previously discussed in Sections 3.11.3 and similar comments apply to AOX compounds and BTX (benzene, toluene, xylene). In all cases, activated carbon and adsorption resins can be used. Their regeneration can be brought about with steam or by thermal treatment.

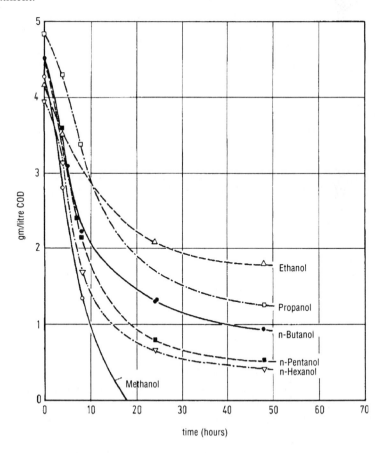

Fig. 3.144   Anodic oxidation of aliphatic alcohols. Current density 2A/sq. dm, platinised Ti mesh anode, pH 12, 35 to 40°C, 15 gm/litre sodium sulphate for conductivity [247].

Selection of an adsorbing material is best based on adsorption isotherm data for
the actual species to be removed from the effluent unless practical data happen
to be already available (for adsorption isotherms see Section 7.7).

In cases where effluent contains BTX type compounds, for example in organic coating plants, these should be removed by air stripping, using the lowest
possible volume of air. This is then passed over an absorber material. Fig. 3.145
shows how this can be done. Table 3.38 summarises the most important parameters for this process. Adsorption proceeds much faster from the gaseous than
the aqueous phase, this being a consequence of the much larger diffusion rates
of BTX molecules in the gas, as opposed to the aqueous phase [293].

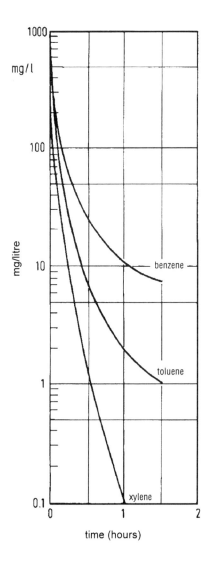

*Fig 3.145   Stripping rate of
aromatic hydrocarbons with air from
aqueous solutions, 60 litres/hour of
air per litre solution [293].*

Though phenol is not commonly found as a constituent of effluent in the metal-finishing and working industries, it might be noted that it can be effectively removed using activated carbon adsorption. Phenols, in this case, are, to a large extent, selectively removed in preference to other substances such as surfactants [294].

TABLE 3.38:
PHYSICAL DATA FOR BTX COMPOUNDS [293]

hydrocarbon	boiling pt. °C	Vapour Press (Torr, 20°C)	Sol'y in water (gm/litre, 20°C)
benzene	80	75	1.8
toluene	110	22	0.5
xylene	144	10	0.2

3.12   NEUTRAL SALTS AND MEANS TO MINIMISE THEIR DISCHARGE

3.12.1   Definitions

By neutral salts is here understood those which, on dissociation in aqueous solution, yield neither hydrogen nor hydroxyl ions. However in the context of treated effluent, this definition must at once be restricted and widened. The restriction limits such salts to those dissolved sodium and calcium salts as also to ammonium salts and those of amines (complexants) which can exist in the pH range 7 to 10. Most of these are very readily soluble, although a few sparingly soluble calcium salts also fall into the same category. There are also small amounts of those alkaline species, not true salts but which are necessary to bring an effluent pH to 7 or above. As a result of the various treatments received, the effluent should not exhibit any particular redox potential with no significant quantities of either reducing or oxidising salt anions present.

As a rule of thumb, the most important salt anions in neutralised effluent, listed in descending order of importance are: sulphate, chloride, nitrate, hydrogen phosphate, bicarbonate, carbonate, dihydrogen borate, tetrafluoborate, hexafluorosilicate. In addition, smaller quantities of non-oxidisable hexacyanoferrate may be present as well as hydroxyl ions as determined by the pH. Toxic anions such as cyanide, cyano-complexes of non-ferrous metals, chromates, nitrites, fluorides and sulphides should not be present at concentrations greater than those permitted by regulations and their total amounts are thus insignificant in reckoning the total dissolved neutral salt concentration. Also to be taken account of, are the anionic organic complexants, when these are present in significant concentrations (e.g. from p.c.b fabrication) and indeed anionic organic compounds in general (such as surfactants).

Sulphates and chlorides together account for the overwhelming majority of total neutral salts, the predominant one of the two depending on the nature of the effluent and the treatment used. Sulphate concentration can be greatly reduced by precipitation, especially from more concentrated solutions (Section 3.10.2). Chlorides predominate mainly in cases where rinsewater is recirculated through an ion exchanger since HCl is the most commonly used regenerant for cationic ion exchangers.

3.12.2   Effluent accumulation and discharge

As long as surface treatments using aqueous solutions are involved, effluent will arise and this will almost invariably contain neutral salts. As will be seen below, the amount of these can readily be reduced though probably never completely eliminated. Leaving aside small quantities of organic compounds, effluent from the metal-working and finishing industries contains mainly neutral inorganic salts, after treatment. If it were ever to be decided that such discharges must cease, the only course of action would be to remove the water, by physical methods such as evaporation, and to collect the soluble salt residues for subterranean disposal. The implications of such a policy would be extremely onerous in financial terms while in environmental terms, it would simply imply a transformation of an effluent disposal problem into one involving solid wastes instead. Furthermore, water recovered from evaporation plants is far from pure and can only be used for certain purposes. In many locations it might well have itself to be treated as effluent and discharged to sewer, quite possibly requiring effluent treatment itself before discharge, during which neutral salts could well be formed. Evaporation of aqueous process solutions is a valid method for recovering and recycling components from solution, and perhaps for water recovery too. If this avoids discharge of neutral salts to effluent, that should be seen as a bonus, but no more. In conclusion, and with a handful of special cases excepted, evaporation is not to be seen as a strategy for avoiding discharge of neutral salts.

Care must be taken to avoid damage which can arise when high concentrations of neutral salts are discharged. These, though not themselves toxic, can affect growth and life-cycles of micro-organisms Tables 3.39 and 3.40 show concentrations of neutral salts at which damage to fish and other aquatic life and micro-organisms occurs. Such damage only sets in at relatively high salt concentrations. Concern exists also lest the concentration of neutral salts in groundwater becomes so high that it precludes their use for drinking water (see Table 3.41).

TABLE 3.39:

HARMFUL CONCENTRATION LEVELS OF NEUTRAL SALTS FOR FISH AND AQUATIC LIFE (AFTER LIEBMANN [295]).

Compound	Harmful concentration (gm/litre)	
	for fish-life	for aquatic life
$NaCl$	7 ... 15	4 ... 11
Na_2SO_4	7 ... 14	0.5 ... 10
$NaNO_3$	ca. 10	2.5 g/l
NaF	0.1 ... 0.56	0.046 ... 0.5
Na_2HPO_4	4 ... 5	3.7 g/l
K^+ salts (KCl)	3 ... 13	3 ... > 8
$CaCl_2$	7 ... 12	1.3 ... 14
$Ca(NO_3)_2$	3.3 ... 5.7	2 ... 10
surfactants		
anionic	3 ... 5	2 ... 10
non-ionic	2.5	5 ... 10

TABLE 3.40:

EFFECT OF NEUTRAL SALTS ON SLUDGE DECOMPOSITION RATE (AFTER McCARTY [296])

Substance	Effect on sludge (gm/litre)		
	rate enhanced	rate inhibited	toxic
Na^+	100 ... 200	3.5 ... 5.5	8
K^+	200 ... 400	2.5 ... 4.5	12
NH_4^+ (as N)	50 ... 200	1.3 ... 3	3
Ca^{++}	100 ... 200	2.5 ... 4.5	8
surfactants	100	150 ... 300	500

At a meeting in 1986 of the Water Utilities from the several European countries which abut the river Rhine, an agreement was reached as to the maximum permitted concentration of neutral salts for discharge (Table 3.41) [297]. Some local authorities prescribed even more severe limits for discharges of neutral salts, not a few calling for such concentrations to be even lower than those found in the mains drinking water at the same location! Neutral salts form part of a natural cycle, originating from the seas and ultimately returning to them. It falls to water users and politicians to avoid a build-up of potentially harmful concentrations in inland waters. Options for achieving this are discussed below.

TABLE 3.41:

MAXIMUM PERMITTED CONCENTRATIONS OF NEUTRAL SALTS AT INTAKE POINTS FOR DRINKING WATER ABSTRACTION FROM [297]

Species	Limiting conc'n (mg/litre) resulting from natural processes	Using additional physico-chem. processes	Mean values for rivers (globally)
chloride (mg/litre)	100	150	7.8
sulphate (mg/litre)	100	150	11.2
bicarbonate (mg/litre)	25	25	
Hydrogencarbonate			58.4
ammonium (mg/litre)	0.2	1.5	
Sodium (mg/litre)	60	90	
electr, cond'y (mS/m)	70	100	

3.12.3   Effluent reduction, avoidance, elimination

The occurrence of neutral salts in effluent is no more avoidable than their ingestion with food and drink and subsequent excretion by all forms of animal life. No public effluent treatment plants in use today are equipped to remove neutral salts although much erroneous material has been published in this respect. It is true that inorganic nitrogen compounds such as ammonia or nitrites can be biologically broken down and phosphates can be precipitated. However the matching cation remains in solution and anionic phosphate must be replaced with some other anion. A sole exception is ammonium nitrate, which can be totally decomposed to gaseous nitrogen and water.

The reduction of neutral salts in water can be simply but most effectively brought about by implementing various measures to reduce drag-out from process baths into rinsewaters (see Section 6.2). Another approach is to use water-economising methods, especially cascaded rinsing (see Section 6.3.2). In the latter case, the same volume of solids is formed in a smaller volume of water, i.e. at higher concentrations. The same principle applies to recirculation of water through ion exchanger loops where water savings are matched by a reduction in volume of effluent to 0.2 to 1% of total circulation. It should be noted that the resin regeneration process requires an excess of the regenerating salt so that the total discharge of neutral salts, using current technology is actually 5% greater than would have been the case without use of ion-exchange. By way of compensation, the volume of metal ions discharged is very much less than would otherwise have been the case. Thus use of ion exchange methods implies a trade-off in which discharge of toxic metal salts is replaced by discharge of a larger amount of neutral salt such as NaCl.

Recycling offers a most effective means of lowering discharges of neutral salt, especially those in which process solutions are recovered by evaporation (Section 7.8.2) or electrodialysis (Section 7.6.3). In theory at least, use of these methods eliminates rinsewater discharge and the dissolved salts which this includes. In addition, such processes largely eliminate the requirement for continuous dosing with fresh chemicals.

Also in this category are those processes which increase the life of a process solution and thus reduce the number of occasions on which such baths must be dumped. Processes in this category include crystallisation (Section 6.4.2), use of active carbon (Section 6.7), selective electrolytic purification methods (Section 6.6.2.2.2), retardation processes (Section 6.2.2.2) and ion exchange (6.2.2.1) although in this last case the reductions in effluent discharge are partly offset by the need to dispose of the spent resin regeneration liquor.

Finally, in addition to the physical recycling methods described above where no chemicals are used, is the approach where reagents which form neutral salts are avoided. Examples of these are hydrogen peroxide, ozone, and "active oxygen" (anodic oxidation, air used with a catalyst). For nitrite reduction, replacement of compounds such as amidosulphonic acid with urea (see Equation 3.94) eliminates salts as reaction products, with gaseous species being formed instead (nitrogen, carbon dioxide). Precipitation reactions are only partly successful in removing neutral salts and only when all ions involved in such precipitations end up as insoluble solids is there an elimination of salts. An example of this is shown below (Equation 3.219).

$$NiSO_4 + Ca(OH)_2 \rightleftharpoons 2\,Ni(OH)_2 + CaSO_4 \qquad\qquad 3.219$$

and in this neutralisation of nickel sulphate with slaked lime, two sparingly soluble products result. In all other cases, equivalent amounts of neutral salts are formed and remain in solution, e.g.

$$Na_2SO_4 + Ca(OH)_2 \rightleftharpoons 2\,NaOH + CaSO_4 \qquad\qquad 3.220$$

The resulting NaOH then again forms a neutral salt. Another example is:

$$2\,NaF + CaCl_2 = CaF_2 + NaCl \qquad\qquad 3.221$$

in which an insoluble fluoride is precipitated. These examples show some options for reducing the volume of neutral salts, the most attractive being reduction of drag-out and recycling.

This section should not end without mentioning the existence of physical processes for removing neutral salts from water with simultaneous formation of concentrated salt solutions or salts. Such methods are widely used for desalination of sea-water and include reverse osmosis (R.O), electrodialysis and evaporation. As previously mentioned, the overall effect of these processes is to "re-cast" the effluent problem as a waste disposal problem. Out of interest, it should be noted that in the Middle-East where these technologies are most extensively used, there has been a substantial increase in the salinity of the seas which feed them, to the point that materials of construction which were satisfactory when the plants were first built, are now corroding thanks to the higher salinity of the waters passing through them. The implications for national economies, were such methods made mandatory for metal-finishing and metal-working industries have yet to be assessed. In any case, however large the amount of neutral salts discharged by such industries, they are dwarfed in comparison with salts formed by actual salt-mining and derived from heavy chemical industries. It would thus be ridiculous to seek to regulate the one source without the other, larger one.

3.13   OILS, GREASES AND EFFLUENTS CARRYING EMULSIONS

A wide range of methods exists for treating effluents in these categories and these, as well as aspects of maintenance and regeneration for such solutions, are covered in Section 8.1 "Mechanical Treatment Methods". Treated below are some basic concepts relating to such effluents.

3.13.1   Origins

Mineral oils and greases as well as non-ionic surfactants are commonly encountered in the surface treatment industry. In the absence of surface-active agents, oils and greases can readily be separated as a second phase from aqueous solutions. When surfactants are present, they tend to form finely-dispersed emulsions which must be broken before the two phases can be separated. In metal finishing, there are two major sources for these substances:

- from emulsion-based coolant liquids used in machining
- cleaning of metal surfaces which have been deliberately or accidentally hydrocarbon coated, using aqueous cleaners.

Coolant liquids, as their name implies, are used to remove heat formed in machining operations while simultaneously lubricating the machining and preventing corrosion. Their normally limited life can be extended by using appropriate measures. Thus particulate matter (swarf, dirt) must be removed as well as additional oil pick-up from the machinery, such oils not usually being emulsifiable.

Cleaning solutions accumulate oils and greases from the surfaces to be cleaned as well as solid soils of widely varying composition. Among these are rust-preventing oils, oils and greases from mechanical working, residues from grinding and polishing compounds and greasy finger-marks. Oils and greases can be taken into solution as emulsions to an extent limited by the type and concentration of surfactant present. The life of such surfactants too, can be increased by removing the second phase of the emulsion using ultrafiltration (see sections 7.5.4.1 and 8.1.2). This now widely used procedure places certain restrictions on the cleaner solutions used (no silicates and surfactants with low cloud-point)

3.13.2 Separation of non-emulsified oils and greases

To the extent that oils and greases are not present as an emulsion, they can be separated from water by flotation, since having a lower density, they float to the surface. This occurs only when dispersing agents (surfactants) are absent from the aqueous solution, e.g. in hot water degreasing. Such separations are also possible when less stable emulsions form an equilibrium with non-emulsified material or when very high concentrations of long-chain hydrocarbons are present. In addition, an oily phase is formed after emulsions are broken, in cases where they are not adsorbed onto a solid phase.

Phase separation can be considered by analogy with sedimentation (Section 3.7.4.3) with Stokes Law governing both processes and because the oily droplets are approximately spherical, experimental data often fits theory more closely than is the case for irregularly shaped solid particles. The equation as previously formulated is modified in that settling rate u_s is replaced by flotation rate u_A, while the density of the lighter phase is subtracted from the value for water while the empirical factor F_A is specific to the properties of the lighter phase:

$$u_A = \frac{g}{18\,\eta}\,(\varrho_w - \varrho_L) \cdot d^2 \cdot F_A \qquad\qquad 3.222$$

The greater the height through which the lighter droplets float upwards, the more likely they are to coalesce. In the absence of turbulence (which serves to break up larger droplets into smaller ones), coalescence works in favour of phase separation. Larger droplet diameter, greater density difference, decreased viscosity and increasing temperature (some of these are interrelated) all operate to shorten coalescence time [299].

Similar parameters hold for the accumulation of oily materials on a separator (simple gravity type or tilted plate) as for solid particle separation (Section 4.2.5.1).

Construction of oil and grease separators as well as their design, installation and operation are described in DIN 1999. Note that these should be installed ahead of sedimenter units (sludge settlers). They should also have adequate capacity for separated oil etc and be of sufficient height for the specific gravity of the oily matter being separated. Neither emulsions nor liquids containing surfactants should be fed to these units. It is advisable and sometimes necessary to embody safety measures as part of industrial oil separators. Typical efficiencies of these devices are around 97% [300] and Fig. 3.146 shows the main features of such a unit.

To further enhance the coalescence process, the oily water is forced upwards, as shown in Fig 3.146, through packing which is a highly effective means of increasing separation efficiency. An alternative approach is to flow the oily water across a series of horizontal parallel plates with a zig-zag profile. This promotes coalescence. Finally, the use of submerged aeration units to accelerate flotation should be mentioned as a method which is used.

Oil and grease which has floated to the surface and coalesced can be skimmed off in various ways (see Section 8.1.2). In spite of the apparent phase separation, the underlying aqueous phase can contain up to 30 mg/litre or sometimes more, of oily mineral matter, even in absence of surfactants. Where this is combined with metal ion containing effluent from a precipitation, these oily residues are removed by adsorption onto the particles of precipitate which can thereby take up significant quantities of hydrocarbon. Studies of sludges from plants with surface cleaning units revealed that from 0.5 to 2% by weight of dried sludge was soluble in petroleum ether [301]. If the residual hydrocarbons cannot be removed in this way, their adsorption onto iron (III) or aluminium salts offers another option. Salting-out by addition of metal compounds is a further approach which can be avoided by addition of iron bicarbonate to the neutral effluent, entraining carbon dioxide at reduced pressure and then adsorbing the organics on the precipitated iron carbonate. Introduction of further carbon dioxide re-dissolves the iron salt, allowing the organic phase to be separated out. The aqueous phase with its dissolved iron bicarbonate is re-used by admitting a new batch of water. The organics produced can be thermally decomposed. [302]. This process is only suitable for certain applications and where other metals, in particular those classified as toxic, are present, it is unsuitable since the solubility of such metals in the buffered pH range of iron bicarbonate is too large.

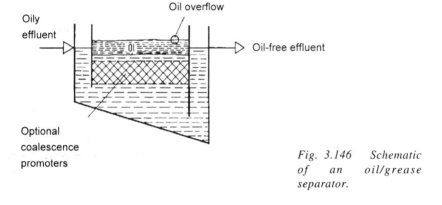

Fig. 3.146   Schematic of an oil/grease separator.

3.13.3 Physico-chemical properties of emulsions

In most circumstances relevant to this book, emulsions are of the type "oil-in-water". They are classified as dispersion colloids and approximate to spherical colloids since the oil droplets are almost perfectly spherical. The size of the oil droplets (disperse phase) in the water (dispersion medium) is usually larger than 0.1 μm in technical situations. The droplet size will not be uniform and for this reason, such systems are referred to as being "polydisperse".

If one were to take a cube of side 1 cm, of the disperse phase material and then divide this into 10^{12} little cubes of size 1 micron, this would require a significant expenditure of energy. The effect would be much more obvious if, instead of taking a cube as above with 6 square cm. surface area, a 1 metre cube were used, corresponding to 6 square metres. The surface energy to break up such a quantity is supplied not by mechanical energy but by mixing two-phase systems with surface active substances which lead to a reduction of surface tension of both phases. Such surfactants which are ideally suitable for emulsifying oils and greases, are sometimes known as emulsifying agents for this reason. In order to function in this way, they must be soluble in both phases. In the case of machining coolant emulsions, the surfactants come already dissolved in self-emulsifying oils. For degreasing baths, they are included in proprietary formulations.

One end of a surfactant molecule is lipophilic (and at the same time hydrophobic) and is soluble in the oily phase. The other end is hydrophilic (and also lipophobic) and imparts water solubility to the compound. Both ionic and non-ionic surfactants share these attributes. In the latter case, ether-acids are widely used to confer water solubility. Surfactants are most effective in lowering surface tension in the phase in which they are most soluble, with the second phase being dispersed. The hydrophilic end of the molecule, oriented towards the water, will usually hydrolyse. The energy thus released provides the surface energy necessary for the dispersion action.

In thermodynamic terms, emulsions are characterised as unstable materials. They cannot exist in an equilibrium form and thus have a tendency to revert to the two-phase state, when surface energy is released. Surface active agents inhibit this reversion and their function is not only to reduce the surface energy to allow mixing but, far more importantly, to stabilise the emulsion once formed.

This stabilisation depends on various effects of the surfactant on the disperse phase and equally on its chemical composition. In spite of the stabilising effect of surfactants, emulsions can coalesce and thicken after extended periods of standing, because of their fundamental instability. Providing this process is not accompanied by chemical or biological degradation of the surfactant, the surface energy released by coalescence can be re-introduced by mechanical means (stirring, shaking etc) thus reversing the coalescence and restoring the emulsion.

Linked to the size of the dispersed phase droplets their thermal motion (Brownian movement) is small and thus also the osmotic pressure of the emulsion. This can still be used to determine the number of particles and their mass, albeit not at larger dilutions. The reason for this stems from interaction between dispersed particles, between these particles and molecules of water and inter-molecular reactions of the water.

The electrical conductivity of oil-in-water emulsions is generally good. The existence of like charges on the droplet surfaces is also one of the main reasons why such emulsions are stable (if only metastable) since like charge repulsion inhibits coalescence. Migration of these charged droplets in an electric field is known as electrophoresis. Their migration rate is largely governed by their potential difference with respect to the surrounding solution, the so-called zeta potential. This is determined by the charge on the surfactant layer formed on the droplet surface. That charge, in turn, depends on the degree of dissociation of the hydrophilic termination of the surfactant molecule and the adsorption of ions from the solvent phase to form an electrical double layer (for further discussion of zeta potential, see below).

Hydrophilic terminations
of an anionic surfactant

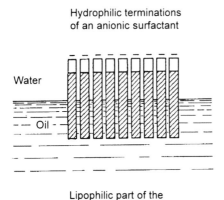

Water

Oil

Lipophilic part of the
surfactant molecule

Fig. 3.147   Idealised model of an interphase with adsorbed anionic surfactant molecules.

Where nonionic surfactants are used, no such strong electrostatic effects can be brought into play. By the adsorption of their lipophilic terminations onto the oil-water interface are formed, as in the case of ionic surfactants, thin layers which, by association, can convert to condensed phase films. With large excesses of surfactant, this can lead to formation of normal micelles capable of firmly binding the oil droplets. Such films, whose thickness in the case of long-chain fatty acids is quoted as 2 to 4×10^{-7} are the main factor in determining the stability of such emulsions. In addition to Van der Waals forces which are the major effect in film formation, further stabilisation can result from hydrogen bonding. These condensed phase films also provide mechanical protection for the oil droplets, though not all surface active compounds are suitable for this. For this reason, it follows that not all surfactants are good emulsifying agents.

From the foregoing, it can be deduced that anionic surfactants (cationic types have no significance for the emulsion types discussed here) are less affected by increased collision forces, thanks to their strongly electrostatic repulsive forces and so better able to function at increased temperatures than nonionic types. Against this, they are less acid-resistant since their anionic groups tend to form undissociated acids, thereby losing their negative charge in the interfacial layer. Conversely, emulsions formed using nonionic surfactants are more temperature sensitive and also more resistant to acids and salts. To obtain good chemical, thermal and mechanical stability for practical applications, numerous formulations based on mixtures of anionic and nonionic surfactants have been proposed as emulsifying agents.

Reference was made above, to the zeta potential. The migration rate caused by this can be expressed as follows:

$$u = DK \cdot \frac{E \cdot \xi}{4\pi \cdot \eta} \qquad\qquad 3.223$$

where u is the migration rate in the electrical field (cm s^{-1}), DK is the dielectric constant of the dispersion phase, E the electrical field strength (potential gradient) in V cm^{-1}, ξ is the zeta potential (V cm^{-1}), η the dynamic viscosity of the dispersion phase (gm/cm/sec).

ξ potential is an interfacial phenomenon, formed as shown in Fig. 3.148. Surrounding the charged particle there is an electrical double layer of counter-charged ions. The thinner element of this double layer is adsorbed directly onto the particle surface. The outer layer is structured as a diffuse cloud of charge-carriers, penetrating more or less deeply into the dispersion phase. Inside the adsorbed layer there is a linear potential gradient which also continues out into the diffuse layer though with decreasing gradient which finally asymptotes to zero. The potential operating at the diffuse layer is known as the zeta potential or electrokinetic potential. It is this external potential which determines the electrokinetic behaviour of the system. In the case of most finely-divided systems it has a value of 50 to 100 mV (disperse colloids require at least 40 mV to be stable). Below 30 mV, a slow coagulation takes place. At 0 mV, the isoelectric point is reached and coagulation occurs instantly [303]. This latter condition can be obtained by addition of one of a number of flocculating agents or polyelectrolytes. In this way, for example, colloidal metal hydroxides whose surface is positively charged by adsorbed hydrogen ions, can be coagulated by adding anionic flocculatingagents or hydroxyl ions. Silicic acids and metal sulphides with negative charges can likewise be coagulated using cationic flocculating agents.

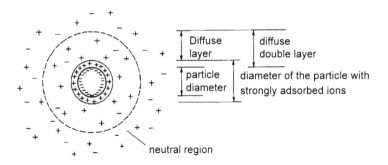

Fig. 3.148   Schematic of a dispersed colloid particle

3.13.4   Emulsion breaking

Breaking of emulsions into their oily and aqueous constituents takes place by interfering with those effects which prevent the disperse particles from coalescing and in this process, the surface energy required to form the emulsion is liberated Phase separation takes place and the system ceases to be metastable, becoming stable instead.

Processes for emulsion breaking as used in the metal-working and -finishing industries can be classified as follows:

- chemical processes.

- physical processes.

- thermal processes.

- mechanical processes.

These are briefly summarised below in terms of process criteria while Section 8.1.2 deals with plant construction and process operation. In many cases, practical installations make use of more than one of the process types listed above,

3.13.4.1 Chemical Processes

The simplest of these involves addition of acid. Degreasing baths containing emulsified material can, almost without exception, be treated by this means. The explanation is that the surfactants used in such baths, though they readily permit emulsification, do not form stable emulsions. Thus, when such baths become so rich in oily matter that they cease to operate, it is wise to de-emulsify, allowing oils to separate before re-emulsification. Separation can be carried out with a

weir system or by skimming. Acidification will, in most cases, result in temperature increase due to release of the heat of neutralisation of the bath and this
itself accelerates the emulsion-breaking process. Where the bath contains silicates, acidification will also result in precipitation of silicic acid and this finely
divided solid further benefits the emulsion-breaking and oil-absorption processes.

As for cutting and cooling oils, very few are used nowadays which allow
emulsion-breaking with acid alone. However acidification remains a cost-effective means for carrying out the first stage here and at least the electrostatic
repulsion forces from anionic surfactants are obviated by this means. Acid breaking of emulsions has one drawback – it can result in an increase in TDS (Total
Dissolved Salt) levels. However if acidic effluent is used, this ceases to be
important since such effluent would in any case have to be neutralised with the
same outcome of increasing TDS levels. In cases where the anions form insoluble calcium salts, slaked lime should be used for neutralisation.

The simultaneous warming of degreasing solutions and/or additions of salts
of multivalent metals, or acidification, warming and using suitable flow regimes
which promote coalescence, are all strategies which have been successfully
used. Salt addition is often the best option. According to the Schulze-Hardy
Rule, the efficacy of salt ions increases as the charge on them. However studies
with salts such as NaCl, calcium and magnesium chlorides and magnesium
sulphate showed that only magnesium salts had demonstrable effects on phase
separation [21]. The good results found using trivalent metals such as Fe(III) or
Al(III) were probably due to adsorption coagulation.

The purpose of adding electrolyte solutions is to neutralise the charge in the
boundary layer which is responsible for the stability of emulsions. According to
the theory of Verwey & Overbeek [304], increased ionic strength in the aqueous
phase modifies the concentration gradient at the double layer and thus reduces
the repulsive forces. Sontheimer [305] confirmed the benefits, as found in practice, of trivalent metals. In cases (as is mostly so in practice) where the colloid is
negatively charged and with positively charged counter-ions, the latter can exchange with trivalent cations, when these are added. This results in increased
potential difference across the double-layer with a reduction in zeta potential. In
consequence, the particles can more easily approach one another. The two theories are in no sense contradictory, indeed they should be seen as complementary.

The increased collision probability resulting from zeta potential reduction
favours a coalescence which is accompanied by hydrolysis of the adsorbed
trivalent metal cations. A flocculated hydroxide can form, onto which oils are
adsorbed and this is known as "adsorption coagulation". To the extent that
electrolyte is added simply for charge neutralisation (e.g.. with magnesium salts),
increase in temperature will significantly accelerate emulsion breaking. However this is often combined with adsorption treatment, e.g. by adding silicic acid
and related species.

The procedure described above, based on metal salt addition coupled with hydrolysis, is specially effective for low-concentration emulsions. In terms of efficiency, concentrated emulsions can also be treated in this way as Ballnus [306] showed, using aluminium salts. However it was found difficult to reduce oil or grease concentrations below the 20 mg/litre level with confidence. The method should rather be seen as a post-treatment stage, when perhaps 50 to 60 mg/litre of oily matter remain to be removed.

It is probably thanks to strong adsorbent properties of metal hydroxides that dilute effluent, after clarification, only rarely contains hydrocarbon residues above the permitted levels thus eliminating a need for a separate emulsion-breaking stage. This is confirmed by solvent extraction studies of effluent sludges showing that they can contain up to 2% oily matter [301].

Finally, some comments on so-called "Demulgators". These are cationic surfactants whose lipophilic grouping is likewise adsorbed at the oily surface. This leads to a charge neutralisation within the phase boundary. Since cationic grouping dissociate most readily in acid, the emulsion to be broken is acidified. For reasons of cost, this approach is usually restricted to smaller loads.

3.13.4.2 Physical processes

The most important and widely used example of this category is the binding of emulsions or their components onto adsorbents. Although in theory, any high surface area solid might be used for this, in practice silicic acid based materials are almost exclusively used.

Typical of these is a finely divided hydrophobic silicic acid with a bulk density of only 150 gm/litre. For emulsion breaking, some 35 to 45% of the weight of oil present are required [307]. On cost grounds, it is recommended that where oily matter exceeds 1%, a pretreatment using magnesium salts is used. According to Knobloch [308], this allows oil concentrations to be brought down to 0.4 – 0.8%. What remains can be broken down and adsorbed on activated silica within 5 to 10 minutes. In the case of more stable emulsions, it is beneficial to raise the temperature during this sequence to 50 to 60°C.

Comparably good results can be obtained with naturally-occurring non-hydrophobic silicas. Depending on the methods used to prepare the silica, a variety of adsorption mechanisms appear to be involved. Whereas the hydrophobic material tends to adsorb the lipophilic component of the emulsion (oils), in the case of Kieselguhr type materials, it is the hydrophilic surfactants which are bound. In both cases, the natural porosity of the material acts to take up the oily phase formed (sponge effect).

Experience shows that simultaneous addition of kieselguhr and magnesium salts at slightly elevated temperatures of 35 to 40°C gives good results. More difficult-to-break emulsions may not succumb to this approach, in terms of reaching permitted discharge concentrations and will certainly not do so by the adsorption stage alone [21].

Currently, there are numerous chemical/adsorbing preparations commercially available for emulsion breaking. They are often complex mixtures including silica derivatives, zeolith, activated carbon, neutral and acid salts, demulgators etc.

Another physical separation method is electrophoresis. Thanks to their surface charges, emulsions are electrically conductive and by imposing an electric field across an emulsion, two processes can take place. The charged droplets migrate to the oppositely charged electrode where, like simple ions, they discharge and phase separation occurs. If very high voltages (1500 V) are used, a shift of charge within the boundary layer takes place. Dipoles are formed which, as oppositely-charged ends approach one another, coalesce [309].

3.13.4.3 Thermal processes

In this approach, increased temperatures are used to remove the more volatile phase − water − so that oil is deposited out as supersaturation occurs. Such oil can be recovered or the now oil-rich residues used as fuel. Using the immersion burner (submerged combustion) method, the emulsion is directly contacted with the flame and its hot gas combustion products [310]. Again, as water is lost, the emulsion breaks as supersaturation occurs. In absence of non-ionic surfactants, an intermediate may form of water-in-oil emulsion which itself can lose water. The residues consist of oil, salts and mechanical impurities. Water from this process can be recovered by condensing and can be re-used, after first removing impurities such as azeotropic mixtures or other volatile species carried over with it. A method based on similar principles but using different equipment is thin film evaporation. The emulsion is fed as a thin film over heated surfaces where, as above, water evaporates and oil forms as a separate phase. Various other designs of evaporator have also been used [311].

It is not necessary to break the emulsion in order to recover the fuel value latent in it and by atomizing the emulsion and injecting into a burner, complete combustion takes place. This option is presently used only in large plants with their own burner units. It is of course the case that some of the energy released by such combustion is effectively wasted in vaporising the water present. The great advantage of all thermal processes is that any emulsion can be so treated, irrespective of its stability. However the capital costs are substantial, thus making the method most attractive on a large scale.

3.13.4.4 Mechanical processes

Depending on their stability, some emulsions can be broken by purely mechanical action. Under suitable hydrodynamic regimes, for example by forced flow through packed beds, meshes, plates or porous materials, the droplets will impinge on solid surfaces or on one another and damage or lose their protective outer layer, whereupon coalescence is likely. Such merged larger droplets will float and phase separation occurs. It is vital that the oily phase is rapidly separated from the aqueous layer, since there may be concentrations of surfactant in the latter which would allow re-emulsification to take place.

It is well-known from work using deep bed sand filters that these can break very dilute and poorly stable emulsions and similar observations have been made using cartridge filters or felts [312]. Another technical method uses impact of oil droplets on solid surfaces with simultaneous acidification and temperature increase [313]. This will break virtually all oil-in-water emulsions. Increasingly used in recent years is ultrafiltration. Because of their relatively large diameter, oil droplets are unable to pass through membrane filters, while water molecules can readily do so. The emulsion is fed, parallel to the membrane, at low overpressure (up to 5 bar). Apart from the oil, some surfactants (micelles) are unable to pass through the membrane and the residue (retained solution) is an oil-rich (up to 50%) emulsion [314]. (see Sections 7.5.1 and 7.5.4.1). On standing, this residue will often "curdle" to give a creamy solid of even higher oil content. As with thermal processes, this method has the advantage that it can treat even the most stable of emulsions. A special advantage of ultrafiltration is that no chemicals are added and there is thus no increase in TDS values. The residue must be disposed of, or burned, in the same way as waste oil.

Another method of great practical importance which combines several principles (adsorption coagulation, electrolysis) is flotation. If aluminium or ion salts are added to an emulsion, the metal ions adsorb at the surface of the colloidal oil droplets as described in Section 3.13.3. and can cause coagulation while themselves hydrolysing (adsorption coagulation). The coarse formed oil-bearing metal hydroxides can bind to small gas bubbles which then float them to the surface. There are two main means for setting up such flotation conditions – compressed air flotation and electroflotation.

In the first case, a portion of the effluent is first saturated with compressed air and then released into the bulk of the effluent. This causes the release of finely-divided bubbles [315].

In electroflotation, electrolysis of the aqueous phase forms finely divided bubbles of hydrogen and oxygen, the electrodes being positioned at the bottom of the tank. In a variant of the method, one electrode is made of iron or aluminium and acts as a sacrificial anode to produce the metal salts referred to above [316]. It has to be recorded that electroflotation was intensively studied in many countries in the 1970's and a few full-scale plants were introduced. It was argued that the electrolytic action did more than produce finely divided bubbles, namely that it acted to neutralise charge on the droplets or particles constituting the colloid or emulsion. It was argued that electrolytically-formed bubbles themselves bore an electrical charge which could act to neutralise charged particles. Without doubt there is a measure of truth in this. Against the method, however, was a need for an aqueous phase with good electrolytic conductivity if very high cell voltages were to be avoided. Production of bubbles by a chemical (electro-chemical) method by definition requires more energy than when they are formed by physical methods (compressed air). Finally the electrodes were expensive and easily fouled. After extensive R & D investment, there are few installations and no indications that this method is of great practical value.

References for Chapter 3

Note The references below have been drawn from major world books and journals. Titles have been translated into English but the language of the original may be in German or French when published in those countries. Photocopies of original articles are available from the Publishers in most cases.

1. Kerber, R., Glamann, "Chemical Kinetics" (in) Ullman Encyclopaedia of Chem. Technol., Vol. 1 Verlag Chemie, Weinheim.
2. Wedler, G.: Textbook of Physical Chemistry (in German), Verlag Weinheim (1982)
3. Schneider, G.M "Principles of Chemical Thermodynamics" (in) Ullman Encyclopaedia of Chem. Technol. Vol 1. 1-53
4. see e.g CRC Handbook of Chemistry & Physics, CRC Press, Boca Raton, Florida (latest edition)
5. Dodge, B.F. Zabban, W.: "Disposal of Plating Room Wastes – III, Cyanide Wastes: Treatment with Hypochlorites and Removal of Cyanates." Plating (1951) 561-586
6. Serota. L. "Science for Electroplaters – 32 Cyanide Waste Treatment – Hypochlorites." Metal Finishing (1958) Jan. 61-67.
7. Lancy, L.E. "Planning a plant for continuous alkaline chlorination of cyanide effluents" Galvanotechnik 49 (1958) 14-19
8. Weber, W. "Hypochlorite for destruction of cyanide effluents" Galvanotechnik 51 (1960) 501-6.
9. Arnold, K. "Effect of free chlorine concentration on reaction for cyanide destruction" Galvanotechnik 57 (1966) 760-765
10. Tricoire, J.-Cl. "Treatment of metal finishing effluents" Galvano (1970 (Jan) 54-60
11. Glayman J. "Cyanide & complex destruction from plating rinse tanks" Galvano (1970) (Oct) 753-757
12. Lohmann, U., Henn, G. "Destruction of toxic species – cyanides" JOT (1973)(Oct) 42-43
13. Delpin, G.D. "Laboratory study of alkaline cyanide destruction with gases" Oberfl. Surfaces. 10 (1969) 476-481
14. Stumm, W., Woker, H., Fischer, H U. "Hypochlorite for cyanide destruction in effluents". Schweiz. Z. f. Hydrologie 16 (1954) 1-21
15. Chapin, R.M.: J. Am. Chem. Soc. 51 (1929) 2112-2117
16. Eden, G.E., Wheatland, A.B.:J. Soc. Chem. Ind. 69 (1950) 166- 169.
17. Price, C.C., Larson, T.E., Beck, K. M., Harrington, F.C., Smith, L.C., Stephanoff, J. "Hydrolyses and chlorinolysis of Cyanogen Chloride". J. Am. Chem. Soc. 69 (1947) 1640-1644.

18. Kieszkowski, M. "Hypochlorite for oxidation of residual cyanide effluents from plating shops"
19. Weiner, R., Leiss C. "Kinetics of complexed cyanide destruction" Metalloberflaeche 25 (1972) 169-204
20. Resnick, J.D., Moore, W.A., Ettinger, M.B. "Behavior of Cyanates in Polluted Water" Ind. Eng'ng. Chem. 50 (1958) 71- 72.
21. Hartinger, L. unpublished work
22. Mirsch, E. "Determination of toxic cyanides in metal finishing effluents" (Lecture). The quoted values of Milne for stability constants of [Cu(CN)3]$^{---}$ were replaced with the more reliable data for [Cu(CN)4]$^{---}$.
23. Schlegel, H. "Residual cyanide in effluent" Galvanotechnik 59 (1968) 52-3
24. Oehme, F., Laube, K.H., Wyden, K. "Methodical comparison of the main classic methods for cyanide destruction" Galvanotechnik + Oberflaechenschutz 7 (1966) 74-84
25. Shashikant, N. "Kinetics of complexed cyanide reaction with hypochlorite in dilute solution" Chem. Technol. 35 (1983) 465
26. Oehme, F. "Trends in effluent technology in the Metals industries" Oberflaeche-Surfaces 9 (1968) 142-145
27. Lörcher, K.P. "Water Projects" Nr 102 06 504 (Germany)
28. Hartinger, L., Scheiffelen, B. unpublished work
29. Koppe, P., Herkelmann, H. "Adsorbent requirements as a criterion for further treatment of municipal effluent" Forum on Urban Hygiene 37 (1986) 194-6
30. Knorre, H. "Observations & experiences in treatment of cyanide effluents" Galvanotechnik 66 (1975) 374-383
31. Peroxid-Chemie GmbH, Hoellriegelskreuth. Techn. Bull. "Cyanide destruction with hydrogen peroxide"
32. DOS 2,352,856 (30.4.75) Degussa
33. Fischer, B., Rüffer, H., Düppers, W., Nagels, G., Knorre, H. "Treatment of cyanide containing gases from blast furnace washings" Z. Wasser. Abwasser. Forsch. 14 (1981) 210-7
34. Schwarzer H. "Hydrogen peroxide for treatment of plating effluent containing cyanides" Galvanotechnik 66 (1975) 22-6
35. DAS 2,219,645 (26.2.1981) Peroxide-Chemie Corp.
36. Lawes, B.C., Fournier, L.B., Mathre, O.B."A Peroxygen System for Destroying Cyanide in Zinc and Cadmium Electroplating Rinse Waters" Plating (1973) 902-909.
37. DOS 2,109,939 (2.3.1971) du Pont
38. Degussa; Techn. Bull. Effluent treatment with formaldehyde and hydrogen peroxide"
39. Hartinger, L., Gülbas, M. unpublished data
40. ELFA, Aarau: "Caro's acid for treatment of simple & complexed cyanides" Oberflaeche-Surfaces. 12 (1971) 149-150
41. Peroxid-Chemie GmbH, Höllriegelskreuth: "CUROX for treatment of cyanide effluent. Techn. Bull. FA 1.4.33 d 0.5 D-289
42. Hoffman, D.C. "Oxidation of Cyanides Adsorbed on Granular Activated Carbon". Plating (1973) Febr. 157-162.
43. Kuhn, A.T., Wilson, C. "Catalytic air oxidation of cyanides" Oberflaeche-Surfaces 18 (1977) 93-6
44. Höke, B., Wittbold, A.H. "Catalytic oxidation of nitrites & cyanides" Galvanotechnik 61 (1970) 468-474

45. Faul, W., Kastening, B. "Treatment of effluents & gas- washings with electrocatalysts" Galvanotechnik 68 (1977) 699-704.

46. Sondak, N.E., Dodge, B.V. The Oxidation of Cyanide-Bearing Plating Wastes by Ozone. Plating (1961) 163-180 und 280-284

47. Fabjan, Ch. "Treatment of metal finishing & other effluent with the ozoniser method" Galvanotechnik 66 (1975) 100-7

48. Stopka, K. "Advanced Ozone Technology for Cyanide Destruction in Electroplating". Plating and Surface Finishing (1980) May 77-80

49. Fabjan, Ch., Davies, R. "Ozone for treatment of cyanide containing effluent" Wasser, Luft, Betrieb 20 (1976) 175-8

50. Bauer, H. "Ozone for treatment of some industrial effluents" Munich Contributions to Effluent, Pisciculture & Freshwater Biology. Vol. 18 181-192. Oldenbourg Publishers, Munich 1977

51. Matsuda, Y., Takasu, Y., Fujisawa, T., Fujikawa, S. "Ozonation of Cyanate Ion In Alkaline Solution." Plating and Surface Finishing (1981) June 86-88

52. Kurz, H., Weber, W. "Cyanide destruction by the CYNOX process" Galvanotechnik + Oberflaechenschutz 3 (1962) 91-7

53. Nagendran, R., Parthasaradhy, N.V., Doss, K.S.G. "An Electrochemical Method of Treatment of Cyanide". Plating (1967) Febr. 179-182.

54. Ruml, V., Topinka, M. "Cyanide destruction from plating shops by continuous chlorination" Metalloberflaeche 23 (1969) 225-231

55. Hillis, M.R. "Treatment of Cyanide Wastes by Electrolysis". Transactions of the Institute of Metal Finishing 53 (1975) 65-73

56. Bustos, S., Kammel, R., Steppke, H.D. "Destruction of cyanide in highly dilute, chloride containing baths". Metall 34 (1980) 843-7

57. Krusenstjern, A. v., Mussinger, W. "Cyanogen formation in cyanide-based silver plating baths" Metalloberflaeche 16 (1962) 263-6

58. Easton, J.K. "Electrolytic Decomposition of Concentrated Cyanide Plating Wastes". Plating (1966) 1340-1342.

59. Jola, M. "Electrolytic cyanide oxidation" Fachber. f. Oberflaechentechnik 11 (1973) 151-3

60. Kuhn, A.T., Biddle, K. "Anodic oxidation of cyanides in aqueous media" Oberflaeche-Surfaces 18 (1977) 182-9

61. Götzelmann, W. "Electrolytic metal recovery – economics & practical aspects" Galvanotechnik 70 (1979) 596-603

62. Götzelmann, W., Spanier, G. "The BW process, a new method for cyanide destruction using iron sulphate" Galvanotechnik 54 (1963) 265-272

63. Weber, W. "Treatment of cyanide containing effluents" Wasser, Luft und Betrieb 7 (1963) 84-6

64. Bortlisz, J. "Methods for cyanide estimation in effluents" Vom Wasser 34 (1967) 196-201

65. Wolfbeiss, E., Schindewolf, U. "Pilot plant studies for treatment of waste cyanides and nitrites" Chem. Ing. Techn. 48 (1976) 63

66. US Patent 4,042,502 (16.8.1977) "Detoxification of cyanide solutions."

67. Tan, T.C., Teo, W.K."Destruction of Cyanides by Thermal Hydrolysis". Plating and Surface Finishing (1987) Apr. 70-73.

68. DOS 3,120,164 (21.5.1981) Dainichi Nippon Cables

69. Jola, M. "Catalytic HCN combustion" Galvanotechnik 61 (1970) 1003-8.

70. Conrad, J., Jola, M. "Cyanide destruction today" Oberflaeche-Surfaces 13 (1972) 143-8

71. Jäckle, H. "Use of acids for cyanide destruction" Galvanotechnik 58 (1967) 470-4
72. Schwarzbach, E. "Cyanide effluent destruction by high temperature hydrolysis" Wasser, Luft u. Betrieb 18 (1974) 26-7
73. Kieszkowski, M. "Studies on thermal decomposition of solid wastes" Electroplating & Metal Fin. (1971) May, 5-10
74. Müller, W., Witzke, L. "Treatment & detoxification of solid wastes from case hardening plants" Chem. Ing. Techn. 45 (1973) 1285-9
75. Fabjan, Ch., Bauer, P. "Treatment of chromic acid containing effluents with hydrogen peroxide" Galvanotechnik 67 (1976) 307-9
76. Klotter, H.E. Müller, W. "Treatment of cyanide & chromic acid containing effluents from small & medium sized plating plants" Galvanotechnik 50 (1959) 245-256
77. Grindley, J. J. Soc. Chem. Ind. 64 (1945) 335-354.
78. Hill, G.B. "Complete Removal of Chromic Acid Waste With the Aid of Instrumentation". Plating (1969) Febr. 172-176.
79. Serota, L."Science for Electroplaters" – 28. "Treatment of Chromate Wastes". Metal Finishing (1957) Sept. 65-71.
80. Channon, H.B. "Reduction of Chromium Wastes by Sulfur Dioxide". Sewage and Industrial Wastes 25 (1953) 923-929.
81. Arnold, K. "Dwell times in effluent treatment" Galvanotechnik 59 (1968) 22-32
82. Bahensky, V., Chalupa, J., Zika, Z. "Effluent treatment in automated plants" Galvanotechnik 55 (1964) 238-242
83. Oehme, F. "Chemical treatment of industrial effluents" Neue Zurich Zeitung 3.3.1965
84. Götzelmann, W. "Planning & scaling of effluent plants" Galvanotechnik 64 (1973) 1091-1107
85. Ibl, N., Frei, A.M. "Electrolytic reduction of chromate containing effluent" Galvanotechnik + Oberflaechenschutz 5 (1964) 117-122
86. Heger, K., Nowack, N., Pechtheiden, C., Schloten, F. "Continuous reduction of chromic acid" Galvanotechnik 73 (1982) 1072-5
87. Pini, G. "Electrolysis of chromic acid containing effluent" Metalloberflaeche (1982) 206-10
88. Schulze, G. "Electrolytic reduction of chromic acid containing effluent" Galvanotechnik 58 (1967) 465-480
89. Schwarzer, H. "Hydrogen peroxide for effluent treatment" Chemie Technik. 8 (1979) 67-70
90. Knorre, H "Hydrogen peroxide treatment of nitrites in effluent". Werkstoffe u. i. Veredlung 3 (1981) 267-9
91. Rodenkirchen, M. "Hydrogen peroxide treatment of nitrite containing effluents" Galvanotechnik 74 (1983) 924-8
92. Farbwerke Hoechst AG: "Amidosulphonic acid for nitrite destruction in dilute solution" Tech. Bull. 1967
93. Oehme, F. "Nitrite destruction - options & problems" Oberflaeche-Surfaces 17 (1976) 102-3
94. Hartinger, L. "Handbook of effluent treatment" Vol. 1, "Chemistry". Hanser Verlag 1976
95. Degussa, Frankfurt: "Hydrogen peroxide for Fe(II) to Fe(III) oxidation in pickle liquors & effluents with hydrogen peroxide"

96. Röver, W., Rüner, H. "Control of neutralisation in acid industrial effluents" Baender Bleche Rohre (1961) 282-9

97. Lohmann, U., Wallat, G. "Neutralisation with lime or NaOH" Chemie Anlagen + Verfahren (1975) 79-81

98. Kobe, K.A., Sheehy, T.M. Ind. Engng. Chem. 40 (1948) 99-102

98a. Louchart, G.W. Howard "Plating Cleans Up Its Act With Magnesium Hydroxide". Plating and Surface Finishing (1987) Nov. 26-30.

99. Ammon, F.v. Viennese Commentaries on Water, Effluent, Rivers 6 (1971) H1-H23

100. Deutsche Babcock Anlagen Aktiengesellchaft. "Babcock Water Handbook" Vulkan Publ. Essen (1982)

101. Schlossberg, L."The Cleaning Phenomenon" Metal Fin. (1965) Dec. 61-8

102. Feitknecht, W. "Equilibrium relationships for sparingly soluble basic salts" (VI. Commun. on Basic Salts). Helv. chim. acta 16 (1933) 1302-1315

103. Feitknecht, W. "Chemistry & morphology of basic salts of divalent metals" (VII Commun. on Basic Salts). Helv. chim. acta, 18 (1935) 28-40

104. Britton, H.T.S. "Electrometric Studies of the Precipitation of Hydroxides". Part I, J. chem. Soc. (London) 127 (1925) 2110-2120.

105. Britton, H.T.S."Electrometric Studies of the Precipitation of Hydroxides". Part II, J. chem. Soc. (London) 127 (1925) 2120-2142.

106. Schwab, G.M. Polydoropoulis, K. "Alkaline precipitation of mixed heavy metals" Z. anorg. allgem. Chemie 274 (253) 234- 249

107. Schwab, G.M., Brennecke, W. Z. Phys. Chem. B24 (1934) 393

108. Schulte-Schrepping, K.H., Deike, A. "Separation of heavy metals from water & effluent" Galvanotechnik 63 (19720 641-3.

109. Dyck, W., Lieser, K.H. "Coprecipitation of Cu, Zn, As, Ag, Cd & Pb by iron hydroxide & iron phosphate precipitation". Vom Wasser 56 (1981) 183-9 Weinheim Publ.

110. Lutter, E. Fachber. f. Oberflaechentechn. 11 (1973) 151-3

111. Krusenstjern, A. v., Axmacher, L. "Neutralisation of plating effluents with calcium & sodium hydroxides" Metalloberflaeche 18 (1964) 65-9

112. Daester, H.H., Jola, M. "Precipitation of heavy metals & compliance with effluent concentration limits" Galvanotechnik 63 (1972) 1117-1125

113. Junghanss, H. "Chemical precipitation processes" Erzmetall 22 (1969) B14-B18

114. Verband der Chemischen Industrie eV. "Physico-chemical effluent treatment processes". Part 7. "Flocculation & precipitation". Frankfurt (1977)

115. Akyel, H., Neven, M. "High M.W. synthetic flocculating agents in modern effluent treatment" Chemie. Ing. Techn. 39 (1967) 172-8

116. Reuter, J. "Synthetic flocculating agents in effluent treatment & sludge dewatering from the metal-working industries". Wasser, Luft u. Betrieb (1969) 129-133

117. Hamboldt, O., Vogel, W. "Assessment of flocculation processes" Wasser, Luft u. Betrieb. (1969) 164-8

118. Jola, M. "Sedimentation & filtration of metal hydroxide sludges – behaviour of nickel, copper & chromium (III) hydroxides". Fachber. f. Oberflaechentechnik 12 (1974) 211-6

119. Reuter, J. "Organic polymer flocculation agents for effluent treatment in the metal industries" Umwelt (1981)(1) 27-35

120. BASF: "Clarification of cloudy liquors" Ludwigshafen (1964)

121. Chemische Fabrik Stockhausen & Cie. "Tables for Prestol Marking", Krefeld (1969)

122. Akyel, H. "Developments in high M.W synthetic polymer flocculating agents for clarifying & sludge dewatering. Glueckauf 102 (1966) 364-8

123. Reuter, J. "Dewatering of industrial & municipal sludges" Wasser, Luft Betrieb 13 (1969) 246-9

124. Jola, M. "Operation & sizing of settling basins" Oberflaeche Surfaces 22 (1981) 7-12

125. Imhoff, K. "Pocketbook of municipal drainage" Oldenbourg Publ. Munich

126. Braha, A. "Compressibility of floc-type sludge in static thickening" Korresp. Abwasser 30 (1983) 612-20, 749-53

127. Wedekind, B. "Effluent treatment in the metal industry" Metall 12 (1958) 515-9

128. Greuter, E. "Practical experiences with automated effluent neutralisation" Wasser Luft U. Betrieb (1959) Vol. 3 57-60

129. Schlegel, H. "Hydroxide precipitation of heavy metals from the plating industry" Metalloberflaeche 17 (1963) 129-133

130. Hartinger, L. "Effluent treatment for the metal industries – precipitation of heavy metals (introduction)" Baender Bleche Rohre (1963) 535-540

131. Hartinger, L. "Heavy metal precipitation from effluents" Wasser, Luft, Betrieb 9 (1965) 303-7

132. Hartinger, L. "Removal of heavy metals from effluent" Wasser u. Abwasserforschung 1 (1968) 30-40

133. Hartinger, L. "Chemistry of metal precipitation from effluent" Interfinish '68 Hannover. Conf. Proc. 279-85

134. Hartinger, L. "Effluent treatment in metal industry – heavy metal precipitation (copper, nickel)" Baender Bleche Rohre (1963) 638-47

135. Näsänen, R. "Solubility of Copper (II) hydroxide & basic Cu (II) salts" Acad. Sci. Fenn. A 59 (1943) (7) 3-15

136. Feitknecht, W. "Solubility product of copper oxide and hydroxide and their solubility in NaOH" Helv. chim. acta 27 (1944) 771-5

137. Hartinger, L. "Effluent treatment from the metal industries – precipitation of heavy metals (zinc, cadmium)" Baender Bleche Rohre (1965) 524-533

138. Feitknecht, W. "Reactions of solids in liquids. 1. Basic zinc salt" Helv. chim. acta 13 (1930) 22-43

139. Feitknecht, W., Häberli, E. "Solubility products of some zinc hydroxy compounds" Helv. chim. acta 33 (1950) 922-36

140. Feitknecht, W. "Structure of basic salts of divalent metals" Helv. chim. acta 16 (1933) 427-454

141. Kiessig, H. Reimers, H. "Reactions of zinc & cadmium sulphates with NaOH" Chem. Ztg-Chem. Apparatur 86 (1962) 488-9

142. Geiger, E., Schnieder, W. Studies of basic zinc sulphates" Chem. Ztg/Chem. Apparatur 89 (1965) 150-1

143. Feitknecht, W., Bucher, H. "Chemistry & morphology of the basic salts of divalent metals. XIII. Hydroxyfluorides of zinc" Helv. chim. acta 26 (1943) 2196-2204

144. Näsänen, R. "Potentiometric determination of solubility product of Ni(II) & Cd(II) hydroxides". Acad. Sci. Fenn. A 59 (2) (1943) 2-9

145. Feitknecht, W., Gerber, W. "Chemistry & morphology of the basic salts of divalent metals VI. Basic cadmium chloride" Helv. chim. acta 20 (1937) 1344-1372

146. Moeller, T., Thymer, P.W. "Observations on the precipitation of hydrous cadmium hydroxide in presence of anions" J. Phys. Chem. 46 (1942) 477-485

147. Feitknecht, W. "Precipitation of hydroxy-salts from cadmium solutions" Helv. chim. acta. 28 (1945) 1444-1454
148. Feitknecht, W., Reinmann, R. "Solubility product of cadmium hydroxychloride & hydroxide" Helv. chim. acta 34 (1951) 2255-2266
149. Hartinger, L. "Precipitation of lead from effluent" Metalloberflaeche 27 (1973) 157-9
150. Heinzelmann, K., Ostermann, A. "Separation of lead from plating effluents & rinsewaters" Galvanotechnik 64 (1973) 366-399
151. Eickershoff, P. "Treatment of lead-containing effluent" Galvanotechnik 73 (1982) 850-2
152. Hartinger, L. "Effluent treatment from the metal industries. Precipitation of heavy metals (iron, aluminium)." Draht Welt 53 (1967) 337-346
153. Guiter, H. "Hydrolysis of chlorides & sulphates of trivalent metals" C.R hebd. Seances Acad. Sci. 226 (1948) 1092-4
154. Van der Giessen, A.A. "Structure of iron (III) oxide hydrate gels" J. inorg. nucl. chem. 28 (1966) 2155-59
155. Feitknecht, W., Keller, G. "On the dark-green hydroxy compounds of iron" Z. anorg. chem. 262 (1950) 61-8
156. Cooper: Proc. Roy. Soc. B 124 (1937) 259
157. Evans, U.R., Pryor, M.J. "Solubility Product of Freshly Precipitated Ferric Hydroxide". J. chem. Soc. Suppl. Issue (1949) 157-160.
158. Feitknecht, W., Michel, K., Buser, H. Helv. chim. Acta 34 (1951) 119.
159. Morkowski, J., Thomaschk, G. "Comments on neutralisation of iron-containing acid effluent" Wasser, Luft, Betrieb (1961) 327-332
160. Jahr, F., Pernoll, I. Ber. Bunsenges. 69 (1965) 221-231
161. Jahr, F., Plaetschke, H. Naturwiss. 38 (1951) 302
162. Heyrovsky, J. "Electroaffinity of Aluminium - Part I - The Ionisation and Hydrolyses of Aluminium Chloride". J. chem. Soc. 117 (1920) 11-26.
163. Kruyt, H.R., Troelstra, S.A. Kolloid Beiheft 59 (1943) 262
164. Lacroix, S. Ann. Chim. (France) 4 (1949) 5
165. Szabo, Z.G., Csanyi, L.J., Kavai, M. Z. Analyt. Chem. 147 (1955) 401.
166. Hartinger, L. "Effluent treatment in the metal industries. Precipitation of heavy metals (chromium). Baender Bleche Rohre (1964) 14-21
167. Kovalenko, P.N. Ukrain. Khim. Zhurn. 22 (1956) 801
168. Hartinger, L. "In-house measures to minimise or avoid inorganic compounds in effluent from the metal industries" Gewaessersch., Wasser, Abwasser Vol. 59 (1983) 179-195
169. Hartinger, L. "Options for heavy metal removal from effluent" Korresp. Abwasser 33 (1986) 394-404
170. DOS 2,154,462 (10.5.71) (Nordstjernan)
171. Schlegel, H. "Nickel hydroxide precipitation from concentrated effluent" Galvanotechnik 57 (1966) 513-7
172. Hartinger, L. "Heavy metal removal from effluents in difficult conditions". Wasser, Luft Betrieb 11 (1967) 353-9
173. DAS 1,262,734 v. 1962. Wieland-Werke AG, Ulm.
174. Saracco, G.B., Prali-Gaglia, P., Genon, G. "Copper removal from effluents by precipitation" Oberflaeche-Surfaces 19 (1978) 249-253
175. Fischer, G. "Decreasing residues in pre-treatment plants" Galvanotechnik 76 (1985) 156-161

176. Rudzki, J. "Sulphide precipitation in effluent treatment" Galvanotechnik 75 (1984) 187-9

177. Hartinger, L. "Effluent & liquid wastes from metal Finishing" Galvanotechnik 77 (1986) 1814-17178. Hartinger, L. "Means for reducing heavy metal concentrations in plating shop effluents" Galvanotechnik 79 (1988) 401-7

179. OS 26 10 637 (15.9.77) Didier-Werke AG, Wiesbaden.

180. DE 36 32 138 Al (31.3.88). Siemens AG, Berlin-Munich.

181. Lauprecht, W.E., Barr, R., Fichte, R., Kuhn, M. "Molybdenum & its compounds" (in) Ullmans Encylopaedia of Chem. Technol.

182. Fichte, R., Retelsdorf, H.J., Ziehl, L. "Vanadium & its compounds" (in) Ullmans Encyclopaedia of Chem. Technol.

183. Lassner, E., Ortner, H., Fichte, R., Wolf, H.U. "Tungsten & its compounds" (in) Ullmans Encyclopaedia of Chem. Technol.

184. Gleu, K. "Complexed Compounds" (in) Ullmans Encyclopaedia of Chem. Technol.

185. Schwarzenbach, G., Flaschka, H. "Complexometric titrations" Enke Publishers, Stuttgart

186. Wagner, R. "Glossary (chelates, hydrolysis, complexes)" Water Calendar, E Schmidt, Berlin (1974)

187. Kober, F. "Principles of Complex Chemistry" Salle Publ. Frankfurt (1979)

188. Schlegel, H. "Complexed heavy metals in effluents" Galvanotechnik 63 (1972) 514-522

189. Hartinger, L. "Complex chemistry in effluent treatment technology" Vom Wasser 44 (1975) 69-117 Verlag Chemie. Publ.

190. Hartinger, L. "Metal complexes in effluent treatment technology" Galvanotechnik 66 (1975) 366-373

191. Doudoroff, P. "Experiments on Fish Toxicity of Complex Cyanides" Sewage & Industrial Wastes 28 (1956) 1020-1040

192. Milne, D; ibid. 22 (1950) 940

193. Kalani, D.K. "Stability constants of cyanides of Ni, Co, Ag" Dept. of Chemistry, Poona Univ. India

194. Hartinger, L. "Precipitation of zinc from cyanide-free bath" Oberflaechentechnik (1973) 441-9

195. Rodenkirchen, M. Removal of zinc from plating rinse water by simple neutralisation" Plating (1973) 698-700

196. Bjerrum, J., Schwarzenbach, G., Sillen G., "Stability Constants of Metal-ion Complexes, Part II, Inorganic Ligands."The Chemical Society, Burlington House, London, 1958.

197. BASF, Techn. Bull. "triethanolamine technology" (1972)

198. Stammler, M. "Metal complexes with triethanolamine" Metall 20 (1966) 841-5

199. Walz, D., Raub, Ch. "Electrodeposition of Pd-Ni alloy from ammonia-free solution with ethylene diamine as complexant" Metalloberflaeche 40 (1986) 199-203

200. Hall, J.L., Jones, F.R., Delchamps, C.E., McWilliams, C.W. "Complexes formed between Cu(II) ion & NNN'N'-tetrakis-(2-hydroxypropyl) ethylene diamine. J.A.C.S 79 (1957) 3361-4

201. Keyworth D.A. "Metal Complexes of NNN'N'-tetrakis- (2-hydroxpropyl)-ethylenediamine" Talanta 2 (1959) 383-384 Pergamon Press Ltd.

202. Rexolin Chemicals AB. Ringbinder, Helsingborg, 1967.

203. Geigy, Division of Geigy Chemical Corporation: Sequestrene, 1967.

204. BASF, Ludwigshafen "BASF Trade names of Trilon" (1971)

205. Guternatsch, H. "Behaviour of nitrilotriacetic acid in clarification & effluent treatment" Wasser/Abwasser 111 (1970) 511-6
206. Bernhardt, H. "Effect of NTA additions to detergents & cleaning compounds on aquatic life". ibid. 125 (1984) 49-56
207. Schoenberg, L.N. "Structure of complex copper species in electroless copper plating solutions". J. Electrochem. Soc. 118 (1971) 1571-1576.
208. Martell, A.E. "Stability Constants of Metal Ions Complexes Part II, Organic including macromolecular Ligands"Special Publication No. 25. The Chemical Society Burlington House London 1971.
209. Jansen, G., Tervoort, J. "Use of gluconates in electroplating" Galvanotechnik 75 (1984) 963-7
210. Anon. "Sodium Gluconate in cleaning processes & plating effluent" Product Fin. (1985) Dec. 17-20
211. Hartinger, L. "Chemical processes for metal removal from complexant solutions in p.c.b manufacture" Galvanotechnik 71 (1980) 797-805
212. Wing, R.E., Rayford, W.E. Doane, W.M. Metal Fin. 75 (1977) May 101-4
213. Wing, R.E., Doane, W.M. Rayford, W.E. "Treatment of rinse waters from electroless copper plating" Plating & Surface Fin. (1977) June 57-62
214. Wagner, R. "Laboratory studies on phosphate recovery from iron phosphates by complex & precipitation reactions" Vom Wasser 44 (1975) 131-172
215. Wing, R.E. Rayford, W.E., Doane, W.M. "Ferrous sulphate treatment for rinse waters from electroless copper plating" Plating & Surface Fin. (19770 Oct. 39-42
216. Wing, R.E., Rayford, W.E. "Use of spent pickle liquor in waste treatment" Metal Fin. (1978) March 31-3
217. Haarst, W.F.M. van, Mulder, R.J., Tervoort, J.L.J. "Removal of metals from gluconate complexes in effluent" Finish Digest (1974) 237-240
218. Ott, D. "Removal of heavy metals from gluconate containing effluents" Galvanotechnik 73 (1982) 339-344, 453-459
219. Srinivasan, B. "Waste treatment of metals in solutions containing gluconates" Plating & Surface Fin. (1984) Apr. 57-60
220. DOS 2,628,649 (26.6.75) Chem. Inst. Schaefer, Basel
221. Degussa AG, Frankfurt: "TMT 15 for heavy metal removal from effluent" (1980, 1987)
222. Stützel K. "TMT 15 for heavy metal removal from effluent" Galvanotechnik 79 (1988) 503-8
223. DOS 2,642,238 (23.3.78) Degussa
224. Wing, R.E. "Process for heavy metal removal from effluent" 32nd Purdue Industrial Waste Conf. Purdue Univ. (1977)
225. Dudenhausen, K.H. private communication
226. C.H. Erbslöh, "Import/export of chemicals & pharmaceuticals raw materials & minerals" Plexon Techn. Bull.
227. Dudenhausen, K.H. "Chemical treatment of effluent containing heavy metal complexes from a user's point of view" Galvanotechnik 73 (1982) 201-2
228. Wing, R.E., Rayford, W.E. "Heavy Metal Removal Using Dithiocarbamates". Plating and Surface Finishing (1982) Jan. 67-71.
229. Abels, C. "Effluent treatment methods" Metall 33 (1979) 961-2
230. Shipley GmbH, Stuttgart: Process guide WTA 2515 (1977)
231. Ehrich, H.J. "Electroless copper for p.c.b production" Galvanotechnik 68 (1977) 960-9

232. DE 25 47 562 C2 (16.7.87). Schering AG, Berlin

233. Hartinger, L. (in) "Effluent Teaching & Handbook" Vol. VII "Metalworking industries" 195-470 publ. Ernst & Sohn, Berlin

234. Fischer, G. "Effluent treatment & metal recovery from copper & alloy etch baths" Draht-Fachzeitschrift 19 (1968) Nov, 2-4

235. DOS 2,242,473 (29.8.72) Dart Industries Inc.,

236. Faul, W., Kastening, B. "Catalytic effluent treatment using electrocatalysts" Galvanotechnik 72 (1981) 808-812; also KFZ Juelich "Treatment of metal-ion containing effluent" June 1982.

237. Schlegel, H., Straub, E. "20 years of effluent treatment in WMF" Metalloberflaeche 36 (1982) 539-550

238. Johnson, J.W., Jiang, H.W., Hanna, S.B., James W.J "Anodic Oxidation of EDTA on Pt in acid sulphates" J. Electrochem. Soc. 119 (1972) 574-580

239. DOS 27 21 994 (12.10.78) BBC AG Brown, Boveri & Cie,

240. Müller, K.J., Bolch. T., Mertz, K. "Packed bed cell for destruction & recovery of organic complexants from effluent" Galvanotechnik 79 (1988) 172-6

241. Lauksminas, W.A., Lauberas, P.J. "Anodic oxidation of ethylene diamine & triethanolamine in dilute solution" Issled v obl. osazd. met. Wilna (1979) 243-6

242. Fabjan, Ch., Marschall, D.K. "Ozone for destruction of complexants in metal finishing & other effluents" Galvanotechnik 67 (1976) 643-5

243. Shambaugh, R.L., Melnyk, P.B. "Removal of heavy metals via ozonisation" J. WPCF 50 (1978) 113-121

243a. Gilbert, E., Hoffmann-Glewe, S. "Ozonisation of NTA in presence of inorganic & organic species found in water" Vom Wasser 62 (1984) 11-23

244. Bober, T.W., Dagon, T.J. "Ozonisation of photographic processing wastes" Journal WPCF 47 (1975) 2114-2129.

245. Ott, D. "Chemical & photochemical oxidation processes for removal of organic residues from effluents" Wasser Kalender Publ. E Schmidt, Berlin (1989)

246. Götzelmann, G. unpubl. work

247. Hartinger, L., Gülbas, M., Scheiffelen, B. unpublished work under BMFT Contract No. 01ZH87116

248. Kauczor, H.W. "Concentrating metals from aqueous solution with solid ion-exchange membranes" Erzmetall 22 (1969) B19- B22.

249. Frick, W. "Treating heavy metal complexes with primary aliphatic diamines" Metalloberflaeche 37 (1983) 55-7

250. Goba Kumar, K., Pavithram, C., Rohatgi, P.K. "Recovery of Tartrate from Electroless Copper Baths" Plating and Surface Finishing (1981) July 70-71.

251. Fraust, C. "Modifying a conventional chemical effluent treatment plant to deal with fluorides & ammonia" Plating & Surface Fin. 62 (1975) 1048-52

252. Czukrasz, G., Hompasz, G. "Treatment of fluoride containing effluent" Galvanotechnik 72 (1981) 832-5

253. DIN 4030 "Estimating concrete-attacking water, soils & gases"

254. Götzelmann, W., Hartinger, L. "Sulphate problems for indirect effluent discharges" Galvanotechnik 71 (1980) 699- 705

255. Morkowski, J., Thomaschk, G. "Neutralisation of iron- containing acid effluents – Pt 1." Wasser Luft Betrieb 5 (1961) 292-4.

256. Lancy, L.E. "Plants for direct effluent treatment" (in) Hartinger "Handbook of Effluent Treatment, Pt. 2, Technology" 114-126 Hanser Verlag, Munich

257. Karnovsky, F. "How can cement-clad materials of construction be protected from corrosion ?" Abwassertechnik 3 (1979) 16-20

258. Christoe, J.R. "Removal of sulfate from industrial wastewaters" Journal of Water Pollution Control Federation 48 (1976) 2804-2808.

259. Keller, H.G., Mehlmann, M., Peschen, N. "Sulphates in acid effluent – concentration reduction by adding lime or calcium aluminate" Galvanotechnik 78 (1987) 392-6

260. EP 250 626 Al, (30.6.86) Walhalla Lime, Regensburg

261. Lohmeyer, S. "Phosphates from surface treatment in effluent" Galvanotechnik 65 (1974) 192-204

262. Kandler, J. "Chemical precipitation" Proc. Seminar "Chemistry & Technology of physico-chemical effluent treatment" TA Wuppertal May 1973.

263. Fischer, E., Merkenich, K., Kandler, J. "Trials to remove phosphates from municipal effluent & detergents" Seifen, Oele, Fette, Wachse 99 (1973) 211-8

264. Pöpel, J. "The third treatment stage - problems & solutions" Wasser Kalender (1969) Publ. Schmidt, Berlin.

265. Dahlqvist, K.I., Hall, L., Bergman, L. "Phosphate removal using divalent iron" Wasser, Luft Betrieb 20 (1976) 107-112

266. ATV-Fachausschuss 2.8: "Phosphate removal" Korresp. Abwasswer 30 (1983) 191-8

267. Schwarzer, H. "Oxidative effluent treatment with hydrogen peroxide" Wasser/Abwasser 129 (1988) 484-491

268. Degussa AG, Frankfurt: "Removal of hydrogen sulphide from exhaust gas with hydrogen peroxide". Umwelt & Degussa (1981)

269. Degussa AG, Frankfurt: "Removal of chlorine & hypochlorite from chlor-alkali electrolysis effluent with hydrogen peroxide before mercury removal with ion exchangers" Umwelt & Degussa (1982)

270. Degussa AG, Frankfurt: "Hydrogen peroxide for hypochlorite removal" Umwelt & Degussa (1980)

271. Bischoff, M. "Ammonia Stripping" Diploma Thesis, Fachhochschule Aalen (1981)

272. Gebrüder Sulzer Aktiengesellschaft: "Cost-effective removal of ammonia from effluent" Metalloberflaeche 39 (1985) 9

273. Gebrüder Sulzer Aktiengesellschaft: "Sulzer column packings for water & effluent treatment" Techn. Bull (1984)

274. DOS 26 04 479 (19.8.76) Sterling Drug Inc. New York, NY.

275. DP 26 24 256 (29.5.76). Degussa, Frankfurt.

276. DGO Working Group on Cleaning. "Planning, implementation & operation of surface treatment plants; volatile chlorocarbons" Galvanotechnik 79 (1988) 3280-6

277. Dürst, J. "Modern methods for recovery of chlorinated hydrocarbons using corrosion-free plant made of plastic" Galvanotechnik 78 (1987) 800-2

278. Nagel, R. "Cost-effective processes for avoiding or treating chlorocarbons in plating shop effluent". Proc. Symp. "Problems in Plating Shop Effluent Treatment" LGA Nuremberg 22.11.88

279. Buss, G. "Treating effluent, exhaust air & solid wastes where chlorinated solvents are used in metal working & finishing industries" Galvanotechnik 70 (1979) 514-524

280. Bauer, I. "Practical experiences in treatment of chlorocarbon containing effluents" Schiftenreihe Praxis Forum (1988)

281. Koth, D. "Activated carbon for treatment of effluent containing chlorinated solvents" GIT Suppl. (1986) (5) 17-22

282. Z-Design, Dipl. Ing. Werner Zyla GmbH. Techn. Bull.

283. Lohmeyer, S. "The Overall AOX parameter in effluent from the metal-working industries" Galvanotechnik 78 (1987) 678-686

284. Stiefel, R. "Halocarbons in industry and the environment" Metalloberflaeche 41 (1987) 109-113

285. Heffels, H.W. "Recycling organic solvents in cleaning & degreasing of metals" Galvanotechnik 69 (1978) 704-714

286. Becherer, E. "AOX reduction in effluent" Galvanotechnik 80 (1989) 4370-4372

287. Staudinger, F. "Some experiences in newly developed plants for halocarbon cleaning" Metalloberflaeche 43 (1989) 105-7

287a. Anon. "Volatile chlorocarbons" Galvanotechnik 79 (1988) 3686-3690

288. Khurshid Ahmad Khan: "Chemical Actinometry of High Intensity 185 nm-200 nm Ultraviolet Radiation" Project Thesis, Univ. of Salford, UK.

289. Peleg, M. "Chemistry of Ozone in the.Treatment of Water". Water Research 10 (1976) 361-365, Pergamon Press, 1976.

290. Bandemer, T. "Raw water treatment with u.v irradiation & hydrogen peroxide" Wasser + Boden 4 (1988) 194-8

291. Verfahrensberichte des Verbandes der Chemischen Industrie. 6th Report "Effluent treatment by wet oxidation" (1976)

292. Randall, T.R. "Wet Oxidation of Toxic and Hazardous Compounds", Zimpro Inc. Technical Bulletin I-610, 1981.

293. Bender, H., Reinartz, D. "Air stripping to treat water with organic contaminants" Galvanotechnik 75 (1984) 958-962

294. Weigand, N., Lieberherr, B. "Phenol removal" cav (1985) Nov. 145 & 150

295. Liebmann, H. "Handbook of freshwater & effluent biology" Vol. 2. Publ. Oldembourg, Munich

296. McCarty, P.L."Anaerobic Waste Treatment Fundamentals, Part Three, Toxic Materials and their Control". Public Works, 1969.

297. 44. Bericht der ARW Arbeitsgemeinschaft Rhein-Wasserwerke eV. (1987) DVGW Forschungstelle, Engler Bunte Institute, Univ. of Karlsruhe

298. Livingstone, D.A. "Chemical Compositions of Rivers and Lakes" (in) Fleischer, M. "Data of Geochemistry", 6th ed. U.S. Ged. Surf. Prof. Paper, 400-G, 1963.

299. Belouschek, P., Weiler, W. "Determination of directly removable volatiles – differential analyses of lipophilic volatiles in waters" Korresp. Abwasser 30 (1983) 635-641

300. DIN 1999 (Pts 1 & 2) "Separators for volatile liquids - gasoline, petrol, heating & diesel oils"

301. Hartinger, L. "Problems with effluent sludges" Galvanotechnik 64 (1973) 582-7

302. DE 31 17 513 Al (18.11.82) AGS. Augsburg

303. France, P., Fuka, T. "Separation of colloidal silicates from effluent by coagulation" Galvanotechnik 76 (1985) 695-8

304. Verwey, E.J.W., Overbeek, J.T.G. "Theory of Stability of Lipophobic Colloids". Elsevier, New York, 1948.

305. Sontheimer, H. "Basics of chemical effluent treatment" IWL Forum 64 Vol. 2 132-147 Cologne (1964)

306. Ballnus, W. "Breaking of drilling oil emulsions" Wasser, Luft Betrieb (1965) 518-520

307. Degussa, Frankfurt: "B20 Separating Agent" Tech. Bull. (1968)
308. Knobloch, H. "Break emulsions, save costs - treating oil emulsions with separating agents based on kieselguhr" Maschinenmarkt 78 (1972) Vol. 45
309. Lohmeyer, S. "Emulsion breaking, Pt 2" Galvanotechnik 64 (1973) 795-805
310. Rathgeber, F. "Plant & processes for treating oily effluent" Wasser, Luft, Betrieb (1973) 69-75
311. Bradke, H.J. "Plant & processes for emulsion breaking & for oil & grease separation" Metall 33. (1979) 267-272
312. Wilke, H. "Emulsion breaking with filtration" CAV (1970) April, 89-90
313. Pohl, K.M. "Breaking industrial oil-in-water emulsions with acids & temperature increase". Munich Reports on effluent, freshwater & river biology 18 (1977) 273-6 Publ. Oldembourg Munich & Vienna
314. Gütling, W. "Processes for breaking oil emulsions & oily effluents" Galvanotechnik 65 (1974) 417-424
315. Köhler, R. "Technology & applications of compressed air flotation in effluent treatment" Wasser, Luft Betrieb 19 (1975) 72-6, 104-9
316. Baer, E.H. "Treatment of oil-emulsions in effluent" Munich Reports on Effluent, Fisheries & River biology" 18 (1977) 265-271 Publ. Oldembourg, Munich & Vienna
317. Christ, W., Cyprian, J., Gasser, A. "New processes for separating used oil emulsions & oily effluents" Wasser, Luft, Betrieb 12 (1968) 755-7.

4   Technology

4.1   EFFECT OF LEGISLATION ON CONVENTIONAL PROCESS TECHNOLOGY

Conventional process technologies in waste water treatment involve both continuous and batch treatment processes. Toxic dipping processes are discussed in Section 6.3.5, as these nowadays are linked more with recycling and rinsing procedures, and have a bearing on waste water treatment proper only in special cases.

The most important effect of recent legislation will be the increasing use of batch treatment at the expense ofcontinuous processing, wherever this is possible. Legislation trends demand that waste water be discharged only if steps have been taken to prevent losses from the process baths and to achieve economy in water consumption, by the use of cascade rinses, ion-exchange circulation plants and other appropriate means. Thus it is that the tighter demands of new legislation can often be met only by means of batch treatments.

Rearranging a factory to accommodate batch-wise waste water treatment presupposes as a rule that water saving measures are in place. This, on the other hand, may depend on alterations made in the course of relocating the production units. A continuous treatment operation, should not be lightly abandoned even if the water consumption is dramatically reduced. More will be said about this later.

In terms of investment it is not too difficult to attain the new and stricter limits on metals. Listed below are suggested modifications, additions and repositioning of equipment, divided into steps to be taken immediately and those to be carried out in the longer term [3,4].

The immediate measures include:

- Conversion of hydroxide to sulphide precipitation (see Tables 3-17). Less soluble metal compounds are thus obtained, enabling the 0.5-mg/l limit, specified for most metals, to be met with greater confidence.

- Post-filtration of waste waters already treated. Any metal excess can generally be traced back to the incomplete removal of metal bearing solid material (see Section 4.4.1).

- Connecting selective ion-exchange units on the outlet side to remove the remaining metal cations and cationic metal complexes (see Section 4.4.2).

- Modifications of this kind can be carried out from within a few days (sulphide precipitation only; as a batch operation!) up to two years.

These measures will be cost-effective if water saving and material economy measures become important and recycling processes are involved, for example:

- Multiple utilisation of rinse waters by circulation through ion exchange units or by cascade rinsing or combinations of these (see Sections 5.3 and 6.3.2). These processes require different alterations to be made and new production equipment to be installed. Normally, they offer the opportunity to introduce batch process systems.

- Construction of batch process plant. In difficult cases, where there is an accumulation of effluents containing high concentrations of complex formers or where expensive precipitating agents must be used, a two-stage plant may be necessary, if circumstances permit (see Section 4.3.3).

- Installation of recycling plant (see Section 7).

This may also require additional or new structural installations. Their implementation can take several years.

In summary, it may be said that new legislation is leading to a change-over in waste water treatment to batch processing and recycling of water, materials and process solutions. Both continuous and batch processes continue to play their roles in effluent treatment technology.

This Section should not be concluded without referring to the conflicts that arise from the legal requirements and their practical consequences. Water saving always leads to less waste water, which however is correspondingly more concentrated. All precipitable species (metals, sulphates fluorides phosphates) readily allow a considerable reduction of TDS. Concentrations of remaining, species are controllable only if additional treatments (of metals) are carried out (see Section 3.7.7.1 and 5.3.4). The breakdown of COD − ammoniacal substances takes place biologically in the municipal sewage plants.

Regarding the problem of AOX − see Section 3.2.1.7. As soon as Caro's

Acid ($KHSO_5$) is available at an acceptable price, it will be possible to prevent increasing AOX formation during cyanide detoxification, and as a rule to maintain a value of 1 mg/litre AOX.

4.2   CONTINUOUS TREATMENT PLANTS

4.2.1   Characteristics

Continuous treatment plants are used where dilute waste water in large quantities, being generated continuously, is to be chemically processed and separated to yield solid phases by appropriate chemical treatment. In line with Section 4.1, however, this should take place only if re-circulation using ion exchange or other water saving measures, which permit the operation of a batch treatment plant, are not possible, and the effluent contains no dangerous substances. But exceptions prove the rule here as elsewhere. A continuous process plant is one in which the same volume flow that enters it untreated leaves it again after treatment at the end of the plant. (The volume of added chemical solutions is normally negligible.) In view of the reaction times, relatively small reactors can be employed for chemical reactions. For this reason continuous process plant has an advantage in relation to detoxification and neutralisation. The thorough mixing of the effluent with the treatment chemicals in the reactor should take place as quickly as possible. That is a prerequisite for the rapid completion of the chemical reactions, but some preconditions concerning the chemical reaction itself must also be stated, if continuous reactions are used.

- High rate of reaction.

- Reaction equilibrium tilted to a large extent towards non-toxic end products, even when there are present only very small excesses of reactant chemicals.

- Ability to measure at least one of the reactant and end products, respectively.

Thus the time within which a reaction is completed is dependent on two parameters: a technical factor, the mixing time, and a kinetic factor, the chemical reaction time. The latter depends on the choice of the reaction employed. Acceleration of the reaction by catalysts or by raising the temperature is not practicable in continuously operating plants, since the reaction parameters are adversely affected. Mixing must be completed as quickly as possible; and indeed, the smaller the reactor, for a given volume flow, the faster this occurs. The effectiveness mixing equipment (stirrer, pump) must be very high in relation to the volume flow-rate of waste water to be treated continuously.

The reactor volume for an average theoretical dwelltime τ depends on the chemical reaction time, in accordance with the relationship shown in Equation 4.1.

$$\tau = \frac{V_N}{Q} \qquad\qquad 4.1$$

VN effective volume of the reactor (l. m³); Q Volume flow rate of the effluent where (l. s⁻¹ m³ min⁻¹, m³ h⁻¹).

In normal continuous process reactors the effective volume should be turned over 90 to 120 times per hour (corresponding to 1.5 to 2 times per minute). Not only is this important in relation to the rate of completion of the reaction, but it is also a control engineering precondition for automatic operation.

The mechanical processes of phase separation (precipitation, sedimentation) are a different matter. Turbulence, such as are generated by the mixing equipment, would disturb the upwards and downwards movements of the materials to be separated, and should be avoided entirely. If there is no disturbing turbulence present, the phases to be separated have a tendency to move upwards or downwards at a rate indicated by Equations 3.222 and 3.132, respectively. They may also, however, be disturbed by the flow of effluent in the separation tanks, whether in a horizontal, inclined or counter-flow direction. Establishing the theoretical residence time of waste water in these separation tanks calls for considerable experience on the part of the plant designer. Calculation of the settling area (see Section 4.2.5.1) and empirical determination of precipitation and sedimentation rates, respectively, for the particular case provide help in fixing the lay-out of the installation.

Theoretical models have been suggested for the steps outlined in exact terms. The first goes beyond the case of a continuous process reactor with turbulence and corresponds to an ideal stirred reactor [5,6]. Complete mixing in the reactor is assumed to occurr instantly. A substance, which enters at concentration c_0, immediately acquires the concentration c_0, as an inevitable consequence of dilution by the existing contents of the reactor. The solution leaves the reactor with concentration c_0.

The second model corresponds to separating tanks without turbulence. It is described as ideal pipe flow [7,8,9]. The water passes through the pipe, not with turbulent, but with laminar flow, and with diffusion hindered by exclusion of axial mixing (plug flow). Therefore only the main flow brings about mass transport.

By submitting both ideal reactors to two theoretical tests, it is possible to evaluate residence time behaviour.

- One injects in to the feed, as a single dose, a quantity of a substance, with concentration c_0 at the inlet (pulse marking) and with concentration c at the outlet; one can then obtain the residence time spectrum.

- One doses the feed constantly at the point of entry with a solution of concentration c_0 (displacement marking); then the value of the concentration at the outlet yields the residence time function.

The characteristics of the ideal reactors with residence time spectrum and residence time function are displayed in Fig. 4.1. The behaviour of all real reactors, such as those used in practice, lies between the limiting behaviours of the two ideal reactors. Examples of these are the stirred tank cascade, the static mixer and the separating tank operating without artificial turbulence. Their residence time spectra and residence time functions are also illustrated in Fig. 4.1, showing clearly that they lie between those of the two ideal reactors. By determining the outlet concentrations as a function of time it is possible to assess them practically in terms of the pulse-marking and displacement-marking theoretical tests. The behaviour of the stirred tank cascade with multiple compartments comes close to that of the ideal pipe flow model, as does the behaviour of the static mixer, which can be thought of as a sequence of many stirred tank cascades [10]. The static mixer is, however, suitable only for spontaneous reactions. The fast reactor with highly turbulent continuous mixing and very low reactor capacity leads to a similar residence time behaviour (see Section 4.2.4.5).

For phase separation it is necessary to strive for the plug flow which is a feature of the ideal pipe flow model, but practical results may deviate greatly from this [7,8]. The reasons for this are unavoidable turbulences, which arise from the devices needed to equalise flow over the large cross-section of the tanks, localised turbulences, which bring about upward or downward particle movement at their edges, as well as convection currents, which arise from temperature differences in reactors, often located out of doors.

Concentration changes at the inlet end of continuous process plant, provided they do not reach excessively high levels, can be counteracted by control technology. Increases in volume flow rate must never exceed the throughput capacity of the reactor. Jerky, spasmodic changes are to be avoided, as they increase turbulence in the phase separation process.

4.2.2 Waste water separation

As seen in Chapter 3, the different substances contained in waste water, and especially their various toxic anionic components, must be subjected to a variety of chemical treatments. So far as metal precipitation is concerned this is normally not difficult, so long as no complexing agents are present The following kinds of waste water should be kept separate and diverted to designated equipment:

- Cyanide-containing effluent to the cyanide detoxification unit.
- Chromate-containing effluent to the chromate detoxification unit.
- Nitrite-containing effluent to the nitrite detoxification unit.

- Acid or alkaline and metal-containing, but complexant-free, effluent to the neutralisation and precipitation unit. The effluents listed earlier also come to this plant after detoxification if they contain metals. The plant is connected to the solid separation unit and observation/sampling sump.

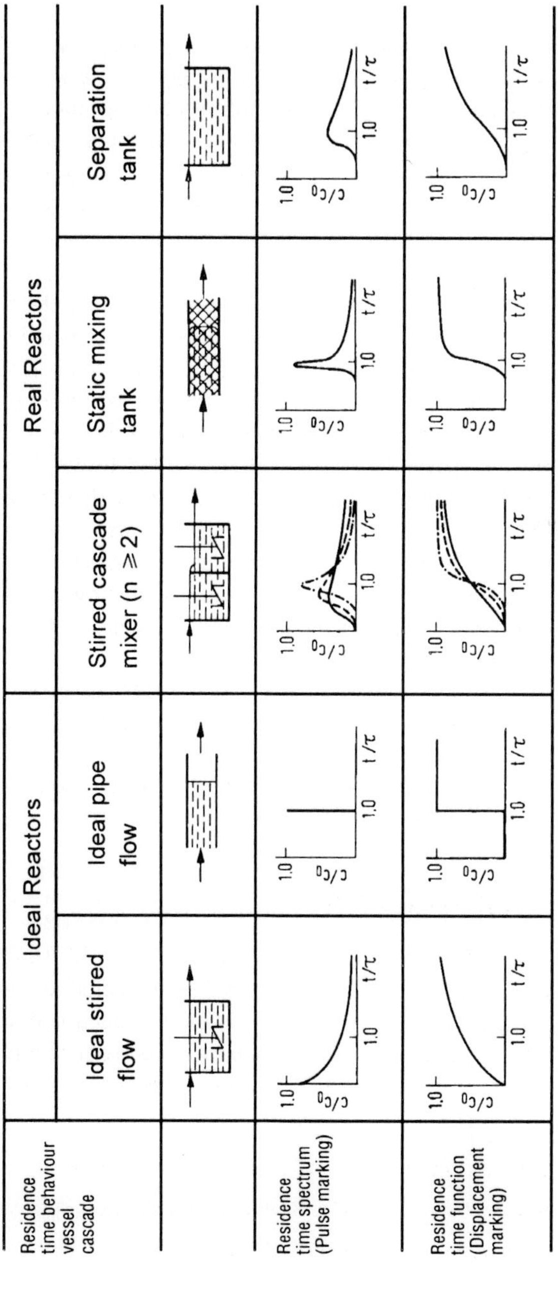

Fig. 4.1 Schematic representation of the residence time behaviour of ideal and real reactors

τ average theoretical residence time

t measured residence time

- Cadmium-containing effluent. This has to be subjected to a complete treatment sequence detoxification neutralisation and precipitation, solid material separation (with independent sludge drainage) and can be led over a separate inspection sump to the general effluent discharge sump. The treatment should, however, be carried out in a batch process unit.

Waste water containing complexing agents should be dealt with in a batch treatment plant. The same goes for effluent which contains containing volatile hydrocarbons. This is treated separately in line with Section 3.11.3. So far as oil- or grease-containing waste is concerned, this is passed through a skimmer and then an emulsion breaking unit, as far as is necessary; except for flotation equipment, these units should always operate on a batch basis. If the aqueous phase contains cyanide, chromate, nitrite or metals, it should be passed to the appropriate plant for their treatment.

If unusable process solutions (concentrates) are discarded they should be passed into the storage corresponding to·the relevant separation as listed above, and from there through a batch treatment plant. After treatment, it is recommended that metered volumes be supplied to the appropriate continuous process detoxification plant. State-of-the-art technology no longer permits the direct dosing of continuous process plants with untreated concentrates. In some countries this has been forbidden for a long time. Fig. 4.2 shows the flow of waste water through a continuous treatment process. It is to be expected that the operation of continuous process plants for detoxification purposes will be approved only in very special cases.

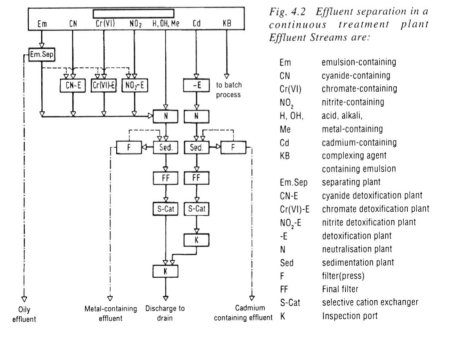

Fig. 4.2 Effluent separation in a continuous treatment plant Effluent Streams are:

Em	emulsion-containing
CN	cyanide-containing
Cr(VI)	chromate-containing
NO$_2$	nitrite-containing
H, OH,	acid, alkali,
Me	metal-containing
Cd	cadmium-containing
KB	complexing agent containing emulsion
Em.Sep	separating plant
CN-E	cyanide detoxification plant
Cr(VI)-E	chromate detoxification plant
NO$_2$-E	nitrite detoxification plant
-E	detoxification plant
N	neutralisation plant
Sed	sedimentation plant
F	filter(press)
FF	Final filter
S-Cat	selective cation exchanger
K	Inspection port

Gas-scrubbing water and floor waste water (effluent from cleaning treatments, emptying and filling operations, leakages, drips etc, as well as the waste rinse water and concentrates, must also be led away for treatment.

Fig.s 4.3 and 4.4 show examples of waste water separation and gullies.

Fig. 4.3   Corrosion resistant floor with gullies for waste water separation (KERAMCHEMIE GmbH, Siersahn/Westerwald)

Fig. 4.4   Waste water separation pipework of an automatic plating plant (LPW-Galvanotechnik GmbH, Neuss)

In connection with the floor waste water, it should be mentioned that floor drains connected directly to the sewerage system are forbidden in wet-process shops. These drains must be linked with the corresponding concentrate storage units, with their respective reservoirs, all being equipped with protection against overflowing. In addition, effluent accidentally spilled on the ground floors, must not find its way either into the sewer or into the soil. The areas occupied by tanks or the entire floor space must be chemically resistant and have bunds to prevent escape of liquid. The capacities of the bunds must be sufficient to accommodate the entire contents of the tanks installed.

4.2.3   Construction of continuous treatment plants for detoxification and neutralisation/precipitation [1-16]

Continuous treatment plants are used mainly when waste water to be neutralised, is generated in large amounts and contains no dangerous substances, and where it is not, or only with difficulty, feasible to remove unwanted materials from circulating rinse water by exchange methods, owing to their concentration being too high. Examples are:

- Pickling of iron or steel.

- Pickling, etching and chemical machining of aluminium.

- Anodising effluent (sometimes).

- Waste water that must be made strongly alkaline to precipitate metals after separation of solid materials.

In special cases, however, continuous treatment of waste waters containing dangerous substances may be used, examples being:

- Chromating, if post-chromating static rinse baths and cascade rinsing systems are used (waste water constituents: zinc, trivalent chromium, chromate).

- Dipping non-ferrous metals in solutions containing nitric acid (waste water constituents: copper, zinc, nitrite).

- Etching of alloy steels (waste water constituents: nickel, chromium, molybdenum, vanadium, tungsten, copper).

- Anodic colouring of aluminium surfaces (waste water constituents: zinc, nickel, cobalt, copper etc).

Continuous treatment plants for detoxification, neutralisation and precipitation are generally built-up from items of equipment described in Section 4.2.3.1.

4.2.3.1   Reactors

Reactors can be made of steel with appropriate linings (hard rubber, PVC and other plastics) or wholly out of plastic (polyethylene, polypropylene, PVC) with the necessary outer steel reinforcement, which may sometimes be protected by a plastic coating. Other continuous process reactors consist of concrete with chemically resistant linings (fillers, ceramic segments) and may be built under or above ground. Their disadvantage is that leakages are hard to detect. They may be round or rectangular in plan, so long as the ratios of diameter to height of liquid or the length and breadth to the height of liquid do not differ very much from 1:1. They contain components such as baffles, immersion pipes, dividing walls, which prevent the waste water from taking the shortest path between the inlet and the outlet. The outlet is as a rule suitable as an overflow, so that the reactor remains permanently full. The inlet is arranged at somewhat over half the height of the water level.

Fig. 4.5 shows one option for the construction of continuous process reactors. Their usable capacity must be greater than is needed to provide the residence time required for the chemical reaction and to accommodate the volume flow rate that is to pass through.

In detoxification plants account has always to be taken of any gas evolution. Reactors should therefore in general be kept enclosed and connected to an extraction system. If these gases are subject to legislation, the associated extracted air must be scrubbed. The scrubbed liquid should be treated as effluent, according to its degree of saturation. Reagents which, occur, and which can be continously analysed, in continuous process plants, include:

- Cyanogen chloride and chlorine in cyanide detoxification.

- Sulphur dioxide in chromate detoxification.

- Nitric oxide in nitrite detoxification.

The formation of cyanogen chloride can largely be avoided by control of the reaction. The remaining gases can be scrubbed out close with caustic soda, and nitric oxide, also with weakly acidic solutions of aminosulphonic acid. In precipitating metals with sulphides care should be taken with hydrogen sulphide evolution as well, especially as these precipitations are carried out in batch treatment plants.

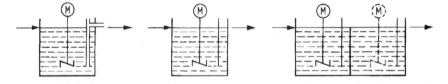

Fig. 4.5   Basic construction of continuous process reactors (diagrammatic)

4.2.3.2 Mixers [17,18]

In continuous treatment plants, stirrers are used as mixers. They must be able to churn the contents of the compartment twice or at least 1.5 times per minute. The number of revolutions per minute in unimportant in relation to this. In this type of plant, however, a mixer with 1500 rev/min is usually installed. Its shaft should be immersed to between two-thirds and three-quarters of the depth of the liquid.

4.2.3.3 Measuring and control equipment [19-23]

Detoxification reactions call for precise pH control, using dosing alkali or acid and must be held constant during the reaction. Measurement of pH is with a robust pH assembly, comprising a combination of glass and reference electrodes. The relationship between of gradient and 59 mV/pH unit potential and the pH value is therefore a straight line (see Fig. 4.6, solid line). pH-meters suffer from two errors that can cause the pH value to diverge from this straight line. The first is the so-called asymmetry potential error. Each glass electrode includes a specification for the pH value of its internal buffer. If the pH measuring assembly is immersed in a buffer solution having the same pH value, then the cell voltage will amount to 0 mV. As a rule the pH value of the internal buffer is near to 7.0.

If a pH measuring assembly is dipped in a buffer solution at pH 7.0, however, a small potential is usually recorded, so that the indicated pH differs from 7.0. This asymmetry potential error is represented in Fig. 4.6 by the broken straight line. If this is compensated for on the potentiometer of the measuring instrument so that it becomes zero, then the line shifts to the parallel dot-and-dash line which intersects the X-axis at pH 7.0 (zero point adjustment).

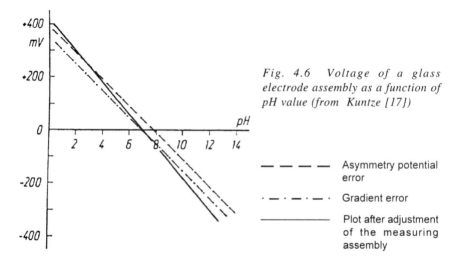

Fig. 4.6 Voltage of a glass electrode assembly as a function of pH value (from Kuntze [17])

— — — — — Asymmetry potential error

· — · — · — · Gradient error

—————— Plot after adjustment of the measuring assembly

The second error is the gradient error. If a measuring assembly calibrated at pH 7.0 is immersed in a buffer solution, whose pH value is as far removed as possible from 7.0, then the reading deviates somewhat more from the theoretical value of 59.15 mV per pH unit. The pH value on a second measuring potentiometer is then set to the known value of the second buffer solution. Thus the dot-and-dash line is appropriately rotated around the zero point at pH 7.0 until it coincides with the theoretically correct solid straight line. Any increase in the gradient always corresponds to an increase in the sensitivity of the measuring electrode.

Regular calibration of electrodes is necessary. In continuous process treatment plants it should be done weekly, in critical cases more frequently.

The measurement of pH also shows a temperature dependency, which is corrected for the temperature of the waste water by third potentiometer. The temperature error can likewise be automatically compensated. pH measurement in continuous process plants normally takes place at between 10 and 20°C, depending on the location and weather. If the temperatures of buffer solutions for which the adjustment is to be made lie in this range, then a single temperature correction can be relied on, as in most cases measurement errors are negligible [24].

At high pH's the alkali error, must be taken into account. If, however, so-called high-alkali electrodes are used, in which the sodium oxide in the glass membrane is replaced by lithium oxide, then it is possible to make correct measurements of pH up to 13 [19].

Detoxification reactions are measured at constant pH values by redox potential measurement using metal electrodes. For reasons of corrosion resistance the electrodes used are precious metals such as platinum, gold and silver, with the latter also in amalgamated form. The reference electrode, is usually silver-silver chloride and calomel electrodes. If possible, measurements not only of the potential of the substance to be detoxified but also that of the detoxifying agent should make a contribution to the redox potential. The two single potentials are thus set one against the other so as to achieve in these measurements relatively large potential changes at the equivalence point of the reactions. This is a prerequisite in continuous process plants as well, as in addition to pH-value fluctuations, other redox systems, such as Fe^{++}/Fe^{+++}, Cu^+/Cu^{++}. Ni^{++}/Ni^{+++}, and the concentrations at which they are present, can interfere. Using established methods of redox measurement, determinations can be made of the redox systems CN^-(concentration dependent)/OCl^-, CN^-/HSO_5^-, $Cr_2O_7^{--}/HSO_3^-$ and NO_2^-/OCl^-. The sudden change from reduction to oxidation potential and vice versa takes place spontaneously and, except in the case of cyanide, is not concentration-dependent (see Fig. 4.7). An excess of reducing agent gives rise to potentials in the more negative voltage region, and an excess of oxidising agent to those in the more positive region. Because of possible potential shifts due to other solution constituents the potential differences for detoxification in continuous process plants must be at least around 200 mV.

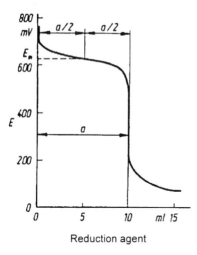

Reduction agent

Fig. 4.7   Change in potential during the detoxification of a chromate-containing solution with sodium hydrogen sulphite solution at pH = 2 with a platinum/silver-silver chloride measurement system (from Kuntze [20]). At a/2 where the reduction agent is half consumed, half the chromium is present in trivalent and half in hexavalent form

Calibration of the redox potential assembly is not carried out with solutions of various concentrations, as for pH measurement assemblies at various hydrogen ion concentrations, but set to the two redox potentials of the corresponding solutions at the start and after completion of the reaction, the one of which contains a small excess of oxidising agent and the other a small excess of reducing agent. Both solutions must be maintained at the pH at which the detoxification reaction takes place.

The electrode assemblies for pH-value and redox potential measurement are installed in fittings permanently immersed in the solutions to be measured. To keep the electrodes active they must be kept clean. Deposits of lime or hydroxides are removed by a periodic acid purge [25]. Gold electrodes are cleaned with motor-driven brushes or, better, kept bright by scratching with a hard abrasive material.

Other potential measurements, eg with ion specific electrodes, are rare in continuous process plants, as test samples have to be conditioned before measurement. Fluoride measurement by automatic analysis is rare. Larger buffer reservoirs must be incorporated in the treatment plant, so that too great a change in concentration does not occur, due to the time elapsing between the withdrawal of the sample and the availability of the measurement results. Optical methods are also used in automatic analysers. Previously conditioned samples are introduced to a photometer previously calibrated for the appropriate measurement range, which converts absorbances into concentrations. These optical methods are mainly used for final control measurements.

The output voltage of the high-impedance pH and redox measurement assemblies is fed to amplifier, (strictly an impedance transformer) which converts the high impedance of the measurement assembly, between 100 and 500 MΩ, to a few ohms. This is achieved on an input current of a few picoamperes, which is amplified to feed 0-20 or 4-20 mA loops [26]. These operate indicating instruments, regulators, recorders, alarm systems, etc.

Previously in effluent controllers, only on-off control devices were used which, in regulating neutralisation, frequently allowed the set pH values to overshoot in one direction then the other. At first this was checked by retarding the dosing using time switches. Today it is usual to work with quasi-proportional controllers. These control the alkali or acid dosing as a function of the deviation of the measured value from the desired value. The larger the deviation is, the longer a solenoid valve is opened and the shorter are the intervening pauses and vice versa. This system is very suitable for small to medium-sized plants and is widely used. For control of redox reactions, however, as a proportional set-up is not possible (see Fig. 4.5), the on-off controller continues to be used.

For rather more difficult tasks and especially in larger neutralisation plants, continuous regulators are installed, which constantly and proportionally convert the measured value deviation into the dosage rate (proportional control). These use servo-motors to vary the flap opening of valves or else adjust the size of stroke or number of strokes of the dosing pumps. Proportional-integral-control regulators are installed for difficult tasks most often in larger continuously operating plants. A further refinement incorporates supplementary differential responses in the regulators (proportional-integral-differential-control).

Developments in measurement technology now make continuously operating processes safer. These should therefore find application wherever it is impossible to avoid the operation of continuous processes in the presence of dangerous substances.

Among these, one of the most important is double measurement, by which the uncertainty resulting from failure of a measurement system or drift in the measured values of a system can be eliminated. For this, using conventional control technology, two complete sets of measuring equipment are required, plus an additional apparatus, consisting of difference measurement amplifier, switch amplifier for difference analysis and difference indicator [27,28]. Double measurement is possible for the measurement of both pH and redox potential. Microprocessor-controlled systems, comprising two measurement elements connected to a single compact unit, simplify this double measurement [24,28]. Automatic calibration of measurement assemblies is also possible nowadays. Microprocessor controlled continuous regulators use software, which allows interchange of the computing functions for the different types of conversion behaviour P-(proportional-), PI-(proportional-integral-), PD-(proportional-differential-) and PID-(proportional-integral-differential-) behaviour [28].

4.2.3.4   Dosing equipment

In continuous process plants in the metal working industry, use is normally made of solenoid valves for adjusting pH values and quasi-proportionally controlling neutralisation. An on-off regulator is sufficient for control of detoxification by means of redox potential measurement. It is a prerequisite that the solutions should be gravity fed to the valves. If this is not possible, then metering pumps must be installed. If the control is constant, then dosing responds to be out-of-balance. This can be achieved by positioning the flaps with the electric motors, pneumatic or hydraulic devices, or by changing the size of stroke or number of strokes of the dosing pumps. Dosing valves and the capacity of dosing pumps must be adapted in terms of area and volume to the quantities to be delivered. To protect against mechanical contamination of the solutions, screens are fitted ahead ofthe dosing pumps.

More detailed descriptions of dosing technology can be found for dosing valves in Pauli [29] and for dosing pumps in Vetter [30].

4.2.3.5 Containers for chemicals

Among various containers for chemicals the following can be distinguished:

- Dosing tanks.
- Chemicals storage containers and silos.
- Preparation vessels.

Dosing tanks are situated as near as possible to the continuous process treatment tanks and connected by the shortest routes to the dosing units. In smaller facilities, where replenishment is from primary chemicals containers, the amounts needed should be taken approximately weekly. The installation of a liquid depth measuring probe to signal the need for replenishment is highly desirable.

In larger plants, dosing tanks usually have a capacity corresponding to one day's requirement, sometimes even less. At minimum liquid level, they must be refilled automatically from the chemicals stores and so need liquid depth measuring instruments with at least two switch contacts (control of pumps for filling, overflow protection). Dosing tanks are usually filled with solutions supplied by the chemicals industry at the concentrations required. These are: caustic soda (up to 50%), hydrochloric acid (normal 32%, maximum 36%), sulphuric acid (various concentrations), sodium hypochlorite solution (commercial, 13-15% active chlorine), sodium hydrogen sulphite solution (mostly 20%), iron(III) chloride (up to 30%), aluminium sulphate (20–30%). If solutions must be diluted for dosing, then this is done in separate preparation vessels (see below) and only rarely extent in the dosing tanks themselves. For that reason the liquid depth gauge for automatic make-up needs an additional switch contact for the supply of diluting water. Dosing tanks for solutions that evolve gases (concentrated hydrochloric acid, sodium hypochlorite, sodium hydrogen sulphite solution) must be enclosed and their ventilation ducts should be connected with an absorption vessel (see chemicals stores) or with the air extraction system.

The material for the dosing vessels is normally polyethylene or polypropylene. Temperature, which can rise when solutions are being diluted, should be monitored. It should never rise above 60°C. and thought should be given to the temperature resistance of the associated pipework and dosing equipment.

Chemical storage containers rank among the most important in terms of process safety and the cost-effective purchase of larger quantities of chemicals. Hence no limits are placed on their capacity. Commercial round trip containers are more and more often installed linked to the tanks. Feeding stored solutions into the dosing tank is via an automatic pump control controlled by the liquid depth measuring device.

To avoid crystallisation of dissolved substances from more concentrated salt and caustic soda solutions, chemicals storage containers should be situated in heated areas (18°C). They are filled by tankers from the outside of the building. The level to which they are filled should preferably be continuously monitored centrally. In addition, switch contacts for overfilling, communications for the punctual ordering of chemicals and dry operating protection for the supply pumps are also necessary (see Fig.s 4.8 and 4.9). For filling, connecting sockets are situated on the outer walls, and designed that filling a storage vessel with the wrong solution cannot occur. There are many possible ways of doing this: colour coding, different couplings, the counterparts of which are kept by the chemical supplier, lockable mechanisms with the keys at the disposal only of authorised personnel, and so on. Nevertheless, a mistake can sometimes occur, if the supplier has filled the tanker with the wrong solution. Therefore, it is basically advisable to test the tanker's contents before filling the storage vessel. This is possible using simple methods, such as for example reagent papers. Under the sockets, collection receptacles should be placed to collect the spillage when the tanker's pipelines are uncoupled; these can be drained to the corresponding chemical store or effluent reservoir (see Fig. 4.10).

Fig. 4.8   Chemical storage containers with exterior drains for recovering of leakages and pumping station for supplying outlets (GOEMA Dr Götzelmann GmbH, Vaihingen/Enz)

*Fig. 4.9   Chemical and concentrate storage containers protected
against leakage by bunded floor (with collection reservoirs)
(GOEMA Dr Götzelmann GmbH, Valhingen/Enz)*

The ventilation ducts of the chemical storage containers for volatile fluids are connected to the air extraction plant or to absorption vessels.

If the qualities of hydrochloric acid and caustic soda ordered are also suitable for regeneration of ion exchangers, then the corresponding storage containers can be connected up to the regeneration points of the recirculation plant as well.

For storing solutions of chemicals, round containers of polyethylene and polypropylene, glass reinforced polyester resins or steel vessels with corresponding plastic or hard rubber linings are employed. For sodium hypochlorite solution an oxidation resistant material has to be used. For concentrated sulphuric acid, tanks of low-carbon steel with walls of considerable thickness are used, protected against air humidity.

In so far as certain solutions are not stored but prepared from their salts or other dry substances, use may be made of preparation vessels, e.g. for sodium hydrogen sulphite, iron(III) chloride, aluminium sulphate and sodium carbonate solution, as well as solutions of flocculating aids. These are equipped with a mixer, and if the substances to be dissolved are dusty or the solutions evolve gas, they are covered and the air extracted from them disposed of as mentioned above. Examples of this are preparation vessels for sodium hydrogen sulphite solution and lime slurry (see Fig. 4.11). Preparation vessels are also used for diluting concentrated solutions. With hydrochloric acid, attention has to be given to the air extraction measures necessary, and with concentrated sulphuric acid to the steep increase in temperature. In the latter case, lead-lined steel tanks are safest. Hard rubber lined steel tanks can be used if it is clear that certain temperatures will not be exceeded.

Fig. 4.10   Filling station for chemicals. For each chemical a separate pipe connection is provided with collection receptacle and "store-full" warning (GOEMA Dr Götzelmann KG, Vaihingen/Enz)

Fig. 4.11   Preparation vessel for lime slurry with screw conveyor to bring in the dry lime (right above) and mixing vessel (centre); pumps for feeding into the dosing pipes (left below) (VERAMCHEMIE GmbH, Siershahn)

If only small amounts of such solutions are to be added to the waste water, then the preparation vessels can function simultaneously as dosing tanks. Where solutions of flocculating aids are added, the solids should first be steeped in water. Mixers running at a slow speed of not more than 50 rev/min are recommended for mixing them with water [31,32] (see Fig. 4.12).

Lime slurry is produced from hydrated calcium oxide, which is stored in silos, and water. In larger plants, silos have a capacity of more than 10 m³. Hydrated calcium oxide, which has an apparent density of only about 0.4kg/l, is transported on screw conveyors or vibrators into the preparation vessel with enough water to produce suspensions containing 5 to 15% of calcium hydroxide (see Section 3.6.3). The entire operation is automated (see Fig. 4.12). If lime slurry is needed in only small amounts, paper sacks full of hydrated calcium oxide can be used. The preparation vessels used in this case have a tube on the upper sealing plate, which, when the sack has been placed there, can be closed with a lid. Paper sacks are such that lime can be emptied into the vessel with the help of a shaking device, without the lime dust getting out into the room. Another way of doing this is using making a water slurry from a commercially available pasty calcium hydroxide.

All chemicals containers must be installed so that no fluids resulting from leakages can escape into the sewerage system or into the subsoil. Storage vessels are therefore located in sunken or built-up basins with chemically resistant linings or inside larger containers, which are equipped with liquid level gauges connected to alarm systems. If several containers are installed in a common vessel, this must be capable of holding the contents of at least the largest of the containers. The contents of storage vessels or other containers held within a common outer vessel must not undergo any chemical reaction with each other (see Figs. 4.8 and 4.9).

Fig. 4.12 Preparation and dosing tank for flocculating aid with geared mixer (GOEMA Dr Götzelmann GmbH, Vaihingen/Enz

4.2.3.6   Concentrate storage vessels

As mentioned in Section 4.2.2, spent process baths must also be drained away and stored. Since they will afterwards have to undergo batch treatment, it simplifies matters if they are segregated as described in Section 4.3.2. If a cyanide-containing concentrate is involved, it can be stored with cyanide-free alkaline concentrate, and so can chromium-containing baths with acidic ones. Concentrates containing complexing agents must always be stored separately, and the whole waste liquid regarded as complex-containing. The same applies to effluents containing oil and grease, cadmium or chlorinated hydrocarbons.

Acidic and alkaline concentrates should never be drained to the same storage vessel. The following hazards or damage could arise from this:

- Increases in temperature due to released heat of neutralisation. The storage vessels, their equipment and the fittings attached to them could be harmed as a result.

- Evolution of carbon dioxide, which in the presence of surface active agents gives rise to frothing.

- Precipitates (metal hydroxide, silicic acid) which sediment during storage and can cause trouble on emptying.

Regarding air extraction, connections to ventilation equipment (when gases are formed), liquid-depth and full-level indicators, collectors of discharges from leaks and the alarm system, the same considerations apply as have been described already in Section 4.2.3.5 for chemicals containers.

Round tanks of polyethylene and polypropylene are mainly used and, for emptying process baths at temperatures above 60°C, rubber-lined steel tanks. The resistance of the storage vessel material in the case of concentrates containing oxidising substances, should be borne in mind especially if they are hot.

4.2.4   Method of operation of continuous treatment plants for detoxification and neutralisation/precipitation

4.2.4.1   Cyanide detoxification plant

Cyanide-containing waste water flows through the reactor or its first compartment in which pH measurements are made continuously with an immersed glass electrode and of the redox potential with a gold or an amalgamated silver electrode against a counter electrode (usually silver-silver chloride). Dosing with caustic soda continues if the pH value is too low for the cyanide detoxification reaction, until a pH value of 10 is exceeded or one of 11 reached. If the value of the redox potential is too low, sodium hypochlorite solution is added, until a potential difference about 100 mV more positive is attained. This occurs as soon as there is an excess of only a few mg/litre of active chlorine. The reaction proceeds according to Equation 3.24 within the detoxification plant. This is followed by Reaction 3.27, which accelerates after the pH is reduced in the neutralisation plant. As the reaction consumes caustic soda, the pH falls and must therefore continually be adjusted to a constant value.

Care must be taken that the redox electrode assembly is correctly set for the desired pH value. As metal electrodes also are influenced by pH, values that are too high produce redox potential values that are too negative. A false cyanide reading thus results, leading to continuous useless dosing with sodium hypochlorite. In the opposite case, it gives rise to no dosing and therefore also no detoxification.

During dosing, the mixer, which operates at a rate coordinated with the throughput of the continuous process and also with the volume of the reactor, ensures that the reaction takes place quickly. The average theoretical residence time τ (see Equation 4.1) may be extended in cyanide detoxification between 30 and 60 minutes, depending on local legislation. This accords less with scientific observations than with heightened safety requirements. The experiences of the writer with about a hundred such plants confirm increasingly that the reaction time of cyanide detoxification, even in the presence of copper, zinc and cadmium, is a function of the reactor technology and not of the reaction kinetics. It is a different matter in the presence of silver and especially of nickel (the latter should not however be allowed into cyanide-containing waste waters). Here the reactions take considerably longer, because of the lower rate of dissociation of the metal-cyanide complexes. In the presence of these metals, detoxification must be carried out as a slow reaction in batch process plants. In the presence of iron, ferricyanide is formed, which cannot in any case be destroyed in the way described (see Section 3.2.1.6).

In an operating facility, the values for the waste water flowing through the neutralisation plant at pH 10-11 are less than 0.1 mg/litre of easily released cyanide, around 0.5 to a few mg/litre of active chlorine, which in the plant as a whole is almost completely converted into chloride by oxidisable material, and ammonia corresponding to the original cyanide content. In addition, the waste water still contains the metals present as hydroxide, with the zinc perhaps partly as zincate. Depending on how the reaction is carried out, it may be that a small amount of copper cyanide (CuCN) is present in a form that is difficult to dissolve, which can be recovered in the settled sludge at the end of the whole plant. Because of possible gaseous emissions (chlorine, cyanogen chloride) the cyanide detoxification plant must be covered and connected up to an air extraction plant. It should also be noted that the copper in the cyanide complex is monovalent and so additional sodium hypochlorite is needed in the detoxification to oxidise it.

If the waste water contains higher temporary hardnesss salts calcareous deposits often form in a short time on the electrodes, the mixer and other plant components. To remove them, it is usually not possible to avoid careful dissolution in dilute acids. Where the electrodes are concerned, this may be necessary daily.

Nowhere is the conflict with respect to AOX, so marked as in cyanide detoxification. In continuous process treatments the diluted waste waters usually more than meet the required value of 1 mg/litre. If water saving measures are employed, so as to make batch treatment feasible, this is often not the case (see Section 3.2.1.7). Hydrogen peroxide is not suitable for continuous process treatment, as the redox potential of cyanide alone is insufficient for control and this oxidation agent makes no contribution itself to the redox potential. Peroxymonosulphuric acid generates a truly distinct oxidation potential, but its cost in relation to its oxidation capacity is too high. With a litre of fresh sodium hypochlorite solution (13 to 15% active chlorine) it is possible to detoxify 96 to 110 g of cyanide and with a kilogram of Caro's acid (4.9% active oxygen) 80 g of cyanide by oxidising it to cyanate.

4.2.4.2   Chromate detoxification plant

The pH and redox potential of chromate-containing waste waters flowing through are measured in the first compartment of the reactor. The pH measurement electrode assembly is used for the first and a combination of a platinum or gold with a counter electrode for the latter. These combinations are generally known in the trade as redox potential measuring assemblies. Dosing with acid continues as long as the pH remains above 2.5. Normally this is done with dilute sulphuric acid (eg 1:3), or, in cases where a limit is set on the effluent sulphate content, with hydrochloric acid (1:1). If measurement of the redox potential indicates the presence of chromate, dosing takes place with sodium hydrogen sulphite solution (usually 20% prepared from sodium pyrosulphate) until a potential drop of some 100 mV in a more negative direction is reached (see Fig. 4.7). As one of the participants in the reaction is acid (see Equation 3.78) the pH value rises. It must be held constant at pH 2.5.

For the same reason as in cyanide detoxification, the redox measurement assembly must be calibrated for a reaction pH of 2.5. At pH 3, the reaction no longer proceeds quantitatively.

Chromate detoxification takes place spontaneously, if the reagents are well mixed. The reactor is designed therefore on the basis of an average theoretical residence time of about 15 minutes.

It is absolutely essential to avoid treatment of non-chromate-containing, eg acid, waste waters in the same plant, as metals present may set up other redox systems, which interfere with the measurement. Those concerned have been named in Section 4.2.3.3 dealing with the measurement of redox potentials. The effect shows itself in the way that large enough potential differences no longer occur, or only a sluggish drift of the potential is observed, similar to what happens when an electrode is poisoned. Over dosing with sodium hydrogen sulphite thus ensues. A small addition of a very strong reducing agent (sodium dithionite) puts a stop to this immediately. Where necessary this periodic addition is carried out automatically, as in the chemical cleaning of electrodes.

If the plant is functioning perfectly, the waste water that flows subsequently into the neutralisation plant has a pH value $\leq$ 2.5, a hexavalent chromium content between zero and less than one-tenth of a mg/litre, a small excess of a few mg/litre of sulphite and a concentration of trivalent chromium corresponding to the original chromate content and possibly other metals.

Because of the occurrence of sulphur dioxide emissions, the chromate detoxification plant must be covered and connected to an air extraction unit.

4.2.4.3   Nitrite detoxification plant

Nitrite detoxification in continuous process plants is possible only by oxidation with sodium hypochlorite (possibly also with Caro's acid, see Section 4.2.4.1) for reasons connected with measurement technology. The pH is adjusted to pH 3 to 5 with acid (sulphuric acid 1:3). By dosing with sodium hypochlorite, the redox potential measured in this pH range is adjusted to give a potential difference in a more positive measurement range. The reaction proceeds in accordance with Equation 3.90. No detectable potential change occurs in the reduction reaction with amido acid.

Furthermore this reaction proceeds spontaneously at a rate influenced only by the reactor technology. The average theoretical residence time of the reactor is designed to be about 15 minutes. If the waste water contains cyanide and nitrite (heat treatment operations) then the waste water is first put through a cyanide then a nitrite detoxification plant. The treated waste that flows into the neutralisation plant has a pH between 3 and 5 and a nitrite content between zero and a few mg/litre. Because of the occurrence of nitric oxide emissions, the nitrite detoxification plant must be covered and connected to an air extraction unit.

4.2.4.4   Plant for neutralisation/precipitation

To this plant are brought the detoxified waste waters and all other different metal-containing waste waters, with the exception of those containing cadmium. The latter must be given separate treatment, including detoxification and removal of solid material. If the waste water shows wide fluctuations in pH value, then it is useful to connect a buffering vessel in series with the neutralisation/precipitation plant. This saves neutralisation chemicals and reduces the amount of salt produced.

The pH of the combined waste water is measured and it is neutralised with a quasi-proportional or continuous-proportional or proportional-integral controller. The pH should be kept within as narrow a range as possible, and this along with the neutralisation agent to be added must be chosen in the light of the metals present (see Section 3.7.6). The control systems used are mostly bi-functional regulators, ie they are capable of neutralising acidic as well as alkaline effluents.

As a rule, since for metals it is generally necessary to increase alkalinity, neutralisation/precipitation is carried out with alkalis (caustic soda, sodium hydroxide-carbonate solutions, lime slurry). When acids are involved, dilute sulphuric acid is used or, if there is a limit on sulphate concentration, hydrochloric acid. Carbonic acid is also used in a few cases for washing machines employing alkaline cleaners. Reactors of different design are then necessary.

Neutralisation and metal precipitation with caustic soda take place very quickly and its rate depends only on the reactor technology. So long as no simultaneous oxidation of divalent iron occurs, it is sufficient here also to use reactors with an average theoretical residence time of 15 minutes. If, however, lime slurry is employed, then double this time should be allowed. Calcium hydroxide dissolves relatively slowly at pH 7 (1.7 g/l, 20°C) and a longer time is necessary to reach solution equilibrium. If the waste water contains fluoride or precipitable sulphate or phosphate, then lime slurry must be used for neutralisation. The need to optimise precipitation with this compound is not always made clear. An equivalent measure of acid is necessary in the effluent, as dosing with lime slurry takes place continuously on the basis of pH value measurement. Additions of calcium salts, e.g. calcium chloride, make up for any deficit of calcium ions which cannot be determined at least by continuous measurement. This is a disadvantage of continuous process treatment plants.

If the waste water contains compounds of divalent iron, then these must be oxidised while still in the neutralisation plant, or problems could arise if pH subsequently falls (see Section 3.5). This oxidation can be performed, with proper measurement and control, using sodium hypochlorite. If the oxidation is carried out in the neutral range by passing in air (aeration pipes, air agitation), then, depending on the concentration of divalent iron, it requires longer, sometimes much longer, reaction times. Design of reactor size must then be determined by experiment.

If the waste water, because of the presence of finely dispersed material, requires flocculation with iron(III) or aluminium salts, then the solutions in the neutralisation reactor must be dosed, as these agents also produce a fall in pH. The particle size of the precipitates must be increased in order to increase the speed of sedimentation, so dosing with flocculating aids is carried out in a compartment preceding the neutralisation reactor. The mixing must be done without vigorous agitation, e.g. with a slowly turning mixer. Dosing with flocculating agents and flocculating aids in continuous process plants can only be carried out in excess, relative to the volume flow of the waste water, not to the calculated need, or according to previous empirical determinations.

After neutralisation, precipitation, coalescence and flocculation, the waste water flows into a sedimentation plant for separation of solid material.

Fig.s 4.13 and 4.14 show continuous process treatment plants of normal construction.

Fig. 4.13. *Continuous process treatment plant for detoxification and neutralisation (KERAMCHEMIE GmbH, Siershahn)*

Fig. 4.14. *Continuous process treatment plant for final neutralisation of metal-free waste water, capacity 10 m³/h (GOEMA Dr Götzelmann GmbH, Vaihingen/Enz)*

4.2.4.5   Special designs for continuous process treatment

In cases where alkaline waste waters are neutralised with carbon dioxide, liquid carbon dioxide is first released under pressure and the gaseous carbon dioxide brought via a gas mixer into an enclosed continuous process reactor. If demand is not called at the outlet, this is closed and the reactor continues in operation until the theoretical value is reached [33]. Flue gases from boiler furnaces and incinerating plants are also used in the neutralisation of such solutions [34].

Some conceptual designs showing the most important features of continuous process plants were described in Sections 4.2.4.1 to 4.2.4.4 There are additional characteristics, which relate to two qualities:

- Improvement in residence time behaviour.

- Greater safety in continuous process treatment.

Cramer [35] describes a fast reactor with minimum reaction space and higher intensity of mixing. The treatment chemicals are already mixed with the waste water in the pump room and this mixture is conveyed at high speed into the reactor through a nozzle with a diffuser. The residence time behaviour, as a result of this reaction technology, approaches that of a stirred tank cascade with several compartments. If the required value is not reached at the outlet, then the valve situated there is closed and the waste water flows back into the receiver. Fig. 4.15 shows the construction of such a reactor. A reactor in which there is a high turbulence stirrer would also be described as a fast-mixer reactor [36].

The same objective, but at greater risk, is achieved if static mixers are installed in reactors of still smaller capacity. Static mixers are tubular with inbuilt rotating propellers to produce turbulent flow, into which the treatment chemicals are injected directly. For fast reactors of this or the previously mentioned type it is a prerequisite that the reactions proceed spontaneously; thus, no excessive fluctuations in concentration should occur in the static mixer. Pini gave a reported on chromate detoxification and neutralisation trials using static mixers [37].

The second type of plant, which guarantees increased security in continuous process treatment, incorporates batch processing capability. Like the fast reactor, it operates as a batch reactor if the required value is not achieved at the outlet. Bahensky [38] describes a plant, which converts to batch operation if there is any abnormal operation in which the outlet valve of the reactor closes so that the level rises to the maximum while dosing is continued. When the correct value is reached, the outlet valve opens again. However, if the theoretical value is not attained quickly enough, then the waste water flows through an overflow into the first vessel or a storage tank. The storage volume of the reactor is such that it can retain the waste water arising in an hour, for example, before it overflows. Within this time, direct action can be taken to overcome the problem.

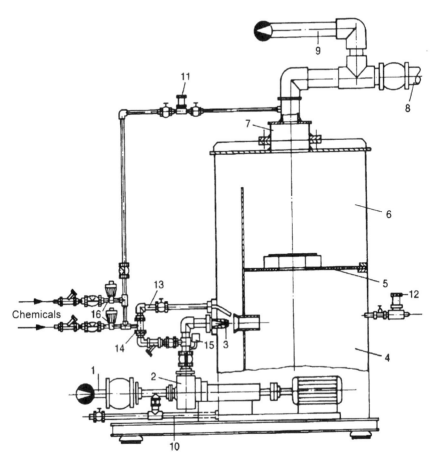

Fig. 4.15   FAV-Fast reactor
(THYSSEN ENGINEERING GmbH, Essen)

1 Intake pipe, 2 Pump, 3 Nozzle and diffuser, 4 Reactor, 5 Partition, 6 Post-reaction space, 7 Venting dome, 8 Outlet, 9 Return pipe to receiver, 10 Back flow, 11 Final control gauge, 12 Measuring electrode for control of chemical dosing, 13 Bypass, 14 Ejector, 15 Pressure gauge, 16 Solenoid valve.

The diagram in Fig. 4.16 shows how this reactor works. Another version operates as a cascade reactor. The waste water flows through two reactors and is dosed automatically with treatment chemicals. The overflow of the first of the two reactors then closes if the required value is not attained in the second. In the meantime, reactor 2 operates as a batch unit. The linking valve opens again when the correct value is obtained in reactor 2. If the malfunction continues for a longer time, then the waste water overflows from the first reactor into a storage tank [39] (see Fig. 4.17).

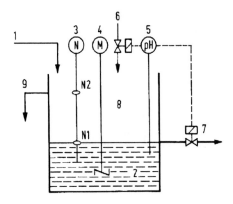

Fig. 4.16. Continuous process reactor with additional safety control (from Bahensky [38]). 1 Waste water inlet, 2 Reactor, 3 Liquid height measuring instrument (N1 normal height, N2 maximum height), 4 Stirrer, 5 Measuring equipment for controlling additions of chemicals and shut-off valves in the outlet, 6 Dosing with chemicals via solenoid valve, 7 Shut-off valve in the outlet, 8 Buffer space, 9 Overflow to storage tank.

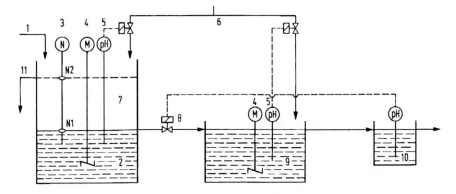

Fig. 4.17 Reactor cascade with additional safety control [39].

1. Waste water inlet, 2 Reactor 1, 3 Liquid height measuring instrument (N1 normal height, N2 maximum height), 4 Stirrer, 5 Measuring equipment for controlling additions of chemicals, 6 Dosing with chemicals via solenoid valve, 7 Buffer volume, 8 Shut-off valve in the outlet from reactor 1, 9 Reactor 2, 10 Final control measurement, 11 Overflow from reactor 1 to a storage tank.

4.2.5 Sedimentation plants for separation of solid materials

4.2.5.1 Fundamentals

In waste water treatment many harmful materials are precipitated in solid form and must then be removed. In the main these are metal hydroxides, basic metal salts, metal sulphides, other fairly insoluble metal compounds with iron and/or calcium, as well as a few sparingly soluble compounds of calcium with various anions. Along with these there are also inert materials of low solubility, such as silicic acid.

The following methods are available for separating solid materials:

- Separation by sedimentation under gravity in continuous process vessels. In a variation of this, the effect of gravity is enhanced by centrifuging.

- Separation by flotation, in which lighter materials, particularly gases, are attached to the particles, so that they are segregated at the surface of the waste water.

- Separation on the basis of particle size through filtration.

In the treatment of waste waters of the metal working industries, sedimentation is employed and, to an increasing extent, filtration, especially in the batch treatment plants that are becoming more and more common. Direct filtration in smaller plants (without preceding sedimentation), as used in the 60s and 70s [40], has become less significant. It is forseeable, however, that to an increasing extent filtration plants will be added later to continuous sedimentation plants as a second stage of solid material separation (see Section 4.4.1).

Flotation is little used for removing solid material but more for separating emulsions after successful flocculation, as oil-laden hydroxide settles only with difficulty if at all (see Section 3.13.4.4).

The parameter on which the rate of sedimentation depends varies according to Equation 3-132 even if its application is scarcely ever possible, because of the lack of knowledge of the constants relating to particle shape and the specific sedimentation properties of the relevant materials.

For separating solid materials in continuous process operations, use is made of sludge settling reservoirs, the useful waste water capacities of which determine the average theoretical residence time. The useful volumes of the different shapes of construction is determined by their geometry. (see Fig. 4.18). In addition, further space is required for the skimmer,the inlet and outlet structures, the space above the water surface and the space for accumulation of the sludge, if this is included in the body of the settling reservoir. In the sludge settling reservoirs installed for sedimentation purposes, the waste water must be given sufficient time, as it flows through them, for the particles of solid material to deposit. For this they need a certain useful volume V. In rectangular reservoirs this is given by:

$$V = A \cdot h \quad [m^3] \hspace{4cm} 4.2$$

A = area of cuboid (m^2); h = height of cuboid (m)

If the waste water stream flows in uniformly over one of the narrow sides at the rate Q, then the smallest particle must have reached the bottom of the cuboid within the time τ taken by the waste water to reach the opposite side, ie it must in this time have covered the vertical distance h:

Part of the reservoir that is important in determining the useful volume		Calculation of volume V [m³]	Example
Description	Form		
Cuboid	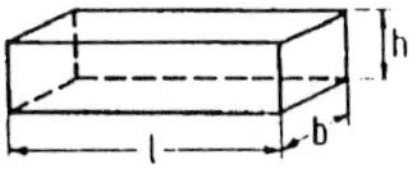	$V = l \cdot b \cdot h$	Longitudinal reservoir
Cylinder	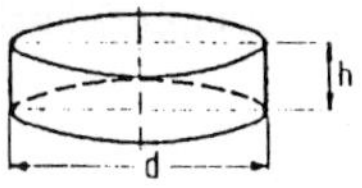	$V = \dfrac{1}{4} d^2 \pi \cdot h$	Round reservoir
Frustum	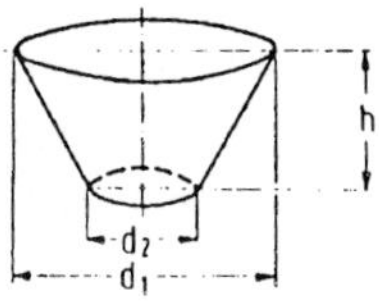	$V = \dfrac{1}{2}(d_1^2 + d_1 d_2 + d_2^2) \pi \cdot h$	Inverted cone container

Fig. 4.18   Geometrical shapes of the useful volumes of important sedimentation units and their calculations

$$\tau = \frac{V}{Q} \quad [h] \hspace{6cm} 4.3$$

τ = average, theoretical residence time (h); Q = volume flowrate of waste water (m³/h)

Equations 4.2 and 4.3 give the distance h travelled by the settling particles:

$$h = \frac{Q \cdot \tau}{A} \quad [m] \hspace{6cm} 4.4$$

the settling rate μs of the smallest particle for the relevant reservoir is:

$$u_s = \frac{h}{\tau} \quad [mh^{-1}] \hspace{6cm} 4.5$$

Inserting Equation 4.4 in Equation 4.5 gives the formula applicable to all clarification processes:

$$u_s = \frac{Q}{A} \quad [mh^{-1}] \hspace{6cm} 4.6$$

Equation 4.6 states that the settling rate μs is identical to the surface loading. From these considerations the size of cuboid for a longitudinal reservoir can be established (see Fig. 4.18). Its bottom floor is an imaginary plane, which separates the space for reception of settled sludge (sludge sump) from the usable space V assumed to be available for sedimentation [41,42 etc.] This model is only correct if water flows continuously through the cuboid over the cross-section b.h. at a constant steady rate (plug flow) without any temperature variation, and any turbulence or diffusion is excluded (see Section 4.2.1).

Such ideal conditions are never realised in practice. There is a whole series of reasons for this [7,43 etc.]:

- Turbulence due to wall effects and at the edges of the necessary components.

- Turbulence due to convection currents: heat of reaction of the incoming water, heat gain or loss to atmosphere at the surface.

- Short-circuit flows. The main water flow takes the path of least resistance. Consequently separate portions of the whole volume do not have the same residence time.

- Temperature dependent solubility of air. The solubility falls as the temperature increases and fine air bubbles are formed, which adhere to the flocs and carry them upwards (flotation).

- Rising of flocs, which are rendered hydrophobic by surface active agents and flocculation aids.

- Acceleration of the settling rate through capture of small, slowly sedimenting flocs by larger ones, as a result of which the latter settle even faster. Further turbulence thus develops at the periphery of the flocs.

It is very important that waste water flowing into the sedimentation plant is already uniformly distributed over its flow cross-section. Its speed of flow is thus simultaneously reduced to a minimum, so that the least possible disturbance is caused to the sedimenting particle on its way downwards. This is not just a matter of distribution over the surface, but also using baffles (longitudinal reservoirs), immersion tubes (conical tanks) and centre manifolds (round reservoirs), in order to make use of the greatest possible proportion of the usable volumes, to minimise short-circuit flows arise, which considerably decrease the efficiency of these plants. This uniform flow can be maintained only if the overflow to the outlet is also uniform.

4.2.5.2 Sludge separation reservoirs

The clarification surface formula can be used in the design of sludge separation reservoirs only in conjunction with considerable practical experience. Many attempts have been made to model sedimentation processes. These provide a theoretical basis for certain phenomena, but do not serve as a substitute for practical knowledge. It is best to make observations in similar instances by experimentally determining residence time behaviour and carrying out sedimentation trials in cylinders, corresponding in effective height with the planned reservoir [8]. Specialist companies have firm ideas about the underlying principles for estimating the dimensions of such reservoirs, which have been developed on the basis of theoretical insights and practical observations.

Gütling and Trapp [44] have suggested for longitudinal reservoirs based on the overall dimension ratio of l:b:h = 3:1:1, the author prefers 3:1:0.66 (see Fig. 4.18). The immersion height of a long-sided retaining wall should be about two-thirds of the usable height and retaining walls at the outlet should be immersed only just enough to hold back the floating sludge. No flow takes place through that part of the volume of the reservoir whose height corresponds to the immersed height of this wall.

The findings concerning clarification surface velocity and the parameters with an adverse effect on this can be applied to other reservoir shapes as well. In round reservoirs, the inlet is centrally located with slow, radiating outward flow. The flow rate is thus always low and facilitates particle settling before the annular overflow is reached. Other reservoirs have radially disposed channels through which the waste water is able to flow in from both sides. The clarification surface formula is applicable here with the same reservations as for longitudinal reservoirs. Given large diameters, the settling direction is substantially perpendicular to the direction of flow, as in rectangular reservoirs.

Some tanks have a conical construction (see Fig. 4.19). Here the settling direction is essentially in opposition to the direction of flow. The settling rate of the particle must therefore be greater than that of the opposing rate of flow of the waste water. As a result of the upwardly expanding conical shape, however, the flow rate of the waste water falls continuously, until eventually the smaller particles stay in balance. This layer forms a filter for still smaller particles. Larger ones thus form, which are able to settle [45,46]. As a rule, for the upward rate a maximum level of 0.4 m.hr^{-1} is required by the authorities [47]; settling rates up to a maximum of 10 m.hr^{-1} can be reached if the hydroxide is well flocculated. Basic dimensional data for building conical tanks can be found in Resch [48,49].

Almost without exception, reservoirs for the separation of hydroxide sludges are designed to have an average theoretical residence time of four hours. This time is insufficient, for sludges such as aluminium hydroxide or chromium hydroxide which settle only with difficulty. The residence time is increased to six hours for these compounds. Depending on the dimensions, the surface loading for separation of metal hydroxides is around 0.3 to 0.7 ms^{-1}. Sludge break-through resulting from only slight overloading of the reservoirs governs this.

In the metal working industries the simple longitudinal or rectangular reservoir is used, where the plate separator has not replaced it. Conical tanks are also installed occasionally. Large clarifying reservoirs, which are shallow to avoid structures of greater depth (round reservoirs and large longitudinal reservoirs), are equipped with sweepers, which continuously transfer the sedimented sludge from the gently sloping floor area into the sludge sump, from where it can be pumped away. The most important types of construction of sludge separating reservoirs are illustrated in Fig. 4.19.

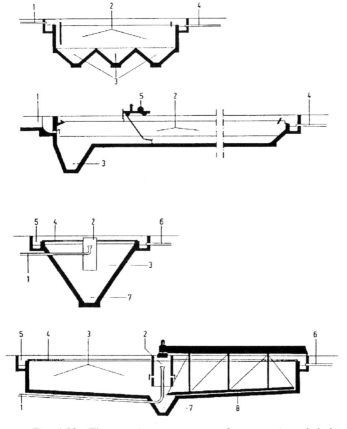

Fig. 4.19   The most important types of construction of sludge separating reservoirs, as employed in the metal working industries. Above rectangular or longitudinal reservoirs, below round-shaped reservoirs (diagrammatic)

(a) Rectangular reservoirs for smaller to middle-sized plants.
(b) Rectangular reservoirs for large plants.
   1 Untreated water inlet, 2 Usable space for sedimentation, 3 Sludge space, 4 Clear water outlet, 5 Sweeper for sedimented and floating sludge.
(c) Conical tanks for small to middle-sized plants.
(d) Round reservoirs for large plants.
   1 Untreated water inlet, 2 Immersed pipe or central structure, 3 Usable space for sedimentation, 4 Clear water overflow, 5 Clear water collecting channel, 6 Clear water oulet, 7 Sludge space, 8 Skimmer.
   (Pipes for removal of sludge not shown).

The central structures of round reservoirs serve not only to achieve uniformity
in water distribution, but also as reactors for flocculation and floc-growth processes. With additional components to improve flow, highly effective arrangements have been developed, of which an example is shown in Fig. 4.20. Finally,
separation systems in reservoir construction were established, which there is
scarcely any chance of bettering in terms of clarification area. In these the
clarification surface loading amounts to around 12 mh^{-1}! The waste water is
introduced tangentially at the side and flows downwards in the form of a spiral
which then moves upwards in an immersed pipe. At the point of flow reversal
most of the solid material present deposits in the sludge cone situated below.
The immersed pipe expands conically in the upper part as far as the outer wall,
so that a post-clarification ensues. Efficient flocculation is a precondition for the
employment of such an arrangement [50,51].

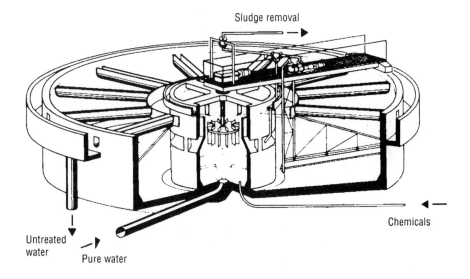

Fig. 4.20 Rapid clarification equipment "Sedimat"
(LURGI Apparate-Technik GmbH, Frankfurt/Main)

It is recognised that there is no sludge separating system that can always be
guaranteed to satisfy the stricter metal limits of new legislation . The reasons are
the break-through of the finest particles or floating flocs. The clarified waste
water must therefore be subsequently filtered (see Section 4.4.1).

The thin sludge leaving the separation systems has a solids content between
1 and 4% depending on the sludge's age. In larger plants with continuous sludge
sweeping it is lower, in smaller plants, sludge is periodically removed and so
has time for further compaction. These sludges are all still fluid and therefore

pumpable. They are transferred into the sludge storage tank, where by ageing, they thicken and thereby lose water. According to Möller [52] and Zimmer [53] settled sludge contains water in the following forms, ranked in order of importance:

- Interstitial water. This is found without any bond between the particles of solid matter.

- Bound- and adhering-water. This is firmly held by long-range molecular forces. Much more water to kind is bound up with the hydroxides with their pores and voluminous structure than with compact particles.

- Capillary water. This is found in interstices of capillary size and is bound by capillary forces.

- Adsorbed water. This is bonded by molecular forces directly on to the particles' surface.

- Internal water. This is a form of cellular liquid, which is trapped in the particles as internal capillary water and water of hydration.

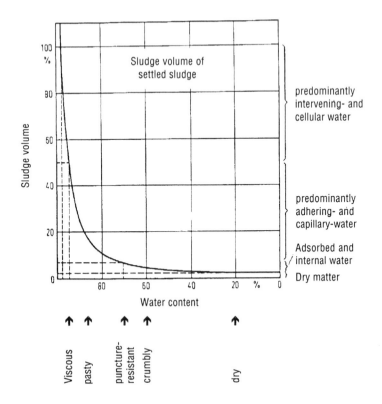

Fig. 4.21 Volumes of hydroxide sludges as a function of water, content [54]

Interstitial- and bound-water alone separate under the effect of gravity compressing the sludge in the storage tank. To separate the capillary- and adhering-water, forces must be applied which overcome the capillary and adhesion forces. This is possible by increasing gravity (centrifuging) or by pressure filtration, hence by measures applied to achieve mechanical dewatering of the sludge (see Section 4.5). Adsorbed and internal water can be driven out only by drying and by thermal treatment.

The relationship between sludge volume and water content displayed in Fig. 4.21 is based on the author's investigations and experiences [54]. Settled sludges, which are removed only periodically from the sedimentation plants, have a solids content from 1 to 4%, which increases to 5 to 6% in the sludge storage tanks, and after dewatering with filter presses is around 30 to 40%. Further dewatering by natural drying is slight. Considerable amounts of water are released by ageing and weathering processes, even if the sludges are exposed to the weather.

4.2.5.3   Plate separators [55]

Sludge separation reservoirs represent the most expensive parts of continuous process treatment plants in terms of cost and space. Considering how to reduce their size leads inevitably to the demand for a shortening of the separation time of the particles of solid matter. Two possible ways of doing this have been known for a long time:

- Increasing gravity by centrifuging. This calls for substantial increase in investment and above all in running costs for energy consumed.

- Increasing the particle size of the solid matter. The use of flocculation aids to produce bigger and faster-settling flocs is practised near universally. Previously, this was often employed to prevent sludge separation reservoirs from being overloaded.

The third possibility, which arises from the clarification surface formula, brings a new solution to bear:

- Reducing the separating distance and thereby accelerating sedimentation.

Imagine several horizontal floors situated in a rectangular reservoir between the water surface and the separation level of the sludge space, and the volume flow of the waste water as being distributed uniformly over the cross-section in the direction of flow; this reservoir would then possess a clarification area corresponding to the number of floors. The separating distance is thus reduced to a fraction of the original and the settling time to an equal extent. The sludge that has settled on the floors might block the space between and the waste water would finally wash it away. If, however, these floors (layers) are set at an angle, so that the sludge can slide down, then one gets a functional sedimentation system, the tilted plate separator. The clarification area of the stack of sloping plates that results corresponds to the product of the area, in plan, of one sloping floor and the number of plates. In spite of the slope, which reduces the clarification area, the latter is substantially increased overall.

Fig. 4.22 shows an illustration of a plate set at an angle α, with a gap h and a length l. The particles to be separated have a maximum distance of h_s to cover. This amounts to:

$$h_s = \frac{h}{\cos \alpha} \quad [m] \qquad\qquad 4.7$$

To cover this at a settling speed of u_s they require time t_s:

$$t_s = \frac{h_s}{\cos \alpha \cdot u_s} \quad [mh^{-1}] \qquad\qquad 4.8$$

At a waste water velocity in the up-flow between the plates of u_F and a plate length l, the particles must reach the floor of the plate, at the latest, in the time:

$$t_F = \frac{l}{u_F} \quad [h] \qquad\qquad 4.9$$

Finally, from Equations 4.8 and 4.9:

$$\frac{l}{u_F} = \frac{h}{\cos \alpha \cdot u_s} \quad \text{or} \quad u_s = \frac{h \cdot u_F}{l \cdot \cos \alpha} \quad [mh^{-1}] \qquad\qquad 4.10$$

If u_s is the settling speed of the smallest particles to be removed (particle-size limit), then it is possible to calculate the surface velocity of the plate separator. It must always be borne in mind that the results will be correct only under conditions of laminar flow between the plates. In practice, however, there are always turbulent currents present near the entrance to the plates, which are smoothed to laminar only after a certain distance. In actual plate separators an allowance has to be made for this transition distance.

Fig. 4.23 shows the method of operation of a plate separator for counter-current use. The waste water flows in in a downward direction, then upwards between the plates, allowing the sludge to settle on their upward-facing surface then slide down these in a direction contrary to the liquid flow to fall into the sludge space. For this to occur, the surface of the plates must be smooth and their angle of inclination for hydroxide separation must be at least 54°. In practice it is usually 60°. As cos 60° = 0.5, for every square metre of surface installed there is available a clarification area of 0.5 m². 4 to 8 cm is recommended as the distance between the plates and 1 to 2.5 m for their length [55]. For high efficiency it is also important that the waste water be introduced between the plates with as little turbulence as possible and be removed equally uniformly at the other side. The space requirement for a plate separator amounts to about 10% of the volume of a sludge separating reservoir of the same capacity. The hold-up time in the separator is around 5 to 10 minutes.

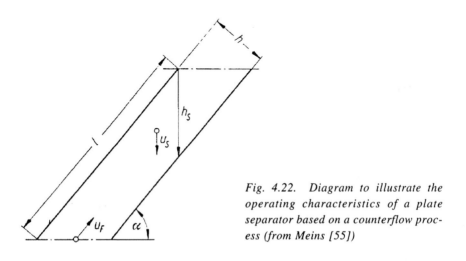

Fig. 4.22.   Diagram to illustrate the operating characteristics of a plate separator based on a counterflow process (from Meins [55])

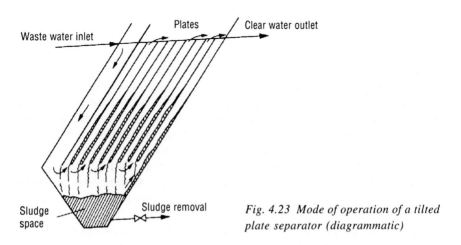

Fig. 4.23 Mode of operation of a tilted plate separator (diagrammatic)

Owing to the short residence time of the waste water in the plate clarifier and the small sedimentation distance it is not to be expected that the same floc growth and consequent accelerated sedimentation will occur as in the sludge separating reservoirs. Therefore it is necessary to carry out formation and/or further growth of flocs in a preceding stage. This is obligatory if plate separators are to be employed. The clarification surface loading is maintained in the range 0.4 to 0.8 mh^{-1}, thus corresponding approximately with that in sludge separating reservoirs.

In practice there are numerous variations possible in plate construction: smooth plates, plates with an undulating or zig-zag-shaped profile, inclined pipes with round, four- or six-sided cross-sections, etc.

In addition to the counter-flow processes there are also co-flow and cross-flow processes in use. In co-flow processes the angle of incidence is actually set at a low value [56,57], yet larger plate gaps are necessary because of the return-flow of the clear water. There is also a distinct disadvantage in this, in that the places at which the clear water is diverted upwards lie in the vicinity of the sludge outlets and there arises a certain danger that small particles will be entrained in the clear water.

As a result of the considerably reduced volume of the plate separator as compared with the equivalent sludge separating reservoir, the sludge space is also substantially smaller than in the latter case. It therefore fills quickly, and the sediment has little time to become compacted. In constructions with conical sludge spaces therefore, sludge break-throughs can very easily occur, but a sludge level indicator warns of their rise to the region below the plates. The operation of these plants stands and falls on as regular as possible, continuous sludge removal. If there are greater amounts of removable material in the waste water, above about 80 to 100 ml/litre, reductions have to be made in the through-put of the equipment, or a simple vessel for pre-clarification removal of the fast settling portion has to be connected in series ahead of it. There are also combined units with optimised pre-separation in use [58,59], which bring about separation of about 90% of the solid material. Difficulties in removing the sludges are avoided by the improvements made in these appliances. They have flat floors in place of the conical sludge spaces and hence also the advantage of a much reduced height. The deposited sludge is taken away continuously, without supervision, by suction through perforated pipes that are moved over the floor. There are systems in which the suction pipes are rotated laterally over the floor [58] or are moved slowly over the length of the vessel [60] (see Figs. 4.24 and 4.25). With these arrangements rather more water than necessary is sucked away with the sludge. Sludge break-throughs are thus avoided. The thin sludge drawn off is conveyed to the sludge storage tank, where it can be thickened. The overlying clear water, which will also contain a smaller amount of solid matter, passes out over an overflow and returns to the flocculation inlet of the plate separator. The underflow from the storage tank is taken to a filter press to remove the water from the sludge.

The clarified water, as in the operation of sludge separating reservoirs, is passed through a filter unit for safety's sake before discharge to drain.

Fig. 4.24   Compact clarifier with pre-clarifying compartment (left) and plate clarifier (right) with rotating sludge suction pipe, capacity 15 m³/h (Gütling GmbH, Fellbach)

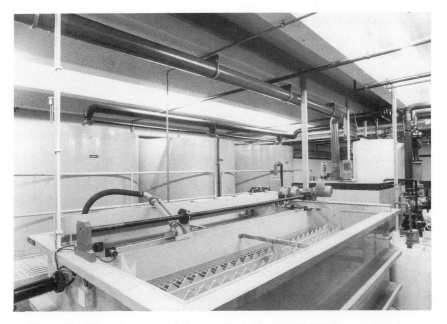

Fig. 4.25   Plate clarifier with longitudinal movement of sludge suction pipe; capacity 12 m³/h (GOEMA Dr Götzelmann GmbH, Vaihingen/Enz)

4.2.6   Safety instrumentation, final checks. The continuous process plant according to current technology

Process technology for continuous plants, wherever the use of batch plants is not possible, requires special safeguards. Possible precautions to take are: dual measurement with difference value averaging, post-treatment of treated water by filtration and selective ion-exchange, treatment of concentrates in batch plants, and so on. To these may be added numerous safety devices with optical and acoustic signals, to indicate departures from normal conditions. In addition to procedures to ensure process safety, automatic action must be taken simultaneously to prevent the escape of waste water or its leakage into the subsoil. The construction of continuous process detoxification and neutralisation plants described in Section 4.2.4.5 allow failures to be rectified in a short time. Storage tanks, which hold the raw effluent in the meantime, generally have capacities equivalent to one hour or so. If the malfunction is not corrected after reaching the maximum filling level of these storage tanks, then in response to repeated alarms the water supply to the production area generating the waste water must be interrupted. This is best done with fail-safe valves that close when current is off. Simultaneously, all the waste water supply pumps within the treatment plant are also switched off. If the continuous process plant has no reserve space or storage tank, then the fresh water inlet and supply pumps are immediately closed and switched off, respectively.

The most important alarm concerned with process safety then operates if an operational fault in the reactor is not corrected within a certain preselected time. There may be various reasons for this:

- Dosing tank and chemical storage tank are empty.

- Dosing valve does not respond or is blocked, or the dosing pump is misbehaving or not working.

- Toxic species concentration increase in the waste water. Dosing units not supplying enough solution.

- Concentration of treatment chemicals too low.

- Mixer has failed. Consequently the required conditions cannot be established quickly enough.

- Measurement electrode or measurement amplifier or control equipment has failed.

- pH values (in detoxification processes) are incorrect or there are foreign substances in the waste water that affect the redox potential.

So as to determine the reasons for failures more quickly, a few of these possible circumstances should be tested initially by reference to the alarms: eg empty dosing tank from minimum level gauge, failed electric motor (mixer, pumps) by electrical checks on their operation (an indication that the unit is switched on is not sufficient here), solution level measurements in the chemicals storage tanks, which give an early signal as to whether there is a sufficient supply of chemicals available for one or two days, and so on.

Finally, controls at the end of the process are vital for supervision of plant functioning. A final pH meter and recorder at the outlet from the entire plant has long been obligatory in many countries. If potentiometric final measurements are used for detoxification processes, they must be installed in compartments that are joined to the outlets from the reactors, so that the redox potential measurement can be made at the pH value appropriate for the detoxification. These compartments should be equipped with mixers. If in the outlet to a plant certain parameters other than pH value are to be determined, then more or less sophisticated automatic analysers are available for this purpose. The samples taken are conditioned in line with usual, and as simple as possible, analysis procedures and then measured. Use is made in this connection of potentiometry, photometry and, where convenient, emission spectroscopy. Numerous measuring electrodes are available for potentiometric determination for analytical purposes, but of these only the fluoride-sensitive electrode has become generally accepted for measurements in waste water [61,62]. Its sensing element consists of lanthanum fluoride. It detects free fluoride ions as well as to some extent the fluoride present in hexafluorosilicate. The measuring range is between 0.02 and 19,000 mg/litre. Just as suitable, but still less established, is a membrane electrode for measuring ammonia in the range of 0.02 to 17,000 mg/litre, which Bellwied and Wilholm [63] recommend for final test measurements. For the same purpose, Oehme [64] describes an electrode for the determination of cyanide, which has as its sensing element a compacted mixture of silver iodide and silver sulphide. This allows cyanide concentrations between 0.026 and 260 mg/litre to be determined.

Continuous process treatment plant according to the present state of technology, including final filtration and selective ion exchange units as described in Section 4.4, is depicted in Fig. 4.26. Additional items not usually found in past years are principally:

- Dual measurement in the continuous process reactors
- Final testing in the compartments or separate vessels linked to the detoxification reactor outlets
- Final filtration of treated waste water
- Post-treatment of filtered waste water with selective ion exchangers
- Treatment of all concentrates and regeneration products in batch process plants.

Continuous process neutralisation plants for waste water from which dangerous components and solid materials are already removed, fall into a separate category; the pH value, which may result from metal elimination at very high pH values, must be corrected.

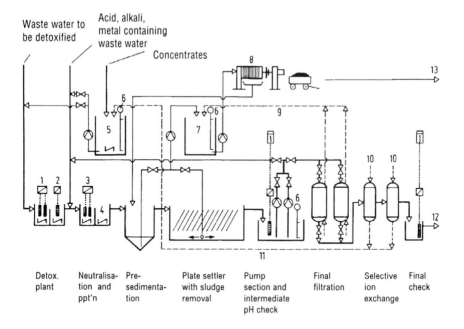

Waste water to be detoxified

Acid, alkali, metal containing waste water

Concentrates

Detox. plant	Neutralisa-tion and ppt'n	Pre-sedimenta-tion	Plate settler with sludge removal	Pump section and intermediate pH check	Final filtration	Selective ion exchange	Final check

Fig. 4.26 Continuous process treatment plant according to current state of technology (diagrammatic), with equipment to increase process safety (dual measurement, intermediate and final testing), more thorough elimination of heavy metals (final filtration, selective ion-exchange unit) and batch treatment for concentrates.

1 Dual measurement with difference value averaging for the detoxification, 2 Final testing for detoxification, 3 Dual measurement with difference value averaging for neutralisation/ precipitation, 4 Reaction compartment with slow-running stirrer for the flocculation, 5 Batch plant for treatment of concentrates, 6 Liquid height measuring gauge, 7 Storage tank for watery sludge, 8 Filter press, 9 Back-rinsing water of the final filter, 10 Inlet for regenera-tion media, 11, Regeneration products, 12 Inlet for waste water to be treated, 13 Sludge disposal.

4.3   BATCH TREATMENT PLANTS

4.3.1   Characteristics

As was pointed out in Section 4.1, a batch treatment plant should be used wherever possible, having regard to the quantities of waste water arising. Water saving measures should also be taken into account in considering this question, ie this Section should be read in conjunction with Sections 5.3 and 6.3.

Batch treatment plants display no residence time behaviour and are not bound by the disadvantages associated with this. They are filled with untreated waste water and emptied after treatment. As in batch treatment waste water is not passed through continuously, it must be retained in the plant until treated and then be transferred to the next treatment stage or to the point of discharge. It is clear then that much larger reactors are necessary than for continuous treatment processes. Generally they can be used only once daily, or at most, twice. The limits on reactor size are determined by the fact that problems of reactor technology arise in larger vessels when attempts are made to achieve good mixing (at high speed also) of the reactor contents with treatment chemicals. Whereas formerly, larger plants of the reservoir type were installed, the upper limits for modern tank plants are as a rule around 20 to 30 m^3. Consequently, the maximum plant capacity is also around 60 m^3/day.

Batch treatment plant is the safest arrangement for waste water treatment, as each step is controllable and critical matters can be dealt with manually. Furthermore, there is the advantage that the entire chemical treatment can be performed in a beaker in the laboratory and is easily scaled-up in the reactor. There are other benefits as well that arise in connection with these treatments, which do not occur in continuous process plants:

- Reactions which cannot be continuously analysed using physical methods. Chemical methods have to be used to monitor these.

- Slow reactions, eg detoxification of cyanide complexes of nickel. Reaction times of 10 to 12 hours are required for these.

- Treatment of waste water containing metal complexes. These often need as a two or treatment stages, which are not all amenable to analysis measurable and in continuous process plants can be carried out only with the resulting uncertainties.

- Carrying out several treatments one after the other in the same reactor.

Carrying out disposal of concentrates in existing reactors. The feeding of concentrated solutions into continuous process reactors has not been permitted for years in some legislations. It is to be expected that the other countries will come to follow this. The superior safety of batch treatment plants as compared with continuous process plants in the treatment of concentrated solutions should be mentioned in particular.

It should be noted here that only batch treatment can be considered for treating separate streams of waste water containing, for example, cadmium or volatile hydrocarbons. Incidentally, a whole range of processes, such as the emulsion treatment processes, (except for flotation), as well as almost all recycling processes, are in principle batch treatment processes. In recycling processes, the reactor should be seen as a loop-system linking over rinse or process baths.

If batch treatment plants are used in conjunction with water-saving measures, however, some new problems arise. Where highly concentrated solutions are to be treated, as opposed to where accumulated rinse water is processed continuously, significant temperature increases, depending on the reaction, result from exothermic processes; the effects of these on materials and in accelerating reactions have to be taken into account. Hydrogen peroxide, the use of which for cyanide detoxification is not possible in continuous process plants, because it makes no contribution to the redox potential necessary for analysis purposes, can be used with reservations in batch operations, where this is not a requirement. The heavy metal ions present (other than iron) catalyse the spontaneous decomposition of hydrogen peroxide. Copper ions, at least, display a marked catalytic effect in the oxidation reaction. The consequence is that a spontaneous reaction can set in if a temperature increase occurs. This happens most often if the greater part or the whole of the hydrogen peroxide is decomposed. As a result of this, there is increased gas evolution, foaming and overflowing of the reactor. In waste waters that contain only iron ions, eg as in the case in heat treatment shops, there is no problem. If non-ferrous metals are present, specialist companies should be consulted.

With certain waste water compositions, conflicts can arise. Owing to the increased concentrations of waste waters it is easy to arrive at excesses of C.O.D. or, if cyanide detoxification with sodium hypochlorite is being carried out, formation of AOX compounds. Further consideration must be given to neutralisation/precipitation at higher pH values, as the increasing influence of neutral salt concentration on the solid phase becomes noticeable (see Section 3.7.7), and may sometimes not be completely counteracted by raising the hydroxyl ion concentration. It is then necessary to precipitate with sulphide. Also the solubilities of materials that are not so very insoluble, such as for example calcium sulphate, are increased on the same grounds as the solubility of the metal hydroxide (see Section 3.10.2).

4.3.2   Waste water separation

Separation of waste waters is easily managed in batch treatment operations, as it is often possible to perform two or more treatments one after the other in the same reactor. This is possible, in contrast to continuous process treatment, because use of physical analytical methods is not necessary. For instance, cyanide detoxification and neutralisation/precipitation, or cyanide detoxification, nitrite detoxification and neutralisation/precipitation, or neutralisation/precipitation and chromate detoxification, or for smaller amounts of waste water, where no detoxifications are required, the treatment of complexant-free and complexant-containing waste waters, can be carried out in a single reactor. The separation of different effluents and the route they take during waste water treatment are shown in Fig. 4.27.

   The waste water produced from treatment in batch plants is usually considerably more concentrated than that from continuous process plants. It is not, as in
the latter cases, separated again into dilute effluents and concentrates, which are
then stored, but, corresponding to composition, the concentrates, regenerates,
semi-concentrates and relatively concentrated rinse water are collected together
in storage tanks, according to their make-up. If particular solutions are to undergo a recycling operation, they must be collected in their own storage tanks.
Regarding the basic types of waste water, the same applies here as was set out in
Section 4.2.2.

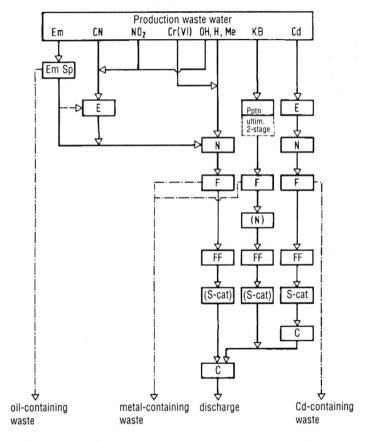

*Fig. 4.27   Waste water separation and passage of waste water through batch treatment
plants. Em-emulsion-containing waste water, CN-cyanide-containing waste water, Cr(VI)-
chromate-containing waste water, NO₂-nitrite-containing waste water, OH-alkaline waste
water, H,-Me-acid and metal-containing waste water, KB-complexing agent-containing
waste water, Cd-cadmium-containing waste water, Em Sp-emulsion separating plant, E-
detoxification plant, N-neutralisation and chromate detoxification plant, PP (Pptn)-
precipitation plant for metals, F-filter press, (FF) final filter, (S-cat)-selective cation
exchange exchanger, (C)-inspection shaft, () not necessary at present.*

4.3.3   Construction of batch treatment plants

The structure of a batch treatment plant reflects to its functions: storage, reacting and filtering. Different variations depending on the quantity of waste water produced, are shown in Fig. 4.28 [16,65]. The simplest version (a) can be employed for waste water that arises sporadically. The reactor is at the same time a storage tank for the waste water as it is generated and for the water requiring filtration after the reaction. If there is enough time available for filtration and the waste water is being produced constantly, then the combination (b) is sufficient. The configurations in (c) and (d) are batch treatment plants with quasi-continuous modes of operation. Correspondingly large storage tanks before and after the reactor are thus necessary. These can operate, however much waste water is produced and whatever the reaction and filtration times, which are usually the slowest processes in the treatment. Version (d) is the most common. If two reactors are used and charged alternately (c), then it is possible to operate for longer reaction times at the same throughput as in (d) and provide a further increase in process reliability [66,67]. In relation to the throughput capacity of batch plants, particular attention must be paid, not only to the reaction and filtration times, but also to the pumping times for filling and emptying the tanks.

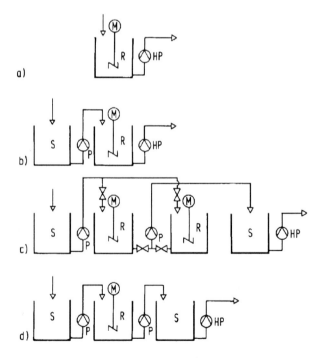

Fig. 4.28 Batch treatment plant variants (diagrammatic). See text for commentary.

Fig. 4.29 shows the arrangement of a quasi-continuous batch plant in its simplest form. In the past this required visual control of the filtrate. If it runs cloudy
initially it may have to be returned to storage tank II. Only when the filtrate runs
clear should it be switched over to final pH stage. In Fig. 4.30 a batch plant is
represented with all the additional equipment for post treatment. These are the
same as have already been described for increasing the safety of continuous
process plants. The final filters coupled to the outlet side are state-of-the-art
technology and so relieve the filter press of the responsibility for turbidity
control. Depending on needs and requirements, the layout of the batch treatment
plants lies between those illustrated in Figs. 4.29 and 4.30.

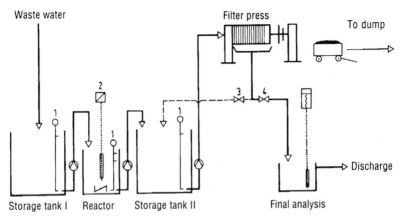

Fig. 4.29 Batch treatment plant with necessary turbidity analysis at the filter press.
1 Liquid height measurement gauge, 2 pH value measurement, 3 Return route for cloudy
filtrate, 4 Outlet for clear filtrate

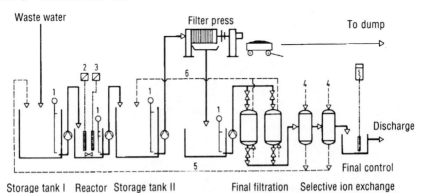

*Fig. 4.30 Batch treatment plant without turbidity control of the filter press
and additional equipment for more thorough elimination of heavy metals.*

1 Liquid height measurement gauge, 2 pH value measurement, 3 Redox potential measurement, 4 Supply of regenerating agents, 5 Regeneration product, 6 Back-rinse water
from final filter.

In particular cases, especially in the presence of very stable metal complex compounds, eg in printed circuit board technology, it is advisable on cost grounds alone to carry out metal precipitation in two stages. For example, in stage 1 that part of the metal is eliminated which is precipitated as hydroxide by increasing the alkalinity or as sulphide, and in the second stage a post-precipitation is performed, eg with dimethyldithiocarbamate and flocculation with iron(III) salts. Filtration is necessary in between, so that the added dimethyldithiocarbamate does not react with the more readily soluble precipitates and so lead to an increase in the consumption of the more expensive precipitating agent. The two-stage treatment is illustrated in Fig. 4.31.

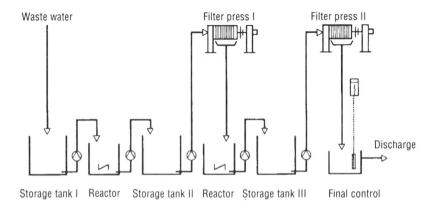

Fig. 4.31 Batch treatment plant with two-stage treatment

Figs. 4.32 and 4.33 show effective batch treatment plants.
    The component parts of the usual batch treatment plants are described below.

Fig. 4.32 Batch treatment plant. In front: cyanide detoxification plant, behind: neutralisation plant. Automatically fillable container for addition chemicals with absorption vessel for gases evolved during filling (GOEMA Dr Götzelmann GmbH, Vaihingen/Enz)

*Fig. 4.33   Batch treatment plant. Right: neutralisation plant, capacity 15 m³,
centre: detoxification plant, capacity 4 m³, left: selective ion exchanger for final
treatment (HAGER + ELSÄSSER GmbH, Stuttgart)*

4.3.3.1 Reactors

The volumes of the reactors should exceed the quantity of waste water produced
each day including the necessary treatment solutions. The latter constitute a not
insignificant proportion for more concentrated waste waters.

The reactor vessel can be round, square or rectangular in plan, but if rectangular its sides should not be too long. Its height should not be greater than 1.5
times the diameter or the width of the vessel. The materials are mostly
polyethylene, polypropylene and rubber-lined steel. The reactors are equipped
with mixers (see Section 4.3.3.2) and liquid level measuring devices (as dry-
running protectors for the pumps, and to set waste water quantity and liquid
level maxima). They are covered and should be connected with an air extraction
plant.

4.3.3.2 Mixers

The mixer, as in a continuous process plant, should be designed for a circulation
of 1.5 to 2 times per minute, depending on the vessel's capacity. The reaction
time for fast reactions is highly dependent on the mixing capability. The

reactants are brought together by the turbulence generated in the main flow, plus the turbulences produced at the edge and finally by diffusion. The sequence of mass transport reveals that the mixing time bringing about the reaction must be larger than the circulation time of the mixer. Voncken and Thoenes [68] found an empirical rule for turbulent mixing, according to which the mixing time to reduce the concentration difference to 0.02 of the initial value is four times as great as the circulation time of the mixer:

$$t_{M(0.02)} \approx 4\, t_U \quad [h] \tag{4.11}$$

This applies equally well, if the circulation efficiency is the same, to fast- or slow-running mixers being used. Slow-running mixers should be employed, therefore, if the neutralisation/precipitation stage is coupled with flocculation and/or floc growth stages. In round reactors the mixer should be located off-centre, as otherwise air vortices are formed, which lead to loading on the main bearing of the mixer.

4.3.3.3 Measurement and control equipment

Reactors should always be equipped with a pH measuring device, which provides an on-the-spot reading for manual control of the dosing units. In automatic operation, this must be achieved with a proportional controller.

In detoxification processes, an on-off control device cannot be used, as in continuous processes, because the degrees of alkalinity and acidity, the metal concentrations and their proportions, depending on the storage capacity, are so completely variable. The measurement potentials and the potential changes from treatment to treatment can also be very variable, thus calling automatic detoxification into question. Either manual dosing is undertaken and the occurrence of potential changes observed on the spot, or a so-called titration-regulator or tracking-regulator is installed [27,28,69]. This operates independently of the above-mentioned effects (see Fig. 4.34). The instrument first registers the measured value (M). There follows a dosing with a certain amount of detoxification solution (D) and afterwards a pause for adjustment of the potential (P). The measured value that follows is compared each time with the previous one, and so on. According to the potential change required the dosing is terminated, as soon as the movement in potential after a dosing addition is less than a predetermined difference value (E). The titration curve is then followed to the end of the potential change and the detoxification is concluded.

The instruments installed for the measurement of pH value and redox potential correspond to those in a continuous process plant. The immersed assemblies are equipped with a device, which guarantees that the electrodes remain immersed in aqueous solution even in an emptied reactor, so as to prevent their drying out. Externally situated flow-through types of measurement cell are rarely used. They have, it is true, the advantage that they can be automatically calibrated in a simple manner, but they are readily subject to mechanical interference by solid matter in the waste water.

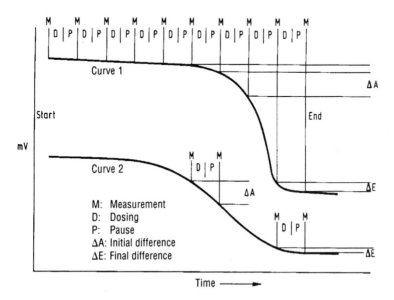

Fig. 4.34 Diagram of the mode of operation of a titration-regulator.
For commentary see text (Wilholm [26])

Finally, the end-point of a reaction can also be determined by chemical analysis. It is in this that batch process treatment has a direct major advantage, as only one part of the reaction taking place is followed by potential measurements. The means employed are therefore often very simple. They range from reagent paper and testing probes of measuring instruments, which have been developed solely for such a purpose, to simple chemical tests. Complicated analyses however, with an atomic absorption spectrophotometer for instance, may also be necessary.

4.3.3.4 Dosing units

The same dosing units can be installed as for continuous process treatment (solenoid valves, dosing pumps). Where no automatic treatment is involved, manually operated valves are necessary on the spot. As a rule manually operated valves should also be available in the bypass to the solenoid valves. Manual dosing can also be carried out via solenoid valves with the aid of switches at the main control panel.

The waste water that is treated in batch plants, is generally much more concentrated than the rinse water fromcontinuous process treatment. It thus has a higher density and viscosity. Consequently, and on account of the greater reactor contents as well, it is often advisable to mix the treatment solutions, not in a single tank, but with branched supply lines in several places. On the same grounds it is also advantageous to introduce the feeds under the surface of the liquid. Thus, faster mixing is achieved and reaction gases, which form are absorbed by the solution.

In batch treatment, hydrated calcium oxide is frequently added in solid form. This is done by a screw conveyor, which raises the powdery material from the bottom end of a silo over the top edge of the reactor, where it falls downwards through a short vertical length of pipe. Small amounts of chemicals, which are added in excess, such as for example sodium sulphide, organosulphide or iron salts, can also be added as solids, so long as they are fine grained or crystalline in nature.

4.3.3.5   Chemical containers and concentrate storage tanks

Modes of construction, material, accessories and installation have already been described in Sections 4.2.3.5 and 4.2.3.6.

As it is mainly concentrates that are treated in batch plants, their storage is facilitated if solutions that do not react with each other are stored adjacent to one another. After detoxification or if they contain no substances requiring detoxification, such solutions can be can be used for neutralisation, thus saving chemicals and reducing discharged salts to of to a minimum.

4.3.4   Operation of batch treatment plants for detoxification and neutralisation/precipitation

Depending on the different treatments required in a plant, several batch plants may be necessary. More detail on waste water separation is given in Fig. 4.27. In many cases it is possible to get by on one reactor, e.g. if chromic acid, acid, alkaline and metal-containing effluents are involved. If the air extraction plant of the reactor is connected to a scrubber that absorbs nitric oxide, then nitrite-containing waste water can be included with it in the following treatment. Cyanide-containing waste water always requires, a separate reactor coupled with a preceding storage tank; in that case, the whole of the waste water should be regarded as cyanide-containing. Complexant-containing or cadmium-containing waste waters also always need completely separate batch treatment plants up to and including filtration and post-treatment.

Cyanide detoxification. Concentrated waste water does not usually have to be made alkaline. It should already be so. Its oxidation can be carried out with sodium hypochlorite, with hydrogen peroxide or Caro's acid, if costs allow. Attention should be paid to the details in Section 3.2.1.7 for sodium hypochlorite and those in Sections 3.2.2.1 and 4.3.1 for hydrogen peroxide. Where there is sufficient time available for detoxification (several days), anodic oxidation (see Section 3.2.2.5) is the ideal process. If, however, the waste water contains chloride, then chlorine is evolved anodically and, if sodium hypochlorite is used, AOX compounds are formed.

The detoxified, alkaline, waste water can subsequently be used for neutralisation of acidic or perhaps chomate-containing effluents. The latter can either be pumped into the reactor in which the detoxification was carried out, or the detoxified waste water can be pumped into the neutralisation reactor. In the first case, the detoxification reactor must be equipped with a mixer, having a shaft with a triple support, as the detoxification should be performed in an only partly full tank.

Neutralisation/precipitation and chromate reduction. The neutralisation/precipitation of the acidic or acidic and chromate-containing waste water is effected, as far as possible, with the treated water from the cyanide detoxification plant or with alkaline (complex-free!) waste water. The treatment of the alkaline waste water can be carried out with acidic and also chromate-containing effluents. The rules in Section 3.7.8 apply in this case. If hexavalent chromium is present, then care must be taken that the whole of the waste water is alkaline (pH · 10). Chromate detoxification is carried out with iron(II) compounds and monitored with a test probe. The use of sodium dithionite is also possible. The reaction in an alkaline medium corresponds to Equation 3.86.

Nitrite detoxification. If nitrite is detoxified by oxidation, then the waste water can either be stored with the cyanide-containing solution or added to the cyanide-containing solution after cyanide detoxification. Then it is acidified to a pH between 3 and 5 and finally oxidised. In this way it is possible to make use of the neutralising capability of the originally alkaline waste water. It is better to collect the nitrite-containing waste water by itself and detoxify it in a smaller reactor, also at pH 3 to 5, with an amido acid, e.g. amidosulphonic acid. The reaction then proceeds while the pH is being measured (see Section 3.4.2).

Fluoride and sulphate precipitation. If fluoride is present, the neutralisation/precipitation must be performed using lime slurry. A test must be made, however, to confirm that the quantity of lime slurry needed for this is sufficient (see Section 3.10.1). If this is not the case, then additional dosing with calcium chloride solution may be required. Testing for the fluoride still present is carried out using test paper, or by observing whether in a filtered sample after addition of some calcium chloride solution, any turbidity still occurs, or best of all by potential measurement with a fluoride-sensitive electrode (see Section 4.2.6).

Calcium sulphate also is precipitated using the same procedures. Although it has a relatively high solubility in concentrated solutions, a much greater quantity can be eliminated than is possible in continuous processes (see Section 3.10.2).

In addition, this treatment brings about the co-precipitation of any phosphate present, so long as it is not already there in the form of relatively insoluble basic metal phosphate.

Treatment of complex-containing waste waters. Here a completely separate batch treatment plant is necessary in each case. If shared use is made of other reactors and filter presses, more soluble metal complex compounds are formed in the solutions left in the never completely emptied reactors and the hydrolysed metal salts on the filter cloths and the complexing agent that still remains. It is essential therefore to maintain the vertical separation shown in Fig. 4.27.

Precipitation, where stable metal complexes are concerned, is carried out with sodium sulphide or organo-sulphides at pH 7 to 8 (see Section 3.9.5.1.2). It is possible to use for this redox potential measurement combined with visual observation or automatic treatment with a titration-regulator. Oxidation of the complexing agent, except for cyanide, is not possible or only incomplete. So far nothing useful has emerged from research on processes, such as anodic oxidation or oxidation with ozone or with hydrogen peroxide under the influence of ultra-violet light (see Section 3.9.5.3). Incidentally, the equipment necessary for these includes batch plants as well, though of a completely different construction.

Treatment of cadmium-containing waste water. In many legislations, cadmium-containing waste water must be handled and treated separately (see Fig. 4.27). The same considerations apply concerning detoxification as for other waste waters. Neutralisation/precipitation is carried out at a pH value of 10.4 or higher, if caustic soda or lime slurry is used, or at pH 8.5 at least, if sodium carbonate is added. Furthermore, filtration to dewater the sludge, the final filtration and the treatment with selective cation exchangers must be carried out separately from those for all other waste waters.

Treatment of waste waters that contain hydrocarbons or volatile organics. These waste waters also are treated separately (hydrocarbons, see Section 3.13). For volatiles, measures in accordance with Section 3.11.3 should be adopted.

4.3.5   Sedimentation and separation of solids

After detoxification and neutralisation/precipitation, the complexing agent-free, complexing agent-containing and cadmium-containing waste waters are separated from precipitated solid matter. For this, each effluent is first transferred to a storage tank, which at the same time serves as the pumping vessel for a sedimentation unit or a filter (see Figs. 4.29 and 4.30, storage tank II). The sludge then thickens to some extent, depending on the time it spends in the tank. Normally the contents of the storage tank are conveyed through a high-pressure pump to a filter press. With larger quantities it is recommended that the suction side of the pump be supported by a float, thus drawing the waste water from the surface so that the surfaces of the filter are not loaded at the start with excessive amounts of solid matter, which can lead to a premature decline in capacity due to the resulting fall in filter pressure. In such cases, however, if no final filtration is to follow, special attention should be paid to any initial cloudiness of the filtrate which would necessitate its return to storage tank II. If a final filter is installed (see Section 4.4), these checks can be omitted.

If larger quantities of waste water are to be separated from solid matter, then it is quite permissible to insert, between the storage tank and the final filter, a continuously operating plate clarifier, which provides the highest possible rate of continuous sludge removal under given conditions, as explained in Section 4.2.5.3. This thin sludge is passed into a separate storage tank and from there to the filter press for dewatering (see Section 4.5.1.1).

The separation of solid matter in batch treatment thus takes place by compression in the storage tank or by separation in the plate clarifier. In subsequent filtration through the filter press, sludge is simultaneously dewatered. A final filtration unit is coupled to the outlet from the plate clarifier. As filtrate from the filter press has to be fed to the inlet of the layered clarifier, a turbidity check on the latter is not necessary. That is so, if the filtrate from the filter press in a plant in accordance with Fig. 4.30 passes through a final filtration step.

4.3.6   Safety equipment, final checks

In so far as the detoxification and neutralisation/precipitation processes are carried out automatically, the same arrangements are required for the dosing tanks and the chemical storage tanks as in continuous process plants. Safeguards at filling points and arrangements to guard against leakages are likewise obligatory here.

Final checks, except pH metering and recording, are not necessary, as these are made after each reaction. In automatically operated plants, it is an advantage to record the values measured during treatment, as by this means it is possible to determine the times and the frequencies of treatments carried out. Measurement of pH in the reactor is essential.

All reactors, storage tanks and other vessels should be equipped with liquid level measurement gauges (dry-running protection for supply pumps, pre-set filling level for reactors, overflow protection).

For automatic operation, it is preferable that plants as shown in Figs. 4.29 and 4.30 work on a quasi-continuous basis. In each case, however, the program should be automatically interrupted by an end point signal after completion of the last reaction and before transferring the reactor contents into storage tank II, so that the operating personnel can satisfy themselves that the waste water has been properly treated.

The continuation of the program must subsequently be triggered manually. It is precisely this interruption, allowing checks to be made by all possible means, which confers on the batch treatment plant its outstanding safety. There is also however a certain staff cost involved. That is especially so if reactions are undertaken which can be monitored only by chemical analysis and these cannot be caried out at all possible in continuous process plants.

For the determination of free cyanide, chromate, nitrite, fluoride, total metals and, in many cases, individual metals, a range of simple test methods are available. However because of the ever-present possibility of interference, checks should be carried out where appropriate.

4.4   POST-TREATMENT BY FINAL FILTRATION AND SELECTIVELY OPERATING ION-EXCHANGERS

4.4.1   Final filtration

Units of this type were earlier described in comments on continuous and batch process plant (see Figs. 4.26 and 4.30). There are no settling plants used in the metal finishing and metal-working industries which guarantee the capture of all solid particulates. The reasons for this are discussed in Section 4.2.5.1. Even when the solids content consists only of particles scarcely visible to the naked eye, these can suffice, if metal-containing, to bring the metal concentration over the critical 0.5 mg/litre level. Even when settling plants are greatly over-specified and not over-loaded in any way, this remains true. For complete assurance that excess metal concentrations are not exceeded, there is no alternative to a final filtration and this is so irrespective whether the precipitate is a hydroxide, a sulphide or organosulphide.

Final filtration can be carried out with a gravel bed filter, the operation of which as a deep bed filter, acts to the same effect as when it is used in recirculation of rinsewater, i.e. to remove small amounts of hydrolysis products (Section 5.3.2.1). In continuous plants, two such units operate on an alternating basis and the same practice is recommended for batch plants, not least because quite small units will often suffice. They are typically rated at 20 to 30 cubic metres/hour. As in the case of recirculating plants, the gravel bed should be overlaid with granulated hydroanthracite. The latter is not for adsorption but serves as a primary filtration medium which retains larger solids particles but is more readily cleaning by water and air back-flushing. The back-flush water is fed to the filter-press entry manifold. Fig. 4.35 shows a gravel bed filter used for final filtration.

A further possibility is the use of membrane filters, the pores of which are larger than those of an ultrafiltration membrane, but considerably smaller than those in a cloth filter. Such filters are known as Microfilters or cross-flow microfilters. The structure of the modular units corresponds to that for ultrafiltration (see Section 7.5.4). This is true also of the membrane materials employed. Microfiltration functions like a sieve. The waste water to be treated flows through the membranes in parallel, at velocities between 1.5 and 3 ms⁻¹. Thus the deposited residues are partly washed off again, so that significant back reaction occurs at the beginning, and constant filtration conditions then follow after a certain time. The pressure difference on the two sides of the membrane amount to 0.5 to 3 bar. In order to avoid an increase in the deposits, which are conditional upon the material and the concentration, the modules are back-washed at regular intervals [70-73].

Fig. 4.35   Gravel bed filter for final filtration of treated effluent (GOEMA).

Microfiltration enables particles to be separated that have a size over 0.05 μm. Because of the larger diameters of the pores compared with ultrafiltration the filtrate throughput is also higher. Ripperberger [70,72] recommends the use of this process in electroplating just for the final filtration. In a practical case, a filtrate flow of between 200 and 300 lm⁻²h⁻¹ was found in removing chromium hydroxide residues, so that final concentrations of less than 0.2 mg/litre of chromium were attained [70]. Tran [74a] even reports, for a membrane with pore sizes over 1 μm, throughputs of 340-680 lm⁻²h⁻¹ for filtrate in which the remaining metal concentration was less than 0.1 mg/litre. Nevertheless, considerably lower filtrate flow rates have also been mentioned elsewhere. Generally however, depending on the amount of solid matter, they tend to be above 60 lm⁻²h⁻¹. Achieving final concentration levels which lie below, and at time even well below, 0.5 mg/litre of metal, has been reported many times in the technical literature [70-72,74]. Difficulties seem to be encountered only with tin [71]. The initial concentrations lay in these cases at a few mg/litre, but in a couple of instances also at levels above 100 mg/litre [71]. Fig. 4.36 shows a microfilter, as used for these purposes (see also Fig. 8.10).

For filtration with an acceptable filtrate throughput the so-called continuous mode of operation is of interest. In this, to keep the deposition low and the filtrate throughput high, the waste water is circulated at great speed (e.g. 2 ms⁻¹) over the membrane and through a storage tank. Thus as much waste water is supplied continuously to the system as is removed as filtrate and passed as retained concentrate into the storage tank of the filter press. The solid matter concentration in the latter may reach levels up to 50 g/l [70]. In a two-stage treatment the filtration throughput is increased. Pilot trials are indispensable in individual cases.

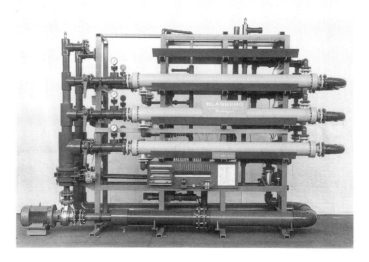

Fig. 4.36   Microfilter for final filtration of treated waste waters, membrane area 48 m², the plant is suitable also for direct filtration after neutralisation/ precipitation (BLASBERG Anlagentechnik GmbH, Solingen)

A prerequisite for maintaining filtrate output is regular back washing of the membranes, e.g. with filtrate, in the opposite direction to that normally taken, e.g. from the outer side to the inner side of the membrane. This should be done after every 3 to 10 minutes of filtration time. A back-washing operation lasts from 1 to 4 seconds, during which time there should be a throughput of about 0.3 to 1 lm⁻². The energy expended in microfiltration is around 2 to 20 kWh/m³ [72].

In principle, a pressure filter can also be used for the final filtration, if its filter medium possesses sufficiently small pores, as for example in the layers of filter aid in a precoated filter. As these operate work as surface filters, the filtration cycle tends to be short as a rule, because of the increasing pressure loss, which is caused by the layers of retained hydroxide.

4.4.2   Selective ion exchangers for post-treatment

If metal concentrations exceed the required levels, not because of solid matter but on account of dissolved species, then, provided it is present in cationic form, it can be retained by use of selective ion exchangers. In these, metals are present either as cations or as cationic metal complexes. In the latter case, the complexing agents themselves are partly cationic, so that they too can be absorbed (see Section 3.9.5.4). It should be noted that the regenerates contain further metal complexes, at this point in more concentrated form, which must be put through an appropriate treatment.

Furthermore, in the presence of tartrate and gluconate, metals can be removed by selective ion exchangers, if the pH value is kept at about 7, so that the complexing agent remains in the eluate and does not accumulate in the regenerate. If copper is present, the pH value must be lowered to 3.

Normally, dissolved metals in known forms are removed by selective ion exchange to a level below 0.5 mg/litre, but this is not certain in all cases. In some kinds of waste water, it is possible only to guarantee a value below 1 mg/ litre of metal. The reasons are not understood in detail. Prior tests should therefore be carried out on the waste waters in question. Details of the elimination of complete metal complex compounds can be found in Section 3.9.5.4 and on the properties of selective cation exchangers in Section 5.2.3.3.2.

In practice, two of these ion exchangers are installed, connected together one behind the other. When the first one is fully loaded, it is regenerated, while the following one continues to operate. Subsequently the regenerated ion exchanger is connected up again behind the other. This manner of working has the advantage that the second of the ion exchangers makes up for any bleed through by the first. As the resin employed here must be operated only at a relatively low specific impingement rate ($10 \ l \cdot l^{-1}h^{-1}$), large columns are needed in continuous process treatment plants. For impingment rate and regeneration see Section 5.2.3.3.2). Through the conditioning carried out after regeneration it is possible to ensure that the outflow from the exchanger lies in the neutral range, so that no subsequent pH correction is necessary. Fig. 4.37 shows a plant for post-treatment of waste waters with selective cation exchangers.

Comparable selectivity cannot be achieved with anion exchangers. Actually they have a high affinity towards anionic metal complexes, especially the strongly basic anion exchangers, but this is in the presence of numerous other anions, which would be bound only weakly, and not sufficiently to achieve the required metal elimination.

Fig. 4.37   Selective ion exchange plant for more thorough elimination of cationic heavy metals and heavy metal compounds present in already treated waste water, with gravel filter (GOEMA Dr Götzelmann GmbH, Vaihingen/Enz)

4.5   SEPARATION OF SOLID MATTER BY FILTRATION, AND SLUDGE DEWATERING

4.5.1   Filter systems with possible applications in the metal working industries

Filtration is one of the most important unit processes in waste water and recycling technology. The service requirements and the filter systems installed are of many different kinds, such as:

- Separation of solid matter. Many filter systems can be used for this. If removal of solid matter and dewatering of the solid product are needed at the same time (batch treatment), then chamber filter presses are used.

- Dewatering of solid matter. Because of their considerable dewatering capabilities, chamber filter presses are almost exclusively employed (see Section 4.5.1.1).

- Final filtration. This is performed using sandfilters and microfilters (see Section 4.4). Other systems are used in special cases.

- Filtration of recirculated water. This is done with sand filters, gravel-coke filters and very rarely with precoated filters.

- Separation of finely dispersed systems. Ultrafiltration (see Section 7.5.4) is used for emulsion separation (see Section 8.1.2) and maintenance of lacquer electrodeposition (electrophoretic) baths (see Section 8.6.2.2).

- Separation of solid matter that is very difficult to filter. There is a range of materials, which block filter media, are themselves largely impervious to water and as residues are difficult or impossible to remove from the filter medium. For these, fibre- or paper-band filters are used (see Section 4.5.1.2).

As this list already shows, there is a preference for specific filter systems for individual tasks. Even so, there are many other filter systems, which are employed in particular cases and which are briefly mentioned below:

- *Gravity filters.* Filter bags made of textiles and filter paper, as well as perforated cases lined with filter paper. Such very effective devices tend to be used in small and very small plants. Gravity filters also include filter drums, which are loaded from the inside, the deposited residue being withdrawn and taken out at the top end. Filters of this kind are well known for the removal of coagulated lacquer residues. Dewatering containers for hard-to-filter waste waters and sludges should also be mentioned here [76].

- *Vacuum drum filters.* These filters have the advantage that they operate continuously. At reduced pressures up to 0.7 bar, hydroxide residues are usually very thin (1-3 mm). The dry material content is about 20%. If any difficulties are encountered in removing the filter cake by doctor blades or rubber rollers, it may be necessary to work with a precoat layer, which is stripped off by a knife in a thin spiral layer along with the residue [77, 78]. Monitoring the drum and stripping of the precoat layer have an adverse effect on the costs of this unit.

- *Pressure filters.* Use is made of vertical and horizontal plate filters, candle filters and bag filters. In contrast to filter presses, these operate at low pressures up to about 6 bar. After opening out the set of plates horizontally or after opening a cover on the underside of vertical filters, the filter residues are ejected by operation of a shaking device or vibrator. A special design is the self-cleaning centrifugal-cleaning filter [79]. All pressure filters are operated with fabric filter media with or without precoating material. The dry material content is about 18 to 20%.

Descriptions of the different filter units can be found in many places in the technical literature [80-83].

Centrifugal appliances and hydro-cyclones should also be mentioned. They are used in special cases where large amounts of solid matter arise at the first stage of the separation. The thin sludge produced is subsequently dewatered further with the help of filter presses [84-86].

For the dewatering of sludges, the chamber filter press has proved itself the best over the course of the last three decades, so far as waste water from the metal working industries is concerned. It is almost exclusively employed for sludge dewatering. For solid matter that is very difficult to filter, band filters are relied upon to an increasing extent. The following descriptions refer therefore to these filter units only.

4.5.1.1 Chamber filter press

Filtration with the chamber filter press is a surface filtration with pressures applied amounting to 10 or 15 bar for hydroxide, depending on the high-pressure pump chosen. The filter medium also has an effect, for reasons ranging from the properties of the filter cloth to those of the sludge that is already deposited. People therefore refer to so-called cake filtration, which also constitutes a surface filtration, if the properties that are typical of deep filtration (see Section 5.3.2.1) still cannot be fully attained, even as the cake thickness increases. Finally, pore sizes and porosity change very considerably during the filtering process. Thus, the filter press operates with a throughput that changes during filtration. It is switched off and emptied after the slow rate drops below a certain level. Thus the filter press operates discontinuously.

The mode of operation of the chamber filter press is best described in terms of a Sectional view (see Fig. 4.38) and a diagram from Hoffmann [87] (see Fig. 4.39), which shows the filter flow and the build-up of pressure.

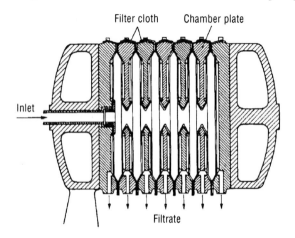

Fig. 4.38 Sectional view of a chamber filter press (RITTERSHAUS & BLECHER, Wuppertal)

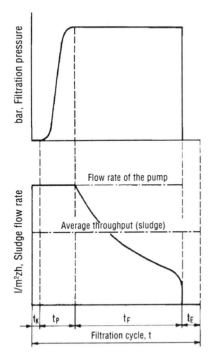

Fig. 4.39   Variation in throughput and pressure as a function of time in one filtration cycle (filter run) of a chamber filter press.

t_K Chamber filling time
t_P Pump filtration time
t_F Cake filtration time
t_E Emptying time [87]

The construction of the chamber plates, in contrast to that of the frames of the frame filter press, allows high working pressures. As shown in Fig. 4.38 they are covered with special fitting filter cloths. Complete with filter medium, these plates are pressed and self-sealed at high pressures between the thrust- and head-pieces of the filter frame. Filtration takes place as the spaces formed between each pair of chamber plates are loaded, until they are completely full of residue cake. After the press is opened the filter cake usually falls out spontaneously.

The time-dependent course of filtration is shown in Fig. 4.39 [87]. First the empty chambers are filled, at a flow rate determined by the pump, without increase in filter pressure and without yielding any significant amount of filtrate. The time taken is called the chamber filling time, t_K. This is followed by the pump filtration time t_P with a constant flow of filtrate corresponding to the pump's supply rate. During this phase the filtration pressure rises up to a value as high as the high pressure pump is able to go. As soon as this is reached, the cake filtration time t_F begins. At a pressure that now stays constant, the sludge flow rate falls and with it the filtrate flow. In practice the cake filtration time is continued so long that the filter chambers are filled with filter cake. Only then by achieving acceptable dewatering is it possible to ensure trouble-free removal of the cake when the filter is opened. The cake filtration time depends on the properties of the sludge and is established by trial and error. With hydroxide sludges it is terminated if the sludge flow rate has fallen over the filter run to 25 or 15% of the average flow rate. There follows the emptying time of the filter t_E.

This takes an average of about 15 minutes. The sum of the times quoted ($t_K + t_P + t_F + t_E$) gives the time for a filtration cycle (t), also known as a filter run.

The average sludge flow rate, taken over the cycle time, can be calculated by integration. It is the equivalent of a continuous sludge flow rate for normal hydroxide sludges of around 60 to 65 $lm^{-2}h^{-1}$ as a rule, depending on a filterability between 40 and 80 $lm^{-2}h^{-1}$. At the end of the filter run it falls to about 10 $lm^{-2}h^{-1}$ [87]. If the sludges are difficult to dewater, then the filtration time is increased and the sludge flow rate at the end of the cycle is lower. The use of lime slurry for neutralisation/precipitation improves the filterability in general, as the sludge is more porous owing to the insoluble lime component.

The filter cakes have solids contents between 25 and 40%, depending on the filterability and the atomic weight of the metal on which the insoluble compounds are based. If the precipitation is with lime, the solid matter contents are nearer to the upper limit, but the amounts of residue obtained are greater.

If the sludge is thin with a low solid matter content (about 1%), the cake thickness has a large effect on the fall in sludge throughput, while at higher solid matter contents (over 3%) the effect is much less noticeable [87].

In order to guarantee trouble-free operation of the sludge dewatering plant, the sludge storage tank, which serves as collector for the chamber filter press, must be made sufficiently large.

Piston diaphragm pumps (up to 15 bar) and eccentric helical pumps (up to 10 bar) are used as high pressure pumps. Their feed capacity is based on double the filter capacity. Too large a pumping rate, however, produces no corresponding additional filter capacity. It does shorten the the chamber filling time and the pump filtration time, but makes only a small contribution to increased filter capacity. In larger plants, however, it is worthwhile to install a second pump to shorten the two first-named times [88].

The plates of the filter chambers are nowadays manufactured almost without exception, of polypropylene and have edge lengths between 300 and 2000 mm. Only the largest with edge lengths over 1200 mm are manufactured from spheroidal graphite cast iron. The chamber depths amount to 10 mm (up to 1200 mm edge lengths) and 15 mm above that. The cake thicknesses are correspondingly 20 or 30 mm, respectively. Chamber filter presses are loaded with up to 140 chamber plates at the maximum. Each filter chamber has a separate outlet into the filtrate collection channel, so that any turbidity resulting for example from damage to a filter cloth can be detected immediately.

As filter media, cloths made from polypropylene, polyamide and polyester have superseded those made from natural fibres. The use of monofilamentary cloths is beginning to be preferred. They have a high permeability and the filtercakes fall away easily after the press is opened. The high permeability does, however, often lead to turbidity at the start of the filter cycle, but this rapidly ceases as the filter cake is built up [89-91]. Because of this property of the filter medium, it must always be possible to return the filtrate to the pump collector of the press, as in the sedimentation system.

The closing of the press takes place at pressures of 300 to 350 bar. At plate sizes up to 800 mm this is possible with manually operated hydraulics. Above this electro-hydraulic closure must be used [87,88]. Larger press units are today automatically driven. The shutting off of the filter takes place after attaining a minimum output of filtrate. If the the volume flow in the filtrate collecting channels is so low that the filtrate can no longer reach a certain height above an in-built weir, then the conductivity measurement at this point is interrupted and the high pressure pump switched off. Thereupon, the chamber plates, which hang from rollers running on rails, are pulled apart by a conveyor mechanism and the filter residues fall into the container placed below. After this, the automatic process should be interrupted, so that a safety check can be made that the whole of the residue has fallen out. Then the automatic process can be continued with the closure of the filter and, after attaining the closing pressure, the switching on of the high pressure pump. For the protection of the operating personnel, automatically controlled filter presses are fitted with photo-cells, which immediately stop the automatic plate transport if a light beam passing along the equipment is interrupted.

Figs. 4.40 to 4.42 show filter presses of different sizes and types.

Fig. 4.40 Filter presses for small amounts of waste water,
plate size 40 × 40 cm, 5 m² filter area. Right: piston diaphragm pump
(BLASBERG Anlagentechnik GmbH, Solingen)

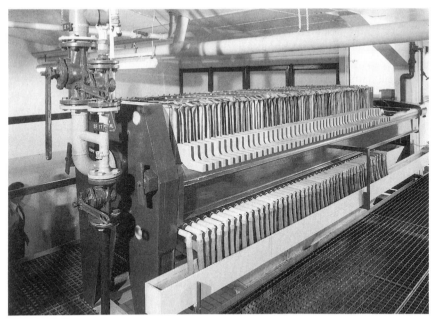

Fig. 4.41   Filter press for moderate amounts of waste water,
plate size 80 × 80 cm, 40 m² filter area, with automatic cut-off
(GOEMA Dr Götzelmann GmbH, Vaihingen/Enz)

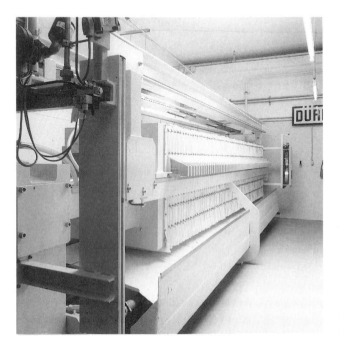

Fig. 4.42   Automatically operated chamber filter press for larger amounts of waste water, plate size 100 × 100 cm, 116 m² filter area, with safety light beam (Dürr GmbH, Stuttgart).

4.5.1.2   Paper and non-woven fibre band filter

Solid materials, which are finely dispersed, gelatinous or sticky, can be filtered through filter presses only with difficulty or not at all. Either a turbid flow is maintained for too long, or too impermeable a residue is formed, which very rapidly reduces the filtrate flow to an insignificant level, or too great a difficulty is encountered in removing the cake. In all cases the filter cloth is so obstructed that after the cake has been removed it needs to be cleaned, if this is at all possible. Also, if pressure filters are being used with a precoating of filter aids, only short filter cycles can be achieved. Satisfactory filtration of solid matter with such properties can be carried out in two different ways:

- The characteristics of a more easily filtered material is conferred on the solid matter by mixing in metal salts, e.g. iron or aluminium salts, which are hydrolysed by the addition of alkali. The metal hydroxides permeate the solid matter and enclose them in their voluminous flocs. This method, a kind of flocculation, in fact increases the proportion of solid matter, but contributes to its better filterability. Another possibility is to mix the waste water containing solid matter with inert materials such as kieselguhr. These are incorporated in the filter cake and make it more porous. In many cases after such a treatment it is even possible to separate and dewater it in filter presses.

- The poor filtrability is accepted and the still thin cake is discarded together with the filter medium, much as is done using vacuum barrel filters operated with precoat layers.

The principle of the latter method is best seen in the diagram of Fig. 4.43. In the simplest case the filtration takes place by gravity through a band of filter paper, which lies on an endless trough-shaped conveyor belt, made for example of wire netting.

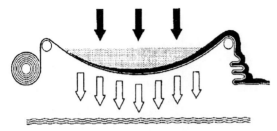

Fig. 4.43 Principle of paper- or fibre-band filtration (Deutsche Filterbau GmbH, Siegen)

If the separated residue of increasing thickness is so compact that the outflow of filtrate falls, the liquid level above it is raised. The conveyor belt supported on a float is then put into operation as the transport mechanism, so that new filter band is available and the filtration process can be continued. The filter band laden with solid matter falls into a container. Before filtration there is generally further flocculation or addition of a filter aid, to improve the porosity of the residue and increase the filter performance. The mode of construction and of working of these filter machines has been described many times, mostly however for other application areas, such as the purification of cold lubricating agents and phosphating baths [92-95]. Because of the mixing of the filter media, they are employed in the metal working industries for the following tasks [96, 97]:

- Solid matter separation in chemical separation of emulsions using flocculation.

- Solid matter separation from waste waters arising in machining and grinding.

- Separation of polymeric residues in the treatment of resist waste water in printed circuit board manufacture.

Normally band filters operate under gravity and so at a static pressure of only 10 to 20 mbar, resulting from the filter's stationary head of liquid. But the possibility exists also of accelerating this type of filtration by raising the pressure difference. Fig. 4.44 illustrates, as well as the example of gravity filtration, both filtration with increase of pressure and with reduced pressure. Fig. 4.45 shows a band filter in operation.

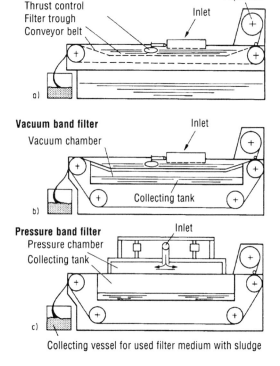

Gravity band filter
Thrust control
Filter trough
Conveyor belt
Inlet
Paper roll
a)

Vacuum band filter
Vacuum chamber
Inlet
Collecting tank
b)

Pressure band filter
Pressure chamber
Collecting tank
Inlet
c)

Collecting vessel for used filter medium with sludge

Fig. 4.44 Paper or fibre band filters for different modes of operation.

a   Gravity filtration
b   Vacuum filtration
c   Pressure filtration
(FAUDI Feinbau GmbH, Oberursel)

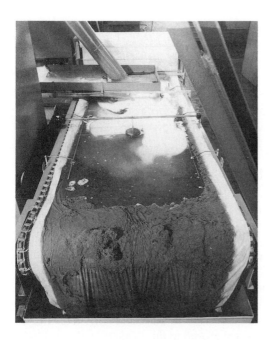

Fig. 4.45 Band filter of non-woven fibre fabric in operation (FAUDI Feinbau GmbH, Oberursel)

4.5.2   Dumping and further processing of sludges

Sludges for disposal are the solid residues arising from waste water treatment. As they consist partly of hazardous materials in insoluble form, they must generally be taken to special waste dumps. As the latter have to comply with numerous requirements, locations for such special waste dumps are limited. Thus there is a need to avoid as far as possible the accumulation of residues for disposal, and, where it is unavoidable, to reduce their amount by further measures (reprocessing and recycling), and in so far as such measures may not be available at the present moment, not to exclude their introduction in the future.

A whole range of measures are known for avoiding or diminishing the generation of sludge:

- All possible ways of reducing the material carried off in the waste water (see Section 6.2).

- All measures that lead to a reduction in the quantity of waste water and so allow harmful substances to be precipitated from concentrated solutions (see Sections 5.3, 6.3.2 to 6.3.5).

- All recycling processes for Reclamation of complete metal-containing process baths from rinse water (see Sections 7.6.3.1, 7.8.2.1.2 and 7.8.2.2.2).

- Removal (recovery) of metals from rinse water (see Sections 7.2.1, 7.4.1, 7.6.2.2 and 7.6.3.1).

- Lengthening the service life of short-lived metal-containing process baths (see Sections 7.2.2, 7.4.2, 7.6.2.2, 7.6.4 and 7.7.2).

To these may be added, where possible, the use of mechanical surface treatment processes in the place of chemical and electrolytic processes.

4.5.2.1 Composition of sludges

A knowledge of the composition of the sludges produced is required for their disposal and for their further processing as well. It also affects, at least partly, their behaviour after dumping. In so far as sludges arise from the maintenance of process baths in the course of production, e.g. in the filtration of nickel electrolytes with active carbon additions or of iron phosphate sludges from phosphating baths or of carbonate sludges from cyanide-containing electrolytes, it should be noted that the water incorporated in them contains all the dissolved substances in the process baths. These sludges must be mixed in waste water containing the same harmful substances, then subjected as well as this to appropriate treatment. This leads to generation of more of the usual waste water sludges. In many cases it may be sufficient to wash the sludge residues in the filter.

The residues generally called hydroxide sludges, which as a rule after dewatering in chamber filter presses have a water content between 60 and 75%, can contain solid matter in various forms [98]:

- Hydroxide, hydrated oxide, basic salts of each of the metals that have got into the waste water.

- Relatively insoluble iron salts, e.g. molybdate, vanadate, tungstate (alloy steel pickling plant).

- Relatively insoluble calcium compounds, e.g. fluoride, sulphate, phosphate and carbonate, if neutralised with lime slurry. If there are metals present, phosphate can occur as highly insoluble metal phosphate or basic metal phosphate.

- Inert material, above all silicic acid. It can come from the alkali silicates of cleaning baths and the insoluble components of hydrated calcium oxide deriving from lime slurry and silicate-containing impurities, which get from ground water into the effluent.

- Relatively insoluble metal cyanides in very small amounts, if the cyanide detoxification is carried out in continuous process plants (see Table 3.3 [99]). In incorporated water they can also dissociate into cyanide that may easily be set free. During the longer reaction times possible in batch plants, metal cyanides are also oxidised and do not find their way into the sludge. Furthermore, cyanides can in addition be present as ferricyanide, though in non-oxidisable form.

- Materials which are absorptively bonded to the active surfaces of hydroxides and basic salts. These are mainly organic substances. Where degreasing baths are in use, an average 0.5 to 2% petroleum ether extractable compounds can be detected in the dry material. It is mainly because of the adsorbing capacity of hydroxides that the limits for hydrocarbons are satisfied, especially if no other measures to eliminate them are adopted. Also if mineral oils in waste water are excluded, 0.2 to 0.6% of extractable petroleum ether compounds can be detected in many sludges, which stem from surface active agents and organic electrolyte additions [54,98], even inorganic substances, used as neutralisation reagents, are adsorbed (see Section 3.7.2). The adsorption of neutral salts has often led in the past to an assumption that basic salts have been formed.

Hydroxide sludges from plating companies are differentiated basically according to the types of materials and their quantitative relationships. This is apparent from a comparison of numerous analyses, as in the compositions reported by Lohmeyer [100], Ott and Raub [101-103] as well as the author [98]. Various sludge compositions arise even in the same factory, as the simultaneous use of different process baths, the losses resulting from exchange of workpieces and the time in service of short-lived process baths during their lifetime, and the surface area throughput of product are subject to continuous fluctuations with time. Generally this is true even from one filling of the chamber filter press to the next. As the sludges of an individual factory are collected together before being taken away in containers to a dump or to an intermediate storage area within the factory, it is scarcely ever possible to provide actual sludge analyses for a load of sludge transported to the dump. So long as no change is made in the factory's production or recycling procedures, it is not only easier but safer and more correct as well, for the average values of many sludge analyses to form the basis for disposal of the sludges. Hydroxide sludges with one predominant metal are produced in the pickling of unalloyed steel and in anodising shops, if no heavy metal-containing electrolytes for colouring or heavy metal-containing sealing baths are included. The sludges arising can then be regarded as harmless and should not occupy expensive space on special waste dumps. Furthermore, because of the new waste legislation, more hydroxide sludges are being obtained, each which are "mono-metal sludges" contains only one of the metals copper, nickel, zinc and chromium. They are particularly likely to arise if separate rinse systems are installed (see Sections 6.3.2 and 6.3.5).

The production of sludges containing metal sulphides is tending to increase because of the stricter requirements with regard to heavy metals. In printed circuit technology it is common for copper sulphide-containing sludges to be generated. Sulphide-containing constituents can also occur in the hydroxide sludges discussed above. Sludges that result from the use of sulphides are also always characterised by the presence of iron sulphide and iron hydroxide, which are a consequence of the addition of iron salts to combine with excess sulphide.

Silicate-containing sludges are of a special kind, and may also contain other inert materials. Among these are the residues from enamelling works and grinding plants. The sludge from the latter, however, can contain metal hydroxides and metal sulphides as well as metallic and abrasive debris. Generally there are hydroxides of flocculating metals (iron(III) compounds and aluminium compounds) and silicate-based materials added to improve filtrability. Sludges containing large amounts of organic materials result from particular processes or manufacturing procedures:

- Printed circuit board manufacture. In the treatment of waste waters containing water soluble resists. In this case polymers and adhesives may be present, which are normally mixed with iron or aluminium hydroxide to stop them sticking together and to improve their filtrability.

- Mechanical processing. In the adsorption of oil resulting from chemical emulsion breaking on to adsorbents such as kieselguhr. The residues after gravity filtration contain up to 60% of oil.

- Lacquering operations. Lacquer residues, which are produced by the coagulation of lacquer-containing waste water and from the stripping of lacquers.

Sludges of this kind should be put through an incinerating plant to reduce their volume.

4.5.2.2   Behaviour of sludges at the disposal site

Care must be taken over the behaviour of solid residues if they are to be deposited on waste dumps. A range of physical, chemical and biological effects are involved there. However, it is not only a knowledge of the properties of the fresh sludges, such as for example their solubility in rain water and the initial influence of pH value and neutral salt content, that is important, but also that of those of the modified sludges produced as the result of ageing and weathering. Even the properties of the unaged sludges do not closely approach those of the freshly precipitated material. At the time when the residues reach the dump they have already changed in behaviour, which can in the main be put down to ageing processes and above all to the dewatering operation. These changes result principally from the loss of their large surface area. As this also implies a loss of energy, their solubility in water is reduced still more. With the loss in surface area, however, goes a fall in adsorptivity. This means that adsorbed substances may sometimes be desorbed again.

The changes that take place on the dump should, however, be considered in terms of the influence of the longer periods of exposure and especially also of the effects of the climatic conditions. Wolters [104] and Bucksteeg [105] have made a systematic study of these and distinguished between ageing and weathering.

Ageing is principally the result of an anaerobic process occurring in the presence of moisture. It starts immediately the sludge is dewatered in the factory and continues with its deposition, overlaying and compacting on the dump. The interfacial energy of the hydroxide produces a reversion beyond the formation of hydrated oxide to that of oxide, and thus the desorption mentioned above occurs, followed by the release of adsorption and capillary water. That is the reason why dumped sludges after longer outdoor exposure appear to have a higher water content (pasty consistency, thixotropic properties), although their solid matter content has in fact risen. The pH value is higher because of the desorption of adsorbed neutralising chemicals. Consequently, and because metal sulphide is formed as a result of the anaerobic production of hydrogen sulphide from sulphate, the metal compounds become more insoluble. These processes completely or partly inhibit the tendency of carbon dioxide to produce the more readily soluble metal hydrogen carbonates (bicarbonates).

Weathering in contrast to ageing, is an aerobic process. Its consequences are a disintegration of the deposited wastes and an increase in their solubility through reaction with the components of the atmosphere dissolved in rain water. It always starts at the surface and progresses inwards according to climatic conditions. Through weathering effects, such as the formation of crevices due to drying out or frost, erosion, displacement of washed out material, etc (physical weathering) new surfaces are formed, which become subject to chemical action. The reactive gases from the air and various combustion products (oxygen, carbon dioxide, sulphur dioxide, nitrogen oxide and so on) which are dissolved in rain water lower its pH value and confer oxidising properties upon it. Sulphuric acid and nitric acid can also be generated from the sulphur dioxide and nitric oxide. The final products of weathering are therefore more readily soluble compounds such, such as bicarbonates and basic metal salts. If organic substances are present, they can be converted to carbon dioxide, nitrate, sulphate and phosphate under the influence of microorganisms, so contributing to the effects of weathering.

Ageing and weathering are interrelated one to the other. At a certain depth, which depends on the degree of compression of the deposits and the climatic conditions, a boundary layer forms, which is subjected alternately to both processes. As a result of repeated dissolution and precipitation a thicker cementation layer is produced, which ultimately impedes or prevents further penetration of the effects of weathering.

An area of a special waste disposal site, which is loaded with sludges from the metal working industries, contains materials whose properties sometimes show many stepwise variations depending on the weather and stratification levels, right up to the those of the freshly deposited sludges. The percolating waters are derived from precipitation and their components depend on the neutral salts present and on the solubilities of the solid materials under different conditions. In addition, the pH of rain water and the substances dissolved in it have an influence. The compositions of the percolating waters are therefore only qualitatively determined by the water leached from fresh sludges.

Numerous leaching investigations by different authors [98,100,102 and 103] with various old sludges demonstrated that the metal values over long periods, if extracted by the waters described, had concentrations of only fractions of a mg/litre up to a maximum of a few mg/litre, as experienced in extreme cases on disposal sites occupied by inorganic residues. Readily soluble materials such as neutral salts are always removed after three extraction steps (in accordance with DIN 38414, Part 4). If metal-containing sludges are disposed of separately on regulated sealed dumps, the percolating waters that are formed can easily be treated by conventional methods and because of their low metal concentration with supplementary flocculation based on iron(III) salts. Acutely toxic substances, which may be produced when disposing of organic residues, do not occur here. Investigations undertaken by Kayser and his co-workers [106] on percolating water at a German central disposal site in Gelsenkirchen revealed metal contents which for copper, chromium, zinc and cadmium were under 0.1 mg/litre and only for lead were about 0.1 mg/litre.

4.5.2.3   Reprocessing of deposited sludges

There has been little inclination to reduce dumping by the reprocessing of metal-containing sludges. Partial success is achieved however only if this is done at the start of the recycling process sequence, and if unadulterated, concentrated solutions and sludges can be transported to the reprocessing facility. Small proportions of calcium, chromium, iron and also nickel in non-ferrous metal sludges upset hydrometallurgical separation processes, and so sludges arising in non-ferrous metal works are not processed, especially as only small amounts may be available in a single place.

The first thought regarding the reduction of discharges or lessening of the harmfulness of metal-containing sludges is thermal treatment. Distinctions should be made between drying at temperatures under 200°C and incinerating at temperatures over 800°C.

A reduction in volume is produced by drying sludges, in that a sludge leaving a chamber filter press with about 30% dry matter loses about 60% of its original volume if the aggregates formed on heating are broken up. From a dry matter content of about 80%, sludges are described as dried. For metal-containing sludges, drying is rarely undertaken inside the factory. Simple cylindrical rotary kilns or tunnel kilns with travelling grates, in which the hot air moves in a direction opposite to that of the sludge, would be most appropriate for this. It is advantageous if use can be made of waste heat, e.g. from an incinerating plant. Schmidt-Burr [107] describes a drier for metal hydroxide sludges, which operates on the principle of thin layer evaporation. The material is spread out 1 to 2 mm thick on a rotating body, which can be heated by steam, hot water or heat-transfer oil.

With the exception of the sludges discussed in Section 4.5.2.1 which have a high organic content ratio, incineration of sludges from the metal working industries is not possible, as they contain scarcely any combustible substances. The sludges mentioned that do contain organic material occur, however, in such small amounts that separate incinerating plants are not worthwhile, even though direct incineration would be ideal for their disposal. The best solution is to burn them along with other industrial organic waste or with refuse.

Metal-containing sludges may contain very small amounts of organic substances (oil, surface-active agents, brightening additions, etc), which make scarcely any contribution to combustion. Their thermal treatment at high temperatures is more of a calcination. The metal compounds are converted into oxides whose solubility becomes lower and lower as the temperature increases. As the sludges contain sulphides, but also basic salts and adsorbed neutral salts, sulphur dioxide, hydrogen chloride, hydrogen fluoride and nitric oxide may also be emitted. Furthermore, vapours of metals and their salts may be produced, especially at very high temperature, and cause dusts to be generated. Appropriate waste gas purification methods therefore require to be adopted. Further volume reductions compared to the dried sludge are practically undetectable.

Because of incorporated alkaline water, adsorbed alkalis and undissolved calcium oxide, metal-containing sludges are, as a rule, alkaline. If they contain chromium(III) compounds, these are oxidised to chromate under the influence of the air supplied for combustion. De Jong [108] demonstrated chromate even in chromium-containing sludges dried at about 105°C. In combustion furnaces, oxidation in the presence of an excess of oxygen can even be quantitative. As Rehn and Nietz [109] as well as Nietz [110) established, there is at combustion temperatures of about 700 up to 800°C a maximum for the formation of water-soluble chromate. At higher temperatures acid-soluble chromates are formed increasingly. This must be taken into account when disposing of sludges that have been treated in combustion furnaces.

Incineration of metal-containing sludges, together with organic industrial wastes or house refuse, is carried out only in existing large plants on special disposal sites, large industrial facilities and sometimes also in municipal incinerating plants. Established incineration plants are described by Reh [111], Shin [112] and Bradke [113].

Experiments should be performed to maximise the insolubility of metal-containing sludges. Thus Brown [114] determined in sintering trials that the solubility of the sinter product changed with the treatment temperature and that practically insoluble residues were obtained at between 1100 and 1200°C. Additions of glass waste did not produce the benefits hoped for.

Ruml and Soukop [115] submitted sludges, which had been adjusted with lime slurry to an alkaline pH of 8.3 to 10, to a chromate reduction with silicate (sand or clay) and iron(II) sulphate, thickened the mixture with coal dust and combusted it at 1200°C. After cooling a glassy product is formed, practically insoluble in water.

In the first half of the 1970s, special consideration was given to the reprocessing of plating shop sludges, and in some places a record was kept of the sludges produced. Because of the presence of unwanted chromium and iron, an oxidising leach was proposed, with subsequent chromate separation by ion exchange. Finally, copper and then zinc were to be removed by selective liquid-liquid extraction. After the precipitation of aluminium it was planned to recover the nickel also by liquid-liquid extraction [116]. Wackernagel [117] in the mid-seventies reported a sludge yield of 37,257 t/a, in one part of Germany from which a metal content of about 0.9 t per registered factory was calculated. A projection for the electroplating shops of the whole Federal German Republic, based on average metal contents for sludges and bearing in mind the author's reservations concerning such an estimate, indicates annual amounts of about 8580 t each of copper and nickel, about 14,580 t each of chromium and zinc and about 216 t of cadmium. Müller [118] calculated from the results of a study of 460 factories in the area of Southern Germany having suitable sludges with a water content of two thirds relative to the dry matter 5.85% copper, 1.92% nickel, 6.72% zinc and 4.59% chromium. Hydrometallurgical recovery is attractive only if there are sufficient amounts of recoverable metal and stable metal prices. According to the information available however they would have increased from a maximum of 8 million DM in 1976 [116] to about 13 million DM in 1980 [118].

In these considerations it should not be forgotten that the proportion of suitable sludges with appropriate levels of the metals mentioned is constantly decreasing. The reasons for this are the increasing measures taken to reduce drag-out losses, to encourage the separation of purer, more concentrated metal solutions by special rinsing procedures and by recycling processes. This development is continuing. It can be seen also from sludge investigations over the years 1985 to 1987, as reported by Dietl [119]. The metal content of the dry matter, compared with the values quoted above, were about 2.8% copper, 2.1% nickel, 7.5% zinc and 2.2% chromium, i.e. the concentrations of the metals copper and chromium have already fallen by a half while those of nickel and zinc remain relatively stable.

Metal recovery from hydroxide sludges using liquid-liquid extraction appears to have been gaining in importance recently [119]. This is largely connected with the growing shortage of suitable disposal sites and of financing hydrometallurgical metal recovery by offsetting increases the costs of disposal. A helpful trend is the production of sludges containing predominantly one metal and especially of separate pure sludges. In order to avoid the difficulties that chromium causes in reprocessing, Böhm and Bischof [120] suggest that chromium-containing waste waters also, after separate detoxification and neutralisation/precipitation procedures, should be submitted to a solids separation process and sludge dewatering. The recovery of cadmium sludges is already no longer difficult, since regulations require cadmium-containing streams to be treated separately. In any case, sludges containing mainly a single metal which are acknowledged to be suitable for reprocessing, should be disposed of separately, so as not to impede their possible reprocessing at a later date. The stimulus for this was apparent as early as the year 1966 in the scheme described in a pamphlet on the organisation and control of industrial and commercial waste disposal [121].

A further consideration regarding relief of the burden of disposal of hydroxide sludges concerns their use in brickworks [122]. In this application the solid component of the sludge can be regarded as an additive and its water component can contribute to water saving in brick manufacture. Care must be taken that:

- A water content of 20% must not be exceeded in the processing of the brick raw materials.

- The solid matter ratio of sludge/clay must not exceed 1:25.

Brick manufacture using hydroxide sludges has already been piloted [123]. Difficulties are caused mainly by variations in sludge composition. Further investigations are necessary before implementation. Fewest objections are likely in the production of inner wall components, where an excess of more than 2.5% by weight of dry matter from sludge should be avoided. Several waste gas investigations in brick manufacture have revealed that the emissions of sulphur dioxide and hydrogen fluoride are lowered when sludge additions are made, while those of hydrogen chloride are distinctly increased [124].

References for Chapter 4

1.   German legislation WHG. 5th Ed'n BGBl. I Nr 50 (1986)
2.   German legislation 40th App. 8.9.89
3.   Götzelmann, W.: "Implementing new effluent treatment legislation" Metalloberflaeche 40 (1986) (Aug)
4.   Hartinger, L.: "Means for minimising heavy metals in electroplating effluent" Galvanotechnik 79 (1988) 401-7
5.   Weikard, J.: "Reactors for carrying out liquid-phase reactions" (in) Ullmans Technical Encyclopaedia Vol. 3
6.   Thoenes, D.: "Basics of chemical reaction technology" (in) Ullman's Technical Encyclopaedia Vol. 1
7.   Jola, M.: "Operation & Planning of Effluent Holding Tanks" Pt. 2 Oberflaeche-Surface 22 (1981) 37-45
8.   Villemonte, J.R., Rohlich, G.A., Wallace, A.T.:"Hydraulic and Removal Efficiencies in Sedimentation Basins."Third International Conference on Water Pollution Research. Munich Section II, Paper 16 (1966)

9.   Kraft, A.: "Technical aspects of water & effluent treatment using flocculating agents" Wasser, Luft Betrieb (1966) 809-816

10.   Pahl, M., Muschelknautz, E.: "Installation & lay-out of static mixers" Chem. Ing. Techn. 51 (1979) 347-364

11.   Bradke, H.J.: "Classical Detoxification Processes" Galvanotechnik 64 (1973) 556-573

12.   Hitzemann, G.: "Effluent treatment from the metal-working & finishing industries using batch & continuous methods" Galvanotechnik 61 (1970) 556-573

13.   Friedr. Blasberg GmbH & Co. KG, "Effluent treatment" (Techn. Bulletin)

14.   Hartinger, L.: "Effluent treatment for metal finishing" (Techn. Note, LPW Co)

15.   Hartinger, L.: "Batch & continuous processes prove themselves for metal finishing effluent treatment" Metall. (Austria) Nr 9. (1973)

16.   Hasler, J.: "Batch, continuous & direct effluent treatment processes" Proc. Training Semins. TAE Esslingen (1977 to 1988)

17.   Thoenes, D.: see ref. 6. pp 213-291

18.   Polland, K.O.: "Stirring mechanisms" (in) Hartinger, L.: "Pocketbook of effluent treatment" Vol. 2 pp 477-488 (Hanser, Munich, 1977)

19.   Kuntze, A.: "pH measurement" (in) Hartinger, L.: (ibid)

20.   Oehme, F.: "Applied potentiometry – a Guide to Measurement of pH, Redox & Ion concentrations" Polymetron AG Zurich (1969)

21.   Wilholm, G.: "Measurement & Control Technology" Proc. Training Semins. TAE Esslingen (1979-1988)

22.   Kuntze, A.: "Redox Potential Measurements" (in) Hartinger, L.: "Pocketbook of Effluent Treatment" Vol. 2 (Hanser Verlag Munich)

23.   Gablenz, E., Michel, E.: "Control technologies" (in) Hartinger, L.: ibid.

24.   Jola, M.: "Monec – a computer program for industrial pH, redox & conductivity measurement" Chemie-Technik 12 (1983) 10 pp 30-39

25.   Berg, F.: "Analytical methods in metal finishing effluent treatment" Chem. Rundschau 34 (1981) 3. 1-2

26.   Wilholm, G.: "Measurement & control techniques in effluent treatment & recycling" Galvanotechnik 71 (1980) 1102-1107

27.   Wilholm, G.: "Process control in demanding situations" Galvanotechnik 79 (1988) 1136-1140

28.   Steinert, G., Wilholm, G.: "Greater safety in process monitoring of effluent treatment plants" Chemie-Technik 11 (1982) (1) 49-51

29.   Pauli, W.: "Dosing valves" (in) Hartinger, L.: "Pocketbook of effluent treatment" Vol 2. pp 571-587.

30.   Vetter, G.: "Dosing pumps" (in) Hartinger L; ibid

31.   Reuter, J.M.: "Use of organic polymer flocculating agents in effluent treatment from the metal industries" Umwelt (1981) (1) 27-35

32.   Scordialo, A., Reuter, J.: "Continuous effluent clarification with synthetic organic flocculants" Maschinenmarkt 81 (1975) 1137-1138, 1143.

33.   Thomas, P.: "Neutralising alkaline effluents with carbonic acid" Chemie-Technik 8 (1979) 73-74

34.   Fuchs, H.: "Plants for neutralising combustion gases" Wasser Luft Betrieb 19 (1975) 224-228

35.   Cramer, F.: "Chemical treatment of industrial effluents" Wasser Luft Betrieb 13 (1969) 422-429

36. "Turbulent reactor for treatment of industrial effluent" (Ges. f. Umwelttechnik mbH, Essen)

37. Pini, G.: "Using static mixers for treatment of plating shop effluents" Metalloberflaeche 33 (1979) 190-195

38. Bahensky, V.: "An automated flow-through effluent treatment plant" Oberflaeche-Surface 11 (1970) 222-226

39. anon. "Effluent neutralisation" Wasser Luft Betrieb 18 (1974) 693

40. Hartinger, L.: "Use of pressure filters for effluent treatment in the metal working industries" IWL Forum 69/III-IV "Mechanical dewatering of industrial sludges". Cologne 1970

41. Gundelach, W.: "Sedimentation" (in) Ullmans Technical Encyclopaedia

42. Schramek, H.: "Flocculation & sedimentation" Wasser Luft Betrieb 19 (1975) 219-224

43. Jola, M.: "The functioning & sizing of effluent holding basins. Pt 1" Oberflaeche Surf. 22 (1981) 7-12

44. Gütling, W., Trapp, W.G.: "Studies on horizontal lengthwise flow-through basins" Wasser Luft Betrieb 15 (1971) 407-410

45. Neumann, H.G.: "Improving performance of effluent settling tanks. Post settling in fully biological treatment systems" Letters, Abwasser 18 (1971) 236-240

46. Jaisle, J.: "Calculations for sludge thickeners" Chemie. Ing. Techn. 36 (1964) 1127-1133

47. Reuther, H.W.: "Modern settling tanks" Techn. Note 1. VDI Working Party, Iserlohn 1977

48. Resch, H.: "New basis for calculations on vertical flow-through post settling reactors in biological treatment plants" Wasser/Abwasser 122 (1981) 236-242

49. Resch, H.: "Construction & sizing of vertical flow-through post-settling reactors in biological treatment plants" Wasser/Abwasser 29 (1982) 468-475

50. Kronenberger, G.: "High performance sludge thickener for treating mineral materials" Aufbereitungstechn. 23 (1982) 648-9

51. Didier-Werke AG, Königswinter: "Settling cyclone" (Techn. Note)

52. Möller, U.: "Dewatering trade & industrial sludges" Industrie Abwasser 1963, pp 14-20

53. Zimmer, E.: "Theory & practice of sludge dewatering" Chemiker Ztg 95 (1971) 224-236

54. Hartinger, L.: unpublished work

55. Meins, W.: "Plate settlers" (in) "Pocketbook of effluent treatment," L Hartinger, Vol 2 (1977)

56. Schade, H., Sapulak, A.: "Parallel plate settlers" Letters, Abwasser 31 (1984) 104-111

58. Schüssler, R.: "Compact settlers" cav (1983) March, 56-64

59. Oswald, E.: "Compact settlers" Galvanotechnik 75 (1984) 956- 957

60. Götzelmann, G.: "Feasibility of reaching new discharge limits with new effluent treatment plant" Metalloberflaeche 40 (1968) (Aug)

61. Oehme, F.: "Direct potentiometric concentration measurement with ion-selective electrodes in process solutions" Galvanotechnik 72 (1981) 373-383

62. Metrohm AG, Herisau: "Specific ion electrode EA 306F" Techn. Note (1971)

63. Bellwied, U., Wilholm, G.: "Ammonia determination using an autoanalyser to monitor effluent discharges" Galvanotechnik 72 (1981) 823-826

64. Oehme, F.: "Need for and methods for cyanide determination" JOT (1974) May 44-51, Sept 30-34

65. Hartinger, L.: "Industrial effluent with inorganic contents" Lehr u. Handbuch Abwassertechnik. Vol VII, pp 276-279 Verlag Ernst & Sohn Berlin (1985)

66. Agfa Co. "Fully automated effluent treatment plant" Galvanotechnik 53 (1962) 194-196

67. Schuster, H., Hartinger, L.: "Automated batch treatment plant "Siemens System" – a little known effluent treatment method" Industrieabwasser (1968) 49-54

68. Voncken R.: "Basics of chemical reactor technology" (in) Ullmans Technical Encyclopaedia

69. Jola, M.: "Tracking Regulator for control of batch treatment effluent plants" Oberflaeche-Surface 18 (1977) 173-181

70. Ripperger, S.: "Microfiltration in metal finishing" Galvanotechnik 75 (1984) 566-569

71. Hartmann, M.W.: "Microfiltration as an alternative to sedimentation for neutralised plating wastes" Deutscher Kommunal Verlag GmbH Dusseldorf (1984)

72. Ripperger, S.: "Crossflow microfiltration of process solutions" Chemie-Techn. 16 (1987) Nr 4. 36-44

73. Szarafinski, D.: "Sizing of crossflow plants" Chemische Produktion (1988) May 93-8

74. anon; "Cost-effective treatment of plating effluent" Oberflaeche + JOT (1988) Nr 4. 94-96

74a. Tran, T.V.: "Advanced Membrane Filtration Process Treats Industrial Wastewater Efficiently" CEP (1985) March, 29-33.

75. Spanier, G.: "Residual metal removal by ion exchange after effluent electrolysis" Galvanotechnik 68 (1977) 705-709

76. Manzei, J.: "Criteria for sludge dewatering containers" Wasser Luft Betrieb (1985) Nr 3. 14-15

77. Rüb, F.: "Thickening & dewatering of effluent sludges" Wasser Luft Betrieb (1964) 463-467

78. Schwalbach, W.: "Importance of filtrate pipe system in vacuum drum filters" Chemie-Techn. 9 (1980) 569-571

79. Trommelschläger, G.: "Use of centrifugal cleaning filters in chemical applications" cav (1983) Apr. 15-18

80. Alt, Chr; "Filtration" (in) Ullmans Technical Encyclopaedia

81. Werner, S., Litzenberger, F.: "Filtration equipment" Chemie Ing. Techn. 36 (1964) 1285-1290

82. Rüb, F.: "Filters for sludge dewatering" Wasser Luft Betrieb 11 (1967) 651-566

83. Benninghoff, H.: "Filtration – equipment & technology" Chem. Rundschau 37 (1984) Nr 1/2 13-18

84. Hemfort, H.: "Overview of separators with nozzles – self-cleaning & fully shrouded barrels" Chem. Ing. Techn 51 (1979) 479-484

85. Brunner, K.H.: "Use of separators for treatment of industrial effluent" Chemie-Techn. 12 (1983) Nr 2. 55-60

86. Trawinski, H.F.; "Latest developments in processes for mechanical phase separation by centrifugal forces" Chemie-Technik 12 (1983) 15-21

87. Hoffmann, E.: "Sludge dewatering plants" (in) Hartinger, "Pocketbook of Effluent Treatment," Vol. 2

88. Merz, A.: "Netzsch chamber filterpresses for industrial & municipal sludge dewatering" Fachber. f. Metallverarbeitung (1979) Nr 1/2 26-28

89. Ammermann, K.: "Comments on selection of filter cloths" Wasser Luft Betrieb 10 (1966) 748-753

90. Anon; "Use of technical filter textiles" Wasser Luft Betrieb 15 (1971) 229-235

91. Bremus, H.J.: "Criteria for use of cloth filters" Chem. Ing. Techn. 53 (1981) 433-438

92. Faudi Feinbau GmbH, Oberursel "Filters for water treatment" Wasser Luft Betrieb 21 (1977) 600-601

93. Wenzel, K.H.: "Belt filters for cleaning cutting & cooling oil emulsions" Werkstatt u. Betrieb 112 (1979) 557-558

94. Schmid, H.P.: "Belt filters for dewatering gypsum sludges" Die Chemisches Produktion (1985) March 14-18

95. anon; "Automatic fibre fleeces as belt filters for solution purification" Chemie-Technik 15 (1986) 86

96. Hinz, H.E.: "Vibratory finishing and its effluents" Galvanotechnik 74 (1983) 151-155

97. Prüller, H.: "Treatment of effluent from vibratory finishing in suitable stages" Maschinenmarkt 89 (1983)

98. Hartinger, L.: "Problems with effluent sludges" Galvanotechnik 64 (1973) 582-588

99. Weiner, R., Leiss, C.: "Kinetics of cyanide destruction" Metalloberflaeche 25 (1972) 169-204

100. Lohmeyer, S.: "Reaction of electroplating & other sludges with the atmosphere at industrial dumping sites" Galvanotechnik 65 (1974) 759-768

101. Ott, D., Raub, C.J.: "Studies of sludges from the metal-working industries in terms of their dumping or recycling" Metall 31 (1977) 862-868

102. Ott, D., Raub, C.J.: "Elution studies on hydroxide sludges from metal-working industries" Metall 31 (1977) 1307-1310

103. Ott, D., Raub, C.J.: "Behaviour of plating sludges after dumping" Galvanotechnik 72 (1981) 1288-1301

104. Wolters, N.: "Ageing, weathering & leaching of dumped solids" Wasser Luft Betrieb (1965) 154-156

105. Bucksteeg, W.: "Dumped solid wastes and their effect on surface & groundwaters" gwf 110 (1969) 529-537

106. Kayser, R., Knoch, J., Stegmann, R., Geissler, Ch.: "Studies on seepage from rubbish tips" Wasser Abwasser Vol. 10. Ges. z. FIS, TH Aachen (1973)

107. Schmidt-Burr, P.: "Sludge disposal" Galvanotechnik 78 (1987) 1246-1247

108. De Jong, B.: "Standardising laboratory tests for solids determination" ref. 106 pp 619-634

109. Rehn, F., Nietz, S.R.: "Critical study of solid waste treatment by incineration" Galvanotechnik 56 (1965) 462-466

110. Nietz, S.R.: "Study of chromate formation in incineration of chromium containing species" Galvanotechnik 64 (1973) 998-1006

111. Reh, L.: "Incineration & pyrolysis of liquid & sludges from industrial wastes" Chemie Ing. Techn. 39 (1967) 165-171

112. Shin, K.C.: "Development of municipal incinerators in Germany" Wasser Luft Betrieb 20 (1976) 63-69

113. Bradke, H.J.: "Incineration plants" (in) Hartinger L, Pocketbook of Effluent Treatment, Vol. 2

114. Braun, R.: "Problems in treating of inorganic industrial sludges" DECHEMA Monograph Vol. 64 (1970)

115. Ruml, V., Soukop, M.: "Sludge treatment after effluent neutralisation in plating plants" Metalloberflaeche 38 (1984) 101-102

116. Müller, W.: "Options in recycling of plating sludges" Galvanotechnik 67 (1976) 381-383

117. Wackernagel, K.: "Review of the solid waste disposal position in the metal finishing industry" Metalloberflaeche 31 (1977) 43

118. Müller, W.: "Processes for recovery of non-ferrous metals and their compounds from plating sludges" Galvanotechnik 69 (1978) 450-452. Also (in) Seng H J et al. "Waste Disposal by Recycling" Vol. 42 expert verlag, Grafenau (1980)

119. Dietl, F.: "Reprocessing plating sludges" Galvanotechnik 78 (1987) 2797-2802

120. Böhm, E., Bischof, J.: "Options for producing more easily recoverable plating sludges" Galvanotechnik 78 (1987) 3564-3567

121. AflA Techn. Note G7 "Regulations for storage of industrial & trade wastes"

122. Seng, H.J.: "Options for recycling metal hydroxide sludges in brick works and implications for their status as hazardous wastes" (in) Waste Disposal by Recycling" expert verlag, Grafenau (1980)

123. Hilger, E.: "Effect of metal hydroxide sludges on the manufacture & properties of bricks" (in) ref. 122. pp 191-200

124. Quellmalz, E.: "Health & safety in use of metal hydroxide sludges in brick works" (in) ref. 122 pp 201-5

5 Ion-exchange processes

5.1 POSSIBLE APPLICATIONS IN THE METAL WORKING INDUSTRIES

The high performance ion-exchange units generally available today make it possible to shift the equilibrium of ionic reactions to the right or to the left as desired. This combined with the option often adopted of carrying out the forward and back reactions repeatedly, creates numerous opportunities for the application of ion exchangers in waste-water and recycling technology, but most particularly in the latter.

Ion exchange methods are used for the following processes:

- Recirculation of rinsing water (see Section 5.3)

- Recovery of materials from rinsing water (see Section 7.2.1)

- Regeneration of process solutions (see Section 7.2.2)

- Final purification of treated waste waters using selective ion exchangers (see Section 4.4.2)

These possibilities are shown diagrammatically in Fig. 5.1.

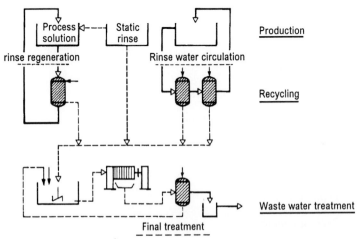

Fig. 5.1 Applications of ion exchange processes in the metal working industries

5.2   ION EXCHANGE RESINS

5.2.1   Definition

Ion exchangers used today are based on synthetic resins with substituted active ionic groups, which swell, but are insoluble, in water. They thus comprise electrolytes in solid form. Ions of particular kinds diffuse out of the aqueous phase into the water-containing ion-exchange phase where they are exchanged at the active exchange groups for equivalent quantities of similarly charged ions. Detailed descriptions of the properties and functions of ion exchange resins are to be found in Griessbach [1], Nachod and Schubert [2], Kunin [3], Buser, Graf and Grütter [4], Martinola and Naumann [5], G Kühne [6] and F Helfferich et al. [7].

The synthetic polymers that confer insolubility on the ion exchangers consist of copolymers of resins suitable for the purpose, self-polymerisable organic compounds such as styrene, styrene derivatives, acrylate, acrylamide, and their reaction products with divinylbenzene (DVB), the agent that produces a zeolitic structure. Fig. 5.2 shows the structures of two ion exchange resins based on polystyrene and polyacrylate. In polystyrene resins the active ion groups may be present in the original monomer, but may also, as is most often the case today, be substituted after polymerisation. In polyacrylates and polyamides the ion exchange groups are reported to be present in the initial monomers [5].

Use is made of 'pearl-polymerisation' in the production of ion exchange resins, the polymerisation is performed in suspensions of the starting materials. As a result the ion exchange materials acquire a spherical form, particle sizes between 0.3 and 0.5 mm being usual commercially. If liquid hydrocarbons or higher alcohols are added to the reaction mixture, then these can be taken out again after polymerisation by vaporisation or extraction. The pores that their removal produces are responsible for the macroporosity of many resins.

Three criteria are of major importance in relation to the physical properties of ion exchange materials:

- Immobilised-ion concentration.

- Zeolitic structure (microporosity).

- Macroporosity.

The chemical properties depend on the following criteria:

- Type of immobilised ions.

- Type of counter ions.

- Type of resin matrix..

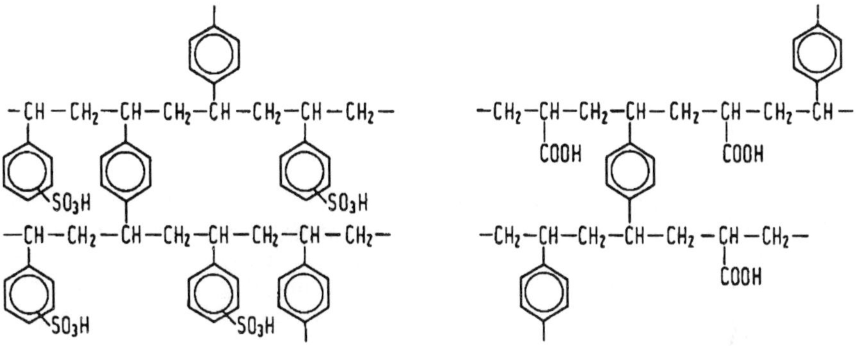

weak acid cation exchanger
(zeolite-structured polyacrylic acid)

strong acid cation exchanger
(zeolite-structured polystyrene sulphonic acid)

*Fig. 5.2   Chemical structure of ion exchange resins. In anion exchangers,
instead of acid groups, there are tertiary or quaternary ammonium
groups as shown in Table 5.1*

5.2.2   Physical properties

Swellability, i.e. a capacity to absorb water, is a prerequisite of an ion exchange
reaction. So that an ion can reach the active exchange sites in the resin, it must
diffuse into them and then the exchanged ion must diffuse out again. Also, so
that the chemical reaction can proceed, the ion to be exchanged must be present
as such, i.e. the reaction components and end products must be able to dissociate
in the exchanger phase. By taking up water the resin makes diffusion and
dissociation possible. Physically, it constitutes a gel.

In the swollen condition, the ion exchange resin is thus a solid electrolyte,
whose phase-boundary surface can be regarded as a semi-permeable membrane.
As the concentration of immobilised fixed ions (active exchange groups) is
relatively large (about 0.5 to 4 litre of wet resin), an osmotic force develops if
the external solution is less concentrated, which tends to dilute it. This force
increases, the higher the fixed concentration (the capacity of the resin) and the
lower the concentration of the external solution (see also Section 7.5.1). The
resin expands to an extent depending on the osmotic pressure developed within
it (swelling). Its expansion is a function of the elasticity of the zeolite structure,
and of course increases with the degree of zeolitic structure. The pressure developed in this state of equilibrium is called the swelling pressure. It determines the
operating volume of the resin and is dependent on the following dimensions and
conditions:

- Extent of zeolitic structure of the resin.

- Capacity of the resin (concentration of fixed and counter ions).

- Valency of the counter ions. The higher this is for a given resin capacity,
  the lower is the swelling pressure.

- Volume of the hydrated counter ions.

- Tendency to association. If the number of particles becomes lower owing to the formation of undissociated groups, e.g. undissociated carboxyl groups in weakly acidic cation exchangers in an acid medium, then the swelling decreases along with the osmotic pressure.

- Concentration in the aqueous solution. As this increases, the osmotic pressure in the resin falls as a result, and along with it the swelling.

Knowledge of the osmotic properties of ion exchange materials is just as important in choosing the right resin as knowledge of their chemical properties. The osmotic properties in particular have to be borne in mind in devising special regeneration processes and in designing ion exchange columns.

The water content of swollen resin is reported to be 40 to 90%. The volume increases, which occur in charging and regenerating, lie between 5 and 60%, depending in each case on the relevant charged species with the smallest volume. The swelling pressures generated can exceed 100 bar.

The swellabillity of exchange resins is controlled in their manufacture by the choice of degree of zeolitic structure. This is determined by the percentage of added divinylbenzene, which usually lies between 6 and 10% DVB. With too low a zeolitic structure the resin swells too strongly and might in extreme cases, if it contained many polar, active exchange groups, tend to disintegrate. Too high a zeolitic structure on the other hand gives rise to embrittlement of the resin and to insufficient swellability. This can lead to a tearing apart of the resin particles by osmotic pressure. In contrast, resins with increased zeolitic structure are resistant to oxidation. Depending on the application of the exchange material, the resin manufacturer has thus to weigh up the influences of the fixed ion concentration and of the degree of zeolitic structure, especially as the diffusion controlled reaction rate is also controlled by these. The zeolitic structure is furthermore responsible for the microporosity of the resin within the gel. Its dimensions lie between 10 and 30, A° and is thus of the order of magnitude of many ions and molecules. Its sieving effect can be distinguished by determining dye absorption at different degrees of zeolitic structure as illustrated in Fig. 5.3.

Another important criterion for the physical behaviour of exchange resins is the macroporosity. In contrast to the microporosity, it is characterised by the passages and cavities that can be occupied by the aqueous phase. According to Martinola and Richter [8] the microporosity dimensions fall within the size range from 300 to 500 A° (3 to 5 $\times$ 10^{-2} μm).

Macroporosity confers further advantages on exchange resins:

- Diffusion conditions for ions are improved. They arrive more rapidly at the active groups situated in the interior of the beads of exchange resin, i.e. the diffusion controlled exchange rate is increased.

- Because of the larger pore diameters, bigger ions can be admitted as well, and an important point can also be recovered in the regeneration process.

- Reversible adsorption processes are made possible. Because of the porosity which has a value of 25 to 50 volume percent [8] the exchanger has large surface area, in the range of 45 to 85 m²/gm so that it behaves like active carbon with respect to non-ionic, large-molecule, organic compounds (see Section 7.7.1 for physical data on adsorption). It even has advantages in comparison with it, as the greater part of the adsorbed matter can be desorbed again during regeneration.

- Increase in mechanical stability and resistance to osmotic and thermal stresses.

The much better adsorption properties of a macroporous resin compared to a gel-resin are illustrated in Fig. 5.3 for strongly basic anion exchangers [5]. The surface-enriched structure of a macroporous resin can be seen in the electron microscope picture in Fig. 5.4. At the same magnification, a gel-resin displays only a homogeneous mass.

Macroporous resins are today preferred for waste water treatment in the metal working industries, so as to prevent irreversible processes due to the bonding of organic materials, which are always present in such waste waters. This is considered in more detail later.

The particle size of spherically-shaped exchanger particles lies between 0.3 and 1.5, at most 2 mm. Because of the variation in the length of diffusion paths, it affects the exchange rate and, when in continuous use, the size of the exchange zone and hence the useful capacity. In practical applications however this particle-size spectrum has scarcely any detrimental effects, as it is tightly specified for individual batches of resin. Too low a range of particle size might lead to too great a drop in pressure. Granules that are too large cause detrimental boundary effects in smaller-diameter columns, which lead to enlarged reaction zones and/or impaired selectivities. These effects can be eliminated at the manufacturing stage. The spherical shape has the advantage of very uniform packing and low mechanical losses during operation and particularly in back-flushing.

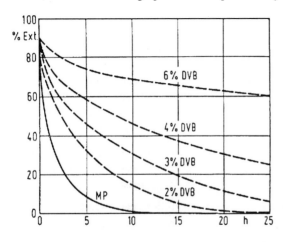

Fig. 5.3 Adsorption of a dyestuff (acilan naphthol red G) by strongly basic anion exchangers with various degrees of zeolitic structure (Lewatit 500) and by a macroporous resin (Lewatit MP 500). (Martinola, & Naumann [5])

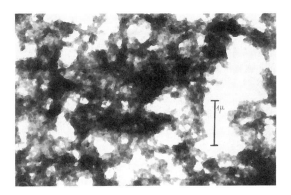

Fig. 5.4 Electron microscope image of an ultra-microtome section of a macroporous exchange resin. Magnification 100,000 times (Bayer AG, Leverkusen)

The temperature resistance of the resin, so far as continuous rinse water recirculation is concerned, is well in excess of requirements. In the treatment of more concentrated solutions, care must be taken with elevated temperatures especially because of possible chemical effects. In general, precautions need to be taken wherever such circumstances arise.

5.2.3   Exchange reactions

This section is concerned primarily with the ion exchange reaction equilibrium in the static system, which has to be changed for use in column technology (see Section 5.2.4). Reaction rates are in the main dependent on the diffusion processes at the interfaces between phases and in the exchanger resin gel. Affinities displayed by the individual types of resin in reacting with the ions have an influence on the selectivity and therefore on the competing equilibria as well. These concepts, which are important for the chemical behaviour of the ion exchangers, are dealt with in the following sections.

5.2.3.1   Reaction equilibrium

If an ion exchanger is loaded with ions of type $A^{(\pm)}$ in a vessel with a solution containing ions of type $B^{(\pm)}$ having the same charge, then the following reaction occurs, irrespective of whether it is a system with a cation exchanger and cations or an anion exchanger and anions:

$$R - A + B^{(\pm)} \rightleftharpoons R - B + A^{(\pm)} \qquad\qquad 5.1$$

R = resin component including active exchange groups

   An equilibrium is thus established, in which all four reaction participants are present simultaneously but in different proportions.

To describe this equilibrium two aspects have to be borne in mind, one physical (the Donnan equilibrium) and one chemical (the law of mass action.). Both have their advantages in particular circumstances, but neither is completely appropriate in the present context. While in the law of mass action only the cations or anions participating in the exchange reaction are covered, in the Donnan equilibrium the charges and the dynamic properties of the ions situated on both sides of a membrane are involved.

Donnan equilibrium

If on two sides of a membrane there are solutions with different types of ion and if the membrane is not permeable to one type, then the ions of the other kind will endeavour to set up an equilibrium on both sides.

In ion exchange materials immersed in a dilute electrolyte, the phase boundary constitutes a membrane. If it is a cation exchanger, this membrane is not penetrable by the negatively charged immobilised ions, which are present in relatively high concentration. The counter ions A^+ associated with these (cations) are freely mobile within the gel-resin. In the aqueous solution there are cations present of the type B^+ (counter ions) and the anions belonging to them Y^- (co-ions). The latter are not present in the resin phase and seek to diffuse into it to set up an equilibrium. As there is also a tendency operating to achieve electroneutrality in both phases, a particular equilibrium (Donnan equilibrium) is also brought about by the diffusion of co-ions into the resin. If there is a high fixed ion concentration and a low electrolyte concentration in the solution, the diffusion of co-ions in the exchanger phase remains low (Donnan exclusion). The higher the valency of the co- ions Y^- and the lower the valency of the counter ions B^+, the lower this is. To preserve electroneutrality in the two phases, inward diffusion of counter ions of one kind must be offset by outward diffusion of the other kind. In anion exchangers the behaviour is similar with the charges reversed.

If the concentration in solution increases, the Donnan exclusion loses its effect and electrolyte invasion of the exchange resin occurs.

Law of mass action

If the law of mass action is written for the reaction Equation 5.1, the equilbrium constant K is obtained, if ions A and B are both assumed to be monovalent:

$$\frac{(a_{A^+})_F \cdot (a_{B^+})_L}{(a_{B^+})_F \cdot (a_{A^+})_L} = K \qquad\qquad 5.3$$

$(a_{A^+})_F \ (a_{B^+})_F$ -activities of ions A^+ and B^+ in the exchanger phase (F);
$(a_{A^+})_L \ (a_{B^+})_L$ -activities of ions A^+ and B^+ in the solution (L)

K may also be described as the selectivity constant. The law of mass action in this form is valid to only a limited extent, and if abnormal changes in the activation coefficients due to activity $a = f_a \cdot c$ are already compensated for, then the volume changes in the resin during the reaction are not taken into account either. The following equation for the exchange of monovalent ions enables the calculation to be made [9] (see also [5]):

$$\ln K = \frac{\pi}{RT} \cdot (V_{A^+} - V_{B^+}) + \ln \left(\frac{f_{B^+}}{f_{A^+}} \right)_L - \ln \left(\frac{f_{B^+}}{f_{A^+}} \right)_F \qquad 5.4$$

π = swelling pressure; V_{A^+}, V_{B^+} = partial molar volumes of the hydrated ions; f_{A^+}, f_{B^+} = activity coefficients of the ions A^+ and B^+ (F in the resin phase, S in the solution)

In very dilute solutions $a = c$ and therefore $(f_{B^+}/f_{A^+}) = 1$, so that Equation 5.4 is simplified by the elimination of these terms. The equilibrium of the reaction 5.1 lies further over on the right hand side, the smaller the partial molar volume of B^+ is and the smaller $(f_{B^+})_F$ is in comparison with $(f_{A^+})_F$.

Because of the difficulty of ascertaining the value of the unknowns in Equation 5.4, it is usual however in practice to select an empirical fixed partition coefficient ':

$$\frac{(c_{A^+})_F}{(c_{A^+})_L} \cdot \frac{V}{m} = \alpha_{A^+} \quad \text{and} \quad \frac{(c_{B^+})_F}{(c_{B^+})_F} \cdot \frac{V}{m} = \alpha_{B^+} \qquad 5.5$$

V = volume of the solution (ml); m = weight of exchange resin (g)

From the relationship between the partition coefficients established for individual ions the actual selectivity coefficient a can be determined;

$$\frac{\alpha_{A^+}}{\alpha_{A^+}} = \beta_{A^+/B^+} \qquad 5.6$$

If, instead of the concentrations used for the partition coefficients, the equivalent fractions X_F (for the resin phase) and X_L (for the solution) are employed, and X_F is treated as a function of X_L, then mutually comparable isotherms are obtained as a representation of selectivity, as shown diagrammatically in Fig. 5.5. For ion exchange between monovalent ions they are substantially symmetrical, as in the illustration. With ions of higher valency there is a deviation from this symmetry (see Fig. 5.7). A linear function is purely coincidental, but was observed by Kressman and Kitchener [10] for the exchange of NH_4^+ for K^+ ions on a phenylsulphonic acid resin. The selectivities shown by the isotherms are due to a range of properties of the substances and exchange resins:

- Valency. As a rule the ion with the higher valency is more strongly bonded to the exchange resin.

- Ionic radius. Of two ions with the same valency, the one preferentially absorbed by the exchange resin is that whose hydrated ion has the smaller diameter and hence produces the lower swelling pressure. The resulting selectivity so generated increases with the magnitude of the swelling pressure. On the other hand, this increases with increasing zeolitic structure of the exchanger and decreasing concentration of the solution.

- Specific effect of active exchange groups. A pronounced specificity should not be expected here. Thus weakly acidic cation exchangers in acid solutions preferentially absorb H^+ ions, since they form undissociated carbonic acid groups with them. The selective effect of chelate-forming resins on metals must also be mentioned here (see Section 5.2.3.3.2).

- Complex formation. An indirect selectivity arises in the exchange resins owing to the formation of complexes of particular metals in the solution.

- Sieve effect. Resins with a strong network structure can hinder the penetration of larger ions into the microporous part. They therefore preferentially absorb smaller ions.

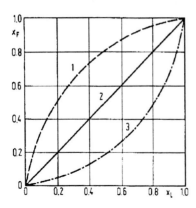

Fig. 5.5 Equilibrium isotherms for exchange reactions (diagrammatic).

X_F equivalent fractions in the exchange phase, X_L equivalent fractions in the solution, 1 large affinity between ion and exchange resin, 2 the affinity between ion and exchange resin in balance with the tendecy to remain in the aqueous phase (chance governs what happens in practice), 3 small affinity between ion and exchange resin.

5.2.3.2   Reaction rate

The process of ion exchange is a chemical reaction so far as material changes are concerned. Its progress as a function of time is subject to physical influences. For the ion to be exchanged there are the following steps:

- Diffusion of the ion in solution to the surface of the exchanger.

- Diffusion through the diffusion boundary layer to the phase interface (film diffusion).

- Diffusion in the gel-resin up to the active exchange groups (grain diffusion).

- Chemical reaction.

There follow in reverse order the diffusion processes for the ion passing out of the exchange resin into solution.

Diffusion in solution to the surface of the resin depends on the degree of turbulence. This is a function of the specific type of impingement associated with the column technology employed. If it is assumed that the concentration of the solution is always maintained because of the turbulence on the outer side of the laminar films surrounding the exchanger particles, then the exchange rate is determined by the film and/or the grain diffusion. Detailed descriptions are given by Boyd, Adamson and Myers [11], Boyd and Soldano [12], Becker-Boost [13] and Kühne [6].

So long as the concentration in solution is high, the quantities of ions diffusing through the film are also large and the exchange rate is determined by grain diffusion. It is proportional to the diffusion constant of the ion in the resin and inversely proportional to the radius of the exchanger particle. The diffusion constant however depends on the valency and diameter of the hydrated ion, the material properties of the resin (porosity, network structure, swelling) and naturally on the temperature. In addition, according to Boyd and Soldano, there is a certain retarding effect, which can be traced back to electrostatic influences, in which the valency of the ion plays a role. In practice it means that the removal of polyvalent ions takes longer during regeneration than that of monovalent ions.

If, on the other hand, concentration in solution is low as is the case in continuous recirculation of rinse water then film diffusion is the rate-determining process. It is proportional to the diffusion coefficient in the solution medium, the concentration of ions at the phase boundary layer as well as its area and inversely proportional to the thickness of the diffusion boundary layer. The magnitude of the latter is estimated to be 10-2 to 10-3 cm. Diffusion of ions in the macropores of the resin is also attributed to a film diffusion action.

There are exchange reactions, which, because of the concentration of the solution in the boundary region, proceed by both mechanisms or pass from one to the other after a certain induction period. In the case of processes employed in the metal working industries, the diffusion kinetics correspond fairly clearly to one or other of the two theories, depending on whether it is a case either of very dilute or more concentrated solutions. Film diffusion takes place in dilute solutions below 0.01 M and grain diffusion in more concentrated ones above 0.1 M.

From factors known to influence the diffusion processes it is also possible to assess the possibilities of accelerating the reaction. Grain diffusion can be influenced only by the properties of the exchanger: good swellability, high porosity, high fixed ion concentration and low particle size (Haagen [14]), and film diffusion only through good mechanical flow conditions at the phase interface. Raising the temperature has a major influence in accelerating both processes. This is because of its effect on diffusion coefficients, not on the ion exchange processes themselves.

The exchange rate is easily determined empirically by mixing (air-dried) exchange resin in fixed amounts with a quantity of electrolyte greater than its exchange capacity, terminating the reaction after different times and determining the number of equivalents exchanged. A reaction-time curve obtained in this way is shown in Fig. 5.6. As suggested by Stach [15] the half-life is a valid criterion of the reaction rate, i.e. the time within which the counter ions corresponding to half of the fixed ions are exchanged. Modern high-performance exchange resins all have small half-value periods. For strongly acidic cation exchangers in the acid condition it is always less than 0.5 and usually around 0.1 min. The procedure for determining the half-value period can also be used specifically to explore the whole of the exchange reaction. It is useful in testing of used resins for irreversible reactions or mechanical contamination, along with measurement of capacity and determination of the wash water requirement after regeneration.

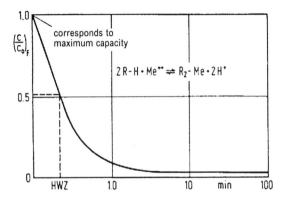

Fig. 5.6   Schematic reaction curve for determining the half-life of an exchange reaction.

c_0 *usable exchange capacity (corresponding to the H^+ concentration in the regenerated cation exchanger),   c H^+ concentration in the cation exchanger during the reaction*

5.2.3.3   Properties of common resins

According to the active exchange groups which they contain, resins are divided first of all into cation and anion exchangers. These groups are further differentiated, in that cation exchangers exist as strongly and weakly acidic, types and the latter may also exist in a complex-forming version. Among anion exchangers too, there are weakly and strongly basic resins. Exchange reactions can also be classified as strong and weak acid or strong and weak bases types or as complex forming processes. Ion exchange resins common in waste water and recycling technology are listed, along with their active exchange groups, in Table 5.1.

5.2.3.3.1   Strongly acidic cation exchangers

Strongly acidic cation exchangers are resins based on polystyrene with core sulphonate groups ($-SO_3H$). They are used both as gel-resins and as macroporous resins, and are available commercially with particle sizes between 0.3 and 1.3 mm. Owing to a balanced zeolitic structure they are highly resistant to oxidation, the microporous resins in particular. They show relatively slight variability in volume in the separate charged forms and good bead stability. The latter property depends on the mechanical stability of the beads of resin which has the effect of reducing particle fracture and wear. The losses of resin by mechanical breakdown depend in practice on various influences such as impact, number of regenerations and back rinses per year and so on. Taken together with the losses due to chemical effects they amount to about 1 to 5% per year.

TABLE 5.1:
COMMON ION EXCHANGE RESINS

Description of the ion exchanger	Active exchange groups
Cation exchangers	
Strongly acidic cation exchanger Polystyrene-based, sulphonate groups	$-CH-CH_2-$; phenyl ring with $SO_3^- H^+$
Weakly acidic cation exchanger Polyacrylate-based, carboxyl groups	$-CH-CH_2-$; $COO^- H^+$
Weakly acidic cation exchangers Polystyrene-based Iminodiacetate groups	$-CH-CH_2-$; phenyl ring with $CH_2 \cdot N \langle \begin{array}{l} CH_2 \cdot COO^- H^+ \\ CH_2 \cdot COO^- H^+ \end{array}$
Anion exchangers	
Strongly basic anion exchanger Type I, polystyrene-based Quaternary ammonium groups	$-CH-CH_2-$; phenyl ring with $CH_2 \cdot N \langle \begin{array}{l} CH_3 \\ CH_3 \\ CH_3 \end{array} \rangle^+ OH^-$
Stongly basic anion exchanger Type II, polystyrene-based Quaternary ammonium groups	$-CH-CH_2-$; phenyl ring with $CH_2 \cdot N \langle \begin{array}{l} CH_3 \\ CH_2 \cdot CH_2OH \\ CH_3 \end{array} \rangle^+ OH^-$
Stongly basic anion exchanger polyacrylamide-based Quaternary ammonium groups	$-CH-CH_2-$; CO ; $NH \cdot CH_2 \cdot CH_2 \cdot CH_2 \cdot N \langle \begin{array}{l} CH_3 \\ CH_3 \\ CH_3 \end{array} \rangle^+ OH^-$
Weakly basic anion exchanger Polystyrene-based Tertiary ammonium groups	$-CH-CH_2$; phenyl ring with $CH_2 \cdot N \langle \begin{array}{l} H \quad CH_3 \\ CH_3 \end{array} \rangle^+ OH^-$
Weakly basic anion exchangers Polyacrylamide-based Tertiary ammonium groups	$-CH-CH_2-$; CO ; $NH \cdot CH_2 \cdot CH_2 \cdot CH_2 \cdot N \langle \begin{array}{l} H \quad CH_3 \\ CH_3 \end{array} \rangle^+ OH^-$

Strongly acidic cation exchangers are the most trouble- free and robust resins. They are used in the pH range of 1 to 14 and are resistant up to 120°C. Typical average data for strongly acid cation exchange resins are:

- Total capacity: 2 to 2.3 Eq./litre wet gel-resin. In macroporous resins it is 1.4 to 1.8 Eq./litre lower by volume because of the lower resin mass.

- Usable capacity: 1.5 to 1.7 Eq./litre wet gel-resin and 1.1 to 1.3 Eq./litre wet macroporous resin.

- Volume decrease on charging the exchanger in the H- condition: ca 7 to 8% in gel-resin and ca 4 to 5% in macroporous resins.

- Liquid velocity during charging: 50 mh⁻¹ maximum.

- Mass of regeneration medium: 80 g HCl (100%) per litre for all resins. This is supplied at a concentration between 5 and 10% and mostly between 7 and 8%. The liquid velocity at this stage is ca 5 mh⁻¹.

- Consumption of wash water: ca 5 litre/litre of resin at liquid velocity of 5 to 10 mh⁻¹.

Special resins with particularly high exchange rate and grain stability can tolerate velocities up to 120 mh⁻¹. Macroporous resins resist pressure drops up to 2.5 bar, while with gel-resins these should not exceed 1.5 bar.

Strongly acidic cation exchangers in waste water and recycling technologies used in the metal working industries are invariably operated in the H-condition (the acid-regenerated condition).

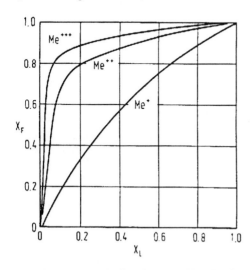

Fig. 5.7 Equilibrium isotherms (diagrammatic) for the reaction of a cation exchanger (H-condition) with metal ions of different valencies [5].

X_F *equivalent fractions of the metal ions in the exchanger phase,* X_L *equivalent fractions of the metal ions in the solution.*

Individual ions have different affinities for the exchange resin, which above all depend very strongly on their valencies and on aspects of the exchanger shown here to have most influence on their selectivity towards particular ions. Fig. 5.7

shows this diagrammatically in the form of equilibrium isotherms. A note should be made here also of the increase in asymmetry with valency, which was mentioned above. Oehme [16] quotes the following order of affinities for strongly acid cation exchangers (H-condition):

$$Ti^{4+} > Cr^{3+} > Al^{3+} > Ba^{2+} > Pb^{2+} > Fe^{2+} > Ca^{2+} > Ni^{2+} > Cd^{2+} >$$

$$> Cu^{2+} > Zn^{2+} > Mg^{2+} > Ag^+ > Cu^+ > K^+ > NH_4^+ > Na^+ > H^+$$

The more strongly the ions are bonded the less they are inclined to dissociate in the resin gel and the more difficult they are to remove during regeneration.

5.2.3.3.2   Weakly acidic cation exchangers

The ordinary weakly acidic cation exchanger is a polyacrylate with a zeolitic structure produced by divinylbenzene, which has its own active exchange groups (–COOH groups). Macroporous resins are commercially available with particle sizes from 0.3 to 1.3 mm. Oxidation resistance and particle stability are good. The temperature resistance extends up to 100°C.

Typical average data for weakly acidic cation exchangers of the carboxyl type are:

- Total capacity: 4 to 5 Eq./litre wet resin.

- Usable capacity: 1 to 3 Eq./litre wet resin, depending on operating conditions and solution pH.

- Volume increase in going from the H- to the Na-condition: ca 60%.

- Liquid velocity during charging: up to 40 mh^{-1}. As used in the metal working industries usually only 10 mh^{-1} (metal recovery, final purification, removal of surface active agents).

- Amount of regeneration medium: 70 gm HCl (100%) as a 7 to 8% solution. So long as the exchanger is not being used to recover surface active agent (H-condition), it can be conditioned afterwards with excess caustic soda (ca 60 to 70 g NaOH/litre resin). Liquid velocity in each case 5 mh^{-1}.

- Consumption of wash water: after regeneration with acid, this has to be displaced; after conditioning, 5 litre/litre resin at a liquid velocity of 5 to 10 mh^{-1}.

The resin can be operated at a maximum pressure drop of 2.5 bar.

Weakly acidic cation exchangers behave as weak acids, and as such are largely undissociated in the H-condition. In this state they absorb scarcely any other cations. Also, because of this high selectivity towards hydrogen ions, they should be exposed only to solutions with as high a pH value as possible, up to almost the point of hydrolysis of the metals present, generally not less than pH 4.

Because of their high capacity, enabling them to be used connected in series of two or three columns practically up to their maximum capacity, carboxyl resins are widely used for recovering metals from waste water.

The order of affinity of these resins is [16]:

$$H^+ > Cu^{2+} > Pb^{2+} > Fe^{2+} > Zn^{2+} > Ni^{2+} > Cd^{2+} > Ca^{2+} > Mg^{2+} >$$
$$> NH_4^+ > K^+ > Na^+$$

These resins are slightly more attractive for final purification (selective exchangers), since there are weakly acidic cation exchangers with iminodiacetate groups. These have not the same high capacity, but they have a considerably better selectivity. They enable divalent metal ions to be absorbed even in the presence of high concentrations of calcium, e.g. in waste water that has been neutralised with lime slurry.

Weakly acidic cation exchangers with iminodiacetate groups are based on polystyrene; they also have particle sizes from 0.3 to 1.3 mm, can be used in the pH range of 1 to 14 and are resistant to temperatures up to 80°C,
Their specific properties are:

- Total capacity: 2.5 to 2.8 Eq./litre wet resin (H-condition).

- Usable capacity: 2 to 2.3 Eq./litre wet resin depending on the pH of the solution.

- Volume increase going from the H- to the Na-condition: 25 to 30%.

- Liquid velocity during charging: 10 mh⁻¹ operating as a selective resin.

- Amount of regeneration medium: 140 g HCl (100%) per litre resin as 5 to 10% acid. Sulphuric acid (200 g/l resin) up to 20% concentration is used for the recovery of metals from sulphate. As with carboxyl resins, this also is followed by conditioning with excess caustic soda (ca 50 gm/litre as a 5% solution). Both solutions are supplied at a liquid velocity of 5 mh⁻¹.

- Consumption of wash water: after regeneration with acid this has to be displaced. After the conditioning ca 5 litre/litre resin at a liquid velocity of 5 mh⁻¹.

The iminodiacetate resin does not react quite so selectively with hydrogen ions as the carboxyl resin. Copper, for example, is able to be absorbed from acidic solutions, but otherwise the behaviour with respect to the pH value is similar to that of carboxyl resins.

The order of affinity of these resins is as follows [17, 18]:

$$Cu^{2+} > Pb^{2+} > Ni^{2+} > Co^{2+} > Zn^{2+} > Cd^{2+} > Fe^{2+} > Ca^{2+} > Na^+$$

This order is in agreement with the stability constants of the iminodiacetic acids (see Table 3.36). This accounts for the good selectivity of these exchange resins. The constants differ by two or three orders of magnitude from metal to metal, and indeed with calcium by six orders of magnitude. This explains also why divalent metals can be separated from calcium-containing waste water. The ion exchanger is capable of binding divalent heavy metals even in the Ca-condition.

5.2.3.3.3 Weakly basic anion exchangers

Weakly basic anion exchangers based on polystyrene and polyacrylamide are known. They have tertiary ammonium groups as active exchange substituents. As a rule, commercial resins are based on polystyrene. They are macroporous and their particle size is about 0.3 to 1.3 mm. Because of their good oxidation resistance they are used in waste water treatment to absorb oxidising anions in series with strongly basic resins which are susceptible to oxidation.

With their properties as weak bases they preferentially absorb the anions of stronger acids, and easily release them again, in slight excess, to the regenerating solution. Their behaviour is comparable to that of the weakly acidic cation exchangers in that, as weak bases, they form only slightly dissociated compounds with OH ions, and so act in a particularly selective way. Because of their relatively large capacity and good oxidation resistance, they are especially suitable for the continuous recirculation of industrial waste water.

Their specific properties are:

- Total capacity: ca 1.5 Eq./litre.

- Usable capacity: ca 1.1 Eq./litre. Thus they surpass, in this respect, the strongly basic resins by about 100%. Therefore, and because of the protection against oxidising anions, they are always used up stream of the strongly basic exchangers, wherever these are used.

- Volume increase in passing from the OH-condition to the charged condition: 20 to 30%.

- Liquid velocity during charging: up to 40 mh^{-1}.

- Amount of regenerating medium: 60 g caustic soda per litre of resin as a 4 to 5% solution. A liquid velocity of 5 mh^{-1}. can be employed.

- Consumption of wash water: this is given as ca 10 litre/litre resin, but in practice it is usually considerably less. The liquid velocity may be 5 to 10 mh^{-1}.

Mainly on account of their macroporosity, the weakly basic anion exchangers are able to bond both with anionic surface active agents chemically and also with non-ionogenic ones adsorptively, and to give them up again during regeneration, even if only incompletely in the latter case.

The affinity, as well as the selectivity, of weakly basic anion exchangers increases with the valency of the anions and with the degree of acidity of the corresponding free acids. As such, they rank above strongly acidic cation exchangers after neutralisation. In the alkaline region they would be unable to function as weakly basic exchangers because of the presence of OH⁻ ions which have a greater affinity towards the resins. The following order of affinities is valid for a weakly basic anion exchanger in the OH-condition in contact with de-alkalified water [16]:

$$OH^- > [Fe(CN)_6]^{4-(3-)} > [Cu(CN)_4]^{3-} > [Ni(CN)_4]^{2-} > \text{anionic surface}$$

$$\text{active agents } > CrO_4^{2-} > SO_4^{2-} > HPO_4^{2-} > NO_3^- > NO_2^- > SCN^- >$$

$$> Cl^- > \text{formate} > \text{complex anion} > \text{citrate} > \text{tartrate} > \text{oxalate} > F^-$$

Relevant practical experience indicates that anionic metal complexes fall close to the ferricyanide ion in the series of affinities.

There are also moderately strongly basic anion exchangers, which contain 10 to 20% of strongly basic groups alongside the tertiary ammonium groups. These allow anions of weak acids to bond, but only temporarily.

5.2.3.3.4   Strongly basic anion exchangers

These are zeolitic polystyrene or polyacrylate with quaternary ammonium groups as well. They fall into two types depending on the nature of these groups (see Table 5.1).

Type II resins are rather weakly basic like type I resins, have higher capacities, but are susceptible to oxidation because of the ethanolamine groups. Oxidation agents must be kept away from them completely, just as from type I resins, by preceding them with a weakly basic anion exchanger. This applies also to metals with a catalytic effect on the oxidation of these groups by dissolved oxygen.

Because of their strongly basic character, these resins are also capable of absorbing the anions of weak acids, if less strongly than those of the stronger acids. In line with their affinities, absorption displacement occurs during the charging process, so that the anions of the weakest acids are in the end removed. The trend is understandably stronger with anion exchangers of type II than with those of type I. The former are therefore more easily regenerated. If anions such as free cyanide (CN^-), dihydrogen borate ($H_2BO_3^-$), hydrogen silicate ($HSiO_3^-$) and bicarbonate (HCO_3^-) are to be removed from water then strongly basic anion exchangers should be employed.

Strongly basic anion exchangers are generally sensitive towards strong alkalis, which can cause the active exchange groups to decompose with formation of ammonia. Their alkali resistance is therefore less that than of the other resins. The working range of type I strongly basic anion exchangers is from pH 1 to 12, and that of type II from pH 1 to 11. Their temperature resistances also differs. For type I resin (polystyrene type) it goes up to 100°C (for the polyacrylamide type only as far as 40°C!), while type II resins are less stable with resistance up to 70°C.

A criterion of weak acid removal is the loss of silicic acid, which increases with increasing pH value, increasing temperature and the presence of smaller amounts in the regenerating medium.

The most important specific properties of strongly basic anion exchangers are:

- Total capacity: in polystyrene in gel form 1.4 to 1.6 Eq./litre, in macroporous resin somewhat lower around about 1.2 Eq./litre, and similar in macroporous resin based on polyacrylamide.

- Useful capacity: type I gel-resins about 0.6 to 0.7 Eq./litre, macroporous ones somewhat less; type II resins generally have rather higher capacity of ca 0.9 Eq./litre, or 0.8 for macroporous resins. In practice they are more likely to be a little less (0.5 to 0.7 Eq./litre. Macroporous resins based on polyacrylamide have useful capacities of barely 1.

- Volume change on passing from the OH-condition to the charged condition: in contrast to the weakly basic anion exchangers, the strongly basic types contract by about 10 to 15%.

- Liquid velocity in charging: type I polystyrene resins 50 mh^{-1}, type II and polyacrylamide resins 40 mh^{-1}.

- Amount of regenerating medium: for all resins 80 gm caustic soda/litre resin, 4 to 5% solution with a liquid velocity of 5 to 10 mh^{-1}.

- Consumption of wash water: ca litre/litre of resin for all exchangers with a liquid velocity of 5 to 10 mh^{-1}.

Although the usable capacity of the strongly basic resins is considerably lower than that of the strongly basic, type consumption of the regenerating medium is distinctly higher. This is associated with their strongly basic character and their consequently distinctly lower affinity for OH ions. Their regeneration with caustic soda is therefore carried out in practice in conjuction with weakly basic anion exchangers (see Section 5.3.2.8.1).

The order of affinities of strongly basic anion exchangers is reported by Oehme [16] as follows:

$$NO_3^- > CrO_4^{--} > PO_3^{---} > \text{ oxalate } > NO_2^- > Cl^- > \text{ formate } >$$
$$\text{citrate} > \text{tartrate} > \text{phenolate} > F^- > \text{ acetate } > HCO_3^- > HSiO_3^- >$$
$$> CN^- > H_2BO_3^- > OH^-$$

Cyano-complexes, metal chelates and anionic surface active agents are so firmly bonded to strongly basic anion exchangers that it is scarcely possible in practice to remove them again by regeneration. They should be eliminated in each case by passing them first through weakly basic anion exchangers.

5.2.4   Reactor technology in ion exchanger applications

The considerations dealt with in Section 5.2.3 relate to steady state reactions, which are well known in batch process waste water treatments. This kind of reactor technology may not be usable for ion exchangers in waste-water and recycling processes, as it does not result in appropriate equilibria.

   If however, the solid reaction participant of an ion exchanger is loaded into vertical columns and a solution that contains the other participant in the reaction is fed into them from the top or the bottom, then the reaction product that passes into the solution is removed from the reaction zone of the ion exchanger. According to Equation 5.1 this comprises ions of the type $A^{(\pm)}$. As they are removed, the equilibrium is displaced completely to the right. The ion exchanger is thus charged with ions of type $B^{(\pm)}$ in the direction of fluid flow.

   In column technology applications, the reversibility of the exchange resin also allows the reaction according to Equation 5.1 to be shifted from the right to the left if an exchange resin in the B-condition is subjected to a flow of solution which contains ions of the type $A^{(\pm)}$. This process occurs usually in regenerating the exchanger. Thus it is possible to determine which type of ion added during charging will exchange with the ions to be removed. As the amounts exchanged are equivalent, it is possible to use ion exchangers to desalinate water or waste water and to regenerate process solutions.

   Column ion exchangers are fed with volumes of solution (m^3_L) per hour, which have to be related to the amount of resin (m^3_{AT}). The liquid replacement rate (b) is:

$$b = \frac{m^3_L}{m^3_{AT} \cdot h} \quad [h^{-1}] \qquad \qquad 5.7$$

Assuming the height of the bed of resin to be l, the velocity of flow (u) in the empty columns is

$$u = \frac{m^3_L}{m^3_{AT} \cdot h} \cdot 1 \quad [h^{-1}] \qquad \qquad 5.8$$

Liquid velocities for solutions, regeneration media and wash water were provided in the sections above. In full columns, however, the solution flows only through the interstices between packed globules. These interstices, for a normal commercial particle size range, amount to about 35% of bulk volume, i.e. the actual mean velocity of flow is almost three times as great as that calculated for the empty volume. Average residence time τ of solution in contact with the resin in minutes [19a] is:

$$\tau_m = \frac{60 \cdot 0.35}{b} \approx \frac{20}{b} \quad [\text{Min.}] \qquad\qquad 5.9$$

That is, with liquid replacement rates such as are usual in this technology, resin contact times are obtained similar to those given in Table 5.2.

TABLE 5.2:
CONTACT TIMES OF SOLUTIONS IN VARIOUS
PROCESSES AND ION EXCHANGERS

Liquid feed rate b [h⁻¹]	Contact time τ_m [min.]	Process
25–50	0.8–0.4	Charging (rinse water circulation)
10	2	Charging (final exchanger)
		Charging (regeneration processes)
5	4	Regeneration medium delivery
5–10	4–2	Washing

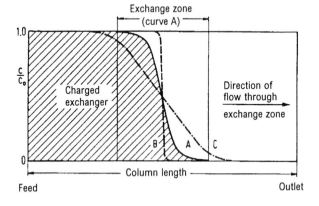

Fig. 5.8 Exchange zones in flow-through columns. The curves become flatter if the diffusion rate is lowered or the velocity of flow is increased, i.e. the zones become larger.

c_0 *concentration in the feed.*

c local concentration in the solution

The exchange processes must of course be completed in considerably shorter times, as, dependent upon the fluid velocity and the reaction rate (see Section 5.2.3.2), they have to take place in fractions of the height that is filled with the resin. The portion of the column in which the reaction proceeds up to a certain time is called the reaction zone or exchange zone. It is represented by a concentration profile, which can be ascertained, for example, by analysis of the exchanged ions. Examples of reaction zones are illustrated in Fig. 5.8.

The rate at which the reaction zones advance at their point of inflection depends only on the amount of material to undergo exchange (concentration, liquid velocity). The gradient of the profile depends on many factors. It increases as the fluid velocity and the particle size of the exchange resin fall and as the diffusion rate (film and/or grain diffusion) and the affinity of the exchange reaction increase. According to Becker-Boost the size of the exchange zone is dependent on a factor that is influenced by the diffusion properties and is proportional to the square root of the flow rate (in cms^{-1}) [20].

5.2.5   Use of particular types of ion exchanger

As the general applications of ion exchangers have been reported in Section 5.1, in this section, after the properties of particular resins have been described, reference will be made to their specific applications. The most important criteria in this respect are usable capacity, selectivity and oxidation resistance. The charged condition in which it is used is a decisive factor, however. The ions given up by the resins in the course of the reaction must create no problems in the solutions that are to be treated.

Table 5.3 surveys the purposes for which particular resins are used and the charged conditions necessary for these applications [21]. Further details are given, according to service requirements, in the sections that follow.

TABLE 5.3:
ION EXCHANGERS AND THEIR POSSIBLE APPLICATIONS

Type of exchanger	Active exchange groups	Possible applications	Corresponding charged condition
Cation exchanger strongly acidic	$- SO_3H$	Rinse water circulation	H^+
		Purification of acidic process solutions	H^+
Cation exchanger weakly acidic	$- COOH$	Recovery of non-ferrous metals	Na^+
		Recovery of non-ferrous metals	Na^+
(complex forming)	$-N{<}^{CH_2 \cdot COOH}_{CH_2 \cdot COOH}$	Final purification of treated metal-containing waste water	Na^+
Anion exchanger weakly basic	Tertiary amino groups	Rinse water circulation	OH^-
		Recovery of CrO_3	OH^-
		Recovery of precious metals (reversible)	Cl^-, OH^-
Anion exchanger	$- N(CH_3)_3OH$ (type I) $- N[(CH_3)_2(C_2H_4OH)]$ OH (type II)	Rinse water circulation	OH^-
		Recovery of precious metals (irreversible)	OH^-, Cl^-
		Separation of free acids from their common ion	Corresponding to common anion
Special exchanger	$- SH$	Final purification of treated mercury-containing waste water	Na^+

5.3   RINSE WATER CIRCULATION

5.3.1   Conditions of use and principle of process

Rinse water circulation with the help of ion exchangers is one of the most important methods of drastically reducing the consumption of water and consequently the amount of waste water produced. Hence it has become an indispensable means of fulfilling the requirements of recent legislation concerning waste water treatment.

The process was introduced in the early 1960s, mainly in Switzerland and in Germany. Since then it has come into use in all industrial countries. Originally, economics, shortage of water and poor water quality were the principal reasons for rinse water recirculation. The main factors today are the need to save water and the ability to treat substantially smaller amounts of waste water in safely operated batch plants (see Section 4.3.1). Economical operation is without question essential, in view of the high cost of water at the present time and that of dealing with and disposing of waste water, even where there is a possibility of self sufficiency thanks to the availability of well-water. The costs of continuous water circulation depend mainly on the concentration of the water to be treated (collected rinse water). This is decisive for the service life of the ion exchanger between regenerations, and has a major effect on the cost of regeneration chemicals and also on the labour costs especially for manually operated plants. Every attempt should also be made to keep the concentrations in the water to be treated as low as possible. This is accomplished by following high-concentration process baths with static rinse tanks, so that the greater part of the drag-out can be returned (see Section 6.3.4). Their concentration is allowed to increase until it corresponds to 10 to 20% of that of the process bath. Their contents can be returned to process baths which are operated hot, to make up for evaporation losses, or transferred to a recycling plant, or if further utilisation is not possible, passed on for waste water treatment. By choosing the period of accumulation in static rinse tanks and the way in which they are used it is possible to exercise an important influence on the concentration of the water requiring treatment. More static rinse tanks or slowly flowing cascade rinsing units (see Section 6.3.2.2) after a process bath can also be employed.

It is the drag-out from these static rinse baths that primarily determines the concentration of the water to be treated and hence the cycle time of the ion exchangers. Typical rinsing technology employed in a rinse-water circulation system is illustrated schematically in Fig. 5.9.

On economic grounds, a maximum concentration of 2m Eq./litre in the water to be treated should not be exceeded, and the cycle times for fixed-bed plants should be not less than two shifts. On the evidence of practical experience, such conditions can be attained after installing static rinse tanks, where previously without this kind of rinsing technology the rinse water concentration might have reached 20m Eq./litre.

Under unfavourable conditions, the circulation can also be adjusted if additional special measures in terms of pre-rinsing are introduced (see Section 8.3.2). Nowadays the water concentration in most continuous circulation plants is below 1 m Eq./litre. The effect of the concentration of the water needing treatment on the cycle time of the ion exchanger is shown in Fig. 5.10 [22].

After the water concentration, the throughput capacity of continuous circulation plants also influences the economics. Both effects can be seen in Fig. 5.11. The labour, energy and wear- and-tear fractions of the costs fall with increase in throughput capacity. Differences in re-circulated water costs compared with those for mains water purchase and effluent disposal can be applied to amortisation of the plant [23].

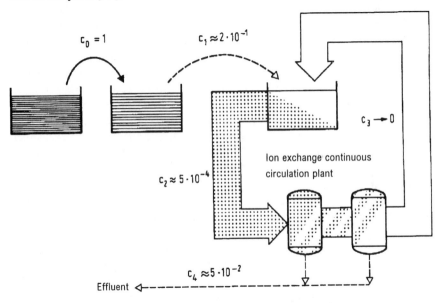

Fig. 5.9   Rinsing system for continuous circulation of rinse water over ion exchangers (diagrammatic) showing orders of magnitude of the concentrations.

c_0 *process bath,* c_1 *drag-out from the static rinse tank,* c_2 *flowing rinse water,* c_3 *purified water,* c_4 *regeneration product (waste water)*

As well as the concentration of the water to be treated, its pH value must also receive attention. It should be weakly acidic, around pH 4 to 6, so that hydrolysis of metal compounds is entirely prevented. There is then no need to couple up filters with the exchangers. Because of the need to adjust the water to a weakly acidic condition, no static rinse tanks should follow very concentrated acid process solutions (e.g. acid pickle baths). On the other hand, care must be taken in doing this, so as to avoid regeneration effects in the cation exchangers and overloading of the anion exchangers [24].

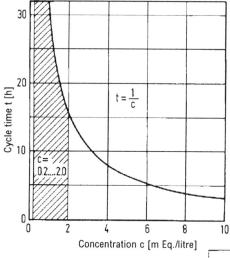

Fig. 5.10   Dependence of the cycle time of ion exchangers on the concentration of ions in continuously circulated rinse water. The area corresponding to usually found concentrations is cross-hatched [22]

Fig. 5.11   Dependence of costs of re-circulated water on concentration of the water to be treated and the size of the plant (without capital charges) [23]. Total height of the ordinate corresponds to US $ 4-5 (1990).

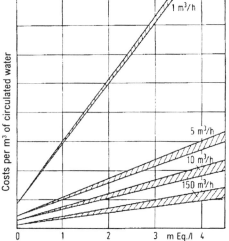

A range of compounds sometimes found in water can chemically or mechanically damage the exchange resins. Rinse waters of this kind need pretreatment or must be excluded from the circulatory system. This will be treated later (see Section 5.3.2.7).

If the foregoing conditions are observed, rinse water can be desalinated by passing over cation and anion exchange resins, and then led back again to the flowing rinse tanks. This can be done both with recirculated rinse waters purified of all their components and with rinse waters containing just one metal which it is desired to recover. Individual steps necessary for this will be discussed in the sections that follow. Descriptions of continuous rinse water circulation systems can be found in Furrer [25], Götzelmann [24,26], Marquardt [27,29,31,32,33], Schlegel [28], Kunin [30] and Hartinger [23,24].

5.3.2   Chemical and physical processes

5.3.2.1   Untreated water, filtration of untreated water

Rinse water suitable for continuous circulation according to the previously stated criteria (concentration, exclusion of detrimental constituents, pH value) is usually collected together in the untreated water storage tank. Its concentration should so far as possible be well under 2 m Eq./litre, its pH value weakly acidic, and its surface tension over $60 \cdot 10^{-3}$ Nm^{-1} (= 60 dyne cm^{-1}). The latter is ensured by excluding process baths with a high concentration of surface active agents or by appropriate pre-rinsing after such baths. The pH value is monitored frequently and the effect of intervening changes in concentration noted by electrical conductivity measurements. The concentration of ions loaded in the exchange resin is ascertained by determination of the m-value.

If the untreated water has a pH value under 7, then a filtered 100-ml sample is passed over a strongly acidic cation exchange resin in the H-condition, which is washed with desalinated water afterwards, the eluate (inclusive of the wash water) being back-titrated with 0.1 M NaOH to the measured pH value of the rinse water. The number of ml required then gives directly the litre of the ions loaded in the cation exchanger. If the same sample is titrated further to pH = 7.0, a measure is obtained of the hydrogen ions not loaded into the cation exchanger. The total volume of 0.1 M NaOH gives an indication of all cations present, the concentration of which is identical to that of all the anions loaded in the anion exchanger.

If the pH value of the rinse water is above 7, then a titration with 0.1 M HCl is performed first to pH = 7.0. The number of ml required indicates directly the hydroxyl ions not loaded in the anion exchange resin. From pH 7.0 on, the procedure is as described above. The consumption of 0.1 M NaOH in returning to pH 7 corresponds to the m Eq./litre of the anions loaded in the anion exchanger. The sum of acid and alkali consumption then corresponds to the cations loaded in the cation exchanger.

Before reaching the ion exchangers the untreated water is pumped through a filter. Gravel filters have proved best for removing the solid contaminants visible to the eye, and they are used almost exclusively today. The medium used to fill them consist of purified, round-edged quartz grit, whose height in the filter columns is in the range 0.8-1.5 m and whose particle size extends from 0.5 to 3 mm. A more uniform particle size is also often employed, e.g. preferably between 1 and 2 mm.

Gravel filters do not work like surface filters, on which the residues build up as cakes, but as so-called in-depth filters, in which the residue accumulates three dimensionally. The pores of the filter medium are determined by the intermediate spaces between the grains of quartz. They are considerably larger than those of the particles to be removed. Nevertheless, filtration takes place which can reach down to a particle size of 10 μm. According to Alt [35] various processes are responsible for this, ranging from (flow-rate-dependent) sedimentation, through deposition in the pores, to adsorption. The last-named arises as the result of surface interactions (Van der Waals' and electrostatic forces).

Gravel filters are principally employed where a clear filtrate is required at high throughput rates and low concentrations of solid matter. In continuous circulation technology they are fairly highly loaded at a velocity of 20 mh^{-1}. The following have an effect on in-depth filtration: concentration of solid matter, particle size and particle characteristics such as size, shape and surface condition of the filter medium.

It is common nowadays to over-lay on the quartz gravel a layer of hard carbon granules (hydro-anthracite). This filter carbon, however, should under no circumstances be assigned the task of removing organic material by adsorption, as is often attempted. Principally, it holds back the larger particles and allows faster and better purification during back washing because of its comparatively low specific gravity.

Back washing is performed on reaching a predetermined pressure difference across the filter and is assisted by compressed air. It is continued until the back-washed water no longer exhibits any turbidity. The back-washed water is transferred to the waste water treatment plant. Only very slightly turbid water should be passed to the storage tank preceding the ion exchange units.

5.3.2.2 Cation exchange

Filtered water to be treated goes first to the strongly acidic cation exchange resin in the H-condition, where in accordance with the following equation all cations (metal ions, ammonium ions, cationic organics) are exchanged for hydrogen ions. The exchange reaction for metals is generally:

$$zR - H + Me^{z+} \rightleftharpoons R_z - Me + zH^+ \qquad\qquad 5.10$$

The filtrate leaving the exchanger is known as demineralised water. It contains the acids of all the anions present in the untreated water, e.g:

$$R - H + NaCl \qquad \rightleftharpoons R - Na + HCl \qquad\qquad 5.11$$

$$2R - H + CuSO_4 \qquad \rightleftharpoons R_2 - Cu + H_2SO_4 \qquad\qquad 5.12$$

$$3R - H + Fe(NO_3)_3 \rightleftharpoons R_3 - Fe + 3HNO_3 \qquad\qquad 5.13$$

The determination described above of the ions loaded in the exchange resin is also based on these reactions. The treated water is therefore markedly more acid than the untreated water. As the concentrations in the untreated water lie between 0.2 and 2 Eq./litre, theoretical pH values between 2.7 and 3.7 can be calculated for the treated water (see Fig. 5.12).

Break-through in a cation exchange column, when the cations are no longer all being exchanged at a constant untreated water concentration, makes itself seen as a fall in the m-value. This is especially easy to determine for two cation exchangers connected in series, by differential titration or measurement of the conductivity difference of both eluates, as the conductivity of hydrogen ions is considerably higher than that of other cations.

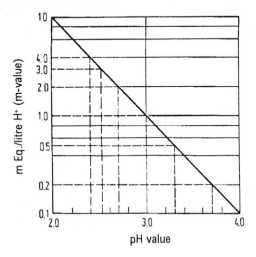

Fig. 5.12 pH value of water after cation removal as a function of the concentration of exchangeable cations in the untreated water

When breakthough of the ion exchanger occurs, the cation with the lowest affinity for the exchange resin accumulates in increasing amounts in the eluate. It therefore also has the broadest exchange zone and is the first to reach the end of the column as a result of absorption displacement. In view of the substances present in the rinse water, this is the sodium ion for ion exchangers in the H-condition.

Also for freshly regenerated resins, there are always small quantities of sodium ions present in the eluate. The reason for this is that the sodium ions are not completely removed in the regeneration. The absorption of ions with greater affinity for the exchange resin liberates hydrogen ions, which on their way through the bed of resin themselves elutriate small quantities of sodium ions. This phenomenon is known as sodium leakage. It is at its most significant right at the start of the charging phase, then falls, only to rise steeply again at the breakthrough. Sodium leakage increases with increase in the ratio of the sodium ion concentration to the concentration of higher valency cations in the untreated water, and with decrease in the amount of acid used in regeneration (see Fig. 5.13). It gives rise to the formation of caustic soda at the subsequent anion exchanger and, depending on the method of operation, is responsible for the low conductivity of the purified water.

In addition to the adsorption of certain amounts of non-ionogenic surface active agents, secondary reactions also occur at the cation exchanger. Because of the high acidity inside the exchange zone, compounds are decomposed that are not stable in an acid medium. Included among these are the metal coordination compounds of low stability. Decomposition of very strong complexes, such as ferricyanide, tetracyanonickelate and metal-EDTA complexes among others, has not been detected, but that of less strong complexes, such as the cyanide complexes of other metals, is observed. After protonation of the part of the cyanide forming the complex, the metal cyanide remaining loses its solubility and is precipitated out, as shown for the example of tetracyanozincate in Equation 5.14:

$$[Zn(CN)_4]^{--} + 2\,H^+ \rightleftharpoons Zn(CN)_2 + 2\,HCN \qquad\qquad 5.14$$

As blockages do not occur in the columns it must be assumed that the zinc cyanide is redissolved by the free cyanide that follows, eventually leaving as tetracyanozincate and becoming bonded to the anion exchange resin. The remainder of the poorly soluble salts are removed by back-rinsing the resin before regenerating it. Low cyanide contents in the regeneration product from the cation exchanger are also among other effects attributable to acid decomposition of cyanide complexes.

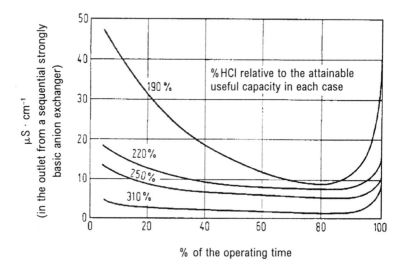

Fig. 5.13 Dependence of the sodium leakage (as specific conductivity of the desalinated water) on the operating time of a strongly acidic cation exchanger and the amounts of hydrochloric acid added for regeneration. Concentration of the water before treatment 3.5 to 4 m Eq./litre; sodium ion content 55 to 60% (Bayer AG [36]).

5.3.2.3 Anion exchange

Cation-exchanged water is afterwards passed over a weakly basic anion exchange resin in the hydroxyl-condition, where anions are absorbed in accordance with the general formula:

$$yR - OH + X^{y-} \rightleftharpoons R_y - X + yOH^-$$

5.15

The equilibrium of the reaction is favourably influenced by the acids contained in the cation-exchanged water, as the hydroxyl ions released combine with them to form water molecules and are thus removed from the equilibrium. Hence the second part of the desalination, representing in effect a neutralisation, is completed. Continuing from Equations 5.11 to 5.13 this proceeds in accordance with the following reaction equations:

$$R - OH + HCl \rightleftharpoons R - Cl + H_2O$$

5.16

$$2R - OH + H_2SO_4 \rightleftharpoons R_2 - SO_4 + 2H_2O$$

5.17

$$3R - OH + 3HNO_3 \rightleftharpoons 3R - NO_3 + 3H_2O$$

5.18

Apart from the inorganic ions, the weakly basic anion exchanger also removes anionic surface active agents and, because of its absorptive macroporosity, a certain portion of the non-ionic surface active agents as well. Larger amounts of anionic surface active agents can lead to a temporary reduction in capacity, which is not always rectified by a single, normal regeneration. Steady-state concentrations of up to 0.3 mg/l are considered to be acceptable for weakly basic anion exchangers. 1 gm/litre, even intermittently, always leads to difficulties.

In the presence of anions whose acids are of higher basicity, a considerable increase in the useful capacity can be observed as compared with monovalent anions. This is dependent upon the degree of dissociation of the corresponding hydrogen ions. Thus some sulphuric acid at the prevailing pH value dissociates, not to SO_4^{--} ions, but only to HSO_4^- ions, which only use up a single equivalent of immobilised ions when they are absorbed.

Specific electrical conductivity is a valid criterion for the breakthrough of the anion exchanger and also for the quality of desalinated water. It is a function of the sum of all ions present in the eluate from the anion exchanger. Before breakthrough, these derive essentially from the caustic soda produced by the sodium leakage of the cation exchanger. Anions of weak acids not absorbed by the weakly basic exchange resin have scarcely any effect on the conductivity. The strongest acid amongst them is carbonic acid. This is also the reason why rinse water desalinated by passing over a weakly basic anion exchanger always has a pH value below 7. The carbonic acid is produced from carbonate that gets into the untreated water from cleaning baths and alkaline electrolytes.

In particular cases it is noticeable that completely desalinated, continuously circulated water is consequently also free from the anions of the weak acids (CN^-, $H_2BO_3^-$, $HSiO_3^-$, HCO_3^-). Then, but above all where the untreated water has a high concentration of free cyanide, it is essential to connect a strongly basic anion exchanger in series after the weakly basic one. In principle the exchange of weak acids takes place in the same way as that of strong acids with weakly basic exchangers. The following equation, in which R–OH represents a strongly basic anion exchange resin, is valid for cyanide absorption:

$$R - OH + HCN \rightleftharpoons R - CN + H_2O \qquad\qquad 5.19$$

Breakthrough is determined here also by the anions with the lowest affinity for the exchange resin. These are the dihydrogen borate and cyanide ions. They can be detected only by analytical methods.

As carbonic acid also is removed by the strongly basic anion exchanger, there is no buffering substance available for the caustic soda produced as a result of sodium leakage in the cation exchanger. The pH value of the desalinated water, when a strongly basic ion exchanger is employed, thus lies in the range 7-9 (see also Fig. 5.16).

In many factories with cyanide-containing process baths, no particular difficulties are caused by free cyanide, provided the concentrations remain low, not exceeding a few mg/litre CN^-. The reason is that free cyanide in the untreated water storage tank reacts with metal ions to form cyano-complexes, forming anions of stronger acids and bonded with the weakly basic anion exchange resins.

More detailed information on the behaviour of weak acids in ion exchange as part of water treatment is available in relation to the removal of carbonic acid and silicic acid. Boiler feed water for high pressure plants has to be very pure in terms of silicic acid. The capacity of strongly basic anion exchangers depends on the CO_2/SiO_2 ratio in the water. It is almost twice as high for water that contains carbonic acid only, because of the greater affinity of this for the exchange resin, as is so for water that contains only silicic acid. Thus silicic acid also undergoes leakage and breakthrough in water purification. Its leakage increases with pH and temperature. The former is affected by the sodium leakage in the cation exchanger. According to Arnold and Martinola [37] sodium leakage between 0.01 and 0.25 m Eq./litre in type I resin produces a silicic acid leakage of 0.01-0.04 mg/litre SiO_2 and in type II resin one of 0.03-0.08 mg/litre SiO_2. Silicic acid leakage can be reduced by the use of larger quantities of regenerating medium and also by several technical measures.

5.3.2.4   Purified water

Desalinated water, leaving the final column of a continuous circulation plant, is returned to the surface treatment plant as rinse water. Its quality must be able to satisfy the requirements of the production process involved.

The most important criterion of the purity of continuously circulated water is the specific electrical conductivity. It is measured simply and continuously. It is significant, however, only in respect of desalination. Conductivity values below 50 or perhaps 30 μS cm^{-1} are normally satisfactory for surface treatment purposes. In plants operating at normal loads they are usually even below 10 μS cm^{-1}. This is equally true, whether the desalination is concluded with a weakly or strongly basic anion exchanger, as the weak acids, because of their low dissociation, make scarcely any contribution to the conductivity of the water.

As well as providing constant quality control, conductivity measurement has the advantage also of indicating breakthough in an exchange plant or individual exchanger. The equivalents of the ions loaded into the cation and anion exchangers are normally not equal. Furthermore, the useful capacity of the two resins is different. That means that one of the two exchangers reaches the breakthrough point first. If this is the cation exchanger, then the NaOH concentration rises, if it is the anion exchanger, then the acid concentration in the purified water increases. The breakthroughs have a particularly strong effect on the conductivity, as hydrogen and hydroxyl ions have an ionic conductivity that is many times higher than that of other ions. Hence the equivalent conductance of hydrochloric acid is over 3.5 times, and that of caustic soda double, that of sodium chloride. Fig. 5.14 shows for comparison the conductances of several acids, bases and salts at low concentrations. If the conductivity rises above a pre-set value for the factory water quality usually 30 or 50 μS cm^{-1} then the ion exchangers must be regenerated. Certain techniques allow only the exchanger that is truly exhausted to be regenerated, while the others can continue to be operated until they reach the limit of their useful capacity (see Section 5.3.3.2.1).

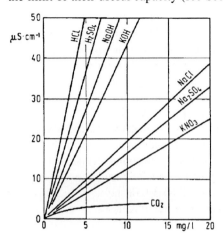

Fig. 5.14   Specific conductivity of a number of substances at low ionic concentrations (18°C) [38]

A low conductance usually below 10 μS cm⁻¹ can always be measured during the desalination of rinse water. It is caused by sodium leakage in the strongly acid cation exchanger. Weak acids that pass through the weakly basic anion exchanger, carbonic acid in particular, make scarcely any contribution towards increased conductivity (see conductivity of carbonic acid in Fig. 5.14).

The pH value of water purified by passing through a strongly acidic cation exchanger and a weakly basic anion exchanger is always below 7 because of its carbon dioxide content. As Fig. 5.15 shows, pure water with only 10 mg/litre dissolved carbon dioxide has a pH value as low as 5.0. The concentration in treated water is often higher than this. pH values of 4.5 are not uncommon. Pure water saturated with carbon dioxide at 15°C (ca 1900 mg/litre CO_2) has a pH value of about 3.8. That it is not lower than this is simply due to the fact that the major part of it is undissociated or present as dissolved gas. By heating up a sample and measuring the pH value again it can be established whether the lowering of the pH value is attributable mainly to the carbon dioxide. The carbon dioxide content of treated water is no disadvantage in surface treatment so long as no highly susceptible materials are involved. The rinsing effect after treatment in alkaline baths is even improved. The carbonic acid is able to compensate to some extent for sodium leakage, as has already been mentioned. This is especially so if a strongly basic anion exchanger is coupled into the circuit. 1 mg/litre NaOH arising from sodium leakage always increases the pH value of a purified water to 9.4 (see Fig. 5.16). With a strongly basic anion exchanger connected to the outlet the pH value is always over 7. A sodium leakage of only 0.03 m Eq./litre results in a pH value of 9.5, one of 0.1 m Eq./litre in a pH value of 10.

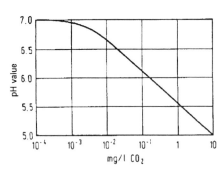

Fig. 5.15   Dependence of pH value on concentration of carbon dioxide [38]

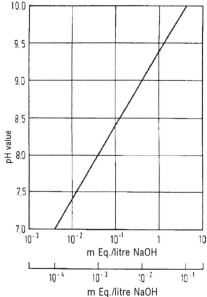

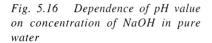

Fig. 5.16   Dependence of pH value on concentration of NaOH in pure water

The surface tension of purified water depends on the concentration of non-ionogenic surface active agents in the untreated water. So long as its value in the treated water is not less than $60 \cdot 10^{-3}$ Nm^{-1}, processes will not in general be adversely affected; below this value care must be taken. If it does not fall below $65 \cdot 10^{-3}$ Nm^{-1} over the long term, it is a sign that the ion exchangers are coping adsorptively with the input of surface active agent. These indications should be guiding principles only, as various surface active agents customarily used affect the surface tension to widely different extents at the same concentration. As a rule of thumb it can be accepted that if the surface tension of the untreated water is equal to or greater than $60 \cdot 10^{-3}$ Nm^{-1}, passage through the ion exchangers will improve it by about $5 \cdot 10^{-3}$ Nm^{-1}.

In addition to surface active agents, rinse water in the surface treatment industry contains other organic compounds as well, such as brightening additives, levellers, inhibitors and so on, which cannot be completely removed by ion-exchange treatment. These may cause problems in production both in the rinsing processes themselves and as a result of their feeding along with rinse water into high performance electrolytes. Best known are the inhibition effects and formation of blemishes in bright copper and bright nickel coatings, as well as the blue-tinting of oxide films on aluminium. The latter can be attributed with reasonable certainty to the adsorption of organic materials by the very active aluminium oxide layer. Between the anodising baths and subsequent sealing, no recirculated water should be used that has a low surface tension in any case, not unless additional purification has been carried out, as for example by active carbon. The problem is most likely to occur if rinse waters from anodising and electroplating plants are combined in a single circulation system.

If the water to be treated contains cyanide and if the plant has no strongly basic anion exchanger as the final stage, then free cyanide can get through into the purified water. Its concentration however, assuming a rinse technology appropriate for a circulatory system, does not normally exceed a maximum value of about 15 mg/litre. Such concentrations in no way interfere with production, provided neither plating on plastics nor zinc plating with final chromate passivation is involved. The reason for adjusting the free cyanide to a constant concentration is its reaction with metal ions in the untreated water, as already mentioned.

5.3.2.5   Improving the quality of purified water by additional measures

In normal operations, complete desalination by using a strongly basic anion exchanger at end of line always brings about an improvement in the quality of the purified water. As already mentioned, this leads to pH values exceeding 7 in the purified water. This can cause problems in difficult production operations. If however a so-called buffer-exchanger is connected in as well, then neutral, fully desalinated water can be produced. For this, use is made of weakly acidic cation exchangers in the H-condition. They react with caustic soda as follows:

$$R - H + NaOH \rightleftharpoons R - Na + H_2O \qquad\qquad 5.20$$

The eluate from this exchanger is neutral. By analogous means, weakly acidic purified water can also be neutralised using the same resin in the Na-condition, though a further increase in conductivity is produced by the sodium chloride formed.

Anionic surface active agents can be partly and reversibly removed from untreated water by weakly basic anion exchangers. The macroporous structure of weakly basic resins enables anionic surface active agents to be bonded partly by adsorption and partly by ion exchange. During regeneration, if loading is not too high, the surface active agent can be removed to an adequate degree if not quantititatively. Anionic surface active agents have a relatively high affinity for weakly basic anion exchangers, while they are practically impossible to remove from strongly basic resins. The part that is held adsorptively, probably because of osmotic effects is largely released again (see Section 5.3.2.6).

Non-ionic surface active agents, in accordance with physical equilibria, can be partially removed from continuously circulated water by ion exchangers. This occurs on the one hand through adsorptive bonding to macroporous resins and on the other through hydrogen bonding to cation exchangers in the H-condition. Weakly acidic cation exchangers in the H-condition are particularly suitable for trapping surface active agents if non-ionic types are to be removed completely. As investigations have been carried out specifically to clarify the elimination of non-ionic surface active agents using ion exchangers, reference should be made to the next section for a more detailed interpretation.

Weakly acidic cation exchangers (H-condition) can successfully use active carbon and adsorbing resins as traps for surface active agent. With the latter these are macroporous resins based on polystyrene without active ion groups. A porosity of 40-45%, an inner surface area of 450-500 m^2/g and a pore diameter of 40-60 A° are suitable for a resin of this type [39]. Martinola [40] describes their use in comparison with ion exchangers. As these resins work adsorptively only, they also bond with anionic surface active agents as well as organic dyestuffs, e.g. such as are used in the dyeing of anodised aluminium surfaces. In contrast to the weakly acidic cation exchangers (H-condition) they have the advantage that they can also remove surface active agents from salt-containing water without loss in capacity. This is important if the removal of surface active agents must take place before desalination by ion exchangers, e.g. in order to protect these from damage. Normally however active carbon and adsorbing resins are inserted into the stream of purified water only after desalination, before it is passed on to surface treatment processes that need rinse water with a particularly high surface tension. This is because of the low rinse water flows that are generally possible if only small amounts enter the adsorbers, so that they need not be regenerated.

They are replaced by fresh materials when they are exhausted. Under certain conditions, high capacities can be achieved, e.g. 190 gm/litre nonylphenol (with 10 mol ethylene oxide) or 125 gm/litre alkylbenzenesulphonate [39]. Regeneration of the adsorbing resin is possible, but must be carried out using methanol or iso-propanol. This presupposes the availability of specially resistant plant and a means of disposing of the regeneration product (e.g. incineration).

In comparison to the adsorbing resins, weakly acidic cation exchange resins (H-condition) are easy to regenerate, but they should generally be inserted only in the desalinated rinse water stream, where they operate also as buffer exchangers. Their regeneration to remove surface active agents is performed with dilute caustic soda and then with dilute hydrochloric acid, so that they can then be used again in the H-condition.

Cationic surface active agents are no longer found any more in most metal working industry rinse water; they can no longer be used because they are so very difficult to break down biologically in accordance with legislation relating to detergents [41]. Anionic and non-ionogenic surface active agents, however, must also, according to some legislations concerning washing and cleaning media [42], be at least 90% biodegradable.

If high purity rinse water is necessary at particular points, then the purified water is passed finally through 1-2 mixed bed exchangers (strongly acidic cation and strongly basic anion exchange resins mixed together). The electrical conductivity of the water should then be well below 1 μS cm^{-1} .

In all cases where the quality of the water has to be improved, it is, as already noted concerning the elimination of surface active agents and as described by Götzelmann [43] for the removal of free cyanide, sufficient for part of the flow of purified water to be so treated.

5.3.2.6   Behaviour of ion exchangers with respect to non-ionic surface active agents

In discussing macroporous ion exchange resins reference has already been made several times to their ability to adsorb non-ionic organic compounds as well. If the circulatory sytem is carefully planned, water needing treatment contains concentrations up to 0.1 mg/litre of predominantly non-ionic surface active agents, giving surface tensions of 60 to 65 $\cdot$ 10^{-3} Nm^{-1} .

In the metal working industries the most important non-ionic surface active agents are derivatives of the following compounds:

a) fatty alcohol polyglycolether.

$$RO - (CH_2 - CH_2 - O -)_n \ -CH_2 - CH_2 - OH$$

b) alkylphenol polyglycolether.

$$R -\!\!\bigcirc\!\!- O - (CH_2 - CH_2 - O -)_n - CH_2 - CH_2 - OH$$

c) addition product of polypropyleneglycol and ethylene oxide

$$HO - CH_2 - CH_2 - (CH_2 - CH_2 - O -)_a - (CH_2 - CH_2 - O -)_b -$$
$$- CH_2 - CH_2 - O -)_c - CH_2 - CH_2 - OH$$

The chain length and molecular weight depend on the number of ethylene oxide molecules added. 10-20 ethylene oxides at the most are incorporated in units such as a) or b). The ethylene oxide groups, through their polar ether oxygens, are responsible for the water solubility achieved in spite of the hydrophobic parts of the molecule.

Götzelmann [44] has thoroughly investigated the conditions of adsorption of non-ionic surface active agents by cation and anion exchange resins, as they occur in continuous circulation systems

The adsorption capability of variously charged cation and anion exchangers is illustrated in Fig. 5.17. A weakly basic anion exchanger is capable of taking up considerably more surface active agent than a strongly acidic cation exchanger. At the same time the results for the sulphate condition are rather better than those for the hydroxyl condition. In comparison with this the strongly acidic cation exchanger shows only a weak adsorption capability, which falls still further after conversion to the salt condition. The latter was clarified by Wolf and Wurster in studies with weakly acidic cation exchangers (see below for further information).

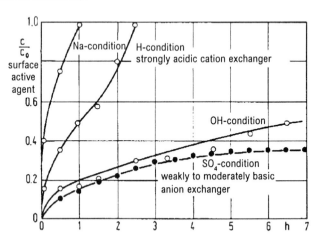

Fig. 5.17   Time dependence of the relative surface active agent concentration in the eluate from macroporous ion exchangers in various charged conditions.
$c_0 = 10$ *mg/litre nonylphenol with 1 litre ethylene oxide (Götzelmann [44])*

Interesting results were obtained as to changes in capacity and reversibility ocurring with an increase in the number of regenerations. It can be inferred from Figures 5.18 and 5.19 that in regenerating exchangers, a constant amount of surface active agent is removed independent of the degree of loading. Quantitatively in this regard the strongly acidic cation exchanger and the weakly basic anion exchanger behave very differently there too, the cation exchanger displays a smaller adsorption capacity, but also a relatively superior regenerability. (see Fig. 5.17). The anion exchanger takes out about three times as much non-ionic surface active agent as the cation exchanger, but gives up no more than this on being regenerated.

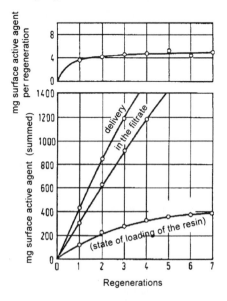
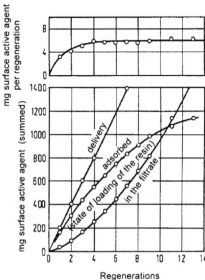

Fig. 5.18   Take-up and release of surface active agent by a strongly acidic cation exchanger (H-condition) as a function of the number of regenerations. Loading solution: 1m Eq./litre H₂SO₄, 1m Eq./litre Na₂SO₄, 2 mg/litre nonylphenol with 1 litre ethylene oxide (Götzelmann [44])

Fig. 5.19   Take-up and release of surface active agent by a weakly to moderately basic anion exchanger (OH-condition) as a function of number of regenerations. Loading solution: 1m Eq./litre H₂SO₄, 1m Eq./litre Na₂SO₄, 2 mg/litre nonylphenol with 1 litre ethylene oxide (Götzelmann [44])

These examples show that exchange resins normally used for continuous circulation systems are able to maintain the precise surface tension of the purified water necessary for trouble-free operation. After longer periods of operation the non-ionogenic surface active agents can, however, by exceeding their micelle-formation concentration, accumulate so much in the resin that no further adsorption takes place, and the concentration of surface active agent in the untreated water may even be increased by back flow from the exchange resin. The surface tension of the purified water can then be worse than that of the water before treatment [45]. In order to avoid this, special regenerations of the resins have to be performed at regular intervals (see Section 5.3.2.8.2).

In further investigations, Götzelmann [44] coupled two similar ion exchangers one behind the other, so that the freshly regenerated one was always last. It was found that the final cation exchanger which is completely in the H-condition absorbs surface active agent, while the cation-charged one which because of its pre-positioning is already loaded with surface active substances releases surface active agent. The amount of adsorption and desorption depends on the supply of surface active agent in the untreated water. In contrast, the pre-coupled anion exchanger that already contains surface active agents does not appear to give them up in any considerable quantities. The adsorption forces operating here are so strong that non-ionic surface active agents accumulate relatively quickly in the exchanger.

The behaviour in relation to non-ionic surface active agents of macroporous, strongly acidic and weakly basic exchangers can be summarised as follows:

- Macroporous ion exchangers adsorb non-ionogenic surface active agents

- The adsorption decreases with the length of the polyglycol chain

- Anion exchangers adsorb more strongly than cation exchangers

- Cation exchangers in the H-condition adsorb much more strongly than in the salt-condition

- Desorption during (normal) regeneration is incomplete, and so the reversible behaviour of the cation exchangers is considerably better than that of the anion exchangers.

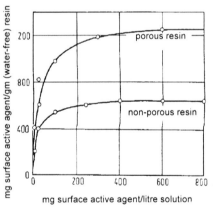

Fig. 5.20 Isotherms for the adsorption of nonylpolyglycolether with 12 ethylene oxide on weakly acidic polyacrylic acid resin in the H-condition, with and without a porous structure (Wolf, Wurster [46])

Fig. 5.21 Isotherms for the adsorption of nonylpolyglycolether with 12 ethylene oxide on a weakly acidic polyacrylic resin with a porous structure in the H-condition at different pH values (Wolf, Wurster [46])

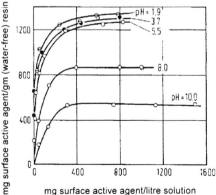

Wolf and Wurster [46] discovered that non-ionic surface active agents are particularly well adsorbed by weakly acidic cation exchangers in the H-condition, and indeed the more so, the more porous the resin is (see Fig. 5.20), the shorter the polglycol chains are, and the lower the pH value is (see Fig. 5.21). From these observations, which were also confirmed in studies by Saito and colleagues [47], it is concluded that the adsorption of non-ionic surface active agents occurs as a result of hydrogen bonding. This is conceivable only if the surface active agent molecules accumulate alongside one another on the resin surface. The observed adsorption on strongly acidic cation exchangers in the salt-condition can be attributed to hydrogen bonding with the phenolic hydroxyl groups of the condensation resins used. The formation of hydrogen bonds arises from the electrostatic attraction between the proton of a weak acid, undissociated groups and the ether oxygen atom of the ethylene oxide with its polar electron pair. The bonding between a nonylphenolpolyglycolether and the proton of an acidic group as well as the dissociation processes that remain in equilibrium with it can be represented as follows:

$$
R \atop |
$$

$$
H
$$

$$
C_9H_{19}-\!\!\bigcirc\!\!-O-(CH_2-CH_2-O)_n-CH_2-CH_2-\overset{\cdots}{O}-CH_2-CH_2-OH
$$

Association

5.21

$$
\uparrow\downarrow
$$

$$
R^-
$$

$$
H^+
$$

$$
C_9H_{19}-\!\!\bigcirc\!\!-O-(CH_2-CH_2-O-)_n-CH_2-CH_2-O-CH_2-CH_2-OH
$$

Dissociation

With increase in pH value of the solution, the exchange resin transforms into the salt-condition, in which it is no longer able to form hydrogen bonds, so the adsorption decreases, as shown clearly by the isotherms in Fig. 5.21. The way to regenerate the resin can also be deduced from this behaviour. It is done with caustic soda, to convert the exchange resin to the salt-condition, thus destroying the bonding of the surface active agents. In order to use it again it must once more be conditioned with acid. Even more successful is its regeneration with methanol [46,47]. It must therefore be assumed that the non-ionic surface active agents associate to form micelles as they accumulate on the resin, an observation that Richter [48] also made in relation to the adsorption of cationic soaps on polyacrylic resins. In alcoholic solution the micelles formed immediately break up again into monomers, which are easily elutriated. The great adsorbability of non-ionic surface active agents on strongly acidic resins can only be explained in terms of micelle formation in the resin. The following values are reported for a resin of this kind:

- Useful capacity on regeneration with caustic soda and conditioning with hydrochloric acid: up to 80 gm/litre surface active agent/litre resin;

  recommended loading: 30 gm surface active agent/litre resin

- Useful capacity on regeneration with methanol solution of hydrochloric acid: up to 150 g surface active agent/litre resin: recommended loading: 80 g surface active agent/litre resin

Where necessary, the surface active agent adsorber is coupled behind the weakly basic anion exchanger in a continuous circulation plant. Its capacity, as may be gathered from Fig. 5.21, is quite large enough at pH values below 7. If surface tensions above $70 \cdot 10^{-3}$ Nm^{-1} are to be maintained, the useful capacity is only about 10-20% of those cited above.

5.3.2.7   Influences and chemicals damaging to the resins or impairing their mode of operation

The negative effects on exchange resins of particular influences and chemicals can vary widely. They range from simple temporary blocking to irreversible damage. In many cases it is possible by chemical treatments differing from those involved in normal regeneration to restore the full performance of the resins. Harmful influences on the exchange resins and instructions for their avoidance are to be found in a leaflet from Bayer AG [50], in Oehme [16,51,52] as well as Kühne and Martinola [53].

Excessive mechanical stress, due for example to too high a throughput, leads to increased pressures, whereby the plastic beads of resin are squeezed together. The same sort of thing occurs also if the periods between back rinsing processes and regenerations become too long. The result is formation of resin conglomerates, into which the regeneration medium penetrates only with difficulty and from which it is equally difficult to wash out again. Purified water of lower quality and increased consumption of washing water are the consequences. To rectify this, stronger and longer back rinsing is required, with the help where necessary of compressed air, giving rise to increased breakage and wear of the resin. If the products of this, together with the mechanical contaminants always present in small quantities in the untreated water, are not regularly rinsed out, the pressure drops are likely to increase further and bring about additional resin damage. The normal impurities in the untreated water arise from small amounts of mechanical contaminants, unavoidable leakage from pre-coupled filters, and hydrolysis products produced between the filter and the following ion exchanger. They may thus be present even if the filtrate seems to be completely clear.

Mechanical damage is also caused by severe temperature variations. Multiple freezing and thawing during storage or operational shut-downs leads to cracking of the resin particles and as the result of consequent osmotic stresses to resin fracture [53]. Excessively high temperatures combined with osmotic stresses arise in regeneration, particularly in special regeneration procedures. In these the resins come into contact with relatively concentrated solutions. It is especially important to ensure that, in special regeneration processes, cation exchangers in the H-condition are not treated directly with caustic soda nor anion exchangers in the OH-condition directly with acid. These regenerations should where possible be carried out on the resin in its loaded condition. Exchange resins loaded with cyanides constitute an exception to this [50]. Here the acid treatment should be performed carefully after normal regeneration. Temperatures then should not exceed 45°C. Experience shows that these cause no damage to the resin. Particular care concerning temperature increases is advisable with hydroxyl-charged anion exchangers, especially type II resins (maximum 60°C). There they can lead to the splitting-off of amines and hence to a reduction in capacity. Incidentally, a moderate temperature increase has a more favourable effect on the rate of the reaction.

Major fluctuations in concentration, such as occur in regeneration or special regeneration and in washing, affect the shrinking and swelling of the resins owing to osmotic processes. They do not in fact lead directly to resin fracture, but they can cause cracks to develop [53]. It is therefore important to ensure that the low flow rates prescribed during regeneration are maintained. Organic substances, if present in excessively large amounts in the water before treatment, lead mainly to blocking of the resins, as they are deposited on their surface and in their macropores. This has already been described in the previous section for surface active agents, which eventually associate to form micelles. Concentrations up to 0.3 mg/litre cause no problems in general. It should be noted however that no anionic surface active agents should come into contact with strongly basic anion exchangers, as they would be almost irreversibly adsorbed.

Apart from surface active agents, other ionic and non-ionic organic substances are also more or less reversibly bonded to the ion exchange resins. They usually behave in a similar way to surface active agents. A generally valid statement cannot be made, however. Non-ionic organic compounds, which do not adsorb, necessarily accumulate in recirculated water and ultimately cause production defects.

Cation exchangers can be blocked by oil. This can happen over long periods even with the lowest concentrations as determined by their solubility in water or with very small quantities of emulsified oil not removed by coalescence on the filter. The emulsified oil undergoes acid breakdown in the acidic exchange zone of the cation exchanger, is adsorbed as oil, obstructs the micropores and makes the surface of the resin particles water-repellent. By special regeneration using appropriate non-ionic surface active agents it can be emulsified and removed. In all events, visible amounts of oil or emulsion in water to be treated should be avoided.

Resin damage due to effects of organic solvents has not been observed. These are much more likely if they remain in the recirculated water to make their presence felt by damaging the non-solvent-resistant materials of the equipment and generating defects in the production operation.

Damage of exchange resins is most frequently caused by inorganic substances. Most dangerous are the oxidising agents such as chromic acid, chromate, chlorine, oxygen (catalysed) and above all peroxides (persulphate, hydrogen peroxide). Ion exchange resins are more or less oxidisable as organic substances, depending on the strength of an oxidising agent, its concentration and the temperature. The danger in this regard is greater for anion than for cation exchangers, for strongly basic than for weakly basic anion exchangers, for gel resins than for macroporous resins and for less zeolitic than for the more highly zeolitic grades.

In cation exchangers oxidation affects the structure of the resins. They are weakened and thereby become soft and yielding, so that the resulting pressure drop makes their further use impossible. In addition, dissolved short-chain molecules with intact sulpho-groups reach the anion exchanger, where they are absorbed and can be removed during regeneration only with very great difficulty.

In anion exchangers on the other hand, oxidation takes place at the active exchange groups. As a result tertiary and quaternary nitrogen groups are split off, forming carboxyl groups that remain attached to the core. The latter constitute active exchange groups of weakly acidic cation exchange resins. During regeneration they are changed into the sodium-condition and in the ensuing wash generate, by hydrolysis, a constant small amount of caustic soda, so that the conductivity value specified for the purified water can no longer be attained. If oxidation is due to chromic acid, then as well as the loss in capacity there is also contamination due to precipitated chromium(III) hydroxide. Because of its amphoteric nature this is partly dissolved in the ion-exchanged water as well as in the regeneration liquor. It can be removed, however, by special regeneration using hydrochloric acid.

Most damaging to the exchangers, especially strongly basic resins, are peroxides such as persulphate and hydrogen peroxide. Type II resins are the most susceptible. Particular care is necessary in continuous circulation systems in the printed circuit board industry.

Oxygen itself can cause oxidative damage to slightly zeolitic resins in the presence of metal compounds acting as catalysts. Care must even be taken concerning the iron content of the regenerating solution [53]. Cation exchangers have good resistance up to 0.05 mg/litre of chlorine, but for anion exchangers the concentration of chlorine must not exceed 0.05 mg/litre. Reducing substances already present in the untreated water often render oxidising agents harmless. If peroxide cannot be excluded, it is usual to dose the water before treatment with a small excess of sodium bisulphite, a practice followed in many printed circuit board factories.

In contact with nitric acid the tertiary and quaternary amino groups of anion exchangers form nitrates. These are chemically similar to ammonium nitrate, which tends to decompose in accordance with Equation 5.22a:

$$2\ NH_4NO_3 \longrightarrow 2\ N_2 + 4\ H_2O + O_2 \qquad\qquad 5.22a$$

This exothermic reaction leads to heat evolution (fire) or explosion. In the USA, as Calmon [53a] reports, accidents have occurred attributed to this reaction. Care must always be taken therefore that nitrate-loaded resins are not exposed to high temperatures and pressures, that they are not stored in this form and that they are not brought into contact with oxidising agents. Nitrate-loaded resins must always be kept under water as is in any case essential during operation and must not be dried out.

If the water to be treated contains cyano-complexes of the metals copper, zinc and cadmium, decomposition can occur in the acidic exchange zone of the cation exchanger, with formation of metal cyanides of low solubility. Because of free cyanide in the water following, these salts dissolve again as complex compounds and so move about in the exchange zone as far as the end of the column. The extremely stable complexes of iron and nickel do not undergo this decomposition and are absorbed by the weakly basic anion exchanger. They must in no circumstances get through to the strongly basic anion exchanger, however, as their bonding there would be almost irreversible.

Metal hydroxides represent the most frequent form of contamination in anion exchangers. The most common cause is an unnoticed breakthrough in the cation exchanger. The metal ions are precipitated out at once in the alkaline exchange zone as hydroxide. They can be removed again by special regeneration with hydrochloric acid. If the continuous circulation plant is working in series (both cation exchangers operating with one coupled behind the other), metal breakthroughs can be avoided. If the water for treatment contains chromium(III) ions, then these also will penetrate in the course of time to the strongly acidic cation exchanger and must be removed after a certain period of time by special regeneration.

If the untreated water contains only, or predominantly, the metals zinc or lead, there is a danger that their colloidal hydrolysis products will pass through the gravel filter and reach the ion exchangers. They can be removed only incompletely or with great difficulty. Similar considerations apply to all other colloidal substances with low acid or alkali solubility.

Clogging of cation exchangers can also arise if poorly soluble compounds are formed during regeneration, e.g. lead chloride (solubility 9.7 mg/litre) or silver chloride (solubility 1.51 mg/litre) in a regeneration with hydrochloric acid. One must turn in such cases to other regenerating acids.

The damage caused to ion exchangers by the processes that have been described can be determined by three types of investigation:

Microscopical examination

- Formation of cracks and resin fracture can be determined. These can be traced back to all the defects caused by osmotic, temperature or mechanical stresses. Weakening by oxidation can be ascertained by crushing the beads of resin between the fingers.

Capacity

- Reductions in capacity can be attributed to oxidative destruction of the active exchange groups in anion exchange resins. Everything that causes an increase in wash water consumption (in column systems) gives the impression in practice of a fall in capacity. Lowering the diffusion rates makes the exchange zones larger and breakthrough more likely.

Wash water consumption

- All resin blockages due to contaminants (oil, hydroxides, surface active agents and so on) have the effect of increasing the wash water consumption. The increase can arise from lower diffusion rates and from the release of conductive substances. If an appropriate special regeneration does not produce an improvement in the wash water consumption, the blockage must be considered irreversible, if successful, the regeneration volumes will be reduced and the capacity increased.

5.3.2.8   Regeneration of ion exchangers

5.3.2.8.1   Normal regeneration

Normal regeneration of ion exchangers is carried out after they have been loaded. This reverses Equations 5.10 and 5.15 for the cation and anion exchanger, converting them again to the H⁻ and OH⁻ condition, respectively. The semi-concentrates produced i.e. the solutions of elutriated substances, the excesses of the regeneration media and the first concentrated wash waters are called regenerates. On average, they have 100 times the concentration of the circulated rinse water and accordingly about 1% of its volume. The times between regenerations depend on the useable capacity of the exchanger and the concentration of the untreated water under the operating conditions prevailing.

Especially for strongly acidic cation exchangers and strongly basic anion exchangers there is also the effect of the amount of regeneration medium employed. If an appropriate choice of regeneration conditions is made, the operating conditions (concentration of regenerating medium, its quantity and rate of flow) ensure:

- optimum operating time
- constant high quality of purified water
- long lifetime of the resins.

Just as the exchange reaction on loading requires a certain reaction time, so also does the opposite reaction of regeneration. As however in regeneration substantially larger amounts of material have to undergo reaction in a shorter time, and as the speed is controlled by the rate of grain diffusion, sufficient time must be allowed, or in other words, a throughput velocity must be selected that is lower than that for loading. The solution velocity recommended by the resin manufacturers for regeneration is for most resins around 5 mh^{-1}, corresponding according to Equation 5.9 to a residence time of ca 4 minutes. The total contact time depends on the time it takes to feed in the regenerating medium and amounts to at least 30 minutes. If regeneration could be accomplished under ideal conditions small and uniform resin particles, very small amounts of substance to be removed, low solution velocity then it would be possible to get a narrow elution curve, with a shape approximating to that of the Gaussian error curve. In practice, the elution curves naturally depart from this ideal form, because of the wide particle size spectrum, the high concentrations and above all because of the simultaneous removal of various substances. The regeneration of ion exchangers consists of three steps, if back rinsing is ignored:

- Supply of regenerating medium (5 mh^{-1})

- Displacement with water. The regenerating medium not yet consumed, which is still present mainly in the water situated between the beads of exchanger resin, should be passed through the exchange column at about the same rate as the regenerating medium was supplied. The displacement of the concentrated regenerating medium at the rate of flow of the dilution water occurs in a normal manner if carried out during regeneration after the diluting and dosing processes (see Section 5.3.3.2.3).

- Washing. The velocity of the wash water is increased to ca 10 mh^{-1}.

The eluate produced while the regenerating medium is being supplied, while it is being displaced, and while washing occurs until the conductivity falls to that of the untreated water, constitutes the actual regenerate and is transferred to a storage vessel, which should also contain other factory waste water that is to undergo the same waste water treatment. The final wash water is transferred to the untreated water storage tank.

Resins used only with down flow (loading and regeneration) must be back rinsed before regeneration to remove mechanical contaminants and to loosen them up with water. So far as rinse water recirculation is concerned this applies only to the older plants. In plants operated with up-flow loading and down-flow regeneration (counterflow), back rinsing is necessary only at long intervals of time (usually externally, in a special back-rinse vessel). Untreated water can be used for back rinsing and passed on afterwards to waste water storage.

Regeneration of strongly acidic cation exchangers used in continuous rinse water circulation is carried out as a rule with 7-8% hydrochloric acid solution. The conversion proceeds largely quantitatively, provided as in all chemical reactions there is a certain excess of acid. 150 to 200% of the theoretical quantitiy of acid ought to be sufficient to maintain a satisfactory useful capacity. The latter increases with the amount of regenerating acid employed. In this connection an increase in the elutriation of sodium ions is important, as it leads to less sodium leakage during circulation and a better quality of purified water (see Fig. 5.22). Counter-flow regeneration is common nowadays, in which the fresh acid comes into contact first with the less highly loaded sodium-containing parts of the resin and, also produces a reduction in sodium leakage. Of the cations involved, sodium ion has the lowest affinity for the strongly acidic cation exchanger and, depending on its ratio to the total cation concentration, is critical in determining the useful capacity of cation exchangers.

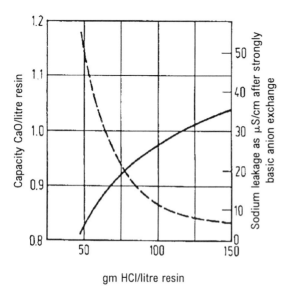

Fig. 5.22 Dependence of capacity (—————) and sodium leakage(– – – – –) on consumption of regenerating medium by a strongly acidic macroporous cation exchanger [54]. Water concentration before treatment 7.5 m Eq./litre; 36% Na content.

Particular regeneration methods allow savings to be made in regenerating acids. Thus, for example, the fraction of eluate with the greatest excess of acid can be collected and used again to do the job of fresh acids in the next regeneration. In regeneration with 200% of the theoretical amount of acid, fractions are produced that still contain 4% acid. Although they also contain metals and surface active agents they can nevertheless be used for pre-regeneration purposes. This kind of economical regeneration has found greater use, however, where ion exchangers are applied in recycling rather than in continuous circulation applications.

Sulphuric acid is used in regeneration only in exceptional cases; e.g., where sparingly soluble chlorides could be formed. Its regenerating effect corresponds anyway only to an equivalent of hydrogen ions, so that there is scarcely any advantage in terms of cost. In particular circumstances, e.g. if the untreated water contains large proportions of silver, dilute nitric acid may be employed.

Acidic regenerates contain alkali, calcium, ammonium and heavy metal chlorides, elutriated non-ionic surface active agents and excess hydrochloric acid. The cyanide and chromate present in the water held between the resin beads in the columns before the start of regeneration are transferred during rinsing and back rinsing to the tanks containing waste water or water requiring ion-exchange treatment. When cyanides are present, the regenerate from the cation exchangers is collected and, after detoxification of the alkaline regenerates, is added to those for neutralisation or also for acidification. In the presence of cyanides the regenerate together with that from anion exchanger or exchangers must be detoxified. There remains a possiblity that the resin contains poorly soluble metal cyanides that have formed in the acidic reaction zones, those of silver and of copper in particular.

Weakly acidic cation exchangers, because of their high selectivity towards hydrogen ion, require scarcely any excess of acid. 100-105% of the theoretical amounts are usually sufficient for regeneration. When they are employed as buffer exchangers, surface active agent adsorbers or selective exchangers, use is made of hydrochloric or sulphuric acid at concentrations between 3 and 10%.

The weakly basic anion exchangers that are always employed in continuous circulation technology require no great excess of caustic for regeneration, because of the relatively slightly dissociated bases formed. Techniques for economising in the use of regenerating media do not therefore have the same advantages as, for example, in the case of strongly acidic cation exchangers. The regeneration is carried out with 4-5% caustic soda, at a liquid velocity of 5 mh^{-1}. Depending on the organic loading, 120-150% of the theoretical amount is used.

The regenerate from weakly basic anion exchangers contains the sodium salts of all the anions contained in the water before treatment, cyanide, cyano-complexes and chromate, if they are present, along with the excess caustic soda. In addition there are the elutriatable anionic and non-ionic surface active agents. Fig. 5.23 shows the surface tension values generated by surface active agents during the regeneration of anion and cation exchangers. If the regenerate contains cyanide, it is collected in a storage tank for detoxification, along with cyanide-containing waste water originating elsewhere.

Strongly basic anion exchangers need a large excess of caustic soda to transform them substantially to the OH-condition. Taking into account their low useful capacity, 200-400% of the theoretical quantity is necessary. The regenerating caustic is supplied at 4-5% concentration and a liquid velocity of 5mh^{-1}. As with strongly acid cation exchangers, the capacity is affected by varying the amount of regenerating medium. This is on the condition that the strong acids are already absorbed by the pre-coupled weakly basic exchanger, depending on the amounts present relative to the weak acids. The higher the proportions of the weak acids, the lower the useful capacity and the more favourable therefore the effect of a larger amount of caustic soda. Fig. 5.24 shows the influence of the quantity of alkali on the capacity and on the silicic acid leakage of a strongly basic anion exchanger of type I. Because of the effects of different proportions of ions, it is possible to plot the curves only in relation to a quantity of regenerating medium of 100 gm NaOH/litre of exchange resin.

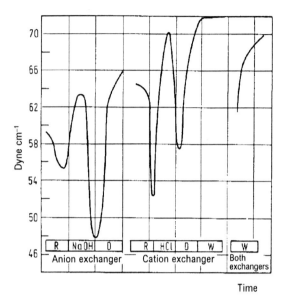

Fig. 5.23   Surface tension during the regeneration of an anion and a cation exchanger [45]

R   Back rinsing

NaOH,   Supply of regenerating

HCl   medium

D   Displacement with water

W   Washing

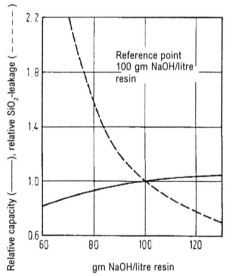

Fig. 5.24   Dependence of the relative capacity (———) and relative SiO₂-leakage (– – – – –) on the consumption of regenerating medium by a strongly basic macroporous anion exchanger of type I [55]; NaOH 4%, 40°C

The regenerates of the strongly basic anion exchangers, because of the high proportion of unconsumed caustic soda, are always used to regenerate weakly basic ones, so that the high excess of alkali required is not wasted. The regenerate leaving the strongly basic exchanger contains all the anions of weak acids including free cyanide as well as a large excess of unused caustic soda. The joint regeneration of strongly and weakly basic anion exchangers coupled one behind the other is called compound regeneration.

Because of the possibility of amines splitting off over a period of time, resulting in a loss of the capacity associated with them, strongly basic anion exchangers should not, so far as possible, be regenerated with more concentrated caustic or at higher temperature or for longer times than necessary [53]. During long interruptions in operation they should be kept in a loaded state or unloaded in the salt-condition.

In regenerating a sequence of exchangers, as is usual in rinse water circulation sytems (strongly acidic cation exchanger weakly basic anion exchanger), the elution of surface active agents is followed by measurement of the surface tension (see Fig. 5.23). The diagram shows that both exchangers produce a distinct lowering of the surface tension of the water during back rinsing. Clearly the resins release adsorbed non-ionic surface active agents when water free from surface active agents is passed through them at a high rate. When the regenerating medium is introduced (in practice, it is noticed first when the displacement water is discharged), further elution of surface active agents occurs during regeneration.

The regeneration of ion exchangers employed in recycling systems sometimes deviates substantially from that described here (see Section 7.2).

5.3.2.8.2   Special regeneration

Given the very varied compositions of continuously circulated waters, over a period of time, substances find their way into the exchangers that can be removed only partly, or even not at all, by normal regeneration. In so far as these substances are bonded ionically, they produce a direct reduction in capacity, and in so far as they are adsorbed or precipitated, an indirect reduction in capacity, which is attributable to a lowering of the exchange rate and which as a rule also causes an increase in wash water consumption. The requisite special regeneration processes are used, depending on which types of exchanger and which substances are involved. Alternating regeneration is a versatile procedure. It employs normal concentrations of regenerating media. The treatment sequence is:

For strongly acidic cation exchangers:

- normal regeneration with 7-8% hydrochloric acid. Wash (metal-free!)

- supply of 4-5% caustic soda (1.5-2 times regeneration amount). Wash

- conversion of the exchanger with 7-8% hydrochloric acid into the H-condition (1.5-2 times regeneration amount). Wash

For weakly basic anion exchangers:

- normal regeneration with 4-5% caustic soda − wash

- supply of 7-8% hydrochloric acid (1.5-2 times regeneration amount) – wash (metal-free!)
- conversion of the exchanger with 4-5% caustic soda into the OH-condition (1.5-2 times regeneration amount) – wash.

It is appropriate to examine the eluate and, if necessary, to extend each of the second treatments or to repeat the alternating processes. During this treatment, alkali-soluble substances are elutriated from the cation exchanger and acid-soluble substances from the anion exchanger. The alternations in osmotic pressure associated with this also bring about the elutriation of non-ionic contaminants, while the temperature increases occurring (40-45°C) make a positive contribution towards the dissolution processes and to the concluding regeneration. It can be affirmed that considerably greater quantities of surface active material can be removed by this means than by normal regeneration.

Non-ionic surface active agents can also be elutriated by methanol, with break-up of the the micelles formed inside the resin, ion-active species are also, more successfully treated with methanolic regenerating media. Because of the methanol, however, these methods give rise to waste disposal difficulties.

Depending on the kind of blocking surface active agent, treatment with alkaline sodium chloride solution (5% each of caustic soda and sodium chloride) is often helpful, first in a batch process and then in a slow continuous operation. This is followed by thorough washing and normal regeneration.

During acid treatment of anion exchangers, dissolution occurs simultaneously of metal hydroxides, which have accumulated there by transfer from the cation exchangers or over an extended period of plant operation. In addition, cyanide complexes, especially the very stable ones of iron and of nickel, are also removed. On the whole, however, if the anion exchanger has taken up trivalent chromium, not much will be achieved with dilute hydrochloric acid.

More severe contamination of anion exchangers with metal hydroxides and anions of the more stable metal complexes their presence is usually indicated by a greenish coloration (copper, nickel) of the upper part of the resin can only be removed by the use of stronger hydrochloric acid (1:1) in a batch treatment. In this, the acid must remain in contact with the resin for several hours, especially in order to dissolve the chromium hydrolysis products that have become less soluble as a result of ageing. This procedure must be repeated as required according to the degree of contamination. In order to avoid osmotic shock and excessive heat of neutralisation, the weakly basic resin should be slowly converted into the hydrogen-condition using normal regenerating acids in a flow-though process before treatment with strong acid. By these means it is often possible to remove more metal/litre of resin. Trivalent chromium can also be produced by reduction of chromate absorbed on the exchange resin. This possibility always exists, if the water to be treated contains anionic reducing agents, e.g. bisulphite.

It is also possible for cation exchangers to become blocked with basic salts of chromium after long processing of water that contains trivalent chromium. A successful special regeneration with hydrochloric acid (1:1) in batch-mode is possible here also.

Cation exchangers have also been known to become blocked with mineral oils. These are formed by acid decomposition in the acidic exchange zones with a scarcely perceptible emulsion content. They attach themselves to the surface and the inside of the micropores, so lowering the exchange rate and the useful capacity. They can be emulsified again and removed by special regeneration with surface active agents. The surface active agents used are those with the highest possible turbidity points (high micelle-formation concentration) and good emulsifying capability. Their 1% solutions should remain clear. The treatment is carried out in batch mode, in which the resins are agitated repeatedly from below by the introduction of compressed air. Repetition once or several times is more often than not required.

Along with these processes there are others as well, which are used specifically on particular substances that have collected in the exchangers, e.g. the removal of silver chloride with dilute ammonia solution.

5.3.2.8.3   Treatment of algal-contamination

As in all continuous water circulation where the water is used many times over, rinse water circulated continuously through the ion exchangers and parts of the plant that remain in contact with it can become contaminated with algae. This affects the storage tank and pumps for the untreated water, the gravel filter, the ion exchange columns including the resins, the rinsing vessel and all pipelines that connect these units together. The algal-contamination develops particularly quickly under the influence of light.

To eliminate algae or bacteria also, all the above-named parts of the plants must be brought into contact with algae-destroying or bactericidal solutions, respectively, after the coarsest contamination has been removed by forceful back rinsing or mechanical cleaning (rinse tanks). In the circumstances it is simplest to replace the filter gravel with new material.

With the exception of the anion exchangers, all pieces of equipment can be treated with the following agents:

- Chloramine T, 200 gm/m^3

- Formaldehyde, 30 litre/m^3 35% formaldehyde solution

  Both of these solutions are poured into the plants and left in contact with them for 5-10 hours. The cation exchangers must already be in the loaded state. Afterwards, back rinsing follows to remove the dead organisms then regeneration of the exchangers [56].

- Peracetic acid, 6 litre/m³ 30% acid. The solution is circulated for 1 hour before or during the regeneration.

- Chromic acid, 200 gm/m³ chromic acid is pumped around the relevant parts of the plant, preferably overnight.

All these chemicals must still be detectable after treatment, otherwise it is necessary to repeat it.

Anion exchangers must not be exposed to chloramine T and chromic acid. These would combine with the ion exchanger and damage the strongly basic resins by oxidising them. The following media can be used on anion exchangers:

- Caustic soda. After regeneration the parts of the plant are filled with 3-6% caustic soda and left in contact for 5-10 hours (taking care with strongly basic resins)

- Formaldehyde (same treatment as for cation exchangers)

- Peracetic acid, but only after consulting the manufacturer.

The instructions of the supply companies should be followed regarding concentrations and contact times. Damage to the hard rubber linings of plant components has been known to be caused by certain commercial preparations. In addition, care must taken with disposal of the used solutions, in so far as organic products are concerned, and with regard to their biodegradability

5.3.3   Process technology of rinse water recirculation

5.3.3.1   Basic construction of continuous circulation plants

The basic equipment of an ion exchange continuous circulation plant, independent of its design, is [57]:

- Raw water storage tank

- Raw water pumps

- Gravel or gravel-carbon filter

- At least two cation exchangers (except in continuous plants)

- At least two anion exchangers (except in continuous plants)

- Conductivity measuring devices

Fig. 5.25 shows this schematically.

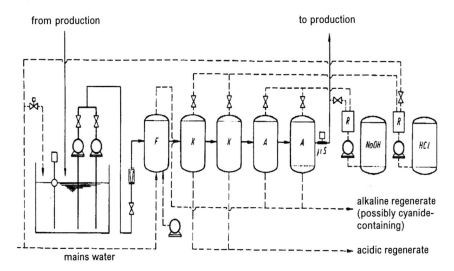

from production   to production

alkaline regenerate
(possibly cyanide-
containing)

acidic regenerate

mains water

untreated water supply filter and exchange columns regeneration equipment

Fig. 5.25 Schematic diagram of an ion exchange continuous circulation plant. F Gravel filter, K Cation exchanger, A Anion exchanger, μS Conductivity measuring device, R Regeneration equipment (various possibilities)

5.3.3.1.1 Raw water storage tank and pumps

Rinse waters, for treatment in a single continuous circulation system, are collected together in the untreated water storage tank. Usually, this means all the circulating rinse waters of a surface treatment plant, without separation into their various types. If certain materials in these rinse waters are to be recovered, after their concentration, from the exchangers, then a separation of rinse waters takes place afterwards, in contrast to waste water separation, which is arranged in accordance with the type of treatment. Species thus recovered in increasing quantities are, in addition to precious metals, non-ferrous metals and chromic acid (see Section 7.2.1).

Untreated water storage tanks can be constructed of concrete with chemically resistant inside linings or they could be hard rubber lined steel containers or plastics (PVC, polyethylene, polypropylene, polyester, glass-fibre reinforced). Their shape is not critical. Nowadays, cost-effective and corrosion-resistant round vessels of polythene or polypropylene are most frequently used.

The capacity of the storage tank should not be too small and where possible should correspond to half the volume of the amount of water continuously circulated per hour. This has the following advantages:

- Possible fluctuations in concentration are evened out

- Low-density materials present in the rinse water (e.g. oil) can separate and be removed by skimming. Where these are a danger from the start, however, a light-material skimmer should be connected to the inlet.

- Fluctuations in the volume of continuously circulated water are evened out and can be allowed for. As a rule problems occur, if the continuous circulation is interrupted, e.g. to back rinse the gravel filter or to wash the cation exchangers, or also as a result of evaporation.

Volume fluctuations in the continuously circulated water always make themselves felt in the untreated water storage tank. It should therefore be fitted with a solution level measuring device having at least two, or still better four, switch contacts. Their functions from bottom up are:

- Dry running protection of the pumps at the minimum water level

- Opening of the intake valve for mains water at a low water level

- Closure of the intake valve for mains water on reaching the required water level

- Sounding the alarm on reaching the maximum permitted level.

In many plants measurement of the conductivity and/or pH value of the untreated water is provided for, to arrest intake of concentrated solutions (arising from unforeseen incidents) and to allow counter measures to be instituted.

In special cases, dosing with small amounts of chemicals may also be carried out in the untreated water storage tank, e.g. addition of sodium bisulphite if peroxide has got into the water.

The raw-water pumps are used to transfer the water to the gravel filters and ion exchangers. As the recirculation plant must operate continuously during production periods, a reserve pump is necessary in addition to the working pump. Both pumps should be operated alternately at intervals, so that the breakdown of one of these important units can be immediately diagnosed and corrected. The output of each of the two pumps must, after allowing for likely pressure drops, be able to maintain the nominal operating rate of the continuous circulation plant, through the exchange units and associated pipework as well as at the water pressure sampling points. Single-stage high-grade steel centrifugal pumps have proved best for this application. As they are normally not self-priming but only work on a suction basis, they are mounted at the same level as the floor of the untreated water storage tank. Priming reservoirs and valves at the end of the suction pipe are required for pumps arranged at higher levels. Inlet strainers have to be fitted to these if mechanical contamination of the untreated water is a possibility.

The volume flow rate at the inlet to the gravel filter is adjustable and readable on a flowmeter. The pressure at which the untreated water is supplied to the plant can be read on a pressure measuring device on the pressure side of the pumps or at the gravel filter intake.

5.3.3.1.2   Filter and exchange columns

By filtration, cation-replacement with strongly acidic cation exchangers in the
H-condition and neutralisation with weakly basic anion exchangers in the OH⁻
condition, the rinse water is desalinated (fully desalinated if strongly basic anion
exchangers are used as well), then returned again to the rinse units.

Almost without exception, gravel filters or gravel filters with an additional
layer of granulated hydro-anthracite [58] are used for filtration. The mechanism
of the depth filtration that occurs and the basis for the use of hydro-anthracite
are described in detail in Section 5.3.2.1. The speed of filtration in continuous
rinse water circulation plants is about 20-30 mh⁻¹ and the height of the bed of
filter gravel about 0.8-1.5 m. The flow in the filters, in contrast to that in the
exchange columns, is downwards. The filters are cleared out using an upflow,
generally of untreated water and with the assistance of compressed air. The
back-flush water is passed on to the waste water treatment plant and, if cyanide
is present, to the cyanide detoxification plant. Back flushing is necessary when a
certain pressure difference develops between the inlet of the filter and its outlet
or the inlet of the following cation exchanger. Its value is usually between 1 and
1.5 bar. Gravel filters are frequently fitted with pressure difference measuring
gauges, which sound an alarm as soon as the the preselected pressure difference
is reached. As the gravel filter columns have to stand the highest pressure, they
should be provided with an excess pressure relief valve. The pressure built up by
the untreated-water pumps against the closed outlet valves at the end of the
plants is about 6 bar, but in plants with longer recirculation pipelines for the
purified water can be up to 8 bar.

Because of the chemical stress and the working pressure, the ion exchange
columns, like the gravel filters, are made from boiler-steel-sheet and usually
coated inside with 3 mm-thick hard rubber. Only plants with small throughputs
are equipped with columns made of plastics, the pressure characteristics of
which need particular care. The inner components consist of hard rubber coated
steel, alloy steel or plastics. Among these there are in particular the lower filter
floors (plates with inset filter nozzles) in all columns, the solid matter capturing
devices (wire baskets, strainers) intended to reduce back rinsing losses in plants
that operate in down-flow, drains (pipes with filter nozzles) for down-flow
columns and upper filter floors for columns employing fluidised and restrained
bed processes (see Fig. 5.35).

The size of the exchanger columns depends on the volume flow rate of the
continuously circulated water. Experience shows that a resin bed height of ca
1000 mm (maximum 1200 mm) is appropriate, because of the pressure losses at
normal throughput and because of the processes of expansion and contraction of
the resins in their various forms. The useful column height (usually the height of
the cylindrical casing) above the lower filter floor should be ca 2000 mm (except
for the fluidised bed process)..For most exchange resins, manufacturers quote a
recommended liquid velocity, with a maximum of 40 mh⁻¹. A well-established

value for this is 25 mh^{-1}. At ca 1000 mm bed height, the equivalent specific flow rate is 25 m^3/m^3 of resin per hour. It follows that 40 litres of resin is necessary for an water flow rate of 1 m^3/h. From this it is possible to calculate the column diameter for the plant throughput required.

A liquid velocity of 25 mh^{-1} also ensures a favourable contact time between the water under treatment and the resin, which because of the diffusion-controlled exchange rate largely dependent in continuous rinse water circulation on the film diffusion is essential. The pressure drop increases linearly with the flow rate and the bed height. Deviations from the data quoted above occur however at the limits. That is true also for unduly low velocities. In the limit, approaching 0 mh^{-1}, the results are similar to those in batch processes. An equilibrium might be established in which absorbed ions return again to the aqueous phase. The consequence is a lower quality of water, which in practice, when bringing continuous circulation plants back into service after a few days shutdown, is indicated also by the observation of poor conductivity values. These results are approached as the flow rates become too low. On the other hand, a certain flow over the surface of the resin is necessary to maintain a favourable exchange rate. The thickness of the boundary diffusion layer falls as the flow rate increases.

The question of a clear increase in the capacity of a quantity of resin because of higher concentrations in the untreated water does not really arise in the case of rinse water. It can be of importance however in recycling plants. There, in some circumstances, larger quantities of resin are provided for a given flow rate, especially as in weakly acidic cation exchangers lower liquid velocities are also favourable. An operating time of ca 20 hours should still be sufficient however at a concentration of 2 m Eq./litre under the conditions stated above.

The columns need couplings for the inlet and outlet of the water to be processed, regeneration media, regenerates, back-rinsing, displacement and wash water, which are designed to link the columns with the associated pipework. In addition, air pipes are required for the upper spaces of the columns as well as for the spaces below the nozzle floors. These can be used also to introduce the air needed for cleaning and loosening up processes. Depending on the size, hand holes or manholes must be provided for repairs, filling and emptying, and through which the upper and lower main connections can also be be made. Only in small plants can these be omitted.

The connecting pipework for all the above-mentioned liquid and air flows must be laid out clearly in front of the columns, depending on the type of circuit in which the plants have to operate (see Section 5.3.3.2.1), including the necessary ancillaries (valves, flow-meters and so on). It must possess the same resistance to chemicals and pressure as the columns. For nominal diameters up to 100 mm, plastic pipes can be used (PE, PP, PVC); for larger diameters, hard rubber lined steel pipes are used.

Ion exchange columns are operated at liquid velocities between 25 and 40 mh^{-1}.

The following devices are required for the control of continuous circulation plants:

- Flow-meters for untreated, diluting and displacement water (mainly at the regeneration station) and wash water

- Pressure measuring device on the inlet side of all the columns

- Sight glasses for the level of the gravel and resin surfaces

- Columns

- Sampling points at the outlet pipes of all columns.

Conductivity measuring instruments.   They are absolutely essential for continuous quality control of the purified water. In plants connected in line, these change over from one to another when breakthrough occurs and in automatically operated plants simultaneously begin regeneration of the disconnected units. Hence a measuring gauge is necessary after the anion exchangers. If the plants are operated in series, then conductivity measuring instruments are necessary after all columns if automatic operation is envisaged. Then the columns made ready for regeneration are only those in which breakthrough has actually occurred (see Section 5.3.3.2.1).

Figs. 5.26 to 5.28 show the mode of construction of ion exchange continuous circulation plants.

*Fig 5.26   Automatically operated ion exchange plant for rinse water.
Rating 3m³/hr. (BLASBERG)*

*Fig 5.27   Automatically operated ion exchange plant for rinse water.
Rating 14 m³/hr. (GÜTLING)*

*Fig. 5.28   Automatically
operating ion exchange
continuous circulation
plant for rinse water in
plumbing ware factory,
capacity   30   m³/hr.
(HAGER   +   ELSSSER
GmbH, Stuttgart)*

If it is specified that complete desalination must be accomplished in a continuous circulation plant, then the additional use of a strongly basic anion exchanger becomes necessary. This means in practice that either an exchanger of this type is connected at the end, in which case the size of the column is chosen to be the same as that of the others, or both the weakly basic and the strongly basic anion exchange resins are put in one column, the so-called two-storey column. In the first case a single stand-by column is usual. The increasing concentrations in the circulated water of anions of weak acids during its regeneration are calculated after its reconnection, so that they never reach a damaging level. With the two-storey column the amounts of both anion exchange resins necessary are adjusted to the requirements, e.g. 2/3 weakly basic and 1/3 strongly basic anion exchangers. The columns have an extra filter floor and are correspondingly taller. The combined regeneration is particularly easy to carry out in such cases. Fig. 5.29 shows a plant with two-storey columns for the anion exchangers.

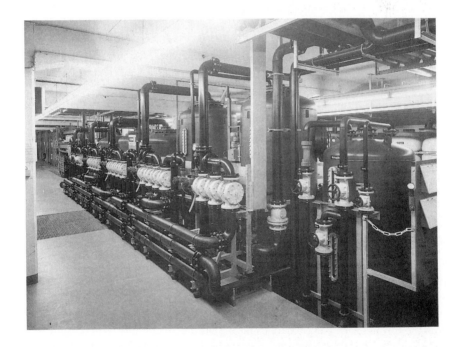

Fig. 5.29   Ion exchange continuous circulation plant with 2 two-storey columns each for strongly basic and weakly basic anion exchangers (centre of photo), capacity 50 m3/hr, automatically operated (GOEMA, Dr Götzelmann GmbH Vaihingen/Enz)

If a further improvement in the quality of the water is to be achieved by incorporating buffer exchangers, surface active agent removers or active carbon in parts of the circuit, this is frequently done downstream with small, simply constructed columns.

If pressure drops occur in continuous circulation plants, so that a satisfactory pressure cannot be guaranteed at sampling points, then pressure-boosting pumps can be provided at the end of the plant or better, at the purified water stations. The latter each consist of a purified water storage tank and two purified water pumps (working and reserve pump). The purified water storage tank has an overflow to the untreated water storage tank and the throughput of the purified water pumps is throttled back to somewhat less than that of the untreated water pumps.

5.3.3.1.3   Regeneration facilities

The normal regeneration of ion exchangers includes, as described in Section 5.3.2.8.1, back flushing, supply of regenerating medium, displacement of the regenerating medium and washing. The facilities required for this are component parts of each continuous circulation plant, except for those plants that are so small that they are regenerated externally (see Section 5.3.3.2.4). Reference should be made to Section 5.3.3.2.3 for further information about these.

Back flushing, being the first step, helps in removing the smallest mechanical contaminants and in loosening up the resin. After long periods of use, resin conglomerates can be produced by adhesion of the small resin particles. By vigorous back flushing with filtered untreated water or mains water of the cation exchanger, or with cation-exchanged water of the weakly basic anion exchanger, or with desalinated water of the strongly basic anion exchanger, the resin beads are loosened up and freed from such solid impurities as may be present. All the back flush water is transferred to the untreated water storage tank. This process can be carried out in all columns with available back-flushing volume.

Ion exchangers operated with upward flow remain loose and do not suffer from the formation of conglomerates during normal operations. Where fluidised bed procedures are concerned, the resin is washed in an external vessel only at long intervals, approximately every 6 months. It is flushed out to this via an overflow at the side of the column (see Fig. 5.35). Back flushing vessels can be open or closed, like a column (see Fig. 5.30).

Dilution of the regenerating medium and the displacement and washing that follow are carried out with the same quality of water as that used for back flushing. The regenerates are transferred to the storage tank for acidic (cation exchanger) or alkaline (anion exchanger) concentrates, respectively, or in the latter case, if cyanide is present, to the storage tank for cyanide-containing concentrates, or cyanide- and alkali-containing concentrates.

*Fig. 5.30 Back flushing vessel for resins in a plant with
fluidised-bed processes
(GOEMA, Dr Götzelmann GmbH, Vaihingen/Enz)*

If the rinse water contains no cyanide, then both regenerates may be collected together in the same storage tank. Regenerates containing complexing agents are transferred to a special storage tank for waste waters containing such materials.

The discharge of wash water is taken to the water-for-treatment reservoir, as soon as the electrical conductivity has been brought down to that of the water already in it. When the wash water attains the conductivity prescribed for the purified water, the regeneration is terminated and the regenerated column disconnected and made ready for re-use (line circuit) or, in accordance with Fig. 5.32, connected into the continuous circulation plant (series circuit).

The regenerating chemicals are taken from storage tanks for concentrated hydrochloric acid (32-36%) or concentrated caustic soda (40-50%) and diluted in various ways (see Section 5.3.3.2.3). These, as well as all the process-related vessels, pumps, water injection pumps, flowmeters, level indicators, control valves and other accessories are component parts of the regeneration facilities, which are usually operated automatically. Fig. 5.31 shows a regeneration facility.

*Fig. 5.31   Regeneration facility of an ion exchange continuous circulation plant
after the dosing processes (dilution water supply with flowmeter).
Acid and caustic tanks are protected against leakage and overfilling
(GOEMA, Dr Götzelmann GmbH, Vaihingen/Enz)*

5.3.3.2   Technical variants

In addition to special designs, there are several variants regarding the arrangement of the exchange columns, the control of liquid velocities within them and the input as well as the discharge of the regeneration media, which are described below.

5.3.3.2.1   Possible ways of connecting up the columns

As ion exchange circulation plants, except for continuously operating plants, always include two or more columns of the same kind, various circuit arrangements are possible. These are:

- In-line circuit
- Parallel circuit
- Series circuit

They are illustrated in Fig. 5.32.

The In-line circuit [59] constitutes the simplest circuit in terms of pipework and automation. A cation exchanger and an anion exchanger, which as a rule are equal in size, are each operated until such time as an increase in conductivity occurs on the purified water side. That means that one of the two exchangers is so fully loaded that it can no longer hold back all the ions that it should absorb. These cause the increase in electrical conductivity and so indicate breakthrough in the line. Both exchangers are shut off and must be regenerated. At the same time the second regenerated line is brought into operation. Because of this alternating mode of operation, these plants are also known as pendulum plants.

If it is necessary to increase the capacity further, a third line can be installed. Operation then takes place as described for plants with parallel circuits.

The strongly acidic cation exchangers and weakly basic anion exchangers used for continuous circulation generally have different capacities. Furthermore, the untreated water because of its weakly acidic nature always contains more exchangeable anions than cations. If the concentration ratios in the untreated water and the capacities of the exchangers were completely constant, it would be possible bring the difference in line with these conditions by adjusting the quantities of resin [59]. In practice this has not been accomplished.

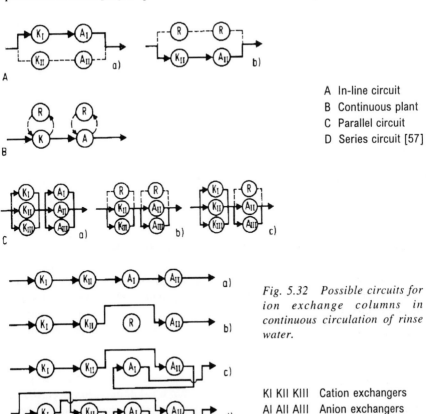

A In-line circuit
B Continuous plant
C Parallel circuit
D Series circuit [57]

Fig. 5.32 Possible circuits for ion exchange columns in continuous circulation of rinse water.

KI KII KIII Cation exchangers
AI AII AIII Anion exchangers
R Exchanger undergoing regeneration

In such cases it is possible to make use of existing cross connections with regenerated lines [59], or simply to take into account that when breakthrough occurs in one line, one of the two exchangers, the cation exchanger as a rule, is still not fully loaded.

The so-called cross circuit of Schüssler [60] has been described in relation to this. According to a particular method the connections of the cation exchangers (K) and anion exchangers (A) are changed and the conductivity measured after a time delay downstream of each of the anion exchangers. Possible couplings are KI-AI, KI-AII, KII-AI and KII-AII. By changing over from one of the two operating exchangers to the other of the same kind, the exchanger in which breakthrough has occurred can be identified, then regenerated and made ready for use again. Fig. 5.27 shows a continuous circulation plant with a cross circuit.

A continuously operating plant is coupled up like a line circuit, but its action ought rather to be compared with that of a plant having a series circuit (see Section 5.3.3.2.4).

Three or more exchangers of the same type are used in a parallel circuit [61,62]. The sequence of the cation and anion exchangers connected in parallel corresponds again, however, to that in the line circuit (see Fig. 5.32c). In the simplest case the electrical conductivity of the purified water is measured at the common outlet from the anion exchangers. This measured value represents a mixed value for each of the exchanger lines in use, which vary in the length of time they have been operating and, because of the related differences in pressure drop, in their liquid flow rate also. When an increase in conductivity occurs, the exchanger pair that has operated for the longest time is uncoupled and regenerated. The capacity of the plant is so designed that the pairs still in operation satisfy the requirements of the plant as a whole. After regeneration the pair of exchangers is reconnected in parallel with the others.

As the exchange zone takes up a greater part of the bed height of the resin in the small columns used here than in large columns, it is particularly important in plants connected in parallel to determine whether the cation exchanger or the anion exchanger of a pair is exhausted. There are two possibilities here.

A conductivity measuring gauge is placed after each exchanger pair, and the reading from it is compared with the overall conductivity at the common purified water outlet. If a cation exchanger is exhausted, then the conductivity rises at all the measuring points and the cation exchanger that has been operating longest is regenerated. If an anion exchanger is exhausted, then this will be indicated by the conductivity gauge situated after the anion exchanger in question.

A small pilot column with a strongly basic anion exchange resin and a conductivity meter are connected in a bypass following the cation exchangers. An increase in the value measured indicates that breakthrough has occurred in the longest operating cation exchanger. So long as this measured value is normal, an increase in conductivity at the gauge in the common purified water outlet is a sign of breakthrough in the longest operating anion exchanger.

Parallel-circuit plants presuppose an automatic mode of operation. Because of the small nominal width of the connections for the individual exchangers this can be achieved at a relatively favourable cost as well. The possibilities for enlargement are straightforward.

Fig. 5.33 shows a plant connected in parallel, in which in particular the smaller exchange columns should be noted.

Fig. 5.33   Automatically operating ion exchange continuous circulation plant connected in parallel; capacity 35 m³/hr; gravel filter and 3 cation and 3 anion exchangers with possible expansion to 4. Pair of exchangers, left: ion exchangers for removing complexed copper ions
(THYSSEN Engineering, Essen)

A series circuit of ion exchangers [24,63] provides for the coupling of the same types of exchange resin one behind the other and thus for the simultaneous use of all the columns (outside the regeneration periods). The connection is K-K-A-A, the last regenerated cation or anion exchanger always being the second of the pair (see Fig. 5.32d). Two similar columns coupled one after the other correspond during loading to one column with double the height of the resin, both halves of which can be individually regenerated however. As the freshly regenerated part is always connected after the other, it corresponds to the lower half (with downflow) or the upper half (with upflow) of a whole column. Thus the series-connected ion exchange plant corresponds to a continuously operating plant with a large charge of resin and correspondingly long cycle time (see Section 5.3.3.2.4), without the resin itself having to be moved.

With this kind of circuit the anion exchangers are better protected against metal breakthrough in the cation exchangers, and expensive special regenerations, which are necessitated by the precipitation of hydroxides on the anion exchange resins, are largely avoided. The quality of the purified water is improved by the reduction in ion leakage.

The series circuit requires more elaborate control technology, however, as each exchange column after it becomes loaded is individually regenerated. Determining when the cation exchangers are loaded is carried out with the help of two conductivity gauges, after each of the columns. In manually operated plants it is also possible to do this by measuring the m-value after each cation exchanger. From the conductivity measurements the difference value can be determined also and used as a triggering signal for the regeneration of the pre-coupled cation exchanger. Because of the considerably higher conductivity of hydrogen ions compared with other cations, the measured value decreases more and more after breakthough in the pre-coupled cation exchanger and becomes constant when the loading is completed. The measured value at the post-coupled cation exchanger remains large, as it exchanges all the cations except for the hydrogen ions. The useful capacity of the pre-coupled exchanger can therefore be fully utilised on the basis of the difference observed between the values measured. This can best be demonstrated by reference to the diagram in Fig. 5.34. This shows the advance in the exchange zone (see Fig. 5.8) through two similar columns connected one after the other [64]. In the case of the series circuit, regeneration of the cation exchanger is accomplished only in stage d, or after loading is completed. Conductivity gauges are also positioned after each of the two anion exchangers. In the line and parallel circuits, the breakthough indicating the need for regeneration occurs as shown in diagram b. The capacity of a quantity of resin m, as represented by Equation 5.22, is not fully utilised. The other type of exchanger involved, if regeneration is carried out in pairs, is at the same time at a stage between a and b.

$$m = \frac{d^2 \pi}{4} \cdot \frac{R_H}{2} \quad [l] \qquad\qquad 5.22$$

R_H – Breadth of the reaction zone [dm], d – Diameter of the column [dm].

Further utilisation of the capacity corresponds to a more effective use of regenerating medium and hence to a lower concentration of salt in the water. If more exchangers of the same type are included for materials recovery, then a series circuit is always used to avoid ion leakage.

The ion exchange continuous circulation plant shown in Fig. 5.29 is operated as a series circuit.

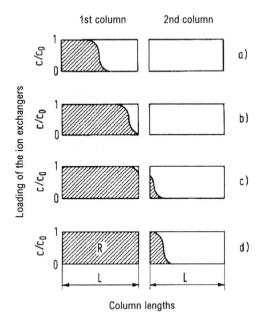

1st column   2nd column

a) Approx 50% loading of the first column.

b)

c)

d)

Loading of the ion exchangers

Column lengths

Fig. 5.34 Schematic representation of the loading of ion exchange plants connected in series.

a) Approx 50% loading of the first column.

b) Breakthrough of the first column. This triggers regeneration in line-circuit and parallel-circuit plants.

c) Changeover to loading of the second column.

d) The first column is fully, the second partly loaded. In series-circuit plants, regeneration is triggered here [64].

5.3.3.2.2 Liquid feed options in columns

Basically, it is possible in vertical columns for solutions to flow from top to bottom or in the opposite direction. This applies to continuously circulated water during loading as well as to the regenerating medium and wash water. The various liquid flow possibilities used in continuous circulation technology are illustrated schematically in Fig. 5.35. They give rise to variations in the consumption of the regenerating medium, impose various mechanical demands on the resin and require different column designs.

The co-current process is the oldest and simplest way of operating an exchange column. The ion exchanger is loaded from the top to the bottom in down-flow and then regenerated and washed in the same direction. Before regeneration it must be back flushed, in order to reduce the pressure drops that have been produced by compression of the resin and to remove particles of dirt. To prevent the resin being flushed out during this process, particle-retaining meshes are mounted at the upper exits of the columns.

Because when the regenerating solutions are diluted by the water already present in the back-flushing space, their consumption is greatest with this liquid-flow arrangement. This system is no longer employed in continuous rinse water circulation facilities. It is frequently used however for recycling processes.

All the other systems illustrated in Fig. 5.35 operate in accordance with the counterflow principle. The type of liquid flow represented as the countercurrent process allows the loading to be carried out in down-flow and the regeneration in up-flow. There are various ways in which this can be done, of which only the simplest version is shown in Fig. 5.35 [66-68]. The regenerate and the wash water are drawn off with the upper outlet closed through a drainage system (e.g. perforated pipes, which are covered with plastic screens), which is situated near the top of the bed of resin. Compared with the co-current process, the consumption of regenerating medium and that of wash water are considerably less, the latter by about 60%. The reason is that no dilution of the regenerating medium takes place and no intermixing of the wash water with whatever liquids lie on top of the resin. Back rinsing before regeneration, as in the co-current process, is carried out with the drainage valve closed. It is an advantage with this system to also include a device to ensure particle retention.

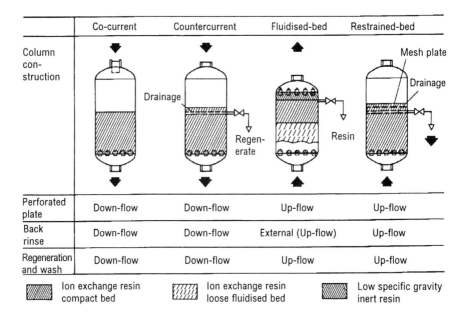

	Co-current	Countercurrent	Fluidised-bed	Restrained-bed
Column con- struction				
Perforated plate	Down-flow	Down-flow	Up-flow	Up-flow
Back rinse	Down-flow	Down-flow	External (Up-flow)	Up-flow
Regeneration and wash	Down-flow	Down-flow	Up-flow	Up-flow

Ion exchange resin compact bed   Ion exchange resin loose fluidised bed   Low specific gravity inert resin

Fig. 5.35   Diagram of the various liquid flow possibilities of ion exchange columns

In the fluidised bed process the flow during loading is from bottom to top. Thus the resin bed is lifted and pressed against an upper perforated plate. So that the holes cannot be blocked, there is an approximately 20 cm-thick layer of more buoyant inert resin of larger particle size on top of the exchange resin [69-71]. The flow rate must be high enough to ensure that at least 25% of the exchange resin is pressed against the inert resin under the upper perforated plate (fixed bed portion). The rest of the resin forms a loosely packed bed beneath this. This has a high exchange rate, and it takes up any leakage from the fixed bed. The resin remains loose, and back flushes have to be carried out only at long intervals (ca every 6 months) to remove impurities. These back flushes are performed in a separate back-flush vessel (see Fig. 5.30). The resin is flushed out into this and after back rinsing is swept back again into the work column. As back-flushing is carried out externally, the exchange column can be lower in height. The free space up to the limiting upper filter plate is designed on the basis of the volume of inert resin and the resin expansion occurring during loading or regeneration. It should however not exceed a height of 30 cm, so that no mixing of the layers occurs during disconnection from the plant. Pressure drops are lower than in down-flow columns.

A process that combines the characteristics of the fluidised bed with the ability to back flush in the column is the restrained bed process. Loading takes place with upward flow, the upper outlet being closed and the water passing out through the drainage outlet, without the resin becoming turbulent at the top. Regeneration and washing are carried out, as in the fluidised bed process, from top to bottom. During these stages the liquids pass first through a layer of resin above the perforated plate and the drainage outlet. It is therefore possible to accomplish the dilution of the regenerating acid in a cation exchanger and its washing with untreated water, or the dilution of the regenerating caustic in the anion exchanger and its washing with cation-exchanged water, without there being any consequent reduction in the capacity of the particular resin beds (below the drainage outlet). The benefit relative to the fluidised bed does not apply though, if the drop in capacity occurring in that case during regeneration is counteracted by a corresponding increase in the amount of resin. It is an advantage, however, that the back flushing of the resin can be carried out in the same column. This is done by opening the upper outlet and closing the drainage oulet.

5.3.3.2.3   Possible ways of producing a flow of regenerating medium

There are various possible ways of preparing the dilute regenerating medium (7-8% hydrochloric acid, 4-5% caustic soda) from more concentrated acid and alkali plus water, as indicated in Fig. 5.36.

In the make-up process a vessel containing a mixer is first supplied with the diluting water and then with the concentrated regenerating medium. The operation of the valve in the water-supply pipe and that of the pump are controlled by reference to a liquid level measuring gauge, so that the desired concentration is achieved. The diluted regenerating medium is transferred to the ion exchange column, at the liquid velocity specified in Section 5.2.3.3, using a second pump.

The dilution process is carried out with a water-jet pump, in which the jetted water functions simultaneously as the dilution water for the regenerating medium. The latter is contained in a small vessel, with a volume depending on the resin and the amount present, and is sucked up by the water-jet pump. The volume flows of the concentrated and diluted regenerating medium are adjusted by valves and measured.

The dosing process constitutes a third possibility. The concentrated regenerating medium is passed through a standard dosing pump for a predetermined time into a mixing zone, which is provided with a standardised flow of water. Another version, as in the dilution process, allows for a certain amount of the concentrated regenerating medium to be present in a vessel and then to be fed in from there to the mixing zone with the help of a dosing pump.

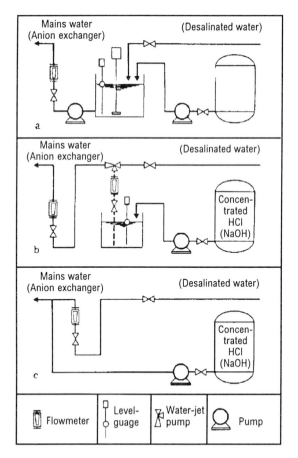

Fig. 5.36 Diagrammatic illustration of the most frequently used regeneration processes.

a   Make-up process
b   Dilution process
c   Dosing process

The preparation and delivery of the regenerating medium can be carried out manually, but in modern plants they are integrated with the automatic operation of continuous circulation systems.

5.3.3.2.4   Continuously operating plants and other designs

Ion exchange plants operated on a continuous basis became prevalent at the end
of the sixties. It is characteristic of continuous exchange processes that loading,
back flushing, regeneration and washing of the resin can be carried out simultaneously in each of the different parts of the plant [74]. Hence the resins are
moved through the plants at intervals or continuously, enabling the various
liquids involved to be passed through them in counterflow.

The Higgins Process has been known longest [75]. Like the Asahi Process
[76] and the Fluicon Process [77] and several others, it has been used only for
water treatment. Kauczor [78] noted the trend towards continuous operation at
the time and evaluated the suitability of these procedures. The processes were
not adopted for continuous circulation of rinse water in the surface treatment
industry, not least because of the large height of the equipment required. Instead,
one of the plants described by Marquardt [79-81], Servo-Kontimat, was adopted,
which has normal height for continuous circulation plants and is particularly
suitable for rinse water recirculation in metal finishing.

The construction and mode of operation of the Servo-Kontimat plant are
illustrated in Fig. 5.37. The two operating columns working in series (cation and
anion exchangers) are operated in up-flow. The resins move in the opposite
direction from top to bottom. Consequently, as there is always freshly regenerated resin present in the upper part of the column, the mode of operation is
similar to that of a line-circuit plant (see Figs 5.32 and 5.34).

The loading condition of each operating column can be determined by continuous conductivity measurement in a measuring stream taken from a point
about half way up the resin bed. When the loading is well enough advanced, the
flow of water is interrupted for 60-90 seconds and the conical space beneath the
lower nozzle floor opened up. It is filled with a certain amount of loaded resin,
while a similar amount of regenerated resin, metered through a dip pipe, is fed in
a transporting stream of water into the top of the operating column from the
uppermost section of the regenerating column. The supply lines for transport,
wash and diluting water in the upper part of the wash columns and for the supply
of back rinse water in the lower part of the back-rinse columns are, for the sake
of clarity, not shown in Fig. 5.37. After this short interruption, the plant returns
to operation. The loaded resin in the cone of the operating column is then pushed
into the lower part of the back-flushing column. A charge of resin that has
already been back flushed is measured off and transported with purified water
through a dip pipe into the U-shaped regenerating and wash column. The resin
that remains in the back-rinsing column is now back flushed. In the regenerating
part of the regenerating and wash column two or more charges are treated in
countercurrent with regenerating medium, while the dilution water already present
in the washing part is used as wash water, also in counterflow. Optimum utilisation is thus achieved of the regenerating medium and the wash water. The next
cycle is then initiated when conductivity measurements show it is due.

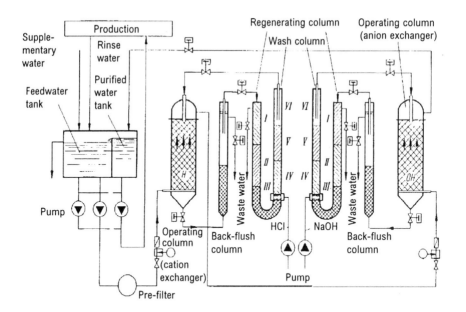

Fig. 5.37   Schematic representation of a continuously working ion exchange continuous circulation plant (HAGER + ELSSSER GmbH, Stuttgart)

Fig. 5.38 shows the construction of a continuously working continuous circulation plant. Automatic operation is essential. The liquid velocities in the operating columns go up to 100 mh^{-1}. Hence resins must be chosen that have particularly good mechanical stability and high exchange rates. The bead size must also be small and its distribution uniform [78].

Filtration in continuously working plants can be handled in a similar way to the treatment of the exchange resins [58,79,81]. The plants consist of operating and back-flush columns. Filtration is carried out in up-flow. The particle size of the filter material is relatively fine at 0.6 mm, and the bed height is chosen to be commensurately large. Macroporous resins (including inactive ion exchange resins) or other polymers are used as filter material. The separation efficiency during filtration can be further improved by the addition of finer materials (e.g. active carbon, kieselguhr and so on) [58].

Transportable compact plants are available for the treatment of small quantities of rinse water and of other solutions as well, their resins being in exchangeable modules or in columns which can themselves be regenerated externally. One of these processes, the RMA system is particularly well known because of its frequent use. Descriptions can be found in numerous publications, from which only a selection is cited [82-86]. Each unit consists of a reception vessel with a capacity of about 45 litres, supply pump, small filter and four exchangeable ion exchange modules placed on top of one another with about 14 litres of resin in each. For continuously circulated rinse water treatment two of these units (cation and anion exchanger) are connected one behind the other. Such a plant has a maximum capacity of 1.2 m³/h. The columns operate in up-flow. Regeneration of the loaded modules is carried out externally in a central plant. A gauge give a visual indication of the loading of the modules. As each loaded module is removed at the bottom, a regenerated one is put back again at the top. The regeneration service covers transport to and from site, of the modules.

Fig. 5.39 shows the construction of a modular type of plant for continuous circulation of rinse water.

The plants operate in line circuit, but their mode of action, like that of the continuous plant, is similar to that of the series circuit because of the post-coupling of the regenerated resin. The so-called progressive process [87] works in the same way, but without exchange of column segments. Here it is a case of a large plant, with beds of resin in the columns divided up into several tiers. They are operated in down-flow and regenerated in counterflow, the tiers being treated individually as required.

Fig. 5.38 Continuously and automatically operated ion exchange continuous circulation plant, capacity 60 m³/h (HAGER + ELSSSER GmbH, Stuttgart

Fig. 5.39   Ion exchange continuous circulation plant of modular design with external regeneration, capacity 1.2 m³/h (INOVANSTRÖBE GmbH & Co KG, Birkenfeld)

Flat bed filters may also be chosen, which have exchange columns of very low height, constituting sometimes only the height of the reaction zone [88]. In order to achieve equivalent loading, ion exchange material with very small particles and uniform particle size, must be employed. The loading-regeneration cycles are correspondingly short. This type of construction is not suitable for the continuous circulation of rinse water in the manner described, as breakthrough occurs in the columns soon after they are put into operation. The process is chiefly used where the resins can be fully loaded during circulation. For example in recycling technology.

As well as the designs mentioned, others are known, expecially for ion exchangers as used in recyling technology (see Section 7.2).

5.3.4   Ecological importance of continuous rinse water circulation

If by the use of static rinse tanks the concentration of the water to be treated is kept below 2 m Eq./litre, the volume of the regenerates arising is as a rule between 0.2 and 1% of the quantity of rinse water in circulation. The contents of the static rinse bath, after reaching a certain concentration, should be put to use, partly to compensate the process solutions for evaporation losses and partly through recycling procedures. This is certainly not possible in all cases. The solutions then qualify as waste water. The cleaning-up and back-rinsing water from the gravel filters must be added to these also, so that the amount of waste water is increased by another 1% over the above percentage. Naturally, this is in terms only of that part of the rinse water that can be continuously circulated. In the regenerates the substances present in the rinse water are found at about 100 times greater concentration. If constituents separate by precipitation, the residual values should be those determined by the solubilities of the poorly soluble substances. The precipitation of metals as hydroxides or basic salts or sulphides may be given as an example. The neutral salt concentration of waste waters from which these are to be precipitated is considerably greater however than that of rinse water; precipitation is therefore more difficult and higher residual concentrations are encountered at the same pH values (see Section 3.7.7.1). But even when these concentrations are doubled, quite large amounts can still be kept out of the discharge.

If, for example, from 100 m^3 of rinse water in continuous circulation, a maximum of 1 m^3 of regenerate arises, and if a metal can be precipitated from the same quantity of rinse water to leave just 0.5 mg/litre, but from the regenerate only to leave 1 mg/litre, then the amount in the first case would be 50 g of metal, but in the second case only 1 g of metal, hence only 2%. The manufacturer as a waste discharger has then, if copper, nickel, chromium or lead is involved, exceeded the specified value, even though they has complied with the requirements of the water economy legislation. Similar circumstances can be expected to arise in applying other water saving measures.

So far as the precipitation of metals from dilute and concentrated solutions is concerned, current German legislation has so addressed only the question of zinc and has fixed on 2 mg/litre as the limiting value. Zinc can certainly not be classified, however, among the more dangerous materials.

Water saving measures, of which the continuous circulation of rinse water is among the most important, give rise to a further ecological advantage, due to the reduction in the quantities of metals involved. It is known that, even though the limiting value requirements may be satisfied, the unprecipitated residual metals can still accumulate in activated sludge by adsorption, absorption and mechanical entrapment. As several investigations have shown, the sludge can bind up between 30 and 95% of the residual metals still present [89-91]. Decomposition of more organic material occurs during the subsequent degeneration of the sludge, causing a further increase in the proportion of metal. If the contents of the individual metals exceed the values specified in the sludge regulations, then the contaminated sludge cannot, or can only with special authorisation, be used for fertilisation purposes Reducing the metal content of contaminated sludge is achieved in almost all cases by continuous circulation of rinse water.

References for Chapter 5

1. Griessbach, R.: "Exchange adsorption in theory and practice" Akad. Verlag Berlin (1957)
2. Nachod, F.C., Schubert, J.:"Ion Exchange Technology". Academic Press Inc., Publishers, New York, 1956.
3. Kunin, R.: "Ion Exchange Resin". John Wiley and Sons, Inc., New York, 1963.
4. Buser, W., Graf, P., Grütter, W.F.: "Physical chemistry of ion exchange" Chimia 9 (1955) 73-93
5. Martinola, F., Naumann, G.: ""Ion exchangers as reactive high polymers" Angew. Makromol. Chem. Vol 4/5 (1968) 185-211 Huethig and Wepf Verlag, Basel
6. Kühne, G.: "Ion exchangers - activated resins for exchange reactions, adsorption and catalysis" Chem. Ztg 96 (1972) 239-247
7. Helfferich, F., Schmidt, E., Lieser, K H., Naumann, G., Martinola, F., Ladendorf, P.: "Ion exchangers" (in) Ullmans Techn. Encyclopaedia
8. Martinola, F., Richter, A.: "Macroporous ion exchangers and adsorbents for treatment of water contaminated with organics" Vom Wasser 37 (1970) 250-264
9. Griessbach, R.: "Anomalous reactions at ion exchangers" Chimia 11 (1957) 29-56
10. Kressman, T.R.E., Kitchener, I.A.: J. Chem. Soc. (London) 259 (1949) 1190.
11. Boyd, G.E., Adamson, L.S., Myers, L.S.: J. Am. Chem. Soc. 69 (1947)2836
12. Boyd, G.E., Soldano, B.A.: J. Am. Chem. Soc. 75(1953)6091.
13. Becker-Boost, E.: "Technology of ion-exchange Pt II." Chem. Ing. Techn. 28 (1956) 411-418
14. Haagen, K.: Z. f. Elektrochemie 57 (1952) 178-183
15. Stach, H.: "Assessing ion-exchangers in terms of their exchange rates" Angew. Chem. 63 (1951) 263-7
16. Oehme, Ch.: "Experiences with ion-exchangers in rinsewater recycling loops" Oberfl. Surf. 12 (1971) 105-110

17. Kauczor, H.W.: "Solid ion exchangers to concentrate metal containing solutions" ERZMETALL 22 (1969) B19-B22

18. Bayer AG, Leverkusen: "Lewatit TP 207" Techn. Inform. Note

19. Bjerrum, J., Schwarzenbach, G., Sillen, G.: "Stability Constants of Metal-ion Complexes, Part II, Inorganic Ligands." The Chemical Society, London, Burlington House W.1, 1957.

19a. Becker-Boost, E.: "Ion exchange technology. I – Planning on empirical guide-lines" Chem. Ing. Techn. 27 (1955) 579-596

20. Becker-Boost, E.: "idem. III – Planning on basis of theory" Chem. Ing. Techn. 28 (1956) 532-542

21. Hartinger, L.: "Effluent and heavy metal discharge reduction in the metal-work-ing industries using ion exchange" Korresp. Abwasser 33 (1986) 895-907

22. Hartinger, L.: "Recycling options for recovery of solvents and chemicals" Metalloberflaeche 29 (1975) 221-280

23. Hartinger, L.: "Rinsewater recirculation" Private publ. LPW Co (1975)

24. Götzelmann, W.: "Planning and defining effluent treatment plants" Galvanotechnik 64 (1973) 1091-1107

25. Furrer, F.: "Ion exchange plant for recirculation of process water" Galvanotechnik 51 (1960) 105-113

26. Götzelmann, W.: "Ion exchangers for effluent treatment" Galvanotechnik 56 (1965) 66-72

27. Marquardt, K.: "Trends in effluent treatment in the metal-working industries using ion exchangers" Galvanotechnik + Oberflaechenschuetz 7 (1966) 231-242

28. Schlegel, H.: "Ion exchange in effluent treatment" Werkstatt u. Betrieb 100 (1967) 19-23

29. Marquardt, K.: "Points to watch out for when installing ion exchange plants in the metal-working industries" Oberflaeche-Surf. 9 (1968) 90-98

30. Kunin, R.: "Ion Exchange for the Metal Products Finisher". Products finishing (1969) Apr. 66-73, May 71-79, June 182- 190.

31. Marquardt, K.: "Recirculating ion-exchange plants" Galvanotechnik 61 (1970) 561-574

32. Marquardt, K.: "Developments in ion exchange technology for effluent treat-ment" Galvanotechnik 62 (1971) 785-798

33. Marquardt, K.: "Ion exchange or conventional effluent treatment plant ?" Galvanotechnik 65 (1974) 33-47

34. Hartinger, L.: "Basics for installing ion exchange plant" Galvanotechnik 61 (1970) 541-7

35. Alt, Ch.: "Filtration" (in) Ullmans Technical Encyclopaedia

36. Bayer AG, "Lewatit S100" Product Note

37. Arnold, K.H., Martinola, F.: "Silicic acid leakage in demineralisation plants" Proc. VGB Water Conf. (1969) 47-52

38. Deutsche Babcock Anlagen AG, "Water Handbook" 7th Ed'n

39. Bayer AG, "Lewatit OC 1031" Techn. Note (19080)

40. Martinola, F.: "Ion exchangers and adsorbers – versatile tools for the chemical industry" Chem. Ing. Techn. 51 (1979) 728-736

41. German legislation WRMG (19.12.86) BGBl I. 2615

42. German legislation (degradability of surfactants) Tens. V. (4.6.86) BGBl I. 851

43. Götzelmann, W.: "Residue-free cyanide removal from rinsewater" Fachber. f. Oberflaechentechn. 12 (1974) 52-4

44. Götzelmann, W.: "Problems with non-ionic surfactants when installing ion-exchangers in the metal-working industry" Kommissionsverlag Oldenbourg (1972)

45. Hartinger, L.: "Surfactants and rinsewater recirculation through ion exchangers" Oberflaeche-Surf. 10 (1969) 157-162

46. Wolf, F., Wurster, S.: "Removing long-chain non-ionic ethylene oxide adducts from aqueous solutions" Tenside 7 (1970) 140-6

47. Hempel, H., Kirschnek, H.: "Using ion-exchangers to remove surface active species, especially cationic and non-ionic types" Fette-Seifen-Anstrichmittel 61 (1959) 369-374

48. Richter, G.: "Behaviour of soaps in ion exchangers" Z. f. phys. Chem. NF 12 (1957) 247-263

49. Bayer AG, "Lewatit ATP 202" Techn. Inf. (1972)

50. Bayer AG, "Plating shop rinsewater - re-use by recirculation through ion exchangers" Techn. Note (1970)

51. Oehme, Ch.: "Macroporous ion exchangers, their properties and prospects" Galvanotechnik 62 (1971) 859-866

52. Oehme, Ch.: "Modern applications of ion exchangers" Galvanotechnik 64 (1973) 233-5

53. Kühne, G., Martinola, F.: "Ion exchangers – their resistance to physical and chemical effects" VGB Kraftwerkstechn. 57 (1977) 173-184

53a. Calmon, C.: "Explosion hazards of using nitric acid in ion- exchange equipment". Chemical Engineering 17 (1980) 271-274.

54. Bayer AG, "Lewatit SP 120" Tech. Note (1965)

55. Bayer AG, "Lewatit MP 600" Techn. Note (1978)

56. Bayer AG, "Disinfection of ion exchangers and filters" (1980) (ref: OC/I20351)

57. Götzelmann, W.: "Recirculating loops with solid bed ion exchangers" (in) Hartinger L; "Pocketbook of Effluent Treatment" Vol. 2. (1977)

58. Marquardt, K.: "Recirculation of plating shop effluent using continuously operating ion-exchanger with on-line treatment and neutralisation" Metalloberfl. 29 (1975) 232-240

59. Marquardt, K.: "Special considerations in planning banked ion exchange recirculating loops" Galvanotechnik 65 (1974) 582-6

60. Schüssler, R.: "Analytical control for optimum ion exchange operation" Galvanotechnik 72 (1981) 827-831

61. Glaser, H., Siegle, K.: "Automating ion exchange recirculation loops for plating effluents" Galvanotechnik 61 (1970) 575-581

62. Berninger, A.: "Advances in ion exchange processes" CZ Chem. Techn. 1 (1972) 463-8.

63. Dia-Prosim, France; "Duolite 20. Series linking of 2 filters". Techn. Note U.N 69.008 (1969)

64. Hartinger, L.: "Recovery of chemicals in liquid form" Galvanotechnik 68 (1977) 721-730

65. Urban H.J.: "Ion exchange plants - different methods and technical considerations" Metalloberfl. 36 (1982) 201-5

66. Limbach, F., Martinola, F.: "What future for counter-current regeneration ?" Proc. Conf. VGB Water supplies (1967) 33-40

67.   Schmidt, D., Wieland, G.: "Counter-current regeneration of ion exchangers" Wasser, Luft Betrieb 12 (1968) 294-9

68.   Wieland, G.: "Developments in counter-current regeneration of ion exchangers" Techn. Ueberwachung 10 (1969) 40-45.

69.   Bayer AG, "The Lewatit restrained bed process" Techn. Information (1977)

70.   Siegers, G., Martinola, F.: "The restrained bed system - simple counter-current regeneration for ion exchangers" Vom Wasser 39 (1972) 377-388

71.   Berninger, A.: "The restrained bed process in ion exchange" Galvanotechnik 65 (1974) 311-318

72.   Siegers, G., Martinola, F.: "Advances in ion exchange technology" VGB Kraftwerkstechn. 61 (1981) 586-7

73.   Kauczor, H.W.: "The Lewatit rinse-bed process" Galvanotechnik 73 (1982) 843-5

74.   Marquardt, K.: "Continuously operating ion exchangers" (in) Hartinger L; "Pocketbook of Effluent Treatment" Vol. 2

75.   Higgins, I.R.: "Continuous Ion Exchange Equipment" Ind. Eng. Chem. 53 (1961) 635-637.

76.   Bucher, H.: "Water treatment using continuous ion exchange" cav (1971) (July) 55-6

77.   DAS 1,280,761 (29.4.67) to M.A.N, Nuremberg

78.   Kauczor, H.W.: "Progress with continuous ion exchange" Chem. Industrie (1979)(11)

79.   Marquardt, K.: "Continuous regeneration of ion exchange resin" Oberflaeche-Surf. 13 (1972) 203-213

79a. Marquardt, K.; Observations on continuously operating ion exchangers" VGB Conf. Proc. Water Supplies (1972) 93-105

80.   Marquardt, K.: "Recirculation of plating effluent with continuously operating ion exchangers, continuous treatment and neutralisation plant" Metalloberflaeche 29 (1975) 188-195

81.   Marquardt, K.: "Continuous ion exchange" Galvanotechnik 74 (1983) 1430-8

82.   Reinhard, F., Gold, M.: "Regional waste treatment and recycling centres for plating shops" Proc. Lecture TA Esslingen 9.12.1980

83.   Wahl, K.L.: "Rinsewater treatment as a service function" Galvanotechnik 71 (1980) 721-8

84.   Reinhard, F.: "Rinsewater treatment and heavy metal recycling in metal finishing shops" Galvanotechnik 73 (1982) 1089-92

85.   Naujoks, R.: "Effluent treatment as a service" Galvanotechnik 74 (1983) 1441-4

86.   Wahl, K.: "Economic and ecological benefits of the Dornier RMA System" Galvanotechnik 75 (1984) 972-6

87.   Pfaudler Werke AG, "The progressive process" Pfaudler-news "Pro-aqua" (1971)

88.   Petzold W.; "Recycling in metal finishing" Galvanotechnik 65 (1975) 301-4

89.   "Report on heavy metal discharge to drain in the Munich area" TU Munich, Dept of Water Quality (1981)

90.   Meixner, G.:"Heavy metals in sewage sludges" Korresp. Abwasser 29 (1982) 260-6

91.   "Origins of heavy metals in sewage sludges and compost at Mura Biel. II Heavy metals in effluents" Schriftenreihe Umweltschutz Nr 9, Bern (1983)

6 Rinsing Technology

6.1 TERMINOLOGY

Rinsing implies little more than dilution. In previous times, use of the word "rinsing" in surface treatment implied the removal of residual species adherent to the surface of the work, dragged out from the process solution and removed with copious amounts of water. This provided assurance that no interfering chemical substances would be carried forward from one stage to the next, or that, after the final treatment, no toxic or corrosive substances would remain on the surface. It was around the 1960's that water ceased to be seen as a "free" commodity and this brought about a new rigour in discussion of rinsing, the so-called "rinsing criterion". This parameter defines the number of times that liquid, dragged out with the work, is diluted by the rinsing process and this provided a quantitative measure of the efficacy of the rinsing process. The parameter can be defined as:

$$\frac{c_0}{c} = \frac{Q'}{V'} \qquad\qquad 6.1$$

where $c(o)$ is the concentration of a given species in the process solution, c the concentration of that species in the rinsewater; Q' is the flowrate of the rinsewater and V' the "flowrate" of drag-out solution, i.e. a function of surface area of the work being plated and its throughput rate.

Fig 6.1 should be examined. The dimensionless ratio c_0/c is the rinsing criterion and Equation 6.1 describes the situation at equilibrium, which sets in after a given time and number of rinse operations, in the rinse tank. The volume of the rinse tank, though it will affect the time to reach equilibrium, does not affect the actual concentration values. Typically this occurs after 1000 to 10,000 rinse operations. The question then arises whether one can in some way make better use of the rinsewater, either re-using it or using less of it. Two technologies have been found practicable and both are widely used, saving both water and thus the volume of effluent. These are:

- rinsewater recirculation using ion exchange (Section 5.3).

- multi-stage (cascaded) rinsing (Section 6.3.2).

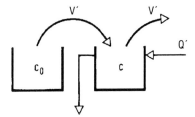

Fig. 6.1   Schematic representation of immersion rinsing (see Equation 6.1)

In the first case, the same water is used many times over for rinsing but with the chemicals dragged-in being removed at each recirculation. In the second case, rinsing takes place in two or more rinsewaters of progressively higher purity. In both cases, acceptable rinsewater criteria can be achieved but with reduced water use. But there is also a further criterion. Even though use of these methods allows improved rinsing criteria to be achieved with less water, there must – overall – be a concentration effect in the effluent. For classical effluent treatment, such concentration effects are broadly neither beneficial nor detrimental. However if recycling or recovery is desired, it is invariably preferable to carry this out from concentrated rather than dilute solutions. Thus rinsewater economising fits in well with a strategy for chemical or metals recovery.

It should be recorded that, as long as 30 years ago, a number of surface treatment specialists have written on the subject of efficient use of rinsewater. Only under legislative pressures did their ideas find use in practice. References [2,3,4,5,6,11] are typical of these.

Nowadays, reference is often made to "zero discharge" metal finishing operation. The author suggests that although this is almost possible in reality, there must inevitably be some effluent wherever surface treatment involving aqueous solutions is carried out. It is true that, of all effluent sources, it is those originated from rinsing which can most easily be avoided. However as part of the overall operations process, it would seem impossible to avoid liquid or solid wastes from being produced, for example when solutions are discarded, in flushing down pipework loops etc. It could be argued that a plant producing solid wastes poses more problems and hazards than one with a reduced effluent output. It would be possible to design and operate a metal finishing operation producing only solid wastes, but in most expert's views, such plants will always be the exception rather than the rule.

6.2 DRAG-OUT AND ITS MINIMISATION

As Equation 6.1 shows, if the drag-out volume V. can be reduced, then so can the amount of rinsewater required, without prejudice to the quality of rinsing and, in any strategy to reduce rinsewater use, this is the first area for action. Indeed, this often requires a simple change in procedure and no capital expenditure. Several approaches recommend themselves.

The first step should be to estimate the quantities of material dragged-out at each stage. This can be done using the actual plant and equipment. However care must be taken to ensure that, when making these measurements, the time for which the racks are out of the process bath is the same as found in normal operating conditions. The work on the racks is transported to stand over an empty tank and is then sprayed or rinsed so that the rinsings fall into the empty tank. In the case of difficult-to-rinse items, immersion in agitated clean water may be necessary. Analysis of the resulting solution allows the amount of drag-out to be calculated, usually in litres/hr as shown in Equation 6.2

$$V = \frac{c}{c_0} \cdot \frac{I}{t} \quad [l/h]$$
6.2

where I is the volume of rinsewater (litres) and t the time (hours) of the experiment. Both Krusenstern [6] & Ochotta [11] described such measurements while similar information can be derived from studies by Jola [7] who followed chemical consumption patterns. (For such a study, a species must be chosen which does not react in the process solution).

Prior to the start-up of a plating plant, drag-out figures can only be estimated. An empirical Equation for their prediction was produced by Kushner [8] who took into account transfer times, viscosity and density to give the following Equation for the thickness of the drag-out film:

$$f = 0,02 \sqrt{\frac{l}{t} \cdot \frac{\eta}{\varrho}} \quad [cm]$$
6.3

where f is the thickness of the adherent drag-out film, (cm), l is the vertical dimension of the work and the rack (cm), t is the transfer time (sec), η the dynamic viscosity of the process solution (gm.-sec/sq. cm) and ϱ the density of solution (gm/cubic cm.)

If f is then multiplied by the total surface area of work and rack, A (sq. cm) one obtains an approximate value for the volume of drag-out per operation, while the hourly rate is given by multiplying this figure by z, the number of operations per hour.

$$V = 2 \cdot 10^{-5} \cdot A \cdot z \sqrt{\frac{l}{t} \cdot \frac{\eta}{\varrho}} \quad [l/h]$$
6.4

From this Equation, a number of useful conclusions can be drawn. Drag-out is proportional to surface area, solution viscosity, inversely proportional to the rate of transfer and solution density. Effects due to surface tension, temperature, shape of the work and its surface roughness are ignored in this treatment and any adjustments to the predicted figures can be made only on the basis of practical experience. A rule-of-thumb suggests that racked work produces drag-out of 150 ml per square meter, with barrelled work giving 1 to 3 litres/barrel load for a normal-sized barrel. The most effective means of reducing drag-out is to mini-mise its extent in the first place. This can be accomplished by the following means:

- Allowing racks or barrels to drain over the process tank. In order to avoid the drying of these surfaces, especially with baths operating at higher temperatures, it is unusual for such drain-off times to exceed 10 sec. Indeed extending this to 15 secs brings little added benefit [2]. The drain-off action is favoured by increased temperature (reduced viscosity), lower surface tension (use of surfactants) and higher density. However increasing the latter will (Equation. 6.4) increase drag-out, other factors being equal.

- Improved drainage results when, after draining over the bath, the rack is raised to the top of the transporter. The resulting impact serves to knock off any persistent drops.

- Alternatively, a light mechanical tap on the rack when dripping has ceased, will achieve similar benefits.

- Use of water to rinse the work while it is still positioned over the process-ing bath, especially for baths operating at elevated temperatures (rack plating). The added water from the washings draining into the bath should not exceed the water lost by evaporation (see Fig. 6.5 for the magnitude of this).

- Use of squeegees (for printed circuit boards).

- Wiping (for wire and pipes).

- Draining of barrels over the process tank. In most cases, the barrel ceases to rotate once it is withdrawn from the bath. Once draining appears to be complete, the barrel should be rotated through 90° so that liquid trapped in cavities etc can be allowed to drain. According to Boger & Kübler [9], overall drainage time should be not less than 20 seconds.

- Blowing (air blast), especially for p.c.b's, strip and barrels. Fig. 6.9 shows how this can be applied to barrels. The approach is applicable both to drainings from process baths and rinsewaters. In recent years, the

- So-called "air- knife" technology has been developed so that even geometrically unfavourable objects can be dried, a further development being the "pulse-blast" in which the action is intermittent. Drainings should be returned to the system where possible.

- Suction (for mass finishing). Application of a small negative, pressure (e.g using a blower) can be applied to a barrel as it rests in a trough and as before, the same approach can be used both for process solution and rinsewaters [10] (see Fig. 6.8).

- Another idea is to use a brush attached to a vacuum and to apply this to the underside of the barrel while it rotates. It has been claimed this gives a 30% better liquid removal than simple draining [11].

- Centrifuging (for mass finishing). There have been numerous proposals aimed, for example, at increasing the rotation speed or for transferring the work from one vessel to another. The rinsing stage can be carried out in the same container [12].

- Most recently, printed circuit boards (held in baskets) have been spray-washed down while the baskets rotated, after which water was removed by centrifuging (see Section 8.9.2)

In reducing drag-out, a vital element is to ensure that both the work itself and the racks are so designed as to minimise liquid retention. In this context, care should be taken to avoid:

- work with unnecessary concavities, blind holes, threaded sections. Likewise work with folds or pleats or configurations where capillary action might operate [13]. Rough (unmachined, as-cast surfaces) which retain more liquid than smooth surfaces. Tube sections must be drained after plating, unless both apertures are first blanked-off. Longer tubular and similar sections should be jigged at an angle so that they can readily drain. Likewise any other shaped work capable of retaining liquid (e.g. cup forms) should be jigged upside-down.

- The same principles as above should be applied to rack design. Fixtures should be configured to avoid work mounted higher up on the jig draining onto work mounted lower down, thereby multiplying the whole drainage sequence [14]. Work should never be horizontally mounted. When racks are provided with an insulating coating by immersion in plastisols or similar, they should be immersed upside-down. This avoids formation of drop-shaped excrescences on the under-side of the rack, these again leading to increased drag-out, each such nodule causing a drop to form.

In designing plating barrels, such as those based on hexagonal construction, it should be recognised that the perforations do not extend to the horizontal edges. If the barrel comes to rest with the edge section (i.e. where two adjacent flats are joined) at the bottom, liquid will be retained in the resulting "V". Perforations in the barrel should be as large as possible. In some designs, very large holes are then covered with plastic mesh. The individual components of a barrel should be welded or glued together, in order to avoid a crevice in which liquid is retained by capillary action [9].

Poorly maintained racks or baskets result in increased drag-out. In the former case, small pockets form at sites where the plastic coating has peeled away from the metal. With both racks and barrels, metals deposit at the electrical contacts which should be chemically, or better, electrolytically, stripped at regular intervals. If this is not done, nodular growths build up which not only increase drag-out but can then only be mechanically removed. This in turn can damage the insulation, leading to yet more drag-out for reasons described above.

Any plating plant contemplating introduction of water-economising measures should only do so once all possible measures for minimising drag-out have been implemented. Then and only then should the unavoidable drag-out be determined so that appropriate rinsing technology can be introduced.

6.3   RINSING SYSTEMS

6.3.1   Simple flow rinses

Simple flow rinses, to the extent these are used with substantial water flows to give an effective rinsing criterion without forming part of a recirculatory system, are now almost completely a thing of the past. However the concept remains as a useful benchmark and a model to test various aspects of rinsing.

Making the assumption that rinsing occurs spontaneously, i.e. that drag-out films mixed immediately with the rinsewater, then increasing the water flowrates affects the equilibrium described in Equation 6.1. In practice, the position is somewhat different. Thus is the rinsewater flow Q' is small, e.g. $\leq = 10\ V'$ then a correction to Equation 6.1 is called for in that the drag-out volume must be adjusted for dilution:

$$\frac{c_0}{c} = \frac{Q' + V'}{V'} \qquad\qquad 6.5$$

This correction will be seen to be important for schemes discussed in Section 6.3.2.2. In considering such equilibria, the actual hold-up volume of the rinse tank does not feature and in practice, it is advisable to use the smallest possible container for given Q' and V' conditions. Obviously equilibrium will be achieved faster, the smaller the rinse tank. Small tanks have a further advantage in that they allow the greatest degree of agitation for a given energy input. The methods employed in this respect are very similar to those used in electroplating. In both cases, efficient mass-transport to the surface is of the essence. Such methods include [1]:

- pumped circulation

- pumped circulation with Venturi nozzles ("Eductors")

- air agitation (compressed air, use of injectors) [15]

- reciprocating rack motion

- rotation of barrel in the rinse

- multiple immersion of barrels or racks.

One should also include spraying as an alternative to immersion. But despite all of these methods, a laminar film of liquid remains on the surface even after rinsing. Its thickness is inversely proportional to the relative velocity of rinsewater and the work, reaching an asymptotic limit at the highest flowrates. The removal of species dissolved in this film occurs only by diffusion, in which it obeys Fick's First Law:

$$\frac{dm}{dt} = -D \cdot A \cdot \frac{\Delta c}{\delta} \qquad\qquad 6.6$$

where dm/dt is the mass-transport (gm/sec); D the coefficient of diffusion (cm^2/sec); A the surface area over which the process acts, Δc the concentration gradient across the diffusion layer (gm/cm^3); δ the thickness of the diffusion layer (cm). The temperature dependence of the process is contained within the diffusion coefficient:

$$D = \frac{k \cdot T}{6\pi \cdot \eta \cdot r} \qquad [cm^2 s^{-1}] \qquad\qquad 6.7$$

where k is Boltzmann's constant [$1.38 \cdot 10E^{-23}$ JK^{-1}]; T the temperature (degrees Kelvin), η the dynamic viscosity (dyne-sec.cm^{-2}); r the radius of the diffusing molecule or ion with its hydration shell (cm).

The denominator corresponds to the frictional resistance opposing the forces of molecular motion kT. Diffusion coefficients of inorganic ions lie in the general range 10^{-6} to 10^{-5} cm^2/sec.

If Δc is small, satisfactory diffusion rates can only be reached by raising the temperature or increasing the flowrate relative to the work. Both approaches are used in practice.

Maintaining a satisfactory degree of rinsing in a simple flow-rinse or in a cascaded rinse system is easily assured by use of electrical conductivity measurements or using high frequency inductive measurement as described by Bahensky et al [16]. Having set the conditions in terms of the required dilution in the rinse operation, increase or decrease in the rate of water flowrate is controlled by the signals from such measurements. The quickest attainment of equilibrium, as defined in Equations 6.1 to 6.5 is assured by use of diagonal flow through the rinse tank, with water entering from the bottom and leaving from a surface outlet at the opposite corner or v.v.

6.3.2   Cascade rinsing

6.3.2.1   High flowrate cascade rinsing

Together with recirculation through an ion-exchange unit, cascade rinsing is the most effective single method for effecting water economies and thus reducing the volume of effluent. The rinsewater flows through a series of containers in the same sense as the work being rinsed so that the first tank contains the most contaminated water, the last tank the cleanest as shown in Fig. 6.2 in which it is assumed the volume of drag-out V' is constant.

In most cases, equilibrium in the final tank can be calculated using Equation 6.8 even though this is mathematically not strictly correct. This can be seen for the case $Q' = V'$ The rinse criterion for all tanks would then be 1.

$$\frac{c_0}{c} = \left(\frac{Q'}{V'}\right)^n \hspace{4cm} 6.8$$

where n is the number of stages in the cascade.

This Equation does not take into account that the drag-out volume V' itself contributes to the dilution (see comments on Equation 6.5) and is transported together with the rinsewater volume Q' to the outlet of the first rinse stage. Thus Q' should be replaced in the denominator with $(Q' + V')$ thus removing at least one source of error. The Equation would then be valid for a series of rinse tanks in which each stage had identical inlet and outlet flowrates [1].

In spite of the remaining error, Equation 6.8 may be safely used where $Q' >$ 10 V' as graphically depicted in 6.3.2.2. However in practice, this condition only obtains when a single stage is used. Fig. 6.3 plots the rinsing criterion as a function of the ratio $Q':V'$ The volumes of water are shown to give rinsing criteria of 1000 and 10,000 per litre drag-out dilution for various numbers of cascades. The reader is reminded that the calculations using Equation 6.8 assume an instantaneous removal of drag-out from the surface with its immediate distribution in the bulk of water in the rinse tank. Means by which such an approximation can be justified are given below. Observations by the author and colleagues [17] showed that in many cases, the ratios derived from Fig. 6.3 were in good agreement with actual observation up to and including the third stage of a cascade rinse.

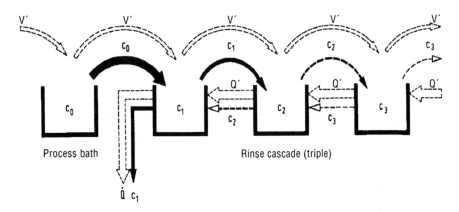

Fig. 6.2   Schematic of a cascade rinse system for
determining volume and mass balances [32].

Numerous publications, some cited below, deal adequately with various aspects
of cascade rinsing [1,2,3,18,19,20]. Also of interest are the publications of Schene
et al. [22,23] who comment on practical aspects.

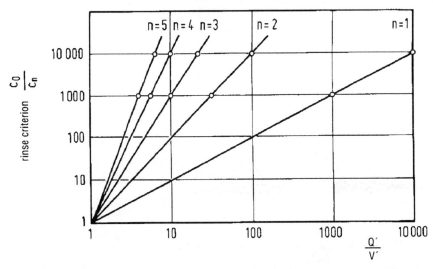

Rinse criteria	Litres rinsewater per litre drag-out				
1000	4	5.6 10	31.6	1000	
10 000		6.3 10	21.9	100	10 000

Fig. 6.3 Rinse criteria as function of rinsewater flowrate Q' and drag-out V'
for rinses with "n" cascade stages, calculated from Equation 6.8 [32].

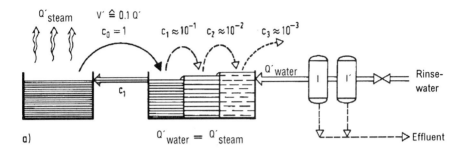

a)

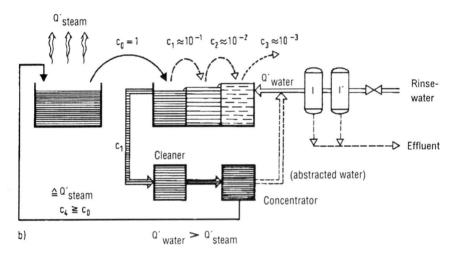

b)

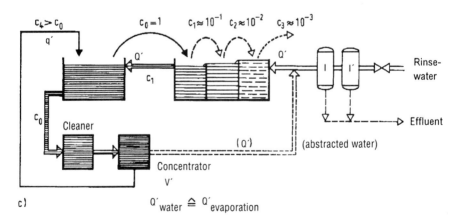

c)

Fig. 6.4　Various configurations of cascade rinses with recycling. [32]

　　a)　High temperature bath operation (ideal case).

　　b)　warm bath operation (normal case)

　　c)　low temperature operation. Process bath is at the left hand side, work moves to the right as shown by the arrows. I & I' are ion exchange units

Applications of cascaded rinsing for various conditions but with $Q' \simeq 10\ V'$ are shown in Fig. 6.4. The simplest case, but also an ideal one arises when the volume of rinsewater required matches or is less than the evaporative loss from the process solution (Fig. 6.4a) and can be fed back to the process tank to make good evaporative losses. The approach is best suited to processes operating at elevated temperatures and where the dwell time of work in the tank is relatively long, i.e. the drag-out volume is small because throughput rate is low. Hard chrome plating is an example of such a process. Any overall water deficiency must be made up with DI water, otherwise hardness will build up with time.

In cases where the volume of rinsewater exceeds evaporative loss, this excess must be removed by evaporation (Fig. 6.4b) either under atmospheric or reduced pressures (see Section 7.8.2). Fig 6.5 shows evaporative losses from the surface of aqueous solutions as a function of their temperature and air velocity above them. It should not be forgotten that, in such cases, while water is removed, impurities and contaminants (including decomposition products) will be progressively enriched until there is no option but to scrap the bath. In broad terms, baths to which no solutions are recycled, are totally replenished between four to eight times a year, as they lose their constituents by drag-out, these being replenished with new chemicals. By the same path, impurities are also removed. The statistic serves to underline the possible dangers in recycling when impurities are present in the return loop.

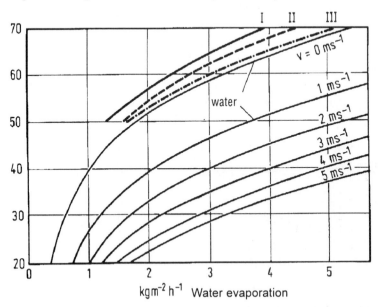

Fig. 6.5 Water evaporation rates from an electrolyte surface with or without air extraction as function of air velocity over the surface (after Sprenger [24]).

Figs 6.4 b) and c) show configurations including a cleaning stage where some of the impurities which build up can be removed. These can be based on any of several methods widely used in metal finishing, depending on whether dilute or concentrated solutions are involved and other factors. Examples are:

- ion exchange for removal of foreign ions from chrome plating baths (see Section 7.2.2.1)

- activated carbon and selective electrolysis for removal of organic break-down products and foreign metals, in nickel plating (see Sections 7.7 and 7.6.2.2.2)

- recrystallisation processes for carbonate removal from cyanide baths (Section 7.4.2).

In cases where virtually no evaporative loss occurs, as with process baths operating at room temperature, evaporators must be used to remove water from rinse solutions before returning the concentrates to the bath. In some cases (e.g. case b) in Fig. 6.4) when cleaning measures are inadequate, the unpurified effluent from rinse tanks or process baths are evaporated and the concentrates returned to the process baths, condensates passed to the rinse feed.

In order to set the correct flowrate for main or DI water to make up any losses in such systems, flowmeters and valves are necessary. Better still is to use a calibrated reservoir upstream of the rinse line as shown in Fig. 6.6. This device is immersed in the rinse tank to a depth depending on the volume of water to be delivered. Water enters through the valve on the floor of the device, rising to the overflow. As the transporter leaves the previous stage, the vessel is pressurised with compressed air so that the water contained acts as a spray-rinse. For further discussion of this and other means of effective rinsing, see Section 6.3.3. Similar arrangements can be beneficially installed in barrel plating lines, the additional spray effects serve at very least to improve rinsing in the barrel.

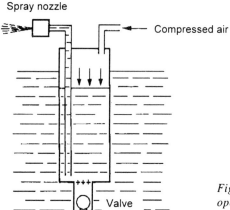

Spray nozzle

Compressed air

Valve

Fig. 6.6 Schematic of a pneumatically operated spray-rinse delivering a set spray- volume (after Schene et al. [22])

Because of the large amounts of water carried over when barrels are immersed in the rinse tank, these should have high partitioning walls, in contrast to racked work which carries less drag-out and which calls for partitioning walls of moderate height in the rinse bath. Use of metering chambers for water feed is usually more cost-effective than using pumps and associated measuring sensors. The drawback of cascade rinsing is that it occupies a large floor area and in most cases, requires additional transporters to be installed. This last can be avoided, for suitably shaped work which can be rinsed by spraying alone. A single spray station may then suffice, with a receiver – which takes the place of the cascade rinse – positioned at a lower level (external cascade). The way this works is seen in Fig. 6.7 [25].

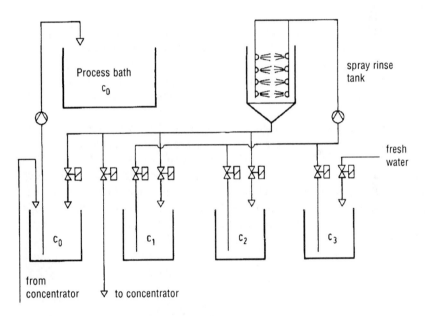

Fig. 6.7 Principle of an external rinse cascade with spray rinsing [33]

The drag-out removed in the spray rinse is returned to the process bath with the concentrate liquor from the concentrator (e.g. an evaporator) via the first tank (c_0). The first spray rinse is carried out using solution from the second tank (c_1) and passes to the concentrator. The second rinse is drawn from tank (c_2) and is returned to tank (c_1) etc.

In principle, a similar set-up may be used for barrel plating automatics. To accelerate liquid removal from the barrel, a partial vacuum may be applied or compressed air is blown through. In the first case, a special barrel is used which drops into a trough to which it is sealed. The drag-out or rinsewater from the various stages is then returned to the system as shown in Fig. 6.7. [26-28]. Fig. 6.8 shows the layout and depicts an actual installation.

Use of compressed air is simpler and allows the drag-out to be returned direct to the process bath. In cases where the water balance allows, the washings from the first rinse can also be returned to the bath. Subsequent rinsings can be recycled upstream in a normal cascade system or via an external cascade is in Fig. 6.7. To implement this, the transporter is fitted with a clam-shell which contains the barrel and liquid draining from it, as it is raised from each stage. The valve providing the compressed air is also mounted on the transporter.

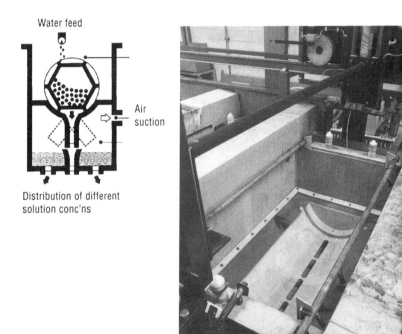

Fig. 6.8   Water-economising barrel rinsing unit with drag-out recovery, using suction. Left − sketch of the basic method. Right − an actual installation showing rinse tank, barrel receiving trough with seals and suction slots. The spray nozzles are seen at the sides. (courtesy Schering AG, Berlin.)

Boger *et al.* [9] describe these rinse configurations, quoting a value of 2 litres per barrel as the volume of drag-out in a zinc plating barrel automatic line. Further details are shown in Fig. 6.9.

Another method of rinsing barrels is to seat the barrel above the rinse tank and rotating it about its lower axis, allowing it to swing to and fro at the same time. The barrel and its contents are thus sprayed. As before, the process can be repeated several times with the rinsings being segregated from first to last, and fed to different reservoirs as shown in Fig. 6.7. When rinsing is complete, the barrel is removed while still rotating [29].

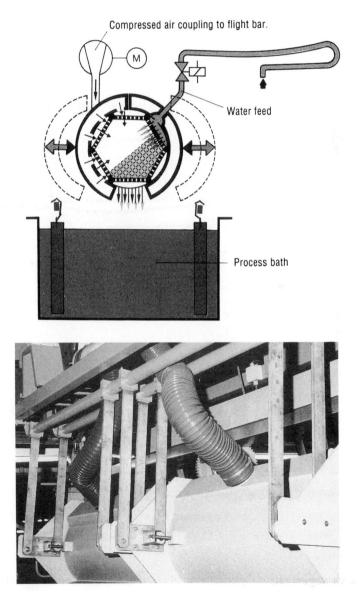

Fig. 6.9 *Water economising barrel rinsing with drag-out recovery by air blowing.*
Top: schematic. Bottom. Clam-shells enclosing the barrel with compressed air feed
(Schering AG, Berlin).

Yet another option for rinsing barrels and their contents uses a spray rinse applied with the barrel located in an empty tank (Fig. 6.10) The sprays rinse the exterior of the barrel while apertures in the cheeks of the barrel allow internal rinsing to take place. Here too, the rinsings can be "fractionated" [30].

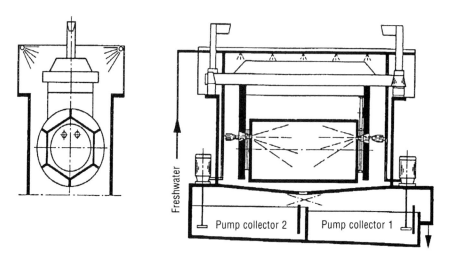

Fig. 6.10   Schematic of flood-spray rinse with recovery of rinsewaters (LPW Galvanotechnik).

6.3.2.2   Cascade rinsing with reduced flow

The two most widely used water-saving rinse set-ups, i.e. recirculation through an ion-exchanger and cascade rinsing as described in the preceding section have advantages and disadvantages, the most important of which are:

Ion-exchange recirculation

- *advantages:* Using high flowrates guarantees good rinse criterion values.

- *disadvantages:* only moderate concentrating effects in the pre-rinse tanks with fluctuating concentrations there.

Cascade rinsing

- *advantages:* relatively concentrated and uniform flow from the first stage.

- *disadvantages:* For a given number of stages in the cascade, water flow must be adjusted if work with different drag-out volume is processed. As drag-out increases, more water is needed or a poorer rinse criterion must be accepted. The system occupies much floor space.

A further option is to combine the principles of the two methods, ignoring minor drawbacks. The aim is to use as little water as possible, so giving the highest possible concentration of chemicals for recycling, but with no detriment to the quality of rinsing. A steady flow of pre-rinse water should be used (with no major fluctuations in concentration) and the system should not take up too much floor space. The rinse quality criterion can be selected using a post-rinse tank with recirculating flow. To obtain the highest possible concentration in the rinsings, the water flow rate in a cascade should be as low as possible. To calculate the condition $Q' \leq 10 V'$, Equation 6.8 can no longer be used. Instead, any calculation should be based on knowledge of the flowrates and the mass-balance, based on equality of water and chemicals entering and leaving the system. For a second stage in a cascade, the following example is given:

$$c_1 \cdot V' + c_3 \cdot Q' = c_2 \cdot V' + c_2 \cdot Q' \qquad\qquad 6.9$$

Similar calculations are shown in [31]. Advantages of a pre-rinse with low flow-rates are shown in publications by Hefele [2], Nohse et al;. [4], Hartinger et al. [17], used in conjunction with several static rinse tanks with intermittent flush-ing.

When rinse criteria for cascade rinsing are calculated on the basis of flow rate and mass balance, one obtains for small Q'/V' ratios, a series of plots as shown in Fig. 6.11. These should be compared with predictions based on Equa-tion 6.8

For small Q'/V'- ratios (< 10), using a linear abscissa, and volume flow and mass balance for the calculation and setting a concentration value for the efflu-ent from the first cascade stage, the data shown in Fig. 6.12 is obtained. The upper diagram clearly shows how, as the number of stages in the cascade in-crease, the greatest increase in recovery comes when a second stage is intro-duced, after which the law of diminishing returns begins to operate. From such analyses, the author was able to arrive at an optimised rinse system [32].

Using rinse cascades alone and taking, for example a Q'/V'-ratio $= 4$, then a five-stage cascade would be required for a rinse criterion of 1000. In this case, the exit concentration of the first stage would be $0.27 c_0$. However five stages occupy a great deal of floor space and constitute a somewhat inflexible system, should the opportunity arise with a different type of work, to reduce the total drag-out volume. The compromise solution would be to use just a two-stage cascade, with exit concentration being $0.25 c_0$. For $Q'/V' = 4$, the rinse criterion would be 20. This can be increased to the desired value by using a recirculation loop. Adjustments reflecting the various volumes of drag-out from different types of work can thus be adjusted using the flowrate, which will normally operate at a larger value than is strictly necessary. As can be seen in Fig. 6.13, for a given value of Q'/V' using a two-stage cascade, no more than 50 V'/hr flowrate is required for a rinse criterion of 1000 [33]. Although the recirculation system makes a major contribution to this, its size would typically be only 20 to 25% of one where no cascade system was installed.

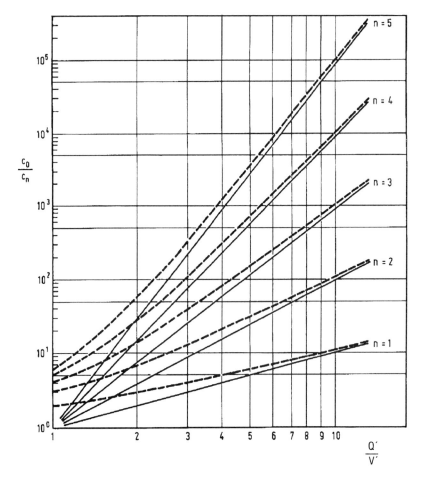

Fig. 6.11 Rinse criteria for cascade rinsing for small Q'/V' -ratios. Solid lines calculated from Equation 6.8, dashed lines using volume and mass-balance data [32].

The same system is schematically shown in Fig. 6.14 and is the most suitable for installation of recycling loops and for maximum economy of water (see Section 6.4). A further advantage is that the pre-rinse tank will normally contain only one metallic species (except where alloy plating is used). In some countries, disposal of sludges based on a single metal is charged at a lower rate, in cases where recycling is not an option.

Graef *et al.* [34] compared a number of rinse systems in terms of the rinse criteria obtainable from each. They concluded that a pre-rinse two stage cascade

linked to a flow rinse with recirculation using ion exchange offered the best
solution in terms of consistent rinse quality and ease of recovery of dragged-out
chemicals. They included in their calculations the price of water and chemicals
used or saved (including regeneration of ion exchangers), finding a definite
minimum for various Q'/V'-ratios for a given water price. In general, it is more
cost-effective to lower the Q'/V'-ratio, compensating this by more frequent
regeneration of the ion-exchangers.

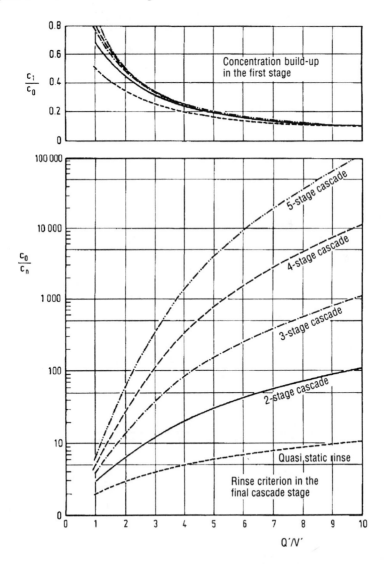

*Fig. 6.12 Concentration in the first rinse stage (above) and rinse criterion in the
last stage of a cascade (below) for small Q'/V'-ratios [32].*

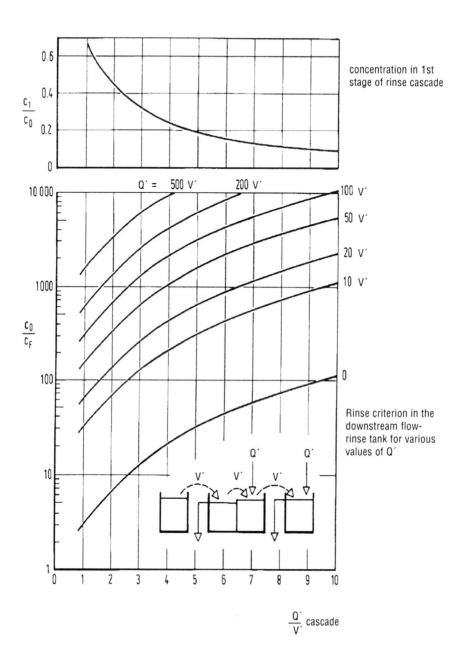

Fig. 6.13 Concentration in the first rinse stage of a pre-rinse two-stage
cascade (above) and in the associated flow rinse with recirculation
for various flowrates expressed as a function of V′ [32].

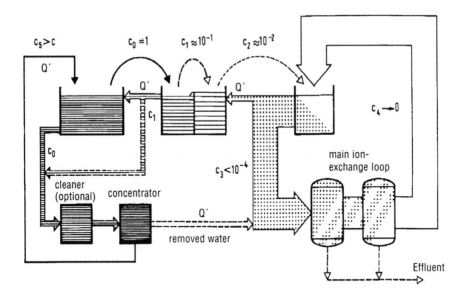

$c_5 > c$ $c_0 = 1$ $c_1 \approx 10^{-1}$ $c_2 \approx 10^{-2}$

Q'

Q' Q' Q'

c_1

$c_4 \rightarrow 0$

c_0

main ion-
exchange loop

$c_3 < 10^{-4}$

cleaner
(optional)   concentrator

Q'

removed water

Effluent

*Fig. 6.14   Schematic of a combination rinsing system based on
2-stage pre-rinse and recirculating flow rinse [32].*

6.3.3   Spray rinsing

Spray rinsing can be conceived as a multiple cascade rinse and so affords, where installed, a comparably high rinse criterion value. Water impinging on the upper part of the work flows downwards, becoming enriched in drag-out as it flows down. Subsequently impinging water further dilutes the drag-out film. Attempts have been made to mathematically model spray rinsing. However these are of limited success unless empirically-determined wetting factors are included [1,2]. In practice it is simplest to treat the whole process on an empirical basis.

In addition to the benefit of acting as a cascade, there is a further factor in favour of spray rinsing. The energy with which the jets strike the surface, create a high degree of turbulence there and effectively reduce the diffusion layer thickness thus reducing the time necessary for rinsing. In this, the water pressure is the critical factor.

Racked work is sometimes, but not always, amenable to spray rinsing, depending on its shape. Work with mainly flat surfaces is suited to spray rinsing. Thus, items which, because of their size and/or shape present only 30% of their exposed area to the spray result in 70% of the water being wasted. Undercut profiles, edged profiles, blind holes are all features inappropriate to spray rinsing and if these are to be effectively spray rinsed, it can easily result in higher water consumption than would have been the case with cascade immersion rinsing. [1]. In addition, racked parts for spray rinsing need to be extremely well secured to the jig.

Successful spray rinsing depends in no small measure on the spray configuration itself. The number and size of nozzles, whether they produce solid cones, hollow cones, fan-shaped jets (the latter for flooding a flat surface) and the water flow-rate – all are critical to ensure optimum rinsing. Horizontally orientated fan jets are usually best, especially when the rinse operates as the rack leaves the solution. Other important parameters include duration of the spray, interval between sprays. In some cases, it may be profitable to operate different spray regimes on the back and front of the work, reflecting perhaps different work geometries. Another useful device is to segregate the contaminated rinsings which drain from the work from the (more or less) uncontaminated "overspray" water impinging on the walls of the tank (see Fig. 6.15).

To prevent blockage of the nozzles, de-ionised water is recommended. This is in any case required for recirculating water systems. Occasionally, problems (e.g. foaming at the nozzles) can arise as a result of surfactant concentration going out of balance in recirculation. In such cases, very simple perforated tubes may be necessary in place of nozzles. In this case, an immersion pre-rinse is required.

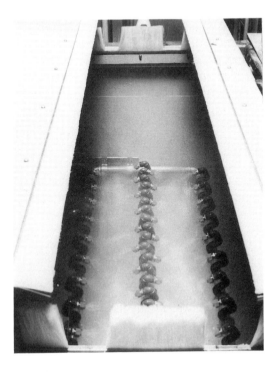

Fig. 6.15 Tank fitted with spray-rinse (LPW Galvanotechnik)

In p.c.b manufacture, spray rinsing is nowadays almost exclusively used e.g. in spray etching machines (Section 8.9.2). It is also widely used after degreasing, phosphating, electrophoretic coating in paint plants and in any operation where large flat surfaces are treated. In cases where spray rinsing alone is used, it should be configured, for efficient water re-use, as a cascade with several spray stations and an external   reservoir. In many cases, spray rinsing is used in conjunction with immersion rinsing and here it should be piped in as additional stages in the cascade (Section 6.3.2.1).

As to the efficacy of a spray system or predicting volumes of water required, little has been published. Reh [29] quotes 2 to 5 litres rinsewater/litre drag-out for a 95% rinsing effect in a particular situation while Schuster *et al* [21] suggest good rinsing is found with 5 to 15 litres/square metre surface. Elsewhere, the very low value of 50 to 150 ml/square metre has been quoted for smooth surfaces using high pressure jets [25].

To some extent, spray rinsing can also be used in barrel processes to rinse the exterior of the barrel after which it is then sucked or blown out, as part of an overall cascade system (as in Section 6.3.2.1).

6.3.4   Static rinsing

One of the simplest rinse methods and one long associated with recovery and recycling, is the simple static rinse tank. Though filled with water, there is no through-flow. The method has been widely used in a recirculatory loop with ion-exchangers. Such tanks can remove up to 90% of drag-out contaminant before rinsing with the recirculated water (see Section 5.3.1). It was only with the use of such static rinses that recirculation became economically attractive. In use, the water becomes progressively richer in dragged-out species. On reaching 10 to 20% of the concentration in the process solution, it is replaced with fresh water. Prior to this, it will be used to replace evaporative losses in the process bath, especially when these operate at elevated temperatures. Where not needed for this purpose, the spent rinse is fed to concentrate holding tanks for process-ing. Without such periodic replacement, the concentrations in the static rinse would approach the values in the process bath itself.

A static rinse will normally be included in any process bath using recirculation, especially for ionic solutions and where recirculated water is then used for rinsing. Exceptions are baths such as acid pickling or etching solutions, which give a relatively acid drag-out. Static rinse tanks are usually installed singly though in cases such as chromic acid where drag-out is greater, two or more static rinses might be used in series. The build-up of contaminants in a static rinse is given by the following Equation:

$$c_t = c_0 \left[1 - \left(\frac{1}{1+V} \right)^t \right]$$

6.10

where c_t is concentration after time t (gm/litre), c_0 is concentration in the process bath (gm/litre), I the volume of the static rinse (litres) and V the drag-out (litre/hr), t = time (hours). Where the drag-out is known, a value for c_t at any given time can be calculated, on the assumption that equilibrium is quite rapidly attained. Some static rinse tanks are fitted with stirrers or pumped circulation or air sparging in order to promote agitation and increase rinse rate (exchange with the surface). In the case of barrel plating, it is advisable to immerse the barrel repeatedly in order to accelerate exchange within the barrel. The concentrations in a static rinse which is periodically emptied and refilled have a saw-tooth shape over time. If two such baths are coupled one behind the other, the first one being periodically emptied and replaced with the content of the second, which is itself filled with fresh water, the effect is that of a pre-rinse double cascade (see Section 6.3.2.2) in which the drawbacks of such saw-tooth profile are largely damped out.

Useful literature discussions include those by Kushner [35], Ochotta & Hefele [1], Schuster et al. [3], Jola [36], Hartinger et al. [37], Lehr & Kutschera [37], Winkel [38].

6.3.5   Chemical rinsing

Chemical rinsing is sometimes known as immersion reaction rinsing, integrated rinsing, the Lancy process or direct drag-out destruction.

All chemical rinse methods follow the principles shown in Fig. 6.16. Initially, they were installed without using a static rinse and served merely to neutralise or detoxify the drag-out. Put another way, they could be better described as static detoxification units than chemical rinses. Thus sodium hypochlorite would be used after a cyanide plating bath or acid sulphite after a hexavalent chromium plating tank. The reactions taking place were those described in Equations 3.24 and 3.78. In addition to any detoxification process, these solutions were acid or alkaline to neutralise the drag-out liquid.

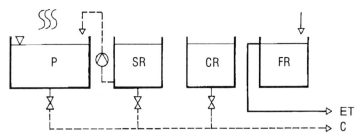

Fig. 6.16   Schematic of a final stage using chemical rinsing.
P = process bath at elevated temperature, SR = static rinse
CR = chemical rinse, FR = flow-rinse, ET = effluent treatment,
C = concentrate hold tank.

Operators of this type of system soon established that rinsing, as well as detoxification was improved. This is due to the fact that not only is mass-transport occurring by diffusion of drag-out through the diffuse layer into the bulk but also because the chemical species present are diffusing in the opposite direction. At the point where toxic species encounter detox reagents, the two react. Local heating (heat of reaction) and formation of faster diffusing species accelerate the overall rinse effect. Clarke *et al.* [15] had earlier found similar effects and practical platers had long made good use of the effects without necessarily understanding the underlying physical chemistry.

The pickling of work in acid media after an alkaline degreasing can readily be seen as a neutralising of a difficult-to-rinse alkaline film to give an acid surface which allows easier rinsing.

Chemical rinsing, as shown in Fig. 6.16 can lead to problems when certain solutions are used at too high a concentration. Thus sodium hypochlorite can passivate some surfaces, while solutions of sodium carbonate can form layers of basic metal carbonate on nickel [15]. No such problems were found using acid bisulphites after chromium plating baths.

Development of rudimentary chemical rinses into a more sophisticated technology stands to the credit of Lancy [39]. The contents of the rinse tanks were pumped into reservoirs where the requisite chemicals were metered in using appropriate dosing equipment so the the overall operation proceeded with only a small excess of reagents, thereby avoiding problems such as those mentioned above. In addition, the reservoirs were configured, and were large enough to allow settled matter to accumulate and be easily removed. The concepts are seen in Fig. 6.17.

By retention of metal-containing species, recycling becomes possible and rinsewaters collected in such an operation can be used several times over. Static rinsing was largely mandatory and contributed to the benefits enumerated in Section 6.3.4.

The widespread use of chemical rinsing is summarised in many publications, of which those of Pinner *et al.* [40] and Lancy [39] could be singled out. Table 6.1 summarises these processes.

In terms of detoxification, chemical rinsing has largely been superseded, sulphate removal apart. It will certainly reduce the load on any final effluent treatment method. More important is its rinsing action as summarised in this section with the associated chemical recovery. Some comments on problems and means of overcoming them are noted below.

For sulphate removal, especially from sulphuric acid pickling baths, a rinse solution containing slaked lime is used. Calcium sulphate precipitates which notwithstanding a solubility of 1.4 gm/litre (as sulphate) results in the removal of substantial amounts of sulphate from the effluent stream. According to Fischer [41], the method reduces sulphate discharges from a acid degreasing plant to 3.5% of the amount where no such treatment is used and so allows effluent discharge well within statutory limits, (see Fig. 6.17).

TABLE 6.1:
EXAMPLES OF CHEMICAL RINSING

Associated process bath	Chemical rinse solution	Reaction products
cyanide baths acid copper baths copper etch baths	OCl⁻, NaOH hydrazine, NaOH	metal hydroxides, oxide (Cu), Cu_2O
nickel baths	soda, NaOH	basic Ni carbonate
acid baths (Zn, Cd, Sn) pickling baths	NaOH	hydroxides, oxides
chromium baths	$NaHSO_3$ (acid, lst stage) hydrazine (neutral, stage 2)	Cr(III) cmpds Cr $(OH)_3$
concentrates with SO_4^{---}, PO_4^{---}, F⁻	$Ca(OH)_2$	Ca salts
concentrates with complexants	reagents as appropriate hydroxide	metals, oxide

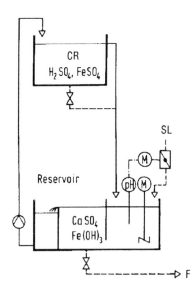

Fig 6.17   Schematic of a chemical rinsing plant with recirculation to a reservoir where dosing is carried out and where sedimentation occurs. In this case, sulphate and ferrous ions are removed after sulphuric acid pickling. CR = chemical rinse, SL = slaked lime dosing, F = filtration. (from Fischer, [41])

Another process worth noting is the use of reductive chemical rinsing to follow non-cyanide copper baths (pyrophosphate or fluoborate) or sulphuric acid pickles contaminated with copper. Chemical rinsing takes place in an alkaline hydrazine solution (danger!!!) where Cu(I) oxide hydrate precipitates, so reducing the solids content by 50% [42]. The same process is used in chemical rinsing of precision-bore copper and brass tubing following a chromic acid pickling in order to ensure rapid removal of the aggressive acid and so "kill" the pickling action at the desired point. Copper precipitates as the Cu(I) oxyhydrate, zinc as hydroxide while chromate is reduced to Cr(III) and precipitated as its hydroxide [43]. Use of a subsequent immersion in slaked lime serves to reduce the sulphate content.

Chemical rinsing is also useful when fine diameter holes are filled with chromic acid. Wellinger [44] describes the chemical rinsing of such holes in multilayer printed circuit boards [44] after their etching in strong chromic acid, again using sodium bisulphite.

For chromium plating, it is nowadays considered better to recover unreacted chromium (VI) salts using cascade rinsing than to react them as described above. Chromium plating lines designed for chemical rinsing can simply and inexpensively be converted to cascade rinsing, as seen in Fig. 6.18.

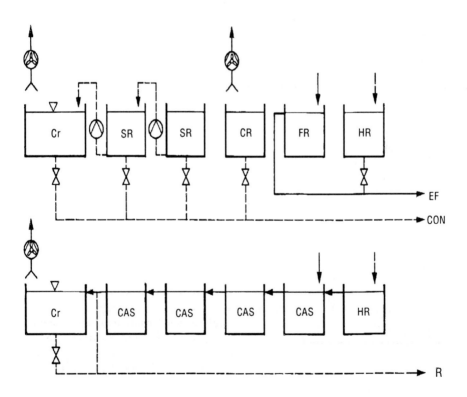

Fig. 6.18 Representation of a chromium plating line with chemical rinsing (above) and its conversion (below) to cascade rinsing with total recovery of the chromic acid. Cr = chromium bath, SR = static rinse, CR = chemical rinse, HR = hot rinse, FR = flow rinse, EF = effluent treatment, CON = concentrate storage, CAS = cascade rinses, R = recovery of chromic salts.

6.4   WATER ECONOMISING SYSTEMS – PRECONDITIONS FOR RECYCLING

Any means for minimising drag-out or water economising in rinsing are useful starting points for recycling systems and examples have already been described, such as recirculation using an ion-exchange plant where both pure water and concentrated eluate are potentially valuable recovery products. Whatever system is used, it can be asserted that the more concentrated the recovered solution, the greater is value. Before illustrating these principles, mention should be made of a rinsing configuration based on cascade rinsing which follows not a single process bath as detailed in Section 6.3.2 but which follows a number of different process solutions. In certain cases, no rinsing at all is required, for example when the species present in the first process bath are similar or identical to those in subsequent tanks. Thus there is no point in rinsing after pickling in dilute acid if the subsequent plating bath is based on the same acid. An example of this would be pickling in dilute sulphuric acid followed by nickel plating in a sulphate solution. With care, the acid concentration of the former can be set so that acid losses from the plating bath are made good.

Again, rinsing is often omitted between hot degreasing baths and electrolytic cleaning where both have similar compositions. Fig. 6.19 shows a further example of these concepts. It shows a barrel zinc plating line with throughput of 10 barrels per hour, each barrel equating to 2 litres of drag-out [33]. After hot degreasing, a slow-flowing three-stage cascade rinse is used, fed by a two-stage pre-rinse after the pickling stage. The water flowrate through both cascades is equivalent to four times the drag-out volume in the same time, i.e. 80 litres/hr. Apart from the fact that the same rinse water is used in two processes, a further benefit accrues. The content, certainly of the third stage after hot degreasing, will be acidic, since pickling baths are usually more acidic than hot degreasing solutions are alkaline. Two benefits result. Alkaline drag-out is more easily rinsed off by acid rinses, and no alkali is carried over into the pickle bath, which would over time neutralise this. The bath life is thus prolonged. Following the two-stage pre-rinse, a recirculating rinse rate of at least 1000 litres/hr is used, corresponding to a rinse criterion of 1000. This is followed by a cyanide zinc plating stage. Because it operates at close to ambient temperature, evaporative losses are minimal. The outflow from the following two-stage pre-rinse cascade is returned to the zinc bath but only after sufficient water is removed to preserve a balance, e.g. using a vacuum evaporator (see Section 7.8.2.1). Prior to this, a chiller crystallising unit is used to remove carbonates (not shown). Condensate is recycled to the mains feed water loop. Further rinsing is carried out using recirculated rinsewater to achieve a rinse criterion of 1000. The zinc deposits are usually brightened by immersion in very dilute nitric acid (0.5%). The subsequent rinsing calls for very small water flowrates or indeed can be entirely omitted if subsequent chromating solutions themselves contain nitrates. The chromating bath, if not too strongly acid, can be regenerated using a strong cationic ion-exchange resin in protonated form (see Section 7.2.2.1). Finally a two-stage cascade is used with sufficient water flow to achieve a rinse criterion of 5000. Fig. 6.14 shows flowrates for each stage. Total recirculation flow rate is 3600 litres/hr but since, depending on the type of work being treated, drag-out volumes could easily be greater, the plant is designed to handle at least twice this volume. Flowrates in the pre-rinse cascades should on no account be altered, since this would affect the volume of effluent generated. However even this can be limited if some of the measures described below for recycling, are incorporated.

At the foot of the diagram, the annual effluent volumes are listed. They include spent process baths (hot degreasing, pickling, brightening, part of the chromating bath), regeneration of the recirculation loop (based on 10 cubic metres/hr) and that of the ion-exchange plant for the chromating bath as well as floor swills. All told, these come to 630 cubic metres/p.a. In practice, this volume is handled on a batch basis at either 2.5 cubic metres/day or 12.5 per week.

It may be that a selective ion exchange unit also needs to be installed, if high neutral salt concentrations are found. Because effluent volumes are relatively small, the size of this is not excessive. Given uniform flowrates, two columns each with 30 to 50 litre resin content will cope with 8 to 10 hours of flow.

An alternative pre-treatment method for this line would be to increase the bath life of the hot degreasing using an oil skimmer and/or ultrafiltration (Section 7.5.4.1)   and to continuously regenerate the pickling bath using a retardation method (Section 7.2.2.2).

A few comments here about metal stripping plant. These operate (for anodic stripping in a warm solution at 30 to 40°C) using cyanide (for copper, zinc, cadmium, silver) or sulphuric acid (copper and nickel) and usually do not incorporate a rinse. If rinsing is deemed necessary, using a five-stage cascade, a water consumption of around 5 litres/hr should suffice, this flowing in at the front end and being removed at the final stage by evaporation. The stripping electrolyte itself is recirculated through an electrolytic regeneration cell from which metals are recovered (Section 8.8.2.1).

Plants corresponding to that seen in Fig. 6.14 are also appropriate in situations with high drag-outs, e.g. etching, pickling, metal colouring and blacking, phosphating). Other examples such as for copper + nickel + chromium plating lines [32] or cadmium plating with rack or barrel lines [45] have also been described.

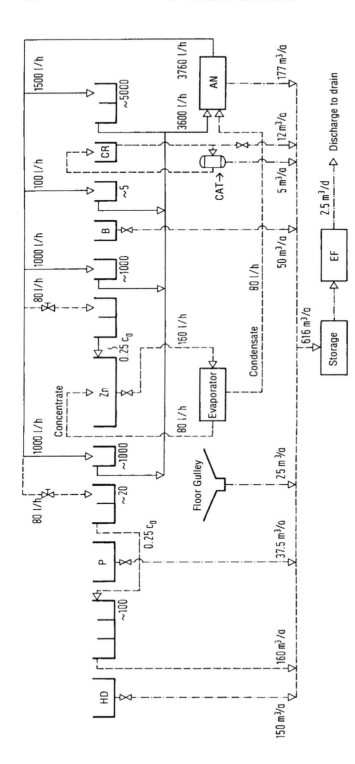

Fig. 6.19 Schematic of a barrel zinc plating line with combined water economising measures and integrated recycling.

HD = hot degreasing, P = pickling Zn = zinc plating bath (cyanide-based), B = brightening bath, CR = chromating bath, CAT = cation exchanger, AN = anion exchanger, EF = effluent treatment

References for Chapter 6

1. Ochotta, G., Hefele, H.: "Rinsing methods in metal finishing" Galvanotechnik + Oberflaechenschutz (1962) 2. pp 19-54
2. Hefele, H.: "Tried & tested rinsing & recovery techniques lower on-costs & ease effluent problems" Galvanotechnik 55 (1964) 589-611
3. Schuster, H.J., Gruhl, W., Winkel, P.: "Effect of rinsing technique on profitability of metal finishing" Galvanik. Metall 20 (1966) 651-6.
4. Nohse, W., Wystrach D.: "Rinsing techniques & recirculation in Plating Shops" Galvanotechnik 58 (1967) 481-491
5. Nohse, W., Wystrach, D.: "Rinsing & recirculation techniques" Proc. IWL Forum Vol. 4 (No. 5) (1966) 61-79 publ. 1968
6. Krusenstjern, A.v., Schmidt, G.: "Drag-out losses from Plating Baths" Metalloberflaeche 15 (1961) B149-153
7. Jola, M.: "Problems in obtaining data for planning an effluent treatment plant for the metal working industry" Fachber. Oberflaechentechn. 11 (1973) (2) 60-64
8. Kushner, J.B.: "Rinsing Pollution & Natural Recycling of Plating Baths" Metal Finishing July (1971) 36-39
9. Boger, K., Kübler, E.: "Rinsing techniques for barrel zinc plating plants" Galvanotechnik 76 (1985) 1659-1666
10. Anon: "Direct electrolytic recovery in barrel plating" Galvanotechnik 70 (1979) 532-3
11. Hunt, J.E.: "Improved Method of Removing and Reclaiming Solution in Barrel Plating" Plating and Surface Finishing (1987) Oct. 38-39.
12. DWP 115,705 "Means for recovery of dragged-out electrolyte" Galgon, H U; Galvanotechnik 67 (1976) 622
13. Cook, T.H., Cubbage, M.L., Fister, L.M.: "Draining Process Solutions from Sheets, Baskets, Pipes, Threads & Fins" Metal Finishing (1984) July 33-39.
14. Bahensky, V.: "Rack plating" Galvanotechnik 68 (1977) 226-8
15. Clarke, M., Kieszkowski, M.: "Rinsing Part IV. - Effects of Agitated Water Rinses and of Chemical Rinses for Nickel and Chromium Solutions". Trans. Inst. Metal Finishing 52 (1974) 79-86.
16. Bahensky, V., Machacek, M., Prusek, J.: "Water management in plating shops" Metalloberflaeche 17 (1963) 301-5
17. Hartinger, L., Götzelmann, W.: "Rinsing & evaporation in recycling technology" Metalloberflaeche 31 (1977) 386-398
18. Hartmann, M.W.: "The Blasberg RS 80 system - an environmentally friendly & cost-effective effluent treatment method" Blasberg Commun. 28 (1975) 15-17
19. Winkel, P.: "Rinsing processes & their implications for effluent" Galvanotechnik 68 (1977) 692-699
20. Jelinek, T.W.: "Rinsing to rationalise effluent disposal" Galvanotechnik 70 (1979) 626-630
21. Schuster, H.J., Winkel, P.: "Rinsing systems" (in) Hartinger L: "Pocketbook of effluent treatment" Vol. 2 pp 22-34.
22. Schene, H., Boger, K.: "Savings & recovery with the right rinsing method" Galvanotechnik 67 (1976) 377-381
23. Schene, H.: "Rational rinsing in metal finishing" Galvanotechnik 72 (1981) 633-640

24.　Sprenger, E.: Heizung. Lueftung 17 (1943) 7-8 (see ref. 21)
25.　Meurer, H.: "Rinsing in surface treatment" Blasberg Comm. No. 33 (1980) 43-7
26.　Kreisel, R.: "Saving & recovery" Oberflaeche + JOT (1978) 224-8
27.　Anon: "Direct electrolytic recovery in barrel plating" Galvanotechnik 70 (1979) 532-3
28.　DE 2947810 Bl (16.7.1981). Friedr. Blasberg GmbH & Co., KG,
29.　Reh, H.J.: "Cost reduction & environmental protection" Galvanotechnik 70 (1979) 720-8
30.　Langbein-Pfanhauser Werke AG, "Static multi-stage spray-flood rinsing as a cascade" Techn. Bulletin
31.　Dr. Ing. Max Schlötter GmbH & Co., KG, "Rinsing in actual conditions – a case study" Galvano-Trommel 32 (1981) 9-15
32.　Götzelmann, W., Hartinger, L.: "Water saving rinse system & ion exchange recycling" Galvanotechnik 73 (1982) 832-842
33.　Hartinger, L.: "Combined water economising measures & recycling in the metal industry" Wasserkalendar 1988, 100- 124, Erich Schmidt Verlag, Berlin
34.　Gräf, R., Schwering, H.U.: "Combined rinse techniques for effluent minimisation or avoidance" Galvanotechnik 77 (1986) 294-300
35.　Kushner, J.B.: "The Plater and Pollution". Plating & Surface Finishing (1977) Aug. 24-30
36.　Jola, M.: "Effect of economy rinsing on operating costs of an effluent treatment plant" Oberflaeche Surface 15 (1974) 355-361
37.　Lehr, K., Kutschera, W.: "An alternative rinsing method replaces a costly effluent treatment process" Galvanotechnik 78 (1987) 378-384
38.　Winkel, P.: "Are static rinse systems still useful today?" Galvanotechnik 78 (1987) 2255-2261
39.　Lancy, L.E.: "Plants for direct detoxification" (in) Hartinger "Pocketbook of effluent treatment" Vol. 2
40.　Pinner, R., Fischer, G.: "Metal recovery combined with the Lancy Process" Galvanotechnik 59 (1968) 430-6
41.　Fischer G: "Dramatic decrease in effluent in a jobbing platers" Galvanotechnik 66 (1975) 638-645
42.　Fischer, G.: "Effluent treatment & metal recovery in pickling of copper & its alloys" Draht Fachzeitschr. 19 (1968), Nov.
43.　Fischer, G.: "The Lancy process for direct effluent treatment" Galvanotechnik 62 (1971) 399-405
44.　Wellinger, B.: "Chromic acid for through-hole cleaning in multilayer boards" Metalloberflaeche 40 (1986) 367-9
45.　Götzelmann, W., Hartinger, L.: "Reducing cadmium discharge" Galvanotechnik 74 (1983) 140-4.

7   Recyling Processes

7.1   DEFINITIONS

A simplistic definition based on materials recovery and its return to the system, is too restrictive to be of use at this point. Included also, should be methods of extending the life of a bath or process solution or converting a short-life process into one of longer life (e.g. by regeneration). The question then arises as to whether any waste products are formed in such a regeneration or whether more, or more noxious products are formed than would otherwise be the case. The aim must be to reduce the amount of waste products formed. The concept of a truly zero-discharge plant, however desirable as an ideal, does not appear to be a realistic goal, as long as aqueous solutions are used for surface treatment. Striving towards this, in the opinion of the author, unattainable goal, could in the end be counter-productive.

Our definition of recycling in the context of surface treatment includes all measures which reduce the volume of effluent or solid waste or avoid its formation, by recovery of useful materials or by extending the life of process solutions. Such a definition forms the basis of several European legislative documents and may become definitive in EU law.

Many processes in the above category were specially developed by or for the metal-working and finishing industries, some of these are already installed or in process of this. Others are under development. Fig. 7.1 attempts an overview of such processes by class or principles involved [1]. They are displayed horizontally in terms of what they seek to achieve, with links to associated operations also shown. Vertical links suggest options for changes in conditions etc which might involve other methods shown above or below. All recycling processes share one attribute – they involve a separation function [2].

In most cases, recycling involves not less than two of the processes shown in Fig. 7.1

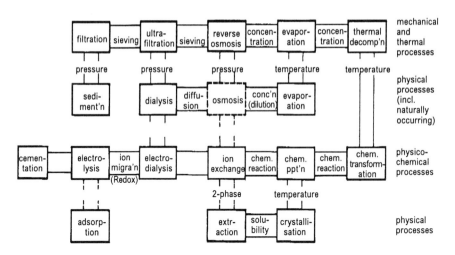

*Fig 7.1 Recycling processes in the metal working industries
and their relation to one another [1].*

The introduction of recycling processes is inevitably linked to the whole question of their cost-effectiveness. In the narrowest sense, it is really only the precious metals and silver, for which a case can be made. In other cases, the attraction of recycling includes indirect savings, e.g. on disposal costs not incurred. Indeed, at least under German legislation, continuing future operations of a company may be predicated on its operating in what is deemed to be an environmentally acceptable manner. The following sections treat processes which, by virtue of their widespread use, can be deemed "state-of-art". For a thorough review, see also Kammel & Lieber [3].

7.2   ION EXCHANGE METHODS

With the exception of selective ion exchangers used for effluent treatment, all other applications of this technology, including the recirculation of rinsewater, qualify as recycling operations. Considered below are somewhat more rigid definitions which relate to recovery of chemicals from rinsewater and the regeneration of acid process baths, aspects of these having been defined by the author [4,5]. For the theory of ion exchange, see Sections 5.2.3 and 5.2.4.

7.2.1   Recovery of materials from rinsewater

7.2.1.1   Precious metals

For understandable reasons, the recovery of precious metals from effluents in surface treatment is one of the oldest forms of recycling. Concentrated solutions are usually recovered in electrolytic cells while more dilute solutions, sometimes with no more than a few mg/litre, are concentrated in an ion-exchange unit. However developments in electrolysis cell design continue to lower the concentrations at which direct electrolytic recovery becomes attractive. Silver and gold have been recovered electrolytically for well over 50 years, the former from spent photographic processing baths [6].

In alkaline solutions, gold is present as the dicyanoaurate. In HCl it exists as the chloro-complex and sometimes again as the dicyanoaurate. The latter is so stable it is not decomposed even by acids. As is the case with all anionic complexes, that of gold is very strongly, almost selectively bound to anion exchange resins and this is specially marked with strongly basic types of anionic resins. Once loaded with this complex, the resin is so stable that regeneration using NaOH is almost impossible (see Section 5.2.3.3.3 – affinity series). Adsorption follows the equation:

$$R - OH + [Au(CN)_2]^- \rightleftharpoons R[Au(CN)_2] + OH^- \qquad\qquad 7.1$$

The use of strongly basic anionic resins for gold recovery is described in [6,7,8]. Resins are supplied in OH⁻ or Cl⁻ forms and would in any case adopt these forms if exposed to alkaline cyanide or chloride (in acid medium). Nor, by the same token, is there any need for them to be de-alkalinised using a cationic exchanger.

Figures quoted for useful capacity are typically 100 gm Au/litre resin [6,9,10], though Weitz [6] quotes 275 gm/litre. As usual with ion exchangers, it is normal to couple two units in series with the front unit taking the main load, the downstream column accepting any excess. To test for signs of over-loading, a sample is acidified and, after heating, tin (II) chloride solution is added. In presence of gold, the solution goes dark. At higher Au concentrations, a dark precipitate consists of finely divided gold [11].

When a resin column is fully loaded with gold, it is sent to a refinery for recovery. Usually this is by incineration in an oxygen-rich atmosphere at 500 to 600°C, the gold is found with the residual ash [4]. Recovery is around 95% efficient. Fig. 7.2 shows a gold recovery system. The eluate may be either cyanide-containing or acidic, depending on the process bath type, and must be fed to the effluent plant. Fig. 7.3 shows a compact unit for this purpose. Each resin column contains around 4 litres.

The resin capacities quoted above are based on the assumption no other metals are present in solution. Because alloys of gold rather than the pure metal are increasingly used in surface coating, such values should be down-rated accordingly and electrolytic recovery may prove more favourable. (see Section 7.6.2.2.1)

In cases where a very large volume of rinsewater is used following gold plating, an option is to retain the small unit (e.g. as in Fig. 7.3) and recirculate the content of the static rinse bath through this. The gold concentration in the static rinse is thus kept so low that further losses from drag-out into the following cascade rinse can be neglected.

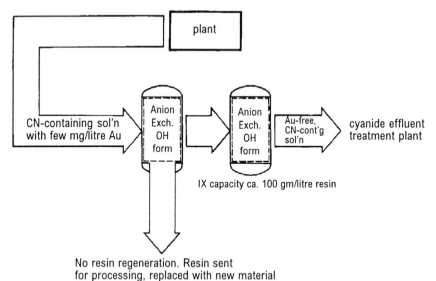

Fig. 7.2 Schematic of a gold recovery system with strongly basic anionic resin (two columns in series) for metal recovery from rinsewater [4].

Fig. 7.3 Ion-exchange resin unit for gold recovery from rinsewater. Rating – 100 litres/ hr (GOEMA)

Silver can be recovered in much the same way as gold. However because of the much lower price of this metal than gold, it is not feasible to recover the metal by incineration of the resin. Instead, it is better to use weakly basic and thus more readily regenerable ion exchange resins. Typical hold-up values in a cyanide-based solution would be 50 to 75 gm/litre of resin, based on silver as dicyanoargentate. In some plants, ion exchange units are built into a recirculatory rinse loop and periodically regenerated using alkaline cyanide solution. In this case, and indeed others, the main recovery method is by electrolysis, whether directly from the rinsing system or using the regenerate from ion exchange beds, as described in [12]. Silver can also be recovered by cementation. Disposable units are commercially available (mainly for the photographic industry) where effluent trickles through steel wool packing. Ionic silver deposits out, ferrous ions pass into solution.

Mercury and its compounds are very rarely found in industry today. As the tetrachloromercurate complex, mercury can be removed with a weakly-basic anionic resin in its chloride form (though regeneration is difficult) to a level of less than 13 ug/litre [13]. Using special resins containing sulphide groupings, it can be stripped from acid solutions down to the 10 µg/litre level [14] though these resins cannot be regenerated. As with silver, cementation should be possible.

Little has been published concerning the recovery of metals of the platinum group. Palladium, used in the p.c.b industry, can be recovered from chloride-containing acids (pH ca 2) where it exists as a chloro- complex, using strongly basic ion exchange resins. The same process readily allows its separation from copper. At inlet concentrations of 5 to 6 mg/litre, an eluate with around 0.1 mg/litre is typical [16]. Typical resin capacities, of the type used for gold, are 30 to 50 gm/litre resin.[10].

7.2.1.2   Copper and its alloys, non-ferrous metals

Ion-exchange methods can be used in these cases, with copper and nickel recovery from acid solutions being technologically the most important. Barrel-plating lines with their high drag-outs are of particular relevance. Because the metals are all in cationic form, they can readily be removed with cationic exchangers. While strongly acid cationic type resins can be used, it should be noted their capacity is smaller than that of of the weakly acid types and they are also more difficult to regenerate. Of the weakly acid cation exchangers, two commercially available types should be distinguished in terms of capacity and ease of regeneration. These are based on carboxyl groups and iminodiacetate groups respectively. Water not containing any hardness salts can be treated using the less expensive carboxyl group type resins. In other cases, the more expensive iminodiacetate resin is preferred. In the former case, any calcium or magnesium ions present reduce the capacity of the resin, in the latter case, metals can be bound in even in presence of calcium (see Section 5.2.3.3.2)

Since weakly acid cation exchangers have a high affinity for protons to the extent that this could lead to reduced capacity or even loss of bound metal ions, the resins are used at the highest possible pH, i.e. just below the hydrolysis pH for the metals in question. Here too, the carboxyl type resin is more sensitive to condition than its more selective imidodiacetate counterpart. Thus the latter can absorb copper even from quite acid solutions when it exists in the protonated form. This is connected with the stability of the complex formed between the metal and the imidodiacetate functional grouping. However below pH 1.5, copper is scarcely absorbed with pH 2.5 and 2.7 being the corresponding limits for nickel and zinc respectively. Alkali metal ions, even at quite high concentrations, have little effect on metal removal. The weakly-acid cationic resins are not suitable for removing trivalent metals since these tend to hydrolyse at low pH.

With the exception of their use for copper, resins are installed in their as-delivered form (the protonated form is not completely converted to the Na-form after regeneration) or else in their Na-form. The exchange reaction is then:

$$2\,R - Na + Me^{++} \rightleftharpoons R_2 - Me + 2\,Na^+ \qquad\qquad 7.2$$

In practice, two or even three weak-acid cation resin columns are coupled in series to ensure that the first is fully loaded before regeneration, without fear of metal loss for the same reason. Since metal recovery (except in special circumstances) requires this be in the form of its sulphate, sulphuric acid, usually fairly concentrated, is used. This ensures that the first flushing yields a concentrated metal salt solution. Regeneration is straightforward, thanks to the high selectivity of the resin to protons. This can also be seen in Fig. 7.4 which compares regeneration of strong and weak acid type cationic resins [15].

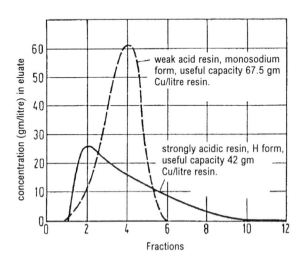

Fig 7.4 Elution curves for regeneration of strongly and weakly acidic cationic resins loaded with copper using 15% sulphuric acid.

The regenerate is recovered and used for its free residual acid value in the pre-regeneration stage of the next cycle. After this, it is replaced with fresh sulphuric acid which again is recycled two or three times. By such cascaded regenerations, copper concentrations of 80 to 100 gm/litre or nickel concentrations of 60 gm/litre can be obtained. Such sulphate salt solutions with modest acid content, are ideally suited to electrolytic metal recovery, unless they are used to make up an actual plating bath. In the p.c.b industry, excess copper chloride HCl solutions are converted, by ion exchange, to sulphates and again used to make good losses in copper plating baths (Section 8.9.4).

After regeneration, it is usual to recondition the ion exchangers using a sub-stoicheometric volume of NaOH. The eluate is then roughly pH neutral and can often be discharged direct to sewer with no further treatment, provided the metal content is less than 0.5 mg/litre. The process flow-sheet is shown in Fig. 7.5

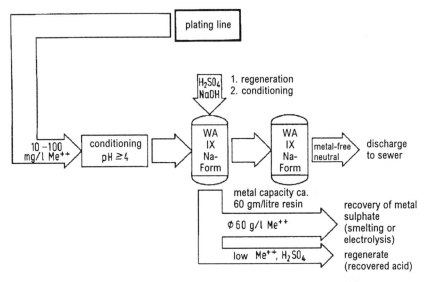

Fig. 7.5   Schematic for recovery of copper and other non-ferrous metals from rinsewater, using weakly acid exchangers [4].

Fig. 7.6   Plant for nickel sulphate recovery from rinsewater (Blasberg).

There are conflicting comments in the literature regarding recovery of nickel and copper from rinsewaters, though no particular problems are highlighted there. Kauczor [16] describes how copper (down to pH 1.5), cadmium (pH 4 to 7.5) and nickel can be removed. In this context, use of resins incorporating the iminodiacetate group is advantageous for sulphuric acid regeneration since there is no scale formation within the exchanger column, in contrast to that found when using carboxylic acid resins. For the recovery of nickel as its sulphate and an associated electrolytic metal recovery from this, see Marquardt [17] and Schwering *et al.* [18]. Fig. 7.6 shows an ion exchange unit for nickel sulphate recovery.

In smaller plants, the best approach is often to use small cation exchanger units on the individual rinse tanks as a means of minimising sludge formation at the effluent plant. Wahl & Reinhardt [19] give details. While weakly-acid cationic resins can be used for removal or recovery of metals from solution, this can also be carried out using flat-bed exchangers although concentrations such as those quoted above, will not be obtained. Thus Nadeau & Dejak [20] quote concentrations of 22 and 35 gm/litre for nickel and copper respectively.

In much the same way, it is possible to recover other divalent metals such as zinc or cadmium from rinsewaters following acidic plating baths although the author is not aware of any actual installations. Such recovery operations are not restricted to sulphate solutions but could equally well be carried out in tetrafluoborate or hexafluosilicate or chloride media.

7.2.1.3 Chromic acid

Chromic acid is still recovered from rinsewaters following chromic acid baths, in some plants, using ion exchange. Most plants use a cascaded rinsing configuration or one which would allow this method of operating following a static rinse, chemical rinsing or flowing rinse tanks (see Fig. 6.18). In such cases, it is more effective to to recover the chromic acid by evaporation (see Section 7.8.2).

In cases where there is insufficient floor-space, the best approach is to use a separate rinsewater circuit with ion exchange columns of strongly acid cationic resin, then weakly basic anionic resin. In such a system, the cation exchanger retains foreign metals (mainly Fe(III) or Cr(III), but also copper & nickel) the latter derived mainly from the substrates being plated. This regeneration effect using a cation exchanger is represented in Equation 5.10.

Chromate and dichromate ions are then retained by the anion exchange unit, as shown in Equation 7.3. The unit is operated until chromate is visible at its outflow, thus assuring that the resin is largely loaded. A better method is to couple two anion exchangers in series, ensuring the first of these is fully loaded.

$$2 R - OH + CrO_4^- \rightleftharpoons R_2 - CrO_4 + 2 OH^-$$

7.3

The demineralised rinsewater is then returned to the flow rinse after the chromium plating bath. The regenerate from the cation exchange unit is piped to a neutralisation precipitation plant.

The anion exchanger, which, apart from chromate, will adsorb only the anions of "foreign" acids (sulphate, silicofluoride, fluoride) from the acid, is regenerated in the usual way with NaOH. The chromate capacity is usually much greater than that listed by the manufacturers, being usually in excess of 100 gms chromium trioxide/litre capacity. This stems from the ability of chromic acid to associate (e.g. dimerise) into larger molecules, resulting in formation of dibasic acids ($Cr_2O_7^{--}$, $Cr_3O_{10}^{--}$). Such associations are favoured by conditions downstream of the cation exchanger unit.

In order to regenerate the anion exchanger, since chromate has a higher affinity for the resin than most other anions, a much longer contact time with NaOH is necessary, implying a higher consumption of this species (2 to 2.5 times the usual amount). The first half of the regenerate solution is reddish yellow and contains 50 to 60 gm/litre of the trioxide (as chromate) at a pH around 7. This portion is used to recover chromic acid. The second portion is paler yellow and contains significant free caustic soda. It should be stored and used in the next cycle, after making up any shortfall in NaOH, as a pre-regenerant solution, thereby providing the greater part of the NaOH requirement for regeneration of the resin. The chromate containing wash-waters are fed to the mains water intake of the recirculation loop, so retaining virtually all the chromic acid within the system. The concentrated portion of the regenerant from the ion exchange unit is converted by passing over a cation exchange resin in its protonated state. The eluate would typically contain 50 gm/litre of chromium trioxide and is returned to the chromium plating bath to make good evaporative or drag-out losses, being concentrated in an evaporator beforehand, if this is appropriate [21]. Equation 7.4 describes the conversion process while Fig. 7.7 shows the process flow-sheet.

This process has been described by Furrer [22], Marquardt [24], Krug [23], Nadeau [20] and others. Marquardt describes a variant in which electrodialysis is also used to concentrate both portions of the regenerant solution from the anion exchanger. Nadeau, drawing on experience with flat-bed exchanger units, suggests that the sulphate content of recovered chromic acid can be up to three times greater than in the electrolyte and recommends barium carbonate to precipitate such excess sulphate.

In cases where only small amounts of rinsewater are produced, there is a simple means for chromic acid recovery by passing the rinsewater over a strongly acid cationic exchanger in the H form. This removes foreign metals and can be passed either direct back to the plating bath to replace loses or returned after partial concentration in an evaporator.

$$2 R - H + Na_2CrO_4 \rightleftharpoons 2 R - Na + H_2CrO_4 \qquad\qquad 7.4$$

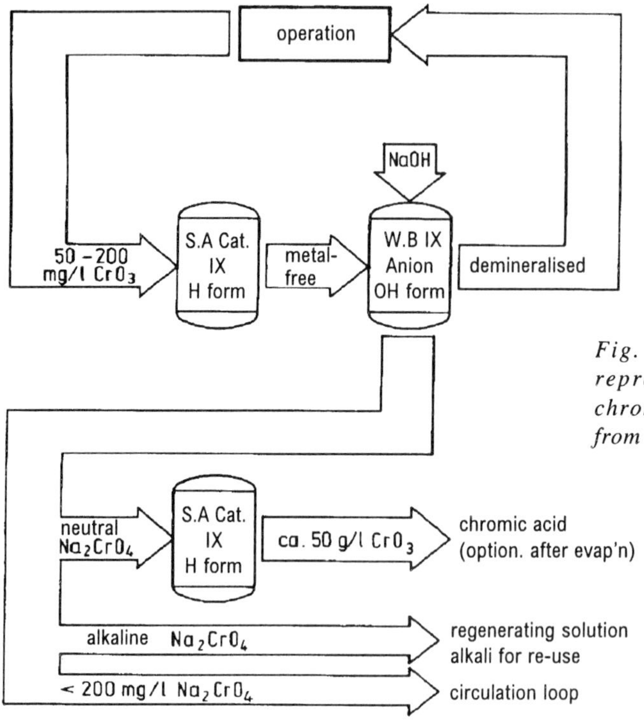

Fig. 7.7 Schematic representation of chromic acid recovery from rinsewater [4].

7.2.2　Regeneration of acid process baths

7.2.2.1　Regeneration using ion exchange

To the extent process solutions are not too strongly acidic, cationic impurities can be removed using strong-acid cation exchangers in their protonated form or, more correctly, the concentration of such foreign species can be maintained at a set level by continuous use of such methods. In many cases, this allows solutions previously considered to be of short life to be used for extended periods.

　　The exchange process is a true regeneration in which foreign metals removed from solution are replaced with the hydrogen ions which would have been lost when the foreign metals were originally ionised.

$$z R - H + Me^{z+} \rightleftharpoons R_z - Me + zH^+ \qquad\qquad 7.5$$

These reactions do not proceed to completion even within the column of the ion exchanger, this being a consequence of the relatively small concentrations of foreign metal ions in the context of a much greater hydrogen ion concentration. It is these H ions which are usually the regenerating species for the strongly acid cation exchanger. The unfavourable equilibrium situation can be envisaged by showing the concentration profiles as liquid passes through the column, drawn as Fig. 7.8 (centre and right hand side) [25]. This type of process is used in conjunction with chromium plating baths., chromating solutions and phosphoric acid etch baths.

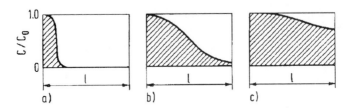

Fig. 7.8   Schematic representation of reaction zones for favourable (a) and unfavourable (b), (c) equilibrium conditions in the removal of metal ions from a strongly acid solution.

Hexavalent chromium plating baths can become contaminated with metal ions derived from the basis metal of work being plated (iron, brass) or from coatings previously electroplated on such work (copper, nickel or zinc) while trivalent chromium can arise by cathodic reduction of Cr(VI) anions in solution. While small amounts of Cr(III) are indeed necessary for good operation of a hexavalent chromium plating bath with 2 to 3 gm/litre iron likewise having a beneficial effect, larger concentrations of foreign metals or Cr(III) serve to reduce solution conductivity, current efficiency and throwing power. In addition, copper impedes formation of a regular cracking pattern in plating micro-cracked chromium [26, 26,28].

To remove such metals, the chromic acid is diluted to 200 or better still, 100 gm/litre. The reason for this is not such much to prevent oxidation of the ion exchange resin (modern resins are quite oxidation resistant) but rather to favour the equilibrium which is disadvantaged by high acid concentrations. In addition, solutions should first be cooled to room temperature. According to Wiedemann [28], a life of 1200 regeneration cycles can be expected for styrene resin cross-linked with 12% divinylbenzene. Where this value falls to 8%, only 400 regenerations can be anticipated, assuming in both cases an exit concentration of 100 gm/litre chromium trioxide. Where this value is doubled to 200 gm/litre, the number of cycles falls to 400 and 100 respectively.

In practice, the diluted chromium solution is pumped over the ion exchange column at a rate which should not exceed 10 litres/litre resin per hour. This should be continued until no further reduction in Cr(III) or Fe(III) concentrations is observed. Fig. 7.9 shows the effect of chromium trioxide concentration on iron removal.

Regeneration calls for prolonged contact times as is the case with all exchangers contaminated by Cr(III). It is often best simply to allow the resin to stand overnight in contact with the regenerant acid, without pumped circulation, after which modest circulation serves to complete the operation. Because the resin is affected by chloride ions, 15% sulphuric acid is the best medium for regeneration.

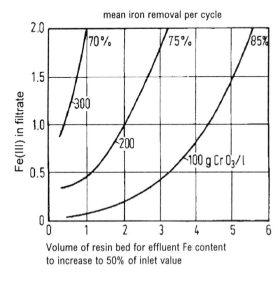

Fig. 7.9   Dependence of iron removal on throughput and Cr(VI) concentration in decontamination of chromic acid. Inlet conc'n 4.0 gm/litre Fe(III)/litre of chromic acid.

The overall process is schematically shown in Fig. 7.10 with an actual plant shown in Fig. 7.11. Only one column is needed for this process and this can be positioned as seen in Fig. 7.10.

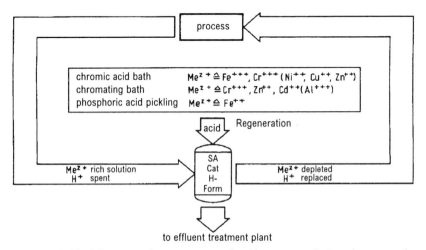

Fig. 7.10 Schematic of regeneration of acid process solutions by removal of foreign ions using a strongly acidic cation exchange column [4].

Fig. 7.11 Ion exchange plant for chromic acid regeneration (GOEMA)

Regeneration of chromic acid is a long-established operation, having been described by Costa in 1950. Since then, the process has become virtually mandatory wherever evaporation is used to make good drag-out losses, since impurities can no longer be discharged to sewer, as was formerly an option.

Recovery of chromic acid is cost-effective both in its own right and in terms of savings in effluent treatment then no longer called for. Indeed, classical effluent treatment of chromic acid is almost unthinkable in some legislations. 1 kg of chromic acid would require 1.5 kg of sulphuric acid, 1.4 kg of sodium pyrosulphite, 0.6 kg of NaOH, all of which would result in formation of 4 kg of soluble neutral salts or 3 kg of compacted sludge (30% dry solids).

In the case of chromating solutions also, foreign metal ions build up. These are most commonly $Cr(III)$, zinc, cadmium, less often silver but also copper (from brass) and aluminium.

Chromating baths usually contain $Cr(VI)$ at much lower concentrations than is the case with plating baths, though some formulations have up to 500 gm/litre as chromium trioxide. Mostly, the concentration increases as one moves up from clear, through blue and green to the olive finish, with 50 gm/litre being exceeded only rarely. Higher concentrations (ca 100 gm/litre) are found where the surface is prepared for organic coating or in plants operating on a continuous, through-flow basis where higher reaction rates are called for (100 to 500 gm/litre).

The regeneration process is substantially the same as described above for chromic acid. However no dilution of the solution is possible, a subsequent re-concentration would make this unrealistically expensive.   Patents apart, the process has scarcely been described in the literature. McGarvey & Fisher [29] describe the regeneration of a very dilute chromating bath (0.3 gm/litre), used after Bonderizing. They confirm that the contact time of solution with the resin, as one might expect, is a function of the concentration of the regenerant acid and increase in pressure. By means of regeneration, zinc and Cr(III) ions could be almost totally removed and the specified pH maintained. Chromium (VI) must be made up to the specified concentration.

Wolotzkow & Hartinger [25] carried out systematic studies of chromic acid over the concentration range 5 to 100 gm/litre with zinc ions in the range 0.5 to 10 gm/litre. Fig. 7.12 shows the contact times in the ion exchange unit as a function of chromic acid strength with 1 gm/litre Zn present. The effect of pH on the exchange capacity of the resin is evident. From Fig. 7.13 it is seen that the useful capacity increases both in terms of zinc (at constant chromic acid strength) and with decreasing chromic acid concentration. For chromating baths with up to 50 gm/litre chromium trioxide, useable capacity values in excess of 1 gm equivalent per litre of resin can be reached.

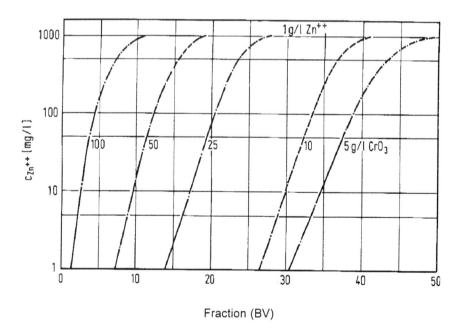

Fig. 7.12   Ion exchange regeneration of chromating solutions using a strongly acid macro-porous cationic resin as a function of chromic acid concentration and constant 1 gm/litre zinc ion . BV = bed volume [25].

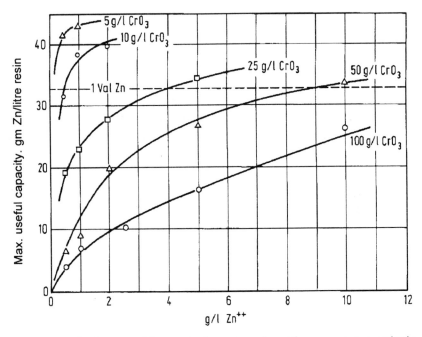

Fig. 7.13   Maximum useful capacity for a strongly acid macro-porous cationic exchange resin for regeneration of chromating baths as function of zinc io and chromic acid concentrations [25].

More recently, Huss [30] has described an automated process for regeneration of chromating baths. A yellow chromating solution is maintained at 0.8 to 1.6 gm/litre Cr(VI), 1-10 gm/litre Zn, less than 1.5 gm/litre Cr(III) & Fe and pH 1.4 to 1.6 with chloride being the major anion. Alkali & ammonium ions are not present.

As seen in Fig. 7.14, the concentrations of dichromate and chromic acid and pH are maintained by HCl additions. Cr(VI) is measured colorimetrically, pH using a meter. The solution then passes over the first cation exchanger which is loaded with zinc ion. An exchange of Zn with Cr(III) and Fe ions occurs with no pH change. Exhaustion of the ion exchange column is indicated colorimetrically by presence of chromium ions and this triggers the regeneration of the first cation exchange column. The solution then passes to the second cation exchange column, which is loaded with hydrogen.. This serves primarily to take up zinc ions. As soon as this unit is exhausted, it is switched over to the function of the now-regenerated first cation exchange column so that the whole cycle can start again. The nett result is a removal of Cr(III) and Fe ions from solution, with the zinc remaining largely in the bath. If the pH value should fall, from the second column, it is put out of circulation until the correct value is restored. Successful operation depends on good mixing of the chromating bath (using compressed air) and proper measurement of the concentrations of key species in the solution.

According to Huss, in the case of clear chromating of zinc, it suffices to make up drag-out losses of the zinc with other minor adjustments of pH etc. Regeneration of blue chromating baths (nitrate + fluoride based) is somewhat different in that the pH electrode is replaced by a fluoride ion sensitive electrode which controls dosing with nitric acid and HF instead of the HCl used in other baths. Similar procedures have been devised for automated regeneration of yellow, clear and olive chromating of cadmium [30].

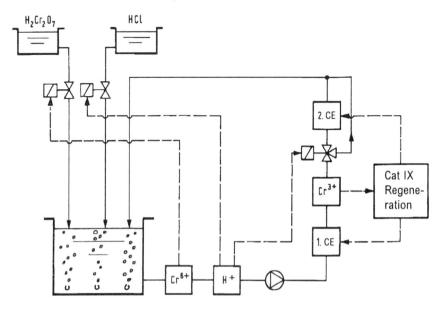

Fig. 7.14   Schematic for automated regeneration of yellow chromating baths (after Huss, [30])

Phosphoric acid etch baths can likewise be regenerated using strongly acid cation exchangers in their protonated form. Such etch baths are often used upstream of the actual phosphating of iron and offer the advantage that acid film adhering to the surface of the work does not result in corrosion, as would be the case with HCl or sulphuric acids. If the iron concentration in the etch acid bath becomes too high, a pre-phosphating action takes place, especially if the time interval before rinsing is too long (in excess of 15 to 20 seconds) and this is most undesirable. This has been described by Heinrich & Busse [31]. The concentration of Fe(II) should not exceed 2 to 3 gm/litre. Control at this level offers a further advantage that less Fe(III) forms. Being relatively insoluble, this leads in due course to crust formation of the tanks and associated equipment.

The useful capacity of strong acid cationic resins for Fe(II) uptake is high. At the typically found phosphoric acid concentrations of 5 to 15%, resin capacities are 30 to 35 gm Fe/litre resin [32,33]. Since phosphoric acid is relatively weak compared with other mineral acids, the process described above can be used for up to 40% acid concentrations. The resin capacity decreases as acid strength increases and increases with increasing iron concentration. Other divalent metal ions such as copper, nickel & zinc can also be readily removed from phosphoric acid [15].

Regeneration of loaded ion exchange resins can be carried out with HCl or sulphuric acid, although greater excess acid is required, 150 gm HCl (as 10% solution) or 200 to 250 gm sulphuric acid (as the 15% solution per litre of column capacity [32]. The supply for re-charging should be 5 to 10 cubic metres/cubic metre capacity/hour and a prolonged contact time is advised, i.e. by standing the regeneration solution overnight in the column. Another approach which seems to have fallen out of favour, is the regeneration of HCl pickle liquors, described by Kraus & Moore in 1950. However this is important in the context of the retardation process (see next section). In chloride or chloride ion containing solutions, a whole range of metal chloride complexes form, such as those with Fe(III), Cu(II), Cu(I), Zn(II) all of which can be adsorbed by anion exchange resins. Use is made of this in the method for regeneration of HCl pickle liquors for iron, described by Borgolte [35-38].

For such regeneration, iron must be present in the Fe(III) form in order to form an anionic complex. Oxidation can be carried out with gaseous chlorine in slightly less than stoichiometric amounts so that no free chlorine is present since this could damage the ion exchange equipment.

$$FeCl_2 + \frac{1}{2} Cl_2 \rightleftharpoons FeCl_3 \qquad\qquad 7.6$$

The Fe(III) chloride, in presence of strong HCl (not less than 160 gm/litre) forms a chloro complex, tetrachloroferrate.

$$FeCl_3 + HCl \rightleftharpoons [FeCl_4]^- + H^+ \qquad\qquad 7.7$$

In the case of copper, Goetzlemann et al. [39] showed that complex formation set in above 20 gm/litre HCl, being complete at 200 gm/litre concentrations. All chloro-complexes adsorb readily from strong acid solutions onto strongly basic anionic resins in their chloride form as:

$$R - Cl + [FeCl_4]^- \rightleftharpoons R - [FeCl_4]^- + Cl^- \qquad\qquad 7.8$$

Regeneration of the ion exchange resin differs, in this case, from normal practice and resembles in some way the Retardation process (see next section). It is carried out using water, whereby the adsorbed complex anion is hydrolysed, releasing a deep brown eluate. This is not the Fe(III) chloride shown in Equation 7.7 which is only stable in acid. It is believed to be a basic ferric chloride, probably Fe(OH)Cl$_2$. The method has also been tested using liquid ion exchange (Section 7.3).

7.2.2.2   Regeneration using the retardation effect

An entirely different means of separating acidic process solutions into acids and salts is the so-called retardation process. Though it utilises strongly basic anion exchange resins, it has nothing to do with ion exchange.

Using first amphoteric ion exchange resins (with cationic and anionic exchange sites) and subsequently anion exchange resins, Hatch & Dillon [40] reported separation into acid and salt. Brown et al. carried the work forward to a technical scale using a flat bed exchanger to regenerate anodising baths [41] and nitric acid pickle liquors [42]. The author has himself and with colleagues worked in the field since the mid-70's in an attempt to broaden its application [39,43,44]. Krausenegger [45] and Rituper [46] have reported on the method for regeneration of mixed acids (e.g. HF + sulphuric) as used for pickling stainless steels.

If a column loaded with strongly-basic anion exchange resin, pre-charged with the common anion from the solution is alternately fed with a solution of mineral acid and its salt and with water. the eluate will alternate between a low-acid salt rich composition and one which is acid-rich and low in salt as shown in Fig. 7.15 which represents a full cycle. The salt flows through the column virtually unaffected and the amount leaving is primarily determined by the specific loading and the void fraction between the resin particles (approx. 35% of gross volume). By contrast, the concentration of acid from the column can be explained only in terms of a physical exchange reaction with the resin in a way not the same as ion exchange itself.

Fig. 7.15 is the best way of visualising a separation between sulphuric acid solution and that of its salts. The continuous if modest acid elution is considered in the following text as a background rate, with the increased acid concentration at the end of the cycle treated as excess acid discharge.

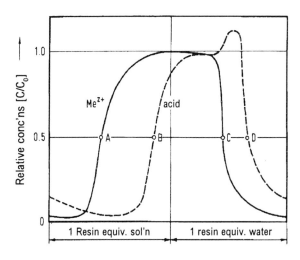

Fig. 7.15   Typical elution curves for acid-salt separations (sulphate & sulphuric acid) by the retardation method [39].

Hatch & Dillon [40] examined four possible theories which could be invoked for acid adsorption on a resin. These included effects such as salting-out of the acid in the resin phase by the salt present, interactions between strongly acid protons with the benzene rings in the resin, adsorption of undissociated acid molecules and entropy effects. None of these were found to be wholly satisfactory.

The observed facts accord most closely with a theory based on the adsorption of associated - and thus uncharged - acid molecules. The positively charged metal and hydrogen cations present in the aqueous phase are kept apart from the solid phase anions in the anion type resin by electrostatic repulsive forces. The anionic resin has cationic exchangeable groups (solid ions) at concentrations of 1 to 1.5 equivalents/litre resin, i.e. in fairly high concentrations. The phase boundary operates as a separating membrane for ions of like charge in the solution phase. The only ions capable of diffusing into the resin are those capable of there being exchanged, namely anions. However since the resin active groupings are already saturated with those anions present in bulk solution, virtually no exchange takes place, mainly because of electroneutrality effects, as seen from mass balance studies. It is thus no more than a possibility that uncharged species such as acid or salt molecules can diffuse into the bulk resin.

An equilibrium exists in aqueous solution, dependent on concentration and temperature, between the dissociated and associated forms of the acid, as shown below for HCl.

$$H^+ + Cl^- \overset{\text{association}}{\underset{\text{dissociation}}{\rightleftharpoons}} HCl \qquad\qquad 7.9$$

The HCl molecule may possibly be capable of diffusing into the anionic resin to constitute the more significant fraction of the equilibrium in Equation 7.9. If HCl is abstracted from the equilibrated state, more will be formed to replace it, as long as the diffusion process suffices to maintain the resin as saturated. It is also possible that the acid molecules form clustered aggregates in the resin phase, since the uptake capacity of the resin is relatively large. Such a diffusion of acid molecules would only be possible as long as salt molecules were excluded. Just why no undissociated salt molecules, which are after all present as part of the equilibrium, appear able to diffuse into the resin is far from clear. Possible the diameter of the unhydrated ions is a factor of importance. Hatch & Dillon [40], in a study of various nitrates, showed that nitric acid exhibited the normal retardation effect while potassium nitrate did not. Lithium nitrate, by contrast, showed a pronounced retardation effect and it is worth noting that the unhydrated lithium ion has the smallest diameter of all mono-valent cations. Other effects cannot be precluded. Fig. 7.16 summarises present theory.

Once the uptake capacity of the anionic resin for undissociated acids is exhausted, elution of these commences. If the column is supplied with water as a regenerating species there follows, with a delay corresponding to the void volume between resin particles, a further elution of acid taken off the resin. This often results in a transiently greater acid concentration than that found in the inlet feed (cf. Fig. 7.15). The effect of water diffusing into the resin causes an internal dilution, resulting in dissociation of the acid. This in turn accelerates transfer of acid to the aqueous phase, in consequences of electrostatic repulsion of the hydrogen ions. Evidence for this is seen in the very steep, often near-vertical gradients observed at the end of the elution process for strong acids.

The retardation process can be said to have its origins in the time taken for undissociated acid to diffuse into the resin and then, in its dissociated form, to diffuse out once more while the salt ions pass through the column unaffected by adsorptive processes. This explains the separative action of the method. If the eluate is optimally fractionated (into two portions), the acid rich, metal salt poor portion can be returned direct to the process bath. However care should be taken that the volume of the returned acid does not exceed that which was initially removed for separation, indeed it should be somewhat less because of minor acid losses associated with the process as well as losses corresponding to anions rinsed out with the metals. Such losses should be made good by topping up with fresh acid. In general, around 85 to 90% of the volume removed will be returned although the acid concentration in this reduced volume will be somewhat higher, so offsetting the apparent loss.

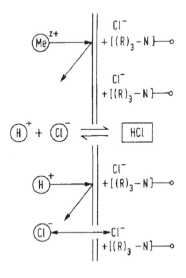

aqueous phase   resin phase

Fig. 7.16 Mechanism of acid-salt separation using an anionic exchange resin in a retardation process [43].

This loss of volume must be compensated in a recycling operation and for such reasons, because of the acid loss which is specially marked when metals dissolve in pickling operations, it is not possible to process solutions based solely on concentrated acids.

The rich-in-metal, low-acid fractions which come first in the elution can be subjected to other processes, e.g. electrolysis, for metal recovery. In the case of iron or aluminium containing solutions, use of flocculating agents in a neutralisation precipitation is recommended. In cases where metal recovery is not practicable, the solution must be treated in an effluent plant.

The distances between points A and B, or C & D (Fig. 7.15) at the intersection point with the 50% line can be used as a guide to the effectiveness of the separation. Systematic studies by Goetzelmann *et al.* [39] of factors affecting the separation gave the following results:

Resin effects

- strongly basic resins gave better results than weakly basic types, even though the chemical capacity of the latter is larger.

- As between the behaviour of Types I & II strongly basic resins (Section 5.2.3.3.4), there is little difference.

- Gel resins are superior to macroporous resins. For the same fill density, they offer a larger fraction of resin by weight. This leads to a greater distance A-B and C-D.

- resins less crosslinked (i.e. with larger micropores) perform best

- more finely divided resins give sharper separations (steeper flanks in the elution curves) but they reduce the amount of acid recovered quite substantially. Because the mean diffusion path length is shorter, the distance A to B, C to D is correspondingly reduced. A corresponding increase in the number of cycles in a given time is still not enough to restore the overall volume.

- The type of resin used has a marked effect. Resins based on polyacrylamide perform much better than those based on gel resins based on polystyrene, even though this is macroporous.

Process parameter effects

- Effect of column size.

- Fig. 7.17 shows the separation curves for different columns containing the same mass of resin and fed with the same liquors but with different height-to-width ratios. What is revealed is not only the drawbacks of flatter columns but also that columns with large diameters perform less well. Since resin fill heights in excess of 1200 mm are best avoided because of pressure losses, it is preferable on larger plants to have more columns of small diameter than a single larger diameter column.

- Effect of specific loading.

- Increasing the throughput leads to a flattening of the "flanks" of the separation plots and thus to poorer separation (as shown in Fig. 7.17).

Nature of load

- As Gulbas [47] showed, counter-current flow gives better performance than co-flow. The elution curves can be somewhat distorted in the former case, but the small amount of acid eluted (which increases with basicity of the acid and the higher losses) decrease. (background acid loss).

Effect of temperature

- Although one might expect increased temperature to favour diffusion of acid into the resin on the one hand while on the other, shifting the equilibrium of 7.8 to the left hand side, Gulbas [47] found no significant temperature effects in the range 20 to 60°C, even when the liquor was fed at at elevated temperature with regeneration at room temperature.

Effect of acid type

- Best results are obtained with monobasic acids such as HCl, nitric or perchloric acids [47] and this applies both to background and to peak outflows. Background effects tend to disappear almost completely as concentration increases, whereas excess concentration effects become larger the lower the acid concentration. HF shows no excess acid effects although it does show substantial background behaviour. This can be linked to the association of HF with the diminution of hydrogen ion concentration. It behaves, in its separation, like a weaker acid. Polybasic acids such as sulphuric, phosphoric, also behave like weak acids because they can form ionic species such as HSO_4^-, $H_2PO_4^-$, HPO_4^{--}). The background increases and the excess acid value decreases. Separation suffers as acid concentration decreases because of high background acid flow and a flattening of the elution curves.

Effects of metal type

- Since metal salts flow through the column virtually unaffected, effects are seen only when complexation with acids occurs. In this case, true ion exchange takes place as described by Borgolte (see previous section) for the regeneration of HCl iron pickle liquor. The metals then follow, rather than precede the acid in the elution sequence, an effect which can sometimes be put to good use. Excess metal ion concentration effects of several hundred percent can arise. In trials, the following anionic complexes have been observed: chloroferrate (III); chlorocuprate; chlorozincate; fluoferrate (III) and fluochromite. The metal ion concentration has very little effect on the separation process. A very small and practically insignificant metal retardation effect is found.

The retardation process for acids finds application for countless pickle and etch solutions but also for anodising solutions. Other applications include recirculation of leach acids in metal winning and recovery as well as other circumstances in which acids used in hydrometallurgy are reprocessed.

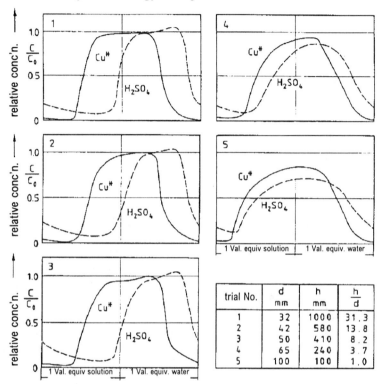

Fig. 7.17   Effect of column dimensions (filled height) on retardation separation for the copper sulphate-sulphuric acid system [39].

In the case of aluminium anodising, apart from maintaining the electrolyte composition, other important factors are reducing the sulphate concentration but most of all, energy saving measures. According to Jenny [48], anodising energy requirements increase with aluminium sulphate content. At constant cell voltage, current density decreases, while if attempts are made to maintain the current, an increased voltage is required. The effect is compounded since the additional energy implicit in this then requires extra cooling to maintain the bath temperature. At an average aluminium concentration, savings of around 0.4 kWhr/ square metre of anodised aluminium surface are possible. Further benefits accrue from maintained quality of product. Figs. 7.18 to 7.22 show examples of salt-acid separation using the retardation process.

The plant required is fairly straightforward in its construction, consisting of exchange columns and two hold-tanks for process liquor and water (see Fig. 7.23).

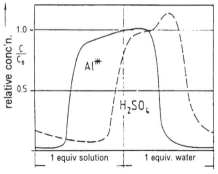

⇐ *Fig. 7.18   Separation of aluminum sulphate and sulphuric acid for regeneration of an anodising bath [43].*

Fig. 7.19   Separation of ⇒ *metal sulphates from leaching acid (sulphuric acid) in working-up of nickel hydroxide sludges and recovery of unused acid [39].*

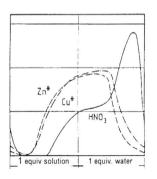

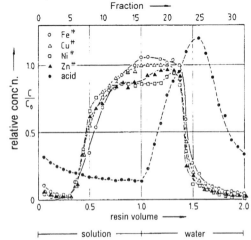

⇐ *Fig. 7.20   Separation of zinc and copper nitrates from nitric acid for pickle liquor regeneration (260 gm/litre nitric acid). [39].*

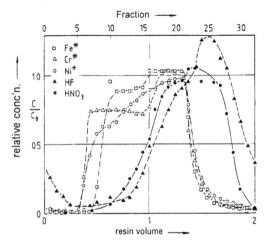

Fig. 7.21   Separation of ⇒ *metal salts from nitric-HF acid mixtures to regenerate pickling acids for stainless steels.*

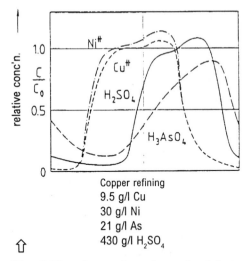

Copper refining
9.5 g/l Cu
30 g/l Ni
21 g/l As
430 g/l H$_2$SO$_4$

⇧

Fig. 7.22   Separation of metal sulphates, sulphuric acid and arsenious acid from spent electrolyte in a copper refinery, for arsenic removal. [39]. Legends as before. Copper refining bath

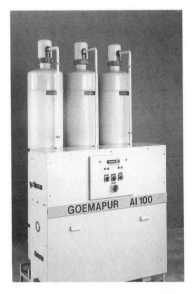

Fig. 7.23 Plant for separation by the retardation process as used in an aluminium anodising plant. Throughput   160   litre/hr (GOEMA).

7.3   LIQUID-LIQUID EXTRACTION

This process as used for recycling, should be described in conjunction with ion exchange since, where it is used in the metal-working industries or might so be, it operates in a basically similar mode. In this, it is irrelevant whether the metals are bound to carboxylic acids or as complexes. In the case of ion exchange, weakly acid resins are those based on carboxylic acid or iminodiacetate groupings. The only difference is that the latter operate in the solid state, the former in the liquid state. This does, however, give rise to major differences in the hardware.

In liquid-liquid extraction, the aqueous phase containing the species to be removed, is contacted with an organic phase, immiscible with water. The species to be extracted are partially transferred to the organic phase. As in the case of ion exchange (see Equation 5.5) an equilibrium is established, corresponding to the partition coefficients shown in Equation 7.10.

$$\frac{(c_A)_O}{(c_A)_L} \cdot \frac{V_L}{V_O} = \alpha_A \quad \text{or} \quad \frac{(c_B)_O}{(c_B)_L} \cdot \frac{V_L}{V_O} = \alpha_B \qquad 7.10$$

where $(C_A)_O$, $(C_B)_O$ are the concentrations of substances A and B in the organic phase; $(C_A)_L$,$(C_B)_L$ are the concentrations of the same species in the aqueous phase while V_L and V_O are the volumes of aqueous and organic phases respectively.

From this equation, and in analogy with ion exchange, one obtains a selectivity coefficient, e.g. for substances A and B:

$$\frac{\alpha_A}{\alpha_B} = \beta_{A/B} \qquad\qquad 7.11$$

For the extraction of a species A, one should use as organic phase, a solvent with the highest possible uptake capacity for A, i.e. such that the partition coefficient α_A or $(c_A)_O$ are large while V_O can be kept small.

In the same way, substance A can be extracted from the organic phase by contacting with fresh acid or alkaline solution and this is known as "stripping". In this way, species can be removed from an aqueous phase into the organic medium and then returned (albeit in more concentrated form) to a new aqueous solution. The process has traditionally been used in the hydrometallurgical branch of the metal-winning industry but is now beginning to find application in metal finishing and working plants.

The organic phase usually consists of two or three components. The most important is the reagent, which is the component reacting with the species to be extracted. In cases where cations are to be bonded, branched chain alkyl substituted carboxylic acids are used, or dialkyl substituted phosphoric acids (most commonly di(2-ethylhexyl)-phosphoric acid [49] or specifically tailored oximes (see below).

$$R_2-\underset{\underset{CH_3}{|}}{\overset{\overset{R_1}{|}}{C}}-COOH$$

branched alkyl-substituted
carboxylic acid

di(2-ethylhexyl) phosphoric acid

Where it is anions requiring bonding, amines are used with the binding affinity of these increasing from primary to quaternary types.

The second component of the organic phase is the solvent or diluent. As this indicates, it acts to dissolve and dilute the reagent and is usually a hydrocarbon in the boiling point range 170 to 260°C. (petroleum, kerosene). Other solvents are based on mainly aliphatic hydrocarbons with 15% aromatics content. However wholly aromatic solvents, e.g. based on xylene [50] have also been reported while perchlorethylene has been proposed [51]. These solvents are broadly speaking inert and their function is largely to distribute the reagent phase in a larger volume and so improve the contact with the aqueous phase [52]. They should have the following properties [51].

- high reagent solubility
- lowest possible solubility in water and no emulsion formation with water
- greatest possible density difference from water [50]
- lowest possible viscosity [50]
- good chemical stability
- lowest possible flammability.

The third component (not in all cases present) is the solvent mediator. On grounds of cost, aliphatic hydrocarbons are preferred to aromatics. However their disadvantage is that they are prone to form highly viscous and insoluble metal complex gels in the organic phase, this leading to complications in phase separation. To minimise this, long-chain aliphatic alcohols such as isodecanol are used at 1 to 10% concentrations [53].Typical reagent concentrations in the solvent are 0.1 to 0.5 Mol [50] or 5 to 30% for copper-specific oximes [54]. Solubilities of the organic phase are around 50 ml/cubic metre which, for a 10% reagent concentration, implies a loss of 5 ppm of this species. [55].

Extraction by cation exchange

In cases where several metals are to be extracted from solution to be concentrated and at the same time separated from one another, the selectivity of the reagent to be used in the process, must be known. Since such reagents are protonated weak acids, they behave like weak acid cationic exchangers in their protonated form, i.e. the pH of the aqueous solution is the main determining factor in their selective behaviour. The actual selectivity can be determined by measurement of the partition coefficients as a function of pH. Typical pH values when using alkylated phosphoric or sulphuric acids are 4 to 7 [50], while for complex forming oximes in strong acids, pH values up to 1 can be used [56]. The choice of anion in the aqueous medium also affects selectivity, e.g. in separation of zinc and cadmium from weakly acid sulphate or chloride containing solutions using alkylated carboxylic acids [57]. In addition, selectivity is affected by the metal ion concentration in aqueous solution [5-56].

In cases where it is simply a question of removing a metal from solution, e.g. in regenerating a process solution, then selectivity is of no importance and extraction can be carried out at a higher pH where the organic phase has a higher capacity.

In such cases care must be taken that the metal is complexed such that hydrolysis will not take place and that the reagent be based on a complexing agent whose stability with the metal is larger than the aqueous phase complex.

The stripping of the metal ions from the organic phase is carried out using a mineral acid, usually sulphuric acid since this lends itself best to subsequent working-up of the solution, e.g. by electrolysis. Because it is usually desirable to have a strong electrolyte for such operations, 20% or even 30% sulphuric acid is used for stripping [54,56].

In terms of applications for liquid cation exchangers, the two most attractive options in the metal-recovery and treatment industries are:

- recovery of metals from their hydroxide sludges (see Section 4.5.2.3)

- regeneration of ammoniacal copper etch baths in the printed circuit board industry.

Mueller [58,59] describes two processes for extracting copper, zinc and nickel from a leach liquor of the metal hydroxide sludge using liquid-liquid extraction. In the Goldschmidt process, sludges are dried and roasted followed by an air oxidation. This produces Cr(VI) and Fe(III) ions. The next stage uses sulphuric acid at 50 to 60°C at pH 1.5 to 2.5 in which over 99% of the copper, zinc and nickel go into solution. At the same time, iron forms jarosite, calcium forms the sulphate, these species and some of the silicic acids are all solids which are retained by filtration. The filtrate is passed over a weak-basic anion exchange resin and this binds chromates and most of the remaining silicates. The resin is regenerated with 6% NaOH from which a sodium chromate liquor containing 30 to 40 gm/litre chromium results (see Section 7.2.1.3). This is then subjected to a first liquid-liquid extraction using a copper-specific organic phase (Reagent LIX 64N). Stripping with sulphuric acid yields a saturated copper sulphate solution which can be used to feed an electrolysis cell. The second extraction stage uses di(2-ethylhexyl) phosphoric acid ester for zinc recovery. Sulphuric and hydro-chloric acids are used for stripping, the product being a zinc salt concentrate. Any aluminium present is precipitated as the hydroxide, using soda. Finally the nickel is extracted at slightly higher pH using the same reagent as for zinc and again stripping with HCl or sulphuric acids to give a nickel concentrate. Fig. 7.24 shows in schematic form the overall process.

The other, so-called MAR Process (MX-Processor, Goeteborg, Sweden) uses weaker sulphuric acid as a leachant so that apart from Cu, Zn, Ni, only small amounts of Fe or Al pass into solution. From this first copper, then zinc are extracted in much the same way as described above. However by using first a weak, then a strong solution of sulphuric acid, the stripping also separates zinc from iron (which emerges only with strong sulphuric acid). If nickel is extracted using a Ni-specific reagent, then only chromium remains in the raffinate and this can be removed with other residual metals by precipitation as its hydroxide. The copper-specific reagent described above is an oxime produced by General Mills [60] or, in another form, by Shell Chemicals [61]. Their structures are shown overleaf.

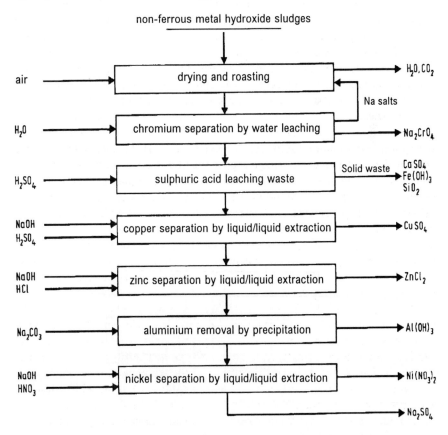

[60]   [60]   [61]

The complete binding of 1 Mole copper requires 2 Moles of oxime, such bonding taking place through the hydrogen atom of the phenylhydroxide and the unpaired electrons of the oxime nitrogen. In the acid stripping, the oximes are again protonated and release the copper ions.

Fig. 7.24 Schematic representation of the various treatment stages for working-up of metal hydroxide sludges by the Goldschmidt process [59].

The development of new, copper-specific, reagents provided a fresh impetus in design and use of liquid-liquid extraction. The result was a process for extraction of copper from spent ammoniacal etch baths used in the printed circuit board industry. Though the copper content in these baths can exceed 150 gm/litre, it is best maintained in the 110 to 140 gm/litre range or even at 80 to 90 gm/litre. The copper after extraction is stripped out with sulphuric acid and passed to an electrolytic recovery cell. The sulphuric acid liberated by this is re-used for further stripping. [62,63] (see Section 8.9.4.2)

The options open to users by making use of selectivity and specificity of reagents in conjunction with liquid-liquid extraction have been reviewed. The terms "capacity of the organic phase", "phase ratio", "reaction time" as well as the actual hardware used, are best described together. To the extent this process is used for recycling in the metal industry, the mixer-settler type system has found most favour. The organic and aqueous phases are first fed into a smaller mixing unit where they are mixed as intimately as possible using a stirrer mechanism. Transfer of the metal to the organic phase occurs within a few minutes – equilibrium is usually attained after three minutes [49,56]. The two-phase mixture then passes to a larger settler section which allows the two phases, after separation, to be separately drained. Because the process is an equilibrium one, some metal remains in the aqueous phase and this is recycled for further organic extraction until the law of diminishing returns is evident. In practice, the requisite number of extraction units are linked in series as a cascade with the metal-containing aqueous solution and the organic phase flowing counter-currently. Fig. 7.25 shows such a set of mixers and settlers. In hardware terms, many designs are used ranging from truly continuous processes on the one hand, to units which act alternately as mixers and settlers within a single vertical column. Muller has surveyed such designs [50].

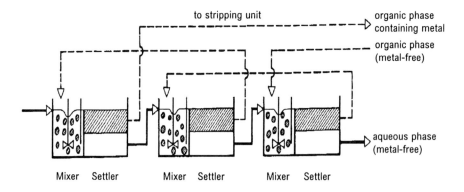

Fig. 7.25   Schematic representation of a three-stage cascade of mixer-settlers for metal recovery by liquid-liquid extraction.

The question of how many extraction stages are best, the capacity of the organic phase and the optimum ratio of phases is best addressed at a semi-empirical level using McCann-Thiele diagrams.

The first step is to determine the equilibrium distribution, starting with aqueous solutions of various metal-ion concentrations and so obtaining a set of extraction isotherms. Then, using the metal ion concentration likely to occur in practice in the aqueous phase, a perpendicular is dropped and also a horizontal line. This should be just below the maximum capacity of the organic reagent. The intersections with the axes are used to draw the "working line". The end-point, E, is determined by the residual concentration in the raffinate and the metal content of the organic phase after stripping. The number of possible stages between the extraction isotherm and the working line A-E also determines the number of mixer-settler stages (see Fig. 7.26) A similar approach is used to determine the optimum number of stripping stages. The gradient of the working line corresponds, in line with Equation 7.10) the existing ratio of flows in the organic & aqueous phases [49,54,56,64].

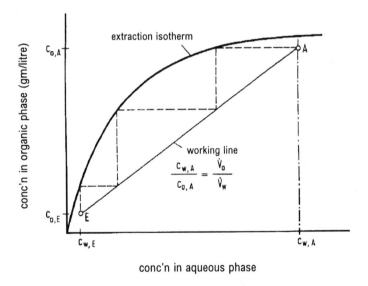

Fig. 7.26 *McCabe-Thiele diagram to determine number of extractive stages for removing species from an aqueous solution.*

$c_{W,A}$ – concentration of substance in the aqueous solution to be treated, $c_{W,E}$ – residual concentration in aqueous solution after extraction. $c_{O,A}$ – useful loading of organic phase, $c_{O,E}$ – residual concentration in organic phase after stripping. The correct mass ratio's of organic phase (V_O) to aqueous phase (V_W) can be derived from the gradient of the working line.

Fig. 7.27 shows a plant for metal recovery by liquid-liquid extraction.

Fig. 7.27   Liquid-liquid extraction plant for metal recovery from plating shop effluent. Left. settler plant for phase separation. (Guetling GmbH)

Extraction by anion exchange

For extraction of anionic species, primary to quaternary amines with aliphatic chain lengths of 8 to 16 carbon atoms are used. Typical examples are tri-iso-octylamine and tri-n-octyl-n-decylamine [53]. In general, they have better selectivity (which is also a function of pH and concentration in the aqueous phase) than their cationic counterparts, with the exception of copper specific types.

There is growing interest in this method for recovery of metals from their oxyanions (chromates, vanadates, molybdates etc) and from their inorganic complexes, such as chloro-complexes. Chromates can be extracted as their sodium salts using long-chain amines (not less than 18 carbon atoms) at 0.1 to 0.2 M in xylene followed by stripping to recovery the sodium dichromate. This can be produced at concentrations up to 350 gm/litre. The content of Cr(III) remains below 150 mg/litre [65].

In Section 7.2.2.1, regeneration of HCl iron pickle liquor using a strongly-basic anion exchanger was described. The same process can be carried out, as described by Wiedemann [66], using a liquid anion exchanger. In this work, a secondary amine loaded with chloride ions was used in 20% concentration in xylene. Iron in the 3-valent state is extracted with chloride as $[FeCl_4]^-$. It can then be stripped out with water to form basic ferric chloride. The regenerated pickle acid should contain at least 150 gm/litre HCl.

The affinity ranking for extraction of metal chloro-complexes using a tertiary amine is: $Zn^{++} > Fe^{+++} > Cu^{++}$ [53]. This agrees with results by the author using a strongly basic anion exchange resin.

Moor & Groenier [67] reported the use of a quarternary amine (methyl-trioctyl-ammonium chloride) in kerosene with tridecyl alcohol for extraction of free cyanide and zinc cyanide complexes from plating effluent. The process is virtually quantitative and stripping is carried out using dilute soda solution.

Yet another process, in which no actual ion exchange occurs, but where an addition compound is formed with the reagent, is described by Muehlberg et al. [67]. This allows free nitric and hydrofluoric acids to be recovered as well as the majority of their salts present in the pickle liquor. The process is schematically shown in Fig. 7.28. The spent pickle acid is taken from a pickle storage tank (1) or reservoir (2) and sulphuric acid is added in order to liberate nitric and hydrofluoric acids from their salts. It then passes through a heat-exchanger (4) to the top of the column (5) for liquid-liquid extraction. A counter-current flow is set up using tributyl phosphate as a reagent and this forms adducts with both acids. At the base of the column, a mixture of sulphuric acid and metal sulphates is withdrawn and passed to a neutralisation stage (6). The organic phase then passes from the column to a reservoir (7) and from there to the base of the stripping column (8) which is fed at the top with water (9). This takes up the acid while the organic phase is returned to reservoir (10) for re-use. The regenerate is passed through an adsorption column (11) to remove organic residues before passing to tank (12) for re-use in the pickle bath.

These examples are intended to indicate that, within the metal working and finishing industries, many potential areas of application remain for the method. Finally, some comments on recent developments which bring together extraction and stripping in a single stage. Marr & Draxler [68], describe what they call a liquid membrane process.

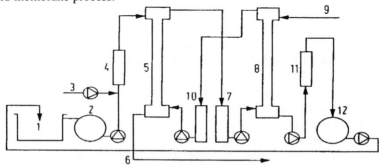

Fig. 7.28 Schematic of liquid-liquid extraction process for regeneration of mixed nitric-hydrofluoric acid pickle baths [67].

1. pickling plant. 2. storage for spent pickle liquor. 3. sulphuric acid dosing. 4 heat-exchanger. 5. extraction column. 6. to neutralisation stage. 7. intermediate hold-tank for charged organic phase. 8. stripping column. 9. water feed. 10. storage for organic phase. 11. adsorption column for organic residues. 12. storage for regenerated pickle acid.

The aqueous stripping solution is dispersed as a finely-divided emulsion in the organic phase and this emulsion is contacted with the emulsion. The organic phase works to separate the two aqueous phases from one another and thus behaves as if it were a membrane between them. Mass transport takes place from the effluent to the organic phase and then, by diffusion, to the internal stripper agent. In a first settler, the emulsion is separated from the purified effluent, A subsequent separation of the organic phase and stripper solution then takes place under the influence of a high frequency electrostatic field. The stripping solution is processed and the organic phase re-used. In an example of the method, nickel is recovered from effluent.

Another different application but still making use of a liquid membrane has been described by Kim [69] and by Schneider & Rintelen [70]. Here the organic phase is immobilised in the pores of a microporous polypropylene and PVDF capillary membrane. The metals are in a weakly acid solution on one side of the membrane, a strongly acid stripping solution is on the other side. The metal ions are complexed by the organic solution and diffuse through to the stripper solution. Meanwhile protons diffuse in the opposite direction to the metal-containing solution. The driving force for the overall process is the pH gradient across the membrane. Since the organic phase has its own, albeit very small, solubility, this too can dissolve. This drawback is eliminated if the membrane modules are used such that the metal-containing solution is held on one side while on the other side, the same organic phase as is immobilised in the pores is pumped counter-currently.

Since the stripping process is inherently faster than the extraction, Scheider & Rintelen propose as optimum, a process in which the metal ion containing aqueous phase is passed over the membrane modules to a mixture of organic phase and stripper solution. Metal ions pass through the organic phase into the stripper. The mixture is flushed out and separated in a settler. The aqueous phase is then treated in the normal way, the organic phase is recirculated for re-use.

7.4   PRECIPITATION & CRYSTALLISATION PROCESSES

Precipitation and crystallisation processes are widely used in recycling either to recover useful materials or to regenerate process solutions by removal of impurities or both together.

In physicochemical terms, crystallisation is classified as a thermal process [71,72], since thermal energy is consumed or formed in this process. Other means for removing solids from solution, such as supersaturation, nucleations and crystal growth, form part of the same process and depend on solubility and solubility product. Once the solubility product is exceeded, formation of the solid phase will, in theory, result. The various aspects are treated as follows, using the same classification as Kammel & Lieber [73]. The following distinction between crystallisation and precipitation can be made:

- precipitation involves addition of a species which forms an insoluble product with the reagent

- crystallisation involves lowering the solubility of the species, either by changing the temperature or by addition of some other species sharing a common ion with the first compound. Alternatively, water can be removed from solution.

For a discussion of solubility when ionic activities are changed, see Section 3.7.3 and for kinetics of formation of insoluble particulates, see Section 3.7.4.1.

7.4.1   Precipitation processes

In terms of recovery strategies, precipitation usually involves hydroxides which lend themselves best to metal segregation. Hydroxides can be dissolved in sulphuric acid and can be worked up by further addition of metal sulphates in acid, to raise the total concentration. Alternatively, electrolytic metal recovery can be carried out. Metal recovery from chemical rinse tanks following metal ion containing process solutions gives readily separable solutions. Recovery processes then proceed via the hydroxide or the basic carbonate. In some cases, such as copper, non-cyanide solutions are reduced to form the hydrated copper oxide (see Section 6.3.5).

Recovery of copper or nickel by the classical hydroxide precipitation method was described by the author in 1972 [74]. The combined recovery of metals and electrolyte regeneration was described by Huss et al. [75,76]. Nickel-containing rinsewater from a cascade rinse is passed to a continuously operating neutralising plant using NaOH. The hydroxide sludges are settled and dewatered in a filter-press to a 20% dry-weight sludge. The residues are dissolved in sulphuric acid to give a solution of pH > 1.8. This is passed to the anolyte chamber of a diaphragm-divided electrolysis cell. Hydrogen evolution in the catholyte compartment results in a solution of pH 2 to 3 which can be used for nickel precipitation. The nickel-depleted catholyte is passed through a filter to the bright nickel plating bath (see Fig. 7.29).

As a result of the precipitation of the nickel, the organic constituents of the bright nickel bath (brighteners, levelling agents, surfactants as well as their chemical and/or anodic oxidation products) pass with the filtrate into the effluent stream, and the bright nickel bath is regenerated. Based on COD values, over 96% of the organics could be removed and hardly any organics were adsorbed onto the nickel hydroxide. It is well known that anodic current efficiencies are larger than cathodic values in nickel plating solutions so that, over time, nickel concentration in solution builds up. Older practice was to remove some of the electrolyte and to replace this with fresh solution containing only the background electrolyte (e.g. acid etc). As described in the literature, metallic nickel was then electrodeposited using a recovery cell. In cases where regeneration via the rinsewater (removal of organics & excess nickel) did not serve, extra background electrolyte would be added. In some cases, the metallic nickel recovered in this way can be recycled as anode metal. The specially designed cell used for this, is further described in Section 7.6.2.2.2 [76]. Fig. 7.30 shows the layout of a complete plant. Variants of the method, virtually identical but not working with nickel, have been described elsewhere [77,78].

Where copper occurs as a contaminant in bright nickel plating baths, the deposit is dark, especially at low current densities (with 2 to 3 mg/litre Cu). At higher Cu concentrations, the corrosion resistance of the nickel is impaired.

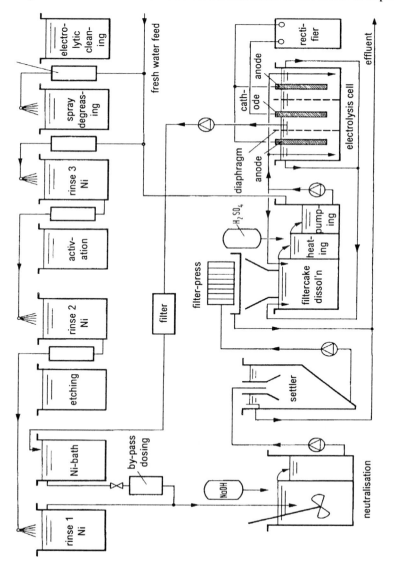

Fig. 7.29   Schematic of a plant for regeneration of bright nickel plating solutions by precipitation of nickel hydroxide & electrolytic recovery of excess nickel (RECON GmbH) [75].

Nawafune *et al.* [79] studied the use of complexing reagents to remove copper impurities from Watts nickel baths. These included compounds such as 2-mercaptobenzothiol and 2-hydroxyphenylmethylene-azanol which can be used to lower copper concentrations below 0.1 mg/litre. These complexants must be removed from solution using activated carbon before the solution is re-used. It is also possible to use specially-treated activated carbon previously impregnated with compounds such as the above.

Fig. 7.30   Plant for regeneration of bright nickel plating baths by hydroxide precipitation with electrolytic recovery of metallic nickel (see Fig. 7.29)

Schoeller *et al.* [80] described a process for removal of divalent metals from effluent, which they named "grain crystallisation" However in view of the definitions offered earlier in this section, their designation should be qualified since it involves addition of soda. In the main, the specially created hydrodynamic environment encourages rapid crystal growth – to this extent the process can be considered as crystallisation. The reactor used is a vertical cylinder, partly filled with filter sand (0.6 mm). It is fed from the bottom and, as long as necessary, operated in a recirculatory mode with concentrated soda liquor being dosed in from below. The partly fluidised sand bed acts to nucleate the basic and simple carbonates which form and grow to a 1 to 1.5 mm size. These then break away from the sand particles on which they first nucleated and are washed out. Typical flow-rates for a 4 metre high column would be 75 cubic metre/hr. In cases where a filter is fitted downstream, metal concentrations (except for Ni) less than 1 mg /litre can be reached. In cases where several metals are present, a slightly higher figure may be found. The crystallites, after being washed out, are treated with HCl or sulphuric acid and the resulting metal salts can then be re-used. As with the basic carbonates, metal sulphides and phosphates can be formed in the same way with sodium sulphide or phosphate being used instead of soda for dosing.

In one plant using large volumes of cyanide silver solution, around 1.5 tonnes of the metal are recovered annually from concentrate and semi-concentrate solutions by precipitation of the sparingly soluble silver cyanide. The metal, complexed as the dicyanoargentate, is decomposed in closed vessels using nitric acid at pH ca. 1, as described by Schlegel [81]. This causes precipitation of the insoluble cyanide as shown in the equation below, down to 0.5 mg/litre

$$[Ag(CN)_2]^- + HNO_3 \rightleftharpoons AgCN + HCN + NO_3^- \qquad\qquad 7.13$$

The precipitate is washed with slighty acidified water (nitric acid at pH 2), then washed to a neutral value with alkaline water. Suction filtration results in a cake with 40% solids. The filtrate is used to treat further silver effluent. HCN formed (see equation) is sucked off, washed with alkali and the resulting NaCN solution can then be re-used.

A similar procedure, described by Peuser [82] is used to recover other metals from cyanides by precipitation as the insoluble cyanide. To the rinsewater is added a metal salt which is identical with the cyano-complex already in the rinsewater. A poorly soluble metal cyanide compound is formed, e.g.:

$$[Zn(CN)_4]^{--} + Zn^{++} \rightleftharpoons 2\,Zn(CN)_2 \qquad\qquad 7.14$$

(See Table 3.3 for solubility values of this type of species)

7.4.2   Crystallisation processes

Perhaps the oldest of all recycling methods in the metal-working industries are those relating to regeneration of pickle liquors for iron and steel, especially for sulphuric acid pickle baths. Unsurprisingly, the solubility of many other metal sulphates is strongly temperature dependant as seen in Fig. 7.31. Apart from cadmium, their solubility increases with temperature until a maximum is reached, when it decreases once more. The implication is that saturation can result both from heating and cooling. Whereas at lower temperatures, the deposit includes significant quantities of water of crystallisation, above the maximum, monohydrates are usually formed. Again cadmium is exceptional – this crystallises with no water of crystallisation [73].

Sulphuric acid pickling baths can contain up to 30% of the acid. They are operated at up to 70°C. The pickling results in formation of ferrous sulphate and the consumption of an equivalent amount of free acid as a result. If the pickling action is to be maintained, fresh acid must be added or spent liquor replaced partially with fresh acid.

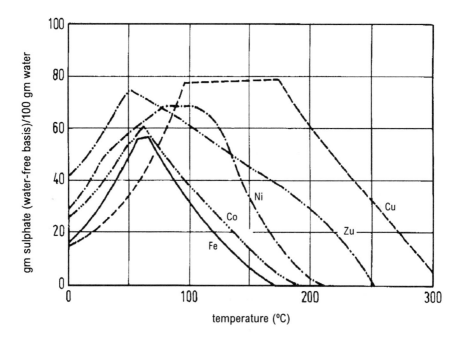

Fig. 7.31   Solubility of metal sulphates as function of temperature [73].

In the first case, the point is reached when no further addition can usefully be made and the entire bath must be discarded. In the second case, the same volume of pickle liquor has likewise to be rejected, except over a period of time in line with the partial replacement of solution. Previously, solutions from either approach would have passed to a traditional effluent treatment plant. However because of the volume of sludge formed as well as the burden on neutral salts discharged, such practice is deemed unacceptable in many locations. The solubility-temperature characteristics of ferrous sulphate can be put to use, as shown in great detail by Dembeck & Meuthen [83], as a means of regenerating the solution. This can be seen more clearly from Fig. 7.32 than Fig. 7.31. It also shows how the solubility of the iron sulphate depends on free sulphuric acid concentration and that three variants of the hydrated salt exist with seven (A), four (B) and one (C) molecules of water of crystallisation. The intermediate form (B) has no practical application. A is the most significant although some plants have been built based on formation of the monohydrate. Iron sulphate concentration can be reduced not just by cooling but also by increasing the sulphuric acid concentration. The latter calls for making good an unavoidable sulphuric acid loss and favours a regeneration cycle. The Figure also shows that increasing free acid from 25 to 30% brings no significant added benefit in relation to cooling.

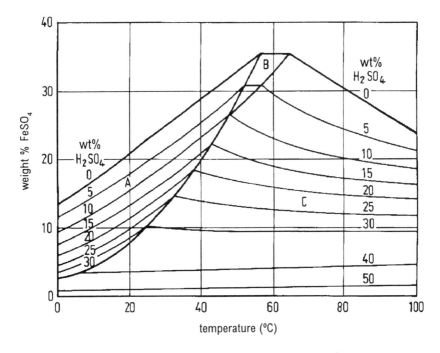

Fig. 7.32   *Solubility of iron (II) sulphate as function of temperature and sulphuric acid concentration [83].*

A – FeSO$_4$ · 7H$_2$O   B – FeSO$_4$ · 4H$_2$O   C – FeSO$_4$ · H$_2$O

Several methods are available for crystallising iron (II) sulphate, based on air, water, brine or vacuum cooling. The hardware configuration differs considerably as between these (see [84,85,86,87,88]).

HCl pickle liquors too can be regenerated by crystallising out iron (II) chloride. These are normally operated with 100 to 200 gm/litre HCl. In their spent form, they typically contain 140 gm/litre iron and 20 to 30 gm/litre HCl [85]. As in the case of sulphate baths, solubility of iron chloride depends on temperature and free acid concentration. However in contrast to sulphuric acid baths, solubility is more sensitive to free acid content, less sensitive to temperature, as seen in Fig. 7.33. Of the various hydration states (6 (A), 4 (B) and 2 (C) Moles of water hydrate), only the tetrahydrate is of industrial importance. To regenerate the bath, fresh HCl is added. However because of the resulting volume increase, only modest amounts can be added. To avoid this difficulty, gaseous HCl is used instead [89,85,83]. For really large scale operations, thermal treatment methods are used (see Section 7.8.1).

Not to be overlooked here, is the regeneration of mixed acid pickling baths containing HF, as used for stainless steels in which fluorides are crystallised out [87,90]. The pickle liquor is filtered and reduced by evaporation to around half its volume. The evaporate is condensed, the concentrate is fed to a crystalliser where metal fluoride crystals drop out. The regenerated liquor is diluted back to its former concentration using the condensate and returned for re-use. Around 90% of unreacted HCl and 55% of unreacted HF can be recovered in this way.

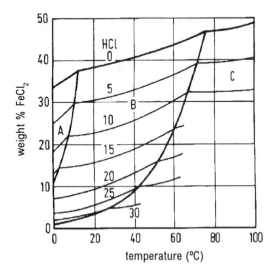

Fig. 7.33 Solubility of Fe(II) chloride as function of temperature and HCl concentration [83].

Sulphuric acid etch baths containing copper sulphate as used in the printed circuit board industry can also be regenerated by crystallisation. The simplest method involves addition of hydrogen peroxide. The spent bath is topped up with lots of 25 litre concentrated sulphuric acid and 10 litres of 35% peroxide and made up to 100 litres, with at least 30 gm/litre copper. The bath operates best at 45°C. When the copper content reaches 70 gm/litre (280 gm/litre of copper sulphate pentahydrate), the etch must be regenerated. By cooling to room temperature, copper solubility will be not more than 40 gm/litre. Crystallisation is allowed to take place overnight [91]. Replacement of lost sulphuric acid further promotes crystallisation (see Fig. 7.34) [92]. Schlitter has described the whole process [93] using a crystalliser as shown in Fig. 7.36. The solution is chilled from its 52°C operating temperature down to 15°C. The crystal nucleation process can be impeded by certain effects while too high an acid content results in formation of very fine crystals which are hard to remove.

In a variant of the etch bath based on sulphuric acid, ammonium sulphate and hydrogen peroxide which operates in the range 30 to 40°C, chilling to 5°C results in crystallisation of the double salt $(NH_4)_2SO_4 \cdot CuSO_4 \cdot 6H_2O$ [94,95].

Cyanide based electrolytes tend to become more concentrated in zinc ions, for the same reason as with nickel plating baths, namely the anodic process is more efficient than the cathodic one. For this reason, some of the zinc anodes should be replaced by those of iron, usually in the ration Zn:Fe = 3:1 [96].

However the downside is that at the iron anodes (which are inert in this medium) cyanide is anodically oxidised to carbonate and the build up of the latter species is detrimental to cathode current efficiency. By chilling, the carbonate content can be maintained at 60 gm/litre so that this problem is held at bay. Fischer [97] quotes a rate of carbonate formation equivalent to 0.15 to 0.2 gm/Amp.hr using 20% insoluble iron anodes and this equates to 0.4 to 0.54 gm of the decahydrated sodium carbonate. It should be noted that the solubility of soda in cyanide solutions is considerably smaller than in water (see Fig. 7.35).

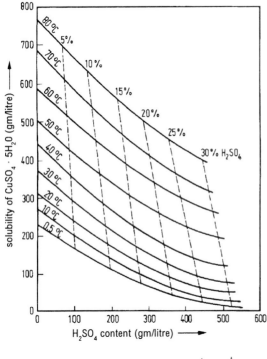

Fig. 7.34 Solubility of copper sulphate pentahydrate as function of sulphuric acid concentration and temperature. [92].

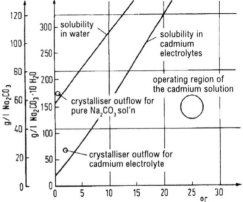

Fig. 7.35 Solubility of soda in water or in cadmium cyanide electrolytes as function of temperature [97].

The sodium carbonate which crystallises out, contaminated with cyanide, can be re-used, for example to make up electrolytic degreasing solutions [98]. Fig. 7.36 shows a chiller crystalliser of a type widely used in plating shops. Other designs are described by Matz [71,99], Schliephake [72] and Woelk & Hofmann [100].

Berry [101] has described a process for regeneration of alkaline aluminium etch baths. It resembles the Bayer process for manufacture of alumina from bauxite. The alkaline bath is supersaturated with aluminates, which hydrolyse to the hydroxide, so liberating NaOH. The aluminium hydroxide is removed by filtration and the filtrate is re-used in the etching process.

$$NaAlO_2 + 2\,H_2O \rightleftharpoons Al(OH)_3 + NaOH \qquad\qquad 7.12$$

It has been reported that the rate of this reaction is unaffected by the concentration of the etch bath and that the reaction shown above is based on an Al:NaOH ratio of 0.36 to 0.5. In some cases, the reaction needs to be initiated by dosing with finely-divided aluminium hydroxide although, as reported by Guelbas [102] at the ratios mentioned above, it is self-starting. The process is a recirculatory one, passing through a settling unit and with dosing to initiate it as required. The benefits are a reduced requirement for NaOH and less solid waste requiring disposal as well as giving an increased constancy of etch rate and thus better quality consistency.

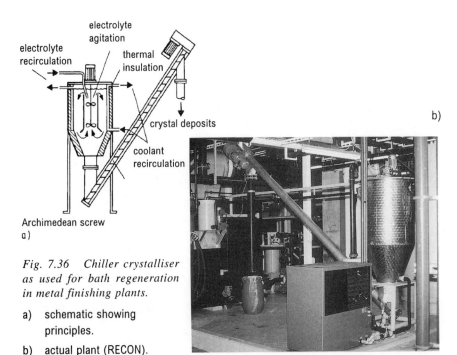

Fig. 7.36   *Chiller crystalliser as used for bath regeneration in metal finishing plants.*

a)   schematic showing principles.

b)   actual plant (RECON).

7.5   MEMBRANE FILTRATION

7.5.1   Theory

Membrane processes used or capable of being used in effluent and recycling include:

- microfiltration
- ultrafiltration
- reverse osmosis (R.O.).

They can be ranked in the same order in terms of their ability to separate finely dispersed, but still macroscopic particles, through colloids and macromolecules down to molecules and ions. Fig. 7.37 summarises the scope of these processes. Within the category "membrane processes" one can include dialysis and, in some instances, electrodialysis. The first of these, in terms of its membrane, can be grouped with ultrafiltration (although it operates without high pressure and uses chemical potential gradients to drive the diffusion) while the latter is more akin to electrochemical processes and is therefore discussed in Section 7.6.3. Processes grouped in this section share the common feature that they are driven by pressure differences. Such pressures vary as a function of the particle size over the range 10^{-8} to 10^{-2} cm. In the case of micro- and ultrafiltration, the pressures are low, and determined mainly by pressure drops arising as particles pass through various forms of membrane (pipe, tube, capillary, between films). Typical pressure drops are 0.5 to 5 bar.

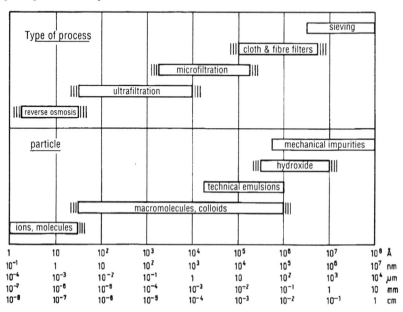

Fig. 7.37 Separation processes as function of size of particle to be separated

In the case of R.O., additional pressure is required to overcome osmotic forces, in this case typically 40 to 100 bar. At the interface between ultrafiltration and R.O., it is hard to distinguish between the two. Likewise there is no major distinction between   micro- and ultrafiltration except that the latter method handles smaller particle sizes, whereas the former copes with finely dispersed, even macroscopic particles.

In order to understand the mechanisms in membrane filtration, two models have been used, embodying the concept of the "ideal pore membrane" and the "ideal solubilising membrane".

The "ideal pore membrane" acts selectively because of the range of pore diameters. Particles too large in diameter cannot pass through and knowing the pore diameter range allows the performance of the membrane to be defined. Pore size distribution determines the "sharpness" of the size range passed or rejected. Membranes with a very narrow range of pore diameters display the sharpest separation behaviour. In other cases, where they have a wider range of pore size distributions, they are known as "diffuse" separating media. Expressed in another way, this type of membrane behaves just like a sieve, albeit (for micro- and ultrafiltration) in the range 10^{-3} to 10^{-6} cm.

The mass flow Q_i for a component i at concentration c_i (in the solution in the pore), is driven by pressure difference and depends on the diffusion constant of the solution molecules. Merken [103] suggests the following formula to describe mass flow through a porous membrane (see also Strathman [104,105]) Saier [106]).

$$Q_i = A \cdot k_i \cdot c_i \cdot \Delta p - D_i \frac{dc_i}{dx} \qquad\qquad 7.13$$

k_i is the permeability coefficient which describes the combined behaviour of membrane and liquid flow through the pores.

$$k_i = \frac{w \cdot r^2}{8\,\eta \cdot \lambda \cdot \Delta x} \qquad\qquad 7.14$$

from which we obtain:

$$Q_i = A \left(\frac{w \cdot r^2 \cdot c_i}{8\,\eta \cdot \lambda \cdot \Delta x} \cdot \Delta p + D_i \frac{dc_i}{dx} \right) \qquad\qquad 7.15$$

where A is the membrane area (square cm), w is membrane porosity (%), r = pore radius (cm), c_i the concentration of component i in the pore (gm/cubic cm); η a correction factor for pore length; λ = membrane thickness (cm), Δp = hydrostatic pressure drop (gm/square cm.), D_i diffusion coefficient for species i (cm$^2 \cdot$ sec^{-1}) and dc_i/dx = concentration gradient across the thickness of the membrane.

In the case of ultrafiltration, it is the first term (pressure difference) which mainly affects convective flow. The second term which embodies the mass-flow rate as affected by concentration gradient is the dominant term in dialysis.

The idealised solubilising membrane is based on the concept of a homogeneous plastic film, whose selectivity, in the way it determines the various diffusion coefficients and concentrations, is governed by the chemistry of the resin matrix. Since one is dealing with aqueous solutions, hydrophilic polymers are used for the membrane. These take up large amounts of water. By contrast, the salt uptake should be as small as possible and this is achieved where the partition coefficient for the salt is as small as possible. Solubilising membranes come closest to those types used in Reverse Osmosis plants and, as this implies, they are used for separation of particles less than 10^{-6} cm in diameter.

The mechanism of transport through a non-porous membrane is determined by other criteria, as described above and these are embodied in the equation derived by Lonsdale *et al.* [107] (see also [104-106]):

$$Q_i = A \cdot \frac{\alpha_i D_i c_i}{\Delta x} \cdot \left(V_i \cdot \Delta p - RT \frac{\Delta c_i}{c_i} \right) \qquad 7.16$$

where a is the partition coefficient of component i between the membrane and the solution, c_i is the concentration of component i in solution (gm/litre); V_i is the partial molar volume of component i (litres), R the gas constant (0.08206 litre bar K^{-1}); π is the absolute temperature (K); Δc_i the concentration gradient across the membrane (gm/litre/unit length). For other parameters see Equation 7.15.

The brackets contain a term which is pressure and concentration dependent. Considering a portion of solution with water as solvent (index = 1), then the concentration of c_1 is high. It follows from Equation 7.16 that for transport of water molecules across the membrane, it is the pressure-dependent term which dominates.

Dissolved low molecular weight molecules or ionic species set up an osmotic pressure as a function of the molar concentration which, assuming ideal gas behaviour, can be represented as follows:

$$p = \frac{n}{V} RT \cdots \pi = c RT \quad [bar] \qquad 7.17$$

where p = pressure (bar), n = number of molecules, V = volume of gas (litres), R = gas constant (as above); π = osmotic pressure (bar); c = molarity (Mole/litre).

The concentration dependent term in Equation 7.16 can be represented by the following simplified equation, assumed dilute solution behaviour and use of a solubilising membrane as found in Reverse Osmosis, which depends on osmotic pressure.

$$\Delta\pi \cdot V_i = RT \, \frac{\Delta c_i}{c_i} \qquad\qquad 7.18$$

This yields an expression for the volume flowrate for this type of membrane:

$$Q_i = A \cdot \frac{\alpha_i \cdot D_i \cdot c_i \cdot V_i}{\Delta x} \cdot (\Delta p - \Delta\pi) \qquad\qquad 7.19$$

The pressure difference set up by osmotic action must be overcome by at least the same value of hydraulic pressure. If, instead, the flow rate of dissolved species (subscript s) is considered, then for dilute solutions the pressure term $V_i \cdot p$ in Equation 7.16 can be neglected and one obtains instead, for component s, an expression based on Fick's first Law for the flowrate:

$$Q_s = A \cdot \alpha_s \cdot D_s \, \frac{\Delta c_s}{\Delta x} \qquad\qquad 7.20$$

It will be seen that at higher pressure differences, the rate of transport of dissolved species becomes approximately a constant. The fact of diffusion controlled transport also explains (see below) why a 100% effective retention is not possible. A few comments on the nature of osmotic pressure might be helpfully added here.

If one considers a hypothetical barrier with water on one side and solution of a species in water on the other, then – on account of Brownian motion – there will be a transfer of the solute molecules into the water, based on their temperature-dependent kinetic energy $\frac{1}{2} mv^2$. Given sufficient time, the concentrations on either side of the membrane will equalise, as predicted by Fick's First Equation (see Equations 6.6 and 6.7). If the hypothetical membrane is replaced with a semi-permeable membrane which permits transport of solvent while retaining solute molecules, then the resulting diffusion is known as osmosis. Salt solution on one side of the membrane is diluted by uptake of water molecules from the other side of the membrane until a given osmotic pressure results (see Figs. 7.38a and b). This is known as equilibrium osmotic pressure.

If the process is reversed in that osmotic pressure is overcome by externally opposed hydraulic pressure, then pure solvent (water) will form on one side of the membrane and concentrated salt solution will build up at the other side. This process (Fig. 7.38c) is known as Reverse Osmosis (R.O.). The equilibrium osmotic pressure is given by Equation 7.17. At a given temperature, osmotic pressure depends only on molarity of the solution and is directly proportional to it. In cases where the same molarities obtain, in the one case for low molecular weight, in the second case for high molecular weight species, there will be a high osmotic pressure in the former case, no significant such pressure in the

latter case. This means that ultrafiltration and microfiltration can be operated
even using low pressures since the pressure serves mainly to compensate for
pressure losses.

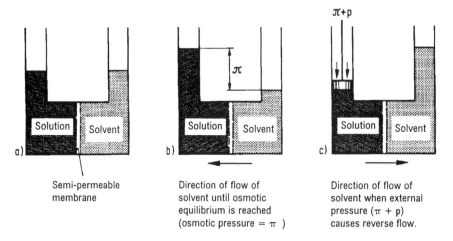

Fig. 7.38 Schematic representation of the basis of reverse osmosis.
a) initial state b) osmosis c) reverse omosis

In the case of compounds which dissociate in solution, this can be seen in terms
of its effect on the osmotic pressure. Thus a solution containing 1M NaCl has an
osmotic pressure not of 22.4 bar but almost twice this value reflecting dissociation of a molecule of NaCl into its two constituent ions. Such effects must be
applied to Equation 7.17 using factors $(v^+ + v^-)$. In cases where dissociation is
incomplete, this is multiplied by the osmotic coefficient for:

$$\pi = \left(v^+ + v^-\right) \cdot f_0 \cdot c\, RT \qquad\qquad 7.21$$

The following example indicates the order of magnitude of osmotic pressures. A
10% solution of a solute of molecular weight 10,000 is only 0.01 expressed in
molarity and sets up, if undissociated, an osmotic pressure of 0.25 bar. A 10%
NaCl solution, by contrast, would be 1.71M and, assuming complete dissociation, to produce 3.42 Mol-ions, would give rise to an osmotic pressure of 83.5
Bar.

A key parameter for membranes is their retention factor R for the dissolved
species. This is expressed below in percent terms:

$$R = \left(1 = \frac{c}{c_0}\right) \cdot 100 \quad [\%] \qquad\qquad 7.22$$

where c_0 is the concentration in the feed, c that of the same solute in the filtrate.

In the case of porous membranes, this is pressure independent so long as no covering layer has been formed as a result of concentration polarisation. In the latter case, these then behave as solubilising membranes in that the properties of the covering layer determine the overall behaviour, indeed the covering layer acts as a membrane in its own right. Outside this case though, the retention factor of pore-type membranes is primarily a function of pore size and its relationship to particle size encountered, as well as the molecular weight of the substance being separated. The first of these is a function of the size of the largest pores, the second depends on the width of the pore size distribution spectrum. This last allows a degree of choice as to sharp or broad type of separation behaviour [105,108].

The retention factor for solubilising membranes is initially a function of pressure though as this increases, a limiting value is reached. Since capacity for a given size increases with pressure while the retention factor remains constant, use of higher pressures gives improved separation rates.

In contrast to porous type membranes with which separation is a function of particle size and/or molecular weight, solubilising type membranes can be used to separate substances with similar molecular weight, provided their partition functions as between membrane and solution are sufficiently different. Though throughput increases with pressure for both types of membrane, the retention behaviour is pressure independent in the case of porous membranes while with solubilising types it increases to a given value. Because of this, low excess pressures are used for ultrafiltration, high pressures are used for R.O., in both cases in terms of excess over osmotic pressure. It should be noted that the specific capacity of an ultrafiltration unit is 10 to 100 times greater than that of the same-sized R.O. plant [104]. Parameters such as retention behaviour, separation limits and the sharpness of separation cut-offs can be empirically determined for a given membrane by using model standard compounds in solution.

Avoiding concentration polarisation is specially important in the case of membrane filtration. If the retained particles are not removed, or inadequately removed from the membrane surface by a suitable hydrodynamic regime, their concentration will build up at the diffusion boundary layer. Above a certain value, a concentration gradient will form such that back-diffusion occurs into bulk solution. In the limit, this reaches the same magnitude as the forward flux to the membrane. The result is a steady-state in which no nett filtration action takes place. By increasing the pressure, filtration resumes − but only for a short time after which saturation again occurs at the surface of the membrane and equilibrium with no nett filtration sets in anew.

In the case of solubilising type membranes, concentration polarisation leads to increase in osmotic pressure linked with excess concentration and retention time, these factors being directly proportional to one another. The end-result is a decrease in filter capacity and poorer quality of filtrate. The last is due to molecular diffusion through the membrane since there are no completely semi-permeable membranes of the solubilising type.

For porous membranes, because of the high molecular weight of dissolved species, osmotic pressure plays no significant part in concentration polarisation. However too high a concentration in the diffusion layer can result in the solubility product value being exceeded and in this case, some dissolved substances can then form a deposited layer. Continuing mass transport to the surface, including that due to increase in pressure, can cause thickening of this layer with the rate of back-diffusion into bulk solution too small to offset the process. Such layers can have a resistance greater than that of the membrane on which they form, indeed they can act as membranes in their own right, such membranes having their own characteristics. In such situations, further increase in applied pressure brings no benefit in filtration rates, a constant value being reached [105,109]. Selecting an optimum pressure value is thus of some importance. Filtrate throughput is only linearly proportional to pressure at very low values (p > 1 bar).

If concentration polarisation is to be avoided, the first priority is to avoid build-up of material at or close to the membrane surface and, of equal importance, to maintain a diffusion layer as thin as possible. In practice both these aims are realised by flowing the liquid across the surface of the membrane at high velocity, so causing turbulence to form there. This implies a difference from classical filtration in which the axes of volume flow and filtrate movement are parallel to one another as opposed to the case described here where the one is normal to the other. Hence the name often used to describe the method – crossflow filtration. These general principles are adopted in the method of construction of filter units used industrially. Most experts suggest the cross-flow velocity should be at least 2 metres/sec.

The effect of pressure on filtration throughput has been considered above. Increased temperature is usually beneficial, reflecting decreased solvent viscosity (see Equation 7.14). In the case of R.O., a 2 to 3% increase in filtration rate per degree C has been quoted [110,111]. However temperatures should not exceed 30°C above which value membrane life is affected by accelerated ageing.

7.5.2   Membranes and Modular Units for Membrane Filtration

Whatever type of membrane is used, filtration rate per unit area is inversely proportional to membrane thickness (see Equation 7.19). However a certain minimum thickness is required for mechanical strength. For this reason, it is almost universal practice to use very thin membranes (0.1 to 1.0 μm thick) with a much thicker (100 to 200 μm) mechanical supporting layer. Where both of these are made from the same material, the membrane is known as "asymmetric type". In cases where two different materials are laminated together, the membrane is designated a "composite type" [110]. Thickness has no effect on the retention factor of a membrane.

The following properties are required of a membrane:

- highest possible specific filtration rate.
- best possible retention factor.
- good mechanical stability.
- good resistance to chemical attack (hydrolysis, resistance to cleaning solutions).
- long life.

Membranes are made of the following materials:

- cellulose acetate (-di and -triacetate, -acetobutyrate, -acetopalmitate, – acetomethacrylate etc).
- polyamide.
- polyimide.
- polyacrylonitrite.
- polypropylene.

These apart, Gaefgen [114] cites the use of fluorinated polymers with their superb acid and chemical resistance (e.g. to solvents) and their high temperature operating capability, for use as tubular membranes in ultrafiltration. Also to be mentioned, likewise mainly for ultrafiltration, are the inorganic, so-called "graphite membranes" with excellent resistance to low pH and high temperatures. In fact only the mechanical support structure is made of graphite, the membrane itself being of thin metal oxide layers (e.g. zirconia) [115]. Glass membranes are also used. However most widely used membranes are those of cellulose acetate, polyamide and polyimide. Table 7.2 lists the pH and temperature limits for specific materials. However it must be noted these two variables are not independent of one another – maximum operating temperature decreases as pH increases.

Filter membranes are formed as tubes, capillaries, hollow fibres and flat sheet depending on the technology being used. Fig. 7.39 shows the most widely used designs. Further information is given by Strathmann [104,209], Saier[106], Rautenbach & Rauch [108] and Marquardt [110].

Tube type design

These are based on membranes in tubular form with an inner membrane film surrounded by a porous mechanical support structure and an outer sheath. For low pressure operation (ultra- and micro-filtration), mechanical support is provided by plastic materials. For higher pressure operation (R.O.) stainless steel tubing is used. Typical pipe diameters range from 1 to 2.5 cm. Units of this design can be operated at up to 100 bar and are thus ideal for all membrane processes (see Fig. 7.39a).

Tube bundle design

This design is based on multiple membrane tubes enclosed within a single pressure casing to give improved space-time yields. Even so, compared with other units, the resulting value which is typically 20 to 200 m^2/cubic metre is modest. In order to achieve high turbulence at the membrane surface, high pumping power is called for. The individual tubes (membrane + reinforcing casing) are tightly sealed into the end plates and incorporated in a pressure casing. Liquid feed is from the inner side of the tubes and the permeate passes from inside outwards, passing through the sheathing casing (see Fig. 7.39b). In some designs this is reversed with the membrane tube outside the support tube, fed from outside [110].

In spite of a somewhat higher cost of construction, the tube bundle is less prone to blockage from concentration polarisation and can readily be cleaned.

Capillary design

This offers a higher space-time yield than the tube bundle design. The membrane elements have a diameter of 0,4 to 1.5 μm thus offering an overlap in terms of space-time yield with hollow fibre designs (see below). The capillary tubes are internally fed and can withstand pressures of up to 10 bar. As such, this design is only suited to ultrafiltration. The multiplicity of capillary tubes are not fixed into end plates as in other designs but are cast in place. Compared with tube bundle designs, the mass-transport process and thus avoiding membrane blocking, is less favourable. For this reason, a pre-filtration stage is advised (see Fig. 7.39c).

Hollow fibre design

This design yields the highest space-time yield so far attained – around 20,000 m^2/m^3. Around 100,000 hollow fibres are shaped as a "U" inside a pressure tube such that their open ends can be cast into an end-plate. The fibres have a diameter of 50 to 200 μm with a membrane wall-thickness of 10 to 25 μm and as such, can operate at pressures of 60 to 80 bar. In comparison with capillary type designs, the mass-transport conditions are less favourable and there is significant risk of blockage, especially from high molecular weight compounds. For this reason, the design is not well-suited for ultrafiltration. The main application is for water desalination or for a first-stage salt removal process. Although their specific filtration rate (based on unit membrane area) is lower than some other designs, in terms of space-time yield, they give the highest value of designs surveyed here (see Fig. 7.39d).

It might be noted here that, in the published literature, the terms "capillary design" and "hollow fibre" design are not always used in the same sense. This can give rise to confusion.

Flat plate or sheet design

Several designs exist within this category. They share in common the use of porous support mats or plates to permit liquid removal – these are placed either side of the membrane sheet and liquid to be filtered passes through their interstices (Fig. 7.39f). Two common configurations are the "juke-box" stack of disks fitted with a central permeate tube or else as a plate-and-frame filterpress design. Another widely-used design is the "Swiss roll". A porous drainage sheet is laid between two membranes and the assembly is rolled-up. This is sealed tight at three side by gluing or welding with the fourth open end facing the filtrate outflow pipe (Fig. 7.39e)and inserted in a pressure housing. Spacers are inserted in both the last two designs to keep the membrane sheets apart. Typical space-time yields are 900 $m^2 \cdot m^{-3}$. Because of the somewhat unfavourable mass-transport regime, blockage can be a problem with this design also, and it is not recommended for use with high molecular weight compounds,being mainly used for R.O. A pre-filtration is invariably recommended.

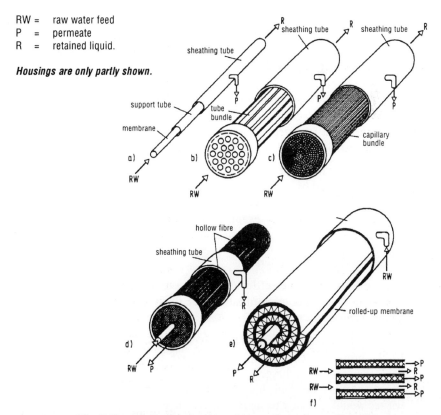

Fig. 7.39   Schematic of various designs of membrane filtration
and the method of operation.
a) tube design b) tube bundle c) capillary d) hollow fibre e) Swiss roll f) flat plate

7.5.3   Reverse osmosis

Designs best suited to R.O. are the tube or tube-bundle types with either inner or outer liquid feed. Swiss roll designs can also be used and, for high pressure operation, hollow-fibre types. As used in the metal-working and – finishing industries, typical ion leakage values are 1 to 10%, caused by the ability of ionic species to permeate solubilising type membranes [118]. In this, ionic radius is the main determining factor, though ionic charge is also significant since the latter can promote interaction with the membrane matrix and the solvent, e.g. via hydrogen bonding [110,117]. Table 7.1 contains data on the retention of various anions and cations, with maximum attainable concentrations also listed [117,118]. Multiply-charged ions are usually easier to separate than single charged types. Ions with the same charge can still be selectively separated from solution because of differences in ionic radius. According to Stelmaszek [119], permeability to a salt is inversely proportional to molecular weight, increases with increasing pressure, decreases with increase in concentration. The permeability of water is greater, the smaller the ionic charge at a given molar concentration (here 0.3M):

$$MeCl_3 < MeCl_2 < MeCl \text{ [119]}$$

This can be mathematically expressed as:

$$P = P_0 \cdot e^{4.75/M} \tag{7.23}$$

where P is the permeation rate of water [$m^3 \ m^{-2} \ s^{-1}$]; P_0 a constant for the solid species [$m^3 \ m^{-2} \ s^{-1}$]; and M the molecular weight of the dissolved species (gm).

The permeability of some species, e.g. metal ions, can be reduced in that larger molecules can form, for example by complexation [118]. The main applications areas where R.O. is cost-effective are:

- concentration of rinsewaters

- final polishing of treated effluent (though use of selective ion exchangers (for ionic species) + microfiltration (for finely dispersed metalliferrous solids) is probably better

- reducing total dissolved salt (TDS) levels (see Section 3.12) R.O. would barely bring effluent concentrations above the 10% level. While this could be legally discharged to sea, in other locations where such discharge was not allowed, use of vacuum evaporators to form solid salt wastes would probably be more cost-effective.

The use of evaporators for concentration of rinsewaters has, in the past, proved the more attractive means of recycling. According to Goetzelmann [120], the same operation using R.O. with a static rinse is equally competitive. At least one plant uses this for treatment of nickel plating rinsewater [121]. By and large, however, use of R.O. to produce concentrations above 10% is not economic. Cartwright [122] reports various applications of R.O. in the USA where zero discharge of effluents and solid wastes are the goals. R.O. alone will not, however, usually permit this and the concentrations produced are not given. However one can be certain that R.O alone could not produce the concentrations used in process baths and the presumption is that further concentration is involved, using evaporators. Incomplete though the description is, some interesting facts emerge. 90 to 97% recovery of solids and additives from bright and sulphamate nickel plating baths is claimed with membrane life of two years using cellulose acetate membranes. In the case of copper sulphate baths, hollow fibre membranes of polyamide and Swiss roll units with cellulose acetate lasted for 1 to 3 years. In the case of sulphuric acid based zinc plating baths, chemicals recovery rate was 87.5%, for cyanide copper or brass plating bath it was 90% with membrane life (cellulose acetate and polyamide) 3 to 4 years. In the case of chromic acid baths using polyamide hollow fibres or Swiss rolls with cellulose acetate, much shorter membrane lives of 1 to 2 months and 6 months respectively, were found. Clearly R.O. is not feasible in these circumstances.

TABLE 7.1:
SUMMARY OF MAXIMUM SEPARATION PERFORMANCE AND DEGREE OF CONCENTRATION POSSIBLE USING R.O. [117,118]

Ions		Max. separation (%)	Max. concentration reached (%)
cations	Na^+	94 ... 96	3 ... 4
	NH_4^+	88 ... 95	3 ... 4
	Ag^+	94 ... 96	–
	Ca^{++}	96 ... 98	–
	Mg^{++}	96 ... 98	–
	Cd^{++}	95 ... 98	8 ... 10
	Fe^{++}	98 ... 99	–
	Ni^{++}	97 ... 99	10 ... 12
	Zn^{++}	97 ... 99	–
	Cu^{++}	96 ... 99	8 ... 10
	Al^{+++}	99*)	5 ... 10
anions	Cl^-	94 ... 95	3 ... 4
	F^-	94 ... 96	3 ... 4
	NO_3^-	93 ... 96	3 ... 4
	CN^-	90 ... 95	4–12 (pH dependent)
	SO_4^{--}	99*)	8 ... 12
	SiO_3^{--}	95 ... 97	–
	CrO_4^{--}	90 ... 98	8 ... 12
	PO_4^{---}	99*)	10 ... 14
	$[Fe(CN)_6]^{---}$	99*)	8 ... 14

*) Presence of other ions may affect max. concentration values if solids form and precipitate.

7.5.4   Ultrafiltration

In contrast to Reverse Osmosis, ultrafiltration offers many possibilities and advantages as compared with alternative processes. In the context of this volume, these include:

- regeneration of alkaline cleaners.

- regeneration of electrophoretic (electrocoat) solutions.

For effluent treatment, the method offers diverse possibilities such as:

- treatment of emulsions and effluent containing emulsions (see Sections 3.13.4.4 and 8.12).

- treatment of effluent containing very finely dispersed solids. This includes effluents after conventional treatment and solids separation which cannot remove the finest particles (Section 4.4.1) or effluent arising from vibratory finishing (Section 8.1.3). Microfiltration is valuable in these cases on account of its greater specific capacity.

- treatment of effluent containing macromolecules such as are found in effluent containing photo-resist residues from the printed circuit board industry (Section 8.9.3) Kock [122a] describes formation of metal complexes from polymeric complexants and their removal using ultrafiltration.

Since this section deals with recycling techniques, the use of ultrafiltration will be discussed only in that context. In principle, all types of cartridge design previously described for R.O. can equally be used for ultrafiltration, provided the correct membrane is used. To maximise space-time yield, capillary type units have been used with success. By contrast, hollow fibre and tube bundle type units with their more restricted flow are prone to concentration polarisation and blocking.

7.5.4.1   Regeneration of alkaline cleaning baths

Degreasing baths can be regenerated for extended life by removal of oil and grease as this builds up. The recovered residues contain up to 50% oily matter content [123-125]. It is possible to operate such that oil or grease concentration in solution is held at 500 to 600 mg/litre and this allows up to 40% of degreasing chemicals to be saved [126]. According to Germscheid [128] and Seyser [126] this allows an up to ten-fold increase in bath life.

The main criteria for successful operation include use of a membrane materials with suitable pH and temperature resistance (Table 7.2), choice of the best cartridge design (see above) and choice of a suitable degreasing solution. Some mechanical means are necessary for removing oil already separated out and other solid impurities. This last is the more important the smaller are the dimensions (fibre diameter, membrane separation) used in the cartridge. Tube units are least prone to blocking – but have the poorest space-time yield. If formation of harmful surface layers and polarisation effects are to be avoided, a flowrate greater than 2 metre/second, up to 5 metre/second is advised.

TABLE 7.2:
pH AND TEMPERATURE RESISTANCE OF
VARIOUS MEMBRANE MATERIALS

membrane material	max. permitted exposure		Reference
	pH value	temp. (°C)	
cellulose acetate	max. 9	max. 40	[116]
	up to 7.2	at 30	[112]
	up to 6	at 40	[112]
polyamide	max. 13.5	max. 57	[116]
	up to 11	at 35	[112]
polyacrylonitrile	max. 10	max. 60	[116]
polysulphone	max. 14	max. 95	[116]
graphit	max. 14	max. 95	[116]
fluorinated polymers	up to 8	at 90	[114]
	up to 10	at 70	
	up to 12	at 50	
	up to 13	at 30	

It is essential to use degreasing formulations which are compatible with ultrafiltration. This applies above all to the most expensive ingredient – the surfactant(s). These make up 0.5 to 15% of solid cleaners in powdered form [127]. They should have the highest possible turbidity value, i.e. their critical micelle concentration (c.m.c) value should be as high as possible. A micelle can be defined as an aggregate large enough to be considered as a colloid, capable of exceeding the separation value for ultrafiltration membranes with a molecular weight of ca 20,000. These are then removed together with emulsified oil and must be replaced with fresh material, so constituting a significant expense. Surfactant monomers with molecular weight 300 to 1000 or so pass through the membrane and are retained in the permeate which is returned to the solution. According to Rossman [127], most suitable are the hydrophilic and also non-ionic types especially those with a high degree of ethoxylation. Less ethoxylated, more oil-soluble and water-dispersible surfactants are less suitable. Like all non-polar molecules, these are preferentially retained by the membrane.

By and large, the presence of surfactants results in lower filtrate volumes than would be the case with pure water by up to 50%. The same is true for dissolved salts and for impurities formed or included in process operation. Thanks to such effects, overall membrane flowrates as low as 10% of the pure water value can result [127,129].

The inorganic component of cleaner baths also has implications for ultrafiltration. Alkali metal silicates have better dispersive action than some other compounds used but are less tolerant of pH changes. At worst, silicic acid can precipitate and block the pores in the membrane [128]. Phosphate based cleaners do not have such problems.

In practice, very variable specific membrane thoughputs are found, depending on membrane material, cartridge design and solution being treated. Typical values quoted range from 20 to 80 litres/hour per metre square. A median value of 40 to 60 is a safe assumption in most cases. Gaefgen [125] quotes relatively high values of 60 to 80 for a tube type unit at 50 to 60°C.

Elsewhere, Schene [124] quotes retention values of 99.7% for oil, 76% (!) for surfactant and for inorganic species 0%, using an inorganic membrane and operating at 70°C.

Fig. 7.40   Ultrafiltration unit with up to 2.5 cubic metre.hr capacity based on the Carbosep process (tube bundles with inorganic membrane). Because of its resistance to pH and elevated temperatures, the method is well-suited to use with degreasing baths (Eisenmann, Boeblingen).

After removal of oily matter and solid impurities, the degreasing solution is normally piped to a a recirculating holder at controlled liquid level, this being then recirculated through the ultrafiltration unit until either the retained solution or the filtrate have reached certain limits. The retained solution is then piped off for disposal with the permeate being returned to the process bath. Certain measures are needed to ensure good membrane life and performance. As time goes on, the filtration throughput rate will decrease, reflecting concentration polarisation at the membrane surface. This can only be reversed by a cleaning operation, usually *in-situ*. Options include mechanical cleaning with a rinsing solution or by using water in the reverse flow direction sense. In the case of tube type units, particles can be deliberately added which will promote a mechanical scrubbing of any deposited layers. Chemical cleaning methods include use of oxidising agents such as hydrogen peroxide or sodium hypochlorite. Increasing the temperature of the circulating rinsewater can also be beneficial [130]. Typical units are shown in Figs. 7.40 and 7.41.

Fig. 7.41   Ultrafiltration unit installed in car body plant for regeneration of degreasing baths. Tube bundle design with 2.5 cubic metre/hr rating (Duerr, Stuttgart).

7.5.4.2   Regeneration of electrophoretic paint baths

There is a major difference between the underlying aim of regenerating an electrocoat paint bath from that involving recycling of degreasing baths. In the latter case, the aim is to remove the useless and valueless organic phase retained by the membrane, while returning the permeate to solution. In the former case, the aim is to remove entrained water and dissolved impurities via the permeate, while the retained solution constitutes the high-value paint bath.

Typical electrocoat solutions contain 1.5 to 2% organic solvents based on the 10 to 15% dry solids weight of the bath. The rate of paint recovery is high, at 90 to 95% [132]. This type of bath consists of special paints (8 to 15%) dispersed in water, made soluble by formation of their salts. Dissociated polymers and pigments are also present. Depending on the type of salt used, such baths are classified as anionic or cationic. In the case of cathodic deposition, hydroxyl ions are discharged, in the anodic version, hydrogen ions are discharged. Following a coagulation process, paint forms at the surface. At the same time, the ionic equivalents are liberated into solution and their concentration builds up accordingly. Were this to become too high, coagulation would occur in bulk solution and this is avoided by continuous removal of such ions by ultrafiltration, allowing a very long bath life if correctly carried out. The permeate contains the ions referred to above and is free from paint. It can be re-used for rinsing, as indicated in Fig. 7.42.

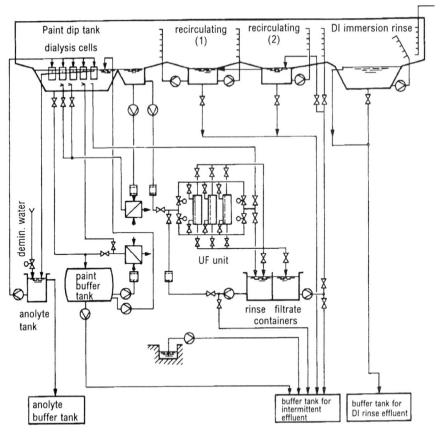

Fig. 7.42   Schematic diagram showing recovery and regeneration of electrophoretic paint baths with re-use of ultrafiltration permeate for rinsing in a cathodic electrocoat plant (Duerr, Stuttgart).

To avoid build-up of salt concentration in the paint bath, the aim should be to establish an equilibrium in which a part of the permeate is discharged to drain. Such a removal of ions entails also a removal of water (of solvation) and this must be replaced with DI water, in some case via the final DI rinse [131].

In the first spray-rinse stage, 1 to 10 litres spray rinse water per square metre of surface (more for complex shaped parts) are used to remove 60 to 90% of the loosely adhering paint and this is returned to the paint tank. The ultrafiltrate is then piped to a cascade-type spray rinse [131].

Kraus [133,134] also describes how, in the case of cathodic systems, organic acids are liberated (e.g. acetic, formic and carboxylics) which are removed from solution by dialysis to avoid a lowering of pH (see Fig. 7.42). Ultrafiltration designs successfully adopted here include flat-plate, "Swiss roll" and tube types. Figs. 7.43 and 7.44 show examples.

Fig. 7.43   Ultra-
filtration unit with two
flat plate modules for
paint recovery
from a cataphoretic
line in a car factory
(E i s e n m a n n ,
Boeblingen).

Fig. 7.44   Ultra-
filtration unit for paint
recovery using Swiss
roll type cell, 6 cubic
metre/hr   rating.
(Duerr, Stuttgart).

7.6   ELECTROLYTIC PROCESSES

Of major importance in any list of recycling strategies are electrolytic processes. These can be classified as:

- cathodic processes, e.g. for metal recovery

- anodic processes, e.g. for regeneration by anodic oxidation of spent oxidants

- electrodialysis, e.g. recovery of bath chemicals from effluent, for partial desalination and for separation of electrically neutral species from charged ions.

All these operate without addition of further reagents, either using electron transfer or (electrodialysis) on the basis of the e.m.f gradient set up. Certain other processes such as electro-osmosis (for sludge dewatering) or electroflotation are only very rarely found and then mainly for simple effluent treatment. Electrophoresis is used only in the laboratory or for preparation of fine chemicals.

In that electrolysis can be used not only for recycling but also for destruction of toxic species, the following section should be read in conjunction with Sections 3.2.2.5 (cyanide destruction) and 3.3.3 (chromate treatment).

7.6.1   Theory

At the core of electrolysis technology lies the concept of an electrochemical reaction. This resembles the classical chemical reaction which takes place homogeneously in solution or in the gas phase. The difference is, however:

a) when written in the formal sense, electrons are transferred to or from the reactants or products

b) these electrons are donated or accepted by the electrodes, and these electrons should themselves be seen as reactants or products.

It will be assumed that readers are familiar with Ohm's Law which, in the main, applies to ionic solutions as it does to solid conductors (but not to the actual processes taking place at electrodes). In Equation 7.24, I = current (Ampere), U = voltage (or potential drop) (Volts) and R = resistance (ohms).

$$I = \frac{U}{R} \quad \text{or} \quad U = I \cdot R \qquad\qquad 7.24$$

A modified form (Equation 7.25) embodies the term $\varkappa$ – specific resistivity, with d = distance between electrodes.

$$U = i \cdot \frac{d}{\varkappa} \qquad\qquad\qquad 7.25$$

In that tables of data exist for values of specific conductivity, the concept is a useful one for predictive purposes. However, at the risk of repetition, an electrolytic cell containing anode and cathode immersed in an electrolyte will not, as a whole, behave ohmically (see below).

Equally important are Faraday's Laws of electrolysis with which readers will be familiar and which describe the relationship between the total charge passed (current x time) and the amount of chemical reaction taking place. The Faraday is 96,487 coulombs (1 coulomb = 1 Amp flowing for 1 second) and this charge will enable 1 gramme equivalent of electrochemical reaction to take place. Faraday's Laws can be summarised as Equation 7.26

$$\frac{dm}{dt} = -\frac{M}{z \cdot F} \cdot I \cdot \gamma \qquad\qquad\qquad 7.26$$

where dm/dt is the rate of mass reacting in time t (gm/second), M = molecular weight (gm), z = valency of the ion or number of electrons required, F = Faraday (96,487 amp-sec), I = current (Amp), γ = current efficiency . This equation is equally valid whether the reaction occurs at the electrode surface or on a membrane.

From a technological point of view, it is desirable that the equipment (in this case a cell) carries the highest possible current. This can also be expressed as current density (amp/unit area of electrode). Attention therefore focusses on the manner in which an electrode reaction (electrochemical reaction) responds to application of a potential to the electrode.

Fig. 7.45 is known as a "polarisation plot". It shows how current (which, thanks to Faraday's Law, can be equated with reaction rate) responds to applied potential. The plots show what occurs when a metal is cathodically deposited. Whenever this is done using an aqueous solution, whether for electroplating or using electrolysis as a means for metal recovery, thought must be given to the major "competitive reaction" which is the cathodic formation of hydrogen from water or hydrogen ions. The competition between metal deposition and hydrogen formation is too complex to be fully treated here. It depends mainly on pH and also the nature of the metal on the cathode surface. Whatever this was originally, after a minute or two of operation, the cathode surface is covered with, and behaves as, the metal being deposited upon it.

Before examining Fig. 7.45 in detail, one further vital point must be made. In electrochemical reactions, the ions or uncharged molecules react at the electrode surface. However first of all, they must reach that surface. A widely held misconception is that ions are attracted to the electrode by electrostatic forces. A moment's thought shows that, though such forces do indeed act in this way, they are not the primary driving force. That uncharged organic molecules, such as brighteners, react at the anode to be oxidatively destroyed or that chromium can be deposited from hexavalent baths where only anionic chromium ions are present, provides an indication that charge attraction alone is not an adequate explanation.

In fact the primary driving force bringing reactants to an electrode is diffusion under a chemical concentration gradient or, expressed another way, driven by differences in chemical potential. This, and in particular Fick's Law of Diffusion, is discussed below. Returning to Fig. 7.45 we see two cases. On the left, a noble metal is deposited. On the right a non-noble metal. In the former case, the solid line represents the increase in deposition rate (i) with voltage (E). The initial portion of the solid line is exponential in shape, becoming linear, then turning over to form a plateau i_G. The significance of this is that increasing voltage brings no commensurate increase in deposition rate. This because the metal ions, driven by diffusion, are then passing at the maximum rate in terms of cell geometry, flowrate etc. After further increase in E, the current increases once more (dashed line) but this increase is almost entirely due to hydrogen evolution and brings no benefits with it. i_G thus represents the highest useful current density for this particular cell design etc. The plateau current is known as "diffusion limiting current" or as "mass transport limited current". E_{Me} and E_{H_2}, though without any theoretical basis, are useful in indicating the approximate potentials at which the react rates of metal and hydrogen deposition respectively attain measurable rates. Much the same picture is seen on the right. In this case however, the metal deposited is less noble than hydrogen and begins to deposit at a more cathodic overpotential that hydrogen evolution. (It should be noted this is a special case of very limited practical interest. In all but more alkaline solutions, hydrogen evolution – because of the high concentration of these ions – fails to reach a limiting value as shown in the Figure.) Recovery of zinc from an alkaline or buffered solution is an example of case b).

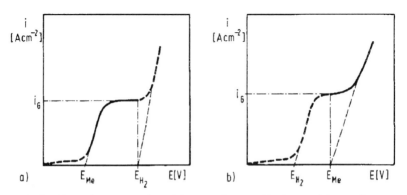

Fig. 7.45   Electrodeposition of a noble (a) and non-noble (b) metal from solution. i_G is the diffusion limiting current for the noble metal (a). In (b) i_G is the diffusion limiting current for hydrogen evolution below which no significant metal deposition occurs.

The foregoing discussion indicates the importance of design and operation of cells such that an adequate diffusion limiting current can be reached. The underlying hydrodynamic phenomena, with their implications, are very similar to those found in membrane cells as discussed in previous sections. As liquid flows past the membrane or electrode surface, there remains a thin layer of electrolyte adjacent to the electrode surface. It is the rate of diffusion across this which determines the magnitude of the limiting current (see also Section 6.3.1). As the flowrate increases, so the thickness of this layer is reduced with corresponding increase in diffusion limiting current. Incidence of turbulent, as opposed to laminar flow is likewise beneficial [135]. The formal embodiment of this is Fick's First Equation (7.27)

$$\frac{dm}{dt} = -D \cdot A \cdot \frac{\Delta c}{\delta} \qquad\qquad 7.27$$

where dm/dt is rate of reaction (gm/second), D = diffusion coefficient ($cm^2 \cdot sec^{-2}$), A = area (cm^2), Δc = concentration gradient across diffusion layer ($gm.cm^{-3}$) δ = diffusion layer thickness (cm). In the diffusion limiting current state, species react at the electrode surface as soon as they reach it, i.e. the steady-state concentration of such species on the surface is zero (which is not otherwise the case). Substituting Faraday's equivalence into Equation 7.27, one obtains 7.28.

$$\frac{M}{zF} \cdot I \cdot \gamma = D \cdot A \cdot \frac{\Delta c}{\delta} \cdot \qquad\qquad 7.28$$

which can be reformulated in terms of limiting current density, with constants lumped into a single parameter as:

$$\frac{I}{A} = i_D = k \cdot \frac{\Delta c}{\delta} \cdot \frac{1}{\gamma} \qquad\qquad 7.29$$

Given values for a particular cell design, this equation can be used to predict at what rate metals can be removed from a solution.

From this point, one can move to a consideration of actual cell design where dozens of different design concepts attempt to maximise diffusion limiting current density and/or space-time yield. A few examples are shown in Fig. 7.46 in which the concept of "three-dimensional" electrodes is also introduced. In all cases, the aim is to minimise δ, maximise electrode surface area and hence space-time yield.

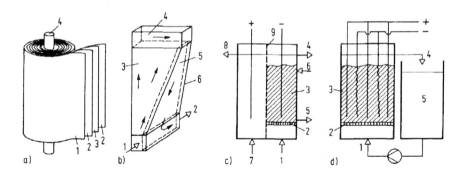

Fig. 7.46   Examples of cell designs suitable for use with dilute solutions.

a) Swiss roll. 1 - anode foil, 2 - textile mesh, 3 - cathode foil, 4 - central electrolyte feed & removal manifold.

b) packed bed cell. 1 - electrolyte feed, 2 - electrolyte exhaust, 3 - particulate cathode, 4 - overflow, 5 - membrane, 6 - anode.

c) fluidised bed cell. 1 - electrolyte feed, 2 - perforated floor, 3 - fluidised bed (particulate cathode), 4 - electrolyte exhaust, 5 - extraction of larger particles (grown by electrodeposition), 6 - addition of new ("seed") particles, 7 - anolyte feed, 8 - anolyte exhaust, 9 - membrane.

d) fluidised bed cell (with non-conducting particles contacting cathode mesh). 1 - electrolyte feed, 2 - perforated floor, 3 - fluidised bed (glass beads), catholyte outflow, 5 - recirculating tank.

The so-called Swiss roll cell developed by Ibl & Robertson (136) consists of a flexible metal anode and cathode foils separated by a plastic mesh as spacer or similar. This "parallel plate" cell is then rolled-up as shown in Fig. 7.46a, with electrolyte feed and exhaust from either side. Conceptually is resembles the membrane filtration unit described in 7.5.2 and both devices benefit from a high degree of turbulence at the electrode or membrane surfaces respectively. The design allows a large electrode surface area operating with small diffusion layer thickness to be packed into a modest volume. However since the inter-electrode gap is so small (a few mm), only a limited metal thickness can be deposited if short-circuiting is to be avoided. Removal of such deposits is most simply brought about by polarity reversal. To avoid damage, during this reversal, cathode materials (which become anodes on reversal) have to be selected with care. A further drawback is that this results in production of a concentrated solution of metal salt, rather than formation of solid metal as is possible with other cell designs.

The packed bed cell (Fig. 7.46b) has been described by several authors including Kreysa [137]. The bed can be of graphite particles, 1 to 1.5 mm diameter, though lead shot, magnetite and other conductive particles have also been reported. The cell design is such that the flowrate remains constant although the bed height increases from entrance to exit. This allows a decrease in current density to match decreasing metal ion concentration as the electrolyte flows through the cell. Conductivity (as metal ions are replaced by hydrogen) and throwing power increase at the same time. The anode is separated from the cathode bed by a diaphragm. This design, which was computer-optimised, aims to maximise space-time yield. As with the Swiss roll design, deposited metal can only be dissolved either chemically or by current reversal. Suitably operated, this design allows metal ion concentrations to be reduced below 1 mg/litre.

In the fluidised bed cell (7.46c), the constant movement between the particles serves to reduce the thickness of the diffusion layer to a very low value. Electrode and particles must be of materials not suffering abrasive wear. Normal practice is to use 0.5 to 1.0 mm particles of the same metal as that being deposited. They are fluidised by an upward electrolyte flow with electrical contact being made by a feeder rod. As in the previous design, the anode is separated from the fluidised bed by a diaphragm or membrane [138]. As the bed particles grow in size, from the electrodeposition of metals onto them, they tend to sink until they reach typically a tenfold increase in mass (not diameter!) at which point they are removed from the bottom of the cell, being replaced with new, smaller beads at the top of the cell. With this technology, copper can be cost-effectively removed from streams flowing at 500 cubic metres/hr at concentrations around 1 gm/litre. Fig. 7.47 shows the effect of flowrate on different cell designs.

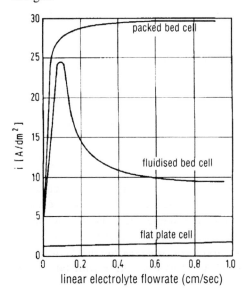

Fig. 7.47   Effect of linear electrolyte flowrate on useful current densities for various cell designs, based on constant cell flux area (after D H Smith)

Another cell design by Betley et al. [139] uses a metal deposition cathode of expanded mesh contacted with a fluidised bed of glass beads up to 500 micron diameter (Fig. 7.46d). The concept goes back to early work by Le Goff and has been successfully exploited by Messrs Bewt in the UK.

The foregoing discussion underlines how important it is to maximise diffusion limited processes and the same principles have been earlier indicated in the context of rinsing and of membrane processing. A distinction should be made between two cases. In the first (see Section 7.6.2.2.1), the aim is to remove as much metal as possible whether because it is a valuable noble metal or to comply with effluent legislation. Electrolytic cells of advanced design can readily lower metal ion concentrations to the 1 ppm level and, if commercial claims are to be believed, at least one order of magnitude lower still. In complete contrast is the situation (as for example with etch bath regeneration) where a relatively high metal ion concentration is acceptable, provided it is held within set limits. In this case (since removal of metals at high concentration is simpler than operating at the ppm level) a simple plate cell design will suffice.

In almost all cases, if other factors allow, increase in operating temperature is beneficial not just because solution viscosity is lowered but in its effect on the activation energy as described by classical reaction kinetic theory.

Together with Faraday's Law and Fick's equation, the third major theoretical basis for electrolysis is the Nernst equation which determines the position of an equilibrium such as:

$$\text{Me} \rightleftharpoons \text{Me}^{z+} + z\text{e} \qquad\qquad 7.30$$

As before, this is described in all physical chemistry textbooks which should be consulted. The major points to be made are firstly that Nernst's Equation applies only to a system in equilibrium, not one which is "working" i.e. through which current is flowing. The terms "at equilibrium" and "at rest" are usually, though not invariably synonymous. Nernst's theory allows us to derive the "electrochemical series" which are a powerful predictor of corrosion behaviour. Equation 7.31 also shows how a more noble or "electropositive" metal such as copper in its ionic form will be displaced by the less noble metal iron.

$$\text{Cu}^{++} + \text{Fe}^0 \rightleftharpoons \text{Cu}^0 + \text{Fe}^{++} \qquad\qquad 7.31$$

The practical significance of this is firstly in production of "displacement coatings" such when aluminium is zincated, and secondly as a means of effluent treatment in which one metal is replaced by another, less harmful, element. Equation 7.31 is sometimes described as a "cementation" reaction.

The equilibrium potential of a metal in solution is a function of its concentration (strictly activity) and temperature and Equation 7.32 shows how these effects operate:

$$E_{Me} = E^0_{Me} + \frac{RT}{zF} \cdot \ln a_{Me^{z+}} \qquad\qquad 7.32$$

where E_{Me} is the half-cell potential (measured potential), E^0_{Me} the same but under standardised conditions, R = gas constant (8.3 Joules/Mol/degree), T = absolute temperature, z = valency of the ion, F = Faraday constant and $a_{Me^{z+}}$ activity of metal ion z.

The half-cell potential, measured under standard conditions (25°C, activity = 1) leads directly to the thermodynamic value for Gibbs Free Energy [140] as shown in 7.33

$$\Delta G^0 = -zFE^0 \qquad\qquad 7.33$$

and the actual equilibrium constant for an electrochemical or ionic reaction can be derived from such thermodynamic data as shown in Equation 7.34.

$$\ln K_c = -\frac{\Delta G}{RT} = \frac{zFE}{RT} \qquad\qquad 7.34$$

where K_c is the actual equilibrium constant.

Table 7.3 lists the so-called "electromotive series" of metals (most noble is gold, least noble is potassium).

TABLE 7.3:
E.M.F SERIES OF METALS AT 25°C

electrode reaction			normal potential (E^0)
K	$\rightleftharpoons$ K$^+$	+ e	− 2.92
Ca	$\rightleftharpoons$ Ca$^{+ +}$	+ 2 e	− 2.84
Na	$\rightleftharpoons$ Na$^+$	+ e	− 2.713
Mg	$\rightleftharpoons$ Mg^{++}	+ 2 e	− 2.38
Al	$\rightleftharpoons$ Al^{+++}	+ 3 e	− 1.66
Zn	$\rightleftharpoons$ Zn^{++}	+ 2 e	− 0.763
Cr	$\rightleftharpoons$ Cr^{+++}	+ 3 e	− 0.71
Fe	$\rightleftharpoons$ Fe^{++}	+ 2 e	− 0.44
Cd	$\rightleftharpoons$ Cd^{++}	+ 2 e	−0.402
Ni	$\rightleftharpoons$ Ni^{++}	+ 2 e	− 0.23
Sn	$\rightleftharpoons$ Sn^{++}	+ 2 e	− 0.140
Pb	$\rightleftharpoons$ Pb^{++}	+ 2 e	− 0.126
Fe	$\rightleftharpoons$ Fe^{+++}	+ 3 e	− 0.036
H$_2$	$\rightleftharpoons$ 2 H$^+$	+ 2 e	± 0.00
Cu	$\rightleftharpoons$ Cu^{++}	+ 2 e	+ 0.34
Cu	$\rightleftharpoons$ Cu$^+$	+ e	+ 0.52
Hg$_2$	$\rightleftharpoons$ Hg$_2^{++}$	+ 2 e	+ 0.798
Ag	$\rightleftharpoons$ Ag$^+$	+ e	+ 0.799
Pb	$\rightleftharpoons$ Pb^{++++}	+ 4 e	+ 0.80
Hg	$\rightleftharpoons$ Hg^{++}	+ 2 e	+ 0.854
Pt	$\rightleftharpoons$ Pt^{++}	+ 2 e	+ 1.2
Au	$\rightleftharpoons$ Au^{+++}	+ 3 e	+ 1.42
Au	$\rightleftharpoons$ Au$^+$	+ e	+ 1.7

In practice, few electrochemical reactions are "reversible", that is additional voltage ("overvoltage") is required to drive them at a given rate [141]. Again, readers are advised to consult a textbook on electrochemistry or electrode kinetics. A useful equation which embodies the concepts summarised earlier is:

$$U_Z = E_A - E_K + \eta_A + \eta_K + I \cdot R \qquad\qquad 7.35$$

This predicts the cell voltage (U_Z) for a reaction driven by a current I. The cell is treated as a separate anode and cathode, each having an equilibrium potential as predicted by the Nernst Equation (E_A, E_K) to which is added the overpotential at each of these (η_A, η_K) and finally the ohmic drop as current flows through the resistive medium of the electrolyte. In cases where oxygen evolution is the anode reaction, this can account for up to 90% of the overvoltage [142].

The golden rule in assessing any electrochemical process is first to examine the equilibrium (theoretical) situation. If the outcome is positive, then there is a good chance that the "kinetics" will be favourable. If the thermodynamic prediction is not encouraging, it is unlikely the reaction will operate under "driven" conditions. However exceptions to this are known, caused by quirks in kinetic and catalytic behaviour. The value of using the Nernst approach cannot be over-emphasised and some examples are given below. Thus wherever one has competitive reactions, such as electrodeposition of copper from a zinc containing solution, and the ever-present possibility of hydrogen evolution, calculation of theoretical (equilibrium) potentials indicates what may be expected. Equation 7.36 allows the hydrogen evolution potential to be predicted as a function of pH (hydrogen ion concentration)

$$E_{H_2} = \frac{RT}{F} \ln a_{H^+} = -0.059 \, pH \qquad\qquad 7.36$$

The same approach is equally valuable in predicting behaviour of redox reactions, such as oxidation of chromium (III) to (VI) as seen in Equation 7.37 where a_{Ox} and a_{Red} are the activities of oxidised and reduced species respectively.

$$E_{Redox} = E^0_{Redox} + \frac{RT}{zF} \cdot \ln \frac{a_{Ox}}{a_{Red}} \qquad\qquad 7.37$$

The equilibrium potentials derived by use of these equations can be ranked, e.g. by reference to the standard hydrogen electrode, and this gives an e.m.f. series, such as that in Table 7.3, which is widely used for predicting corrosion behaviour or other electrochemical reactions. In common parlance, the upper elements are known as "non-noble", the bottom as "noble".

TABLE 7.4:

E.M.F SERIES OF REDOX COUPLES RELEVANT TO EFFLUENT TREATMENT. RANKING RELATES TO ACID CONDITIONS.
(SOURCE: W M LATIMER "OXIDATION STATES OF ELEMENTS & THEIR POTENTIALS IN AQUEOUS SOLUTION", PRENTICE HALL, 1952; "HANDBOOK OF CHEMISTRY & PHYSICS" CRC PRESS, FLORIDA (ANNUAL EDITION).)

| redox reaction in acid solution | | E^0 [V] | redox reaction in alkali | | E^0 [V] |
reduced form	oxidised form		reduced form	oxidised form	
$H_3PO_2 + 2\,H_2O$	$\rightleftharpoons H_3PO_3 + 2\,H^+ + 2\,e$	-0.5	$H_2PO_2^-\ +\ 3\,OH^-$	$\rightleftharpoons HPO_3^{2-} + 2\,H_2O + 2\,e$	-1.57
			$BH_4^-\ +\ 8\,OH^-$	$\rightleftharpoons H_2BO_3^- + 5\,H_2O + 8\,e$	-1.24
$H_3PO_3 +\ H_2O$	$\rightleftharpoons H_3PO_4 + 2\,H^+ + 2\,e$	-0.276	$HPO_3^{2-}\ +\ 3\,OH^-$	$\rightleftharpoons PO_4^{3-} + 2\,H_2O + 2\,e$	-1.12
$S_2O_6^{2-} + 2\,H_2O$	$\rightleftharpoons 2\,SO_4^{2-} + 4\,H^+ + 2\,e$	-0.22	$S_2O_4^{2-}\ +\ 2\,OH^-$	$\rightleftharpoons 2\,SO_3^{2-} + 2\,H_2O + 2\,e$	-1.12
$HS_2O_4^- + 2\,H_2O$	$\rightleftharpoons 2\,H_2SO_3 + H^+ + 2\,e$	-0.08			
$HCHO +\ H_2O$	$\rightleftharpoons HCOOH + 2\,H^+ + 2\,e$	-0.056	$HCHO\ +\ OH^-$	$\rightleftharpoons HCOO^- + H_2O + 2\,e$	
H_2	$\rightleftharpoons 2\,H^+ + 2\,e$	± 0.00	$H_2\ +\ 2\,OH^-$	$\rightleftharpoons 2\,H_2O + 2\,e$	-0.828
H_2S	$\rightleftharpoons S + 2\,H^+ + 2\,e$	$+0.141$	$HS^-\ +\ OH^-$	$\rightleftharpoons S + H_2O + 2\,e$	-0.478
Sn^{2+}	$\rightleftharpoons Sn^{4+} + 2\,e$	$+0.15$	$[Sn(OH)_3]^-\ +\ 3\,OH^-$	$\rightleftharpoons [Sn(OH)_6]^{2-} + 2\,e$	-0.90
Cu^+	$\rightleftharpoons Cu^{2+} + e$	$+0.153$	$Cu_2O + H_2O\ +\ 2\,OH^-$	$\rightleftharpoons 2\,Cu(OH)_2 + 2\,e$	-0.08
$H_2SO_3 +\ H_2O$	$\rightleftharpoons SO_4^{2-} + 4\,H^+ + 2\,e$	$+0.17$	$CN^-\ +\ 2\,OH^-$	$\rightleftharpoons CNO^- + H_2O + 2\,e$	-0.97
$[Fe(CN)_6]^{4-}$	$\rightleftharpoons [Fe(CN)_6]^{3-} + e$	$+0.36$	$SO_3^{2-}\ +\ 2\,OH^-$	$\rightleftharpoons SO_4^{2-} + H_2O + 2\,e$	-0.93
$S_2O_3^{2-} + 3\,H_2O$	$\rightleftharpoons 2\,H_2SO_3 + 2\,H^+ + 4\,e$	$+0.40$	$S_2O_3^{2-}\ +\ 6\,OH^-$	$\rightleftharpoons 2\,SO_3^{2-} + 3\,H_2O + 4\,e$	-0.58
$S_4O_6^{2-} + 6\,H_2O$	$\rightleftharpoons 4\,H_2SO_3 + 4\,H^+ + 6\,e$	$+0.51$			
MnO_4^{2-}	$\rightleftharpoons MnO_4^- + e$	$+0.564$			
$2\,H_2SO_3$	$\rightleftharpoons S_2O_6^{2-} + 4\,H^+ + 2\,e$	$+0.57$			
Fe^{2+}	$\rightleftharpoons Fe^{3+} + e$	$+0.771$	$Fe(OH)_2\ +\ OH^-$	$\rightleftharpoons Fe(OH)_3 + e$	-0.56
Hg_2^{2+}	$\rightleftharpoons 2\,Hg^{2+} + 2\,e$	$+0.920$			
$HNO_2 +\ H_2O$	$\rightleftharpoons NO_3^- + 3\,H^+ + 2\,e$	$+0.94$	$NO_2^-\ +\ 2\,OH^-$	$\rightleftharpoons NO_3^- + H_2O + 2\,e$	$+0.01$
$Mn^{2+} + 2\,H_2O$	$\rightleftharpoons MnO_2 + 4\,H^+ + 2\,e$	$+1.23$	$Mn(OH)_2\ +\ 2\,OH^-$	$\rightleftharpoons MnO_2 + H_2O + 2\,e$	-0.05
			$2\,S_2O_3^{2-}$	$\rightleftharpoons S_4O_6^{2-} + 2\,e$	$+0.08$
$2\,NH_4^+$	$\rightleftharpoons N_2H_5^+ + 3\,H^+ + 2\,e$	$+1.275$	$2\,NH_4OH\ +\ 2\,OH^-$	$\rightleftharpoons N_2H_4 + 4\,H_2O + 2\,e$	$+0.1$
Au^+	$\rightleftharpoons Au^{3+} + 2\,e$	$+1.29$			

Table 7.4 Continued

redox reaction in acid solution reduced form	oxidised form	E⁰ [V]	redox reaction in alkali reduced form	oxidised form	E⁰ [V]
$2\,Cr^{+++} + 7\,H_2O \rightleftharpoons$	$Cr_2O_7^{--} + 14\,H^+ + 6\,e$	+1.33	$Cr(OH)_3 + 5\,OH^- \rightleftharpoons$	$CrO_4^{--} + 4\,H_2O + 3\,e$	−0.13
$Pb^{++} + 2\,H_2O \rightleftharpoons$	$PbO_2 + 4\,H^+ + 2\,e$	+1.46	$PbO + 2\,OH^- \rightleftharpoons$	$PbO_2 + H_2O + 2\,e$	+0.248
$Cl^- + H_2O \rightleftharpoons$	$HClO + H^+ + 2\,e$	+1.495	$Cl^- + 2\,OH^- \rightleftharpoons$	$ClO^- + H_2O + 2\,e$	+0.89
$Mn^{++} + 4\,H_2O \rightleftharpoons$	$MnO_4^- + 8\,H^+ + 5\,e$	+1.51	$MnO_2 + 4\,OH^- \rightleftharpoons$	$MnO_4^{--} + 2\,H_2O + 2\,e$	+0.60
$\tfrac{1}{2}\,Cl_2 + H_2O \rightleftharpoons$	$HClO + H^+ + e$	+1.630	$\tfrac{1}{2}\,Cl_2 + 2\,OH^- \rightleftharpoons$	$ClO^- + H_2O + e$	+0.42
$MnO_2 + H_2O \rightleftharpoons$	$MnO_4^- + 4\,H^+ + 3\,e$	+1.695			
$2\,H_2O \rightleftharpoons$	$H_2O_2 + 2\,H^+ + 2\,e$	+1.776	$3\,OH^- \rightleftharpoons$	$HO_2^- + H_2O + 2\,e$	+0.878
$Ni^{++} + 2\,H_2O \rightleftharpoons$	$NiO_2 + 4\,H^+ + 2\,e$	+1.93	$Ni(OH)_2 + 2\,OH^- \rightleftharpoons$	$NiO_2 + 2\,H_2O + 2\,e$	+0.49
$Ag^+ \rightleftharpoons$	$Ag^{++} + e$	+1.98	$Ag_2O + 2\,OH^- \rightleftharpoons$	$2\,AgO + H_2O + 2\,e$	+0.57
			$2\,SO_4^{--} \rightleftharpoons$	$S_2O_8^{--} + 2\,e$	+2.01
$O_2 + H_2O \rightleftharpoons$	$O_3 + 2\,H^+ + 2\,e$	+2.07	$O_2 + 2\,OH^- \rightleftharpoons$	$O_3 + H_2O + 2\,e$	+1.24

TABLE 7.5:
E.M.F SERIES FOR HYDROGEN-OXYGEN COMPOUNDS ("HANDBOOK OF CHEMISTRY & PHYSICS" CRC PRESS)

redox reaction in acid	E⁰ [V]	redox reaction in alkali	E⁰ [V]
$H_g \rightleftharpoons H^+ + e$	−2.1065	$H_g + OH^- \rightleftharpoons H_2O + e$	−2.9345
$H_2 \rightleftharpoons 2\,H^+ + 2\,e$	± 0.00	$H_2 + 2\,OH^- \rightleftharpoons 2\,H_2O + 2\,e$	−0.8277
$HO_2 \rightleftharpoons O_2 + H^+ + e$	+0.136		
$H_2O_2 \rightleftharpoons O_2 + 2\,H^+ + 2\,e$	+0.682	$ÖH + 2\,OH^- \rightleftharpoons HO_2^- + H_2O + e$	−0.24
$H_2O + ÖH \rightleftharpoons H_2O_2 + H^+ + e$	+0.72	$HO_2^- + OH^- \rightleftharpoons O_2 + H_2O + 2\,e$	−0.146
$H_2O \rightleftharpoons O_g + 2\,H^+ + 2\,e$	+1.229		
$H_2O_2 \rightleftharpoons HO_2 + H^+ + e$	+1.5	$OH^-\,,+ OH^- \rightleftharpoons O_g + H_2O + 2\,e$	+0.401
$2\,H_2O \rightleftharpoons H_2O_2 + 2\,H^+ + 2\,e$	+1.776	$2\,OH^- + OH^- \rightleftharpoons HO_2^- + H_2O + 2\,e$	+0.878
$H_2O + O_2 \rightleftharpoons O_3 + 2\,H^+ + 2\,e$	+2.07	$O_2 + 2\,OH^- \rightleftharpoons O_3 + H_2O + 2\,e$	+1.24
$H_2O \rightleftharpoons O + 2\,H^+ + 2\,e$	+2.422	$OH^- + OH^- \rightleftharpoons O + H_2O + 2\,e$	+1.594
$H_2O \rightleftharpoons ÖH + H^+ + e$	+2.8	$OH^- \rightleftharpoons ÖH + e$	+2.0

Tables 7.4 and 7.5 list some redox values of importance in effluent treatment. Some comments should be made on the use of "reference" or "standard"electrodes. Nernstian theory and the thermodynamic approach is based on the use of the "hydrogen electrode" (Table 7.3) whose redox value is, by definition, 0.000V. In practice a range of secondary standards are used, such as the calomel electrode. These give voltages defined in terms of the hydrogen electrode. It should also be noted that everything that has been stated above, relates to the standard temperature of 25°C. Any deviation from this, will affect the thermodynamics of the system being studied. However it will also affect the "standard" value of reference electrodes such as calomel. As a simple rule, reference electrodes should not be used at temperatures other than 25°C except for purely comparative purposes.

In the case of electrodialysis cells, the same equations used above still apply though with certain provisos. For concentration cells such as shown in Fig. 7.66, total voltage U is the sum of that at each of the "n" cells for current I, as shown by Equation 7.38.

$$U = i \cdot n \cdot d \cdot \left(\frac{1}{\varkappa_K} + \frac{1}{\varkappa_D} \right) + (n + 1 R_{KM} + n R_{AM}) + 2i \cdot d_E \cdot \frac{1}{\varkappa_E} \qquad 7.38$$

where d is the thickness of the concentrate and diluate chambers and $\varkappa_K$, $\varkappa_D$ are the electrical conductivities of the electrolyte in concentrate and diluate chambers respectively; R_{KM}, R_{AM} the electrical resistance of the cation and anion membranes respectively; $\varkappa_E$ the specific electrical conductivity of electrolyte in the main chambers and d_E the thickness of these chambers.

The electrical resistance of ion exchange membranes is very low. Expressed as specific conductivity, it is of the same order of magnitude as strong electrolytes, e.g. a 2M KCl solution [143]. Depending on function (e.g. as in Fig. 7.66) and the conductivity of the diluate, up to 300 cell pairs can be operated in series. Typical current densities are 1 to 1.5 Amp/sq. dm. At this rating, voltage losses should not exceed 1.5V per double cell pair.

In electrodialysis, there exists the same danger of concentration polarisation at the membrane as was discussed in Section 7.5.1 although in the present case, they are caused by too high current densities. As previously described, a laminar boundary layer is formed either side of the membrane with mass transport of species driven across this by diffusion. To minimise polarisation, this laminar layers should be as thin as possible and this can be brought about by using crossflow across the face of the membranes and also by using spacers which promote turbulence. If concentration polarisation does occur, the entire process can be brought to a halt or even reversed. Reflecting their high electrical conductivity, ion transport numbers are often higher in the membrane than in the surrounding electrolyte and this largely governs the limiting current for such membranes. Taking the case where concentration of ionic species at the membrane surface falls to zero (i.e. onset of diffusion limiting current density) this can be described for a given ion by Equation 7.39 [144].

$$i_G = D \frac{zF}{t-t} \cdot \frac{c}{\delta}$$ 7.39

where i_G is the limiting current density at the membrane, D the diffusion constant of the ion and z the ionic valency, F is the Faraday constant and t the transport number of the ion in the membrane, t the transport number in the aqueous phase; c is concentration of ions in solution; δ the thickness of the diffusion layer.

If the limiting current density for the ion in the membrane is exceeded, then hydrogen or hydroxyl ions are driven through instead with the expected pH changes in the electrolytes from which and into which they migrate..

The selectivity of ion exchange membranes is determined by the immobile charged species of which they are formed. Cationic exchangers have large fixed anionically charged species and are permeable only to cations. The converse holds for anion exchange membranes. This is, however, only true for dilute aqueous solutions. If the ionic concentrations approach the values found within the membrane, then the membrane loses its permselectivity. According to Strathmann [105], ion exchange resins contain around 2 to 4 gram-equivalents per litre of immobilised ions. As higher solution concentrations are reached, the Donnan equilibrium (see Section 5.2.3.1) is disturbed, the electrostatic exclusion of co-ions breaks down, leading to the so-called "Donnan invasion". Current efficiency then falls close to zero [105,143]. Broadly speaking, concentrations up to 200 gm/litre can be reached [105].

Performance limits for electrodialysis exist not only on the concentrate but also the diluate side of the cell. As dilution increases, so does the electrical resistance and this is reflected in increased cell voltage, to the point where, as previously stated, cell voltages exceed 1.5V which is not acceptable (phenomena such as bipolar leakages or "shunt currents" then occur). Equation 7.40 shows the theoretical energy required to concentrate a solution (or energy release when it is diluted): [143].

$$\Delta G = 2 \, RT \, \ln \frac{a_{conc}}{a_{dil}}$$ 7.40

where a_{conc}, $a_{dil.}$ are activities of concentrate and diluate respectively. However such calculated values are greatly exceeded by other terms such as concentration overpotential, resistive losses, unavoidable co-transport of hydrogen and hydroxyl ions and also water molecules across the membrane.

7.6.2 Electrolysis

7.6.2.1 Electrodes

At the risk of repetition, one must distinguish between cases where (whether because it is toxic or because it is valuable) the metal(s) are removed down to a very low concentration (1 ppm or less) or, as say with copper etch baths, the aim is simply to maintain a relatively modest metal ion concentration and preventing it rising above set limits. If acceptably high current efficiencies are required, sophisticated electrode design is needed in the former case, simple flat plate cathodes will suffice in the latter. In all cases however, including those involving anodic oxidation, an "insoluble" anode is required. Under this heading can be listed graphite, lead, alloys of lead with antimony, silver or tin, stainless steel, (in a few cases) cast iron, ferrosilicon and the valve metals (titanium, tantalum, tungsten, niobium) coated with noble metals (platinum, iridium) or noble metal oxides (ruthenium oxide). The life of these depends on the solution composition and the operating current density and can range from weeks to years. In almost all cases, these anodes operate at potential above oxygen evolution. Somewhat different, and a spin-off from fuel-cell research is the hydrogen depolarised anode. If hydrogen or hydrogen-containing gases are fed to this, it can sustain anodic currents at very low potentials – typically 0.25V anodic to hydrogen evolution potential. At present, tungsten carbide is used as electrocatalyst in such anodes. However it seems quite possible that novel catalysts will be developed. Attractive though its energy saving possibilities, the higher capital costs and need for hydrogen "fuel" combine to place this idea in the future rather than the present. Against this, it allows operation in strong chloride solutions, for example, without danger of chlorine evolution.

Selection of anode materials is usually a compromise based on:

- overvoltage behaviour for the particular reaction on a given material

- anode corrosion (incl. while at rest)

- mechanical properties and forms in which the material is available, i.e expanded mesh, sheets, rods, pellets, etc.

- price (where noble metal coated Ti and similar anodes command a high premium, reflecting their long life and near-universal usability, as opposed to graphite or stainless steel).

Another relatively new anode material is "Ebonex", a conductive ceramic based on sub-oxides of Ti. A moderately good electrical conductor, it also features a resistance to fluoride ion attack which is near-unique.

For cathodes, sheets, foils or particles can be used, usually of the same metal as is electrodeposited, but also of stainless steel or other metals which allow either a mechanical parting of the deposit from the cathode blank, or its removal by anodic dissolution. Iron, stainless steel. porous carbon, graphite particles, glass or plastic metallised beads, metallised felt or fabric have all been used. The choice of these is largely determined by the nature of the next stage after metal deposition.

In terms of flat-plate cells (monopolar or so-called bipolar filter-press or plate-and-frame designs) and some other designs, the inter-electrode (anode-cathode) gap is critical. Too large a gap leads to excessive energy requirements to overcome the ohmic drop across the electrolyte. Too small a gap leads to problems of short-circuiting by growing dendrites and limits the thickness of metal which can be deposited before its removal is required for the same reason. 3 to 5 cm is a guideline figure. In cases where dendrite formation is found to be a problem, measures must be incorporated for periodic and regular removal of such growths. A similar problem arises where metals are deposited as powders or in friable form. Here too, accumulation of deposits at the bottom of the cell can cause short-circuiting. Incorporating settling volumes or use of high flow-rates to remove such metal and then filtration, are possible solutions. Other approaches are detailed below.

7.6.2.2   Cathodic deposition

7.6.2.2.1   Precious metals

Electrolytic recovery of metals such as silver and gold has a long history, not least (in the former case) in association with the photographic industry where silver has long been recovered from spent fixer baths, at the same time allowing a life extension of solution. From around 1970 on, a renewed emphasis on novel cell designs was seen, in which the aim was to increase the space-time yield and also, by enhancing the diffusion process, to allow their operation at acceptable current densities using solutions of low metal ion concentration. Maximising cathode surface area (Section 7.6.1) formed one part of the strategy. Kammel & Lieber [146-148] have reviewed a range of concepts some of which have been actually constructed and studied, other which have not. Schab & Hein [149] likewise reported various designs including a rotating unit for copper recovery from very dilute solutions. A number of other designs are discussed in Section 7.6.2.2.2 which, though designed for operation in dilute solution, are too large for processing rinsewaters in most plants. So numerous are cell designs for silver or gold recovery that only a few are discussed here, these being designs which have met with commercial success.

Because gold lies so high in the e.m.f series (Table 7.3), it lends itself well to electrolytic recovery even if quite simple flat plate cells are used, provided only a low current density is used and low cathode efficiencies are acceptable. Using so-called "three-dimensional" cathodes, such as a fluidised or restrained or packed bed of graphite particles, high recovery rates can be obtained. Fig. 7.48 shows recovery from gold and gold alloy plating rinsewaters [148,150] using the cell drawn in Fig. 7.49. The final gold concentration is around 0.5 ppm. Electrolytic recovery offers several advantages over ion-exchange. Thus it does not result in increased dissolved salt concentrations and in addition, where other metals are present in similar concentrations, these do not affect the rate of removal of the desired species. That silver and/or copper will co-deposit is no serious drawback. Finally, electrolysis is a true recovery process in contrast to ion exchange which serves only to concentrate the metal in the resin, from which it can only be removed by incineration of the resin or its release again in dissolved form, albeit at higher concentration. In the case of metal recovery from alloy plating bath rinsewaters, the deposits when studied in cross-section, show a stratified structure [147] and this reflects preferential deposition, in some conditions, of the more noble metal.

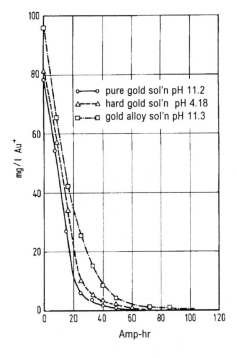

Fig. 7.48 Electrodeposition of gold from rinsewaters of various gold baths using the cell shown in Fig. 7.49 and 7.50 [148].

Fig. 7.49 Schematic of rotating particulate bed cathode cell as used for gold recovery from rinsewaters (GOEMA)

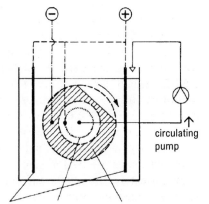

external anodes   internal anodes   particulate bed cathode

Where gold, silver and copper are all present together, they would be expected to deposit in that order, at least until the entire process became diffusion controlled, at which point the three metals would deposit together. However the e.m.f series shown in Table 7.3 applies only to uncomplexed metals. Complexation can drastically change the ranking, as Gal-Or *et al.* [151) showed in their study of cyanide solutions where silver is deposited first, then gold and then copper, i.e. silver behaves as more electropositive than gold. The stability constant of the dicyanoaurate at $2 \cdot 10^{38}$ is substantially greater than that of the dicyanoargentate at 10^{21}. Another implication of this is that the concentration of free gold ions in equilibrium with the Au complex is much lower than is the case for silver. Complexation reduces the standard potential of gold from $+1.4$ to $-0.6V$, the corresponding value for Ag at $-0.31V$ thus being higher and explains why it is preferentially deposited.

The cell design shown in Fig. 7.50 is known as the "rotating tube" type and somewhat resembles a plating barrel. However it features two cylinders allowing use of internal and external barrels (shown in Fig. 7.49) The particles of the bed are roughly spherical with a hole drilled through, thus maximising the surface area. They are made of plastic with electroless copper coating. Typically, these are sent for gold recovery when 1 kg of the metal has deposited. A high rate of mass transport is assured by the particle-electrolyte motion and the pumping action which forces the flow from outwards to the central barrel.

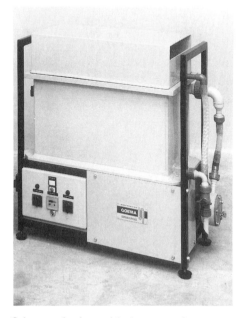

Fig. 7.50 Rotating tube reactor for gold recovery from rinsewater (also shown in previous Fig.) (GOEMA).

Other cathodes with large surface area used for gold recovery are made of carbon fibre or graphite felt. One such cell is described by Farkas *et al.* [152,153] and uses carbon fibres of ca. 6 micron diameter with around 20 micron separation. The authors showed this has a diffusion layer thickness of around as low as 10 microns and typical diffusion limiting current densities of several thousand Ampere/square metre (depending on the solution used). In practice current densities a tenth of this are used. The true cathode surface area is ca 12,000 times greater than the geometrically measured value. Typically, up to 5% of the cathode volume can be loaded with metal. At this point the cathode is removed and put in a specially designed refining cell where it is anodically polarised with two flat sheet stainless cathodes one on either side. This cell also allows final concentrations of 0.5 ppm to be reached. The concentration vs. time plots resemble Fig. 7.48. with zero-order kinetics until quite low metal ion concentrations are reached. Use of a reference electrode at the cathode surface allows cathode polarisation to be measured and suitable values of this can be set for whatever metal is to be deposited. The unit is shown in Fig. 7.51.

Fig. 7.51 Electrolysis cell with graphite fibre cathode (at left), covered with plastic protective mesh. At right – open topped cell for anodic metal stripping (RECON GmbH)

Both these cells can be operated in the batch mode or on a continuous basis. Platinised titanium expanded mesh anodes are used. Though these cells can be used to recover any electrodepositable metal, the economics are most attractive for noble metals and after that for silver. Indeed it has to be said that the decline in world silver prices in recent years has led to a loss of interest in the electrolytic recovery of that metal, even though it is still widely practised.

    A different and more modest approach is shown in Fig. 7.52. This 50 mm thick cell is suspended in the static rinse tank and rinsewater is fed to it and returns by overflowing from the top. It has a central anode with a cathode on either side [154] – a widely adopted approach reflecting the fact that the anode is usually the most expensive single item in a cell and its use must therefore be maximised.

Fig. 7.52   In-tank cell for rinse tanks especially gold with woven copper cathode and expanded mesh platinised Ti anode. (DEGUSSA)

In respect of other metals, Farrer [155] reported on palladium recovery from static rinse tanks. At 3 gm/litre Pd, 0.6 Amp/sq. dm current density was recommended. However using stainless steel cathodes, adhesion of the electrodeposited Pd was so poor that short-circuits resulted. Platinised Ti anodes were used. At low current densities (ca 0.1 Amp/sq. dm), copper impurities in the palladium bath plated out so well that the author suggested instead, this means for purifying the electrolyte and then returning the rinsewater to the plating bath, with evaporation as required for correct water balance.

For precious metal recovery down to 0.5 ppm from very dilute solutions, a patent filing [156] claims pH at least 10, cell voltage = 8V and current density 20 Amp/sq. dm, temperatures >60°C and anode-cathode gap of 8 to 16 cm.

Far more important than gold, in volume terms, is electrolytic silver recovery. A special cell deposits Ag at high current densities (ca 10 Amp/sq. dm and 15V in powder form [157]. An array of tubular stainless steel rod cathodes is positioned round a central anode of graphite or Armco steel. The powdery deposit is removed from the cathode thanks to the high flowrate (8 to 12 cubic metres/hr) and this also drives a plastic scraper. The powder accumulates in a container. At 1.5 to 2.5 gm/litre, silver is recovered with around 75% current efficiency. Deposition of 900 gm Ag requires 4.5 kWHr electrolysis energy, 1.5 kWHr pump energy. While mainly aimed at continuous metal removal from static rinse tanks, the unit is also used for exhaustive electrolysis on a batch basis. It is shown in Fig. 7.53.

A similar cell design used as cathode a basket of stainless steel expanded mesh with central platinised Ti anode. Operating under very similar conditions to the above, silver is again formed as a powder and collected in a bag filter [158].

Fig. 7.53   Electrolytic silver recovery cell for use with rinsewater. Rating – 900 gm Ag/hr (SCHERING).

For electrolytic silver recovery from cyanide solutions, it is difficult to reach concentrations below 50 to 100 mg/litre, using conventional designs. This is also due to the chemical re-dissolution of the metallic silver in the alkaline cyanide. Fig. 7.54 shows the solubility of electrodeposited powdered Ag in the electrolyte after treating a static rinse bath with 3.4 gm/litre cyanide and 3 mg/litre silver [150].

The equilibrium between electrodeposition and re-dissolution makes cell design very difficult in that the high flowrates required to enhance mass-transport also facilitate the re-dissolution. Stainless steel cathodes are used because their passive surfaces largely prevent adhesion of electrodeposited Ag [150, 159, 160]. In some designs, a reciprocating cathode motion is incorporated in which one cathode rod rubs on the adjacent one. A collecting gutter and container are used to catch the scales of metallic silver falling from the cathode in this way. One of these so-called impact-rod cells is seen in Fig. 7.55. At cell voltages of 5 to 7V and currents of 150 Amp, cathode efficiencies are close to 100% at Ag concentrations of 500 mg/litre or above. The unit shown is rated at 750 gm Ag metal/hr although at silver concentrations above 0.5 gm/litre, a higher current can be used with correspondingly increased output. Fig. 7.56 shows a performance curve.

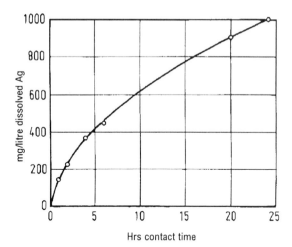

Fig. 7.54 Dissolution of Ag powder in spent cyanide-containing effluent [150].

Fig. 7.55 Impact rod reactor for silver recovery from effluent using continuous mechanical Ag removal from the cathode and collection of the deposit (GOEMA).

As mentioned earlier, the noble metals (at least when uncomplexed) are more readily electrodeposited than their non-noble counterparts, because of their electropositive character and this is true for all cell designs, not least those shown in Fig. 7.46b, practical details of such installations being given by Mueller [162] and in [149] for rolled-up cells.

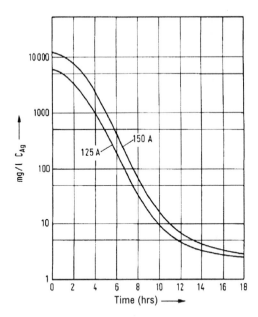

Fig. 7.56   Deposition rate for silver from cyanide solution in the unit shown in Fig. 7.55 (GOEMA).

7.6.2.2.2   Non-ferrous metals

Electrolytic recovery of other metals becomes more difficult the less noble are the metals in question. As they become more electronegative, such recovery becomes possible only at higher pH and (in the case of non electrodepositable metals) is not possible at all. Thus copper is the most important non-noble metal to be electrolytically recovered, followed by nickel. Because of its highly toxic nature, rather than its intrinsic value, cadmium is important, with zinc only occasionally being recovered and lead of no importance outside the deposition of tin-lead alloys (solder) in the p.c.b industry.

Most non-ferrous metals, notwithstanding what was stated above, are recovered from acid solutions. The main aim is not to strip out the last few ppm from solution (although units exist capable of doing this) but rather to avoid formation of large amounts of metalliferrous sludges and the disposal costs associated with these. The value of the recovered metal is often secondary to this. Given these aims, the simple flat-plate electrode cell remains in widespread use. In the case of copper recovery, operating at correct pH is important because of the low overvoltage of the competing reaction – hydrogen evolution. Also, at very low pH, undesirable anode reactions become important, such as chlorine evolution from chloride solutions. In some cases, use of a dividing membrane is then necessary (membrane electrolysis cells). In the following treatment, the most widely-used flat plate cells are first described, followed by more specialised cell designs.

The process flowsheet in Fig. 7.57 for electrolytic copper recovery from pickle liquor goes back to 1967 [163]. 50 litres/hr of copper-containing pickle liquor are dragged out. This volume is first reduced by concentration using an evaporator. The concentrate is returned to the pickling tank, the condensate is used in the cascade rinse system. The heat recovered from condensation is used to heat the pickling tank and the hot rinse tank with corresponding energy savings. In the electrolysis cell, 2 kg/hr of copper and – more important – 6 kg of sulphuric acid are recovered, thereby eliminating the cost of disposing of spent pickle acid (effluent treatment, neutralisation, sludge disposal). The only effluent is the regenerate liquor from the small ion exchange unit, used to treat the recirculating water loop.

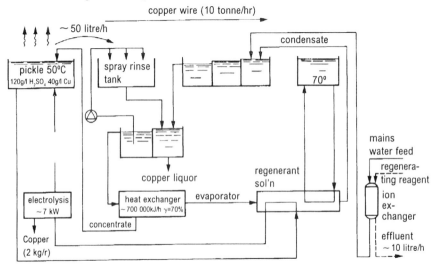

Fig. 7.57   Schematic of electrolytic copper and sulphuric acid recovery from a copper wire pickling plant [163]

One of the earliest descriptions of electrolytic metal recovery from effluent is by Wystrach [164] and the guidelines he laid down for flat plate cells remain equally valid today. They are shown for four most important electrolytes in Table 7.6. Elsewhere, Wiaux reports the use of a modern parallel plate cell with vertical electrolyte feed to the cathode for electrolytic recovery of copper from cyanide solution. In batch operation, copper was reduced from 2000 to 0.2 mg/litre with mean cathode efficiency in excess of 10%. Wiaux established a cause of loss of efficiency which had previously been noted by Schab & Hein [166] using a Swiss roll cell. To their surprise, they found lower cathode efficiencies at high metal concentrations in acid than at lower concentrations. At the copper cathode, the equilibrium:

$$Cu^+ \rightleftharpoons Cu^{++} + Cu^0 \qquad\qquad\qquad 7.40$$

is set up. The equilibrium constant is given by the term:

$$\log K_c = \frac{a_{Cu^{++}}}{a^2_{Cu^{++}}} \cdot \frac{c}{\delta} \qquad\qquad 7.41$$

Both authors realised that loss of cathode efficiency is directly proportional to the concentration of Cu(I) as indicated by Equation 7.41. At higher copper concentrations and with good agitation, the Cu(I) ions are re-oxidised at the anode. This shows that, where ions of a metal can adopt several valence states, efficient agitation can actually be a disadvantage.

TABLE 7.6:
DATA FOR ELECTROLYSIS OF CONCENTRATED METAL-CONTAINING EFFLUENT SOLUTIONS [164].

Electrolyte	nickel	copper		zinc
	static rinse tanks effluent from treatment of nickel baths & filter units	from acid copper baths	from cyanide copper baths	from barrel zinc plating
Electrolysis conditions				
Initial conc'n	120 g/l Ni	57 g/l Cu	20 g/l Cu. 80 g/l NaCN	14 g/l Zn, 33 g/l NaCN
final conc'n	8 g/l Ni	1 g/l Cu	1 g/l Cu. 0.5 g/l CN	1 g/l Zn, 0.5 g/l NaCN
current density – cathodic	1-1.5 A/dm²	2 A/dm²	2 A/dm²	10 A/dm²
– anodic			1 A/dm²	1 A/dm²
cathode material	nickel sheet	copper sheet	copper sheet	zinc sheet
anode material	hard lead	hard lead	graphite	graphite
energy required	5.5 kWh/kg Ni	5.5 kWh/kg Cu	11 kWh/kg NaCN	11 kWh/kg NaCN
misc. details	50-60° C; pH 3-4.5 constant	air agitation for less than 20 gm/litre Cu 5 gm/litre NaNO₃ added	air agitation addition of 20 gm/litre NaCl ... Current efficiency for Cu - 20%	air agitation addition of 20 gm/litre NaCl current efficiency for Zn ...30%

Wiaux established this anomalous current efficiency loss only for cyanide copper solutions. In the case of acid copper solutions, 8000 mg/litre was reduced to 0.8 mg/litre at overall 35% efficiency. In the case of nickel solutions, a pH correction was called for to repress hydrogen evolution. At 2 to 2.5 Amp/sq. dm, 3600 mg/litre was reduced to a level of 5 mg/litre with over 30% efficiency. In the case of sulphamate solutions, a neutral salt had to be added in the same concentration range, to bring about sufficient electrolytic conductivity. When this was not done, cell voltages rose to 7 to 9V and current efficiency fell below 10%.

In order to provide suitable metal ion concentrations for electrolysis of rinsewaters, pre-treatment with a weak acid ion exchanger can be useful, while anions such as chloride, whose anodic reactions are undesirable, can at the same time be excluded. This is described by Schwering & Hasler [18]. The ion exchanger, loaded with Ni, is regenerated with sulphuric acid as described in 7.2.1.2 with the nickel sulphate solution fed to the electrolysis unit. There, Ni is removed to to 1 gm/litre level and the electrolyte is returned to the ion exchanger. This allowed the hourly rate of nickel emission to be lowered from 150 to 6 gms.

An interesting means of obtaining concentrated nickel sulphate solutions is described by Brown [167]. The nickel sulphate solution drawn from the ion exchanger is then treated by the retardation process (Section 7.2.2.2) where it is split into an acid-rich, low nickel and nickel-rich low acid fractions. The latter is treated by electrolysis, the former returned to the ion exchanger.

It should be noted that pure and unbuffered nickel sulphate solutions cannot be easily electrolysed. Electrolysis causes a continuous replacement of Ni with H ions, and the pH drops and must be constantly corrected. The electrodeposition is best carried out at or quite close to pH 2.5 where cathode efficiency and polarisation are minimised, as reported by Goetzelmann [168]. This is shown in Fig. 7.58. At pH values below 2.5, the hydrogen overpotential is low. Above this value, cathodic polarisation is seen. This very strong pH effect is independent of the nickel ion concentration and, at the limits, independent of current density also. In the case of copper, no pH correction is required and the sulphuric acid regenerated by electrolysis can be used to rejuvenate the weakly cationic ion exchangers.

Dejak [169] used strongly cationic exchangers to concentrate copper and nickel from rinsewaters. Thus he obtained copper sulphate containing regenerate with 7 to 15 gm/litre Cu which was fed to a narrow-gap electrolysis cell. The inter-electrode gap was sparged with finely-divided air bubbles. The cathode was stainless steel, the anode a lead-calcium-tin alloy. The spent solution, with ca. 1 gm/litre copper, was returned to the ion exchanger. In similar manner, nickel sulphate solution with 25 to 30 gm/litre Ni was subjected to electrolysis, the spent solution having 3 gm/litre Ni. This was then treated with spent electroless nickel solution and fed to a weakly acid ion exchanger. The regenerate from this contains 5 to 6 gm/litre Ni and is returned to the process.

A similar approach is reported by Marquardt [170] who, however, used a membrane-divided cell with dilute sulphuric acid as anolyte. Using an anion-exchange membrane, the sulphate ions migrate into the anolyte and the acid formed is used to regenerate the ion exchangers. Fischer [171] describes a cell where nickel is electroplated onto Ni pellets from sulphate. As these grow, they are extracted and used to fill anode baskets in the nickel plating tank. Because heavy nickel deposits have a tendency to blister on flat cathode sheets, the author favours the pellet-based approach. Using a plating barrel type unit, pellets of 5 to 6 mm diameter will double in diameter, at which point they are extracted. If a cation exchange membrane is used in nickel recovery cells, simple lead anodes can then be used.

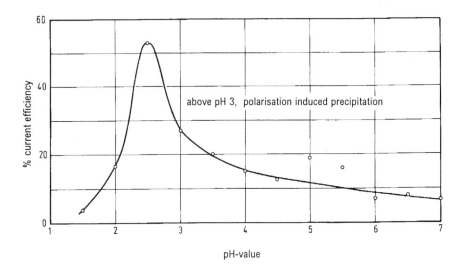

Fig. 7.58   Current efficiencies for electrodeposition of nickel from unbuffered sulphate solution as function of pH. [168].

If nickel is electrolytically recovered with its immediate in-house re-use in mind, the solution may require purification before the electrolysis, for example by precipitation, as described in 7.4.1 [75,76] or using activated carbon to remove organic impurities. Selective pre-electrolysis can be used to remove copper, zinc and some organics at very low current densities (see below). This last procedure can itself be considered as a recycling method since it has been used for decades as a means for extending the life of plating baths and maintaining their condition. Nickel and zinc plating baths have a tendency to increasing their Ni or Zn ion concentration respectively, during operation because the anode process is more efficient than the cathodic one. In addition, metal enters solution by corrosion of the anode even when the plant is not working. In the case of weak-acid zinc baths, a rule of thumb is that zinc concentration would double every 12 months if not treated. The traditional treatment was simply to discard a portion of the bath, replacing it with metal-free electrolyte. There is no need for this, and excess metal can be removed without hydroxide sludge formation, using electrolysis. As before, impurities in solution must first be removed. Meyer [172] describes how this is done (see Fig. 7.59). The first step is to pass the electrolyte and rinsewater over activated carbon to remove organic additives, reaction and break-down products. Next follows a selective electrolysis. The presence of 20 mg/litre or more copper results in nickel deposits with impaired corrosion resistance. 10 mg/litre zinc gives rise to brittle Ni deposits and at 100 mg/litre, the nickel will deposit very dark in colour. In the selective electrolysis, somewhat higher current densities (0.3 to 0.5 Amp/sq. dm) are used at first until the initially dark grey or black deposits become light in colour. At this point, current density is dropped, first to 0.1, then to 0.05 Amp/sq. dm during which stage the organics are removed [173]. Finally, nickel of high purity is deposited on a nickel or stainless starter cathode or on pellet type or other cathodes. Such deposits should be tested and, if satisfactory, can be used as anode metal. In the case of low nickel baths, evaporation losses can be made good using the electrolyte.

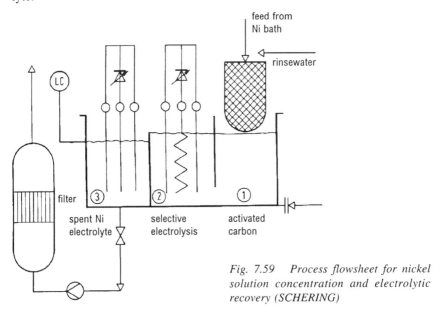

Fig. 7.59   Process flowsheet for nickel solution concentration and electrolytic recovery (SCHERING)

In the case of excess zinc ion concentration in a weak acid bright zinc bath, Schene & Kuebler [174] recommend electrolytic recovery using a membrane cell. This is necessary because of chloride ions in solution, which would otherwise form gaseous chlorine at the anode. The bath is bled into the cathode compartment and a cation exchange membrane is used. After zinc removal by cathodic electrodeposition, the electrolyte is returned to the plating bath.

Because, at the high chloride ion concentrations used, some permeation of these into the anode compartment (to form gaseous chlorine) could take place, a second cation exchange membrane is often used. The the anode compartment and in the space between the two membranes (buffer compartment) dilute (0.5%) sulphuric acid is used. At regular intervals, portions of this (now contaminated with chloride) are withdrawn from the buffer compartment, passing to the effluent treatment plant, and replaced with acid from the anode compartment. The whole process is usually more cost-effective than dumping of metal sludges.

In some cases, the phenomenon of metal enrichment can occur in copper plating baths, for the same reasons as above. Boger & Kuebler [175] describe electrolytic regeneration of an acid bright Cu bath. The cell reduces copper concentration to 1 gm/litre or less, using an insoluble anode. The recovered copper is re-used in a cyanide copper plating bath while the spent electrolyte, containing some organics, is used to neutralise spent alkaline degreasing baths.

In the case of tin, when plated from sulphuric acid baths, static rinse baths after the plating line must be acidified with sulphuric acid to avoid hydrolysis of dragged-over tin salts. [176]. Addition of ca, 1 gm/litre hydroquinone prevents oxidation to tetravalent tin. Electrolysis at current densities of 1 to 2 Amp/sq. dm maintains tin concentrations in the rinse tank at the 2 to 4 gm/litre level. Similar measures are used with fluoborate tin baths [177] where current densities 1 to 1.5 Amp/sq. dm and operation at 40 to 50°C regulate tin concentrations at 1 to 2 gm/litre. In one cell with paddle stirrers which provides linear electrolyte velocities of 1 to 3 metre/sec, current efficiencies up to 85% are obtained. The same author reports on tin recovery from static rinse baths for alkaline stannate plating lines where tin is present in tetravalent form. The rinse tanks here contain sodium sulphate, maintained at pH 12 to 13 [176].

Similar recoveries are possible for zinc and cadmium plating, using parallel plate cells. Although it is a very soft metal, cadmium can be recovered in cells of the type shown in Fig. 7.55. For optimum current efficiency, the pH should be maintained as high as possible, just below the hydrolysis point of the metal. However, as shown by Hedrich & Raub [178], the two metals do behave differently in their electrocrystallisation behaviour, reflecting the greater ease of hydrolysis and amphoteric nature of zinc. Whereas cadmium hydroxide inhibits metal electrodeposition, no such action is found with zinc. While deposits formed at pH > 5 are grey in colour, depolarisation is favoured more by zinc hydroxide than by the hydrated zinc ion.

Vachon et al. [179] used an initial static rinse after cyanide cadmium plating, followed by an 8% NaCl rinse. Solution from the latter is recirculated at high rate (150 turnovers/hr, dwell time = 3.5 sec) through a relatively small electrolysis cell with carbon fibre cathode. This allows a reduction in cadmium levels from 300 to 60 mg/litre in 10 minutes. The anode used is titanium coated with precious metal oxides. Current efficiencies are 98% for cadmium deposition, 93% for anodic oxidation of the cyanide. The cadmium is recovered by re-dissolution of the cadmium in a strong cyanide solution purged with oxygen.

Spent electroless copper and nickel solutions can also be subjected to electrolysis for metal recovery. In the case of copper, this is straightforward, even when EDTA is present in solution as a complexant. After acidification with sulphuric acid, the procedure is as for acid copper baths. The EDTA is simultaneously anodically oxidised [180,181]. In the case of electroless nickel, Meyer [182] reports the use of parallel plate cells with small interelectrode gap, taking nickel concentrations down to 0.5 mg/litre. The cathode was a metallised plastic weave with 2.7 sq. metre surface area. Up to 90 kg nickel could be deposited on this. The amount of electricity required to take the nickel concentration from 7 gm/litre down to 0.5 mg/litre was 55 Amp-hr/litre. Since conductivity decreases as metal ions are removed, cell voltage must be progressively increased, i.e. using a constant current regulator. Figs. 7.60 and 7.61 show two parallel plate cells for metal recovery, with two types of agitation.

One should also mention here the UK-designed "Eco-Cell". This was based on a vertically-rotating cylinder cathode and the novelty for which it claimed patent protection lay in the use of a scraper. At the very high Reynolds Numbers and current densities used, the electrodeposited metal formed as a powder and was thus recovered. The concept was not, with the wisdom of hindsight, a success and the reasons for this are instructive. The finely divided metal formed by electrodeposition is a powerful abrasive, readily capable of damaging bearings. For this reason, the Eco cell design was based on a single bearing above the cylinder cathode and out of solution. This is, as anyone with experience of domestic washing machines will be aware, an inherently weak feature requiring costly engineering and bearing components to mitigate it. The Eco cell was thus an inherently expensive device. Secondly, it can be used to make the following point. As has previously been explained, high flow rates (or their equivalent) are essential to give thin diffuse layers. It does appear that using off-the-shelf pumps, whose costs are low because they are mass-produced, provides the most cost-effective means of producing high flow rates. The secret of a successful electrolysis cell is to minimise the extent to which it has to be custom-built and to maximise the proportion of off-the-shelf components it incorporates.

Only a handful of full-size Eco cells are believed to have been installed and the patents describe their use singly or in a cascade configuration. The cell is described in [183-186] and is seen in Fig. 7.62.

Yet another cell design with a chequered career is the "fluidised bed" cell. Credit for this design is due to Prof. M. Fleischmann (of "cold fusion" fame) and has been many times since been described [187-189]. Its basis is shown in Fig. 7.46c. The cathode is a fluidised bed of small particles, usually made of the same metal that is being recovered. As the metal deposits onto these and they grow in size, they sink lower in the fluidised bed and are in due course removed, to be replaced by fresh "seed" particles. Typical size ranges are 0.5 mm (initial) to 1.0 mm (final). Contact is made to this fluidised bed by feeder bars. In historical terms, the fluidised bed cell marked the first serious attempt to break away from the traditional flat plate cells and it provoked a wave of new thinking, resulting in (to mention but a few) the designs described here. For metal recovery (a process for which it was not in fact designed), it brings certain problems of its own - provision of fresh "seed" particles being one. In addition, there is always a danger that the bed will "freeze" when electrodeposition results in agglomeration of the particles which then cease to be fluidisable. In spite of this, such cells are capable of a remarkably high space-time yield (3,500 sq. metres cathode area per cubic metre) with the corresponding figure for the plate cell being only 16 sq. metres. Fig. 7.63 shows a schematic of the cell which can operate at 10 to 20,000 Amp/cubic metre of cathode. Expressed another way, such cells occupy 200 times smaller volume than a parallel plate cell and can still operate at 2.5% of their current density so that diffusion limiting current conditions are only reached in extreme cases.

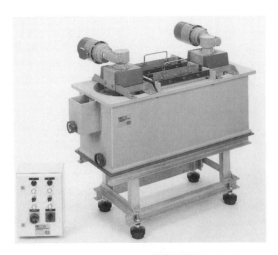

Fig. 7.60   Parallel plate cell for electrolytic metal recovery from static rinse tanks or concentrated solutions. Agitation using paddles. Cathode surface area 0.5 sq. metre (RECON).

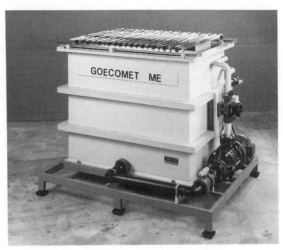

Fig. 7.61 Parallel plate cell for electrolytic metal recovery from static rinse tanks or concentrated solutions. Pumped circulation for agitation. Cathode area ca 10 sq. metre (GOEMA).

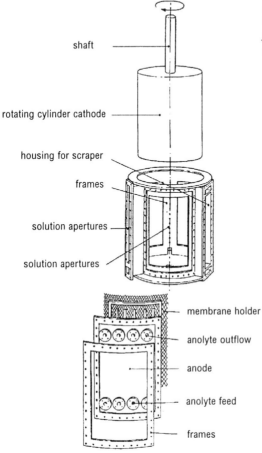

Fig. 7.62   Exploded diagram of the Eco-cell for metal recovery from effluent in powder form [183].

shaft

rotating cylinder cathode

housing for scraper

frames

solution apertures

solution apertures

membrane holder

anolyte outflow

anode

anolyte feed

frames

A typical fluidised bed cell rated at 10 to 20 cubic metres/hr would have a diameter of 0.35 metre, a bed height of 1.2 metre and 7 anodes, with a current rating of 1 to 2 kA. In the case of copper ions, concentrations can be taken from 100 down to 1 mg/litre in a single pass with current efficiency of 70%.

Not to be omitted in this partial listing of cell designs for electrolytic metal recovery from low concentrations, is the Chemelec-Cell, shown in Fig. 7.45d. The cathode (usually of expanded metal mesh) is contacted with small glass beads (up to 500 micron diameter) which are fluidised using pumped flow. The idea goes back to work of LeGoff in the 1960's. According to Bettley *et al.* [190] this type of cell can have diffusion layers around 2 micron thick. Nickel recovered in such cells from solutions of 1500 to 100 mg/litre [Ni] will be ductile. Silman [191] gives an example of a cell with 3.3 square metre cathode surface, contained within external dimensions $0.5 \times 0.6 \times 0.7$ metre (excluding peripherals), operating at 50 Amp/sq. dm. Such cells are suited to metal ion concentrations of 50 mg/litre. This cadmium can be reduced to the 1 mg/litre level or less, while nickel can be maintained at the 500 to 1000 mg/litre level in a static rinse. In the case of Ni or Zn recovery, maintaining the correct pH is vital for optimum performance. Platinised titanium mesh or similar type anodes are normally used.

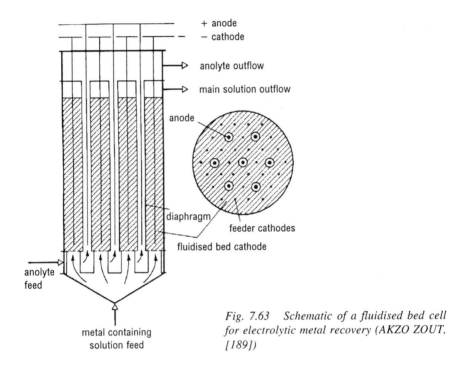

+ anode
− cathode
anolyte outflow
main solution outflow
anode
diaphragm
feeder cathodes
fluidised bed cathode
anolyte feed
metal containing solution feed

Fig. 7.63   Schematic of a fluidised bed cell for electrolytic metal recovery (AKZO ZOUT, [189])

Sections 7.61 and 7.6.2.2 are also relevant to this subject and [192] (Kreysa *et al.*) elaborates on packed bed cells. Section 8.9.4.2 deals with cells for regenerating print circuit board etch baths.

In conclusion, and thinking of the literally hundreds of different cell designs, simplicity of operation (especially recovery the solid metal) and the form in which the metal is recovered, not to mention the cost, are often more important than any one parameter such as space-time-yield.

7.6.2.3   Anodic oxidation

Anodic oxidation of effluent has already been treated (for cyanide) in 3.2.2.5 and (for complexant destruction) in 3.9.5.3.

The main use of anodic oxidation in recycling is for oxidation of chromium (III) to (VI), particularly in the context of chromium plating and etch baths for conditioning plastic for metallising. Presumably the alkaline permanganate solutions now favoured for plastics conditioning will also be regenerated in the same way.

All sources indicate that anodic oxidation of Cr(III) to (VI) is a straightforward process operating at close to 100% efficiency, especially if good mass-transport is assured.

Chromium plating baths build up Cr(III) ions by cathodic reduction of the Cr(VI) species and also as a result of Cr(VI) attack on iron, Fe(III) ions and other solid or ionic species which can be oxidised by the Cr(VI) which is itself then reduced. While some Cr(III) is in fact essential for good operation of most chromium plating baths, it should not exceed 10 gm/litre.

Nohse [193] describes a procedure which has been subsequently adopted by numerous other workers. The chromium solution to be regenerated is passed into the anode compartment of the electrolysis cell fitted with a cation exchange membrane. Cr(III) is then oxidised to Cr(VI) as shown in Equation 3.71. At the same time, cations of other metals migrate through the membrane into the catholyte where they are either deposited on the cathode, or where the catholyte is circulated to a separate deposition cell. In the case of iron, which cannot electrodeposit except at higher pH, the catholyte is removed for subsequent chemical treatment. One drawback is that this cathodic migration takes longer than the anodic oxidation of Cr(III). Anodes are made of lead or lead alloys or platinised titanium expanded mesh. Fig. 7.64 shows the basis of the method. With 1 sq. dm membrane, 1.1 gm iron can be removed per hour while, in this time, 2.2 gms Cr(III) can be re-oxidised. Since Fe and Cr are of almost identical atomic weight, this implies a 2:1 advantage in favour of the redox process.

Since metal deposition is not in general desirable in a membrane cell, whether in powder or adherent sheet form, [194] describes a buffer type cell with dilute sulphuric acid between the anode and cathode compartments, constrained by a second (anion exchange) membrane (Fig. 7.64b). Metal ions pass into the buffer compartment but, thanks to the anion exchange membrane, cannot access the cathode compartment. The electrolyte in the buffer compartment is circulated to a separate electrolytic recovery cell. (It might be mentioned here than metals growing by deposition at a cathode can mechanically puncture a membrane. Though repair procedures exist, it is desirable to avoid this happening).

In most cell designs, a large anode surface area is mated with a smaller cathode. This results in higher current density at the cathode such that metals are more likely to deposit there in powder form, which is easier to remove and recover. In patent [195] an anode:cathode area ratio of from 3 to 20 is recommended with even greater values suggested in [196]. It is claimed that this results in increased current efficiencies, even at low current densities [197]. Self-evidently, current density increases with concentration of Cr(III). Because of their resistance to chromic acid attack, perfluorinated membranes are best-suited for use in such cells, e.g. polyfluoroethylene with sulphone groups substituted. In fact, ion-exchange membranes are not essential and this process can be carried out using a conventional diaphragm (filtering diaphragm) or, with very high anode:cathode area ratios, in an undivided cell. Thus with anode:cathode ratio of 30:1 (using a wire or rod cathode) and current density (cathode) of 60 Amp/sq. dm, satisfactory results can be obtained at 80°C operation [197].

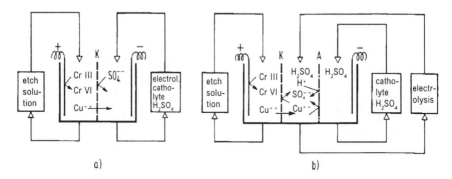

Fig. 7.64 Schematic of anodic oxidation of Cr(III) with simultaneous removal of foreign metal ions. a) simple membrane cell b) membrane cell with foreign metal ion removal from catholyte.

The procedures described above are equally applicable to chromium plating baths, chromic acid etch and chromic-sulphuric pickling baths as well as very concentrated solutions (700 to 950 gm/litre chromic oxide) used to condition plastics for metallising. In the latter case, ion exchange membranes, because of the highly aggressive solution, can be replaced with porous ceramic tile or tube diaphragms. In etching 100 square metre of plastic, some 2 kg of chromic acid will be reduced to ca. 0.5 kg Cr(III) [198]. Heymann [199] has described a cell with concentric lead anode, ceramic diaphragm and cathode. Dilute sulphuric acid is used as catholyte and the cell is suspended in the etch bath. The cell is shown in Fig. 7.65. Operating at a current density of 1 to 3 Amp/sq. dm. this not only re-oxidises Cr(III) to (VI) but also oxidises dissolved organics to carbon dioxide and water. Current efficiency (for chromium) is 50 to 60% but drops off as soon as as Cr(III) concentration falls below 20 gm/litre. Similar information is given by Innes *et al.* [200] who quote 70% current efficiencies for Cr(III) concentrations above 15 gm/litre. At the same time, diffusion of the concentrated chromium solution into the sulphuric acid catholyte was less than 1% over 76 hours at room temperature. In [199], cell design and operation parameters are laid down for maintaining constant Cr(III) concentration in terms of throughput area of plastics to be etched.

Use of a packed bed cell with lead beads for the above reactions is described by Gross *et al.* [201] and brings, as might be expected, benefits in terms of space-time yield. In a patent [202] the anodic regeneration of manganate to permanganate with simultaneous destruction of organics is described. Current

densities of 0.1 to 20 Amp/sq. dm are claimed using a diaphragm cell operating
at up to 25V. A wide range of anode materials is proposed (Ni, stainless steel,
Pt/Ti) with stainless steel or copper cathodes

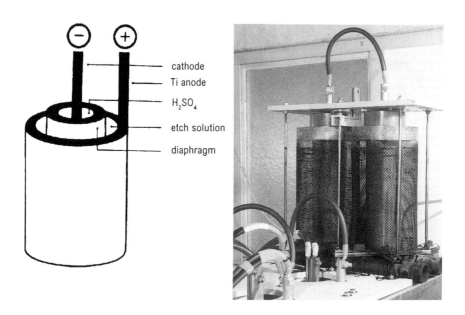

cathode
Ti anode
H_2SO_4
etch solution
diaphragm

Fig. 7.65   Cell for anodic oxidation of Cr(III) to Cr(VI).
Unit can be suspended in a tank.
Left – schematic.   Right – actual unit (Schering)

In electrolysis cells and systems such as those described here, various side-
reactions will take place:

$$SO_4^{--} \quad\rightleftharpoons\quad SO_4 + 2\,e \qquad\qquad 7.42$$

$$SO_4^{--} + H_2O \rightarrow H_2SO_4 + \frac{1}{2}\,O_2 \qquad\qquad 7.43$$

The sulphate radical formed in 7.42 then reacts with water to form oxygen and
hydrogen ions (acidic) 7.43 is an alternative presentation.

At higher sulphate ion concentrations, persulphate can form:

$$2\,SO_4^{--} \quad\rightleftharpoons\quad S_2O_8^{--} + 2\,e \qquad\qquad 7.44$$

in alkaline solution:

$$OH^- \rightleftharpoons Q'H + e \qquad\qquad 7.45$$

$$2Q'H \rightarrow 2\,H_2O + \frac{1}{2}\,O_2 \qquad\qquad 7.46$$

In chloride media:

$$2\,Cl^- \rightleftharpoons Cl_2 + 2\,e \qquad\qquad 7.47$$

The chlorine so formed reacts with water to form hypochlorous acid or hypochlorite:

$$Cl_2 + H_2O \rightleftharpoons HOCl + HCl \qquad\qquad 7.48$$

and subsequently (not shown) chlorates are formed. Anodic reactions of cyanides are shown in Equation 3.54a and 3.54b. In all such cases, the anode must be "inert" i.e. not dissolve to any significant extent. Reactions such as 7.47 can be avoided using ion exchange membranes. Where graphite anodes are used (their life is short except in chloride media), small amounts of carbon monoxide and dioxide will form. Mention is also made here of an undesirable cathode reaction which is reduction of nitrate to form lower oxides of nitrogen:

$$NO_3^- + 3\,H^+ + 2\,e \rightleftharpoons HNO_2 + H_2O \qquad\qquad 7.49$$

$$NO_3^- + 4\,H^+ + 3\,e \rightleftharpoons NO + 2\,H_2O \qquad\qquad 7.50$$

These reactions have low rate constants and, to the extent they take place, do not interfere with the main metal deposition process [142].

7.6.3 Electrodialysis

Electrodialysis is a process which brings together two physical mechanisms − dialysis and electrolysis. Whereas in dialysis, the critical factor is the concentration gradient on either side of the membrane, for electrodialysis it is the rate of ion migration in an electric field. The introduction of ion exchange membranes, especially those made of newer polymeric materials, proved critical to the success of dialysis and electrodialysis, as will be discussed at the end of this section. Both processes had long been known in the laboratory, their practical adoption was postponed until suitable membranes became available. In the case of electrodialysis, the former use of inert membranes allowed only separation of uncharged species (absence of charge) or colloidal or macromolecular species (on account of their size) from ionic species. Use of modern ion exchange membranes allows anions and cations to be separated and only when this became possible was the process of interest to the metal-working and -finishing industries.

Ion exchange membranes, their support mesh apart, and like the resins themselves, consist of artificial resins with exchangeable groupings (see Section 5.2.3.3) in some cases using the same matrix, in other cases a similar matrix to the bead-form resins. In the metal-working industry, only homogeneous membranes are found. By contrast, heterogeneous membranes are constructed of mechanically resistant resin films into which ion-exchange particles are embedded. Electrodialysis is used for two main purposes in the metal-working industry, these requiring two different cell designs:

- concentration processes, e.g. recovery of process bath constituents from rinsewater, with simultaneous de-ionising of water.

- regeneration processes in which interfering ions are removed from process baths or concentration processes to recover individual substances

The main difference is that in the first case, numerous cells can be linked in series using a single electrode pair, in the second case each cell requires its own pair of electrodes.

7.6.3.1   Concentration processes

The cells used for this purpose have their own characteristic design with alternate anion and cation exchanger membranes between an anode and a cathode as shown in Fig. 7.66. The two end compartments containing the electrodes are usually fed with a high conductivity electrolyte such as dilute sulphuric acid. The process start-up involves filling several of the intermediate compartments with the same solution, e.g. the content of a static rinse tank. Under the influence of the potential gradient, anions begin to migrate to the anode, cations to the cathode, until they encounter an ion-exchange membrane which arrests them. In this way, anions and cations come together in every second cell, forming a concentrated solution there, with these same ions being depleted from the remaining cells where the solution becomes progressively lower in ionic strength. For this reason, these cells are known as "concentrate" and "diluate" cell respectively and once the process is under way, the electrolyte streams are divided into "diluate" and "concentrate" circuits. The main features of "concentration-depletion" electrodialysis are, with certain exceptions, based on three separate circuits:

- concentrate circuit
- diluate circuit
- anolyte + catholyte circuit.

In most cases, the diluate is continuously topped-up with rinsewater. The concentrate, however, is returned to the process bath on a continuous or discontinuous basis, in some cases after an intermediate stage such as evaporation, purification, adjustment of composition. Further details are given by Fritsch [203], Deuschle *et al.* [204], Linnhoff [205,206], Nohse [207] and Hartinger [208].

If the energy requirement of the concentration-dilution process is to be optimised, the thickness of each compartment must be held to a minimum, to give the smallest voltage drop. This has the further advantage that migration of ions from the diluate into the concentrate is more rapidly achieved. In modern plants, thickness values of 1 mm or even 0.5 mm are found. Any further reductions would impair the electrolyte flow and so bring no added benefit.

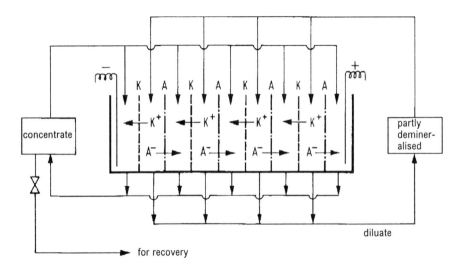

Fig 7.66   Schematic of an electrodialysis cell stack in series array for a concentration-dilution operation. K and A are cation and anion selective membranes respectively

In order to prevent adjacent membranes from contacting one another and impairing the flow, spacers are inserted between the membranes. This serve the following purposes:

- controlling inter-membrane gap.
- promoting turbulent flow by filling the spacer frames with a rough weave material or similar.
- ensuring uniform flow across the membrane surface with feed and exit manifolds on opposite sides.
- sealing the cell stack by compression of membranes and spacer frames between end electrodes.

Fig 7.67 shows in schematic form, the assembly of a typical stack. Fig. 7.68 shows, as example, how four separate liquid streams can be separately fed through a stack, using the spacers. The membranes positioned between the spacers must have sufficient porosity [209].

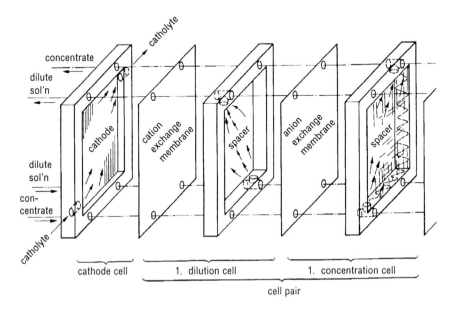

Fig. 7.67 Schematic showing assembly of an electrodialysis stack

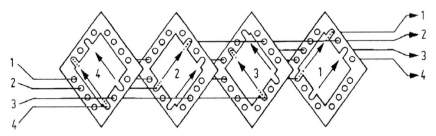

Fig. 7.68 Spacer allowing four different solutions to flow as separated streams
(GKSS Research Centre Geesthacht)

Countless applications of this process could be cited, many (in the context of this book) related to rinse water treatment. However two variants described below are of interest. The use of Electrodialysis for recovery of nickel solutions has been several times described. However many other solutions including cyanide baths, noble metal plating solutions and chemical, rather than electrochemical baths such as phosphating solutions have been recovered.

In the mid-1970's, Sunshin [210] reported his results for nickel recovery. His equipment was as shown in Fig. 7.66. The electrolyte used in the end compartments was sulphuric acid – sodium sulphate. Using a feed with nickel concentration held constant at 9 gm/litre, a concentrate containing 55 gm/litre Ni was obtained after 5 hours at current density of 1.3 amp/sq. dm. The increase in concentrate volume was around 30%, the current efficiency over a range of runs was 85 to 90%. The mean concentration in the outlet solution was 4 to 5 gm/litre Ni and this should never fall below 0.5 gm/litre. Compared with nickel, only 5 to 10% of the boric acid was transported through the membrane because of its low degree of dissociation to form ions. Non-ionic species such as butynediol would not be expected to migrate across the membranes. In another report, Fischer [211] found that transport of boric acid was 50% of the value for nickel. The extent to which various organic constituents and their breakdown products are transported is a function of their structure and properties.

In another series of experiments, Markovac & Heller [212] used a cell of unit size 220 sq. cm × 0.1 cm with 400 volume turnovers per hour, a nickel concentration of 150 mg/litre. This showed a limiting current of 0.18 Amp/sq. dm. Using feed concentrations of 3 gm/litre, corresponding with actual practice, the limiting current increases to 5 Amp/sq. dm, a value of purely academic interest since the membranes are unable to withstand the operating temperatures caused by ohmic heating, which such current densities would imply. In the case of very low nickel ion concentrations, i.e those where comparable concentrations of hydrogen ions exist in solution, (as as pH 4 to 5), the latter will migrate preferentially and this can result in precipitation of nickel hydroxide with corresponding reduction in current efficiency [144].

According to Marquardt & Schmid [213], nickel recovery is best brought about by coupling the first static rinse direct to the ED unit. This allows a steady state concentration of 5 gm/litre Ni to be maintained. The second static rinse is fed by a very slow flow of recirculating water (equivalent to one exchange every 16 hrs). The outflow is held in a buffer storage tank. This too, is fed to an ED unit which reduces the nickel concentration to 50 mg/litre, at which point it can be returned to the feedwater storage tank. At regular intervals, the ED units require cleaning and descaling. This is done using 2% sulphuric acid with polarity reversal and reversed direction of flow.

The same authors also describe the recovery of copper cyanide solutions from a static rinse, where conductivity measurements are used to control the operation, holding the Cu concentration at 5 to 10 mg/litre. From the concentrate stream one obtains a solution with 30 to 40% of the copper concentration used in the process bath. Linnhoff [214] reports on the same application but in this case, the content of the second static rinse is circulated at 8 cubic metres/hr through an ED stack with 250 cell pairs (each measuring 500 × 500 × 1 mm). Fig. 7.69 shows the arrangement, including the 200 litre/hr circulation between a buffer stock tank and the electrolyte. This maintains the concentration in the static rinse tank at less than 1 gm/litre Cu and CN. Current density is 1.8 Amp/sq. dm, stack voltage is 195V (0.75 V drop per cell).

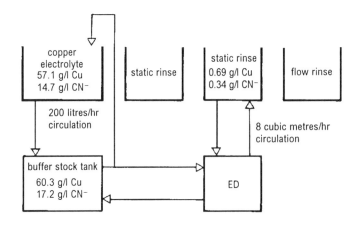

Fig. 7.69   Flowsheet for recovery of cyanide copper by electrodialysis (LINHOFF)

Lindstedt & Millman [214] describe the use of ED linked to a rinsing system for gold recovery. A static rinse tank after the gold plating stage is linked to an ED unit. This is followed by a 3-stage cascade rinse, part of which is circulated through activated carbon filters, part through a selectively acting gold exchanger with its own regeneration loop. The electrodialysis yields a concentrate of 7 to 10 litres/day with 2.5 to 4 gm/litre gold. Figs. 7.70 to 7.72 show the membrane stack and ED unit as used for this application.

As shown by the studies of Sunshin [210] and also Markovac & Heller [144], the concentration found in electrodialysis is also accompanied by a water transport process as the migrating ions bring their hydrate shells with them and this results in a dilution action which can range from 15 to 30% Yet a further process resulting in water transport in the same direction, is due to osmosis which can take place thanks to the water permeability of the hydrophilic ion exchange membranes. It is for this reason that upper limits are found for the degree of concentration possible for some processes.

Fig. 7.70   Membrane stack with 252 cation and anion exchange membranes having 806 square metre total surface area. Spacing = 0.5 mm (GOEMA)

Fig. 7.71 Electrodialysis unit for concentration of dilute solutions (e.g. rinsewater). Membrane area = 1201 square metres (GOEMA)

Fig. 7.72 Electrodialysis unit for concentration of dilute solutions with built-in filter and rectifier (RECON GmbH)

7.6.3.2 Processes for removal of specific substances

A quite different application of electrodialysis is for the extraction of one particular species, be it an anion or a cation, from a solution. It may be that this species is then re-used or otherwise simply disposed of. In this case and in

contrast to the previously shown designs, each cell pair requires its own electrodes. Some economies can be made by using both sides of each anode or
cathode for two cell pairs. Each electrode must be individually connected to the
power source, in order to avoid current leakage through the feed pipes.

The most common application here is for regeneration of pickle liquors and
to remove chromic acid from rinsewaters for re-use. For simplicity's sake, these
processes are discussed in terms of individual cells.

As far back as 1959, Quitmann [215] described how sulphuric acid, used for
pickling of steel, might be regenerated. The spent pickle liquor is fed to the
anode compartment, the Fe(II) ions migrate through the cation membrane to the
catholyte, which consists of alkaline sodium sulphate and there precipitate as
ferrous hydroxide. This is then removed by external filtration. The author showed
that even when saturated with Fe(II) ions, the membrane favoured passage of
these as against hydrogen ions. At 70 to 80°C, 60 to 70 gm iron are removed per
hour per square metre of membrane. Current density was 1.2 to 1.5 Amp/sq. dm,
cell voltage 3.5 to 3.6V. At 70% current efficiency, the energy requirement was
4 to 5 kWhr/kg iron. Fig. 7.73 shows a schematic of the plant.

To avoid problems from undesirable electrode reactions, additional membranes may be used as described in the following examples (see also Section
7.6.2.2.2 where regeneration of weak acid zinc electrolyte is described [174]).

In order to avoid anodic chlorine gas formation in the electrolysis of HCl
pickle liquor, the anode is operated in a sulphuric acid solution as shown in Fig.
7.74 and this is separated from the remainder of the cell with a cation exchange
membrane. Iron is cathodically deposited, preferably externally [216].

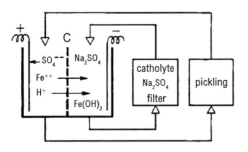

Fig. 7.73   Schematic of ED
process for regeneration of
sulphuric acid steel pickle
liquor. C is cation exchange
membrane [215].

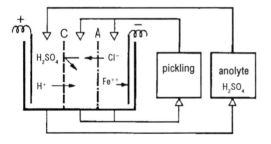

Fig. 7.74   Process flowsheet
for regeneration of HCl steel
pickle liquor using ED. C is
cation exchange membrane. A
is anion exchange membrane
(SOCOMATEN [216]).

In the same way, Lancy avoided contacting of the nitrate anion with the anode in regeneration of nitric acid copper etch baths or zinc etch baths for electrotype plates as shown in Fig. 7.75. Recovery and regeneration of chromic acid from rinsewater in a Japanese plant is reported in [218]. The aim is to recover chromic acid and also to remove foreign metal ions or those of Cr(III). This operation should not be confused with that described in 7.6.2.3 in which chromic acid solutions are regenerated by oxidation of Cr(III) ions. The two processes can, however, be combined.

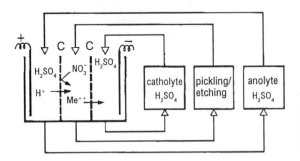

Fig. 7.75 Schematic for regeneration of nitric acid etching or pickling baths (Lancy [217]). C is cation exchanger membrane. [215].

Further details are given in Fig. 7.76. The solutions are fed to the cathode compartment and the bichromate ions migrate through an inert (ceramic) porous membrane to the anolyte. Sodium ions migrate migrate from the anolyte to the catholyte where they form NaOH with the cathodically generated hydroxyl ions. Foreign metal ions precipitate as hydroxides and are removed by external filtration. In order to minimise cathodic reduction to form Cr(III) ions, more concentrated chromic acid liquors should be diluted to at least 150 gm/litre. Energy requirement to recover 1 kg of chromium trioxide is less than 10 KWhr. Lead is used as anode material and, before use, should be anodised in sulphuric acid to form a lead dioxide coating. Trickle coolers are installed in both anode and cathode compartments to hold the temperature below 50°C and at this temperature, evaporation loss results in concentration of the anolyte to the 300 to 400 gm/litre chromic acid level.

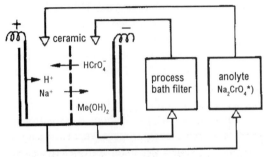

*) Initially, then catholyte.

Fig. 7.76 Schematic for recovery of chromic acid from dilute solution [218].

A similar mode of operation is described in technical literature from a supplier company [219]. Rinsewater for treatment is drawn from the third stage of a rinse cascade at concentration of 50 to 500 mg/litre chromic acid and fed to the cathode chamber of the cell. Feed concentrations greater than 1500 mg/litre should be avoided. The concentrates formed are around 75 to 112 gm/litre chromic acid. Foreign metals precipitate as hydroxides and are flushed out. The diluate stream can be re-used in the rinsing cascade. Platinised titanium expanded mesh anodes are used, with stainless steel cathodes in similar form.

A similar procedure is also reported by Linnhoff [205] but with use of an anion exchange membrane to divide anolyte and catholyte. As before, foreign metals form hydroxides and are externally filtered. Other details are given in Table 7.7. The result is to maintain the chromium trioxide content of the static rinse tank at 20 gm/litre.

TABLE 7.7:
PURIFICATION AND RECOVERY OF CHROMIC ACID
BY ED (LINHOFF [205])

Species	Chromium Electrolyte	
	before treatment	after treatment
chromic acid CrO_3 (gm/litre)	250	270
Cr(III) Cr^{+++}, (mg/litre)	1075	130
Fe(III) Fe^{+++}, (mg/litre)	250	100
Cu(II) Cu^{++}, (mg/litre)	850	210
Ni(II) Ni^{++}, (mg/litre)	2500	875

Energy required to recover 1 kg of chromic acid is around 15 kWhr and the rinsewater from an associated cascade rinse is simultaneously demineralised. The regenerate from the anion exchanger is returned to the ED unit. Fig. 7.77 shows an ED unit with 5 cell units, rated at 10 kg chromic acid/24 hours.

In electroless copper plating of printed circuit boards, copper and NaOH are consumed, sodium sulphate is formed while formaldehyde is oxidised to formate. In order to avoid dumping of these and other products, a treatment procedure was devised by Horn [220] and Krulik et al. [221]. The cell is divided into three compartments by two anion exchange membranes, with a working compartment between the anode and cathode compartments. As shown in Fig. 7.78, NaOH loses hydroxyl ions which migrate out of the cathode compartment. Sulphate and formate ions migrate to the anode compartment where formate is oxidised at the anode to carbon dioxide. Copper sulphate, formaldehyde and various stabilisers for the copper plating bath must be replenished. The authors assert that very little hydroxide ion, no EDTA or copper-EDTA complex enter the anode compartment and this is also true for other anionic complexants and copper complexes. No comment is made on these reports and no volume increase of the copper solution is quoted. The only waste product is dilute sulphuric acid.

Fig. 7.77   ED Unit for recovery of chromic acid from rinsewater. Rating is 10 kg chromium trioxide/ 24 hrs (LINHOFF GmbH)

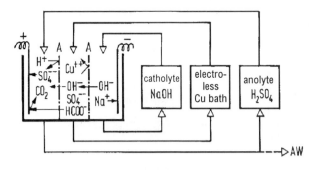

Fig. 7.78   Schematic of an ED unit for regeneration of electroless copper baths. A is anion exchange membrane. [220,221].

7.6.4   Dialysis

Dialysis is not one of the electrolytic processes, nor has it much in common with such widely used regeneration processes as ion exchange or membrane filtration. It is briefly considered here, since it too uses ion-exchange membranes for separating chemical compounds.

The main use of dialysis in the metal-working industry relates to separation of mineral acids and their metal salts, i.e. very much the same function for which the retardation process (7.2.2.2) is used. In the latter case, an anion exchange resin is used while in dialysis, anion exchange membranes are employed. The same underlying theory is applicable in both cases – mineral acids in their undissociated or associated forms can diffuse through the anionic membrane whereas the dissociated metal salts are help back (see theory section in 7.2.2.2).

The most widely used dialysis application uses anionic membranes, with the cell stack being constructed much the same way as series-arrayed cells in electrodialysis concentration processes. The main difference is that only one type of ion exchange membrane (in this case anionic) is used and that there are no end electrodes. A unit is shown in schematic form in Fig. 7.79. Every other cell is supplied with a feed of acid and metal salts. The alternate cells are fed with fresh water, flowing in the opposite direction. The acid then diffuses through the membrane into the freshwater feed. The two outflows thus contain a low acid, metal-rich stream in one case, a high acid metal-lean stream in the other. The latter is returned to the process bath, the former can be subjected to a second concentration, the metal-rich stream is then passed to a metal recovery stage, e.g. electrodeposition.

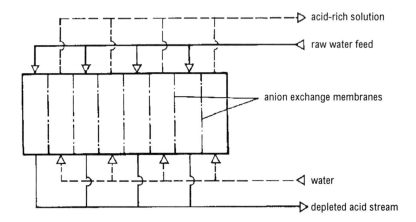

Fig. 7.79   Schematic of a dialysis cell for separating mineral acids from their salts.

Dialysis coefficients for acids are substantially larger than those for salts and this ties in with what is known from the retardation process. In one case, the dialysis coefficient for sulphuric acid is, depending on the temperature, some 50 to 100 times greater than that of nickel sulphate [222]. The units are Mol/square metre/hr. Two examples are summarised in Table 7.8.

TABLE 7.8:
DATA FOR SEPARATION OF MINERAL ACIDS FROM THEIR SALTS BY DIALYSIS [222]

components to be separated	parameter		feed		outflow	
			solu-tion	water	acid-rich	acid-lean
HCl, AlCl$_3$*)	flow rate	[l/h]	830	830	700	960
	HCl	[g/l]	100		85	25
	Al$_2$O$_3$	[g/l]	30		0.7	26
H$_2$SO$_4$, NiSO$_4$		[l/h]	20	20	14	26
	H$_2$SO$_4$	[g/l]	32		26	12
	Ni	[g/l]	1.7		<0.04	1.6

*) For this rating, 1000 sq. metre of membrane are required
(2700 pieces each 43 × 86 cm).

7.7 ADSORPTION

Adsorption is here defined as a separation process. As such, it has a relatively low throughput rate. This can be increased where, for example, the "concentration" of species on the surface of the adsorbent is increased, e.g. by forming a multilayer adsorbate consisting of micelles. The main application of the method in the metal-working industries is therefore not for recovery or recycling but rather to remove contaminants and other undesirable species present in low concentrations, from solution and these can then be recovered. In effluent treatment, its use is limited to adsorption of hydrocarbons, chlorocarbons and phenols where these are present, and also for removal of organic dyestuffs as used in colouring of anodised aluminium.

7.7.1 Theory

Standard textbooks on physical chemistry explain the various short and long-range forces which not only hold together all solids and liquids, but which also bind adsorbing species to a solid. Adsorption onto a solid can take place from the gas phase or from solution. In the latter case, the dissolved species (solute) compete, in the adsorption process, with those of the solvent. It is useful to sub-divide adsorption into "physical" and "chemical" adsorption (sometimes known as "chemisorption"). The forces involved in the former case are much weaker than those in the latter case, which approach in magnitude typical inter-atomic bonding forces. Physically adsorbed species can thus be quite easily removed, i.e by gentle heating, washing etc. Chemisorbed species require much greater energy input before they are desorbed.

There is no more important concept, in any study of adsorption processes, than the use of an "adsorption isotherm". This is an equation (which can be graphically represented as in Fig. 7.80) linking the concentration of species in solution or the gas phase, with that on the surface of the adsorbing material (adsorbent) at a given temperature. The best-known such isotherm is the Langmuir model (Equation 7.51). However equally important and somewhat different isotherms have been formulated by Temkin and by Frumkin.

$$b = \frac{b_\infty \cdot c}{c \cdot a}$$ 7.51

where b is amount adsorbed on the surface (gm/100 gm adsorbent), b_∞ = saturation adsorption (units as before), c = equilibrium concentration of substance being adsorbed (gm/litre) and a is a constant for a particular substance-adsorbent combination (gm/litre). As seen in Fig. 7.80, where c is small, the isotherm is quasi-linear and can be expressed as:

$$c \ll a \quad \text{and} \quad b = \frac{b_\infty \cdot c}{a}$$ 7.52

As c increases, the curve bends over until saturation adsorption is reached. However if multilayer adsorption occurs (association, condensed phase formation, micelle formation), the simple concept of saturation breaks down and isotherms may adopt a "staircase" form, each step corresponding to formation and saturation of a further monolayer.

The classical adsorption model is based on a flat surface onto which species adsorb. The isotherm is then expressed in terms of "coverage" (from 0 to 1). In practice, such flat, smooth catalysts are rarely if ever used. In order to maximise the surface area, finely divided powders are used, and the more finely divided they are, the greater their surface area. This is usually expressed as "specific surface area" (square metre surface area per gm). While such powdered adsorbents have very high specific surface areas (see Table 7.9) the adsorbing species have to diffuse into the bulk and this can take time. The rate of adsorption or desorption is a function of temperature, size of the adsorbing species and the geometry (particle size) of the solid phase. Small species adsorb faster than larger ones. Schwuger [223] gives the following equation in which the particle size of the adsorbent figures:

$$\tau = a \cdot r^n$$ 7.53

where τ is the half-life (time in which the value of c is halved), r = mean radius of particles, n = constant.

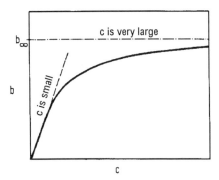

Fig. 7.80 Example of a Langmuir adsorption isotherm

Adsorption from solution is a much slower process than from gases, reflecting the smaller mean free path length and in project planning, contact times from 15 to 60 minutes should be provided for [224].

Adsorptive (bonding) forces increase with molecular weight, at least within a homologous series (Traube's Rule). It also follows that more strongly adsorbing species will displace those already adsorbed, if these form weaker bonds. This is the basis of chromatography as used in preparative and analytical chemistry. pH can also affect adsorption behaviour, mainly in the way it alters the form of the adsorbing species (dissociated or undissociated acids, protonated or unprotonated compounds). Phenols or aniline are two examples of this.

As mentioned above adsorbent can have very high specific surface areas, up to 1500 square metres/gm. However at these high values, stability can suffer and the specific area will normally decrease with age or continuing use. Further details are seen in Table 7.9 where the solids listed can be classified [224] as follows:

- microporous (up to 20Å) (for gases and vapours)
- mesopores   (20 to 500 Å) (for molecules 10 to 30Å in size)
- macropores (over 500 Å) (for macromolecules)

TABLE 7.9:
PHYSICAL DATA FOR SOME ADSORBENTS [225,227]

adsorbent	spec. surf. area m²/gm	micropore vol. ml/gm	macropore vol. ml/gm	fill density gm/litre
granular carbon for water treatment	500–800	0.3–0.6	0.3–0.4	300–550
powdered carbon for dye removal	700–1400	0.45–1.2	0.5–1.9	250–500
closed pore kieselgel	600–850	0.35–0.45	<0.1	700–800
wide pore kieselgel	250–350	0.3–0.45	0.05–0.1	400–800
activated alumina	300–350	0.4	ca. 0.1	700–800
adsorber resin (no active groups)	400–500			650–700

7.7.2   Adsorbents and their use

The most widely used adsorbing materials in effluent and recycling technology are: activated carbon (and charcoal), coke and other powdered carbonaceous materials; adsorbing resins (with and without activated adsorbing substituents); kiesel gel, natural silicates and silicic acid; alumina [226] and also, in a somewhat different sense, metal hydroxides themselves as precipitated in solution. The carbonaceous materials are mainly hydrophobic and are used to adsorb non-polar organics. Alumina and kiesel gel, by contrast, are hydrophilic, indeed so much so that they adsorb polar water molecules and are thus used as drying agents. Adsorbing type resins can be used for both polar and non-polar organics.

From an engineering point of view, a distinction is made between stirred feed reactors and percolation (column) processes. The first is operated batch-wise by addition of pulverised activated carbon or kieselguhr into the stirred reactor. After a given residence time, the solution is filtered and for this reason, a readily filterable absorbent should be selected. In the case of percolation towers, granulated material or pressed pellets are used, 2 to 3 mm in size, packed into vertical columns. The liquid feed can be from top to bottom or v.v. Typical bed heights are 2 to 3 metres, in some cases less [225]. Correct contact times should be maintained and this is done by regulating the feed rates, normally from 0.4 to 1.2 cubic metres/hr/cubic metre of column size [228]. The charging behaviour in this type of percolation column is similar to that of ion exchangers (see Fig. 5.8).

In some cases, activated carbon treatment must be preceded by flocculation with Fe(III) or aluminium salts. This brings with it an adsorptive working of its own and also removed the most finely divided particles which cause turbidity and which would otherwise block the adsorbing sites (see Adsorption coagulation, section 3.7.4.2).

The great strength of adsorptive processes is their ability to remove very low concentrations of absorbable substances either from pure water or concentrated solutions. In the former case, the most important example is organic species in water, including surfactants, brighteners, levelling agents and grain refiners, inhibitors and also entrained oils and greases. Also involved are organics formed by cathodic reduction and anodic oxidation in electrolysis cells. Some of these organics and their reaction products can combine forming, in certain cases, products which are harmful to the deposit properties.

The most important example of adsorption in electroplating relates to maintenance or regeneration of nickel baths and is usually carried out in combination with selective electrolysis (Section 7.6.2.2.2) Brugger [173] discusses in detail, the use of activated carbon and kieselguhr for treating nickel baths. In simple cases, adsorption is coupled with continuous settling filtration, using kieselguhr or other material at 100 to 150 gm per square metre or else kieselguhr with a top layer of activated carbon (300 to 500 gm/square metre). In cases where a very thorough adsorption treatment is required, pulverised activated carbon is stirred into the solution in a separate reactor. Around 2 to 5 gms/litre, in extreme cases

10 to 20 gms/litre of activated carbon are used. Treatments are usually carried out for 30 minutes at 50 to 60°C in a well-stirred reactor. Solids are then removed in a settling filter. When larger quantities of activated carbon are involved, its filterability is improved by addition of kieselguhr. Since thermal energy is released, the solution should be first cooled, but the optimum temperature is presumably a compromise between improved diffusion at higher temperatures and more efficient heat removal at lower values and this is reflected in reduced treatment times. In selecting the activated carbon (of which many different varieties are found), the supplier of the organic addition agents should certainly be consulted, since the aim is to selectively remove the breakdown products and not the remaining additives themselves. Suppliers should be able to advise on the grade or mixture of grades of activated carbon to be used. Pore size distribution and porosity are the most important parameters in this respect. Fig. 7.81 shows a settling filter in use to condition a nickel plating bath.

A second major application of adsorption in metal finishing is for removal of surfactants and other organics from recirculating water loops. In most cases, the porosity of ion exchange units used in such loops serves to maintain sufficient surface tension (see Sections 5.3.2.4 to 5.3.2.6). However where water of the highest specification is required, activated carbon or resin adsorbers should be installed after the ion exchangers.

Fig. 7.81   Plant for maintaining and regenerating nickel plating baths.

Left − settling tank, right − settling filter (Stohrer-Doduco)

In most cases, quite small columns are adequate and these handle water only for particular processes where highest purity is required. In the case of adsorbing resins, suitable types are macroporous weakly acid cationic resins in their hydrogen form [229] or macroporous resins with no functional groupings [227]. The former are well suited to remove non-ionic surfactants, in that they form hydrogen bonds. Taking the case of nonylphenol with 10 Mol ethylene oxide, their useful capacity is 100 gm/litre of resin. Regeneration is carried out with methanolic HCl. The second type of resin adsorbs anionic and non-ionic surfactants and is also well-suited for removing organic residues from liquid-liquid extraction operations (Section 7.3) from the aqueous phase such as tributyl phosphate or di-2-ethyl hexyl phosphate. These resins have a high porosity (40 to 45%), a specific surface area of 450 to 500 sq. metre per gm and mean pore diameters of 40 to 60Å. Their adsorptive capacity, expressed in terms of alkyl benzyl sulphate is around 125 gm/litre resin or 190 gm/litre for nonylphenol with 10 Mol ethylene oxide. In contrast to the case of weak cationic exchange resins, the presence of salts in solution enhances adsorption. Regeneration is carried out with methanol or isopropanol in order to re-monomerise the micelles of surfactant. In certain cases, NaOH, HCl or hot water can be used.

The use on their own of macroporous strongly basic anion exchange resins [230] as "scavengers" for adsorbing dyestuffs and organic ballast materials is not common in metal finishing.

Whereas activated carbon residues from batch stirred adsorption operations may be classified as solid wastes needing special disposal or treatment, comparable materials used in percolation processes can often be regenerated. This is done at around 850°C [226] when the adsorbed organics undergo thermal cracking and pyrolytic carbonation. Adsorbed volatiles (aromatic hydrocarbons, chlorocarbons, phenols etc) can be removed using steam. Typically 1.2 to 1.4 kg steam per litre carbon are required at up to 150°C. Activity of the regenerated material is around 60 to 80% of virgin carbon. There is a roughly 5 to 10% loss of material by combustion or abrasion and this must be disposed of.

Removal of organic such as hydrocarbons, chlorocarbons [231] or phenols [232] by this method is not significant with the exception of effluent which contains only one of the above classes of compound, because otherwise, where there is excess of another adsorbable material, this will "swamp" the adsorbent. Koppe & Herkelmann [233] showed how AOX compounds could be removed from purified mains water from 150 down to 50 ppm (mean concentration in feed = 100 ppm) with 0.1 gm activated carbon required per litre of water. To remove 1 kg of AOX materials, around $3000 worth of activated carbon would be needed, plus the disposal costs for the spent material. Similar costs would be required for treatment of AOX compounds in industrial effluent.

7.7.3   Heterogeneous catalysis

Heterogeneous catalysis is a process almost invariably preceded and followed by adsorption, and as should, it should be mentioned here. The classical definition of a catalyst, due to Ostwald (1891) is:

> "A substance which accelerates the rate of a chemical reaction but does not appear as a product"

and:

> *"A catalyst increases reaction rate but does not affect the equilibrium position".*

If one or more reactants are adsorbed at a solid surface, their effective concentration is increased and this in turn increases the collision number which determines reaction rate. In addition, the energy of adsorption which is released can contribute to the activation energy for the reaction, according to the Arrhenius equation:

$$k = k_{max} \cdot e^{-\frac{\Delta E_A}{RT}} \qquad\qquad 7.54$$

where k is rate constant of the reaction, k_{max} the rate constant for the case where all collisions are fruitful. ΔE_A is the activation energy which at least one partner must possess, if a collision is to be fruitful.

The component $e^{-\frac{\Delta E_A}{RT}}$ of k_{max} thus determines the rate constant for the actual reaction. Examples found in effluent treatment include oxidation of cyanide by air in a packed bed of activated carbon as well as the oxidation of cyanide, nitrite, hydrazine etc by air using dispersed carbon. The incineration of cyanide on platinum (section 3.2.2.3 and 3.2.4) is another example. Homogeneous catalysis (i.e. where the catalyst is a dissolved species) is rarely found in effluent treatment. The main examples are oxidation of cyanide with peroxo and similar compounds where copper ions act as catalysts.

7.8   THERMAL PROCESSES

Two main categories under this heading are important in the metal-working industries:

- thermal decomposition of salts for acid recovery
- water removal from dilute solutions to concentrate these.

The removal of water with formation of more concentrated solutions as part of the strategy of operating an effluent-free process has been discussed at various point in this book. A more appropriate designation might be "zero discharge", since effluent will form wherever aqueous solutions are used for surface treatment. The recovery of water using evaporation is by no means always the answer and the following three points should be borne in mind, where effluent treatment is concerned:

- residues can remain, of very soluble salts either as concentrated solutions or as solids which can no longer be treated as described in this book but which will almost certainly have to be disposed of.

- the energy required for the process will almost certainly have its own environmental implications, e.g. carbon dioxide release
- the evaporated water is by no means pure and can very rarely be used without further treatment, e.g. ion exchange or some form of chemical treatment.

These factors are only seriously affected when the recovered solution can be re-used directly as process solution.

7.8.1   Thermal decomposition

The best-known example here is the recovery of HCl in large steel pickling plants. It is possible that this technology will, in the future, be extended to smaller pickling operations or that it might be used in specialist waste treatment stations. With HCl as pickling acid, the following equations describe the process:

$$Fe + 2\,HCl \; \rightleftharpoons \; FeCl_2 + H_2 \qquad\qquad 7.55$$

$$FeO + 2\,HCl \rightleftharpoons \; FeCl_2 + H_2O \qquad\qquad 7.56$$

The iron (II) chloride is split and oxidised to form Fe(III) oxide and HCl.

$$2\,FeCl_2 + H_2O + \frac{1}{2}\,O_2 \rightleftharpoons 4\,HCl + Fe_2O_3 \qquad\qquad 7.57$$

As this last equation shows, steam and sufficient oxygen are required. The reaction begins at 180°C but only above 850°C is the rate fast enough for a technical process [234]. The ferric oxide is separated out and the HCl gas absorbed into water. The rinsewater from the cascade system can be added to the pickle liquor and treated at the same time. Two main processes are offered by contractors, these being fluidised bed and spray-roasting based.

In the fluidised bed process [235,236] (Fig. 7.82), thermal energy is provided by oil or gas and the process is regulated to form oxide granules 1 to 2 mm diameter and these are withdrawn at the bottom of the bed to form a dust-free oxide (fill density 3.5). The gas formed is cooled to ca. 100°C in a Venturi washer and passed to the absorber tower. The residual gases are cleaned and discharged to air. Fig. 7.83 shows an actual plant.

There are two main spray-roasting processes of which one is seen in Fig. 7.84 [234, 237]. The acid is injected into the roasting furnace at 3 to 5 bar pressure. The resulting iron oxide has a lower fill-density than that from the fluidised bed (0.3 to 0.4 kg/litre) and as before, HCl is absorbed into water to give an 18 to 20% HCl solution.

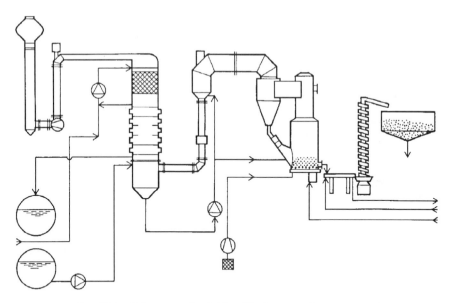

*Fig. 7.82   Schematic of a fluidised bed reactor for recovery of
HCl pickling acid (KERAMCHEMIE)*

*Fig. 7.83   Fluidised bed furnace for HCl recovery from strip steel pickling line.
In the foreground, two cyclones with Venturi separators, centre/rear one of the two
absorbers can be seen. Rating is 2 × 7.5 cubic metres/hr (KERAMCHEMIE).*

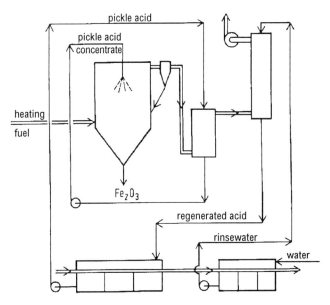

Fig. 7.84 Schematic of a spray-roasting plant for HCl pickle acid recovery (RUTHNER).

In the second spray-roasting process [238] (Fig. 7.85), the pickle liquor is used for cooling in a quencher where it contacts the hot gases. The acid droplets are pyrolysed at 600°C at high turbulence, as shown in Equation 7.57. The combustion gases pass through a heat exchanger and are cooled to 350°C. Iron oxide is separated using an electrostatic precipitator to form a slightly less dense powder (0.2 to 0.3 kg/litre) than the other variant of the process. As before, HCl gas is absorbed in water to form an 18 to 20% acid. Apart from details of construction and contacting, all three processes are based on the same overall chemistry.

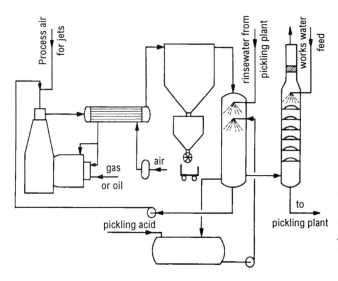

Fig. 7.85 Schematic of a high turbulence spray-roasting plant for HCl pickle acid recovery (DR C OTTO).

Sulphuric acid pickle liquor can also be recovered although for various reasons, this is a more complex operation only viable on a large scale plant. Sulphur dioxide gas is converted to sulphuric acid in a double contacting process.

7.8.2   Removal of water

The two main processes for water removal and recovery are known as vacuum evaporation and evaporation. The former – as will be evident – is better described as distillation under reduced pressure. However it is hardly ever thus described and for better or worse, the terminology "vacuum evaporation" will be used in what follows. For either to be used for rinsewater recovery, a system using the minimum of water must be installed and one with the largest acceptable concentration of dragged-out chemicals. Chapter 6, should be consulted in this respect. Not to be overlooked that a gradual build-up of impurities also occurs in the rinsewater tanks. Water removal serves to further concentrate these and the resulting solutions cannot be returned to the process bath until these impurities are removed or at very least held at a level where they have no damaging effect on the actual deposition process. Section 6.3.2.1 deals with this but the measures called for are usually specific to the type of bath in question. Some are best carried out before, others after the water removal. Thus heavy metal ions are best removed from chromic acid or chromium plating baths before the water removal using strongly acid cation exchange membranes. This is because the competition from hydrogen ions is less marked in the more dilute medium. By contrast, selective treatment of nickel plating baths or carbonate removal from cyanide solutions are better carried out in the more concentrated solutions after water removal.

If these precepts are followed, process solutions can be successfully recovered and the volume of spent rinse water discharged greatly reduced. Evaporation is also the main means for avoiding sludge formation and salting-up of solutions.

In strictly scientific terms, there is no difference between evaporation and vacuum evaporation. However in their practical implementation they differ considerably and will be treated separately in the following sections.

7.8.2.1   Vacuum Evaporation

7.8.2.1.1   Theory

Thermal energy is supplied to a solution which converts it from the liquid to the vapour state. Energy must first be supplied to raise the solution to its boiling point. The energy required is the product of the mass of liquid, its specific heat and the increase in temperature required. For water at 1 bar and 100°C, the

specific heat is 4.18 kJ/kg. To this must be added the heat of vaporisation –
energy required for the change of state from liquid to vapour, without change of
temperature. Total energy requirements are thus given by Equation 7.58:

$$Q' = m'_1 \cdot c_p (T_S - T_A) + m'_2 \cdot h_s \quad [kJ/h] \qquad 7.58$$

where T_A and T_S are the initial and boiling point temperatures respectively, Q' is
the heat flux (kJ/hr), m_1 the initial flowrate of water (litre/hr), m_2 the flowrate of
water to be boiled off (litre/hr), c_p the specific heat of water at constant pressure
(4.18 kJ/kg/degree), h_s the heat of vaporisation of water at its boiling point (kJ/
kg). In the above equation, it should be noted the second term is invariably far
larger than the first.

The heat of vaporisation of water at its boiling point (at 1 bar) is 2256.73 kJ/
kg. It is a pressure dependent term, increasing gently as pressure decreases (e.g.
2394.2 kJ/kg at 0.1 bar). Total energy required for evaporation increases from
2501.88 to 2589.5 kJ/kg as pressure goes from 0.1 to 1 bar. Taking an efficiency
value of 85 to 90% typical of most single-stage plants, change in pressure is not
significant. In practical terms, 2900 kJ or 0.81 kWhr are required to evaporate 1
kg of water, where no energy recuperation is in place. To supply this heat, some
form of heat exchange surface is required with sufficient surface area. The heat
flux Q' shown in 7.58 can then be expressed as:

$$Q' = k \cdot A \cdot \Delta T_m \quad [kJ/h] \qquad 7.59$$

where k is the heat flux transmission of the heat exchanger (kJ/square metre/hr/
degree), A its surface area (square metres) and ΔT_m the mean logarithmic temperature difference. The definition of this last term in the case of liquids (hot
water, heating oils), is complex (see ref. [241]). In the case of steam and neglect-
ing energy required for the temperature change, a simplifying assumption can be
made in which one uses the mean of the condensation temperature of the steam
and the boiling point of the solution. Kreichelt [241] has treated these and other
factors in some detail, as they relate to evaporation of rinsewater.

The heat transmission factor k embodies the transmission at the heating side
(α_H) and the liquid/vapour side (α_L) of the heat exchanger as well as its wall-
thickness δ and thermal conductivity (λ) of the material of its construction as
shown by Equation 7.60.

$$k = \frac{1}{\dfrac{1}{\alpha_H} + \dfrac{\delta}{\lambda} + \dfrac{1}{\alpha_L}} \quad [kJ\, m^{-2}\, h^{-1}\, K^{-1}] \qquad 7.60$$

Typical values for α_H would be 21,000 (steam) but only 8000 (hot water) (units
as above) [241]. Using heat exchanging oils, even smaller values result from
which it follows that use of steam allows the most compact plant.

Thermal conductivities for the most widely used materials in heat exchangers are listed in Table 7.10. Typical wall thicknesses are 1 mm (copper, steel, glass) dropping to 0.4 mm for more expensive materials.

TABLE 7.10:
THERMAL CONDUCTIVITY OF SUBSTANCES
USED FOR EVAPORATOR UNITS

Material	Thermal conductivity λ [kJ m^{-1} h^{-1} K^{-1}] at 20°C
copper	1354
tantalum	195
titanium	135
18/8 stainless	50.4
glass	ca. 4.2
water	2.1
air	0.92

The heat transfer coefficient α_L on the evaporation side is strongly dependent on the hydrodynamic regime and it is at the solid-liquid interface that evaporation actually takes place. By analogy with Equation 7.59, one obtains:

$$Q' = \alpha_L \cdot A \cdot (T_W - T_L) \quad [kJ/h] \qquad 7.61$$

where T_W is the temperature of the heat exchanger surface and T_L that of the liquid being evaporated. Q'/A gives the heat flux q' at the surface. Equations 7.59 and 7.61 are of critical importance in correct sizing of a heat exchanger.

α_L and thus q' show a pronounced maximum for a temperature difference of 28 to 30°C between the heat exchanger surface and the liquid to be evaporated (critical heat exchange area loading) which should not be exceeded. If this difference is too small (only a few °C above boiling point), the resulting convection gives only a low rate of evaporation. For larger temperature differences, vapour bubbles form on the heating surface and these climb the surface and finally become detached. The growth of these vapour bubbles is enhanced by evaporation of the solution, itself a few tenths of a degree above its boiling point. Still further temperature increase leads to growth in number of bubble sites and their area and rate of formation and growth. The climbing action of these bubbles increases turbulence in solution which rips the bubbles from the surface and (for vertical surfaces) leads to natural convection (see below). In addition, the thickness of the diffusion layer decreases and the heat transfer coefficient α_L increases while the solution temperature remains constant. At this

point, optimum heat transfer conditions have been reached. Still further temperature increase is detrimental. A blanketing of the heat exchanger surface by gas (a
good insulator as Table 7.10 shows) occurs and heat transfer rates fall off.
Further temperature increase brings renewed increase in heat transfer – the heat
exchanger surface is now acting as a radiant heat source until finally, at very
high temperatures, the surface burns out [242].

In the optimum region of bubble formation, recommended here, α_L values at
normal pressures are around 80,000 to 125,000 kJ/square metre/hr/degree for
water. For electrolyte solutions, lower values of 16,000 to 25,000 are found
[241].

The foregoing values refer to natural circulation as is found, for example, in
most rinse tanks. Still to be considered is the effect of reduced pressure which
brings two main benefits:

- lowering of boiling point, less danger of breaking down of organic bath
  additives

- increase in mean logarithmic temperature difference and thus the specific
  capacity of the evaporator

The effect of reducing pressure on the boiling point of water is shown in Fig.
7.86 for water and chromic acid at two concentrations [44].

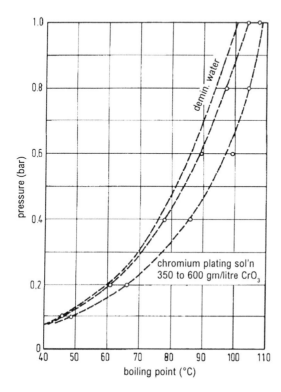

Fig. 7.86 Boiling point as
function of pressure for
demineralised water and
chromic acid [44].

As the boiling point is reduced in this way, so do the temperature differences in Equation 7.59 and 7.61 increase and this in turn leads to increase in the rating of the evaporator even if a reduction in heat transfer coefficient α_L is found [241]. Fig. 7.87 shows how the rating of an evaporator increases with decreasing pressure for various electrolyte solutions and for demin. water [240]. Table 7.11 gives data on boiling points for various salt solutions.

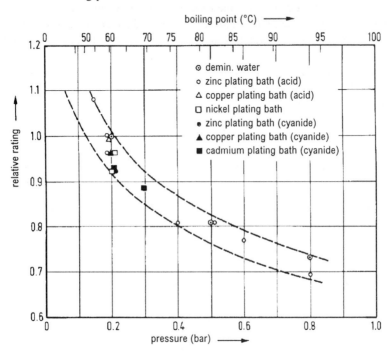

Fig. 7.87   Relative ratings for an evaporator for various electrolytes as a function of operating pressure [240]

TABLE 7.11:
BOILING POINTS OF SOME INORGANIC SOLUTIONS AT 1 BAR [242]

compound solution	boiling points (°C) at various concentrations (gm/100 gm)			
	10	25	50	75
NaOH		108.1	119.5	132.5
NaCl	101.6	104.6		
Na$_2$SO$_4$	100.6	101.6		
ZnSO$_4$		101.0	102.5	
CuSO$_4$		100.6	101.6	103.6
CrO$_3$		101	104	110

7.8.2.1.2   Evaporation of water for recovery of process solution from rinsewaters

In practice (certain exceptions apart) evaporators installed for process solution recovery from rinsewater are rated at 50 to 300 kg water/hr. Because of this relatively low rating, addition of plant for energy recovery was not usually envisaged, although in certain cases, the low grade released could be used elsewhere in the operation (see Fig. 7.57 [163]). Nor were multi-stage evaporators used for this type of application.

In most cases, the plant used is a single stage unit with natural circulation, mostly working at an under-pressure. The construction is vertical with the actual evaporator being based on tube bundle design. The operation is explained with reference to Fig. 7.88.

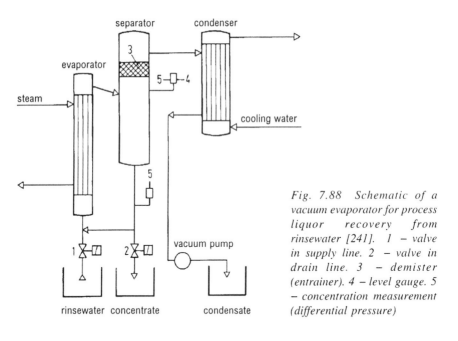

Fig. 7.88 Schematic of a vacuum evaporator for process liquor recovery from rinsewater [241]. 1 – valve in supply line. 2 – valve in drain line. 3 – demister (entrainer). 4 – level gauge. 5 – concentration measurement (differential pressure)

A partial vacuum is set up using the pump. Valve 1 is opened, sucking rinsewater into the system until a certain liquid level is reached. Heating (steam) and cooling (water) is supplied to the evaporator and condenser. Bubble formation causes, in the manner of an air-lift pump, a massive circulation upwards in the evaporator, through the separator unit and back again. Vapour formation reduces the liquid level in the evaporator pipes, and a so-called "apparent liquid level" results which, in the case of water corresponds to 30 to 40% of the column length. The value is affected by the value of α_L, the heat transfer coefficient, obtained and also the tube diameter. In the separator chamber, the condensate is freed from water drops entrained with it and is condensed as water in the condenser. In cases where a liquid ring vacuum pump is used, a portion of the condensate must be returned to the pump to maintain the seal. Once the liquid level has fallen to the lowest acceptable value, fresh rinsewater is admitted and the process continues. At the appropriate time, Valve 2 is opened and concentrate is released and the cycle commences once more. Operation is only possible on an automated basis. Concentration can be determined by measurement of differential pressure or by air bubble injection. In the latter case one or two tubes fed with air dip into solution and the pressure in the feed tubes is measured [241].

The vapour condensation process is the reverse of evaporation. The water vapour surrenders its heat of evaporation to the coolant. Cooling surface areas and heat transfer can be calculated using Equation 7.61 where ΔT is the temperature difference between the condensation temperature of the vapour and that of the cooler wall. For a ΔT of 20°C, cooling water requirement is 33kg/kg of condensate. This highlights one drawback of evaporators. The cooling water, it is true, also be used for other processes and in some cases, the warm water produced can be useful. Alternatively, a small trickle cooler can be used before returning the liquid to the main water lines. The better option is to recover energy and re-use it in the same process (see below). Most liquid ring pumps used for vacuum production can pull a vacuum of 0.02 to 0.04 bar. In most cases, 0.1 bar is used.

Yet another drawback of vacuum evaporators in contrast to simple evaporators, are the expensive materials of construction required. Thus the evaporator tubes, for treating chromic and other acid solutions, have to be made of titanium, tantalum or niobium. In some cases, stainless steel or glass will serve. The separator and circulation tubes can usually be made of borosilicate glass. The outer casing of the evaporator and the pump components in contact with solution, are made of stainless steel, as is the condenser. Glass can also be used in the latter case. Seals are made of ptfe.

Advantages of the vacuum evaporator against direct system are that its capacity is not affected by external factors and that it can be used to process solutions containing surfactants. In the latter case, the separator must be slightly modified to offer an intermediate compartment with its own mist entrainer. Low pressure operation also destabilises any foam formed – the foam cells increase in size as the bubbles move upwards, finally breaking up on contact with the entrainer [44].

With a handful of exceptions, it can be said that any electrolyte capable of withstanding 65°C is suitable for use in vacuum evaporators, be they acid, alkaline cyanide containing or other [44]. As far back as 1965, Culotta [243] describes the use of the method for recovery of cyanide and chromium containing solutions from rinsewaters.

The author has carried out exhaustive studies in which a wide range of electrolytes was evaporated ten times over, with concentration and subsequent re-dilution to the original concentration, at 0.2 bar with no significant changes in overall concentration or results from Hull cell tests. Cyanide losses by hydrolysis are insignificant. The only problems were encountered with bright copper baths where breakdown of the organic additive led to an immediate loss of performance. Apart from chromium plating solutions, no electrolyte exceeded 65°C at 0.2 bar for a doubling of concentration (see Fig. 7.89) [44]. According to Raoult's Law, the boiling point of a solution is raised as the concentration of dissolved species increases and depending also on density of solution:

$$\Delta T = \frac{c}{\varrho} \cdot E \quad [K] \qquad\qquad 7.62$$

where c is Molar concentration and E is elevation of boiling point (°C).

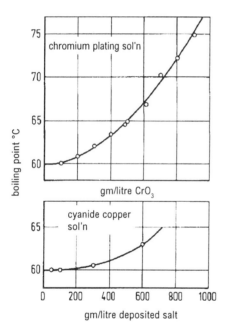

Fig. 7.89   Elevation of boiling point at 0.2 bar as function of electrolyte concentration for chromium plating baths and cyanide copper solution [44]

Problems arise in handling solutions of HF or concentrated phosphoric acid or where such compounds might form. Chromium plating baths often contain fluorides or fluosilicates as stabilisers. Igawa et al. [244] studied the behaviour of such fluoro-compounds in the evaporation of chromium plating baths, establishing their loss rate as function of chromic acid concentration and temperature (see Figs. 7.90, 7.91). The author discovered that such losses are due not only to volatilisation of these acids but also to the formation of gaseous silicon tetrafluoride. In addition, at 0.2 bar, deposits of silicic acid form by hydrolysis on the cooling surfaces where liquid flow is smallest and this is still true at 0.6 bar [44]. Corrosion of glass by HF formed from fluosilicic acid is so slight that the life of equipment with 10 mm thick glass walls is unaffected.

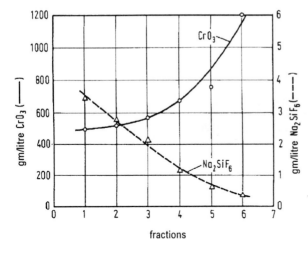

Fig. 7.90 Changes in chromic acid and fluosilicic acid concentrations after evaporation (after Igawa et al. [244])

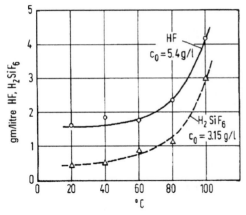

Fig. 7.91 Losses of HF and fluosilicic acid after evaporative concentration of chromium plating baths from 100 to 500 gm/ litre CrO_3 as function of temperature (after Igawa [244]).

Fig. 7.92 shows a vacuum typical recirculatory evaporator. The overall process is significantly improved if, as mentioned above, the heat of condensation can be recovered and returned to the evaporation stage. The additional capital outlay for this is usually justifiable. If the vapours are condensed using a commercially available coolant and this is then compressed in a heat pump, the liquid can then be returned to the process as a heat source for the evaporator. The heat of condensation, together with the work of compression, provides more heat than is required for evaporation and the excess compensates for energy losses in th evaporator. Radke [245,246] has described a unit operating at a boiling point of 70°C at under-pressure. The coolant is heated up during condensation to 40°C and after compression, reaches 90°C when it is used as heating medium. This heating-cooling loop is topped up with minor quantities of heat, the cooling loop has additional air cooling (see Fig. 7.93). All of this saves up to 50% of the energy otherwise required for a single stage evaporator. Further advantages are

absence of cooling water requirement and the ease with which heating energy
shortfall can be made up by electric heating.

*Fig. 7.92 Vacuum
evaporator from chromium
plating solution recovery
from rinsewater. Rating -
120 kg/hr condensate. Left
to right – evaporator,
separator, condenser
(GOEMA)*

Solutions with very low boiling points can also be treated, e.g. at 35°C. However very low pressures are required in such cases – typically 0.043 bar. Under
these conditions, natural circulation is inadequate and units with pumped circulation may be required [246].

An evaporator of this type, resembling in some ways a rotary evaporator
(though without rotating machinery since convective circulation of the solution
to be evaporated fills that function) has been described in a Linhoff publication
[247]. The difference between this and the previously described approach is that
evaporator and condenser share a common container and the vacuum is formed
using a water-jet pump, assisted by a centrifugal pump. The dual-function chamber is made of stainless steel and, for aggressive solutions, PVDF is used where
liquid contact is likely. Fig. 7.94 shows such a unit, made of stainless steel. An
energy requirement for aggressive media of 0.24KWhr/kg condensate is quoted.

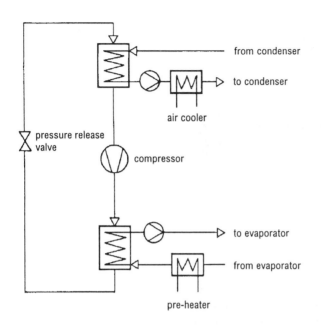

from condenser

to condenser

air cooler

pressure release valve

compressor

to evaporator

from evaporator

pre-heater

Fig. 7.93 Schematic showing evaporator fitted with heat pump for energy saving [246].

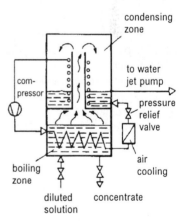

condensing zone

com-pressor

to water jet pump

pressure relief valve

air cooling

boiling zone

diluted solution

concentrate

Fig. 7.94 Vacuum evaporator with integral heat pump. Left – schematic. Right – unit in stainless steel, rating 112 kg condensate/hr. (LINHOFF)

Condensates recovered by evaporation are not to be confused with distilled water and they contain constituents of the original solution which have high vapour pressure. In the case of normal chromium plating baths, this will include chromium chloride (the chloride from make-up water), in the case of stabilised baths, chromium fluoride will be present. In the case of cyanide solutions, cyanide and HCN [15] and various other volatile components will be present, possibly also metallo-organic compounds. Condensates are thus not usually suitable for a final rinsing. They can be used as a pre-rinse in a cascade (see 6.3.2.2) or better still, fed to the inlet of an existing ion exchange unit in a recirculating loop. Specially designed evaporators are used in case-hardening plants for effluent treatment and recovery of heat-treatment baths (see Section 8.2.3)

7.8.2.2   Evaporation

7.8.2.2.1   Theory

At the liquid-air interface, water molecules can form vapour in the air, at a rate depending on various factors. In cases where the vapour pressure of the water is less that the total air pressure, this is known as evaporation. The main difference between this and the previous method is that here, the process operates below the boiling point. In the former case, boiling occurs, even if the boiling point is reduced by use of a vacuum. The reverse process, where water molecules in air are captured to form a liquid, is known as condensation [248]. Evaporation requires that water molecules possess sufficient mean kinetic energy to escape from the bulk. The equilibrium between evaporation and condensation is governed by temperature, humidity of the air above the water and other factors. The moisture content of air, where x (kg water per kg air) is given by:

$$x = \frac{m_D}{m_L} \qquad\qquad 7.63$$

m_w is the mass of water (kg), m_L the mass of air (kg). As long as the water is present in air only in the vapour form, both gases can be treated as ideal gases and the equation:

$$x = \frac{M_D}{M_L} \cdot \frac{p_D}{p_L} = \frac{18}{29} \cdot \frac{p_D}{p - p_D} = 0.622 \, \frac{p_D}{p - p_D} \qquad\qquad 7.64$$

holds. M_D is the molecular weight of water (gm), M_L the molecular weight of air (gm) p_D the partial pressure of the vapour [Nm^{-2}], is the partial pressure of air [Nm^{-2}] p_L the total air pressure [Nm^{-2}].

For dry air, $x = 0$ and for pure water vapour, $x = \infty$. The water vapour content of air can increase only as a function of temperature until it reaches saturation vapour pressure $p_{D,S(T)}$. At this point, the moisture content of the air x_S is:

$$x_S = 0.622 \cdot \frac{p_{D,S(T)}}{p - p_{D,S(T)}} \qquad\qquad 7.65$$

the ratio

$$\varphi = \frac{p_D}{p_{D,S(T)}} \qquad\qquad 7.66$$

designates "Relative Humidity". (R.H.) $\varphi = 0$ is dry, $\varphi = 1$ is water saturated air. The enthalpy (h) of an air-water vapour mixture is:

$$h = h_L + x_D \cdot h_D \quad [kJ / kg] \qquad\qquad 7.67$$

These four variables control the humidity of air and the effect of changes can best be assessed using the h-x diagram (Fig. 7.95) designed by Mollier. This embraces the four variables for a total pressure of 1 bar. Three examples will be used to demonstrate water uptake from air as heat is supplied and which are the basis of the evaporation process. Consider air in condition A (20°C, 7.5 gm/kg moisture content, R.H 0.5, enthalpy 38 kJ/kg). Taking the moisture content as x gm, let us turn to the h-x diagram which normally present moisture content not in kg/kg but in gm/kg of air. I kg of air corresponds to 0.841 φ at 20°C.

If air is contacted with water without supplying heat, it becomes saturated with water. The enthalpy remains constant and temperature drops to point B, around 13°C in an adiabatic process and the moisture content of the air increases from 2.5 to 10 gm/kg. If the process is operated isothermally, at 20°C, then saturation occurs at point C. Water uptake is 7.5 gm water per kg to reach a total moisture content of 15 gm/kg and there is an enthalpy increase to around 57.5 kJ/kg.

If the air is now warmed to 30°C, R.H will now fall to 0.28, i.e. its capacity to take up further moisture is increased. Saturation is reached by uptake of a further 20.5 gm/kg of water to reach a total moisture content of 28 gm/kg and an enthalpy of 102 kJ/kg (Point D).

Model calculations such as the above are required to assess water removal capacity. Below the curve, for $\varphi = 1$, lies the fog zone where moisture contained in the air condenses out. It also shows the curve for the lowest temperature which air can reach before onset of condensation. The mass of water taken up by this process can be expressed as: [249].

$$m_W = \frac{m_{W1} \cdot c_{pW}(T_{W1} - T_{W2})}{\dfrac{(h_2 - h_1)}{(x_2 - x_1)}} = m_{W1} - m_{W2} \qquad\qquad 7.68$$

where m_w is the mass of water evaporated [kg]; m_{w1} is mass of water supplied; m_{w2} is mass of water removed [kg]; c_{pw} the specific heat of water [4.18 kJ/kg/ degree]; T_{w1} is feed water temperature; T_{w2} exit water temperature; h_1, h_2 the enthalpies of inlet (1) and outlet (2) air [kJ/kg]; x_1, x_2 the water content of inlet (1) and outlet (2) air [kg/kg].

Equation 7.68 gives the heat flux as represented by the water feed. In the ideal case, it matches the enthalpy taken from the air flow. In practice, in order to maximise the water evaporation, complete saturation (R.H. =1) is never reached and 0.9 to 0.95 are more commonly found. To this extent, a correction should be applied to Equation 7.68. Actual performance is best derived on an empirical basis (see below).

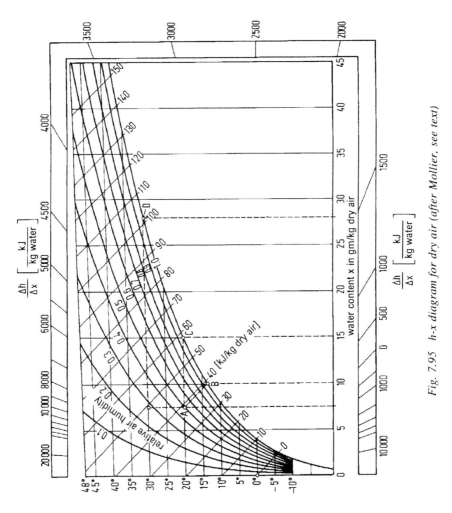

Fig. 7.95   h-x diagram for dry air (after Mollier, see text)

air temp. [°C]

7.8.2.2.2 Evaporation of water for recovery of process solutions from rinsewater

Various designs are found in practice, of which the most common is a trickle cooling tower as used for cooling of hot water. A horizontal configuration, as used in air-conditioning plants, is also found. Figs. 7.96 and 7.97 show diagrams of these. In the case of the trickle cooling tower, the solution to be evaporated is trickled over a bed with packing to maximise surface area, e.g. Raschig rings, various types of hoop etc. Air is drawn upwards in counter-current flow. Moisture saturated air is freed from entrained droplets by a demister and exhausted to atmosphere. At regular intervals, the demister is sprayed with water to wash back into the electrolyte, droplets held there. The concentrated solution returns, now cooled to the holding tank which may be fitted with heating provision, depending on conditions. Optimum results are obtained by balancing air and water flows and contact times in the trickle tower. If the air flow is too large, too much liquid is drawn up in liquid form. If the water flow rate is too great, the interfacial area at which evaporation takes place, is reduced. A term sometimes used is "rain density". A typical value for this is 15 cubic metres/square metre/hr with an airflow of around 11,000 in the same units for a column height of ca. 1 metre. Values for these parameters can also be taken from air conditioning manuals.

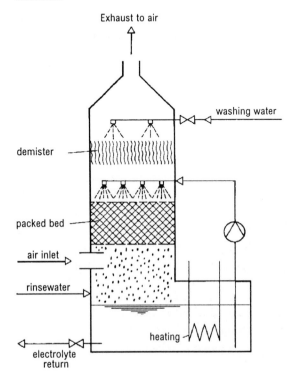

Exhaust to air

washing water

demister

packed bed

air inlet

rinsewater

heating

electrolyte return

Fig. 7.96 Schematic of an evaporator plant, trickle cooling tower type

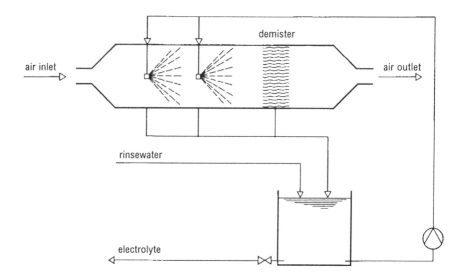

Fig. 7.97 Schematic of an evaporator plant, air conditioning type

Air is usefully pre-heated before use. For water temperatures of 40°C, it should be heated to 30°, for water at 50°C, to 37°C. Uptake capacity for water can be read from the h-x diagram (Fig. 7.95). In the air conditioning design, solution is spray injected into a horizontal airflow. The larger the droplets, the smaller the interfacial contact area. Contacting time is governed by air flowrate and length of the reactor tube. Droplets are captured by a demister as in the trickle tower design to prevent their emission to air.

A patent [250] describes an evaporator in which an exhaust hood is fitted over the electrolyte tank and thence to a packed bed. The concept is a form of trickle tower evaporator. In another variant, a bubble column is used with finely divided air blown up through a column of liquid. A special design for treatment of water from a case hardening or similar plant uses a flat plate evaporator where the air-water interface is irradiated with infra-red. Details are given in Section 8.2.3 while another flat bed evaporator is shown in a patent [252] in which warm air passes of the liquid in flat pans or trays. The foregoing treatment describes the most widely known designs, though many others have been proposed or are used. The main advantages of evaporation are:

- the process simultaneously cools the solutions and this is beneficial for baths operating at ambient temperature where, for example, plating currents tend to heat them. Such thermal energy from passage of current is usefully employed in evaporation plants. The same is true for baths operating at high current densities even if they are designed to operate above ambient, e.g. chromium plating solutions. Further discussion of cooling is given below.

- the air exhaust from process bath evaporators also fulfils the function of air cleaning

- evaporators can be constructed of lower cost materials (plastic, rubberised mild steel).

Drawbacks of evaporators are:

- water containing surfactants may be difficult or impossible to treat because of foam formation which interferes with the process. Only flat plate evaporators can really be used in this case and their capacity is small – 30 to 40% that of the same sized trickle tower [15].

- performance depends on climatic conditions or the previous use of the air. Fig. 7.98 shows the volume of air required to evaporate 1 kg of water and how decrease in temperature or increase in R.H. reduces evaporative capacity. Only at higher temperatures and lower R.H. values is this dependence less marked and this argues for the incorporation of solution heating facilities. Other points to be noted are:

- some species in air build-up in concentration, notably carbon dioxide but also, especially in industrial regions, sulphur dioxide, ammonia, hydrogen sulphide, NO_x and dust if the air is recirculated.

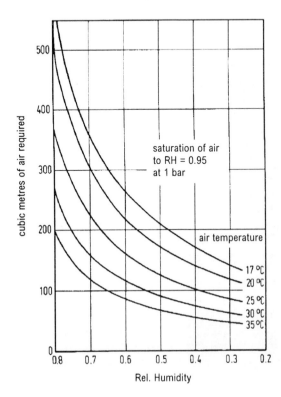

Fig. 7.98   Effect of air temperature and R.H. on volume of air required to evaporate 1 kg of water

Following the resolution of a conflict in the patent field in the 1970's [252], the use of evaporators in Germany became widespread, especially in connection with chromium plating, and in some larger plants, zero-discharge operation of this became possible. Various authors [253,254, 255,256] have described trickle cooling towers. The process can be carried out with solution temperatures of 27°C but operate much better at 50°C [253, 255]. Inlet air will typically have R.H of 0.6 to 0.7 and emerge with R.H at 0.9 to 0.95 [253]. Energy requirements are around 3200 kJ/kg of water, realistically somewhat higher – 3600 kJ or 1 kWh/kg to account for heat losses. Typical efficiencies are around 70%.

For a unit with cross-section around 0.4 sq. metre and air flow of 4200 cubic metres/hr, evaporation rats are 20 litres/hr at 20°C, 200 litres at 50°C [256]. Elsewhere, air volumes from 25 to 150 cubic metres are quoted to evaporate 1 kg water at 50 and 25°C respectively [254]. Since solutions containing dissolved species have lower water vapour pressures than water itself, a correction should be made for this (Equation 7.62).

Though most units have been installed for chromium plating baths, they have also been used for weak acid zinc plating baths and cyanide silver solutions [253-256]. In the former case, a bath composition low in surfactants was specified, in the second case, problems with carbonate build-up were avoided by recirculation of the air. In a first evaporator, rinsewater was heated to 35° before evaporation. The humidified air was chilled in a second unit to remove moisture and then recycled for further use [254].

Fig. 7.99 Evaporator of the trickle tower type for chromic acid recovery from rinsewater (BLASBERG)

Fig. 7.100   Trickle tower for recovery of process solution from rinsewater. Rating is 600 litres/hr for solution at 55°C. At right, the ducting feeding the air to the evaporator or direct to air (GOEMA)

Fig. 7.101   Chromium plating plant with direct mounted evaporators above the baths (LPW Galvanotechnik)

Figs. 7.99 and 7.100 show trickle type units, Fig. 7.101 shows direct mounted evaporators. Where space allows, direct mounting of the evaporator over nickel baths for rack plating is used and these will cope with rinsewater from a 3 or 4-stage cascade. Another application based on barrel plating, evaporates rinsewater which is first heated to 60 to 65°C [257].

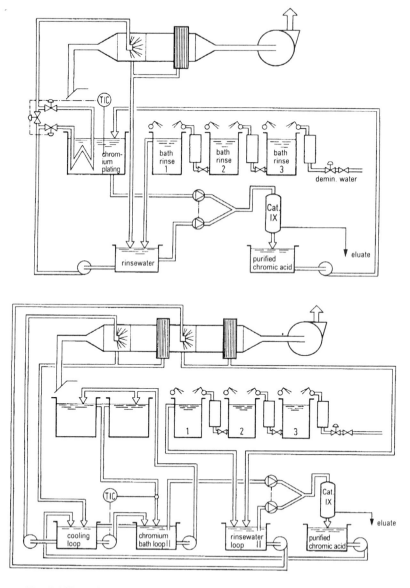

Fig. 7.102   Schematic of chromium plating bath recovery from rinsewater with simultaneous bath cooling, using an evaporator of the air-conditioning type. a) indirect   b) direct cooling

Both designs of evaporator have been used to recover chromium plating rinsings which, because of their toxicity, would otherwise be difficult to treat. Classical chemical treatment of 1 kg of chromic acid requires 1.5 kg concentrated sulphuric acid, 1.4 kg sodium pyrosulphite and 1.8 kg NaOH and this results in formation of 1.03 kg sludge (dry weight) and 2 kg sodium sulphate. Discussions of air-conditioning type evaporators are found in [258-262], these deal mainly with chromic acid treatment. Cr(III) ions and other foreign metals are removed from a 80 to 100 gm/litre chromic acid solution, diluted with rinsewater with a strongly acid cation exchanger (H form). In all cases, the evaporative cooling is utilised while the Joule heating in the plating baths provides heat for the evaporation itself. The current density employed determines the amount of cooling necessary and this is described in [258, 260] and shown in Fig. 7.102.

Indirect cooling uses a heat exchanger in the electrolyte and when cooling is required, rinsewater and the cooled outflow from the evaporator are used. The feed to the cascade rinse matches that produced in the evaporator and, after passing through the ion exchange column, is returned to the process bath.

In direct cooling (b), electrolyte and rinsewater initially have separate streams. The chromic acid loop is piped through the evaporator as soon as electrolyte temperature shows signs of over-heating. Fig. 7.103 shows an evaporator of the air-conditioning design. The process has been used for cyanide zinc baths [259] and another planned use was for a pickling-degreasing tank for steel with the cooling used to promote crystallisation of ferrous sulphate heptahydrate [263].

Fig. 7.103 Evaporator using the air-conditioning design for electrolyte recovery from rinsewater. Rating 50 litre/hr. (RECON).

References for Chapter 7

1. Hartinger, L.: "Recycling in the metal industry promotes environmental protection" Future (1981) 335-336, publ. Ing. Digest Verlag, Mainz

2. Hartinger, L.: "Industrial treatment of metal-containing effluents" Muenchener Beitraege zur Abwasser, Fischerei, Flussbiologie Vol. 34 (1982) 53-76 Oldenbourg Verlag, Munich and Vienna

3. Kammel, R., Lieber, H.W.: "Effluent & recycling technology in metal finishing plants" Galvanotechnik 68 (1977) 57-63, 122-130, 413-418, 710-715, 789-794, 883-6, Vol. 69 (1978) 317-324, 624-630, 687-696

4. Hartinger, L.: "Ion exchangers to reduce effluent volume and metal discharge in the metal-working industries" Korr. Abwasser 33 (1986) 895-907

5. Hartinger, L.: "Effluent treatment & recycling using solid ion exchange resins" Schriftenreihe der GDMB, Vol 53 (1989) 287-311

6. Waitz, Jr., W.A.: "Ion Exchange for Recovery of Precious Metals". Plating & Surface Finishing (1982) Apr.

7. Pritchard, E.J., Pcihoda, W.W.: "Recovery of Gold From Rinse Waters Using Ion-Exchange Techniques". Plating (1969) 1044- 1046

8. Thompson, J., Miller, V.J.: Plating (1971) 809-812.

9. Schumann. R.: "Recovery of precious metals from rinsewater" Oberflaeche (1969) Vol. 11

10. Degussa AG, Frankfurt: "Combined mini-filter and ion-exchanger" Galvanotechnik 72 (1981) 1088

11. DODUCO KG. "Recovery of gold with ion exchangers" Galvanotechnik 72 (1981) 1092-3

12. Marquardt, K.: "Metal recovery in solid form from effluent". II. Ionolysis as extension of the catiolysis process" Galvanotechnik 69 (1978) 115-121

13. Oehme, C.: "Mercury removal from water & effluent using Lewatit ion exchanger" Vom Wasser Vol. 38 (1971) 345-356

14. Hartmann, H.: "Ion exchange for mercury removal from effluent" Muenchener Beitrag zur Abwasser, Fischerei, Flussbiologie Vol. 28, 301-7, Oldenbourg Verlag

15. Hartinger, L.: unpublished work

16. Kauczor, H.W.: "Treatment of metal containing effluents with ion exchange" Metall 32 (1978) 339-342

17. Marquardt, K.: "Recovery of metals from effluent in solid form" Pt I - Electrolysis – a desirable method." Galvanotechnik 69 (1978) 7-15

18. Schwering, H.U., Hasler, J.: "Improving efficiency of production units with optimised disposal and treatment methods" Metalloberflaeche 37 (1983) 48-52

19. Wahl, K., Reinhard, F.: "Eliminating hydroxide sludges in metal finishing" Galvanotechnik 77 (1986) 339-342

20. Nadeau, T., Dejak, M.: "Copper, Nickel and Chromium Recovery in a Jobshop." Plating and Surface Finishing (1986) Apr. 48- 54.

21. Bayer AG, Leverkusen: "Chromic acid recovery from rinsewater using LEWATIT ion-exchange resin" (Techn. Note, 1962)

22. Furrer, F.: "Use of ion exchangers in metal finishing" Metallwarenindustrie u. Galvanotechnik (1957) 72-81

23. Krug, J.: "Ion exchangers – their reliability and economics in treatment of plating shop effluent" Galvanotechnik 55 (1964) 295-300

24. Marquardt, K.: "Recovery of metals in solid form from effluents. III. Recovery of chromic acid from rinsewater and purifying concentrate baths with ion exchangers in combination with the ionodialysis process" Galvanotechnik 69 (1978) 421-429

25. Wolotzkow, F.P., Hartinger, L.: "Recovery of chromating baths using ion exchange" Metalloberflaeche 32 (1978) 104-9

26. Diethelm, C.: "Regeneration of spent chromium plating baths with ion exchangers" Oberflaeche-Surface 13 (1972) 206

27. Meyer, W.: "Regeneration of chromium plating baths with ion exchangers" Tech. Note Schering (1981)

28. Wiedmann, H.: "Regeneration of chromium plating baths with ion exchange for reduced effluent discharge" Metalloberflaeche 14 (1960) B 165-9

29. McGarvey, F.X., Fisher, S.A.: "Effluent Control in Zinc Bonderising Processes. – II. The Recycle of Dilute Chromic Acid Baths Via an Ion Exchange Process." Plating and Surface Finishing (1975) Apr. 335-339.

30. Huss, R.: "Process for automatic regeneration of chromating baths" Galvanotechnik 75 (1984) 550-7

31. Heinrich, K., Busse, W.: "Using ion exchangers in surface treatment of metals" Galvanotechnik 51 (1960) 67-77

32. Bayer AG, Leverkusen: "Lewatit SP 112" Techn. Bull. (1974)

33. Hesler, J.C.: "Recovery and Treatment of Metal Finishing Wastes by Ion Exchange – Part III – Metal Finishing Bath Applications". The DOW Chemical Company, Midland, Michigan.

34. Kraus, K.A., Moore, G.E.J.: J. Am. Chem. Soc. 72 (1950) 5792 et seq.

35. Borgolte, T.: "Process for continuous regeneration of HCl pickling baths" Baender, Bleche, Rohre (1964) (1) 5-8

36. Borgolte, T.: "Iron removal from HCl pickle liquors by ion exchange" Wasser, Luft, Betrieb (1966) 607-9

37. Borgolte, T.: "Treatment of HCl rinsewaters & concentrates by ion exchange as part of an HCl pickling bath regeneration" Baender, Bleche, Rohre 9 (1968) 459-462

38. Borgolte, T.: "Treatment of HCl rinsewaters & concentrate from wire pickling plant by ion exchange" Draht-Welt 55 (1969) 67-70

39. Götzelmann, W., Hartinger, L., Gülbas, M.: "Chemicals separation and recovery by the retardation process". II. Metalloberflaeche 41 (1987) 315-322

40. Hatch, M.J., Dillon, J.A.: "Acid Retardation". Ind. Eng. Chem. 2 (1963), 253-263

41. Brown, C.J., Davy, D., Simmons, P.J.: "Purification of Sulfuric Acid Anodizing Solutions". Plating and Surface Finishing (1979) Heft 1, 54-57.

42. Brown, C.J., Davy, D., Simmons, P.J.: "Recovery of Nitric Acid from Solutions used for Treating Metal Surfaces". Plating and Surface Finishing (1980) (Feb), 60-62.

43. Hartinger, L.: "Recycling process solutions with the retardation method" Proc. Conf. 6th Eur. Effluent and Waste Disp. Munich (1984)

44. Hartinger, L.: "Handbook of effluent techniques" Vol. 7, Verlag Ernst, Berlin.

45. Krausenegger, G.: "Life extension of stainless steel pickling baths using acid retardation and ion exchange". Diploma Study. Aalen Techn, Coll. (1986)

46. Rituper, R.: "Regeneration of mixed acid pickle liquors after retardation" Abstr. Conf. Tri-countries. Innsbruck (1989)

47. Gülbas, M.: PhD thesis. "Separations using the retardation process in metal-working and surface treatment industries" TU Berlin (1990)

48. Jenny H (in) Hübner, K., Speiser, D.: "Practical anodic oxidation of aluminium" Aluminium Verlag (1977)

49. Halwachs, W.: "Liquid ion exchange in solvent extraction" GDMB Series (Germany) 53 (1989) 63-94

50. Müller, E.: "Liquid-liquid extraction" (in) Ullmans Techn, Encyclopaedia

51. Workel, H.A.: "Solvent extraction for metal recovery" Fachber. Huettenprax. Metallverarb. 15 (1977) 265-6

52. Bouvier, A., Gaubinger, W., Marr, W.: "Use of liquid-liquid extraction and liquid membrane permeation in hydrometallurgy" CAV (1982) 43, 46, 51

53. Hoechst, AG, Frankfurt: "Hostarex-A Brands" Techn. Bull.

54. Barthel, G.: Heinrich, R.F.: "Copper recovery by solvent extraction and electrolysis". Techn. Note Krupp Werksber. 32 (1974) (1) 29-41

55. Agers, S.W., De Ment, E.R.: "Evaluation of New LIX- Reagents for the Extraction of Copper and Suggestions for the Design of Commercial Mixer-Settler Plants." TMS Paper Selection. Paper Nr. A, 72-84.

56. Schwab, W., Buchmeier, W.: "Selective liquid-liquid extraction from acid solutions in the zinc manufacturing industry" Schriftenreihe GDMB 53 (1989) 254-6

57. Shell Chemicals: "Versatic acids in metal extraction processes." Technical Bulletin INT:77:M4.

58. Müller, W.: "Options in recycling of plating sludges" Galvanotechnik 67 (1976) 381-3

59. Müller, W.: "Methods for recovery of non-ferrous metals or their compounds from plating sludges" (in) Seng, H J; "Disposal by Recovery" Kontakt u. Studium, 42 (1980) 151-61 Expert Verlag Grafenau

60. General Mills, Chemical Division: "Copper Recovery from Acid Solutions Using Liquid Ion Exchange." Techn. Note.

61. Shell Chemicals: "Shell Metal Extractant 529". Data Sheet FC:PRO:77:6:DS.

62. Transaco MX International AB, Stockholm: "Mecer Process" Techn. Note (1980)

63. Hofmann, H.: "Regeneration of ammoniacal etches and associated copper recovery by liquid-liquid extraction" Galvanotechnik 73 (1982) 311-5

64. Hanson, C.: "Technology of solvent extraction in metallurgical processes." Proc. Int. Symp. Solvent Extraction in metallurgical Processes, Antwerp 1972

65. DAS 2.319.244 (31.5.79) Ugine Kuhlmann

66. Wiedmann, H.: "Regeneration of HCl pickle liquors by liquid anion exchange" Wasser, Luft, Betrieb 15 (1971) 87-94

67. Mühlberg, H; Mensler Th,; "Pickling of alloy steels with regeneration of mixed acids" Stahl u. Eisen 95 (1975) 639-642

68. Marr, R., Draxler, J.: "Liquid membrane permeation" Schriftenreihe GDMB 53 (1989) 327-337

69. Kim, B.M.: "Membrane-based Solvent Extraction for Selective Removal and Recovery of Metals." (in) Symp. "Membrane Technol. for Chem. Process Appl'ns" Denver Colorado 28.8.83. Special offprint GE Co Schenectady No. 9934

70. Schneider, K., Rintelen, T.H.: "Processes for selective separation of metals using supported liquid membrane" Schriftenreihe GDMB 53 (1989) 352-369

71. Matz, G.: "Crystallisation" (in) Ullmans Encyclopaedia

72. Schliephake, D.: "Processes for crystallisation from solution" Chem. Ing. Techn. 52 (1980) 553-561

73. Kammel, R., Lieber, H.W.: "Treatment of plating shop effluents avoiding formation of toxic waste material. Pt 4." Galvanotechnik 68 (1977) 710-5

74. Hartinger, L.: "Effluent treatment in metal finishing" Industrieabwasser (1972) 27-31

75. Huss, R., Fishcher, G., Peters, W.: "Removal of organic breakdown products from bright nickel baths with complete recycling" Galvanotechnik 71 (1980) 712-720

76. Huss, R., Peters, W.: "Process for removing organic breakdown products. II" Galvanotechnik 71 (1980) 818-823

77. Strecke, H.: "Nickel-chromium plating plant with recycling system" Galvanotechnik 70 (1979) 429-434

78. Navazio G., Carbini, M., Contin, R., Schiavon, G., Badan, B.: "Regeneration of nickel plating baths" Oberflaeche Surf. 24 (1983) 378-380

79. Nawafune, H., Mizumoto, S., Kawasaki M.: "Selective Removal of Trace Copper Ion in Nickel Electroplating Bath with Chelating Reagents". Faculty of Science, Konan University.

80. Schöller, M., Dijk, J.C. van, Wilms, D.: "Recovery of Heavy Metals by Crystallization." Metal Finishing (1987) Nov. 31- 34.

81. Schlegel, H.: "Recovery of silver cyanide from plating baths" Jahrbuch Oberflaechentechn. 31 (1975) 169-171 Metall Verlag Berlin

82. Peuser, M.: "Metal and cyanide recovery by the Peuser-Bartolo process" Galvanotechnik 78 (1987) 2803-5

83. Dembeck, H., Meuthen, B.: "Solubility of iron in sulphuric and hydrochloric acid pickling baths" Baender Bleche Rohre 5 (1964) 320-5

84. Mathesius, K.: "New process for treatment & detoxification of pickling effluents" Industrieabwasser (1958) 20-4

85. Borchert, O.: "Effluents from chemical surface treatment" Verlag R Mueller, Koeln (1965)

86. Haubrich, F.G.: "Treatment of spent sulphuric & hydrochloric pickle liquors" Fachber. Huettenprax. Metallverarb. 14 (1976) 1119-1126

87. Jedlicka, H.: "Recovery of pickling acids" (in) Hartinger "Pocketbook of effluent treatment". Vol. 2

88. Hitzemann, G.: "Treatment of pickling effluents and effluent from semi-manufactures" Handbook of Effluent Treatment, Vol. VII (1985) 140-163 Verlag Ernst & Sons, Berlin

89. Hake, A.: "HCl regeneration - experiences in a zinc coating plant" Baender Bleche Rohre 5 (1964) 9-13

90. Ruthner Industrieanlagen AG, "Recovery of nitric and hydrofluoric acids from stainless steel pickling" ACHEMA (1976)

91. Schiller, W.: "New ideas in etching" Galvanotechnik 70 (1979) 924-8

92. Teworte, W., Rabben, H.J.: "Copper compounds" (in) Ullmans Techn. Encyclopaedia

93. Schlitter, F.W.: "Eco-friendly etching with hydrogen peroxide/sulphuric acid" Galvanotechnik 79 (1988) 1631-1634

94. DP 1696137 (12.10.72) FMC Corp USA

95. Bogenschütz, A.F., Jostan, J.L., Merten, A.: "Regeneration of etch baths in the p.c.b industry" Galvanotechnik 70 (1979) 816-823

96. Meyer, W.: "Zinc plating of steel with low effluent discharge and recycling" Techn. Note Schering (1981)

97. Fischer, G.: "Regeneration of process solutions by crystallisation" Galvanotechnik 76 (1985) 691-4

98. Huss R., Peters, W.: "Pilot plant for low effluent cadmium plating" Galvanotechnik 73 (1982) 1208-1216

99. Matz, G.: "Crystallisation – a thermal separation process" Chem. Ing. Techn. 50 (1978) 13-23

100. Wöhlk, W., Hofmann, G.: "Construction of crystallisers" Chem. Ing. Techn. 57 (1985) 318-327

101. Berry, L.: "Cutting Caustic – New Process returns caustic to anodizers etch tank, reducing caustic purchases". Products Finishing (1988) Apr. 92-95.

102. Gülbas, M.: unpublished work

103. Merten, U.: "Desalination by Reverse Osmosis." M.J.T. Press Cambridge, Mass. 1966.

104. Strathmann, H.: "Membrane filtration, a simple process for separation, concentration and fractionation. Pt 1" Galvanotechnik 64 (1973) 664-71

105. Strathmann, H.: "Membrane processes for molecular separation". Chem. Technik 7 (1978) 333-347

106. Saier, H.D.: "Membrane filtration – theory and practice" Chemie f. Labor u. Betr. 28 (1977) 6-14

107. Lonsdale, H.K., Merten, U., Riley, R.L.: J. Appl. Polym. Sci. 9 (1965) 1341.

108. Rautenbach, R., Rauch, K.: "Ultrafiltration and Reverse Osmosis – theory and technology" Chem. Ing. Techn. 49 (1977) 223-231

109. Strathmann, H.: "Membrane filtration – a simple concentration and separation method. Pt 2" Galvanotechnik 64 (1973) 899-905

110. Marquardt, K.: "Membrane processes for fresh water and effluent treatment" Galvanotechnik 78 (1987) 353-362

111. Lohse, U.: "Reverse Osmosis" CAV (1984) (Nov) 107-108, 151

112. Marquardt, K.: "R.O and ultrafiltration for effluent treatment" Korr. Abwasser 31 (1984) 343-357

113. Pusch, W.: "Separation with synthetic membranes by ultra- and hyperfiltration" Galvanotechnik 67 (1976) 975-981

114. Gäfgen, K.: "Emulsion breaking in the metal-working industries" VDMA Conf. "Effluent treatment and avoidance in mechanical engineering industry" (29.4.82)

115. Hansen: "Materials recovery and effluent treatment by the Ucarsep-System" Galvanotechnik 68 (1977) 340-1

116. Wilhelm, F.: "Recirculation in pre-treatment plants" Metalloberfl. 33 (1979) 301-334

117. Marquardt, K.: "Treatment of fresh water & effluent with R.O. and ultrafiltration, compared with ion-exchange" Metalloberfl. 27 (1973) 169-182

118. Marquardt, K.: "Effluent treatment – R.O. and ultrafiltration for chemicals recovery" Vom Wasser 44 (1975) 223-272

119. Stelmaszek, J.: "The efficacy of R.O" Chemie-Techn. 13 (1984) 74-81

120. Götzelmann, W.: "R.O in metal finishing" Galvanotechnik 64 (1973) 588-600

121. Hammers, F.J.: "Reduced metal effluent discharges by adopting new processes

and use of recirculation" Muenchener Beitraege zur Abwasser, Fischerei, Flussbiol. 34 (1982) 77-90

122. Cartwright, P.S.: "Update On Reverse Osmosis For Metal Finishing." Plating and Surface Finishing (1984) Apr. 62-66.

122a. Kock, K.: "Removal of Hg & Cd ions & other heavy metals from effluent with the Berghof process". Techn. Note. Forsch. Inst. Berghof (1977)

123. Rothkegel, J.: "Eco-friendly phosphating for tomorrow" Metalloberflaeche 30 (1976) 449-504124. Schene, H.: "Techniques for reducing plating sludge volumes" Proc. Semin. 1776/9 TA Wuppertal (1976)

125. Gäfgen, K.: "Rationalised degreasing using ultrafiltration". Techn. Note 17, Duerr GmbH (1982)

126. Seyser, H.D.: "Cost-effective recycling of degreasing baths" Galvanotechnik 73 (1982) 720-2

127. Rossmann, Ch.: "Studies on disposal and regeneration of alkaline cleaning baths" Galvanotechnik 71 (1980) 824-833

128. Gernscheid, H.G.:"Trends in industrial cleaning" Metalloberfl. 33 (1979) 229-232

129. Rossmann, Ch.: "Regenerating aqueous degreasing baths" Jahrbuch Oberfl. (1982)

130. Marquardt, K.: "Membrane processes in fresh water and effluent treatment". II. Ultrafiltration" Galvanotechnik 78 (1987) 1279-87

131. Hestermann, G.: "Ultrafiltration for paint savings in electro-coat plants" Oberflaechentechn. (1972) 397-404

132. Saatweber, D.: "Cathodic electro-coating" Metalloberflaeche 31 (1977) 455-467

133. Kraus, H.: "Effluent treatment for car-body painting". Metalloberfl. 36 (1982) 531-8134. Kraus, H.: "Effluent treatment for car-body painting" Duerr GmbH Tech. Note 22.

134a. Ibl, N.: "Role of mass and charge-transport in electrometallurgy" Erzmetall 22 (1969) B 87-98

135. Schmidt, A.: "Applied electrochemistry" Verlag Chemie (1979)

136. Ibl, N., Robertson, P.M.: "Electrolytic metal recovery from dilute solutions" Chem. Ing. Techn. 48 (1976) (Feb)

137. Kreysa, G.: "Packed bed cells for metal removal from effluent" Chem. Ing. Techn. 50 (1978) 332-7

138. Raats, C.M.S., Boon, H.F., Eveleens, W.: "Fluidised bed electrolysis for metal recovery from dilute solutions" Erzmetall 30 (1977) 365-9

139. Bettley, A., Tyson, A., Cotgreave, S.A., Hampson, N.A. "Electrochemistry of Nickel in the Chemelec Cell". Surface Technology 12 (1981) 15-24.

140. Roennefahrt, K.W.: "Electroanalytical methods in water and effluent treatment" Wasser, Luft, Betrieb (1965) 365-370

141. Facsko, G.: "Dissolution and deposition potentials" Galvanotechnik 76 (1985) 1159-1164

142. Jola, M.: "Electrolytic metal recovery" Galvanotechnik 63 (1972) 459-464

143. Wendt, H.: "Electrolysis" (in) Ullmans Technical Encyclopaedia

144. Markovac, V., Heller, H.C.: "Principles of Electrodialysis for Nickel-Plating Rinsewater." Plating and Surface Finishing (1981) Dec. 66-69.

145. Universität Dresden; "Hydrogen anodes" Licensing News (1988)

146. Kammel, R., Lieber, H.W.: "Treatment of plating shop effluents without hazardous waste production. VII" Galvanotechnik 69 (1978) 317-324

147. Kammel, R., Lieber, H.W.: ibid. VIII. Galvanotechnik 69 (1978) 624-630

148. Kammel, R., Lieber, H.W.: ibid IX. Galvanotechnik 69 (1978) 687-696
149. Schab, D., Hein, K.: "Electrolysis with agitated mass electrodes" Neue Huette 27 (1982) 317-322
150. Kammel, R., Lieber, H.W.: "Electrolytic Recovery of Precious Metals From Dilute Solutions". Journal of Metals (1981) Oct. 45-48.
151. Gal-Or, L., Calmanovici, B.: "Gold Recovery from Cyanide Solutions – Part I – Electrochemical Deposition". Metal Finishing (1983) 15-21.
152. Farkas, J.: "An Ecological and Economic Process for Transition Metal Recovery". Journal of Metals 37 (1985) Febr. 72-75.
153. Farkas, J., Mitchell, G.D.: "Prospecting for Gold in Wastewater Pays Off". Electronic Packaging and Production (1985) Jan.
154. Anon: "Treatment and recovery of precious metals" Galvanotechnik 79 (1988) 1871-6
155. Farrar, L.S.: "Recovery of Palladium from Plating -Operations." Plating and Surface Finishing (1987) March, 60-61.
156. OS 2731698 (19.1.78) Matthey Rustenberg Refiners
157. Schering AG, Berlin: "Silver recovery from static rinse baths" Galvanotechnik 68 (1977) 467 and company literature
158. Dr. Dürrwächter, DODUCO KG. Techn. Note 6 (1979); "Savings in precious metal use". Galvanotechnik 70 (1979)
159. Götzelmann, W.: ""Electrolytic metal recovery – cost effectiveness and practical details" Galvanotechnik 70 (1979) 596-603
160. Kammel, R.: "Environmental protection by effluent electrolysis" Lecture N 316 R-W Akademie, West-deutscher Verlag (1983)
161. GOEMA Dr. Götzelmann KG. Techn. Note GEOCOMET
162. Müller, K.J.: "Packed bed electrolysis for electrochemical effluent treatment" Galvanotechnik 74 (1983) 902-6
163. Götzelmann, W., Schuster, H.J., Winkel, P.: "Cost-effective effluent treatment for pickling baths" Metall 21 (1967) 627-633
164. Wystrach, D.: "Recovery of metals" Galvanotechnik 67 (1976) 374-6
165. Wiaux, J.P.: "Electrolytic effluent treatment" Oberflaeche Surface (1988) Oct. 16-21
166. Schaab, D., Hein, K.: "Electrolytic metal recovery in cells with agitated mass cathode" Chem. Techn. 36 (1984) 451-455
167. Brown, C.J.: "A Better Way to Recover Nickel". Products Finishing (1988) Aug. 54-64
168. Götzelmann, G.: unpublished work
169. Dejak, M.: "Ion Exchange + Electrowinning = Recovery at Hewlett Packard". Plating and Surface Finishing (1988) Apr. 35-38.
170. Marquardt, K.: "Recovery of metals in solid form from effluent – I" Galvanotechnik 69 (1978) 7-15
171. Fischer, G.: "Recovery of nickel from plating baths" Galvanotechnik 75 (1984) 563-5
172. Meyer, W.: "Cleaning up nickel solution for return to plating baths" Schering, Techn. Note (1981)
173. Brugger, R.: "Nickel plating" Leuze Verlag, Saulgau (1984)
174. Schene, H., Kübler, E.: "Plant and processes for low waste plating" Galvanotechnik 79 (1988) 3000-3010

175. Boger, K., Kübler, E.: "Water and chemicals saving in metal finishing" Galvanotechnik 76 (1985) 680-690

176. Fischer, G.: "Electrolytic metal recovery in plating processes" Galvanotechnik 74 (1983) 145-150

177. Fischer, G.: "Metal recovery in plating shops" Werkstoff u. ihre Veredlung 2 (1980) 281-5

178. Hedrich, H.D., Raub, E.: "Electrodeposition of cadmium and zinc from acid solutions" Metall 29 (1975) 573-6

179. Vachon, D.T., Bissett, W., Calver, B.A., Dickson, G.C.: "Evaluation of Electro-chemical Recovery of Cadmium at a Metal Finishing Plant." Plating and Surface Finishing (1986) Apr.

180. Müller, K.J., Bolch, T., Mertz, K.: "Packed bed electrolysis for recovery or destruction of organic complexants in effluent" Galvanotechnik 79 (1988) 172-6

181. Hartinger, L., Gülbas, M.: unpublished work

182. Meyer, W.: "Disposal of electroless Cu and Ni baths". Abstr. Tri-country Conf. Duesseldorf (1989)

183. Holland, F.S., Rolskov, H.: "Effluent recycling" Galvanotechnik 70 (1979) 346-7

184. Foster Wheeler Ltd, Reading (UK): "Application of the Eco-Cell- Process to Decopperisation of Copper Refinery Electrolytes, (1978)

185. Walsh, F.C., Gabe, D.R.: "Controlled-Potential Electrodeposition of Metals at a Rotating Cylinder Electrode (The Eco-Cell)". Surface Technology (1981) 25-37.

186. Facsko, G.: "Overvoltage, structure and texture of electrodeposited metals" Galvanotechnik 75 (1984) 1242-5

187. Raats, C.S.M., Boon, H.F., Eveleens, W.: "Fluidised bed cells for electrolytic metal removal from dilute solutions" Erzmetall 30 (1977) 365-9

188. Heiden, G. van der, Raats, C.M.S., Boon, H.F.: "Fluidised bed cell for electrolytic removal of metals from effluent" Chemie Ing. Techn. 51 (1979) 65103

189. Raats, C.M.S., Boon, H.F., Heiden, G. van der: "Fluidized bed electrolysis for the removal or recovery of metals from dilute solutions." Tech. Bull. Akzo Zout Chemie, Netherlands

190. Bettley, A., Tyson, A., Cotgreave, S.A. Hampson, N.A.: "The Electrochemistry of Nickel in the Chemelec-Cell." Surface Technology 12 (1981) 15-24.

191. Silman, H.: "Electrolytic methods for metal recovery from effluent" Galvanotechnik 73 (1982) 589-593

192. Reiche, H., Kreysa, G.: "Effluent treatment and recycling – is packed bed electrolysis the answer ?" Fachber. Rohstoff-Engin'g 104 (1980) (July)

193. Nohse, F.: "Electrodialysis – a new approach to recovery and recycling of process solutions in metal finishing" Fachber. f. Oberflaechentechn. 11 (1973) 262-4

194. DOS 2256280 (2.8.73) Dart Industries

195. DOS 2310622 (12.9.74) BASF

196. Apfelbach, R.D.: "Regeneration of etches for metallising plastics" Galvanotechnik 70 (1979) 144-8

197. Anon: "Re-oxidation of Cr(III) ions in chromic acid baths" Galvanotechnik 77 (1986) 340

198. Schering Galvanotechnik, Berlin: "Oxamate" Techn. Note

199. Heymann, K.: "Recovery of chromic acid from etch baths" Schering Colloq. Sindelfingen (1977)

200. Innes, W.P., Toller, W.H., Tomasello, D.: "Electrolytic Regeneration of Chromic Acid Etchants". Plating and Surface Finishing 65 (1978) 36-40.
201. Gross, W., Heger, K., Nowack, N.: "Anodic oxidation of Cr(III)" Galvanotechnik 72 (1981) 1307-9
202. DE 37 16031 (24.11.88) Schering
203. Fritsch, J.: "Electrodialysis in environmental and recovery technologies" Wasser, Luft, Betrieb (1986) (1/2) 72-3
204. Deuschle, A., Kübler, E.: "Electrodialysis - for chemicals recovery from rinsewater" Galvanotechnik 75 (1984) 968-971
205. Linnhoff, F.: "Electrodialysis for recovery of plating baths" Metalloberflaeche 34 (1982) 1-5
206. Jelinek, T.: "Electrodialysis for recovery of plating solutions" Galvanotechnik 72 (1981) 1076-9
207. Nohse, F.: "Electrodialysis", in Hartinger, L.: "Pocketbook of effluent treatment", Vol. 2 Hanser Verlag.
208. Hartinger, L.: "Recovery of chemicals in liquid form" Galvanotechnik 68 (1977) 721-730
209. GKSS Forschungszentrum Geesthacht GmbH "Membrane stack for multi-chamber processing" Techn. Note (1983)
210. Personal communication; Sunshin MFG, Co Ltd, Nagoya.
211. Fischer, G.: "Electrodialysis for concentration & recovery of metal salt solutions" Metalloberflaeche 43 (1989) 309-313
212. Markovac, V., Heller, H.C.: "Engineering Aspects of Electrodialysis for Nickel Plating Rinsewater". Plating and Surface Finishing (1982) Jan. 84-86.
213. Marquardt, K., Schmid, G.: "Using electrodialysis for effluent treatment" Galvanotechnik 80 (1989) 2300-4
214. Lindstedt, J.S., Millman, W.G.: "Electrodialysis and Gold- Recovery – A Closed-Loop Approach." Plating and Surface Finishing (1982) July, 32-36.
215. Quitmann, H: "Treatment of spent pickle liquors & effluent from the steel industry using ion-exchange membranes" Draht 10 (1959) 225-9
216. SOCOMATEN "The Seprac-Process for the Treatment of Acid Pickling Solutions". Techn. Note (1975)
217. Lancy, L.E.: "The Use of Electrodialysis for Metal Finishing Process Solution Recovery and Regeneration." Proc. Conf. Interfinish Basle (1972)
218. Report "Kehin industrial zone for metal finishers" Galvanotechnik 69 (1978) 391-6
219. Innova, Inc., Clearwater, Florida. Techn. bulletins
220. Horn, R.E.: "Continuous Regeneration of an Electroless Copper Bath". Plating and Surface Finishing (1981) Oct. 50-52.
221. Krulik, G.A., Lipson, M.A., Davidson, J.B., Davis, S.C.: "The copperstat process - a new approach to treatment & process control of electroless copper wastes" Galvanotechnik 76 (1985) 1806-11
222. Asahi Glass Co., Ltd., Tokyo: SELEMION Ion-Exchange Membranes, (1974).
223. Schwuger, M.J.: Dissertation TU Aachen (1966) (see ref. 225)
224. Degussa AG, Frankfurt: "Activated carbon for the environment" Techn. Note (1983)
225. Wirth, H.: "Adsorption" (in) Ullmans Encyclopaedia of Chemical Technology.
226. VCI Process News. "Adsorptive water treatment" Oct. (1975)

227. Bayer AG, Leverkusen: "Lewatit OC 1031" Product Bull. (1980)
228. Kienle, H.v.: "Adsorption on activated carbon" Vom Wasser 37 (1970) 265-275
229. Bayer AG, Leverkusen: "Lewatit 202" Product Bull. (1981)
230. Bayer AG, Leverkusen: "Lewatit OC 1047" Product Bull. (1981)
231. Bender, H.: "Removal of organics from effluent" Umwelt and technik (1983) 36
232. Rohm and Haas, Philadelphia: "Ambersorb carbonaceous adsorbents"
233. Koppe, P., Herkelmann, H.: "Adsorption material requirement as criterion for further treatment of municipal effluent" Forum Staedte Hygiene 37 (1986) 194-6
234. Jedlicka, H.: "Recovery of pickle acids" (in) Hartinger, L; "Pocket-book of effluent treatment" Vol. 2
235. Hitzemann, G.: "Recovery of pickle acids" Industrieabwaesser (1976) 6-9
236. Hitzemann, G.: "Treatment of effluent & spent pickle liquors from manufacture of semi-finishes" Lehr- u. Handbuch der Abwassertechnik Vol VII, Verlag Ernst & Sohn, Berlin (1985)
237. Hake, A.: "HCl pickling with total recovery of acid & rinsewater" Oesterreich. Chem. Ztg. 68 (1967) 180-5
238. Dr. Otto & Comp. GmbH, Bochum: "HCl regeneration plants" Techn. Datasheet (1972)
239. Hartinger, L., Götzelmann, W.: "Rinsing & evaporation in recycling technology" Metalloberflaeche 31 (1977) 386-398
240. Hartinger, L.: "In-house treatment of metal-containing effluents" Munich. Beitraege Abwasser, Fischerei 34 (1982) 53-76
241. Kreichelt, H.J.: "Electrolyte recovery by evaporation" Galvanotechnik 70 (1979) 610-619
242. Hartinger, L.: "Evaporation – a major tool for recovery of process solutions" Galvanotechnik 69 (1978) 408-415
243. Culotta, J.M.: "Treatment of Cyanide and Chromic Acid Plating Wastes". Plating (1965) 545-548.
244. Igawa, S., Uchikoshiki, Y.: "Behaviour of Fluorides in Vacuum Evaporation Recovery of Chromium Plating Solutions." Kinzoku hyomen gijutsu
245. Radke, M.: "Energy saving evaporation" cav (1978) Dec. 55-6
246. Radke, M.: "Evaporation plant for metal finishing" Metalloberflaeche 32 (1978) 459-461
247. Linnhoff GmbH, Staufen: Techn. Data for ECO Vacuum evaporator
248. Hiecke, R., Schubert, M.: "Evaporation processes – theory and technical evaporation" VEB Deutscher Verlag, Leipzig (1976)
249. Spangemacher, K.: "Evaporative cooling and its applications" (in) Ullmans Technical Encyclopaedia.
250. PS 2914722 (25.5.1982) to LPW
251. DE 3632245 A1 (7.4.88) to Siemens
252. DAS 2124864 (8.8.74)
253. Hartmann, M.W.: "Water savings and recycling in metal finishing" Industrieabwaesser (1976) 1-5
254. Hartmann, M.W.: "Rinsewater evaporation technology" Blasberg Communication 32 (1979) and Metalloberflaeche 32 (1978) (May)
255. Meurer, H.: "Economising on water, energy and chemicals" Maschinenmarkt (1977) (Oct).

256. Hasler, J.: "Water saving and electrolyte recovery as alternative to effluent treatment – System RS 80" Lecture No. 4. Colloqu. in Gyoer, Hungary April 1978

257. Wahl, K.: "Reducing recycling costs" Galvanotechnik 76 (1985) 727-8

258. Fischer, G.: "Processes for water and chemicals savings" Galvanotechnik 69 (1978) 416-420

259. Fischer, G.: "Effluent treatment" Drahtwelt (1979) 109-111

260. Fischer, G.: "Improving the economics of chromium plating by chromic acid & energy recovery" Metalloberflaeche 34 (1980) 1-14

261. Fischer, G.: "Recycling in chromium & nickel plating" Metalloberflaeche 40 (1986) 230-4

262. Tscherwitschke, R.: "Effluent free chromium plating in modern automatic nickel-chromium plating lines" Galvanotechnik 70 (1979) 620-5

263. Meissner, K.H., Peters, W.: "A pilot plant for low effluent zinc plating" Galvanotechnik 69 (1978) 431-8

8 Special process technologies for particular production areas

Foreword

Mention was made in previous chapters of special processes, which are applied in particular production areas only, wherever they best satisfy the prevailing conditions. Other process technologies, however, specific for a production area.

The most important are:–

- Mechanical treatment
- Heat treatment (hardening)
- Pickling and removal of surfaces
- Hot dip galvanising
- Enamelling
- Lacquering including pretreatment
- Anodising
- Electroplating
- Manufacture of printed circuit boards
- Manufacture of batteries and primary cells.

8.1 MECHANICAL TREATMENT

8.1.1 Waste water and its constituents

In the mechanical treatment of metals, agents (e.g. oils and emulsions) are used, and cleaning baths (alkaline degreasing baths, chlorinated hydrocarbons) are used between or after the production procedures. These generate waste water during their use (rinse water) or when the time comes to dispose of them. Then there are suspensions of lubricating materials in liquids used predominantly in the deburring, grinding and polishing of articles processed in bulk. The waste water constituents of these production areas are characterised by finely dispersed oils (emulsions) and solid matter.

Although mineral oils and greases may be present in waste water in an unemulsified form, in most cases they were emulsified originally. Phase separation can be brought about by destabilisation of the emulsion, by supersaturation (introduction of foreign oils), by temperature variations or physical and/or chemical influences, as these give rise to emulsion breaking as well and may also to a lesser extent occur during the production operations (eg coalescence). Depending on the origin of the waste water, the aqueous phase can contain water-soluble or solid components of the cooling and lubricating agents or of the cleaning baths. Fig. 8.1 shows a low-density material separator, as it might be set up to remove oil [1].

In shaping metals by cutting methods, such as tapping, milling, planing, broaching, sawing, honing and so on, lubricating coolants are fed to the working areas, which, as their name implies, are intended to take away the heat generated, to lubricate and to remove the wear debris. They also perform all these functions, except for the last, in forming procedures such as rolling, deep drawing, bending and so on [2].

In these operations, the aqueous phase takes care of the cooling, the oil phase the lubrication. Exacting requirements are imposed on these constituents with respect to their tribological characteristics and temperature resistance. See Section 3.13.3 concerning the physical and chemical properties of emulsions.

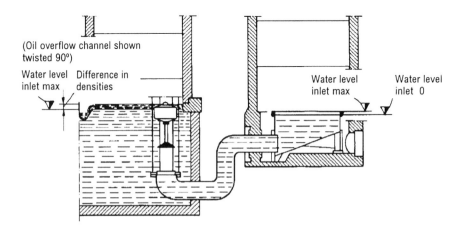

Fig. 8.1 Oil separator with automatic withdrawal facility [1]

Lubricating coolant preparations can contain the following materials: mineral oil (2-20%), synthetic oil (e.g. esters, glycols), emulsifiers, corrosion inhibitors (e.g. nitrites, amines, amides), stabilisers (e.g. alcohols), high-pressure additives (e.g. chlorinated paraffins, organic sulphides), anti-foaming agents, microbiocides and complexing agents. Foreign material can also be encountered in the working process, eg: metal chips and abrasion debris, extraneous oil and grease, drawing lubricants, surface active agents, nitrites, cyanides, phenols, bacteria, yeasts and fungi. The COD of the separated aqueous phase of the emulsions used lies between 5000 and 20,000 mg/litre, the content of dissolved metals, depending on the material processed, up to 100 mg/litre. The accrual of lubricating coolants is intermittent and may be up to 5 m^3 in small central plants and up to 100 m^3 in large ones [2].

Components cleaning plants designed for degreasing give rise to waste waters that can contain cleaning alkalis (sodium hydroxide, carbonate, phosphate, silicate, borate, polyphosphate), surface active agents and complex formers (e.g. gluconate, NTA). To these should be added, where they are present, emulsified oils and greases, dirt (grinding and polishing materials and their binders) and small amounts of metals due to the presence of complexing agents.

If degreasing is carried out in solvents, chlorinated hydrocarbons will be employed, mostly trichloroethylene and tetrachloroethylene. These compounds do not themselves constitute waste water and can not therefore be disposed of as such. In contact with these solvents, however, chlorinated hydrocarbon-containing waste water can be produced, e.g. water of condensation in the water separator of working tanks and in steam regeneration of active carbon. Efforts have recently been made to construct fully enclosed degreasing plants with their own solvent, water and ventilation circuits and with double bottom-shells as a precaution against ground contamination [6] (see Section 3.11.3). Practically no emissions can then escape. The alternative is to replace the chlorinated hydrocarbons by aqueous cleaning solutions [7].

Mass finishing is carried out by producing a relative motion between workpieces and grinding materials (media) in the presence of suitable dissolved chemicals (compounds). The media are made of ceramic materials or plastics with embedded grinding and polishing materials, in various shapes, sizes and weights, chosen according to the desired surface condition and the nature and delicateness of the work. The sliding and grinding movement is produced with the help of rotating vessels (drums, barrels) or vibrating machines, which are filled with the articles to be treated in bulk, the chips and the compound solution.

Waste water with the following constituents is produced during the process [3]:

- Solid matter. Fine debris from the chips, such as corundum, pumice powder etc as well as binding material and abraded metal

- Dissolved components of the compound (estimated amount 5-20 g/litre).

Cleaning alkalis, surface active agents, soaps, complexing agents (carboxylic acids, aminopolyacetates, amines), anti-foaming agents (alcohols), corrosion inhibitors (e.g. nitrites)

- Working-in materials. Emulsified oil and grease, metal complex compounds, soil.

Care must be taken here in relation to fish-toxicity due to the added biocides.

Prüller [4] and Förtsch [5] quote concentration ranges for some characteristic constituents. The lower and upper limits of both authors are as follows: pH value 7.5-10; 1000-10,000 mg/litre COD; 10-100 ml/litre removeable materials; 5-100 mg/litre hydrocarbons; approximately up to 200 mg/litre metals, and according to [4] up to 2000 mg/litre nitrite and 50-500 mg/litre amino nitrogen, e.g. triethanolamine. If there are soluble and stable coordination compounds of iron present, they undergo further complexation with these amines.

The volume of waste water produced is typically 30 to 2000 litre/hr. Waste water generation may be continuous or discontinuous (up to 1 m^3 for each batch emptied out) [2]. It consists of the used contents of the machines and the rinse water.

8.1.2   Treatment of lubricating coolants and alkaline cleaning baths

These two working materials have a great similarity in so far as the elimination of oil and grease is concerned:

- Lubricating coolants are oil-in-water emulsions with relatively high oil concentrations (2-20%) as initially prepared.

- Cleaning baths, generally alkaline, although neutral baths are also known, accumulate increasing amounts of emulsified oil and grease as they are used, but concentrations around 10 g/litre should not as a rule be exceeded.

Emulsified oil and grease can be removed from both liquids using the same processes, although in practice some have been more successful with lubricating coolants and others with cleaning baths.

Before thinking in terms of the disposal of lubricating coolants, it is important to take care to maintain its working life as long as possible. This is done by separating demulsified oils, removing solid matter and taking preventative measures to preserve the emulsion.

The oil that has completely dropped out of the emulsion can be removed from the true emulsion by gravity (see Section 3.13.2) with the help of low-density-liquid separators [8] or laminar separators. The laminar separators operate in inverse manner to that applying when they are removing solid matter. That is, the inlet and outlet have to be interchanged (cf Fig. 4.23). Devices to encourage coalescence, which can precede or, as in the case of low-density separators, form an integral part of the equipment (see Fig. 3.147), improve the oil extraction or can be directly applied to separate the oil [9,10,11]. A further possibility is the use of centrifugal separators. These increase the gravitational force and, appropriately constructed, are able simultaneously to separate light fluids and solid matter from water or emulsion (see also cleaning baths). They are not, however, able to separate genuine emulsions into oil and water.

The oil removal step generally precedes the separation of mechanical pollutants. For this, solid matter separators, sedimentation units of various design, with and without floor sweepers, plate separators, hydrocyclones, magnetic separators for iron particles and filters of all sorts are employed [10,11]. Information on band filters (see Section 4.5.1.2), fine filters (with metal or textile fabrics, filter paper, wrappings of various materials) up to a maximum retention capacity of 1 μm, coated filters with a retention capacity of 0.5-20 μm and others is available. In larger plants use is most often made of a combination of two solids removal processes, a separating step for larger contaminants followed by a filter.

Preventative maintenance of lubricating coolants necessitates the monitoring of their composition, their aeration to suppress anaerobic reactions (putrefaction) and cooling. About 60-70% of the heat generated at the worked surface is transmitted to the lubricating coolant. A rise in temperature over 30°C should be avoided [10], since it encourages the multiplication of bacteria and hence putrefaction.

If it becomes necessary to dispose of lubricating coolants, then reference should be made to the treatment processes described in Sections 3.13.4.1 to 3.13.4.4. Because of the salt production that occurs, chemical processes should only be resorted to if the acids and/or salts needed for emulsion separation are also due for disposal.

Among the physical processes is that of adsorption of the emulsified oil, usually on preparations based on silicic acid. These frequently also contain salts (calcium, magnesium, aluminium, de-emulsifiers, flocculating agents and so on, and are referred to as so-called reactive separating media. In these processes therefore it is mainly a matter of combining chemical and physical actions. Adsorbed oils are removed along with the adsorption medium mainly by gravity filtration. Simple filter bags of paper or cloth are suitable for this. Fig. 8.2 shows a plant for this application. The process is suitable for the treatment of small batches.

Vaporisation in the extraction of oil and its normal separation as well as incineration are relatively costly and can, if special measures are not adopted, lead to other environmental problems (air contamination).

After the well-known adsorption process, the ultrafiltration method is most commonly used for disposal purposes. It has the advantage that it is not affected by the stability of the emulsions. It enables concentrates with an oil content of up to 50% to be achieved along with permeates containing less than 10 mg/litre of oil (see Sections 7.5.1, 7.5.2 and 7.5.4.1 on this). Fig. 8.3 shows an ultrafiltration plant for the treatment of lubricating and cooling media. Each of the modules shown contains 660 membrane capillaries with a surface area of 1.39 m². The output of permeate amounts to ca 21 litre m^{-2} h^{-1} for emulsions and ca 30 litre m^{-2} h^{-1} for wash water [12]. Further information on the ultrafiltration of lubricating coolants can be found in [12,13], [14], and in various reports and technical summaries [15,16].

Fig. 8.2 Batch plant for emulsion separation using adsorbents (CARL DITTMANN & Co, Karlsruhe)

If the emulsion-containing waste water is pretreated with an organic emulsion-splitting agent (demulsifier) and subjected afterwards to microfiltration (using membranes with pore sizes larger than those in the ultrafiltration range), then it is possible, as trials have shown, to obtain specific outputs of permeate that are about a factor of 10 greater than those of normal ultrafiltration without pretreatment [17].

Flotation plants are used for very large quantities of this type of waste water (see Section 3.13.4.4). Following adsorption coagulation with iron or alumiumium salts, emulsion droplets adsorbed on the flocs are floated upwards with air and removed as sludge (swept or sucked off). A distinction is made between conventional flotation [18] and electroflotation [19]. Fig. 8.4 shows a flotation plant for lubricating coolant. The amounts of metal salts added to produce flocculation are large (4-10 kg salt/m^3) in the case of concentrated waste waters. If the oil-laden flocs of hydroxide are dissolved again in acid, then the coagulated oil remaining is easily separated. In a second stage the hydrolysis pH value of the metal can be restored in order to remove the residual oil. The sludge now floated contains only the small quantities of oil that still remain and so can be used again directly for a first-stage separation [19].

Fig. 8.3   Ultrafiltration plant with capillary bundle modules for treatment of emulsion-containing waste water in an engineering works, permeate output: ca 2 m^3/h (Guetling GmbH).

Another process, which like ultrafiltration can split any emulsion, works in extreme conditions, and consists of a combination of acid splitting and coalescence separation. Waste water is heated to ca 90°C, acidified to pH 1.5-2 and then passed through a column containing monocrystalline quartz. The separated oil rises to the top and can be removed there. Flocculation with iron salts is then carried out in order to remove the last remaining oil [20]. Because of the acidification, the process generates a certain amount of salt.

Comprehensive reviews of the methods available for removing oil and grease, for maintaining lubrication coolants and disposing of them, can be found in [21], [10], [11,12] and [23]. The disposal of neutral and alkaline cleaning baths is basically similar to that of lubricating coolants.

Fig. 8.4 Low-tension flotation plant for the treatment of waste water containing lubricating coolant produced in engine manufacture (EISENMANN, Boeblingen).

Degreasing baths generally become unusable at oil contents between 1 and 10 g/litre. In degreasing baths which precede phosphating followed by lacquering, the values are still lower by 1-2 g/litre [24]. Earlier, when this level of contamination was reached, baths were drained and replaced by new solutions. As this could have occured after times ranging from 1 day to 2 months, depending on the surface area processed and the extent to which it was soiled, waste waters that were awkward to treat were generated at short intervals. It is a matter not only of their oil and grease content, but also of the increased difficulty of waste water processing, if solutions such as these are combined with other waste waters for treatment. There may at times be complexes formed with metals (polyphosphate, NTA), but also increased difficulty in eliminating them owing to the formation of colloidal hydrolysis products. Extending the service life of degreasing baths is thus important for the surface treatment industry, especially where large volumes are concerned. Enhancing the service life or converting to long-life process baths (similar to electrolytic) can be achieved by:

- Mechanical separation of oils and solid materials.

- Hot water pre-degreasing.

- Use of degreasing cascades.

- Use of centrifugal separators.

- Adsorption of oils.

- Ultrafiltration.

- Minimising surface soiling of the work pieces.

A variety of often very simple methods is available for mechanical cleaning. Surface active agents should be used first, as they are very effective in removing oil and grease by emulsification, yet without forming really stable emulsions.

In this way oil and grease can be self-removing and then floated off from the surface of the baths over a weir. From here they flow into an intermediate vessel or an oil separator, where they can be removed from the surface by skimming, while the aqueous phase is returned to the degreasing bath. Fig. 8.5 shows devices for skimming using overflow gutters, suction floats and oleophilic bands [11,12]. The last-named are swept over the liquid surface. Disc and roll skimmers, which are immersed below their axes in the aqueous solution with its floating oil layer, are also employed. Ehrler and Janitza [25] describe a band skimmer which takes up oil not only by adhesion but also to a considerable extent by the capillary action of the materials used (needlefelt of polyester, polyethylene, polypropylene). The band is capable of taking up 40 to 50 times its own weight of oil. This is then removed by squeezing between rolls. With such a device it is possible, if the oil layers are more than 1mm thick, to withdraw 0.5-1 tonne of oil per minute using a 5 m² band.

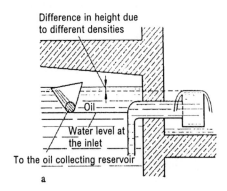

a

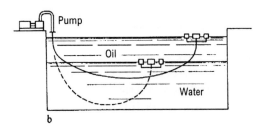

b

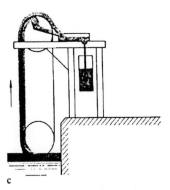

c

Fig. 8.5 Examples of devices for skimming oil separated at the surfaces of aqueous solutions [22]

(a) Adjustable overflow channel (PASSAVANT-WERKE, Aarbergen),
(b) Floating skimmer (MEGATOR Ltd, London).
(c) Band skimmer (SANDVIK CONVEYOR GmbH, Fellbach)

If it is necessary to accelerate oil separation, devices such as laminar separators or coalescing plate units [9] can be connected in series. The latter are plastic plates with a zigzag profile, which are arranged one above the other with little space between and which constantly divert the direction of the waste water passing through. Their upper edges are provided with bore-holes, enabling the oil to rise. An acceleration in oil separation relative to the usual separators (Fig. 8.1) can be achieved by use of so-called annular chamber de-oilers [26,27]. The waste water is pumped tangentially at high speed in these devices and is diverted in its circulation via concentric cylindrical walls, which allow passage alternately over and under, so that the heavier aqueous phase is driven outwards against the inner cylindrical walls and the oil rises upwards in contact with the outer cylindical walls.

Separation of solid material impurities from degreasing baths can be undertaken with all types of sedimentation equipment. It is simplest as a rule to pre-couple heavy material separators as component parts of the plant in the same way as light material separators. Flow-through vessels are also often employed, in which the oil separates above and overflows or is skimmed off, and the solid matter settles out on the flat bottom, from which it can be brought out by a drag belt over the sloping front wall of the tank [31]. Laminar separators are also used successfully.

Efficient mechanical removal of oil and heavy materials brings about an increase in bath service life of two to four times [31].

Using special centrifuges, known as centrifugal separators, it is possible at high speeds of rotation (ca 7000 rpm) to increase the force of gravity by a hundred times. A separation time of hours in a static system can thus be reduced to seconds [26]. Inside the ontrifuge, in order to increase the clarifying area, there is a set of conical plates, which have spacings of only 0.3 to 2 mm, so as to shorten the sedimentation path coresponding to each of them [28]. The effect achieved is similar to that of the laminar separator. The heavier aqueous phase settles towards the outside, whilst the oil is driven towards the inside. Fig. 8.6 shows the mode of operation of a centifugal separator. While the aqueous and the oil phases, as the diagram shows, can be continuously withdrawn from the separator, it is possible at the same time to separate the heavy materials as a solid dirt phase. It can be discharged at intervals through a valve provided at the periphery of the barrel. The (non-emulsified) oil content of the regenerated degreasing bath amounts to ca 0.5 g/litre, the water content in the oil phase 5-10%. Separation of solid matter is sucessful up to particle sizes of ca 5 μm. The process enables degreasing baths to be cleaned up so long as no solid separation, e.g. silicic acid from silicate, takes place and provided oil contents of 0.5 g/litre are acceptable. It has been known for service lives to be lengthened by up to 16 times [28]. For practical guidance concerning the regeneration of degreasing baths with centrifugal separators see also [29,30]. An effective increase in the service life of degreasing baths can be achieved, as Bessey [32] reports, by

simple pre-degreasing using hot water (at 60-80°C), especially if the work pieces
are heavily oiled or greased. The oil dispersed in the hot water can be continuously removed as it circulates using coalescing or other systems as mentioned
above. The degreasing bath following the hot water bath is thus protected from
the greater part of the oil and grease. Up to 99% of the oil may be excluded from
it. It can additionally be treated according to the procedures previously described
in this section.

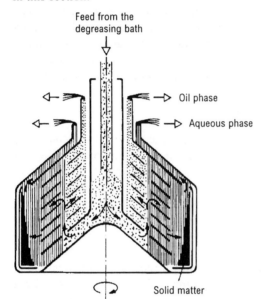

Feed from the
degreasing bath

Oil phase

Aqueous phase

Solid matter

*Fig. 8.6   Schematic illustration of a
centrifugal separator and its mode
of operation [28] (WESTFALIA-
SEPARATOR AG, Oelde)*

Wilhelm [31] commented on the combination of mechanical purification and
ultrafiltration. The requisite output of permeate, beginning with a single stage
degreasing process, diminished, as soon as a second degreasing step was added,
the first step being a mechanical and the second an ultrafiltration purification
treatment. A further improvement, i.e. reduction in the necessary permeate output (at a constant degreasing efficiency), can then be achieved, if part of the flow
of the first stage after mechanical purification is passed on to ultrafiltration and
the permeate obtained after the second stage is used for preliminary rinsing. The
overflow of the second degreasing bath returns to the first, and that from the first
back to the regenerating plant. This basically represents a degreasing cascade, in
which the intake of the purified degreasing bath constitutes the first stage. In the
rinsing areas following this, however, recycling by cascade rinsing is used only
in exceptional circumstances, as there is a danger of rusting with longer rinsing
times and drag-out into the following process baths can be too high.

Ultrafiltration is also, as in this last case, widely used in combination with
mechanical purification. Rothkegel [33] gives a further example of degreasing
bath regeneration in a phosphating plant, in which mechanical purification and

ultrafiltration using tube bundle modules with inorganic membranes are carried out in separate circulation systems. The supplementary chemicals required, surface active agents in particular, are added subsequently on a continuous basis. The rinse water is again supplied to the degreasing bath partly directly and partly via the ultrafiltration unit.

In ultrafiltration modules with larger diameter tubes (25 mm) it is possible, according to experiences reported by Gaefgen [24], to do without pre-filtration in the absence of coarse solid matter. The service life of degreasing baths can be raised by ultrafiltration by ca 12 times, the regenerating efficiency of ultrafiltration plants and centrifugal separators being similar, if allowance is made for the degree of recirculation and relative size. The effective oil and grease content of the permeates is less for ultrafiltration than centrifugal separation because of far better removal of emulsified oil.

For regeneration of degreasing baths by ultrafiltration reference should be made to Section 7.5.4.1.

Ultrafiltration is also a reliable process for disposing of emulsion-containing waste water from degreasing plants and lubricating coolant centres. Fig. 8.7 shows a plant for treating such waste waters in combination.

Rossmann describes a further adsorptive regeneration process for degreasing baths [26]. This is based on the observation that surface active agents and oils are adsorbed on precipitated sediments, such as for example silicic acid, calcium and magnesium phosphates. When using this effect it is recommended that surface active agents used in preparation of degreasing baths be non-ionic, so as to avoid excessive losses. Calcium or magnesium salts are added in the course of the treatment (ca 4-8 g/litre in baths with 50 g/litre concentration). The oil-laden sediment is filtered off. It is recommended that the treatment be repeated at intervals of 10 working days. Surface active agent losses must be replaced. 1-2 g/litre phosphate or silicate should be added previously to neutral cleaners that do not contain these ions.

Operating with degreasing-bath and rinsing-bath cascades should also be considered. With these arrangements, part of the rinse water should be passed to the degreasing-bath cascade and part to an evaporator. For the sake of regeneration, and to make room for the rinse water coming in, the degreasing solution also is fed to the evaporator. There the oil phase separates owing to supersaturation and the de-oiled solution produced is returned to the last degreasing bath. The condensate is pumped into the rinse cascade [34].

Lengthening the service life of degreasing baths should be achieved mainly by avoidance measures. In many cases far too heavily oiled and greased parts have to be cleaned. This particularly affects job shops, whose clients give too much consideration to protection against corrosion and think too little about protection of the environment. Attention must be given, not only to the amounts but also to the types of oil and grease to be removed in preliminary cleaning. Not uncommonly, these may be such as to make it impossible to do without degreasing with chlorinated hydrocarbons.

It should be noted too that oil-, grease- and emulsion-containing waste waters can also contain harmful and dangerous substances such as cyanides (cyanide-containing electrolytic degreasing baths), nitrites (washed off corrosion preventatives), metals (from the complexation of dissolved corrosion products) and so on. After oil and grease have been eliminated, such waste waters must then be passed on to the appropriate chemical treatment process.

Fig. 8.7 Ultrafiltration plant for treatment of lubricating coolants and rinse baths of degreasing plants in an automobile factory; capacity 3 m³/hr; tube bundle modules (Duerr GmbH, Stuttgart)

8.1.3   Treatment of mass finishing waste water

Waste water generated in mass finishing is distinguished by the constituents described in Section 8.1.1. It is most particularly characterised however by the very finely dispersed solid matter comprising debris on the one hand of grinding and polishing media and on the other of the material being processed. As the compounds not only exert degreasing and rust protection effects, but also tend to keep the solid materials dispersed, separation of the latter is particularly difficult. This is connected also with their surface charges, which are usually negative, as well as with their small particle size. Rodenkirchen [3] provides the data quoted in Table 8.1 on the particle size distribution.

TABLE 8.1:
PARTICLE SIZES OF SOLID MATERIALS IN
MASS FINISHING WASTE WATER [3]

Particle sizes	Percentage
0.0– 3.0 µm	66.76%
3.0– 5.0 µm	19.66%
5.0–10.0 µm	11.73%
10.0–20.0 µm	1.64%
20.0–50.0 µm	0.20%

Because of these properties, use has to be made in solid matter separation of coagulation (adsorption coagulation), flocculation, adsorption and flocculation (see Section 3.7.4.2), as Hinz [35], Prueller [4], Fritsch [5] and Rodenkirchen [3] have also pointed out. In what follows the processes mentioned are for brevity's sake as precipitations.

There are mass finishing shops with oil- and grease-free waste water, which, if it is also free from dissolved metals, can be treated solely by precipitation measures with final sedimentation and/or filtration. The media involved are electrolytes, flocculation agents (iron or aluminium salts) and flocculation aids (for negatively surface-charged particles usually cationic polyelectrolytes). If the electrolytes added to assist coagulation are acidic, a final neutralisation is necessary with caustic soda or lime slurry. If dissolved metals (not complexed) are present, then the pH must be adjusted to the optimum value for their precipitation. If metals are present as complexes, then they can be precipitated in the neutral range with an organosulphide (see Section 3.9.5.1.2). The NTA complexes of non-copper metals are, however excluded. In some countries, EDTA-containing waste water may no longer be discharged after such treatment.

Iron forms relatively stable complexes with any amines present (e.g. triethanolamine). These can however be precipitated with organosulphides.

If the waste water contains larger amounts of oil, which on flocculation cannot all be adsorptively bonded to the hydroxides, installing an ultrafiltration unit (see below) may be unavoidable. If the waste water contains nitrite or other harmful materials, then it must also be treated for these following the treatments described below. If the waste water is free from complexing agents, then it can be treated jointly with other waste water from the factory. If it contains complexing agents, however, it must be treated separately. It is not possible to eliminate the causes of COD and amines, if they are not bonded at least to some extent adsorptively to the flocculation media.

As mentioned above, waste water can be treated by precipitation, sedimentation and/or filtration simply by using appropriate re-agents. Precipitation media consist of preparations which contain electrolytes, metal salts and flocculation agents, and which, as with emulsion breaking, are known as reactive separating media. If there are only small quantities of waste water, they can be treated in a batch plant, as illustrated in Fig. 8.8. Filtration is by means of filter bags. After an appropriate draining time, residues are obtained with up to 40% dry matter. These can be further de-watered in intermediate plants in the factory or by drying with infra-red radiation. Facilities of this kind are suitable for waste water yields of 1-2 m³/day.

If larger amounts of waste water are generated, they are passed first to a storage tank. However, it is important that the solid materials should not settle there, as it would otherwise then be very expensive to remove them. Hence a storage tank of this kind is equipped with a slow stirrer (see Fig. 8.9). In medium-sized plants the waste water is pumped out from here into the precipitation-reactor. In large installations it is fed first through a band filter (see Section 4.5.1.2) to remove most of the solid materials and subsequently, in the presence of emulsified oil, through an ultrafiltration plant. After precipitation it passes into a sedimentation tank, where the sludge settles at a rate of a few cm/min, yielding a dry-material content of 5-10% [4]. This can be de-watered in a filter press to a dry-material content of 50-60% [5]. For safety's sake the filtrate is passed through a bag filter before being discharged (see Fig. 8.9).

Fig. 8.8 Batch treatment plant with bag filters for the treatment of smaller amounts of waste water from mass finishing shops (WALTHER GmbH & Co, KG, Wuppertal

Fig. 8.9 Batch treatment plant for waste water from mass finishing shops;
capacity 1.5 m³/batch, left: storage tank and stirrer; centre: reactor with settlin
tank below it; right: filter press with a bag filter to follow it
(Roesler GLEITSCHLEIFTECHNIK GmbH & Co, KG, Untermerzbach)

For intermediate quantities of waste water the appropriate equipment lies between the two described. A gravel filter can also be used as a safety filter for post-treatment following the filter press, and if the waste water still contains metals that are not bound as anionic complexes, a selective cation exchanger can be inserted [3].

Treated waste water, when all solid materials have been removed, can sometimes be supplemented with mains water then returned to the rinse zone of vibratory machines [35]. In doing this however, care must be taken that chemicals added in waste water treatment and accumulations of compound constituents do not give rise to any deleterious effects [36]. The recycleable proportion of the water is approx. to 40-60% [37].

If the mass finishing water can be clear-filtered before chemical treatment, then it is also possible to return it directly to the production plant. This can be done if coarser contaminants are first removed with a band filter and the solution is then subjected to membrane filtration with a microfilter (pore sizes 0.1-0.2 mm) [3]. The residue must be treated as described and the permeate supplemented with mains water by adding this in appropriate quantities. The fact that dissolved metals are returned does not disturb the process. So that settled dirt particles can be removed periodically, it must be possible to back flush the tube modules used, on their permeate side. As asymmetrical membranes cannot tolerate this, mechanically stable self-supporting polypropylene membranes polypropylene are employed [3]. In other circumstances precleaning may be carried out with filter cartridges followed by filtration through polysulphone ultrafiltration membranes [37].

Fig. 8.10 shows a microfiltration plant for the treatment of mass finishing waste water. The requirements concerning the discharge of the waste water from mechanical production units come from Tables 1.1a and 1.1b.

Fig. 8.10 Microfiltration plant for removing solid materials from mass finishing waste water, permeate output 3 m³/h, separation limit 0.1 mm(GEORG SCHOENEMANN & Co (GmbH & Co), Bremen)

8.2   HEAT TREATMENT

Common heat treatments are in addition to actual hardening, tempering, annealing, carburising, carbonitriding, nitriding, and, more rarely, aluminising, boriding, siliconising and chromising.

8.2.1   Waste water and its constituents [39]

The amount of waste water generated in this production area has been declining for years. The trend is due mainly to the growth in gas-phase methods. Waste water arises if the operations involve fused salt baths, aqueous quenching baths, oil baths and cleaning solutions and if exhaust gas washing plants are employed. In general, the quantities of waste water produced in heat treatment shops are small and can be treated in batch plants.

The characteristic constituents of the waste water are cyanides, nitrites, hexacyanoferrates and barium compounds. Cyanates should also be added to this list. Kunst [40] shows the concentrations and materials produced over several months in waste water from a heat treatment facility (see Table 8.2). Additional waste water constituents are oils and oil-in-water emulsions, deriving from hardening, tempering and rust-protecting oils which find their way into degreasing baths.

In waste water containing cyanides, nitrites and iron, prussiates i.e. hexacyanoferrates, can form in which cyanide ions are partly replaced by nitrite ions, although substitution by nitrate and other ions is also known. These prussiates, because of their properties, have an adverse effect on waste water treatment. In contrast to hexacyanoferrates, they form stable complexes in alkalis, but are only slightly resistant in a weakly acidic medium, in which they dissociate to form free cyanide (see below).

TABLE 8.2:
WASTE WATER CONSTITUENTS AND THEIR CONCENTRATIONS, MEASURED OVER SEVERAL MONTHS IN A HEAT TREATMENT FACILITY (FROM KUNST [40])

Waste water constituent	Concentration range [g/litre]	Average concentration [g/litre]
CN^-	0.05–1.70	0.25
CNO^-	0.20–0.45	0.30
NO_2^-	0.13–7.35	1.30
NO_3^-	0.15–8.00	1.50
Cl^-	0.30–3.70	0.55
CO_3^{--}	0.20–2.20	0.50
Ba^{++}	0.13–3.40	0.25
Na^+, K^+	0.34–5.30	1.35

8.2.2   Waste water treatment

The classical treatment process consists in the oxidative processing of cyanide and nitrite, the cyanide being first oxidised at pH values over 10 and then the nitrite at pH values between 3 and 4. This is done in accordance with the processes described for cyanide in Section 3.2 and for nitrite in Section 3.4.1. It is advantageous to employ the same oxidising agent for both. In waste water from heat treatment plants hydrogen peroxide is particularly suitable for this purpose, as it produces no further salts. Decomposition reactions catalysed by non-ferrous metal ions (see Section 3.2.2.1) are rare in almost all cases. Care should be taken, however, if the heat treatment is being performed on highly alloyed steels or non-ferrous metals. The use of hydrogen peroxide is even more favourable in view of the fact that it is also unnecessary to monitor or control the reaction by constant testing in the case of batch processes which are appropriate here. Free cyanide displays a true redox potential, which however, as hydrogen peroxide has no significant countervailing effect on the oxidation potential, gives rise to no substantial movement of potential in the positive direction, as is the case with dissociating oxidation agents. According to Knorre [41], however, a redox indicator can be added to the waste water, and will itself indicate the potential change likely to occur.

Nitrite detoxification is coupled together with cyanide detoxification but requires acidification to pH of 3-4. Among other things, this gives rise in the presence of nitrite, to the dissociation of prussiate complexes, such as for example the pentacyano-mononitrite-ferrates ($[Fe(CN)_5NO_2]^{---}$ and $[Fe(CN)_5NO_2]^{---(-)}$). The consequence is that the freshly alkalified and liberated cyanide is oxidised. While the prussiates display a similar stability to hexacyanoferrates in an alkaline medium, they lose this in an acid medium. Hexacyanoferrates are so stable that the very low quantities of cyanide with which they are in equilibrium neither have a toxic effect nor allow any detoxification of the complex (see Section 3.2.1.6). Their cyanide component is therefore not included with the very small amount of free cyanide and does not contribute to the total cyanide, as defined in most effluent legislation. It should be emphasised in this connection that prussiate complexes during analysis liberate small amounts of free cyanide while the cyanides themselves dissociate, and this can lead to difficulties, even though the prussiate introduced is no longer brought into contact with acid.

If after the cyanide detoxification, sulphuric acid is used for acidification prior to nitrite detoxification, then any barium present precipitates out as highly insoluble sulphate. Without acidification barium carbonate is formed from the alkali carbonate always present. The insolubility of this is not sufficient however to get below the specified discharge limit of 2 mg/litre Ba (see Table 8.3). If barium is present in waste water but no nitrite, then, if the sulphuric acid for neutralisation is not sufficient, an appropriate amount of sodium sulphate must be added.

TABLE 8.3:
DATA FOR SOLUBILITY OF SULPHATE AND
BARIUM CARBONATE (FROM D'ANS-LAX)

Compound	Solubility	Corresp Ba concentration	Solubility product
$BaSO_4$	2.5 mg/l (20 °C)	1.47 mg/l	$1,08 \cdot 10^{-10}$ (25 °C)
$BaCO_3$	17.2 mg/l (18 °C)	11.97 mg/l	$7 \cdot 10^{-9}$ (16 °C)

Nitrite detoxification, whether by oxidation or reduction, is performed in the pH range 3-4. In oxidation with hydrogen peroxide or in reduction with amidosulphonic acid, the changes in pH are measured, and recorded [42] (see Equations 3.92 and 3.93). If nitrite concentrations are high, the required amounts of hydrogen peroxide are added first, acidification then taking place slowly [41]. In this way generation of nitrogen oxides is avoided. If sodium hypochlorite were used as the oxidising agent, chlorine would be evolved.

Final neutralisation takes place at a pH which may allow for the metals present. Finally, the waste water is filtered. After the precipitation of barium sulphate, it should if possible be prevented from settling, since if the high specific gravity precipitate is allowed to stand for a significant time it becomes difficult to remove from the settling tank. Continuous treatment plants in heat treatment facilities are rare. Batch plants have the advantage that all the treatment steps can be performed in the same reactor.

Fig. 8.11 Batch treatment plants for the treatment of waste water from heat treatment shops (GUETLING GmbH, Fellbach)

If waste water is produced from exhaust gas washing plants, it can be added to the production waste water before treatment. If the waste water contains oil or oil emusions, then these should be removed or separated before the detoxification process (see Section 8.1.2).

Fig. 8.11 shows a batch plant, as installed for treatment of waste water in heat treatment shops.

8.2.3   Effluent avoidance possibilities and recycling processes

By the use of gas-phase methods for heat treatment, generation of waste water can be entirely avoided in many cases and partly so in others. The processes can be employed for carburising, carbonitriding and tempering. Both exothermic gas (exothermically cracked town and natural gas) and endothermic gas from pure propane, with ammonia additions in carbonitriding, are used.

In a process developed by Mueller and Witzke [43] the noxious materials produced as by-products of heat treatment can be thermally detoxified, as described in Section 3.2.4. At temperatures above the melting points of the mixtures (580-620°C), which contain cyanides, cyanates, nitrites, nitrates, barium and other salts, cyanides and cyanates are oxidised by nitrites and nitrates, nitrogen oxides being generated and alkali carbonate produced. The sodium cyanide and potassium cyanate percentages of the mixture should not in this case exceed 5%, so as to avoid an explosive reaction. Barium carbonate is formed in the reaction; this can be separated and washed then converted again to barium chloride.

The detoxification outlined above can make use also of another process, which was described by Mohr [44], and which is suitable principally for factories with a large heat treatment throughput. In order to attain high rinse water concentrations, several rinse baths are installed one behind the other. The first, most concentrated (40-50% salt), is slowly introduced into a melt of alkali nitrites and nitrates at 400-500°C. The water is thus evaporated and the cyanides and cyanates present are oxidised. Depending on the amount of salt introduced, the melt becomes exhausted, reaches its collection limit and then begins to solidify. The recovered nitrite-nitrate-salt is added again to the heat treatment bath. At an evaporation rate of 60 litre/hr. up to 25 kg/hr. salt can be recovered.

Another process described by Kunst [40] also involves detoxification in a salt melt, but is intended to reduce the supply of waste water. The waste water is freed from solid materials by sedimentation and filtration, evaporated at low pressure and the condensate returned to the production area. The concentrate (40-50%), sediments and filter residues are transferred to a melt based on alkali hydrides, nitrates and sulphates, where detoxification takes place in the usual way.

Similarly, a surface evaporator may serve also to reduce waste water generation in heat treatment facilities that operate with salt baths and protective gas. The surface of the salt solution is irradiated with infra-red lamps and the water evaporated at 70-80°C. The water vapour is withdrawn continously through a ventilator and passed out into the open air. Salt solution is added corresponding to the amounts evaporated. At the bottom, a drawer-like compartment is pulled out to empty the salt, which, gathered together as a coherent mass, is either added again to the salt baths or disposed of as used salt [45]. A surface evaporator is shown in Fig. 8.12. It has an evaporation capacity of 25-30 l/h, at an air throughput of 600 m³/h.

The requirements relating to the discharge of waste water from heat treatment facilities can be taken from Tables 1.1a and 1.1b.

Fig. 8.12   Surface evaporator for waste-water-free heat treatment facilities; the drawer compartment for removal of the salt residues is open (DEGUSSA AG, Frankfurt)

8.3   PICKLING AND DESCALING PROCESSES

The following belong to this group of processes:

- Pickling of metals in mineral acids and their mixtures

- Treatment of non-ferrous metals in oxidising acids and acid mixtures

- Pickling in alkaline solutions, e.g. aluminium

- Chemical etching in acid and alkaline solutions

- Electrolytic machining processes (countersinking, drilling, turning, deburring, smoothing, honing, lapping etc.) [46]

- Electrolytic processes for surface treatment (polishing, brightening, deburring, metal stripping, etc) [46]

- Chemical dissolution processes (brightening, countersinking, metal stripping, forming, etc).

Regarding chemical etching, reference should be made to printed circuit board manufacture (Section 8.9), chemical and electrolytic brightening prior to anodising of aluminium (Section 8.7) and the chemical and electrolytic stripping of electroplate (Section 8.8).

8.3.1 Amount of waste water and its constituents

Where it is not carried out as an intermediate treatment to remove scale, rust and corrosion products in the manufacture of semi-finished products, pickling is an important part of the pretreatment in almost all production areas involving further surface treatment, such as hot dip galvanising, enamelling, lacquering, anodising and electroplating. It is normally carried out in these instances after degreasing. Its role is to remove from the surface, oxidation and other corrosion products, and inorganic or organic residues from mechanical working operations whether or not affected by temperature. As the surface is attacked, both acid-soluble and acid-insoluble substances are removed. In addition, the work should acquire a uniform shiny appearance and the active surface necessary for the following process stages. By the choice of pickling medium, its concentration, ratio of constituents and any additions, it is possible to obtain bright, matt or rough surfaces, whether these are required for technical purposes or for decorative reasons.

The waste water arising in pickling is characterised by the presence of pickling media and dissolved metals coming from the pickled work.

Pickling media include the well-known mineral acids, hydrochloric acid, sulphuric acid, nitric acid, hydrofluoric acid, phosphoric acid, chromic acid and their mixtures, as well as caustic soda for aluminium. The acids can contain organic inhibitors to prevent acid-induced embrittlement, or surface active agents (pickling degreasants) to achieve an additional degreasing effect. Caustic soda solutions may contain additions of sodium gluconate (as an anti-scaling agent) and other complex formers, trisodium phosphate, as well as nitrate and nitrite (as accelerators).

When iron and other base metals are pickled in acids, metal salts corresponding to the acids (chlorides, sulphates, nitrates, etc.) are produced in accordance with Equation 8.1.

$$Me^0 + zH^+ \rightleftharpoons Me^{z+} + \frac{z}{2} H_2 \qquad\qquad 8.1$$

The pickling of nobler metals, which because of the range of their electrochemical potentials cannot liberate hydrogen from acids, must be accomplished with the help of an oxidising acid or an non-oxidising acid containing an added oxidising agent. If this is done with nitric acid or mixtures containing it, nitrates and other salts of the metals are produced as well as nitrogen oxides, which are scrubbed out in a gas cleaning plant (see Section 8.8.2.5). Pickling of copper proceeds in accordance with the following equation:

$$3\,Cu^0 + 8\,H^+ + 2\,NO_3^- \rightleftharpoons 3\,Cu^{++} + 2\,NO + 4\,H_2O \qquad\qquad 8.2$$

The same occurs also with copper alloys, and silver and its alloys. If sulphuric acid is used for pickling copper, with chromic acid or hydrogen peroxide as oxidising agent or with air passed into the hot acid, the waste gas problem can be avoided.

Burkhardt [47] reports on the pickling of copper and other non-ferrous metals such as zinc, tin, aluminium and their alloys. According to him, sulphuric acid or hydrochloric acid or sulphuric acid-nitric acid mixtures are preferred for zinc and its alloys, and hydrochloric acid or nitric acid for tin and its alloys.

Iron is normally pickled in hydrochloric or sulphuric acid and, alloy steel in nitric acid-hydrofluoric acid mixtures, but in special cases also in alkali melts with oxidising or reducing additions (caustic soda with sodium nitrate or sodium hydride), and silicon-containing steel in pickles with hydrofluoric acid additions.

The range of possible waste water constituents is considerably increased by the chemical and electrolytic material-removal processes. Table 8.4 lists reagents used for chemical dissolution of different metals [48]. For chemical removal the mineral acids are used, especially phosphoric acid, also higher alcohols, perchloric acid, and acetic acid as well as other organic acids. For electrolytic machining, sodium nitrate and sodium chloride solutions are employed. While the metals removed are present in the acidic solutions as ions, they are present in the salt solutions as hydrolysis products.

8.3.2   Waste water treatment [49,50]

Waste water arising from the pickling of metals in acids is based on a neutralisation/precipitation process in which the pH value is adjusted in accordance with the metals present (see Sections 3.7.5.10 and 3.7.6). As there is usually a rather large quantity of rinse water to deal with, the neutralisation/precipitation process is carried out almost exclusively with lime slurry. Continuous process treatment plants are also employed, so long as only pickling waste water is to be treated.

Metal hydroxides are removed from the neutralised waste water by sedimentation, then dewatered with the help of a filter press. Fig. 8.13 shows a continuous process plant for the treatment of pickling waste water. In the presence of iron, however, care must be taken that, if it is in divalent form, its oxidation can nevertheless be accomplished during neutralisation so that no lowering of the pH value takes place afterwards (see Section 3.5). Pickling waste arising should be passed on to a regeneration process (see next section) or utilised in another way (eg as a flocculation material). If this is not possible, it should be treated along with the rinse water, but in a batch process plant.

TABLE 8.4:
SOLUTIONS AND SOLUTION COMBINATIONS FOR CHEMICAL DISSOLUTION OF DIFFERENT METALS [48]

Nature of work piece	Solutions for chemical dissolution											
	NaOH	NaCN	H_2O_2	HF	HCl	HNO_3	H_2SO_4	H_3PO_4	CrO_3	$FeCl_3$	$Fe(NO_3)_3$	NH_4HF
Mg						▒						
Al	▒			▒	▒	▒		▒	▒			
Ti				▒	▒	▒	▒					
Fe, FeCr					▒	▒	▒	▒				
FeCrNi				▒	▒	▒	▒	▒	▒			
Ni, Co					▒	▒	▒					
Zn					▒	▒						
Cu						▒			▒	▒		
Ag						▒					▒	
Au		▒			▒	▒						
Nb, Mo, Ta, W				▒		▒						
Be				▒	▒		▒					▒

The treatment of all other waste waters produced as a result of surface removal processes is similar. It invariably comprises neutralisation/precipitation, sedimentation and sludge dewatering, as fully described in Sections 3.7, 4.2, 4.3 and 4.5. If the waste water contains fluoride, then this is precipitated by neutralising with lime slurry, but care must also be taken in this case, as explained in Section 3.10.1, that sufficient calcium is provided. With alloyed steels, reference should be made to Sections 3.8.2.2 to 3.8.2.5, if the requirements concerning the metals vanadium, molybdenum and tungsten apply. If one of the process solutions contains phases immiscible with water, chromate, nitrite or other dangerous materials, then these must detoxified or eliminated beforehand [50].

In some facilities for electrolytic machining, smaller quantities of waste water are generated, and can be treated in batch plants.

Fig. 8.13   Continuous process treatment plant for neutralisation and settling of solid matter in a wire manufacturing factory (KERAMCHEMIE GmbH, Siershahn)

8.3.3   Recycling processes

In general the high rinse water concentrations in pickling shops militate against continuous rinse water circulation. If, however, an appropriate rinse technology should then be developed it could be effective here, as indicated in the example of a pickle for strip material. The original double rinsing cascade was enlarged to a 4-stage cascade and the post-rinse zone converted to a single rinse bath. The latter was connected to an ion exchange circulation plant. Discharge from the rinse cascade was returned to the hot pickle. The same amount of fresh rinse water was introduced at the end of the cascade. It was possible by these methods to lower the rinse water concentration from 30 m Eq./litre H^+ and the corresponding iron content to 0.6 m Eq./litre Fe^{++}. As the pickle and the regenerate from the cation exchanger are processed in a vacuum crystallisation plant for separation of iron(II) sulphate heptahydrate, only the regenerate from the anion exchanger and the purified water are discharged as effluent. [51].

A further example of rinse water economy is provided by the case of a pickle plant for aluminium with the sequence: pretreatment in nitric acid pickle in caustic soda brightening in nitric acid [52]. The rinsing takes place in double pre-rinse cascades and a 4-stage cascade. The rinse waters are combined, neutralised and freed from aluminium hydroxide by means of a filter press. The neutral salt-containing filtrate can be used again for rinsing, a small amount of fresh water being introduced at the end of the cascade so as to meet the rinse criteria. The water surplus corresponding to this is evaporated.

Moller [53] reports on an electropolishing plant with similar rinse water saving. A pickle and an electropolishing bath are connected to a 3-stage rinse cascade with low water throughput at each stage and a double cascade, the water from the latter being being separately passed through an ion exchanger and circulated continuously. The overflows from the 3-stage cascades are brought together and concentrated. The concentrate is utilised again in the pickle. Given volumes are withdrawn daily from both process baths and disposed of and new additions are made to keep the concentrations constant. At a throughput of 600 m^3/day of alloy steel, 400 litre/day of rinse water was consumed and 300 litre/day passed back to the process solutions.

An increase in the service life of a sulphuric acid copper etch can be obtained as follows. Pickling takes place in sulphuric acid at 60-70°C and the pickle is circulated through an electrolysis cell to keep the copper concentration constant at 30-35 g/litre. A bright pickle is afterwards carried out in sulphuric acid with a hydrogen peroxide addition. There follows a chemical rinse in a reducing solution (see Section 6.3.5). The copper(I) oxide sludge filtered from this last solution is recovered electrolytically. The bright pickle is circulated through a storage tank, from which part of the flow is returned to the first pickling bath [54]. Some sulphuric acid copper pickle must be recovered by electrolysis and evaporation in a simultaneous recirculation of the rinse water, as already described in Section 7.6.2.2.2. In addition, reference should be made to the numerous other regeneration and recovery processes for pickling acids already described by thermal decomposition (Section 7.8.1), crystallisation (Section 7.4.2), retardation processes (Section 7.2.2.2), dialysis (Section 7.6.4) and liquid-liquid extraction (Section 7.3).

Sturm [55] describes a process for the recovery of pure pickling acids, which is more fully explained in other technical reports [56]. The method makes use of two different dialysis processes and can be illustrated by the example of a nitric acid-hydrofluoric acid mixed pickle. Part of the pickle is neutralised with caustic potash and the hydrolysed metals are filtered off. The filtrate passes, together with neutralised rinse water that has been clarified by sedimentation, into an electrodialysis cell, the design of which is the same as that in Fig. 7.66. This generates (see Section 7.6.3.1,) a concentrate and a diluate from the potassium salt mixtures. The diluate is used to wash out the filter or is recirculated within the pickling plant. The concentrate, which contains dissolved potassium nitrate and potassium fluoride, passes to a special dialysis cell, in which ion exchange membranes are arranged between the electrodes in the order: bipolar membrane cation exchange membrane anion exchange membrane bipolar membrane. Bipolar membranes are thus situated on one side of a cation exchange material and on the other side of an anion exchange material. The concentrate is introduced in the gap between the normal cation exchange and anion exchange membranes, the potassium ions then migrating through the cation exchange membrane in the direction of the cathode and the nitrate and fluoride ions through the anion exchange membrane in the direction of the anode. On their way they encounter the bipolar membranes, which they are unable to pass through however, as each of them has its exchanger sides oriented in the opposite direction to that which would allow penetration to occur. At the boundary surface of the bipolar membranes there is present, as in the bulk of the exchanger, water within the resin which dissociates in the electric field into H^+ and OH^- ions. From here the OH^- ions migrate out through the anion exchanger side of the bipolar membrane in the direction of the anode and the H^+ ions in the opposite direction. Thus they come together in the enclosed compartment of the cell together with the potassium ions and the nitrate and fluoride ions, respectively, so that caustic potash and nitric acid-hydrofluoric acid mixture are produced on the two sides of the bipolar membranes. The mixed acids are fed to the pickle, the caustic potash to the neutralisation/precipitation process. Fig. 8.14 shows the lay-out of such a plant with its electrodialyser stack.

The regeneration and recovery of sodium chloride- or sodium nitrate-containing electrolytes used in electromachining are carried out differently. First to accumulate are the hydrolysis products of dissolved material that can block the narrow spaces between the cathodes (tools) and anodes (workpieces). These can be removed by sedimentation, centrifuging and filtering. Whilst the sodium chloride electrolytes are chemically stable, nitrite is formed at the cathode in sodium nitrate electrolytes and then, from it, ammonia. In the processing of chromium-alloyed steels, chromate is produced anodically. Simon [57] described the regeneration of electrolytes with the impurities that result. The chromate present is reduced at pH 8.5-9 with iron(II) sulphate and the nitrite is later reduced with amidosulphonic acid at pH 3.5 to nitrogen. A patent [58] provides for complete regeneration, without allowing foreign matter into the electrolyte. First chromate is precipitated by the addition of barium nitrate as poorly soluble barium chromate (see Equation 3.87). This can be separated by sedimentation, filtering or centrifuging. After establishing a pH value between 4 and 5 with nitric acid, nitrite is reduced to nitrogen with urea (compare Equation 3.94). By the addition of iron(III) nitrate as a flocculation agent and the raising of the pH value to 11-12, the metal hydroxides are precipitated in an easily filterable form. If any ammonia present creates production difficulties, it can be stripped out with air at this pH value. Finally, if no other neutral salt such as sodium nitrate is present, the electrolyte is restored to its working pH value with nitric acid. New developments in this procedure enable chromate to be reduced and nitrite to be simultaneously oxidised by electrochemical treatment [58].

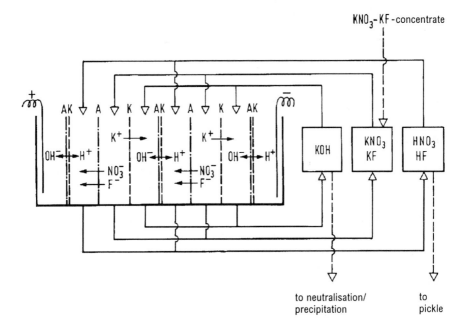

Fig. 8.14   Schematic diagram of an electrodialysis cell with bipolar ion exchange membranes for the recovery of alkalis and acids from salts (example: caustic potash and HNO$_3$-HF mixed acids from KNO$_3$ and KF).

K cation exchange membrane, A anion exchange membrane, AK bipolar membrane (from [56], STEULER INDUSTRIEWERKE GmbH, Hoehr Grenzhausen)

The recycling technologies mentioned can only be examples, given the very varied process stages in the surface removal procedures. This applies also to waste water treatments. Wherever the treatment has to deviate from the examples given because of the use of particular process solutions, a procedure can be derived from those described in Chapters 3 and 4.

8.4   HOT DIP GALVANISING

8.4.1   Amount of waste water and its constituents [59]

Parts made of iron and steel (occasionally other metals) are hot dip galvanised, mainly in order to protect them against corrosion. A distinction must be made here between general galvanising or the galvanising of prefabricated parts (steel components, posts, crash-barriers, screens, stands, refuse containers, railings, engineering products, small parts such as screws for example) and continuous galvanising, especially of wide strip (for further fabrication into roof and wall panelling, garage doors, refrigerator cabinets, transport vessels, drawn, rolled and pressed parts, lightweight steel constructions, etc.) but also of pipes and wires.

Although the number of plants doing zinc electroplating is considerably larger than that of those doing hot dip galvanising, the latter consume by far the larger tonnage.

The normal operating procedure in hot dip galvanising is straightforward: (degrease, rinse) pickle rinse dip in flux solution immerse in molten zinc (quench in water). The degreasing is often dispensed with, but this can have a deleterious effect on the purity of the waste gas (see below).

The following types of galvanising can be distinguished [60]:

- Wet galvanising. Here, after pickling and rinsing the wet workpieces, they are introduced into the molten zinc via a compartment separated by a baffle. Before entering the melt, the parts pass through a foaming fluxing medium containing a large proportion of glycerol on the surface of this compartment. Finally, they are taken further past the baffle into the molten metal bath proper.

- Dry galvanising. After pickling and rinsing, the parts are dried in an oven, then dipped in the fluxing medium and finally in the molten zinc. This process is preferred for almost all bulk galvanising of piece goods.

- Sendzimir process. This process is used for the galvanising of wide strip (600-1500 mm wide, 0.4-3 mm thick); no wet surface pretreatment is used in this, hence no waste water is produced. The strip is first oxidatively annealed at 450-600°C (oven temperature 1000-1200°C), during which oil and grease are burnt off, and afterwards annealed in a reducing, ammonia-containing atmosphere at 900-960°C (strip temperature), during which the oxides formed are reduced again [60,61]. Finally, the strip is cooled somewhat and passed into the zinc bath under a protective cover of gas.

Degreasing is carried out hot in the usual alkaline cleaning baths (see Section 8.1.1). For pickling, hydrochloric acid at ca 15% is used almost exclusively at room temperature, sometimes with inhibitors; iron may accumulate in this up to 100 g/litre or more and zinc up to 15 g/litre. In the processing of silicon-alloyed cast iron, hydrochloric acid-hydrofluoric acid mixed pickles are used with ca 10 vol% hydrochloric acid (32%) and 4 vol% hydrofluoric acid (40-50%). Black-heart malleable iron with graphite inclusions can be galvanised only if it is oxidatively annealed [63]. The fluxing solution is prepared with ca 60% zinc chloride and 40% ammonium chloride at concentrations of ca 200 g/litre, but also up to 500 g/litre [59, 62] and is used at 50°C and pH of 3.5-4.5.

In galvanising in melts heated to 455-465°C, iron-zinc alloys are present in the boundary phases, which carry a final layer of zinc on the outside. The total coating thickness amounts to 50-100 mm in general galvanising, 20 mm on wide strip and 5-45 mm on wire.

The waste water produced in hot dip galvanising derives principally from the pickles used and the related rinse water. In addition there are: degreasing baths, quenching water, waste gas wash water, if a wet process is used for gas cleaning, and cleaning water. As rinsing is carried out as a rule in static rinse tanks, the quantities of waste water are not large, but the waste water is correspondingly concentrated.

If chromating or phosphating follows the galvanising, e.g. if a coating of paint is to be applied finally, then appropriate care must be taken over the waste waters produced. In a post-treatment in wax- or oil-emulsions to prevent white rusting, no waste waters need arise if care is taken. The waste water from galvanising shops is characterised mainly by high concentrations of iron, zinc and ammonium chloride.

8.4.2   Waste water treatment and recycling processes

If hot degreasing is used, then from time to time waste water containing emulsion or oil will be generated. In these circumstances, acid separation is used for treatment, since there are generally pickles to dispose of as well. After decanting off the separated oils and neutralising, further oil will adhere to the surface of the iron hydroxide. After filtration, the filtrate can be combined with the rest of the waste water and during its treatment the rest of the oil can be removed by adsorption on the precipitated hydroxide. Other methods of emulsion separation, e.g. ultrafiltration, can naturally be employed, especially in the larger facilities. Hot water pre-degreasing can considerably lengthen the service life of the degreasing bath if heavily oiled or greased parts are involved.

The treatment of the waste water produced should, wherever possible, be carried out in batch plants (see Section 4.1). Any iron from a hydrochloric acid pickle that is present in divalent form must be oxidised during neutralisation with caustic soda or lime slurry. This is best done by introducing air. If chromate is present, then its reduction can be carried out with pickling solution.

If phosphate-containing waste waters are produced, lime slurry is always used to neutralise them.

In hot dip galvanising, the fluxing agent bath can easily become contaminated by large amounts of iron being carried over from the concentrated rinse bath following the pickle or as a result of inadequate rinsing. Further carry over into the zinc bath increases the formation of hard zinc alloy (1 g of iron produces 25-30 g of hard zinc). Durez [64] and Torres and co-workers [65] describe a method of avoiding this. The rinse water after pickling is circulated through a continuous treatment unit. First it is neutralised with ammonia to pH 5.5-9 and aerated. The precipitated iron(III) hydroxide is settled out or filtered off and the filtrate is returned to the rinse bath. In this way the drag-in of iron into the flux bath is lowered to 30-50% and its ammonium chloride content increased [64]. In the same way the flux bath itself can also be circulated [60]. Losses made good by adding zinc chloride and ammonium chloride. By installing a cascade rinsing system after the pickle the carry over of iron into the flux bath can be reduced still further. As an ancillary treatment in large pickling plants, a thermal decomposition acid recovery process (see Section 7.8.1) can prove worthwhile, in the context of the increasing cost of disposal. Depending on the capacity available, it is also possible for this to be carried out on a contract basis in a different plant.

When parts coated with dried-out fluxing agents are immersed in the molten zinc, fume is produced which consists of hydrochloric acid, zinc chloride, zinc oxide and ammonium chloride and has a particle size of less than 1 μm. This must be extracted and the waste gas cleaned [66]. Enclosure of the zinc bath is therefore recommended.

Although gas cleaning by washing is perfectly possible in galvanising, dry waste gas treatment using filters is more usual so that the generation of waste water is avoided. Cracked products of oil and grease cause problems in this however, as they solidify and block the filters [62]. This can be prevented by the use of degreasing baths, which are being introduced to an increasing extent in galvanising shops [67]. Smoke emissions are also reduced by the use of low-fume fluxing agents. Reductions to 25% or even to less than 10% have been reported [62].

8.5   ENAMELLING

Among other things, metals can be protected from chemical attack and corrosion also by non-metallic materials, e.g. by organic materials in painting, oxides in anodising or, as here, by ceramic materials in enamelling.

Enamelling is defined as follows in German standard RAL 529A:

"A material produced by melting or sintering (without complete fusion), which has an inorganic mainly oxide composition, and which is to be or has been fused on to metallic workpieces in one or more layers, sometimes with the aid of fluxes".

8.5.1 Amount of waste water and its constituents [68]

So-called frits are used as the raw material for enamelling; these may contain the following substances: alkali carbonates, potassium feldspar (in base enamel), potassium nitrate, alkaline earth carbonates, borax or boric acid, aluminium oxide, quartz sand, alkali phosphates, fluospar (for high-value white enamels, cryolite, alkali fluosilicates, sodium fluoride), nickel and cobalt oxides and inorganic opacifiers and pigments (titanium dioxide as white pigment and numerous mineral colouring agents and colour pigments [69]). The frits are fine ground in ball mills and worked up with water into so-called slip, which can be applied to the workpieces by dipping, spraying, electrostatic spraying or electrophoretic deposition under immersed conditions. It can also be applied as a powder by means of electrostatic powder coating. In all electrostatic and electrophoretic processes the particles have a negative charge, so that the work must be connected anodically. Current densities between 0.5 and 5 A/dm^2 are employed in electrophoretic deposition. The coating period is between 10 and 25 seconds.

The coating is subsequently dried then fused at 750-880°C. On chemical equipment higher temperatures are sometimes used, 880-1000°C. Glassy layers are formed during firing which consist of the oxides of the metals and non-metals contained in the frits. The waste gas produced, contains water, carbon dioxide and hydrogen fluoride [69].

The thicknesses of the enamel coatings per layer applied, as in direct enamelling for example, amount to ca 100 μm after firing, there being double thicknesses in two-layer enamelling with enamel primer (bond coat) plus top coat, and many times this on chemical equipment with up to eight top coats.

Enamel qualifies as a true composite material. When it is fired, a roughness develops on the sheet metal surface, which is due to the saturation of the enamel in the boundary layer with a lower oxide of the substrate material. In direct enamelling processes, nickel oxide (cobalt oxide also) serves to improve the adhesion capability. It does not directly affect the adhesion itself however. The mechanism responsible is still not fully understood. It is easiest to apply the nickel by immersion in a solution of nickel sulphate, but it can also be done by electrodeposition and more recently by electroless deposition of nickel. The pretreatment for enamelling in this instance is fundamentally different from the surface preparation otherwise employed.

The surface of articles to be enamelled must be pretreated to remove oil and grease as well as corrosion products. Degreasing is carried out in hot alkaline cleaning solutions. The composition of the baths is in accordance with that given in Section 8.1.1, and their regeneration or other treatment is as described in Section 8.1.2. Mueller [70] particularly recommends for the purification of degreasing baths in enamelling shops the use of centrifugal separators to simultaneously separate solid matter and oil. For this, preparations based on phosphate and without sodium silicate (because of sludge formation) are especially suitable. Also, anionic surface active agents, whose losses remain small, are preferred to non-ionic surface active agents.

Pickling in this industrial sector is mainly carried out with 5-10% sulphuric acid at 60-75°C. The loss where an enamel primer is applied will be 2-5 g of iron/m^2 and in direct white enamelling 20 g/m^2 or more [69,71,72]. In a few cases, hydrochloric acid also is used at room temperature or sulphuric acid with additions of phosphoric acid and nitric acid as accelerators.

In direct white enamelling, nickel-plating is used as an aid to adhesion, and is generally carried out by simply dipping in a nickel sulphate solution. A cementation coating comprising ca 1 g of nickel/m2 of surface is produced on the substrate [69]. These immersion nickel baths have a lifetime of 6 months up to 2 years [73]. Maintenance of the baths requires the replenishment of nickel sulphate and continuous filtration to remove hydrolysed iron [74].

In direct enamelling using electro-immersion processes, an additional zinc coating must be applied for 10-25 seconds [72] to prevent the substrate steel from corroding anodically. The coating should be thin enough to oxidise quickly on firing, without the white opacifying agent being discoloured by reduction.

Rinsing takes place after all these pretreatment operations. The rinse water consumption per process stage is reported to be 30-50 litre/m^2 in conventional enamelling and 50-70 litre/m2 in direct enamelling [72]. These quantities can be reduced still further by the introduction of the cascade rinsing technique with supplementary spray rinsing [75,76].

After the immersion and spray processes there follows as a rule a passivation treatment with nitrite-containing or sometimes with nitrite-free solutions (soda and borax). While on the one hand it is good to dispense with the nitrite-containing solutions, caution is nevertheless necessary, as the nitrite-free solutions may contain alkanolamines. If nitrite is used as a stabilising salt in enamelling, then there is a danger of nitrosamine formation [70].

All the treatments mentioned so far and the materials they contribute to the effluent do not make this waste water greatly different from the others that arise in the metal working industries. A new waste water constituent is however, generated in the enamelling process itself. This is a finely dispersed insoluble inorganic material of similar composition to the frits described above. This arises at the following points:

- Milling of the frits

- Application of enamel by dipping, spraying and electro-immersion enamelling. It appears in the rinse water in dipping and immersed electrolytic processes, and as overspray in spraying and electrostatic spraying.

- Waste gas washing units.

- Manufacturing plants and auxiliary units, in the cleaning water.

8.5.2   Waste water treatment and recycling processes

The waste waters generated during pretreatment, so long as they contain no chromates or nitrites, can be treated by neutralisation/precipitation (oxidising divalent iron!), sedimentation and sludge dewatering. Because of the large quantities of rinse water and absence (in most cases) of dangerous materials, this can usually be done in continuous process plants. Attention should always be paid however to the possibility of reducing water consumption further and hence the quantity of waste water produced. Treated slip-water or waste water containing only low levels of slip can be combined with metal-containing water for neutralisation/ precipitation and flocculation, then freed from solid matter by sedimentation. These processes were dealt with in Chapters 3 and 4, and have been described by Wilhelm [77] and Schraeder [76].

So long as there is no more than 5 mg/litre of nitrogen present as nitrite, as often happens, no detoxification plant is necessary. Chromium-containing waste water can be circulated through an ion exchange unit. In direct enamelling processes this is also possible with the rinse waters following nickel and zinc plating. As it is not worthwhile to recover materials from these solutions, they are put through a common continuous circulation plant. The regenerates must be treated in a batch process plant.

Process baths needing treatment must be processed in batch plants however, so long as no regenerations follow. The service lives of degreasing baths can be extended by centrifuging (see above) and ultrafiltration, and that of pickles for example by retardation methods.

Recovery should be undertaken in waste waters that contain larger amounts of slip, if the type of slip is not changed too often. This is particularly easy to do with overspray in the spray cabinets, which collects in a pasty form. This can be returned to the end of the slip milling process. Care must be taken however that not more than 20% of recycled slip is added to the fresh enamel, so that its properties are not changed [78].

Waste water that is heavily contaminated with slip is mixed with iron salts in a batch process plant, aerated to oxidise the divalent iron, neutralised with caustic soda or lime slurry and flocculated along with a filter aid. Settling takes about 1 to 1.5 hours. Dewatering is carried out with a filter press or with a sieve pan. A solid matter content of ca 70% can be achieved with a filter press and 40-60% with a sieve pan. Fig. 8.15 shows an enamel separator with filter press.

If the enamel sludge contains no dangerous materials, it can be disposed of without risk on a household rubbish dump. It can thus be an advantage to treat slip waste water separately. If however the slip contains cadmium-bearing pigments (red, orange, yellow), then, despite the insolubility of these materials, the waste water arising must be kept separate from other waste waters that contain slip, and treated separately.

The variety of enamelling processes available also has itself an effect on the wastes produced. In normal spray enamelling the overspray amounts to 55-65%.

Fig. 8.15   Separator (right) and chamber filter press for the separation and dewatering of enamel slip (EISENMANN, Boeblingen)

This quantity is halved simply by using electrostatic spraying. A similar effect is observed using the electro-immersion process in contrast to the conventional dip process. A further process is known, which uses specially prepared priming and top-coat enamels, and which can dispense with pickling and nickel plating. The sheet steel must be free however, from grease, rust and scale [71]. In many cases also waste water treatment can be simplified by the frit manufacturer's supplying ready-pulverised frit materials.

8.6   ORGANIC FINISHING INCLUDING PRETREATMENT

Articles of unalloyed or lighly alloyed steels, are often phosphated for increased corrosion resistance and to increase the adhesion of the organic coating.This process is performed in solutions containing inorganic chemicals, in which organic additions, such as surface active agents for example, exert a beneficial effect by improving the wetting and emulsifying properties. Painting takes place subsequently, using materials principally organic in nature.

In painting, two kinds of waste water are produced, which can always be distinguished from each other, and which should generally be treated separately:

- Waste water arising during pretreatment, with mainly inorganic constituents.

- Waste water arising in painting and stripping, with mainly organic constituents..

If phosphating is carried out to produce a lubricating layer for non-cutting shaping operations, the waste water falls into the first category.

8.6.1 Pretreatment

8.6.1.1 Amount of waste water and its constituents

Pretreatment in aqueous solutions includes cleaning;rinsing; phosphating; rinsing; (rinsing in desalinated water). The cleaning takes place in alkaline degreasing baths, such as were described in Section 8.1.1.

Phosphating is carried out in solutions containing phosphoric acid, zinc phosphate, or zinc and calcium or magnesium phosphates. Their pH value is in the range 2.8-3.4, or in alkali-phosphating baths up to 5.6. They contain as well, as accelerators, oxidising agents such as nitrites, nitrates, chlorates, hydrogen peroxide [79,80] or organic nitrogen compounds, surface active agents, and in special cases metal compounds also, polyphosphates, fluorides, borates and so on. In their presence, iron(II) phosphate goes into solution and is partly oxidised to iron(III) phosphate, and as such gives rise to relatively insoluble, so-called phosphate sludge.

Like all pretreatment processes, phosphating can be carried out by dipping and/or spraying. A finely crystalline layer is thereby formed on the iron surface, which consists of iron(II) phosphate and the phosphates of other divalent metals present in the solution. It provides excellent adhesion for the paint layers to be applied to it subsequently. Conversion coatingsof this kind have a preferred range of surface weights between 1.2 and 2 g/m^2 [33, 81], sometimes to 4 g/m^2 [82], They derive ca 25% from the substrate material and are built up on to it to ca 75% [83]. According to Schene and Kuebler [84], of the chemicals used in the phosphating of vehicle bodywork, ca 25% remains on the surface, ca 50% precipitates out as phosphating sludge and ca 25% passes into the rinse water.

After phosphating, chromating in very dilute chromic acid solutions (0.1%) often follows [85]. The pores in the phosphate layer are thereby closed and the corrosion resistance further increased. If electrophoretic paints are applied, a final rinse in desalinated water is carried out after chromating, or if this stage is not included, after the normal rinse that follows phosphating.

Among the special additional process steps that should be mentioned are rinses with additions of activating materials (titanium compounds) in the last rinse bath before the phosphate treatment and the pickling of more severely corroded or scaled workpieces after degreasing. Hydrochloric acid, sulphuric acid, often with phosphoric acid and sometimes organic acids, are used in the pickle [80].

Degreasing baths, pickles and chromating baths are short-lived process solutions and need treatment as concentrates at intervals of time, although these can be lengthened by the use of regenerating procedures. The phosphating sludge characteristic of this pretreatment is also produced. It consists of iron(II) phosphate, retaining within it, however, liquid that contains all the constituents of the phosphating solution; this is why it must be transferred to the waste water treatment facility.

The rinse waters contain all the materials in the process baths used, in more dilute form. Their quantity, which varies according to the types of workpiece, is reported to be from 1 to 15 litre/m^2 per rinsing stage [80]. As will be seen in the following section, it is possible to reduce rinse water consumption, especially in the phosphating processes. In an automobile factory, carry-over amounts of 6 litre/car-body were measured. 80 litre of water was necessary in a simple rinse to meet the specified criteria, but only 20 litre in a two-stage cascade [85]. Gotta [86] quotes for a continuous circulation system (see below) a reduction from the usual 10-15 litre/m^2 to 10% of it. Under certain conditions it may even be possible to avoid producing waste water.

The amount of sludge amounts to about 0.3 litre/m^2 as wet sludge (ca 3% dry matter) [87], which can be thoroughly dewatered to ca 40% dry matter.

When phosphating for cold forming, phosphate coatings are applied with higher weight per unit area between 2 and 30 g/m^2. The process baths used have the same composition as quoted above with respect to metal phosphates and accelerators. The phosphate coating does not itself constitute a solid lubricant, such as for example graphite or molybdenum disulphide. The hard phosphate crystals are partly crushed in the working process and form glassy coatings at the high pressures and the temperatures generated. While still remaining firmly bonded to the surface, these form a very pressure resistant paste in combination with added lubricant (oils, emulsions, greases, soaps, but also graphite and molybdenum disulphide). On wires it is often possible to carry out more than 15 drawing operations with a single coating. If soaps (e.g. stearate) are applied, reaction products (zinc stearate) are formed with the zinc phosphate. The customary pretreatment, degreasing and pickling in inhibited acids, can serve also as an intermediate treatment after annealing and before the next cold-forming process [89].

Another metal that is also often painted is aluminium. It is pretreated by chromating. Treatment comprises the operating sequence: degreasing; rinsing; deoxidising; rinsing; chromating; rinsing. Degreasing is carried out in solutions like those used for anodising (see Section 8.7.1), the deoxidising in nitric acid (with hydrofluoric acid additions for silicon-alloyed material) and the chromating in chromate solutions. The coating is formed partly from the substrate material and partly from the chromate. It has a thickness of ca 0.5 mm and is resistant only up to 150°C [83].

In certain cases the bond coat on aluminium may also be produced by anodising. (see Section 8.7.1). It is a question here of corrosion protective layers, which are 5-25μm thick.

8.6.1.2 Waste water treatment and recycling processes

The possibilities of extending the service life by regeneration of degreasing baths are described in Section 8.1.2, and by ultrafiltration in particular in Section 7.5.4.1. If pickling is carried out, other than with phosphoric acid, the recycling processes for the solutions used can be found in Section 8.3.3.

The accumulation of phosphate sludge cannot be avoided. It can be reduced by about 30% by the use of so-called low-temperature phosphating. This process comprises zinc manganese phosphating at 35°C in contrast to normal zinc phosphating at 55-60°C [82]. As it cannot be operated with hydrogen peroxide as accelerator, however, the continuous circulation system described below is not possible with this process.

Recycling in phosphating depends mainly on maintenance of the phosphating baths (sludge removal and replenishment of the chemicals consumed) and on the special rinse water circulation systems necessary in this industry, and on the inclusion perhaps, of additional rinse stations.

First it is necessary to remove the phosphate sludge. In smaller plants this is possible with paper or felt band filters, but also with normal pressure filters. In larger plants the removal is accomplished in stages, e.g. by first circulating through a laminated separator. The sediment from this (with 1-2% dry matter) can be further dewatered (5-8% dry matter) in a thickener; the clear overflow from this, along with that from the layered separator, is returned to the phosphating bath. The thickened sludge is finally dewatered in a filter press up to 40% dry matter. The filtrate also returns into the phosphating bath [32,88].

After phosphating, rinsing is carried out in a cascade, the outflow from which is normally transferred to a waste water treatment plant. If care is taken, all the constituents of the phosphate bath can be precipitated with lime slurry (metals, phosphate, fluoride), then after neutralisation of the phosphating sludge and rinse water, and after removal of the solid materials, a clear water is obtained, whose electrical conductivity is about 200-400 μScm^{-1} similar in terms of conductivity to mains water [87]. A condition for this is that hydrogen peroxide be used as accelerator, as all other accelerators generate conducting salts. Water recovered in this way can be returned again as rinse water to the second cascade after phosphating. Fig. 8.16 shows the principle of continuous circulation with removal of the poorly soluble iron(III) phosphate and the other precipitated products. The rinse after degreasing can also be supplied with the treated water, if additions are made to the water before the spray heads of surface active agents (ca 0.4 g/litre) that have been lost from the degreasing bath during its purification by ultrafiltration [87]. In this way 90-95% of the rinse water can be saved.

If water treated in this way is continuously circulated through an ion exchange plant before feeding it to the spray heads, good quality water is thus made available for rinsing and replenishment, and it also becomes possible to operate with ionic accelerators (see Fig. 8.17).

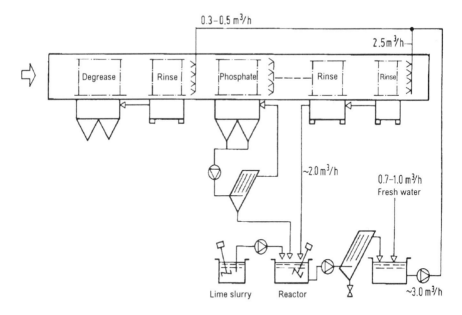

Fig. 8.16   Schematic diagram of continuous water circulation in a pretreatment plant with a surface throughput of 1000 m²/h (Duerr GmbH, Stuttgart)

Fig. 8.17   Plant for continuous circulation of phosphating waste water in the automobile industry, capacity 15 m³/h, right: neutralisation/precipitation and sedimentation, right: ion exchange unit for desalination of the clarified water produced (Duerr GmbH, Stuttgart)

With a simple double cascade after phosphating, with lime slurry precipitation and solid matter separation, it is possible to achieve a rinse water reduction of around 70-80% without any other special measures [31]; with a triple cascade about 80-90% can be attained [86]. By adding at the end a rinsing area using fully demin. water, which has been circulated separately through an ion exchange unit, electrophoretic painting can also be carried out afterwards. If a post-chromating treatment is installed, then it is advantageous to circulate the following rinse water as well through an ion exchanger, because the chromate ions can thus be held back and detoxification need only be carried out on the small amounts of regenerate [31].

Whatever water saving measures are adopted, a waste water treatment plant cannot be dispensed with. The waste water produced in cleaning or emptying the plant and during the regeneration of desalination units must be neutralised and possibly also detoxified and filtered. It is possible to do this in a batch treatment plant.

Another possible way of treating continuously circulated phosphating waste water is electrodialysis (see Section 7.6.3.2). The diluate is added again to the rinse water and the concentrate is returned to the process bath [84].

Unaccelerated zinc phosphating baths can be maintained by oxidising the divalent iron to trivalent iron in part of the flow and filtering off the hydrolysis products. In this way around 70% less zinc is lost than in accelerated baths [90].

If phosphoric acid is used for pickling there is a danger of unwelcome pre-phosphating if the iron content rises to more than 3%. The relatively expensive phosphoric acid can be regenerated with the help of strongly acidic cation exchangers (see Section 7.2.2.1).

There are processes for phosphating without degreasing or without production of waste water, such as that reported on by Lohmeyer [91] among others. In the first case however there are conditions to be met concerning the rust preventing oils used on the sheet, the lacquer that is to form the coating and its mode of application as well. Different procedures allow phosphating and priming to be performed in one operation, others again, degreasing and phosphating. The preparations used for these are dissolved in non-aqueous solutions.

Kraemer [92] describes a non-aqueous process using hydrocarbons and alcohols as solvent, which are recovered by distillation, for the production of 1-3 mm thick conversion coatings. The solvent losses amount to ca 1 kg on 30-35 m² of treated surface. No waste water is produced in this, but a waste gas treatment with active carbon or by post-incineration may be necessary. A process, which operates with methylene chloride as solvent [93], must have scarcely any prospects of application, because of the solvent's high volatility and limited adsorption on active carbon.

In the chromating of aluminium, chromium-containing waste water is generated, which must be detoxified. Afterwards it can, with other aluminium-containing water, be subjected to neutralisation/precipitation, followed by the removal of solid materials. The following can be used as abatement and avoidance measures:

- Water saving measures based on continuous rinse water circulation through ion exchangers or cascade rinsing after chromating

- Multiple use of rinse water by installing counter-flow cascades after various process solutions

- Regeneration of chromating baths using strongly acidic cation exchangers (see Section 7.2.2.1)

- Regeneration of degreasing baths, e.g. by ultrafiltration (see Section 7.5.4.1)

- Regeneration of deoxidising baths by retardation processes (see Section 7.2.2.2).

For waste water generated in anodising aluminium see Section 8.7.1.

8.6.2   Painting and paint stripping

8.6.2.1   Amount of waste water and its constituents

In industrial painting, waste water arises at the following points:

- Paint spray cabinets.

- Electro-immersion painting plants with and without ultrafiltration.

- Paint stripping plants.

- Waste gas washing plants.

In spray painting the following processes can be distinguished:

- Spraying with compressed air.

- Spraying without compressed air (airless spray).

- Electrostatic spraying.

- (Electrostatic) powder spraying.

All these paint spraying processes are carried out in spray booths to protect the staff. Depending on the spray process, varying proportions of the paint do not reach the product but settle on the walls of the booth that lie behind it (overspray). The losses that thus arise, depending on the surface areas of the articles being painted, are in compressed-air spraying between 50 and 80%, in airless spraying between 20 and 40%, in electrostatic spraying between 10 and 20% and in powder spraying only ca 2% [94].

In all processes material recovery measures can be employed (see below), but powder spraying remains the most environmentally friendly, compared not only with spray processes, but with all industrial painting processes.

So that the paint should not get stuck to the booth walls, these are protected by a curtain of water, which takes up the paint. This water is circulated for a long time, often months, in the circulation system of the booth. Hence, to remove the paint components, it is continuously cleaned by coagulating them and separating the conglomerates formed. This always constitutes one part of the waste water treatment (see below). The waste water that is emptied out at intervals contains not only the remaining paint constituents, but also the coagulation medium, so long as it has not been removed in an insoluble form with the paint residues. The constituents of the waste water can be: the remains of the solvents, paint binders, pigments and coagulation media, such as alkalis, calcium chloride and aluminium sulphate. Because of the thorough air extraction of the booths, there is not very much solvent in the booth water. It may however contain bacteriocide, added to extend service life.

Another industrially important kind of painting is electro-immersion painting. The process of anodic paint deposition devised in about 1960 has been almost superseded by cathodic paint deposition developed ten years later. Polymeric acids, which were made water-soluble by salt formation with amines, were used as binding agents in anodic paint deposition. The poly-anions migrate in the direction of the anode in the electric field. Electrolytic water decomposition produces hydrogen ions at the anode, which coagulate the salt by re-formation of the insoluble polymeric acids, which therefore deposit on the surface. The salt-forming amines are thus set free again.

The reverse occurs in cathodic paint deposition. Polymeric amines are made water-soluble by salt formation with organic acids (acetic acid, formic acid), migrate as polymeric cations in the electric field in the direction of the cathode, and coagulate on this by reforming insoluble polymeric amines under the influence of the hydroxyl ions produced there, thus setting free the organic acids again.

Compared with the anodic process, cathodic paint deposition has a series of technical advantages, which result in better corrosion protection, while exhibiting an ecological benefit in lower lacquer consumption. For the same corrosion protection, the optimum coating thickness of 22 mm for the anodically deposited paint can be reduced to 15 mm using cathodic deposition [95].

The electrophoretically deposited paints have a solid material content of 10-20% (cathodic) and 8-18% (anodic) and solvent contents of 1.5-2% (corresponding to 10-15% of the solid matter content). The solid material in the cathodic paints consists for example of epoxide or polyurethane resins with substituted cationic groups, within which are bound water-insoluble, usually mineral, pigments. The proportion of pigment to total solid matter lies between 10 and 80% [96]. Ethyl glycol, butyl glycol, hexyl glycol and higher primary alcohols among others are used as solvents in cathodic paints, iso-propanol, iso-butanol, propylene glycol and higher primary alcohols among others in anodic paints [95].

The constituents may finally get into the waste water, including the salt formers liberated by electrophoretic deposition; organic acids (formic acid, acetic acid) at cathodes and amines at anodes, and ions of different kinds (by drag-out). Waste water arises from the remains of immersion baths, the contents of immersed rinse baths (where recirculated), cleaning after colour change, waste water from the floor, discharged ultrafiltrate and dialysate. The last two contain mainly carbon dioxide or amines and inorganic salts.

In the case of spray and electro-immersion painting the application itself generates waste water typical of industrial painting. In all cases however waste water can also be generated from the washing of waste gases and the stripping of paint.

The air from spray booths, baking ovens and pyrolysis plants for removing organic coatings must be extracted and cleaned. If this is carried out in waste gas washing plants, then equipment of various designs, e.g. counter-flow scrubbing towers, can be employed, using wash-water made alkaline with soda. Whilst the air in the spraying booths contains dilute solvents and small residues of the overspray, the waste gas from the stoving ovens and pyrolysis plants may also contain cracking- and breakdown-products, plasticiser condensates, phenols, phthalic acid derivatives, methacrylates, aromatics and so on [79,97].

Other possibilities for waste-water-free waste gas treatment, which may also be used in combination with gas washing, are active carbon treatment, post-incineration (800°C) and catalytic incineration (200-400°C).

Many processes areused for paint stripping. Organic film removal in molten salt baths at 480-500°C presupposes thermally resistant material (no soldered joints). The paint stripping process lasts only 1-3 minutes. The melts consist of sodium hydroxide and sodium nitrate and become enriched during their period of use with sodium carbonate, nitrite and mineral pigments, while the organic paint constituents are burned off. The rinse water contains the same materials and must be passed on to a nitrite detoxification plant. The pigments that settle from the melt are removed daily. The mass thus obtained contains constituents of the melt with ca 10% pigments and cools, then freezes to a solid. The sodium compounds present can be dissolved out of this (they still contain almost 50% of sodium hydroxide) for return to the hot paint stripping bath, or must all be disposed of as hazardous waste.

Another method of removing paint at elevated temperatures is by heating in the absence of air (pyrolysis). The incompletely burned gas produced can be used to help meet energy requirements. The fully combusted gas must probably then be scrubbed. The inorganic paint constituents remain on the stripped components as pyrolysis ash, which can be removed by sand blasting.

If the parts to be stripped are strong enough, the lacquer can also be removed by high-pressure water treatment at 400-750 bar. Only shredded lacquer is found in the waste water, which after filtering through a sieve can be put back into circulation.

Hot paint-stripping baths are formulated with caustic soda of various concentrations depending on the lacquer and operated at just 100°C. The treatment time may be anything from one hour to several hours. The loosened lacquer is then removed mechanically or by water blasting. The waste water contains caustic soda and shreds of peeled-off lacquer.

The media for cold lacquer stripping are as a rule chlorinated hydrocarbons, such as dichloromethane, trichloroethane and so on. They may also contain hydrocarbon derivatives, such as phenol or cresol. Wherever such media are used, containment measures with closed-cycle circulation must be adopted, such as those that are described in Section 3.11.4. No waste water is then produced.

8.6.2.2 Waste water treatment and recycling processes

Treatment of waste water generated in spray booths is identical to that of waste water from electro-immersion painting plants, in so far as this is carried out conventionally by coagulation of the paint constituents.

Overspray in the booth water is present in a colloidal to finely dispersed condition. It can be coagulated by chemical additions, such as are sometimes used in treatment of oil-in-water emulsions. Both alkaline and acidic coagulating agents can be used for this. As the former, caustic soda or other alkalis, e.g. sodium silicate, are used, pH values up to between 10 and 12.5 being reached. The coagulation medium can also contain other emulsion splitting agents or additions that promote the sedimentation or flotation of the coagulated material, and anti-foaming agents. If the paint contains pigments in large proportions and/ or particularly heavy pigments, then sedimentation will always be preferred [98].

Solutions of calcium chloride and aluminium sulphate act as acidic coagulating media. Pulverulent substances are also well-known. All coagulating media have the additional task of reducing the stickiness of the coagulated material so far as possible, so that no problem is encountered in separating it. Because of the danger of sticking, all piping after coagulation and up to the filtration stage should be as short, open and hard to block as possible. Also, mixing the chemicals in an air powered jet can be used, for example, to keep the solids in suspension. The dense sludge layer is smooth at the top so that the waste water can be cleared of solid matter to less than 20 mg/litre. This process can be used for all electro-immersion paints. 0.5-2 g of coagulating agent is needed per gramme of solid material in the paint [99]. If only small amounts of water are circulated in the spray booths, flotation units can be used for separating the paint sludge; in these the coagulated waste water is mixed with fine bubbles of air from a compressed air cylinder and passed into a tank, where solid matter floats upwards, then is skimmed off and falls into a bag filter. The tank has a conical bottom on the inlet side with a strainer fitted to it, which is able to remove the larger, sedimenting particles. The overflowing clear water is passed back to the spray booths [100].

Usually however the coagulated paint sludge is sedimented and removed from the aqueous phase by various methods. Lohmeyer [97] describes a simple apparatus. Spray booth water flows after coagulation into a vessel subdivided by several partitions. Each partition consists of two layers of wire netting with wood fibre between them. The coarser particles settle in the chambers formed between these partitions, the finer ones being filtered out by the wood fibre. The purified water is circulated back to the booths. Sieves or filter baskets, coke layers, settling tanks and sand filters are also arranged one behind the other to achieve a mechanical separation [101].

Larger spray plants use vessels in the circuit in which the sludge sediments before being led through powered jets on the suction side of the pumps, then conveyed from there into a proper settling tank. The sludge that settles here is dewatered in a decanting unit up to ca 60% dry matter. The overflow from the settling equipment and the decanted water are returned to the circulation tanks [103]. Fig. 8.1 shows a decanting plant for dewatering paint sludge.

Fig. 8.18   Plant for treatment of water circulated through spray booths.
Right: storage tank, centre: coagulating medium, left: decanting unit (above)
and container for paint sludge (EISENMANN, Boeblingen)

A smaller plant operates similarly. The booth water being circulated is mixed with a neutral, powdered coagulating agent and is returned after sedimentation back to the booths. The sludge from the sump of the settling equipment is pumped out through bag filters [104]

Haltinger [105] describes paint coagulation without addition of chemicals by electrostatic treatment of the waste water. This is passed through a tubular module, where a DC potential of 3000-10,000 V is applied between two electrodes, causing coagulation to take place.

Various proposals have been made for the recovery of paints from spraying booths without coagulation. The last overspray residues in powder spraying can be sucked out and recovered in a separating chamber [94]. Overspray of wet paints can be recovered, either by skimming it continuously from a rotating disc in front of the booth's back wall or periodically from a rectangular wall. In the process some solvent is lost around the skimming device. The paint obtained is restored to its working viscosity by addition of solvent and used again [94,106]. The rate of recovery of overspray is around 60 to 80%. There are problems with colour changes, however, in all these procedures. It is understandable that attempts are made to achieve a reduction in overspray, bearing in mind the estimate that around 110,000 t/year of paint is lost through overspray in Western Germany alone [107].

In comparison to spray painting (except for powder spraying), electro-immersion painting is ecologically very advantageous. The effectiveness of paint utilisation is well over 80% compared with only around 40% in spray painting [96]. Further savings can also be made, because it is possible to apply relatively thin but uniform coatings that provide excellent protection against corrosion.

Nowadays all larger electro-immersion painting plants make use of recycling systems for the recovery and regeneration of paints. These are described in Section 7.5.4.2. Electro-immersion painting plants with a high surface area throughput generate only small amounts of waste water. (see the schematic diagram in Fig. 7.42.) Acids and bases liberated during the deposition as well as electrolytes carried over from the pretreatment can be removed by dialysis and ultrafiltration. The permeate from ultrafiltration is returned cascade-style into the rinse (spray-immerse rinse), just a small part of it being discharged to avoid electrolyte enrichment. The retained material is passed back into the electro-immersion painting bath. Kraus [108,109] describes the waste water situation in a large electrophoresis plant in the automobile industry. In it, 1-2 litre of waste water per m^2 of painted surface is generated. Of this, 0.03 litre/m^2 is derived from the dialysis overflow and 0.05-0.07 litre/m^2 from the permeate discharge from ultrafiltration. The remainder comes from the dip tank filled with fully desalinated water. Fig.s 7.43 and 7.44 show ultrafiltration plants such as are used for electro-immersion painting. Reel, pipe and plate modules are coming into use.

All the waste water mentioned above (1-2 litre/m^2) is subjected to coagulation. The same processes are used for this as were described in connection with waste water from spray booths. Fig. 8.19 shows a coagulation plant for electrophoretic paint. After coagulation, waste water, without removing solid matter, is passed to the neutralisation/precipitation stage, where it is neutralised and settled along with the pretreatment waste water. After dewatering in filter presses, sludge has ca 60% dry matter. Fig. 8.20 shows a continuous process treatment plant for the waste water from a painting shop.

Fig. 8.19   Coagulation plant for electrophoretic immersion painting (with alkaline solutions or lime slurry and simultaneous precipitation of lead by addition of phosphate-containing waste water), capacity: 10 m³/h (DÜRR GmbH, Stuttgart)

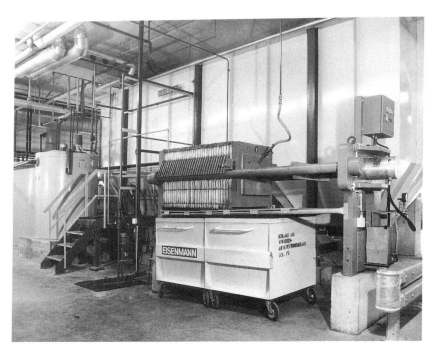

Fig. 8.20   Continuous process treatment plant for waste water from a painting shop. Left: chromate detoxification and neutralisation, behind: settling tank, in front: chamber filter press (EISENMANN, Boeblingen)

It is best to transfer waste water from the waste gas washing plant to the coagulation plant, and that from the paint stripping unit after removal of paint shreds to nitrite detoxification, if nitrite is present, and otherwise to neutralisation/precipitation. Waste water containing chlorinated hydrocarbons must not be discharged. Its production should be prevented or else it must remain in circulation within the confines of the plant.

If phenol- or cresol-containing waste water is generated in paint stripping, it should be examined whether it can be cleaned up by circulating it over adsorbents after removing solid materials. Otherwise, the only sensible treatment is by biological means.

As several references have already made clear, the choice of the type of paint to use can be decisive in satisfying ecological requirements. Application by dipping or spreading using solvent-containing paint (50% solvent) is now of minor importance in the metal working industries. The same is of true environmentally friendly high solids paint with ca 30% solvent, production of which in Western Germany is, incidentally, declining sharply.

The environmentally friendly systems are water-based paints, including electro-immersion paints. Emulsion paints, which also belong in this group, are of no significance in the metal working industries. Far and away the most environmentally friendly system is powder coating. Sprayed powder coatings achieve the highest recovery rate of ca 98% and contain only 2-3% of solvent, which derives from the manufacture of the binding agent. It is astonishing therefore that in Western Germany electro-immersion paints represent only 4.1% and powder coatings only 2% of the total paints used [1987]. Both however showed a tendency to rapid growth in the period 1982-1987: increase in powder coatings around 112%, in electro-immersion paints around 60%. Requirements for discharge of waste water from paint shops can be found in Chapters 1 and 2.

8.7 ANODISING OF ALUMINIUM

As well as the anodising of aluminium there are other production operations for this metal, such as mass finishing (see Section 8.1.1), chemical milling in alkaline solutions (principally used for airframe construction in the aircraft industry), etching of aluminium foils (usually in hydrochloric acid) in the manufacture of condensers, chromating prior to painting (see Section 8.6.1), phosphating and chemical oxidation. The treatment of waste water produced in these is very similar to that of waste water generated in the individual stages of anodising. Chemical and electrolytic brightening of aluminium as surface removal processes are described together with anodising.

8.7.1   Amount of waste water and its constituents

Anodising of aluminium is a surface conversion processes involving oxidation of the basis material, where the coating may also contain foreign matter, e.g. due to colouring.

Cleaning and degreasing for anodising is carried out in alkaline solutions, whose alkalinity must not be too high (pH 9-12), as aluminium can be attacked by alkali solutions. The constituents include those already mentioned in Section 8.1.1. Chlorinated hydrocarbons are also used for pre-degreasing or degreasing, but to a decreasing extent because of new legislation. Articles can also be degreased in bulk in combination with barrel polishing or mass finishing.

Alkaline etching follows the cleaning process. The preceding degrease can be omitted if the article are only very lightly soiled. The etch contains 40-100 g/litre (sometimes even more) caustic soda. The effect is to produce a uniformly active surface and particular surface characteristics (eg a matt finish). The solution usually contains sodium gluconate as a complexing agent (see also Section 3.9.3.4.3) to prevent scale formation (deposition of sodium aluminate on heating surfaces and walls). Sodium nitrate and nitrite are often used as accelerators. These are reduced by the hydrogen generated during the etch, sometimes as far as ammonia, so attention must be paid to the possibility of there being nitrite and ammonia in the waste water.

Deoxidising, or cleaning, by dipping in nitric acid as previously mentioned, follows the etch. The concentrated acid and water are usually mixed in the ratio 1:1. If the articles need to be bright (reflectors, trim strip, jewellery items), an additional process step is inserted between deoxidising and anodising, i.e. brightening. In chemical brightening, ammonium bifluoride, combinations of nitric acid and ammonium bifluoride, and phosphoric acid are used, but also other preparations containing chromic acid or acetic acid. Alkaline brightening baths are also known, which may contain nitrite, nitrate and phosphate as well as sodium hydroxide. For electrolytic brightening, both acid process solutions are known, based on phosphoric acid, sulphuric acid and chromic acid, and strongly alkaline electrolytes also, which contain sodium carbonate and sodium phosphate.

Anodising then takes place. Aluminium accumulates in the electrolytes, and they become unusable at a concentration of about 20 g/litre aluminium. Formerly, portions of them were replaced at short intervals by fresh electrolyte or they were completely discarded after longer periods. Today, to an increasing extent, they are regenerated (see next section).

If the articles are to be coloured, another process, colouring, is used before sealing, unless integral colour-anodising is carried out. This can be done by AC treatment in metal-containing electrolytes to produce colours or by immersion in solutions containing 10-50 g/litre iron(III) oxalate (yellow, gold and bronze colours) or by immersion in rather dilute solutions of organic dyes (all colours plus black). The colour fastness of organic dyes, which are bonded adsorptively to the oxide coating, is less than that of inorganically produced colorations.

Finally comes the sealing of the oxide coating by its conversion to boehmite using hot water (96°C) or steam (over 98°C). Afterwards, there is another rinse.

In normal circumstances the constituents of the waste water are aluminium, caustic soda, sulphuric acid, nitric acid and their neutral salts as well as gluconate and possibly also nitrite and ammonia. From brightening there may be fluoride, phosphate, chromate and other substances present, and from colouring, heavy metals, particularly tin, iron(III) oxalate, and from sealing, nickel.

The quantities of rinse water are generally large, especially from the rinsing stages following anodising, as the diffusion outwards of the constituents of the anodising bath from the pores in the oxide coating takes a relatively long time.

In principle, all process solutions in anodising are short-lived, and become unusable after a certain time, though after longer intervals in the case of the deoxidising and anodising baths. Their service life can however be increased by regeneration. The volumes that have to be dealt with are frequently very large. This is particularly so if parts are being processed for the building and construction industry ($10 \ m^3$ and more).

8.7.2   Waste water treatment and recycling processes [112]

In normal circumstances, separation of waste waters in not necessary in aluminium anodising, unless chromic acid-containing brightening and/or anodising baths are being used. After chromic acid-containing brightening baths, a chemical rinse in acidic sodium bisulphite solution may even be possible (see Section 6.3.5), so that a separation is still unnecessary. In these cases, however, the brightening bath, the chemical rinse bath and the flowing rinse bath are installed within a separate enclosure, so that leakages and washing waste water do not mix with the other waste water. If the alkaline etch contains nitrite, or if nitrite is produced from nitrate, then it must be determined whether the nitrogen content of the rinse water exceeds 5 mg/litre (corresponding to 16.4 mg/litre nitrite) in any circumstances. If not, it is unnecessary to have a detoxification plant, which would moreover involve the generation of additional salt. If the etch contains an accelerator it will be necessary to carry out nitrite detoxification as a separate treatment. As all nitrite detoxification reactions take place in the weakly acidic range, acidification to pH 4 to 5 is essential. This can be used also for acidic process baths (deoxidising and anodising baths) that come up for disposal. The provisions described in Section 3.4 should be observed in this connection.

Especially in anodising plants with very large process baths it is important to provide correspondingly large storage tanks for the solutions to be drained off, which can then be utilised for treatment in batch plants. Reuther and co-workers [113] suggest that the sludge produced be filtered off and only the filtrate passed to the continuous neutralisation plant.

If the waste water contains fluoride, e.g. from a chemical brightening bath, then neutralisation is carried out with lime slurry. Care must always be taken however that the calcium supplied in the neutralising medium is sufficient also for the precipitation of calcium fluoride. Otherwise it will be necessary to make a further addition of calcium chloride solution. Fluoride precipitated is only with great difficulty from hexafluoroaluminate (cryolite), and so the 50 g/litre fluoride specified in normal cases should be easily satisfied.

Waste water from anodising shops is characterised principally by its aluminium and sulphate contents. In line with its solubility in the waste water, the latter can be precipitated as well during neutralisation with lime slurry, but a reduction below ca 2000 mg/litre sulphate should not be expected. The precipitation pH for aluminium lies between 4.8 and 8.7 in terms of waste water technology, but between 7 and 8.7 if a residual aluminium concentration of 3 mg/litre must not be exceeded [114] (see also Fig. 3.84). Beyond pH 8.7, the aluminium begins to dissolve again as aluminate. If chromium is present, this also is completely precipitated in the range quoted above. If other metals are present, e.g. from electrolytic colouring baths, then tests must be made to see whether they can be precipitated in adequate amounts in the presence of larger quantities of aluminium. If this is not the case, then with the consent of the appropriate authorities the pH value should be raised and a somewhat higher aluminium content tolerated. It should also be considered however whether it would not even be worthwhile to arrange for separate discharge and neutralisation/precipitation of the metals responsible for the hazardous substances. Their precipitation residues are classified as special waste, while aluminium hydroxide is usually considered as harmless waste that can be deposited along with domestic rubbish. It must of course be demonstrated that the aluminium hydroxide does not still contain other metals deriving from alloying constituents of the aluminium.

The only complexing agents expected in this waste water are polyphosphate and gluconate. The former is precipitated out with lime slurry, the latter causes no problems below 200 mg/litre at pH 7 or is precipitated out at pH 9 in the presence of an excess of aluminium as poorly soluble metal compounds when neutralised with lime slurry [115]. See also Section 3.9.5.1.1.

If organic dyestuffs get into the rinse water, special precautions are not usually taken to remove them, and with a little care they are not necessary. Freshly precipitated aluminium hydroxide is able to adsorb the dyes, so that after removal of the solid matter the waste water is colourless. If exhausted dye baths have to be dealt with, they must be treated together with aluminium-containing concentrates (alkaline etches, anodising baths).

Colour can also be removed from dye baths with the help of active carbon. According to Grossmann [115a] adsorption takes a relatively long time, as diffusion of the dyestuff molecules has to take place if they are to reach the inner surfaces of the granulated active carbon. In a column process a contact time of ca 45 minutes is required. The volume of the active carbon is therefore spread out to yield a throughput of 10 $m^3 m^{-3} h^{-1}$. For the same reason, the quantity of rinse water should be kept as low as possible by using cascade rinsing. The best adsorption is achieved at the isoelectric point of the dye (under pH 2 for anionic dyes). 1 kg of dyestuff then requires up to 5 kg of active carbon.

Precipitated aluminium hydroxide has a slow sedimentation rate. The usable volumes of sludge separation reservoirs of continuous process plants must therefore be designed with at least a 6-hour theoretical residence time. For other metals, except chromium, 4 hours is sufficient. The clarifying area in plate separators must therefore be around 1.5 times greater than is usual for other metals.

Within limits, cascade rinsing systems and continuous circulation through ion exchange units can also be used as water-saving measures. Cascade rinsing is possible after degreasing, etching, deoxidising and anodising. In each case, two or perhaps even three cascade rinses are again connected one behind the other in reverse sequence, so that rinsing after anodising is always in the cleanest water. In this way, correspondingly little concentrated rinse water is generated. The quantity of sulphate is also considerably reduced in these circumstances. With ion exchangers the rinse baths after degreasing and etching are as a rule excluded from the circuit. They can however still be connected as a series of cascade rinses. Care is necessary, if rinse baths from other facilities, e.g. for electroplating, are linked with them. Organics can then get into the circulated water, and are not completely removed by the ion exchangers. They are adsorbed during the rinsing operations by the active oxide film formed in anodising and can give rise to slight discoloration if the coating is not coloured subsequently. It is possible, to pass the circulating purified water through a small column containing active carbon or an adsorbing resin before transferring it to the post-anodising rinse baths. The regenerates from the ion exchangers have a very small volume in comparison with the circulated water and can be dealt with in a batch-treatment plant. In strip and wire plants water saving measures can be put into effect in a similar manner. Drag-out losses can be reduced in the former case by pinch rolls and in the latter by wiping apertures.

Water saving measures in this branch of industry should also aim to use batch-treatment plants where possible. Continuous process plants should not however be replaced in very large facilities, as large amounts of rinse water are necessary after the alkaline etch because of the viscosity of the solution and after anodising because of the diffusion processes, and these cannot be dealt with in batch plants using the procedures described. The latter should be restricted therefore to the waste water that contains harmful materials (metal-containing electrolytic colouring baths, nickel-containing sealing baths).

A further water saving measure should also be mentioned, based on the possibility of eliminating major waste water constituents, such as aluminium, metals, sulphate, fluoride and phosphate by precipitation [116]. The alkaline rinse waters are combined, acidified to pH 4-6, then mixed with the acidic ones, adjusted to pH 4.7-7 with lime slurry and heated to a temperature of at least

38°C. Solid matter, which can now settle easily, is separated or filtered off [117] and clear water is returned to the rinsing tanks. Through occasional residue washing and replacement of evaporation losses, fresh water make-up is needed, which is best circulated to the rinse bath following anodising, so that the neutral salts (up to 2.3 g/litre calcium sulphate) are kept at a constant value. Water savings up to 90% should be possible.

Exhausted degreasing baths, if there is no acidic waste for simultaneous diposal, can be treated by acid splitting of the emulsified oil followed by its separation (see Section 3.13.4.1). All processes described in Section 8.1.2 can be used for regeneration or reduction in the amount requiring treatment, especially if large volumes are mostly involved.

Regeneration processes for alkaline etches would be very desirable. It has been observed in the USA that sodium aluminate formed, after seeding for example, stimulates hydrolysis with precipitation of aluminium hydroxide and liberation of caustic soda. This hydrolysis should therefore occur at all concentrations at aluminium/caustic soda ratios from 0.36 to 0.5 [118] (see also Section 7.4.2). If no complexing agents are present to disturb this process, the recovery rate for caustic soda after separation of solid matter should amount to ca 70% [119].

In an investigation of a process for recovery of phosphoric acid from brightening baths, Nelson and co-workers showed that the most favourable way, bearing in mind the relative investment and running costs of evaporation or ion exchange plus evaporation processes, is to reclaim 30% of the acid (process bath concentration 75%) with the help of cascade rinsing and to sell the phosphoric acid.

A report was given in Section 7.2.2.2 of the regeneration of anodising baths with the help of the retardation process. The procedure is very advantageous for this process bath in particular, since as well as recovering acid and reducing waste water contamination, especially with respect to sulphate, considerable savings can be made in energy costs (corresponding to ca 0.4 kW/m² of anodised surface) [1221,122]. According to Jenny [123], either the voltage increases with the aluminium content at constant current density or the current density falls at constant voltage (see Fig. 8.21). The same process can also be used to regenerate nitric acid solutions, such as those comprising deoxidising baths [124] (see Section 7.2.2.2).

Exhausted anodising baths or alkaline aluminium etch baths can often be made use of for flocculation or phosphate removal in municipal clarification plants. A process has been described [125] for producing aluminium sulphate or aluminium-containing concentrates for this purpose from anodising waste waters. For this, aluminium hydroxide is first precipitated by mixing alkaline etch and anodising solution. The precipitated aluminium hydroxide is filtered off and part of the residue-cake is just dissolved in the etch. The other part of the residue is just dissolved in the anodising bath and further concentrated by evaporation.

Both products are sold. The sodium sulphate-containing filtrate is crystallised out after evaporation and separated as $Na_2SO_4 \cdot 10\ H_2O$ by centrifuging.

The discharge values for waste water from anodising shops are indicated in Chapters 1 and 2.

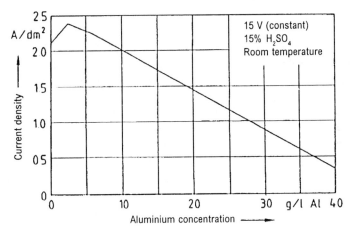

Fig. 8.21 *Dependence of current density on the aluminium concentration (at constant voltage) in the anodising of aluminium (from Jenny [123])*

8.8   ELECTROPLATING

In the narrow sense, electroplating is understood to be the electrolytic deposition of metals on metals and also on non-metallic surfaces that have been made electrically conducting, including all the pre-, intermediate- and post-treatments required. In the wider sense, it can be taken to include metal deposition without external sources of current on metallic and non-metallic substrates, metal colouring and metal stripping (chemical and electrolytic). Procedures or parts of them, such as were described in Sections 8.1, 8.3, 8.6.1 and 8.7, can also be integrated as pre-, intermediate- or post-treatments in surface finishing processes involving electroplating.

A review of the waste water produced and the waste water treatment and recycling processes used can be found in the leaflet ATV-Note M 765 [126].

8.8.1   Amount of waste water and its contents

Like anodising, electroplating can be considered in terms of the following principal treatments:

- Pretreatment: Removal from oily and greasy components of dirt and mechanical residues of all kinds, removal of corrosion products and activation of the surface by dissolution processes.

- Electroplating and intermediate treatments: Besides the electrolytic deposition of individual metals, different metals may also be deposited one after the other (multiple layers) or simultaneously (alloys). Any necessary etches or activations are carried out intermediately.

- Post-treatment: This is performed to increase corrosion protection (e.g. by chromating) or to produce a decorative effect (e.g. colouring).

- Metal stripping: This applies to the contacts of product carriers (racks, barrels) and rejected parts.

In pre-degreasing a distinction must be made between halogenated hydrocarbons, such as trichloroethylene (Tri), tetrachloroethylene (Per), or also fluorinated CHCs, such as 1,1,2-trichloro-1,2,2-trifluoroethane (R113), and degreasing in hot aqueous solutions. For halogenated hydrocarbons, particular care should be taken to avoid generating waste water by using fully enclosed facilities. Where waste water nevertheless does arise it must be kept separate in accordance with the relevant legislation (see Chapters 1 and 2) and treated appropriately (see Section 3.11.4).

Pre-degreasing or cleaning in alkaline, less often also in neutral, hot solutions and the waste water constituents that then arise have already been comprehensively described in Sections 8.1.1 and 8.1.2. This treatment is frequently sufficient for articles in bulk. Attention must always be paid, if non-ferrous metals are being degreased, not only to the oil, grease and dirt that accumulate, but also to the presence of metals, owing to the likely existence in the waste of complexing agents such as polyphosphates, NTA or gluconates.

If there are corrosion products on the surface, then these as a rule are removed by pickling after degreasing (see Section 8.3). The pickles contain mineral acids or mixtures of them, and may also contain inhibitors, to reduce the severity of attack or hydrogen embrittlement, or surface active agent additions to exert a degreasing effect (pickle-degreasers).

For racked articles, the obligatory final cleaning step is an electrolytic degrease, and this is also used to an increasing extent for articles processed in bulk. Its formulation is similar to that of hot soak cleaning baths, except that to avoid foaming they have no, or only very small, additions of wetting agents and that they may contain sodium cyanide. For cleaning, the articles are connected so as to be either cathodic or anodic, or sometimes, by reversing the polarity, cathodic and anodic in turn. In rare cases, treatment with polarity reversal, in solutions with a high cyanide content (formerly also EDTA), is also used for the alkaline descaling of iron and steel components. This gives rise to ferro- and ferricyanides.

Before being finally electroplated, work is usually dipped in a solution to etch it, i.e. in a dilute acid (sulphuric, hydrochloric, fluoboric or hydrofluoric acid). Before plating in cyanide electrolytes it may also be dipped in a sodium cyanide solution. These procedures are also sometimes used between one electroplating process and another. For activation purposes an intermediate degrease may even be necessary.

On the cleaned and activated surface other metals can now be electrodeposited. For this, acidic or alkaline metal-containing solutions are used. The latter require complexing agents to keep the metals in solution, usually as cyanide- or hydroxy-complexes.

The following notes include comments concerning the substances that can get into rinse water following electroplating in the electrolytes indicated. In contrast to the process solutions for pre-, intermediate- and post-treatments, these electrolytes could be described as long-lasting process solutions.

Lead electrolytes: The acidic electrolytes contain divalent lead and are usually based on fluoborate or fluosilicate.

Lead-tin electrolytes: The strongly acidic electrolytes contain the metals lead and tin in the divalent state as fluoborates and free fluoboric acid.

Bronze electrolytes: The alkaline baths contain the copper as a cyano-complex and the zinc as a hydroxy-complex, both dissolved as anions. In addition, the baths contain free cyanide.

Cadmium electrolytes: The acidic electrolytes contain the cadmium as sulphate or fluoborate together with the corresponding free acid. More common are cyanide electrolytes, which contain the cadmium as a cyano-complex and free cyanide.

Chromium electrolytes: The simplest chromium electrolytes consist of chromic acid and some sulphuric acid. In the mixed acid chromium electrolytes, part of the sulphuric acid is replaced by hexafluorosilicic acid. Chromium electrolytes are also known that contain hydrofluoric acid additions. Electrolytes with trivalent chromium compounds (in the presence of ammonium salts) have not gained wide acceptance.

Gold electrolytes: Most electrolytes are cyanide-based, the gold being present as a gold-cyanide complex. Baths based on sulphite or chloride frequently contain additions of gold cyanide as well.

Copper electrolytes: Acid copper electrolytes consist of copper sulphate or copper fluoborate and the corresponding free acid. In cyanide electrolytes the copper is present in monovalent form as a cyano-complex along with free cyanide. There may also be additions of thiocyanate and tartrate. Alkaline cyanide-free copper electrolytes are based on pyrophosphate. They contain ammonium salts.

Brass electrolytes: They contain copper and zinc as cyanide complexes together with free cyanide.

Nickel electrolytes: These electrolytes are always weakly acidic and formulated on the basis of sulphate and chloride; they contain boric acid as a buffer. Other nickel electrolytes are based on sulphamic acid and fluoboric acid.

Silver electrolytes: Most electrolytes are based on cyanide and contain silver as a silver-cyanide complex.

Zinc electrolytes: Cyanide electrolytes contain zinc as a cyanide complex along with free cyanide. The equally common weakly acidic zinc electrolytes are prepared on a basis of chloride or sulphate. They may contain high concentrations of ammonium salts if they are not buffered with boric acid. There are also acidic zinc baths based on fluoboric acid and cyanide-free alkaline baths, which contain zinc as a hydroxy-complex. Organic amines may also be present as weak complexing agents.

Tin electrolytes: Alkaline electrolytes contain the tin as stannate. Acidic electrolytes may contain sulphate, chloride, fluoborate or cresolsulphonic acid.

Frequently, in electroplating the current efficiency in anodic dissolution is greater than that in cathodic deposition, so that the metal concentration can increase with the service life of the electrolytes. A further contribution to this may be made by chemical dissolution of the anodes in the electrolyte on standing. Recycling processes are capable of restoring the metal balances in the systems (see Section 7.4.1 and 7.6.2.2.2). Conversely, if for example inert, i.e. insoluble, anodes are used, as in chromium and gold electrolytes, the strength of the electrolytes can be restored by metal salt additions.

Post-treatment in the electroplating industry consists in the majority of cases of chromating of zinc and cadmium coatings, or occasionally the anodic chromating of silver coatings. In all cases, chromating contributes additional corrosion protection to the coated article, and may also affect its colour as well. Chromated films may be: transparent, blue, yellow, olive-drab or coloured. Then there is the colourless, strongly leached, bright chromated finish and black chromating achieved with the help of silver-containing solutions. The constituents of these process solutions are mainly chromic acid or chromate, various acids and salts of the cations sodium, potassium, ammonium, and the anions nitrate, fluoride, sulphate, and always the metal ions formed or generated during operation, such as those of chromium, zinc and cadmium. Chromating baths operate at various acidities and, without regeneration, have only short service lives. Chromated metal coatings can easily be coloured in solutions of organic dyes to improve the identification of small components.

Dipping in phosphate- or nitrite-containing solutions, as well as in oil-containing solvents, can also be used to produce a slight increase in corrosion protection. To ensure spot-free drying, the treated article may be immersed in hot, fully-desalinated water or in non-aqueous preparations, the so-called dewatering fluids. The latter must not be allowed into the waste water.

All the above constituents may be found in the waste water, which to assist in its treatment is kept in separate streams (see waste water separation, Sections 4.2.2 and 4.3.2), but which may also derive from different facilities and be generated in widely varying concentrations and amounts. In the electroplating industry, waste water is obtained from:

- Short-lived process baths that have become unusable

- Interchange of the contents of special rinsing systems

- Rinse water

- Regenerates from ion exchange plants used in the continuous circulation of rinse water and the recovery of materials and of process solutions

- Waste solutions from various recycling plants

- Operation of ancillary equipment used for filtration or specific electrolyte maintenance, waste gas scrubbing plants, chemical storage tanks, emptying stations, etc

- Cleaning and floor water of electroplating plants, ancillary equipment, as well as the waste water and recycling plants themselves.

Separate consideration is given in Section 8.8.2 below to the waste water constituents of some specific processes.

8.8.2   Specific Electroplating Processes

8.8.2.1   Cadmium plating

Since the mid-70s cadmium has been classified as an especially toxic material, which is why recent legislation requires the segregation and separate treatment and disposal of cadmium-containing waste waters a procedure which was previously prescribed only for waste water containing mercury. In the spirit of this regulation, cadmium-containing waste water should, first, not be diluted with other kinds of waste water, but should be treated in the smallest volumes possible, so as to minimise the amount present. The result has been that numerous companies have given up cadmium plating, and other metals or materials have, wherever possible, been substituted for cadmium.

In the year 1976-77, cadmium consumption in the electroplating industry in Western Germany was about 400 t/year. By the end of 1982 it had already fallen by a half, according to official estimates. Corresponding to this, only about 0.2 t/year was discharged to waste waters. In comparison, the amount for the year 1972 had been around 5 t [128]. This trend will have continued since then.

Low-emission cadmium plating is not conceivable without recycling measures. To satisfy statutory requirements, several firms have built low-waste cadmium plating plants or adapted existing ones. The procedures used and options taken are described below:

Koehler [129] reports on a cadmium plating plant, which was modified so that pretreatment, cadmium plating and chromating are separated from each other spatially and in terms of waste water technology. Transporters for racked products reach the loading stations already stripped of cadmium, so that pretreatment stages hot degrease, pickle-degrease, electrolytic degrease and the etch are kept free from cadmium. Several barrels are used for pretreatment of bulk articles, the work being transferred to others for cadmium plating. Rinse water is passed through a continuous circulation ion exchange plant. Cadmium plating follows in a cyanide electrolyte. Pre-rinsing is then carried out in a cyanide electrolyte similar to that used for cadmium stripping, followed by rinsing in a four-stage cascade. The cadmium-containing rinse water, which is already very small in amount, is further reduced by evaporation. The same situation applies in the cadmium stripping bath, whose volume losses can be replaced with cadmium-containing rinse water. All the cadmium thus finally arrives in the cadmium stripping bath. This solution is passed through an electrolysis cell, which enables a given concentration to be maintained. The waste water from chromating is first oxidised by hydrogen peroxide, which converts the trivalent chromium to hexavalent. Cadmium is precipitated as hydroxide, the filter residue put into the cadmium stripping bath, and the cadmium-free chromate solution transferred to the effluent treatment plant.

Huss and Peters [130] report on a cadmium plating plant, in which the recovery of other process solutions also takes place. For hot degreasing, two baths are connected in cascade. Solution is taken out continuously from the first bath and freed from oil in a gravity separator then passed back to the second bath. From time to time a small proportion of the degreasing bath is removed and replaced by new solution. Rinsing takes place in a cascade, the rinse water being reduced in volume by an evaporator, then passed into the degreasing baths to make good the evaporation losses there. The pickle-degrease containing sulphuric acid that follows operates at about 60°C. By constantly circulating it through a cooling crystalliser to separate iron(II) sulphate heptahydrate, the iron content is kept steady at 50-55 gm/litre (see Section 7.4.2). The rinse water discharge is processed through the accompanying cascade, as was described for degreasing. The subsequent electrolytic degrease operates at about 50°C, and this solution is also passed through an evaporator, so that the rinse water can be incorporated. A small part is withdrawn, passed to the cyanide detoxification plant, and replaced with new solution, to keep the sodium carbonate concentration within the limits set. The cyanide-based cadmium electrolyte is also regulated by means of an evaporator, in which the Joulean heat is used for evaporation. A cooling crystalliser in the circuit ensures the sodium carbonate content remains constant. The cadmium- and cyanide-containing sodium carbonate obtained is added to the cadmium stripping bath.

The chromating baths are subject to drag-in and drag-out, formation of trivalent chromium compounds and changes in cadmium content. Cadmium and chromium concentrations can be kept steady by circulation through a strongly acidic cation exchanger in the H-condition (see Section 7.2.2.1). Chromic acid consumed must be replenished. This regeneration can be performed automatically [131]. The rinse water after chromating is circulated continuously through a small ion exchange plant. Cadmium stripping is carried out in a cyanide electrolyte, kept low in volume by passing through its own evaporator. Because of the simultaneous cathodic deposition of cadmium and the removal of a small part of it in a cyanide detoxifier, the cadmium concentration also remains low. The detoxified portion and the regenerates from the cation exchanger are freed from cadmium by neutralisation/precipitation, and the cadmium is afterwards deposited from a sulphuric acid solution in an electrolysing cell. Finally, detoxified chromium and iron are precipitated as hydroxide and separated by filtration.

Some interesting observations were made following the commissioning of such a plant [131]. At a throughput of 16,200 m^2 of bulk articles (corresponding to 108 t by weight) and 8700 m^2 of racked articles per month, 45% of the investment goes to the production operation and 55% to ensuring low-emission operation. The space demands for these requirements are in the ratio of 1:1, and the staff requirement up to 85% for manufacturing and around 15% for environmental protection considerations. The quantity of waste water amounts to 25.8 m^3/month, the amount of cadmium in the waste water 2.6 g/month and the waste gas pollution 0.9 g/month.

A suggestion of Götzelmann and Hartinger [133] for low-emission cadmium plating is to abandon the attempt to keep the pretreatment stages free from cadmium. Concentrate leakage due to the capillary effect and damaged insulation on racks etc can always cause undesirable carry over. Hence all the waste water in a cadmium plant including floor water and waste water from ancillary equipment is considered to contain cadmium. Using the combined rinsing technique described in Section 6.3.2.2 (pre-rinse cascade continuous circulation of rinses), with good rinsing standards, only very small amounts of waste water are generated, of which those portions containing valuable or dangerous materials are returned via recycling plants to the process baths. Waste water whose contents are worthless, and which contain only the smallest amounts of cadmium, are treated under optimised conditions (cadmium precipitation with soda, use of selective cation exchangers). The example in Fig. 8.22 shows a barrel cadmium plating plant with an hourly carry-over of ca 12 litre/bath for an average amount of waste water produced of only 1.6 m^3/day. The cyanide cadmium plating bath is recovered or regenerated by evaporation, the chromating bath by ion exchange, and the metal stripping bath by electrolysis. The rinse water after the hot degrease and the pickle is used twice. The consumption is only about 48 litre/hr. The rinsing criteria are maintained by means of a continuous circulation ion exchange plant. This has a capacity of 10 m^3/h, so that the actual requirement in this example is only about 2.7 m^3/h. The functioning of the plant is thus guaranteed, even if there is a change of production involving for example greater dragout, without any increased water throughput being necessary in the pre-rinse cascades, which would incidentally have necessitated a higher recycling plant capacity.

Vachon and co-workers report on a special electrolysis technique for eliminating cadmium, which brings about a considerable reduction in cadmium emission while achieving cyanide detoxification at the same time (see Section 7.6.2.2.2 for details).

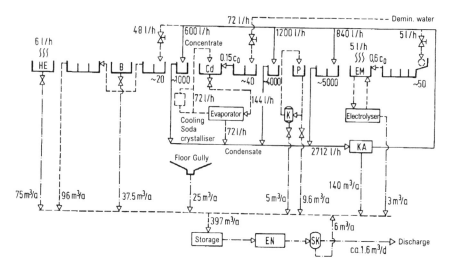

Fig. 8.22. *Schematic diagram of a cadmium barrel plant with optional electrolytic recovery and simultaneous minimisation of the amount of waste water produced.*

HE hot degreasing, B pickling, Cd cadmium electrolyte (cyanide), P Chromating, EM metal stripping, K cation exchanger, KA continuous circulation ion exchange plant, EN detoxification and neutralisation plant, SK selective cation exchanger [133]

These examples, in conjunction with the fact that cadmium has become restricted to small operations specialising in it, and that cadmium can easily be dispensed with, ought to not to dispel apprehensions that this metal once aroused.

8.8.2.2   Electroless nickel plating

8.8.2.2.1   Waste water constituents

If nickel plating is to be carried out without an external source of current, a very strong reducing agent is necessary. So that deposition should not occur spontaneously, the baths contain complexing agents, which lower the nickel ion concentration to an extent determined by the stability of the complexed nickel compounds. In addition, the baths may contain buffering substances to keep their pH value constant. The waste water associated with electroless (chemical) nickel plating contains substances present in the baths as made up, along with oxidation products of the reducing agents. As the baths may sometimes be operated in the weakly acidic range (pH 4-5) and sometimes in the alkaline range (pH 8-12), the pH value of the waste water produced is affected accordingly.

In addition to nickel, present in concentrations around 7-8 gm/litre, the bath constituents may be:

Reducing agents: Sodium hypophosphite (NaH_2PO_2), sodium borohydride ($NaBH_4$), di-alkylamine borane ($R_2 \cdot NH \cdot BH_3$) principally di-methylamine or di-ethylamine borane   and hydrazine hydrate ($N_2H_4 \cdot H_2O$). The concentration at which these are used varies considerably. For hypophosphite it is 10-30 gm/litre at the highest, for sodium borohride 0.6-0.7 gm/litre at the lowest, while for other compounds it is ca 2 gm/litre [134-137].

Complexing agents (sometimes acting also as buffering substances) are:

- Mono- and di-carboxylic acids, such as acetic, propionic and succinic acids.

- Hydroxy-carboxylic acids, such as glycolic, hydroxy-propionic, hydroxy-succinic, tartaric, citric acids, and their salts.

- Amines: glycine, ethylenediamine, EDTA (now prohibited in some countries).

- Inorganic compounds: ammonium salts, pyrophosphate.

Their concentrations are between 10 and 90 gm/litre.

Then there are the oxidation products of the reducing agents, which are produced in accordance with the following fundamental equations:

$$Ni^{++} + H_2PO_2^- + H_2O \rightleftharpoons Ni^0 + H_2PO_3^- + 2\,H^+ \qquad 8.3$$

$$4\,Ni^{++} + BH_4^- + 8\,OH^- + H_2O \rightleftharpoons 4\,Ni^0 + H_2BO_3^- + 6\,H_2O \qquad 8.4$$

Phosphite and borate are also present in used baths. In di-alkylamine borane-containing baths the corresponding di-alkylamine is present as well.

8.8.2.2.2   Waste water treatment and recycling processes

The problematical materials in these waste waters are the complex nickel compounds, wherever the complexes involved are stable, ammonia and phosphorus in hypophosphite-containing baths, which is sometimes present as phosphite. Then there is the reducing agent present, which contributes to the COD.

The COD of the waste water is due not only to the reducing agents, but also to the organic complexing agents (see Table 8.5). COD. After the hypophosphite has reacted, a half of the COD still remains in the form of further oxidisable phosphite, which increases in concentration as the necessary additions are made continually to regenerate the bath. The relatively high COD of the borohydride is not so critical, bearing in mind that the concentration needed is only about 5% of that of the hypophosphite and that its reaction product, borate, has no further effect on the. None of the processes usual in this area leads to the decomposition of the latter or of ammonia. In most cases the COD due to organic compounds can be decomposed biologically only in a public waste water plant. The situation with respect to ammonia is similar. The limiting value of 100 mg/litre N from ammonium compounds can as a rule be met. On passing into the public waste water plant, it is oxidised to nitrate by nitrification or broken down to nitrogen by subsequent denitrification.

TABLE 8.5:
COD OF THE CONTENTS OF A TYPICAL
ELECTROLESS NICKEL BATHS*

Estimated Compound	Formula	Molecular weight	g COD Mol^{-1}	g^{-1}
Acetic acid	$CH_3 \cdot COOH$	60	64	1.07
Propionic acid	$CH_3 \cdot CH_2 \cdot COOH$	74	112	1.51
Succinic acid	$COOH \cdot CH_2 \cdot CH_2 \cdot COOH$	118	112	0.95
Glycolic acid	$CH_2OH \cdot COOH$	76	48	0.63
Hydroxypropionic acid	$CH_2OH \cdot CH_2 \cdot COOH$	90	96	1.07
Hydroxysuccinic acid	$COOH \cdot CHOH \cdot CH_2 \cdot COOH$	134	96	0.72
Tartaric acid	$COOH \cdot CHOH \cdot CHOH \cdot COOH$	150	80	0.53
Citric acid	$COOH \cdot CH_2 \cdot \overset{\displaystyle OH}{\underset{\displaystyle COOH}{C}} \cdot CH_2 \cdot COOH$	192	144	0.75
Amino acetic acid	$CH_2NH_2 \cdot COOH$	75	72	0.96
Ethylenediamine	$NH_2 \cdot CH_2 \cdot CH_2 \cdot NH_2$	60	128	2.13
Sodium hypophosphite	$NaHPO_2$	88	32	0.36
Sodium phosphite	Na_2HPO_3	126	16	0.13
Sodium borohydride	$NaBH_4$	37.8	64	1.69

* Calculated.

If the nickel bath contains complexes of low stability, as with acetic, lactic, propionic and succinic acid, the nickel can be precipitated as hydroxide at higher pH values (10-12) [138]. This is truer still in relation to lime slurry if the carboxylic acid forms a poorly soluble calcium salt as well, as in the case of tartaric acid, even though this forms a relatively stable complex. In this instance nickel can be precipitated from a solution containing 2-3 gm/litre to as low as ca 2 mg/litre, the calcium tartrate also precipitating to ca 1 gm/litre [139].

With the exception of EDTA-containing baths, which have now been banned for chemical nickel plating in some countries, this metal can be directly precipitated at pH 7-8 from the complexes referred to, using sodium sulphide, or better still an organosulphide. The procedures mentioned in Section 3.9.5.1.2 should be followed. Then even the complex with citric acid, which is the stablest of those used in nickel plating, presents no difficulties.

If the stability of a complex is not too great, nickel can also be removed using selectively operating cation exchangers with iminodiacetate groups (see Section 3.9.5.4). This works if:

- the nickel complex is cationic, as for complexes with ammonia and ethylenediamine. Genuine ion exchange occurs here.

- the stability of the nickel complex is distinctly less than that of the nickel iminodiacetate complex. Here breakdown of the complex takes place.

Table 8.6 contains details of the capacity of the ion exchanger for nickel in the presence of a double stoichiometric excess of complexing agent (in respect of the number of bonding sites and the most favourable pH range [140]. It is possible either to pass the rinse water directly over the exchanger or to eliminate the residual nickel at the end of the waste water treatment (final ion exchanger).

TABLE 8.6:
FEASIBILITY OF ADSORBING NICKEL FROM SOLUTIONS CONTAINING COMPLEXING AGENTS ON TO A WEAKLY ACIDIC CATION EXCHANGER WITH IMINODIACETATE GROUPS [140]

Complexing agent	Stability constant	Charge-sign	Capacity [g metal/litre resin]	Operating pH value
Ammonia	$3.1 \cdot 10^8$	+	at least 30	8.5
Ethylenediamine	$2.4 \cdot 10^{19}$	+	at least 20	9
Triethanolamine	–	+	at least 20	7
Quadrol	$7.1 \cdot 10^6$	+	at least 20	7
Citric acid	–	–	0	7
Tartaric acid	$6.3 \cdot 10^{16}$	–	at least 15	7
Gluconic acid	–	–	at least 15	7
NTA	$3.2 \cdot 10^{11}$	–	0	all
EDTA	$4 \cdot 10^{18}$	–	0	all

Since the most recent West German legislation restricting phosphorus to 2 mg/litre, there is the new problem of getting rid of hypophosphite and phosphite. A good part of the phosphite can indeed be precipitated as calcium hydrogen phosphite, but not in satisfactory proportions. Only if both phosphorus compounds are oxidised as far as phosphate can they be precipitated as such with iron, aluminium or calcium salts by treatment within the plating shop or in the third purification stage of the clarifying plant. It is not known to what extent phosphite can be oxidised sufficiently quickly in a modernised plant.

Wei-chi-Ying and Bonk [138] described the elimination of phosphorus by oxidative treatment. To reduce the high costs of oxidising agents, they first precipitated phosphite with lime slurry so far as possible at pH 10. The filtrate was oxidised with potassium permanganate at room temperature and the phosphate formed finally precipitated again with lime slurry. Because of the large amount of potassium permanganate consumed, the process is expensive. If there are readily oxidisable complexing agents present, then even larger amounts of oxidising agent are used. With the usual formulations they amount to 40-60 gm/litre $KMnO_4$. Furthermore, the quantities of lime slurry are very large (150-200 gm/litre); hence, because of the large quantities of solid matter arising for example from the precipitation of manganese dioxide, substantial amounts of sludge are produced, about which no details were given.

It is generally recommended that nickel should be eliminated from baths that have become unusable, by "controlled reductive deposition" on scrap iron or aluminium [141]. For this it is necessary, however, to use extra reducing agent to establish similar conditions as for nickel plating. The baths are heated and agitated with air. In this way, nickel concentrations below 10 mg/litre can be attained, but all other problems, including the precipitation of the residual nickel, still remain.

There has also been no shortage of processes for regenerating electroless nickel baths by precipitation reactions and with the use of ion exchangers. There always remained however the question of eliminating the phosphite formed. Attempts by Parker [142] to precipitate it with iron(II) salts resulted in the precipitation, along with the phosphite, of all the hypophosphite and half of the nickel. The phosphite was more successfully precipitated if trivalent iron was used, but half of the hypophosphite was co-precipitated, whilst the nickel remained in solution. The most successful treatment was with a suspension of calcium hydroxide and calcium sulphate at pH 5, resulting in the precipitation of a large part of the phosphite, very little of the hypophosphite and none of the nickel. In all cases however foreign salts remained in the solution. Parker next investigated the regeneration of the bath with a weakly basic anion exchanger in the OH-condition. It proved possible to remove 136 g phosphite per litre of the resin.

$$R - OH + H_2PO_3^- \rightleftharpoons R - H_2PO_3 + OH^- \qquad\qquad 8.5$$

Anions of weak acids, such as hypophosphite, are not absorbed by weakly basic resins (see Section 5.2.3.3.3). Displacement of the water held within the exchange resin before the treatment and of the bath solution after the treatment results in an approximately 20% dilution of the bath, so that the nickel and hypophosphite contents must again be replenished and the pH value restored. A sodium phosphite solution is then obtained during regeneration with caustic soda.

Levy and Doss [143] carried out further investigations with a strongly basic anion exchanger in the $H_2PO_2^-$-condition. As anions of weak acids are absorbed by this exchanger, the following reaction occurs, which is very useful in these circumstances:

$$R - H_2PO_2 + H_2PO_3^- \rightleftharpoons R - H_2PO_3 + H_2PO_2^- \qquad\qquad 8.6$$

222 g hypophosphite in 5.5 litre of water was necessary however to convert 1 litre of the exchange resin into the $H_2PO_2^-$-condition. At an average capacity for a strongly basic anion exchanger of 0.6 equivalents/litre resin, this is 750% of the theoretical, with a considerably moderated volume of regenerate. The capacity of the ion exchanger is also reduced by any anions of strong acids present, such as chloride or sulphate. The process is therefore not technically practicable.

It appears that the best means at present of reducing waste is to look after the nickel plating baths in such a way as to lengthen their service life. This can be achieved by maintaining the working conditions at their optimum, which requires continuous measurement during operation of the ever-changing parameters, pH value and nickel concentration. The other chemicals required are added as required by their stoichiometric reactions in accordance with Equations 8.3 and 8.4 [144].

Water saving measures, such as continuous circulation ion exchange or cascade rinsing technology, can no doubt be used in electroless nickel plating.

8.8.2.3   Bronzing and colouring of metal surfaces

The most important metal colouring processes are:

- Bronzing (or blackening) of iron and steel parts.

- Colouring of oxide coatings produced by anodising. Reference should be made to Section 8.7.1 concerning this.

- Colouring by chromating, especially of electrodeposited coatings.

- Colouring of chromated films by organic dyes.

- Colouring by solutions of chemicals.

Alkaline oxidising solutions are used at high temperature for bronzing (or blackening). Depending on the concentrations, compositions, exposure times and temperatures, brown to black iron oxide films are formed with coating weights that in multi-stage bronzing reach up to 6.5 g/m^2. The concentrations of bronzing baths are about 70-80 kg salt/100 litre. Boiling temperatures are consequently 135-145°C [145]. The principal components of the salts present are caustic soda, sodium nitrite and nitrate. Additions of other alkaline salts may also be made, e.g. trisodium phosphate. After bronzing, the main area of application being in arms manufacture, the parts are oiled. The chemical treatment of the rinse water produced is carried out as described in Section 3.4. In principle, it is possible to treat exhausted process baths in the same way also. Special attention must be paid, however, to the generation of nitrogen oxides at reaction pH values in the weakly acidic region. It is better in fact to pass these wastes over to a waste disposal specialist.

Rinse water from small units can be treated without difficulty in batch plants. In large bronzing establishments, in spite of the extremely high concentrations of the process baths, it is still possible, if the rinsing system is suitable, to install continuous process plants with ion exchangers, so that batch treatment becomes possible in this case also. After multiple stage pre-rinsing cascades with low water throughput, rinsing is finally completed in each case in continuously circulated water. At a maximum throughput capacity of up to 2.5 t/h and with 12 m³ of water in continuous circulation, the weekly amount of waste water produced in the plant amounts to only 70 m³ [146]. The high evaporation losses of the bronzing bath are replaced by water from the final static rinse baths. Circumstances permitting, bronzing can also be carried out without generating rinse water, if the bronzing baths are so formulated that they are insensitive to drag-in from the pretreatment and if the discharge from the rinse cascades can be returned to each of the process baths. Disposal of the process baths then takes place externally [147].

Figure 8.23 shows a batch treatment plant for oil-containing bronzing waste water for a plant generating waste water in small quantities.

Figure 8.23 Batch treatment plant for oil-containing bronzing water, capacity 1 m³/day (CARL DITTMANN GmbH & Co, KG, Karlsruhe)

Colour effects, and full colouring also, can be achieved by chromating, especially of electrolytically deposited zinc and cadmium coatings. Depending on the concentration of acid, chromic acid or chromate, and the kind of chemical additions made, it is possible to produce, not just bright surfaces, but also bluish, iridescent, yellow and olive-drab colours, and with silver additions black as well. The rinse water generated can be circulated continuously together with other rinse waters over ion exchange resins or treated in accordance with Section

3.3. Many chromate bath preparations can also be regenerated using ion exchangers (see Section 7.2.2.1). Chromated films, if they are produced in a colourless form, can also be coloured with solutions of organic dyes. Waste water that contains dyes, wherever necessary, is treated as described in Section 8.7.2.

Mention should also be made of the electrodeposition of black coatings, such as for example black nickel plating from nickel electrolytes with zinc additions and black chromium plating [145].

There is also a series of chemical colouring methods, especially for copper and its alloys, in which sulphide-containing and also copper-containing preparations have a particular importance. Other procedures for colouring iron (apart from bronzing), tin, lead and zinc exist for many shades of colour, but can be categorised as requiring exotic process baths (see [148] in this connection). The waste water treatment must then be adapted to the constituents known to be present.

8.8.2.4 Metal stripping

Metal stripping is necessary on:

- Rejected parts, in so far as it is worthwhile.

- Contacts of product racks and electroplating barrels.

The metal growing at these points leads to high drag-out by the work carriers and, in an advanced condition, to damage of the insulation on adjacent surfaces, which further increases drag-out (see Section 6.2).

On automatic plants, metal stripping after product unloading is practically obligatory.

According to the bath formulations, metal stripping takes place in acid, seldom in neutral, and in alkaline solutions.

A distinction is made on the basis of the process technology between chemical (most common) and electrolytic metal stripping.

The process solutions are required to:

- Dissolve the coating as quickly as possible.

- Avoid attack on the substrate material.

- Keep the dissolved metal in solution as long as possible, and what frequently conflicts with this –

- Cause no difficulties in waste water technology.

The main constituents of metal stripping baths are therefore:

- Oxidising agents, also known in this context as depolarisers. They are used principally in baths operating chemically. Suitable for this purpose are nitroaromatics, peroxide, nitric acid, nitrate, chromic acid.

- Substances which make it possible for the metals to dissolve and remain in solution. That is, mineral acids and mixtures of them and, in alkaline baths, complexing agents such as sodium cyanide, ammonium salts, ethylenediamine, triethanolamine, diethylenetriamine, NTA and EDTA (in Germany prohibited for metal stripping baths since January 1990,) and, for amphoteric metals, caustic soda also. Sometimes there will also be buffering agents, activators and inhibitors.

Although it gives rise to considerably fewer difficulties in terms of waste water technology, sodium cyanide should for the most part be replaced by amines as the complexing agent. Table 8.7 contains details of some of the most common metal stripping baths [149,150].

TABLE 8.7:
DETAILS OF THE MOST COMMONLY USED
METAL STRIPPING BATHS

Metal stripping bath		Used for removing the following metal coatings	Remarks
Mode of operation	Important constituents		
Chemical	Amine (cyanide), nitro-aromatics	Cu, Ni, Zn, Cd	
	Ammonium nitrate	Cd, Zn	Watch out for nitrite formation
	Caustic soda	Zn	
	Caustic soda, nitro-aromatics	Zn, Sn, Pb-Sn	
	Dilute hydrochloric acid	Cr	
	Dilute hydrochloric or sulphuric acid	Zn	
	Chromic-sulphuric acid or nitric acid	Cu	
	Sodium chlorite, ammonia	Cu	PCB manu-facture
	Ammonium persulphate	Cu, brass	
Anodic	Amine or esterified amine Phenols, nitric acid (pH 4.5-5.5)	Cu, Ni, Cr, Zn, Cd Sn, Ag	Iron basis metal
	Ammonium nitrate	Cu, Ni, Cr, Zn, Cd Sn, Ag	Non-ferrous basis metal
	Caustic soda	Cr, Zn, Pb, Pb-Sn	Watch out for chromate formation
	Hydrofluoboric acid, peroxide	Pb, Pb-Sn, Sn	PCB manu-facture

Provided that the metal stripping baths are free from complexing agents their treatment poses no difficulties. If they contain complexing agents but no nitroaromatics, they can be treated as described in Section 3.9.5. If nitroaromatics are present, then it is best to pass them over to a waste disposal operator. They can be treated together in a separate plant not suitable for treatment of comprehensive batches of waste water. Where possible, electrochemical depolarisation should be employed. This can be done with all nitroaromatic-free alkaline metal stripping baths and those based on sulphuric acid. The metal can in this way be recovered and the bath regenerated. Fig. 8.22 incorporates a hot electrolytic cadmium strip, through which the rinse water can be circulated as well.

With nitroaromatic-containing rinse water it is necessary to investigate whether all the metals present can be satisfactorily eliminated if it is treated along with other complex-containing rinse waters.

8.8.2.5   Waste water from waste gas washing plants

Waste gases are sometimes produced in the surface finishing industry, which must be extracted at the place where they are generated and freed from their constituents by scrubbing. The wash liquids so produced must be regarded and treated as waste water.

According to the type of waste gas, acidic constituents are given an alkaline, and alkaline an acidic wash. Where it happens that a constituent can be stably absorbed in water in an oxidised state, the wash should be made oxidising. If this is done using hydrogen peroxide, then the washing liquid should be acidified, so as to prevent too rapid decomposition of the hydrogen peroxide.

The equipment used for gas washing is similar to the facilities described in Section 7.8.2.2.2 for water evaporation based on the principle of the open-surface cooling tower and the air conditioning plant. Most devices are constructed as counter-flow scrubbers like the one illustrated in Fig. 8.24. The waste gas is sucked into the tower from below and brought into contact with the washing liquid flowing from the top over the packing material. The chemicals for oxidising or absorbing the materials contained in the waste gas are dosed into a tank included in the circuit. The dosing may also be controlled by analysis. As well as this, jet washers, vortex washers, rotating atomisers and venturi washers have also come into use. Three stages may be necessary for much gas washing: e.g. acid wash oxidative wash alkaline wash [151].

Extracted mist, e.g. from chromium plating electrolytes, may be washed out as well, so that the wash water contains process bath constituents also. Care has to be taken concerning this in treating the wash water.

In Table 8.8 the gaseous substances arising in the surface finishing industry are listed, along with the possible places where they may occur, and the types of washing liquid and their subsequent treatment.

Figure 8.24 Waste gas washing equipment for waste gases of a waste water treatment plant (GOEMA Dr Götzelmann GmbH, Vaihingen/Enz)

Waste air containing nitrogen oxides can be washed with chromium-containing waste water, if there is any available. Thus the nitrogen oxide is oxidised to nitrate and chromate is reduced in accordance with Equations 8.7 and 8.8 [152].

$$2\ NO + Cr_2O_7^{--} + 6\ H^+\ \rightleftharpoons\ 2\ NO_3^- + 2\ Cr^{+++} + 3\ H_2O \qquad 8.7$$

$$6\ NO_2 + Cr_2O_7^{--} + 2\ H^+ \rightleftharpoons\ 6\ NO_3^- + 2\ Cr^{+++} + H_2O \qquad 8.8$$

With hydrogen peroxide as oxidising agent, nitrogen dioxide and nitric oxide, which according to Equations 8.9 and 8.10 are absorbed as nitrous acid, are oxidised to nitric acid or, in alkaline solution, to sodium nitrate [153].

$$2\ NO_2 + H_2O\ \ \ \ \rightleftharpoons\ HNO_2 + HNO_3 \qquad 8.9$$

$$NO + NO_2 + H_2O \rightleftharpoons 2\ HNO_2 \qquad 8.10$$

$$HNO_2 + H_2O_2\ \ \ \ \rightleftharpoons\ HNO_3 + H_2O \qquad 8.11$$

Sulphur dioxide, which is absorbed in water as sulphurous acid or in caustic soda as sodium sulphite, can also be oxidised by additions of hydrogen peroxide to sulphuric acid or sodium sulphate.

$$SO_2 + H_2O + H_2O_2 \rightleftharpoons H_2SO_4 + H_2O \qquad\qquad 8.12$$

In a similar way, formaldehyde in an oxidising medium is oxidised to formic acid or sodium formate.

$$HCHO + H_2O + H_2O_2 \rightleftharpoons HCOOH + 2\,H_2O \qquad\qquad 8.13$$

TABLE 8.8:
DETAILS OF WASTE GAS WASHING IN THE SURFACE FINISHING INDUSTRY

Waste gas	Formula	Possible Sources	Washing	Products	Discharge to:
Hydrogen chloride	HCl	Hydrochloric acid pickles and storage tanks	NaOH	NaCl	NA
Hydrogen fluoride	HF	Hydrofluoric acid and mixed acid pickles. Storage tanks for these acids	NaOH	NaF	NA
Sulphur dioxide	SO_2	Chromate detoxification plants	H_2O_2	H_2SO_4	NA
Nitrogen oxides	NO_x	Pickles for non-ferrous metals. Nitrite detoxification plants	$H_2O_2(HNO_3)$,	HNO_3 N_2	NA
Hydrogen cyanide	HCN	Cyanide solutions	NaOH HCHO	NaCN $CH_2OH \cdot CN$	CyA
Cyanogen chloride	ClCN	Cyanide detoxification plants	NaOH	$Na_2CO_3 + NH_3$	CyA
Chlorine	Cl_2	Storage tanks for sodium hypochlorite. Electrolysis plants (anodic)	NaOH	$NaCl + NaOCl$	NA
Hydrogen sulphide	H_2S	Reactors for sulphide precipitation	H_2O_2	H_2SO_4	NA
Ammonia	NH_3	Production and waste water treatment plants in PCB manufacture	H_2SO_4	$(NH_4)_2SO_4$	NA
Formaldehyde	HCHO	Production and waste water treatment plants in PCB manufacture	H_2O_2	HCOOH	NA
Bath mist	Div.	Various Electroplating baths and electrolysis plants with insoluble anodes; pickles, chromium plating baths, etc	Water	Solns	Corresponding detoxification plant or NA

CyA = cyanide detoxification plant; NA neutralisation plant

All other substance are absorbed by forming salts. Bath mist produces an aqueous solution of the corresponding bath constituents. When they have reached a certain concentration, gas wash waters must be treated as waste water. They are discharged as indicated in Table 8.8.

8.9   PRINTED CIRCUIT BOARD MANUFACTURE

Ref 154 provides an overview, inclusive of waste water and recycling techniques, while the Handbook of Printed Circuit Board Technology [155] gives a detailed description of the manufacturing processes.

8.9.1   Amount of waste water and its constituents [154]

Depending on the manufacturing process selected, the conducting tracks are produced by etching away the residual copper foil (subtractive process) or by building up the tracks on uncoated material (addition process) or chemical copper plating of uncoated boards, building up the tracks by electroplating and etching away the copper undercoat (semi-additive process). The subtractive process is most frequently employed in Western Germany.

The preparation and pretreatment in printed circuit board manufacture, especially that for through-hole-plated and multilayer boards, is very complex and subject to variation.

First the required size of board is cut from the basis material. On boards coated on both sides, the holes needed for through contacts are provided by drilling. The coated boards are deburred by a brushing or blasting process, which simultaneously produces the depth of roughness necessary for satisfactory coating by screen printing or photo-resist methods. Rinsing is carried out after brushing, sometimes at high pressure. In the blasting process, quartz or pumice powder suspensions are sprayed on to the boards at high pressure. These media may also be used when brushing. Waste water is produced that contains metallic copper together with the added abrasives and blasting media as solid matter. This is usually removed in small centrifuges associated with the blasting units or in a gravel filter, and the clear water recirculated.

Deoxidising follows, using solutions containing sulphuric acid and hydrogen peroxide or sodium persulphate (often described as conditioners). The waste water generated contains these substances along with dissolved copper. If there is a post-treatment with an organic inhibitor (e.g. benzotriazole), then this also will get into the waste water, though in very low concentrations.

So that the area subsequently needed for the conducting tracks cannot be etched away it must be protected by so-called etch-resists. These are either put on as silk-screened etch-resistant paints or applied as overall coatings which are exposed to UV radiation through a negative film. The unexposed parts remain soluble in chlorinated hydrocarbons (1,1,1-trichloroethane is most commonly used) or alkaline developing solutions and are thus dissolved (developed) while the exposed areas are left behind. After etching the unprotected areas with acidic copper chloride solutions, the remainder of the etch resist is dissolved off in chlorinated hydrocarbons (usually methylene chloride) or stronger alkaline solutions. In recent years, water-soluble resists have been developed for this application, which have to a large extent superseded those that are soluble in chlorinated hydrocarbons, in anticipation of legal restrictions. Wherever it is necessary for resists soluble in chlorinated hydrocarbons to continue in use, developing and stripping machines must be employed that operate within enclosed containers as described in Section 3.11.4. The waste waters produced when using soluble resists have an alkaline reaction and because of their polymeric content a high COD, which can be reduced by appropriate treatment (see below).

The usual photo-chemicals are employed for making the film patterns required for silk-screen and photo-printing, such as developers (metol, hydroquinone, amidol) and fixing baths (sodium and ammonium thiosulphate, sodium bisulphite, dithiosulphatoargentate, and ammonium-containing solutions if diazo films are used. Only small quantities of waste water are generated.

The silk-screen fabric can be utilised over and over again. Removal of the coating from the used screen is carried out after degreasing in 20% caustic soda by a two-stage treatment with potassium permanganate followed by reduction of the oxidising agent in sodium bisulphite solution (both saturated). Washing is then carried out at high pressure. Residues remaining after drying can be removed with butyl acetate, toluene or xylene. Recoating then takes place with a photosensitive copy-coating on a basis of polyvinyl alcohol or polyvinyl alcohol-polyvinyl acetate, with sensitisers based on chromate or diazo compounds. Final curing is with aldehydes and ammonium dichromate solution.

In the course of producing the film underlayers and preparing and cleaning up the screens used in silk-screen printing, the following can find their way into the waste water: caustic soda, aldehydes (formaldehyde, acetaldehyde), chromate, dispersions of polyvinyl alcohol (polyvinyl acetate), diazo compounds, potassium permanganate, bisulphite, butyl acetate, toluene, xylene and some addition agents in low concentrations.

Low-cost, one-sided printed circuit boards without through-contacts for the consumer electronics industry have the copper layer that is not to form conducting tracks etched away in acidified copper chloride solution before the part of the etch-resist that has been cured by radiation is stripped, as has already been described above. This completes the process with this kind of PCB.

For multilayers and some other special circuits, it is necessary before through-hole plating to remove the smeared film (of decomposed epoxide resin) produced by drilling, in a so-called desmearing stage. For this, concentrated sulphuric acid or sulphuric-phosphoric acid mixtures or concentrated chromic acid or alkaline potassium permanganate solutions are used. Simultaneously, a desirable back-etching occurs of the basis material over the drilled area. If chromic acid and potassium permanganate are used, a chemical rinse in acidified sodium bisulphite or hydrogen peroxide solution is carried out afterwards. The waste water will contain these particular chemicals, depending on which are used, together with organic breakdown products.

Some multilayer circuits have up to 30 layers pressed together with intervening insulating coatings. So that these layers should not for some reason become detached from one another, the copper foils of the multilayer are given a treatment in hot alkaline solutions containing strong oxidising agents, so that a 0.5-5 μm-thick oxide layer is produced (black or brown oxide process). These solutions contain caustic soda and sodium chlorite or sodium chlorate.

After these preliminary steps, the drilled boards requiring through-hole contacts are then prepared in several stages for electroless (chemical) copper plating. There is first a dip in a conditioner, which ensures that, with the most commonly used glass fabric reinforced epoxide resin, palladium nuclei are evenly distributed over the epoxide resin and the glass fabric. Additionally, it has a degreasing effect on the copper surface and and a slight swelling effect on the epoxide resin. These solutions may contain ethanolamine, triethanolamine, glycol ether, etc. Deoxidising follows in sodium persulphate or sulphuric acid-hydrogen peroxide solution and then pickling in a solution which resembles the activating solution without palladium, so as to prevent dilution of the activating solution with water. The next step is activation in a solution containing palladium chloride and tin(II) compounds, plus organic substances that give rise to the formation of protective colloids. The tin is then redissolved by dipping in a solution of hydrochloric or fluoboric acid or ammonium bifluoride, which may also contain reducing agent (borane).

It is now possible to bring about through-hole contact by means of electroless copper plating. This is done in an alkaline bath (pH 11.5-12) at 20-60°C. A typical solution composition is: 7 gm/litre copper sulphate pentahydrate (corresponding almost to 2 gm/litre copper), 25 gm/litre EDTA, 6.5 ml/litre formaldehyde (37%, corresponding to 2.5 gm/litre formaldehyde), 14.5 gm/litre sodium hydroxide and milligrammme quantities of a stabiliser (e.g. sodium cyanide). In place of EDTA, other complexing agents are also used, in Western Germany chiefly quadrol and tartrate. As the reduction by formaldehyde of copper and of hydroxyl ions (the latter being catalysed by copper) proceed alongside each other, the overall equation can be written as follows [156]:

$$Cu[EDTA] + 2\ HCHO + 4\ OH^- \rightarrow$$

8.14

$$\rightarrow Cu^0 + 2\ HCOO^- + H_2 + H_2O + [EDTA]^{4-}$$

In the so-called basket process, in which the circuit boards are held parallel within a basket frame, a coating of 3-6 μm copper is deposited, whereas using the rack technique only 0.5 μm is plated, which is thickened up to 5 μm by electrodeposition in acid sulphate baths.

As well as the bath constituents mentioned, the process also causes neutral salts to be present formate and sodium sulphate (due to copper sulphate replenishment).

In the fully additive process an adhesion promoter for the copper layer that is to be deposited is applied by pouring, dipping or transfer coating. It is formulated on the basis of acrylonitrile-butadiene-phenol. With regard to the application of the resist, the procedure is the same as in subtractive processes. Before activation there is an etch in chromic acid solution and chemical rinsing, and then the conducting tracks are built up to their full thickness (30-50 μm) by electroless copper plating. Finally, the portion of resist remaining is stripped off.

In the semi-additive process the boards are prepared with adhesion promoter, etched, activated, coated with ca 3-5 μm-thick electroless copper, then the photo-resist is applied, exposed and developed. An electrodeposited copper coating is next applied to the conducting tracks, which need to be about 5 μm thicker than the final conductors. The remaining portions of resist are stripped off and the spaces between the conducting tracks are etched away to the basis material in acidified copper chloride solution. The advantage of the process is that a maximum of 5 μm only has to be etched away between the tracks compared with at least 17.5 μm in the subtractive process.

Printed circuit boards produced in accordance with the subtractive process are further treated after electroless copper plating as follows: degrease, deoxidise in a solution of the composition already described, etch in dilute sulphuric acid and thicken up the conducting tracks in an acid copper electrolyte to their full thickness. The thickening of the tracks can also be carried out on an intermediate layer of electrodeposited nickel.

Before etching away the spaces between the conducting tracks the tracks must be protected by an electrodeposited lead-tin or tin coating (etch resist). Then the photo-resist between the conducting tracks is stripped off and the copper thus exposed etched away to the base. An ammoniacal etch solution is used for this, containing ammonium chloride or ammonium sulphate plus ammonia. The oxidising agent is atmospheric oxygen.

The etches based on acidified copper chloride or on ammoniacal solutions are used in special spray etching machines (see below). Sulphuric acid-hydrogen peroxide solutions are used for immersion etching in tanks.

The lead-tin or tin coatings present on circuit boards are either dissolved away in solutions based on fluoboric acid and hydrogen peroxide or fused on.

Waste water is generated as well wherever contact strips (pads or fingers) are gold plated, and in fact during stripping of lead-tin or tin deposits, deoxidising, pickling and nickel plating in well-established solutions. Gold is recovered, almost entirely, from the rinse water.

Racks and baskets are stripped of deposited metal before being re-used. In the case of electroless copper this is done with sodium persulphate solutions, in that of electrodeposits (copper, lead, tin) mainly with nitric acid or by anodic treatment.

All the electrolytes used (lead, lead-zinc, gold, copper, nickel, silver) correspond in composition to those described in Section 8.8.1.

It has been possible to give details only of the most important operations in this industry and the waste water constituents that result from them. In any particular case, these must be established individually for each printed circuit board factory.

In general the waste water produced in the circuit board industry is characterised by the following constituents:

- Copper content.
- Complexing agents, which form very stable compounds with copper, their stability increasing in the following order: tartrate, quadrol, EDTA (see also Table 8.9).
- Higher COD content due to waste water from the electroless copper baths (complexing agents, formaldehyde) as well as the developers and strippers for the water soluble resists.
- High concentration, copper-containing waste water from the etching machines.

The waste water comes from:

- No longer usable, short-lived process baths. All the process baths associated with this technology, up to the electroplating baths, are in this category.
- Regenerates from continuous circulation ion exchange plants.
- Concentrated rinse water (from static rinse baths and cascade rinsing systems).
- Flowing rinse water (metal-free in particular).
- Waste water from regenerating equipment for etching solutions, electroless copper baths, deoxidising baths, etc, filtering units, connecting and filling stations, chemical stores.
- Cleaning and floor waste from the production and waste water plants.

8.9.2   Waste water separation and production-specific rinse processes

Cyanide-containing waste water is produced in the printed circuit board industry only to the extent that precious metal (gold, silver) electrolytes are used.

Chromate-containing and potassium permanganate-containing waste water can be detoxified by means of chemical rinsing, so that they do not have to be kept separate. Wherever floor and cleaning waste waters arise, however, they should be put into a ground-level reservoir, so that the collected waste water of the batch treatment being used can be submitted to the reduction processes described in Section 3.3.

Incidentally, care must always be taken on cost grounds to collect complexing agent-containing and complexing agent-free waste water separately, as only the latter can be given a simple neutralisation/precipitation treatment. Complexing agent-containing waste water should be circulated through an ion exchanger to reduce the amount present to such a level that processing can be undertaken in a batch treatment plant [157]. Depending on whether cationic or anionic metal complexes and complexing agents are present, passing through cationic and anionic exchangers will itself bring about a separation, which in certain cases will enable the treatment of complexing agents to be brought down to that of the regenerate from the anion exchanger. Beyond this, it can be said that metal complexes of low stability, such as ammines and amines, are decomposed in the acidic exchange zone of the cationic exchanger, and the cations and anions thus generated are absorbed by the corresponding exchange resins. The regenerate from the cation exchanger may contain complexing agent, however, if waste water with high concentrations of free EDTA or waste water containing quadrol are passed through it. In the former case, protonated EDTA (at concentrations of ca 100 mg/litre or more) may be precipitated in sparingly soluble form; in the latter, the copper-quadrol complex and also quadrol itself may be absorbed in the exchanger. All anionic complex compounds and complexing agents, such as EDTA, tartrate and their copper complexes, become bonded to the anion exchanger. The treated cation exchanger regenerate may, if it is metal-free after its treatment, be discharged, or, if it contains cationic complexing agents, be passed through a selectively operating cation exchanger and then discharged. If it contains anionic complexes (particularly from protonated EDTA), then it is mixed with the anion exchanger regenerate, which may already have added to it the regenerate from the selectively operating cation exchanger (see Fig. 8.25). In doubtful cases all exchanger regenerates should be considered likely to contain complexing agents.

Figure 8.26 shows the waste water separation in a complete printed circuit board factory, in which a distinction is preserved between metal-containing and metal-free waste waters, which cannot put into circulation. In the first category are the rinse waters in the palladium nucleation (activation) area and possibly rinse waters from peroxide-containing deoxidisers, if there is no guarantee that the peroxides are reduced by reducing agents already present or deliberately added to the treated water. Peroxides cause irreversible damage to anion exchangers (see Section 5.3.2.7).

The geometrical form of the circuit boards also permits methods of support and of rinsing, especially spray rinsing techniques, that are not otherwise customary in the surface finishing industry. The boards that are held vertically for chemical pretreatment, electroless copper plating and electroplating also allow adhering fluids to drain away easily. Drops that collect on the bottom edge can readily be blown off the racked products. A brief spray with water above the process baths is also feasible [158]. Blowing and spraying devices for these purposes are integrated within the transporter. Fig. 8.27 shows such an arrangement.

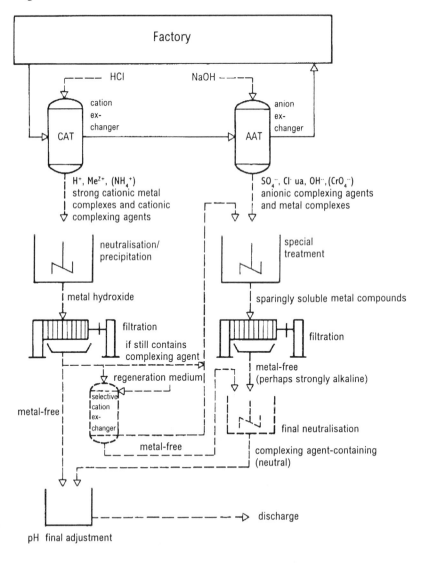

Fig. 8.25 Waste water separation of continuously circulated rinse water in a printedcircuit board factory [157]

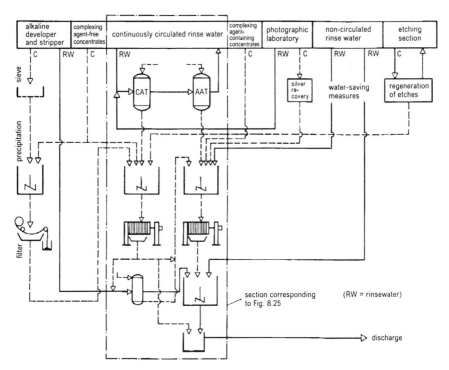

Fig. 8.26 Waste water separation possibilities in a printed circuit board factory [157]

Fig. 8.27 Transporter for chemical and electrolytic treatment of printed circuit boards (racked products) with blowing device and swing pan for reducing bath drag-out (LPW-Galvanotechnik GmbH, Neuss)

Water-saving spray-rinsing systems can also be used when treatment is carried out in baskets, according to a description supplied by Emde [158]. The circuit boards are arranged at an angle in a double basket on both sides of the axis of rotation of a work carrier as shown in Fig. 8.28, so that the adhering liquid can rapidly be drained off diagonally. Cuts of different sizes can be made at 12 mm intervals in the same baskets. In the diagram shown in Fig. 8.28, the product carrier is introduced from above at an angle of 90° in a central rotary rinsing position and set in rotation (50 rpm), while from the same axis it is simultaneously sprayed from the centre to the outside (pre-rinse). It is then centrifuged at 150 rpm. After stopping the rotation, the same operation is performed in the opposite direction of rotation, this constituting post-rinsing. The rinse water that is drained or centrifuged off may be separated into the drainage boxes situated below, in accordance with the nature of the process bath constituents and their concentrations. With this device, of which there will only be one in an automatic machine, the quantities of rinse water can be reduced to 6% compared with cascade immersion rinsing. Fig. 8.29 shows the design of a central rotary rinse in a basket-type automatic machine.

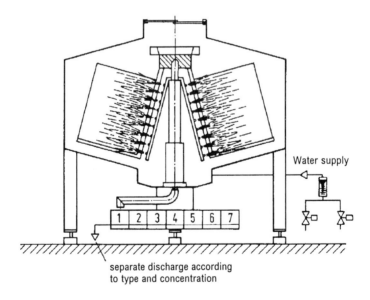

Fig. 8.28   Schematic diagram of a rotary rinsing module
and its mode of operation
(STOHRER-DODUCO GmbH & Co, Rutesheim)

Different rinsing procedures are possible for other circuit board processes, especially for etching. There, the boards are transported horizontally through enclosed machines. Etching is carried out by spraying on the etch solution from above and below. Following the etch machine there are 2-4 rinse modules, connected as a cascade. In each of these, the rinse water from the lower part of the corresponding tank is sprayed on to both sides. As the circuit boards are transferred to the next rinse module or from the etch machine to the first rinse module, they are squeezed through rubber rolls or blasted with air to keep back the solutions. Fig. 8.30 illustrates the design of a plant of this kind for acid etches. The concentrated rinse water produced is either utilised directly or returned again to the etch solution (see Section 8.9.4).

Fig. 8.29 Central rotary rinsing in an automatic machine for the treatment of basket products (STOHRER-DODUCO GmbH, Rutesheim)

Fig. 8.30 Etch machine for acidified copper chloride solutions (right) and rinse modules connected in cascade (left) (HANS HOELLMUELLER Machinenbau GmbH, Herrenberg)

Like the immersion bath cascades now coming into use, all these production-specific cascade-type rinses are prerequisites for the recycling systems they are coupled with, but also for the operation of batch treatment plants.

8.9.3   Waste water treatment

The continuous circulation plants already mentioned, with some effluents treated separately, have long been an essential component of waste water treatment in printed circuit board factories. Fig. 8.31 shows the design of such a plant.

So long as the waste water requiring treatment is metal-containing but free from complexing agents, there are no difficulties in neutralisation/precipitation, provided the amounts of copper and nickel exceed those of lead and tin. The latter precipitate out in a finely dispersed form and need other metals to be present for satisfactory flocculation. Copper serves this purpose as a rule, otherwise iron(III) salts must be added. The treatment takes place in batch plants. All acidic, metal-containing semi-concentrates and usually also regenerates from cation exchangers belong to this category of waste water.

Fig. 8.31 Automatically operating continuous-circulation ion
exchanger plants in printed circuit board manufacture
(GOEMA Dr Goetzelmann GmbH, Vaihingen/Enz)

If there are complexing agents present, as may occur in the circuit board industry, then the metal precipitation is carried out using sodium sulphide or an
organosulphide. Copper can be precipitated out from all the complex compounds
encountered using dimethyldithiocarbamate (DMDTC) (see Section 3.9.5.1.2).
Nickel, lead and tin must not, however, be mixed in with EDTA-containing
waste water. In contrast to copper, these metals cannot be satisfactorily precipitated out. The treatment of complexing agent-containing waste water has to be
performed in batch plants. Waste waters that contain complexing agents, but
also other substances that interfere with the circulation process, are exceptions.
Precipitation with sodium sulphide and DMDTC in batch plants enables the
process to be accomplished satisfactorily and also to be readily controlled by
redox potential measurement. A titration regulator is necessary though, if this is
to be done automatically (see Section 4.3.3.3). Fig. 8.32 shows a batch plant
for complexing agent-containing waste water, which is equipped with such a
regulator.

*Fig. 8.32   Automatically
operating batch treatment
plant for complexing agent-
containing waste water in
printed circuit board
manufacture with titration
regulator (GOEMA Dr
Goetzelmann   GmbH,
Vaihingen/Enz)*

As the complexing agent remains intact when waste water containing it is treated,
care must be taken that, after the metal has been precipitated, the complexing
agent-free and complexant-containing waste waters are still treated separately,
right up to the final stage as shown in Fig. 8.26. This means that each stream
must also be filtered separately. Complexing agents dissolve metal hydroxides
from filter residues, if they come into contact with them, and the conditions
required in the aqueous phase to precipitate the metals then no longer obtain.
Fig. 8.33 shows a double filter station for complexant-containing and complexant-
free waste water.

Fig. 8.33   Filter presses for the separate treatment of complexing agent-containing and complexing agent-free waste waters in printed circuit board manufacture (GOEMA Dr Goetzelmann   GmbH, Vaihingen/Enz)

Where waste water contains high concentrations of complexing agent it is often advisable to precipitate the copper in two stages. In this way copper, can be precipitated from EDTA-containing, but also from tartrate-containing solutions with a greater excess of lime slurry, down to a few mg/litre. While the reason for this in the first case should be sought in the re-complexing of the EDTA by calcium, in the second case it lies in the precipitation of calcium tartrate, which occurs approximately down to 1 g/litre tartrate [139]. If precipitated copper hydroxide is filtered off, then the small amount remaining can be precipitated in a second step with a little DMDTC, and with the help of iron flocculation at low chemicals cost, to some 10 µg/litre. Further filtration is necessary of course, but if the solid matter content remains low this can be carried out jointly with the obligatory final filtration. As in the course of the precipitation procedures metal-free waste waters with a high pH value frequently arise. it may become necessary to carry out a final neutralisation. There is no reason at all why a continuous process treatment should not be used for this. Fig. 8.34 shows a continuous process plant for the final neutralisation of metal-free waste waters.

It is desirable to destroy complexing agents such as EDTA for example by oxidation, and yet, as already explained in Section 3.9.5.3, there is still no process that could be described as state-of-the-art technology. The more advanced studies of the oxidation of EDTA and its complex compounds with hydrogen peroxide under UV radiation have revealed that practically every bond in EDTA can be split. The following were identified as breakdown and oxidation products [159]: ethylenediaminediacetic acid, nitrilotriacetic acid, iminodiacetic acid, ethylenediamine, glycine, glycolic acid, acetic acid, oxalic acid, formic acid, acetaldehyde, formaldehyde, ammonia, nitrite, nitrate, carbon dioxide. Intermediate products therefore are formed as well, which are equally capable of forming complex compounds with metals, sometimes even stable ones. But what is important, except for the end products nitrate, carbon dioxide and water, they can all be further oxidised in this process as far as the final products and they are also biologically decomposable, once the EDTA has lost one of its acetate groups. Such decomposable breakdown products can be obtained when the exposure to radiation is relatively short.

*Fig. 8.34 Automatic continuous process plant for the final neutralisation
of metal-free waste waters in printed circuit board manufacture
(GOEMA Dr Goetzelmann GmbH, Vaihingen/Enz)*

Since, with one exception, all printed circuit board manufacturers are secondary producers, COD and ammonia elimination must be considered from the point of view of coordinating the procedures in the factory (abatement, avoidance, recycling) with the operations of municipal purification plants in accordance with the prevailing regulations. The greatest COD loading derives from from the electroless copper baths (organic complexing agents, formaldehyde) and from the waste waters associated with water-soluble resists. Given an electroless copper bath composition of 25 g/litre EDTA and 2.5 g/litre formaldehyde, 27.5 g/litre COD arises from the EDTA and only 2.68 g/litre from the formaldehyde (compare Table 8.9). The data quoted here are calculated COD values. Measured values are somewhat smaller as a rule.

TABLE 8.9:
ORIGINS OF COD IN ELECTROLESS COPPER BATHS

Complexing agent or compound	Mol wt	g COD*		1 g copper		Mol-ratio
		Mol^{-1}	g^{-1}	g CA	g COD	Cu : CA
EDTA	292	320	1.10	4.60	5.06	1 : 1
Quadrol	292	640	2.19	4.60	10.07	1 : 1
Tartaric acid	150	80	0.53	4.72	2.50	1 : 2
Formaldehyde	30	32	1.07	–	–	–

* Calculated; the measured values are lower. CA = complexing agent

Since, according to latest European statutory requirements, EDTA must be re-covered, when the copper has been plated out from the copper baths no longer usable, the COD should be reduced at least as far as 110-220 mg/litre COD (corresponding to 100-200 mg/litre EDTA) (see below). The COD due to the formaldehyde should be lowered by exactly 50% by oxidation with hydrogen peroxide in an alkaline medium, as the oxidation proceeds only as far as the formate. Thus, the COD due to EDTA and formaldehyde in concentrates can be reduced to ca 5% of the starting concentration. The recovery of EDTA from the rinse water, as current legislation requires, is not possible however under reason-able conditions. As the solubility of protonated EDTA after the removal of all metals must be around 200 mg/litre, this would mean the evaporation of all rinse waters and of anion exchanger regenerates also. It is much more sensible there-fore to considerably reduce the carry-over of EDTA-containing solutions into the rinse water by blowing or squeeze rolling the printed circuit boards.

If the copper baths contain quadrol in concentrations similar to those of EDTA, the COD is increased as Table 8.9 shows quite considerably namely in comparison with EDTA (1.10 g COD/g) by 1.09 g/g (2.19 g COD/g quadrol. Both complexing agents form 1:1 complexes with copper of identical molecular weights. A recovery process for quadrol is as remote a possibility as a biological breakdown process. According to present knowledge, chemical oxidation is best carried out with potassium permanganate in an acid medium (pH<2) [159]. The use of hydrogen peroxide, Caro's acid, chromate and hypochlorite at various pH values is not successful. Using Fenton's reagent in presence of a very large excess of sodium persulphate (14 g/g quadrol and with long reaction times, a satisfactory oxidising action can be achieved.

The outlook seems very favourable where tartrate-containing copper baths are concerned. The COD is considerably lower for equivalent amounts of tar-trate (see Table 8.9). Furthermore, tartrate is readily broken down biologically. As it can also be precipitated to ca 1 g/litre, if necessary, using an excess of lime slurry [139], the treatment of electroless copper baths can reduce the COD due to tartrate to 500-600 mg/litre.

Waste water generated in the photographic laboratory likewise contributes to the COD, but not to any considerable extent. It is treated together with complexing agent-containing waste water after the silver has been recovered from the fixing baths.

The alkaline waste water from resist developing and stripping has a high COD level in concentrate form. The process solutions used are easily broken down. The developers consist of 0.8-1.2% sodium carbonate solution and the strippers of 1-3% potassium hydroxide. Both solutions contain additions of 1-3 ml/litre of anti-foaming agents. Their COD contribution is increased by the dissolved resist components that accumulate in the developers and strippers. These resume their original state again, alongside organic polymers, organic monomers that polymerise on exposure in the presence of photoactive initiators, organic solvents (polyhydric alcohols), dyestuffs, adhesion promoters and sometimes other additions [160]. The waste waters from the developers and strippers usually arise in the relative proportions of 3:1, as the developers are always made up with fresh solution, while the strippers are freed from undissolved resist constituents by filtration and then recirculated [161]. Replenishment takes account also of the evaporation losses from the warm developer at 40°C and the hot stripper at 55-80°C. The concentrations of dissolved organics in the spent solutions are ca 5-10 g/litre, corresponding to a COD of ca 10-20 g/litre. Then, in the developer there is still 1-2 mg/litre copper, and in the stripper still ca 5 mg/litre copper (in some cases up to 200 mg/litre), ca 20 mg/litre lead and ca 60 mg/litre tin. In volume terms, the amount of waste water produced is not large. 1 litre developer is sufficient for up to 0.15 m2 of photoresist and 1 litre stripper for up to 0.5 m^2 [162].

The treatment of the resist waste waters produced is carried out by:

- acid treatment,

  or

- ultrafiltration, so that adjustment of concentration is all that is needed and the residue is further processed by acid treatment.

The acid treatment is most often performed by adding hydrochloric acid, the technical literature quoting pH values from below 1 up to 5. Thus, organic constituents polymerised by acid or only sparing soluble in acids are precipitated out, and can then be separated either from the acid solutions or after neutralising them. The COD reductions achieved by these precipitations can vary a good deal. The author has found that [159] in studies of 3:1 mixtures (developer: stripper), most usual in Germany, acidification to a pH below 1 produced a COD corresponding to 25-30% of the initial value. The rest of the COD was obviously due to low molecular weight organic products, which have a large influence on the properties of the resist, especially its water solubility. Various publications indicate that, even with additional active carbon treatment, a level corresponding to 400 mg/litre COD is not reached. The reason is that the resists do not by any means contain a large proportion of adsorbable higher molecular weight substances. That does not mean however that the result with active carbon cannot be improved upon.

It should be realised that the adsorbing properties of acid precipitates improve with the amounts of acid added, as Suessmuth [163] demonstrated, and they are further enhanced if compounds of trivalent iron are added before acidification, so that their hydrolysis products are incorporated in the precipitate. An addition of 1-3 g/litre iron in the trivalent form is sufficient for this as a rule, depending on the initial COD concentration. The acidification must be carried out slowly, as carbon dioxide is generated due to the soda content of the developer and causes foaming. It is preferable also to neutralise before separating the solid matter. The COD actually rises somewhat if there is slight redissolution, but there is no need to use acid-resistant filtration equipment. It has also been shown that there are preparations that ensure dyestuff is fully incorporated in the polymeric sediment, so that it is unnecessary to include a post-treatment with active carbon after it has been separated.

A few observations and remarks concerning acid treatment, may be helpful: Precipitates from developers have a tendency to settle, and those from strippers to float [161]. When treating with acid in the reactor, the resist solution goes straight into the strongly acidic medium, this having a beneficial effect on the stickiness, while raising the temperature accelerates the formation of a floating cake and final neutralisation with calcium carbonate prevents its redissolution.

Fig. 8.35 Batch plant for the treatment of resist-containing waste water;
batch size: 2 × 5 m³
(HAGER + ELSAESSER GmbH, Stuttgart)

Figure 8.35 shows the construction of a plant for treating resist waste waters. The process provides for dosing with calcium chloride solution in a tube reactor prior to acidification. Separation of solid matter by pressure-release-flotation at a surface loading of 3 mh^{-1} is carried out finally [164]. Precipitation with calcium chloride in the neutral range is possible for the developer also, but not the stripper [163]. The higher molecular weight components of the resist waste water can also be separated by ultrafiltration. Increasing the concentration of residue by a ratio of 1:25 enables the COD of the filtrate to be reduced by 80%. The residue can then undergo acid treatment or must be disposed of [165,166].

8.9.4   RECYCLING PROCESSES

A distinction must be made here between the electroless copper and electroplating sector on the one hand and that of etches on the other.

8.9.4.1   Pretreatment, electroless copper, electroplating

Possible processes in the pretreatment area are the removal of solid matter by brushing and pumicing with continuous circulation of the water (see Section 8.9.1), as well as the reoxidation of trivalent chromium to chromic acid or of manganate to permanganate in the etches and of adhesion promoters used before activation in additive and semi-additive processes. Regeneration of the last named solutions is described in Section 7.6.2.3 (Anodic oxidation). See Fig. 7.75 also in this connection.

Sodium persulphate-containing copper-etching solutions, known as conditioners in a variety of intermediate treatments before activation of the copper surface, can be partly regenerated electrolytically. At current densities of 1.5-2 A/dm^2 and concentrations of sodium persulphate below 50 g/litre, the copper can be deposited at current efficiencies over 60% down to 2 g/litre, whilst almost 30% of the persulphate ion is reduced cathodically to sulphate. The copper deposit is ductile and bright [167]. The persulphate that is used up or cathodically reduced must be replaced. Fig. 8.36 shows the deposition curves for copper from such a solution.

Above 90 g/litre sodium persulphate, no copper deposition is possible. By carrying out diaphragm electrolysis using a cation exchange membrane and a sulphuric acid-containing catholyte, the reduction of persulphate can be prevented if a degree of sulphuric acid enrichment is accepted.

If ammonium persulphate solutions are used, which are more difficult to treat in terms of waste water technology, then it is possible to regenerate them by crystallising out $CuSO_4 \cdot (NH_4)_2SO_4 \cdot 6H_2O$ using cold crystallisation, as described in Section 7.4.2 for acidified ammonium sulphate- and hydrogen peroxide-containing solutions.

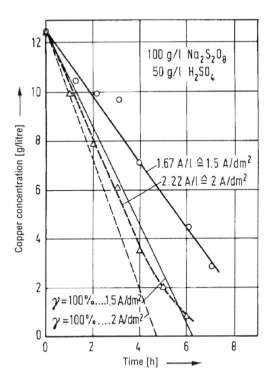

Fig. 8.36   *Deposition curves for copper in the electrolysis of sodium persulphate-containing etch solutions for various current densities [167]*

The electrodialysis process proposed by Horn [168] and Krulik [169], among others, for the regeneration of electroless copper baths is described in Section 7.6.3.2.

Hummel [170] reports on the recovery of material from electroless copper baths by evaporation at reduced pressure (0.1-0.2 bar). As soon as a specified bath density is reached, the concentrate is cooled down, with the result that sodium sulphate and sodium formate crystallise out in accordance with their solubilities. After the crystals have been removed the solution contains ca 70% EDTA (in concentrations of 70-100 g/litre), ca 60% of the copper and ca 7% of the formaldehyde, as well as all of the caustic soda still present. This concentrate is used again as the starting solution for new electroless copper baths, which are adjusted to the desired values by appropriate dilution.

Analytical monitoring and control make an important contribution towards increasing the service life of electroless copper baths. In this, alkalinity, temperature, bath density, the concentrations of copper, EDTA, formaldehyde and stabilisers are measured and adjusted accordingly. For this purpose, temperature-, pH value-, redox potential-measuring devices, photometric measurements, automatic titrations, density measurements, as well as empirical estimates based on the measurements, are employed [171-173].

A recycling procedure, namely the recovery of EDTA from the waste water from printed circuit board manufacture, is specified even under the terms of recent legislation (c.f. comments in Section 8.9.3 on COD). Investigations by the author [174] revealed that it is best to crystallise out EDTA as the sparingly soluble protonated acid in the pH range 1.6-1.8. The process takes ca 48 hours. The residual solubility of the acid is about 100-200 mg/litre (see Fig. 8.37). It is of course only the free EDTA, not that forming part of a complex compound, that is extracted in this way. Fig. 8.38 shows that the whole of the EDTA equivalent to the metals present remains in solution. If it is wished to recover the EDTA, all the complex-bonded metals must be removed. This can be done in two possible ways:

- removal of the copper by catalytic reduction.

- electrolytic deposition of the copper.

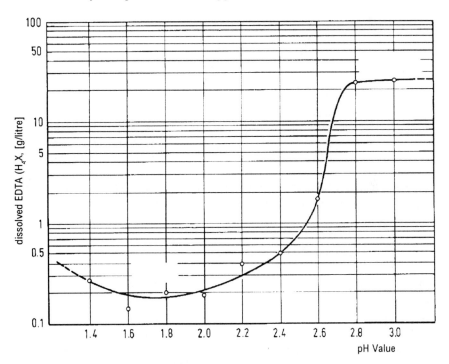

Fig. 8.37   Dependence of the solubility of EDTA
on the pH value [174]

In the first case the same process takes place as in electroless copper deposition. The bath must first, if necessary, be adjusted to the proportions required for electroless copper plating, by caustic soda and formaldehyde additions. Then the solution is brought into contact with nucleated granules of active carbon or another suitable substrate. The nucleation (activation) is carried out in a precious metal solution (palladium, silver). With the help of an appropriate looped reactor the copper can fairly quickly be brought down to less than 0.5 mg/litre [175] (see also Section 3.9.5.2 and Fig. 8.38a). Stirred reactors can also be used for this process [176].

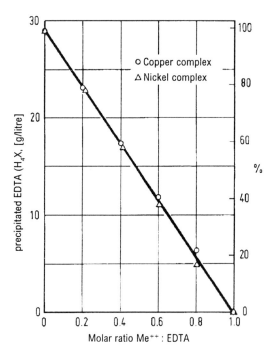

Fig. 8.38   Precipitability of EDTA as protonated acid (pH 1.6) from its complex compounds with copper and nickel at various metal-EDTA ratios [140]

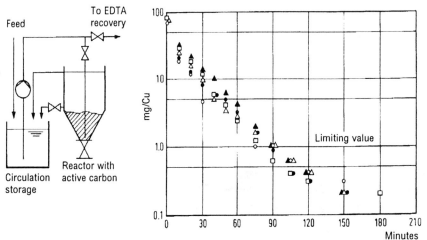

Fig. 8.38a   Schematic diagram of a looped reactor for catalytic reduction of copper (left) with which the deposition rates illustrated on the right can be achieved in multiple reactions [175]

The second possibility for removing copper is electrolysis in baths acidified with sulphuric acid. Copper deposition proceeds as from an acid copper electrolyte. Very low concentrations can also be achieved, though over a long period of time, which allow the EDTA to be precipitated out (see Fig. 8.39). The EDTA can be removed after about 48 hours, preferably using a suction filter (see Fig. 8.40). After washing with a small amount of water it can be used again for bath additions.

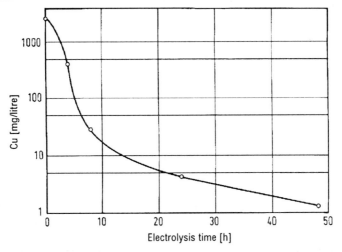

Fig. 8.39   Electrolytic deposition of copper from an EDTA-containing waste water acidified with sulphuric acid [174]

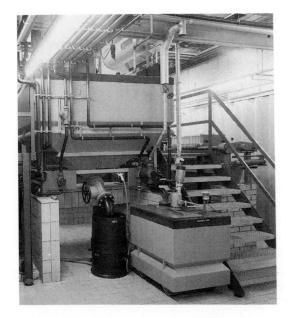

Fig. 8.40 Equipment for the recovery of EDTA from spent electroless copper baths. Above: precipitation reactor; below: suction   filter (GOEMA Dr Goetzelmann GmbH, Vaihingen/Enz)

If in addition to the acidified copper electrolyte there are other electrolytes in use, such as nickel, tin, lead-tin, gold and silver solutions, then maintenance is the same as is usual in electroplating. Recovery measures for metal compounds, so far as nickel, gold and silver electrolytes are concerned, can be found in many places in this volume (precipitation Section 7.4.1, electrolysis Section 7.6.2). Tin baths can only be partly recovered by electrolysis. What remains can be precipitated with caustic soda. Lime slurry has no advantage here, especially if it is planned to process the precipitate further. Lead can readily be recovered electrolytically from fluoboric acid electrolytes down to a few mg/litre.

8.9.4.2   Etching

Three preparations are used for etching in the printed circuit board industry:

- Copper(II) chloride solutions acidified with hydrochloric acid, principally used for subtractive processes without through-hole plating. They contain 90-120 g/litre copper (maximum concentration 160 g/litre) and 60-100 g/litre hydrochloric acid.

- Ammoniacal solutions of ammonium salts. These find use in the production of through-hole-plated printed circuit boards and do not attack etch resists such as tin or lead-tin. They are ammoniacal solutions of ammonium chloride or ammonium sulphate and contain ca 150 g/litre copper as the tetrammine complex (maximum concentration 200 g/litre).

- Sulphuric acid- and hydrogen peroxide-containing solutions. They are used in subtractive processes, mainly for immersion etching, and contain ca 160 g/litre sulphuric acid, 35-40 g/litre hydrogen peroxide and ca 40-80 g/litre copper (maximum concentration 80 g/litre).

Whatever the process used to recycle the etch solution, it can be classified as belonging to one of the three basic schemes determining the way in which the excess volume that is generated in the systems is disposed of, with the copper concentration falling in sequence (down to 0) from the etching machine, from the rinsing equipment and from the regenerating equipment [177] (see Fig. 8.41).

It is sometimes the case that etching in printed circuit board manufacture is carried out exclusively in purpose-built machines with rinsing modules connected to them, like those shown in Figs. 8.30 and 8.44. Rinse water is generated in small amounts in concentrated form, and thus satisfies the conditions for its further exploitation. At the same time, the drag-out losses are reduced by in-built squeeze rollers to 40-70 ml/m^2, depending on the number of drill-holes [178].

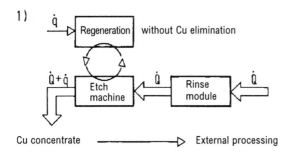

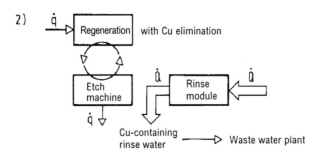

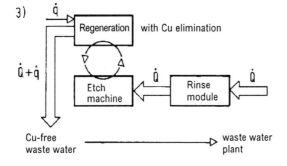

Figure 8.41 Basic possibilities for regenerating etches in the printed circuit industry,

1) with generation of copper concentrates,

2) with generation of copper-containing rinse water.

3) with generation of copper-free waste water [177]

Etch solutions of cupric chloride acidified with hydrochloric acid

In etches comprising cupric chloride solutions acidified with hydrochloric acid, cuprous chloride is generated, which then forms a chloro-complex in the acidified high chloride content solution:

$$Cu^0 + CuCl_2 \rightarrow 2\ CuCl \qquad\qquad 8.15$$

$$2\ CuCl + 2\ Cl^- \rightarrow 2\ [CuCl_2]^- \qquad\qquad 8.16$$

While cuprous chloride has only a low solubility of 0.12 g/litre (at 20°C), corresponding to 77 mg/litre copper, the chloro-complex is readily soluble and has a stability constant of at least $5 \cdot 10^4$. With even low quantities of monovalent copper present (ca 2 g/litre), the etching rates fall appreciably. The etches must therefore be regenerated. This is done by oxidation of the monovalent copper, replacement of the hydrochloric acid consumed and dilution with water to keep the concentration of cupric chloride constant. This concentration is controlled by measuring the density. The regeneration can be represented by the following equation:

$$2 \; CuCl + 2 \; HCl + H_2O_2 \rightarrow 2 \; CuCl_2 + H_2O \qquad\qquad 8.17$$

At the same time a volume corresponding to the additions made, of hydrogen peroxide, hydrochloric acid and water, must be removed from the etch solution.

Monovalent copper in cupric chloride solutions acidified with hydrochloric acid can also be satisfactorily determined by redox potential measurement. The variation in the potential with cuprous ion concentration is particularly sensitive just in the relevant range of concentration up to 2 g/litre. A potential of −500 mV at 0.2 g/litre Cu+ falls to −410 mV at 2 g/litre, while the influence of the hydrochloric acid concentration is only very slight [140].

If water obtained from the cascade rinse is used in place of fresh dilution water, then the excess volume due to the regeneration process is reduced. Using three rinse modules with squeeze rollers, only 10 litre/hr of rinse water, for example, is produced in etching ca 20 m² of printed circuit boards [178]. Fig. 8.42 shows the mode of operation and the circuit for regenerating hydrochloric acid-acidified cupric chloride etches if the rinse water is re-used. Monitoring when the rinse water is used in this way is carried out just the same as in normal regeneration. The concentrated copper solutions arising from the excess volume generated would in the past have been turned over to disposal companies.

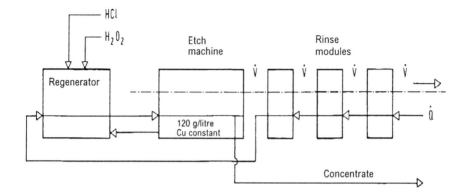

Fig. 8.42 Schematic diagram of cascade rinse and etch regeneration using acidic (HCl) copper chloride

Other oxidising agents have also been used to oxidise monovalent copper, such as oxygen, chlorine, sodium hypochlorite and sodium chlorate, but hydrogen peroxide, owing to a number of advantages, has gained general acceptance for chemical regeneration. Certainly, because its decomposition is catalysed by copper ions, it does not tend to accumulate, as some of the other oxidising agents do [179].

The important reactions in the regeneration of acidified cupric chloride etches are the elimination of dissolved copper produced by the etching reaction and the oxidation of the monovalent copper generated in the solution. Both reactions can be effected together, mainly by electrolysis: metallic deposition of copper by cathodic reduction and oxidation of monovalent copper by anodically generated chlorine. In the past two decades therefore many attempts have been made to establish electrolysis as a means of regenerating etches, especially as it offers the possibility of achieving this without the addition of chemicals. There are however a few difficulties, which are characteristic of electrolysis, that must be overcome as well:

- Cathodic copper deposition tends to be incompatible with the high hydrogen ion concentration and the chemical etching capability, and so the current efficiency is poor.

- Large amounts of monovalent copper are formed at the cathode, and this must be directly re-oxidised in the regeneration process as well as that resulting from the etching process itself.

- The dissolution of anodically generated chlorine gas in the etching solution, bearing in mind the relatively small amounts there are in this industry, can cause problems.

Because of the make-up of the electrolyte, it contains several redox systems:

$$Cu^0 \rightleftharpoons Cu^+ + e \qquad\qquad +0.52\ V \qquad\qquad 8.18$$

$$Cu^0 \rightleftharpoons Cu^{++} + 2e \qquad\qquad +0.34\ V \qquad\qquad 8.19$$

$$Cu^+ \rightleftharpoons Cu^{++} + e \qquad\qquad +0.15\ V \qquad\qquad 8.20$$

$$2\ Cl^- \rightleftharpoons Cl_2 + 2e \qquad\qquad\qquad\qquad\qquad 8.21$$

$$H_2O \rightleftharpoons \frac{1}{2}O_2 + 2\ H^+ + 2e \qquad +2.42\ V \qquad\qquad 8.22$$

In addition to these, there are the redox systems of the chloro-complexes of monovalent and divalent copper. The interactions possible at the electrodes as well as the chemical reactions depend on cathodic reduction and anodic oxidation taking place one after the other in a particular sequence or in separate volumes of electrolyte. The consequences that result from this may be indicated by a few examples.

A patent specification [180] describes a process that prevents the anodic generation of chlorine, the anodes being placed in diaphragm compartments that contain dilute sulphuric acid as anolyte. The copper is deposited on the unenclosed cathodes of this cell. The anodically produced oxygen is brought into contact with the electrolyte in a bubble column afterwrds, to oxidise the monovalent copper. The current passed during electrolysis is controlled on the basis of density measurement. After the hydrochloric acid has been recovered by depositing copper, only that portion of it that has been lost by drag-out needs to be replenished.

Bringmann [118] proceeds in the same way, that is, with the help of anode boxes containing sulphuric acid anolytes, but he oxidises afterwards with hydrogen peroxide in accordance with Equation 8.17.

Another patent application [182] outlines a process in which an etching solution based on potassium chloride (240 g/litre KCl, 12 g/litre HCl) is passed first to the cathode chamber of a diaphragm cell to deposit the copper and then into the anode chamber from which the rising chlorine gas stream flows out in a counterflow direction. The current densities preferred are given as 25-40 A/dm^2, at which a current efficiency for copper deposition of 72% was recorded.

A patent application [183] has been made, for circumstances however in which the average copper concentration of the etch solution amounts to no more than 50 g/litre. The solution is passed between the anode and a porous diaphragm (porosity 20%, pore size 60-100 μm). Anode, diaphragm and cathode are arranged parallel, at an angle. The cathode is longer and projects over the brim of the cell, so that scrapers attached to a chain, which are passed around the cathode, carry the loosely deposited copper from the top side of the cathode into a receptacle placed outside. The anodically generated chlorine rises up to the top over and around horizontal baffle plates, so being brought in contact with the etch solution before it passes out. The electrolysing current is controlled by redox potential measurement at the outlet from the plant. The current density is 60 A/dm^2 on average.

A variant on this is described by Knipps [184]. The exhausted etch solution is transferred from a storage vessel to the electrolysis plant, which has anodes supported in diaphragm compartments. The chlorine generated there is passed to a jet-washer, through which the etching solution from the etch machine is circulated by pumping. Production of the amount of chlorine necessary to oxidise the monovalent copper is regulated by measurement of the redox potential and control of the DC source.

A development, which further reduces the difficulties that still appertain to the systems described and which largely solves the problem of a higher rate of copper elimination, can be gathered from a patent specification [185], as well as from reports by Hillis [186] and Filor [187]. Cathode and anode regions are separated by a cation exchange resin membrane (based on perfluorosulphonic acid). Graphite is used as the anode material and titanium as the cathode. The etch solution is pumped at very high speed through the anode compartment, in which monovalent copper is oxidised to divalent. A relatively high hydrochloric acid concentration of 100-160 g/litre [185] or 125-175 g/litre [187] is required for the etch solutions used here if they have a copper content at the normal level of 120 g/litre. The acid is present in the catholyte as well, even if the copper content there is only 10-20 g/litre. This concentration is kept constant by the migration of copper ions from the anolyte into the catholyte, and as this alone is not sufficient by using the anolyte to replenish the etch solution. At an average current density of 35 A/dm² [185], and temperatures around 50°C, ca 1 kg of copper can be deposited per kAh. This corresponds to a current efficiency of ca 85%.

Under these conditions of concentration and current density, the copper deposits on the cathode in a powdery to dendritic form, and falls to the bottom of the cell. Given the very low copper(II) chloride concentration in this cell compartment, chemical re-dissolution is negligible. The operation of the electrolysis is controlled by measuring the redox potential in the anolyte so that the concentration of monovalent copper stays at less than 1g/litre. In order to avoid chlorine generation at the anode, the electrolysing current is switched off if the potential falls below a preset value. With equipment of this kind, relatively high rates of copper deposition can be attained. Individual units have been made with capacities up to 6 kg copper/h. Another process, which really has nothing to do with the regeneration of etches, but is a means of utilising surplus hydrochloric acid etch solutions, is the recovery of copper sulphate from copper(II) chloride. For this, etch solutions are diluted to ca 10 g/litre copper and the pH value set at just below the hydrolysis point, then this solution is passed through a weakly acidic cation exchanger in the semi-sodium conditions (see Section 7.2.1.2). The exchanger absorbs the copper, while a weakly acidic sodium chloride solution flows out of it. If the regeneration of the exchanger is carried out with sulphuric acid (15%) and the regenerate obtained is employed in the next regeneration to make use of the sulphuric acid still contained in it, and then again for a third time in the following one, copper sulphate solutions are finally obtained with 80-100 g/litre copper and pH values $\geq$ 2.5. Because of the high purity of the copper sulphate, these are suitable for use in making up electroless copper baths. A particularly large demand exists wherever the additive process is in use [174]. The process is illustrated schematically in Fig. 8.43, along with the recovery of EDTA.

Ammoniacal etch solutions

As in etches based on copper chloride solutions, monovalent copper is also produced here and, just like the divalent copper in the initial solution, is present as an ammine complex:

$$Cu^0 + [Cu(NH_3)_4]^{++} \rightarrow 2\ [Cu(NH_3)_2]^+ \qquad\qquad 8.23$$

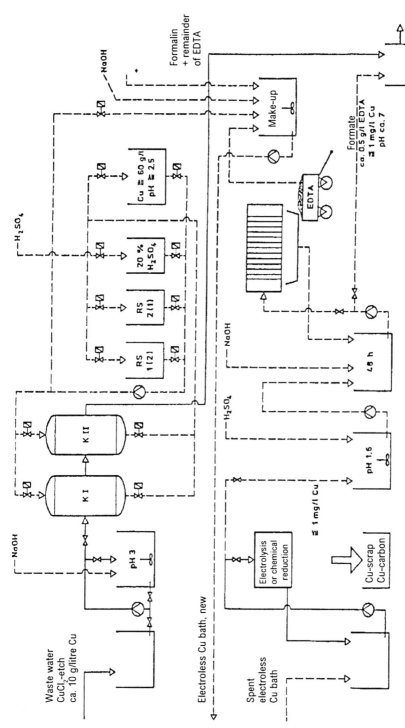

Fig. 8.43 Schematic diagram of the production of copper sulphate concentrates from surplus acid copper chloride solutions and the recovery of EDTA from spent electroless copper baths for use in making up new electroless copper baths [174]

The diammine-copper(I) formed is easily oxidised with atmospheric oxygen, which it combines with in spray etches:

$$2\,[Cu(NH_3)_2]^+ + \frac{1}{2}\,O_2 + 2\,NH_3 \rightarrow 2\,[Cu(NH_3)_4]^{++} + H_2O \qquad 8.24$$

It is necessary therefore only to replenish the ammonium or ammonia.

The drag-out from these alkaline solutions is greater than from copper chloride etches because of their higher concentrations. For squeeze-rolled boards, 70 ml/m² is a reasonable estimate [188]. The higher etching rate also allows a greater hourly throughput of boards, so that as a result more waste water is produced than in acid etching.

With ammoniacal etch solutions, pre-rinse cascades, which contain fresh copper-free etch medium, can be connected up before the rinse cascades proper. These solutions are called replenishers. They are topped up en masse with fresh solution, as they are withdrawn from the pre-rinse cascades to replenish the etch machine. The pH value is corrected using ammonia gas, so as to increase the volume in the system as little as possible. The mode of operation and the circuit for regenerating ammoniacal etch solutions, incorporating replenisher pre-rinsing, are illustrated in Fig. 8.44. Fig. 8.45 shows the construction of an etch machine using this rinsing technique.

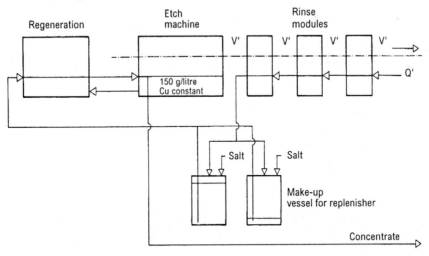

Fig. 8.44 Schematic diagram of cascade rinsing and etch regeneration using the replenisher process for etching with ammoniacal solutions

A further reduction in the amount of rinse water can be achieved, if rinse water is used, with the appropriate ammonium salts, to make up the replenisher. Two make-up vessels, which are filled alternately, are needed for this. Fig. 8.44 shows the way in which it is done. The solution is transferred as required to the regenerator, where the pH is corrected with ammonia gas. The contents are circulated through the etch machine. The surplus volume generated is drawn off from the etch machine and sold to firms equipped to utilise or recover copper solutions [188-190].

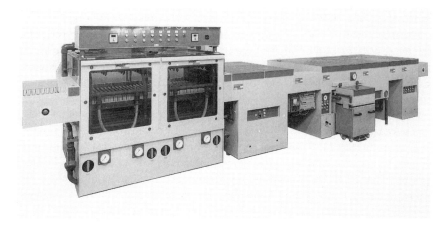

*Fig. 8.45 Etch machine for ammoniacal etching solutions
(left), replenisher/pre-rinsing (centre) and rinse modules connected in cascade
(right) (HANS MOELLMUELLER, Maschinenbau GmbH, Herrenberg)*

Ammoniacal etching solutions can also be regenerated electrolytically, the copper therefore being removed from them by cathodic deposition. Regeneration by this means is considerably simpler than for copper chloride etches, as no precautions have to be taken with regard to oxidation. As Hillis [186] reported, the etch solution is transferred to the cathode compartment of a cell, which is separated from the anode compartment by a cation exchange membrane. At a high current density (almost 40 A/dm^2) ca 1 kg of copper per kAh is deposited in dendritic form. Dilute sulphuric acid is used as anolyte. If chloride etches are involved, platinised titanium anodes have to be used on corrosion grounds, as some chloride is bound to diffuse into the anolyte. Graphite has been shown to be suitable as the cathode material.

According to a patent application [191], the pH of the etch solution in the cathode compartment is adjusted to a value below 8, while as anolyte, separated from it by a cation exchange membrane, an ammonium salt solution is used with 20-80 g/litre salt (preferably ammonium sulphate) at pH 4-10, in order to avoid loss of ammonium ion from the etch. The copper concentration is reduced to 5-80 g/litre and the pH adjusted again to the required value above pH 8 as it is returned to the machine. Lead is used as the anode material, and alloy steel, steel or copper for the cathodes. If the etches are based on ammonium sulphate, they can be regenerated by circulating them through normal electrolysis cells that are not subdivided with diaphragms [192]. Bipolar electrodes can be employed, on the cathodic side of which copper is removed by electrodeposition. The electrolysis is controlled on the basis of density measurement. Ammonia content is adjusted by introducing ammonia gas. Other constituents are replenished only to compensate for drag-out. The rinse water from the cascade rinses is circulated continuously through ion exchangers. The deposition curves in Fig. 8.46 show that the copper is deposited at reasonable current efficiencies provided the temperatures do not exceed 40°C. Certainly, current densities over 7 A/dm² can be applied. Cooling is generally necessary because of the heat generated by the flow of current. Fig. 8.47 shows the construction of an electrolysis plant for regenerating ammonium sulphate-containing etch solutions.

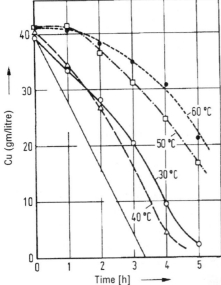

Fig. 8.46 Electrodeposition of Cu from ammoniacal etch bath (ammonium sulphate) [167]

Fig. 8.47 Cell for Cu recovery from solution in 8.46 (HANS HÖLLMÜELLER Maschinenbau GmbH Herrenberg)

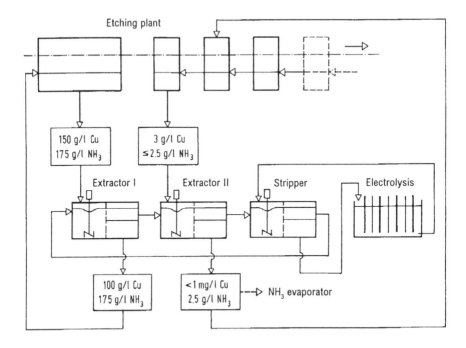

*Fig. 8.48   Schematic diagram of the recovery of ammoniacal etch solutions
by liquid-liquid extraction after production processes [Mercer process]*

At correspondingly high cost, a process that is otherwise little used in the metal
working industries is also suitable for regenerating ammoniacal etch solutions
liquid-liquid extraction. Using copper-specific reagents, which when dissolved
in water-immiscible organic solvents act as liquid ion exchangers, copper can
easily be removed from alkaline and neutral aqueous solutions (see Section 7.3).
Hoffmann [193] describes the operation of a liquid-liquid extraction plant for
reclamation from ammoniacal etch solutions. As illustrated in Fig. 8.48, etch
solution is transferred from the machine to a mixer-settler (Extractor I) and is
thoroughly intermixed with the organic phase there. The two phases then separate out. The aqueous one with a very low copper content is returned to the etch
machine and the organic one is passed to a second mixer-settler (Extractor II),
where it comes into contact once more with copper-containing rinse water from
the first cascade rinse. After separating out again, the organic phase then passes
into the so-called Stripper, which is constructed in a similar way to a mixer-
settler plant. By mixing with dilute sulphuric acid, the organic phase gives up its
copper sulphate again to the aqueous phase. The stripped organic phase is passed
back to the extraction mixer-settlers and the recovered acid copper sulphate
solution, containing about 60 g/litre copper, passes on, for the copper to be
deposited in an electrolysis plant. The sulphuric acid regenerated there is used
again for stripping. The copper is deposited on titanium cathodes, from which it

can easily be removed, at a current density of 1.5 A/dm^2 in layers ca 2 mm thick. A lead-antimony alloy is used as anode material. Part of the aqueous phase separated in the mixer-settler used for rinse water treatment can be transferred to the second cascade rinse, and indeed in such a quantity that the ammonia concentration entering Extractor II does not exceed 2.5 g/litre. The remainder can be used as make-up for the replenisher solution. If such a replenisher is not employed, the ammonia should be removed using an evaporator. Once this problem is resolved, the etching can be performed without generating waste water with the help of this process.

Sulphuric acid-hydrogen peroxide etch solutions

The etching reaction is very simple, in that the sulphuric acid dissolves the copper, while hydrogen peroxide is the oxidising agent necessary for this:

$$Cu^0 + H_2SO_4 + H_2O_2 \rightarrow CuSO_4 + 2\,H_2O \qquad\qquad 8.25$$

The solution works at ca 50°C and, in contrast to the other etchants, contains no copper when newly made up. Its regeneration comprises removal of the etched copper and replenishment of the chemicals consumed, especially hydrogen peroxide. Electrolysis and cool crystallisation are suitable in this case.

At certain concentrations, hydrogen peroxide creates difficulties in the electrolysis. Fischer [194] reports that at concentrations below 5 g/litre hydrogen peroxide the current efficiencies are around 40-50%. As soon as the oxidising agent is reduced, they increase to 90-95%. Gülbas [167] determined efficiencies that were still ca 60% at a current density of 2 A/dm^2 and up to 37 g/litre hydrogen peroxide, but these fell if the current density was reduced. Copper deposition is not possible at all above 50 g/litre hydrogen peroxide. It can only take place when the oxidising agent is decomposed again. Stupar [195] suspects that 18 g/litre hydrogen peroxide might prevent electrodeposition, and so destroys it by heating before electrolysis. At 60-65°C this takes 4-6 hours, while at 88°C the concentration falls as low as 4 g/litre after 1 hour. The copper can then be deposited out on to alloy-steel cylindrical cathodes, which are submerged horizontally in the solution and which are rotated slowly. At 2.5-3.5 A/dm^2, deposited copper layers can be detached as foils above the surface of the liquid.

The removal of copper by crystallisation of copper sulphate pentahydrate is carried out by cooling down the etch solution, which is operated hot. At 20°C the concentration falls to ca 40 g/litre [179], at 7.2°C to 26-28 g/litre [195]. Cool crystallisers, like those described in Section 7.4.2, are suitable for this purpose; they are integrated into the process using the technical devices shown in Fig. 8.49. Whereas in electrolysis, however, the sulphuric acid is set free again, in crystallisation it is lost along with the copper sulphate and must be replenished. On the other hand, the recovered copper sulphate can be utilised for making up electroless copper baths. In this regard, Schlitter [196] points out that the copper sulphate crystals must be neither too fine (as when the sulphuric acid concentration is too high) nor too coarse. In the first case, too much liquid is carried out with them, and in the second, difficulties arise owing to the deposition of crystals on the moving parts. Care must be taken therefore to achieve the relationship between nucleus-formation and crystal-growth rates that favours the production of medium-sized crystals (see Section 3.7.4.1).

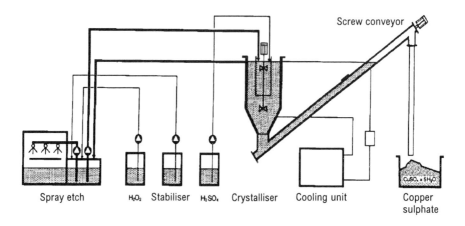

Spray etch H_2O_2 Stabiliser H_2SO_4 Crystalliser   Cooling unit   Copper
                                                                                            sulphate

*Fig. 8.49   Schematic diagram of the application of a cool crystalliser to the
regeneration of sulphuric acid-hydrogen peroxide-containing etch solutions
(RECON Verfahrenstechnik GmbH, Waldenbuch)*

Finally, reference should be made to an etching process that has been described, which can be used with either a sulphuric-acid or an ammoniacal salt solution, in which the etching effect is brought about, however, by positively charging the surface of graphite particles suspended in it. [197]. The etch solution is a suspension containing 25-30% by weight of graphite particles with diameters of about 50 mm. As shown in the diagram in Fig. 8.50, these circulate through an anode chamber where they come in contact with the graphite anode, which confers a positive surface charge on them. When they afterwards contact the surface to be etched they transmit their positive charge to it, thus allowing copper ions to pass into solution. The discharged particles later contact the anode again and are charged once more, while the copper ions migrate through a diaphragm that is not permeable to the particles into the cathode compartment, where they are deposited on to the cathode. Rinsing of the etched printed circuit boards takes place first in particle-free salt solution, which is obtained by filtration from the circulation system (not illustrated in Fig. 8.50) and which afterwards flows back into the suspension. Rinsing with water follows in a second stage, in which the salt solution is washed off; this solution also is then passed into the suspension to compensate for evaporation losses.

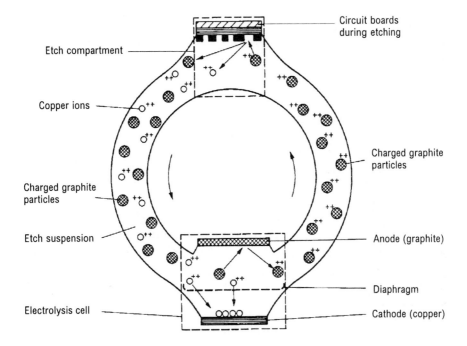

*Fig. 8.50 Schematic diagram of etching with suspensions of charged
graphite particles (from Faul and Kastening [197])*

8.10   BATTERY MANUFACTURE

The battery industry is the only one in the metal working industries that has no
direct connection with surface treatment, except that the steel strips used for
electrode production in the fabrication of nickel-cadmium storage batteries or
for the cases of primary cells are nickel plated as a separate operation. Regarding
these production areas, reference should be made to the sections of this volume
dealing with waste water treatment and recycling technology, in so far as they
apply to nickel-containing waste water, and in particular to the continuous circulation of rinse water (Section 5.3) as well as the other water-saving measures in
rinsing (Sections 6.2 and 6.3.2).

Separate consideration must be given to the waste water and recycling
technologies associated with the manufacture of:

- lead accumulators.

- nickel-cadmium accumulators *and*

- primary cells.

Full details concerning waste-water and recycling processes in the battery industry are reported by Hiller [198,199] and a briefer account can be found in the
ATV-Merkblatt M 752 leaflet [200].

8.10.1   Production of lead accumulators

8.10.1.1   Amount of waste water and its constituents

On average, 160,000 tons of lead per year are consumed in Western Germany in
the manufacture of lead accumulators (average value 1980-1989). That is about
50% of the total consumption. The proportion of lead recycled is 55-60%, but
about 90% of this can be attributed to lead accumulators alone [201].

All the lead compounds used in production or arising during manufacture or
in waste water treatment are more or less insoluble. The water can come into
contact only with metallic lead, lead oxide (PbO), lead dioxide (PbO_2), lead
hydroxide ($Pb(OH)_2$) and lead sulphate ($PbSO_4$). Of these substances, only lead
sulphate has a significant solubility of 42 mg/litre, and lead hydroxide above pH
10.4 has such a low solubility that the concentration of lead can be less than 0.5
mg/litre (see also Figures 3.59 and 3.84).

Production and transport systems are areas where waste water can arise,
generally wherever lead oxide powder or paste and electrodes are involved, and
especially in the cleaning treatments involved in paste preparation. When the
electrodes of filled charged batteries are activated before their first use, small
amounts of acidic, lead-containing waste water can arise as they are filled with
forming acid (dilute sulphuric acid), as this is replaced by fresh acid and as the
batteries are washed. The same applies also when the electrodes of dry charged
batteries are activated, but considerably larger quantities of waste water are
generated in the washing of these electrodes. If no other application exists for
spent forming acid, it must be discharged as waste water. Desalinated water is
used for diluting the sulphuric acid. In regenerating the ion exchangers employed for this purpose, regenerates are produced containing surplus sulphuric
acid or caustic soda. Finally, the water used to wash extracted air must be
discharged from time to time.

The extracted air contains principally lead-bearing dust, and has a lead content that is often only a few mg/m^3, and is always well below 100 mg/m^3.
Particles up to 1 µm can be removed with an efficiency of 98-99% by wet
washing. These come mainly from the paste mixers, the pasting machines and
driers, as well as from the formation and post-treatments of the electrodes. The
extracted air is mixed with water in venturi tubes to make sure that the particles
are wetted. They settle in the lower part of the washer and can be removed from
there as sludge, while the water is kept in circulation [202].

In addition, there is the cleaning water from the floors and machines. The lead emissions in the waste water amount to about 7 g per tonne of lead processed. For Western Germany that corresponds to an annual quantity of 1.12 t lead from accumulator manufacture.

The waste water constituents fluctuate widely. The following values are the approximate upper limits [198]:

- settleable material   up to 10 ml/litre
- lead, dissolved and undissolved   up to 60 mg/litre
- iron, dissolved and undissolved   up to 20 mg/litre
- sulphate   up to 20,000 mg/litre.

The pH value is about 3.

The amounts of waste water produced in the manufacture of starter batteries amount on average to 3.5 m³ per tonne of lead processed, which can fall in the most favourable cases to 1 m³/t of lead. The specific quantity of waste water increases the smaller the product line, e.g. for industrial batteries in this case it is about 5 m³/t, but can rise as high as 20 m³/t.

8.10.1.2 Waste water treatment and recycling processes

The waste water from lead accumulator manufacture is subjected to neutralisation/precipitation. Where the production capacity allows, this should take place in batch plants. Continuous process plants are operated as a rule in two stages, neutralisation and post-neutralisation. Because of the high sulphate content, neutralisation with lime slurry is preferable, but, because of the high pH required to precipitate the dissolved part of the lead as lead hydroxide, the pH value must be at least 10.4. Given the poor solubility of hydrated calcium oxide in alkali (high sludge ratio), this is more easily achieved with caustic soda. Where a raised sulphate content is not critical, dissolved lead can be precipitated using caustic soda containing sodium carbonate at pH values just above 7 as very poorly soluble basic lead carbonate (see Fig. 3.59) [203, 204]. As lead hydroxide and the basic bicarbonate both precipitate out in finely dispersed form, the addition of iron(III) salts is usually necessary for flocculation, to ensure improved sedimentation and/or filtration (see Fig. 3.60). The sludge produced by neutralisation/precipitation with lime slurry consists mainly of calcium sulphate containing ca 1% lead. It is therefore classified as special waste.

As with all plants in which metals are precipitated, a final filtration has to be carried out at the end (see Section 4.4.1). In the lead accumulator industy, partly for the reasons mentioned, this has been done for a very long time using gravel filters. If dissolved lead is still present in the discharge, then this can sometimes be adsorbed on selectively operating, weakly acidic cation exchangers. The feasibility of this must be tested first however.

Fig. 8.51 shows in diagrammatic form the construction of a continuous process plant employing state-of-the-art technology. If it is not possible to reduce the lead to less than 0.5 mg/litre by neutralisation/precipitation, then precipitation with sodium sulphide is recommended. The excess sulphide must be bound up again finally, however, with iron(II) or iron(III) salts. Any excess of the latter helps at the same time to flocculate the lead sulphide, which also tends to be finely dispersed.

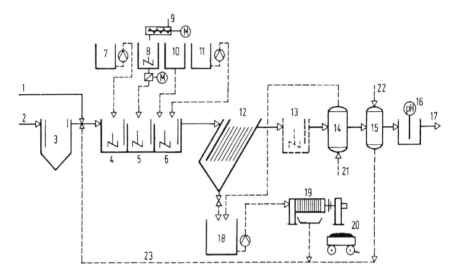

Fig. 8.51 Schematic diagram of a continuous process plant for the treatment of waste water from a lead accumulator factory

1 Inlet for waste water free of solid matter, 2 Inlet for waste water containing solid matter, 3 Solid matter separator, 4 Dosing compartment for iron (III) chloride solution, 5 Neutralisation compartment, 6 Post-neutralisation and flocculation compartment, 7 Iron(III) chloride, 8 Lime slurry make-up and dosing tank, 9 Conveyor for hydrated calcium oxide, 10 Caustic soda, 11 Flocculation medium, 12 Settling equipment, 13 Final neutralisation, if necessary, 14 Final filtration (various possibilities), 15 Selective cation exchanger, 16 Final control cell, 17 Waste water discharge, 18 Sludge storage tank, 19 Filter press, 20 Sludge for disposal, 21 Back-rinsing water (gravel filter), 22 Regenerating medium, 23 Return pipe for filtrates and regenerates.

Except for the recycling of lead from old, spent accumulators, the possibilities of recycling in lead accumulator manufacture are slight. The following recommendations, however, to the extent that they have not already been acted upon, can ease matters by reducing the amount of effluent and other waste:

- Separation of a large part of the rapidly settling lead compounds in dense-material separators, which are connected to the floor drains at the main collection points. Where it is feasible, the collected residues should be partly dewatered in small centrifuges and passed back again to the pasting units. Another and larger dense-material separator should be placed before the inlet to the waste water plant [200].

- The acid that is no longer usable for formation purposes should be collected and made available for production again as blended acid, so as to reduce the sulphate content and the quantity of neutral salt [200].

- Washing processes, e.g. that of the electrodes of dry charged cells, can be carried out in cascade systems, so achieving considerable economies in water use (see Section 6.3.2.1) [200].

- Avoidance of mists by improved enclosures for the machines [200].

- Re-use of waste water that has previously been fully desalinated and is contaminated only with sulphuric acid as dilution water for forming or as battery acid.

Any further increase in the service life of lead accumulators also improves environmental protection in an indirect way. Over the last two decades, this process in conjunction with the increase in energy efficiency has made it possible to manage on a consumption of ca 160,000 t lead/year instead of 400,000 t/year [201].

The proportion of lead recovered from old accumulators is the highest achieved among all non-ferrous metals.

8.10.2   Production of nickel-cadmium accumulators

8.10.2.1 Amount of waste water and its constituents

In Western Germany 50,000 t nickel and 1800 t cadmium are processed annually, of which 400 t nickel (representing 0.8%) and 300 t cadmium (representing 16.7%) are for nickel-cadmium accumulators (average for 1980–1989) [201].

Waste water arises in production mainly where nickel hydroxide and cadmium hydroxide are precipitated from their nitrates using caustic soda. These are required in all types of electrode (pocket plate, tubular plate, sintered plate), nickel hydroxide being used for the positive and cadmium hydroxide for the negative electrodes. The waste water is generated principally during washing of the precipitated hydroxides. In the positive tubular plate electrodes, nickel flakes are also incorporated, which are produced from composite electrodeposited foils of alternating copper and nickel layers by dissolving out the copper with acid. This consequently gives rise to an acidic, copper-containing effluent. Perforated steel strip and tubes are used to strengthen the electrodes. Wherever their nickel plating is carried out in a separate operation, waste water is produced that derives from the drag-out of degreasing solutions, etching baths (sulphuric acid) and nickel electrolytes (see Section 8.8.1).

In addition, waste waters are generated in regenerating the ion exchangers that produce desalinated water, which contain excess hydrochloric acid and caustic soda; finally, alkaline waste water arises in the making up the potassium hydroxide electrolytes. In this industry the exhaust gas is generally cleaned using dry methods, so no waste water is generated in gas-washing. The concentration ranges of the waste waters undergo wide fluctuations. The following values are the upper limits observed:

- settleable material   up to 3 ml/litre
- cadmium, dissolved and   up to 100 mg/litre
  undissolved   (separately handled waste water)
- nickel, dissolved and undissolved   up to 50 mg/litre
- iron, dissolved and undissolved   up to 60 mg/litre
- nitrate   up to 16,000 mg/litre
- sulphate   up to 250 mg/litre.

The pH value is over 9, and the proportions of the dissolved metals are determined by this and by the effect of the amounts of neutral salt present.

Recent legislation prescribes that cadmium-containing waste water be handled and treated separately.

The quantities of waste water per tonne of processed nickel or cadmium can be as much as 200 m³ in the manufacture of sintered plates, but the amounts are considerably less in the manufacture of other electrodes.

8.10.2.2 Waste water treatment and recycling processes

Since cadmium-containing waste water must be treated separately, two treatment plants are necessary. The waste waters that do not contain cadmium can be treated together, if there is no intention of reclaiming the nickel.

Both plants must be operated at relatively high pH values if neutralisation/ precipitation is carried out with caustic soda. At least pH 9.8 is required for the nickel-containing, and at least 10.4 for the cadmium-containing solutions (see Fig. 3.84).

If sodium carbonate or sodium carbonate-containing sodium hydroxide solution is used to precipitate cadmium, then it can be precipitated as the very low-solubility carbonate at a pH of only 8.5. For this, however, it must be pre-conditioned with acid at a pH value below 8.5. The finely dispersed precipitate needs to be flocculated with iron(III) salts of course to assist separation. As the waste water in nickel-cadmium accumulator manufacture contains no complexing agents, there are no problems in precipitating the metals out. Hydrochloric acid is recommended for neutralisation, so as not to increase the sulphate content (in the secondary stages).

To economise in the use of water, batch treatments should be aimed at, especially for the separate cadmium-containing stream. If a continuous process must be used, the neutralisation plants should consist of two compartments (neutralisation/precipitation and flocculation). After the settling units, final filtrations are obligatory (see Section 4.4.1) and, for safety's sake, selectively operating cation exchangers should be provided.

So long as metals other than nickel and cadmium can be excluded, weakly acidic cation exchangers can also be used for metal removal in the treatment of waste water, allowing both metals to be recovered as hydroxides by precipitating them from the regenerates. If these are dissolved in nitric acid, then they can be used again in electrode manufacture. Fig. 8.52 illustrates the circuit and functioning of a waste water treatment plant with metal nitrate recovery. The following measures and modes of operation are feasible for water and material recycling in the manufacture of nickel-cadmium accumulators:

The cascade method of circulating wash water is an appropriate water-saving measure for nickel and cadmium hydroxides. In a new procedure for manufacturing anodes the electrodeposition process the cadmium is deposited electrolytically, and as a result the wash water consumption in this area is considerably reduced.

The processing method as shown in Fig. 8.52, with duplicated installations for the separate nickel and cadmium streams, permits simultaneous waste water treatment and reclamation of nickel and cadmium nitrate (see Section 7.2.1.2 also in this connection). As the metals are precipitated in this process, concentrating the regenerate, as in the recovery of copper sulphate, is not necessary.

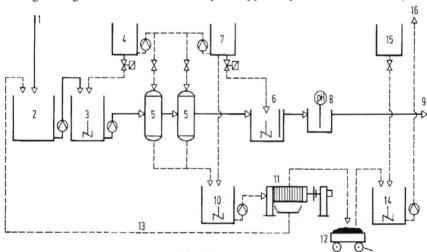

Fig. 8.52 Schematic diagram of a waste water treatment plant for the production of nickel-cadmium accumulators with simultaneous recovery of metal nitrate (showing the waste water route for one metal)

1 Inlet for nickel- or cadmium-containing waste water, 2 Storage tank, 3 Reactor (batch treatment), 4 Acid for neutralisation and regeneration, 5 Weakly acidic cation exchanger, 6 Final neutralisation, 7 Caustic for neutralisation and conditioning, 8 Final control, 9 Waste water discharge, 10 Sludge storage tank, 11 Filter press, 12 Nickel or cadmium hydroxide, 13 Return line for filtrate, 14 Dissolving vessel for hydroxide, 15 Nitric acid, 16 Return line for concentrated nickel nitrate or cadmium nitrate solution.

The nitrate in the waste water cannot of course be avoided, so long as production depends expressly on the use of metal nitrates. The dissolution of the hydroxides incidentally allows each of the metal salts to be produced.

Where the components necessary for the strength of the electrodes are manufactured in a separate operation, recirculation of the rinse water through ion exchangers is feasible. Nickel electrolyte leakages and surpluses can sometimes be transferred direct to the precipitation plant for nickel (Figure 8.52).

Further development of the nickel-hydride accumulator, so that cadmium-containing waste water is no longer generated is under way [205].

Application of the newly developed electrolytic manufacture of cadmium anodes (electrodeposition process), also enables a considerable reduction to be made in the amount of washing water needed.

Nickel-cadmium accumulators contain ca 20% nickel and cadmium. Larger units would always be returned to the manufacturer. For the smaller ones the recycling percentage is difficult to determine because of their long life [201]. Since, however, in the last five years, the consumption of cadmium for small accumulators has almost doubled, an agreement has been reached between governments nature-protection and reactor safety bodies and the battery industry, concerning the labelling of cadmium-containing accumulators and their return free of charge to the points of sale for disposal [205].

8.10.3   Production of primary cells

8.10.3.1 Amount of waste water and its constituents

Annually, ca 460 million primary cells in various forms cylindrical cells, cells with approximately rectangular bases (prismatic cells) and button cells are manufactured, of which the cylindical ones are of the greatest significance. Table 8.10 contains a list of the most important cells classified according to the electrochemical systems, with information about their shapes, mercury contents and production figures. The importance of mercury is that it restrains the strong electrochemical tendency of zinc to bond with oxygen, and inhibits the liberation of hydrogen from the electrolyte. Otherwise, the cells would very rapidly be discharged. Consequently, the zinc anodes of primary cells are amalgamated to some extent with mercury [206]. The amount of mercury used for this purpose has already been reduced from 52 t (1985) to 11 t (1989).

The most important electrochemical systems, with their cathode and anode materials and their electrolytes, are described below. The constituents of the waste water arising during their manufacture can also be gathered from this.

- *Zinc/carbon cells and zinc/chloride cells*
Cathode:	Compact of brown (manganese) ore, carbon black and/or graphite and some zinc oxide.
Anode:	Zinc.

Electrolyte: Ammonium chloride and/or zinc chloride solution with small additions of mercuric chloride to lightly amalgamate the zinc anode, thickening agent (e.g. starch). Because of the very small amount of mercury the system can be described as practically mercury-free.

● *Alkaline zinc/brown ore cells (alkaline manganese cells)*
Cathode: Brown ore compact.
Anode: Amalgamated zinc powder.
Electrolyte: Potassium hydroxide solution.

● *Zinc/mercury oxide cells*
Cathode: Mercuric oxide (tablet).
Anode: Amalgamated zinc powder.
Electrolyte: Potassium hydroxide solution.

Silver oxide also occurs as cathode material in place of mercuric oxide.

TABLE 8.10:
COMPOSITION OF PRIMARY BATTERY SYSTEMS AND THEIR SALES IN WESTERN GERMANY (1989) [206]

Battery system anode/cathode	Geometrical shape*	Anode	Cathode	Hg-content wt%	Sales million items/a	Weight t/a	Hg-content t/a
Zn/MnO_2	R, P	Case	Compact	0	230	9 200	–
Zn/MnO_2	R, K, P	Powder	Compact	0.025	180	6 100	1.5
Zn/HgO	K, (R)	Powder	Tablet	30	10	30	9
Zn/Ag_2O	K	Powder	Tablet				
Zn/O_2	K	Powder	Film (carbon)	1	40	70	0.5
Li/X	R, K	Film	X**)	0	Low		–
			Total		460	ca. 15 400	11

* R = Round cell; P = Prismatic battery (cell); K = Button cell
** Various alternative cathodes: MnO_2, CF_x etc

There are other cells as well with zinc anodes, and, as a new, certainly not yet fully established system, the lithium cell. This has a lithium anode, which can be combined with a brown-ore cathode, but also with cathodes made from other materials. An organic solvent containing lithium salt is used as the electrolyte.

Places where waste water is generated are the mixing plants for preparing the cathode active masses and those for making up the electrolytes, desalination plants, and, if included, nickel plating plants and equipment for spray painting the steel cases. Because of the mercury, the washing water from the floors and protective clothing should also be included in the waste water.

Table 8.11 lists the measured upper limits of substances in the waste water from primary cell production. About 5000 t brown ore and 3000 t zinc are consumed annually. The quantities of waste water produced are very small and amount to only a few m³/day.

TABLE 8.11:
SPECIFIC WASTE WATER CONSTITUENTS IN
PRIMARY CELL PRODUCTION [200]

Parameter	Concentrations Zinc/carbon cells	Mercuric oxide cells Alkaline manganese cells
Settleable materials	5ml/litre	5ml/litre
Cadmium, total	Traces	–
Iron, total	Up to 5 mg/litre	Up to 5 mg/litre
Copper, total	Up to 5 mg/litre	–
Manganese, total	Up to 5 mg/litre	–
Nickel, total	Up to 40 mg/litre	Up to 40 mg/litre
Mercury, total	Traces	Up to 5 mg/litre
Zinc, total	Up to 800 mg/litre	Up to 50 mg/litre
Ammonia as nitrogen	Up to 200 mg/litre	–

8.10.3.2 Waste water treatment and recycling processes

The small quantities of waste water can generally be treated in batch plants. They are collected together and regarded as mercury-containing.

Mercury in the divalent state in the presence of iron is precipitated, along with zinc, using caustic soda. For this, the pH value should be maintained between 8.7 and 9.5 (maximum 10). Quantities of ammonium salts with up to 200 mg/litre ammonium-nitrogen do not interfere with this. Monovalent mercury must first be oxidised with sodium hypochlorite, which may cause any metallic mercury present to dissolve in the divalent state as well. Surplus hypochlorite is reduced using an excess of iron(II) salt, allowing the trivalent iron formed and the surplus divalent iron to act as welcome flocculation media. Final neutralisation is with caustic soda. If satisfactory results are not achieved, then precipitation in the same solution is carried out subsequently with sodium sulphide or an organo-sulphide, both of which form compounds of exceptionally low solubility with mercury, any excess sulphide being combined again with iron salts.

The waste water plant is constructed accordingly. The waste water runs into a storage tank, which can be designed to function also as a dense material separator for rapid sedimentation of insoluble substances (e.g. brown ore). It is pumped out of here under manual or automatic control into the batch plant reactor and chemically treated as described above. If the mercury concentration of a filtered sample is less than 50 µg/litre, the contents can then be pumped into a second storage tank, which functions also as the receiver for the filter press.

The filtrate is again subjected to the obligatory final filtration, which here, because of the small waste water quantities involved, can be carried out with a gravel filter. The treated waste water is passed through a final control cell. In essence the plant is similar to that illustrated in Fig. 4.30, incorporating chemical vessels for dosing with caustic coda, hydrochloric acid, sodium hypochlorite solution, iron(II) sulphate solution and sodium sulphide solution. Since the required final concentration of mercury is easy to achieve, not more than 200-300 g mercury per year should be need to be discharged from the whole of Western Germany's primary battery production.

In the circumstances described, recycling would not be sensible, even if it were possible. The amount of waste water can be reduced somewhat, if the machine cleaning water with the same composition is simply used again for the washing and making up of fresh cathode active mass.

As critical as waste water treatment is the disposal of discharged mercury-containing primary cells. As in the case of nickel-cadmium accumulators, there is an obligation to label and to return them. In addition, the efforts of the manufacturers have been and are directed towards dramatically reducing the mercury content of alkaline manganese cells [207]. Zinc-carbon cells are practically mercury-free. The development and marketing of mercury-free cells (lithium cells) is being driven forward. The annual turnover at present is around 10 million units.

The requirements relating to the discharge of waste water from battery manufacture can be found in Chapters 1 and 2.

References for Chapter 8

1.   Weiler, W.: "Deposition of light fractions" Lehr u. Handbuch der Abwasser Techn. III pp 117-136 Verlag Ernst & Sohn, Berlin

2.   ATV-Notice M765 "Effluent arising from the metal-working industries" Sheet 5 "Mechanical Working" Ges. z. Foerderung der Abwassertechnik e.V; St Augustin (1989)

3.   Rodenkirchen, M.: "Effluent treatment from polishing & vibratory finishing" Galvanotechnik 79 (1988) 2173-2181

4.   Prüller,H.: "Effluent treatment from vibratory finishing using correct sequences" Maschinenmarkt 89 (1983) No 59/60

5.   Förtsch, A.: "Cost-effective effluent treatment from vibratory finishing by flocculation & filtration". Wasser, Luft Betrieb (1988) (3) 26-31

6.   Anon; "New methods for solvent cleaning" Galvanotechnik 79 (1988) 2871

7.   Zange, B.: "Replacing halocarbon cleaners with aqueous solutions" Galvanotechnik 80 (1989) 2288-2291

8.   Passavant-Werke. (Germany) "Deposition Methods" (1975)

9.   Anon; "Oil-skimmers for less than 10 ppm residual levels" Chemie Techn. 12 (1983) (10) p. 73; also Chemie-Technik 12 (1983) (2) 12

10. Knobloch, H.: "Maintenance & cleaning of cooling oils − separating spent emulsions" Techn. Mitt. 68 (1975) Nos 7/8

11. Bradke, H.J.: "Plant & processes for emulsion breaking and deposition of oils & greases" Metall 33 (1979) 267-172

12. Gütling, W.: "Ultrafiltration in industrial effluent treatment" Fachber. Huettenprax. Metallverarb. (1978) 969-972

13. Gütling, W.: "Methods for splitting emulsions & oily effluent" Galvanotechnik 65 (1974) 417-424

14. Seyser, H.D.: "Reverse flow – new generation of ultrafiltration" Galvanotechnik 76 (1985) 960-2

15. Anon; "Treating & cleaning oil-water emulsions with ultrafiltration" Metalloberflaeche 39 (1985) 155-6

16. Metallgesellschaft AG, Frankfurt: "Treating cooling oils with the UCARSEP system" Techn. Bull. (1978)

17. Sedelies, R., Müller, U., Spei, B.: "Combination processes for emulsion breaking" Wasser, Luft, Boden (1989) (9) 19-24

18. Köhler, R.: "Technology & applications of compressed air flotation in effluent treatment. Pt 1" Wasser Luft Betrieb (1975) 72-7. ibid. Pt 2. (1975) 104-9

19. Baer, E.H.: "Treatment of oily emulsions". Munich Beitrag z. Abwasser, Fischerei, Flussbiol. 28 (1977) 265-271

20. Pohl, K.M.: "Breaking industrial oil emulsions with acid & raised temperature" Beitrage z. Abwasser, Fischerei, Flussbiol. 28 (1977) 273-6

21. Lohmeyer, S.: "Emulsion breaking" Galvanotechnik 64 (1973) 601-610; 795-805; 911-921

22. Bradke, H.J.: "Emulsion breaking" (in) Hartinger, L; "Pocketbook of effluent treatment" Vol. 2

23. Neul, H.: "Plant & processes for treating oily effluent". Werkstoff u. ihre Veredlung. 3 (1981) 255-261

24. Gäfgen, K.: "Ultrafiltration for rationalised treatment of degreasing baths". Techn. Guide. No 17. Duerr GmbH (1982)

25. Ehrler, P., Janitza, J.: "Oil removal with "Reutlinger Mats" in industry" Metalloberflaeche 31 (1977) 193-5

26. Rossmann, Ch.: "Processes for regenerating aqueous degreasing baths" Jahrb. Oberflaechentechn. (1982) 29-44 Metall Verlag, Berlin

27. Winkel, P.: "Treating effluent from degreasing baths" Galvanotechnik 80 (1989) 4362-9

28. Schulze-Schwieking, T.: "Centrifuge separators to clean washing liquor" Galvanotechnik 73 (1982) 731-3. Idem cav (1987) April, 88-92

29. Dahlheimer, H.G.: "Practical aspects in separating alkaline liquors" Galvanotechnik 73 (1882) 713-9

30. Sonnenberg, H.: "Reducing effluent volume and cost-saving by continuous treatment of degreasing baths" Galvanotechnik 79 (1988) 1531-2

31. Wilhelm, F.: "Recirculation of pre-treatment solutions" Metalloberflaeche 33 (1979) 301-7

32. Bessey, H.: "Modern methods for processing pre-treatment solutions for reduced environmental loading" Metalloberflaeche 33 (1979) 372-4

33. Rothkegel, J.: "Eco-friendly phosphating for tomorrow" Metalloberflaeche 30 (1976) 449-452

34. Dürr GmbH, Stuttgart: "Aquaclean system for aqueous cleaning" Techn. Note

35. Hinz, H.E.: "Effluent treatment and water purification for vibratory finishing". Vol. 65 Kontakt und Studium. "Gleitschleifen" Expert Verlag (1980)

36. Hinz, H.E.: "Vibratory finishing and its effluent" Galvanotechnik 74 (1983) 151-5

37. Osmota Membrantechnik GmbH, Sigmaringen: "Treatment of effluent from vibratory finishing". Techn. Note.

38. DIN 17014 "Heat treatment of ferrous materials"

39. ATV Notice M 758 "Effluent from heat treatment plants". Ges. f. Foerderung der Abwassertechnik e.V (1986)

40. Kunst, H.: "Zero discharge metal treatment plant – experiences & developments" Degussa News.

41. Knorre, H.: "Hydrogen peroxide for treatment of metal hardening effluent" Haerterei technische Mitt. 34 (1979) (1)

42. Knorre, H.: "Hydrogen peroxide to treat nitrite containing waste water" Chemie Technik 10 (1981) 553-5

43. Müller, W., Witzke, L.: "Complete treatment of waste from molten salt metal hardening" Chem. Ing. Techn. 45 (1973) 1285-9

44. Mohr, E.: "New method for evaporating rinsewater from metal hardening plant with salts recovery" Z. f. wirtschaftliche Fertigung 72 (1977) 534-5

45. Degussa AG, Frankfurt: "New design of flat-plate evaporator for zero discharge metal hardening plant" Galvanotechnik 80 (1989) 1328

46. VDI-Guidelines: "Electrolytic treatment – anodic stripping with and without external current source" VDI 3401, Sheets 1, 2, 3 (electrolytic metal removal & etching) Beuth Verlag GmbH Germany

47. Burkhardt, W.: "Plant & process for pickling & etching etc. Pt 2" Galvanotechnik 79 (1988) 2518-2528

48. Hiermacher, M., Simon, H.: "Chemical etching, chemical & electrolytic deburring" Galvanotechnik 74 (1983) 401-9

49. ATV Note M 765; "Effluent arising in the metal-working industries. Sheet 3. "Pickling & etching, chemical & electrolytic methods". Ges. f. Foerderung der Abwassertechnik (1989)

50. VDI Guidelines: "Electrochemical processing – treatment of electrolytes & etch solutions, effluent & sludges" VDI 3401, Sheet 4 (1982) Beuth Verlag, Berlin

51. Plümer, L.: "Ion exchangers for treating pickling effluent" Vom Wasser 34 (1967) 281-296

52. Gebrüder Decker GmbH & Co., K.G, Berg/Opf.: "Automatic pickling machine with zero effluent discharge" Galvanotechnik 80 (1989) 3422-8

53. Moller, P." Effluent free electropolishing plant" Galvanotechnik 79 (1988) 1817-1820

54. Anon; "Effluent treatment plants for pickling lines" Metalloberflaeche 30 (1976) 230-2

55. Sturm, W.: "Regeneration, not neutralisation" Industrie-Anzeiger (1989) 61/2

56. Steuler-Industriewerke GmbH, Höhr-Grenzhausen: Tech. Note.

57. DE 3542820 C2 (16.6.88) AEG-Elotherm

58. Work at RW Tech. Hochschule, Aachen

59. ATV Notice M 765; "Effluent arising from metal-working industries. Sheet 6. Hot dip galvanising

60. Böhm, A.: "Effluent treatment in hot dip galvanising" Fachber. f.

Oberflaechentechn. 12 (1974) 235-9

61. Oeteren, K.A. van: "Hot dip zinc galvanising as corrosion protection" Jahrbuch Oberflaechentechn. (1976) 279-298

62. Sattler, H.P.: "Emission from general galvanizing as a function of pretreatment, flux and zinc bath composition". Proc. 14th Int'l Zinc Coating Conf. Munich (1985)

63. Renner, M.: "Hot dip galvanising of cast iron products" Metalloberflaeche 32 (1978) 114-7

64. Durez, M.: "A system for treatment of liquid wastes and gaseous discharges." Proc. 14th Int'l Zinc Coating Conf. Munich (1985)

65. Torres, J., Silvent, B., Ferry, J.:"Design of a neutralizer with recycling of water from rinsing after pickling." Proc. 13th Int'l Zinc Coating Conf. London (1982)

66. Köhler, R.: "Exhaust systems - practical experiences based on a hot dip galvanising plant" Umwelt (1984) (1) 49-53

67. Koblenzer, H.D.: "Parts cleaning – aqueous or solvent based systems ?" Metalloberflaeche 42 (1988) 183-190

68. ATV-Notice M 765 "Effluent arising from the metal working industry" Sheet 7 "Enamelling" (1989)

69. Kyri, H.: "Vitreous Enamel" (in) Ullmans Technical Encyclopaedia

70. Müller, K.P.: "Surface cleaning & pre-treatment for iron & steel in vitreous enamelling" Mitt. Ver. Deutsche Emailfachl. 35 (1987) 117-122

71. Noth, H.: "Enamel 1981 – Status of applications and test methods" Jahrbuch Oberflaechentechn. (1981) 269-277

72. George, H.: "Operations in Vitreous Enamelling" Work Not No. 19 Duerr GmbH (1980)

73. Deutsches Email-Zentrum e.V., "Environment and Enamel" Jahrbuch Oberflaechentechnik. (1985) 352-366

74. Wratil, J.: "How Enamellers solve problems of sludge & effluent formation" Mitt. Ver. Deutsche Emailfachl. Special Edition. Oct 1983, 33-34

75. Anon; "Dip enamelling" Oberflaeche + JOT (1988) 107-9

76. Schräder, R.: "Effluent treatment in the enamelling industry". as ref. 74. pp 20-26

77. Wilhelm, F.: "Effluent in the Enamelling Industry" Umweltmagazin (1980) (Apr) 24-7

78. Joseph, W.: "Waste from Enamelling and its Treatment". as ref. 74. pp27-32

79. Lohmeyer, S.: "Effluent from paint coating" Industrieabwaesser (1982) 36-44

80. as ref. 68. Sheet 8 "Painting & special pre-treatment".

81. Blum, H.: "Low temperature spray phosphating as pre-treatment for painting" Jahrbuch Oberflaechentechn. (1981) 171-7

82. Blum, H.: "Phosphating for improved automobile corrosion resistance" Jahrbuch Oberflaechentechn. (1973) 203-225

83. Simon, H.: "Pre-treatment of metal surfaces for organic coating or dry lubricant coating" Galvanotechnik 73 (1983) 1320-9

84. Schenem, H., Kübler, E.: "Report on plant & processes for minimum hazardous waste production in metal finishing" Galvanotechnik 79 (1988) 3000-3010

85. Blum, H.: "Spray & Dip Phosphating of Auto Bodies" Jahrbuch Oberflaechentechn. (1982) 313-324

86. Gotta, H.: "Triple cascade in spray and phosphating plants" Jahrbuch Oberflaechentechn. (1980) 186-93

87. Prein, H.D.: "Spray phosphating with total water recirculation". Duerr GmbH Techn. Note No. 20 (1980)
88. Bessey, H.: "Measures for economising on water, chemicals and energy in pre-treatment plants" Galvanotechnik 73 (1982) 846-850
89. Blum, H.: "Phosphating as pre-treatment for cold forming of iron and steel" Jahrbuch Oberflaechentechn. (1983) 326-334
90. Gerhard Collardin GmbH, Cologne "Regeneration of zinc phosphating baths" Galvanotechnik 79 (1988) 4112
91. Lohmeyer, S.: "Economies in pre-treatment plants" Galvanotechnik 74 (1983) 665-675
92. Krämer, K.: "Organic phosphating free from chlorocarbons". Jahrbuch Oberflaechentechn. (1985) 296-300
93. Adams, K.H.: "Room temperature iron phosphating" Jahrbuch Oberflaechentechn. (1985) 302-6
94. Strauss, E.: "Industrial painting technology - re-use and reprocessing of coating materials" Jahrbuch Oberflaechentechn. (1987) 320-336
95. Strauss, E.: "Industrial paint coating" Jahrbuch Oberflaechentechn. (1984) 338-359
96. Saatweber, D.: "Cathodic electrocoating" Metalloberflaeche 31 (1977) 455-9
97. Lohmeyer, S.: "Treatment of effluent from paint shops" Galvanotechnik 63 (1972) 1130-1140
98. Adams, K.H.: "Effluent technology in spray booths with trickle drain facilities" Jahrbuch Oberflaechentechn. (1984) 165-9
99. Dürr GmbH, Stuttgart: "Effluent flotation with air bubbles" JOT (1970) (Dec) 52-3
100. Hydrochem GmbH, Renchen: "Solids separation with air flotation" Galvanotechnik 79 (1988) 513.
101. Bruggen, B.v.d.: "Multi-function effluent plant for paint wastes" Metalloberflaeche 31 (1977) 460-7
102. Bruggen, B.v.d.: "Useful measures for treating paint shop waste" Metalloberflaeche 28 (1974) 473-9
103. Eisenmann, Böblingen: "Continuous and effluent-free disposal of spray booth wastes" Abwassertechnik (1984) 62-4
104. Anon; "Settling plant for paint waste" cav (1989) (Oct) 132
105. Haltiner, E.: "Electrostatic paint coagulation" Oberflaeche Surf. 26 (1985) 453-6
106. German Ministry for R & D: "Solid wastes; recovery & reprocessing of paint waste"
107. Sutter, H. in Straub, H., Hösel, G., Schenkel, W.: "Avoiding and reprocessing paint sludges" Erich Schmidt Verlag, Berlin (1986)
108. Kraus, H.: "Effluent treatment for auto-body finishing" Techn. Note 22. Duerr GmbH (1982)
109. Kraus, H.: idem Metalloberflaeche 36 (1982) 531-8
110. Aluminium-Zentrale, Duesseldorf; "Anodised aluminium for decorative purposes" 13th (1981), 14th (1982) editions
111. Wernick, S., Pinner,: "Surface Treatment of Aluminium & its Alloys" 5th Ed. Finishing Publications Ltd, Stevenage UK
112. as ref. 80 Sheet 4. "Anodising"
113. Reuther, H., Schippert, E., Hitzemann, G.: "Planning effluent plant for anodising

lines for minimum operating costs" Galvanotechnik 71 (1980) 733-9

114. Hartinger, L.: "Precipitation of heavy metals (Iron, Aluminium) Baender, Bleche, Rohre 8 (1967) 509-518

115. Haarst, W.F.M. van, Mulder, R.J., Tervoort, J.L.J.: "Study of metal removal from metal-gluconate complexes in effluent" Finish Digest (1974) 237-240

115a. Grossmann, H.: "Decolorising effluent from aluminium anodising lines" Galvanotechnik 69 (1978) 16-24

116. DAS 20 48 445 (13.7.78) Lancy Laboratories

117. Anon; "Rinsewater recycling in anodising plants" Fachber. f. Metallbearbeitung (1981) No 3/4 137

118. Berry, L.: "Cutting Caustic". Products Finishing (1988) Apr., 92-95.

119. Dejak, M.: "Aluminium Anodizer Caustic Etch Solution". Plating and Surface Fin.(1984) Apr. 30-32.

120. Nelson, G.D., Schumacher, C.P., Shortreed, W.J., Sommerfeld, J.T.: "Quality and Cost in the Aluminium Bright Anodizing Process-III. The Recovery of Phosphoric Acid from Aluminium Bright Dip Rinse Solution". Plating (1966) Jan., 72-77.

121. Brown, C.J., Davy, D., Simmons, P.J.: "Purification of Sulfuric Acid Anodizing Solutions". Plating and Surface Fin. (1979) Jan., 54-57.

122. Hartinger, L.: "Recycling acid process solutions with the retardation process" (in) Proc. Conf. 6th Eur. Effluent & Waste Symp. Munich (1984)

123. Jenny L; (in) Hübner, F. und Speiser, C.: "Practical aspects of the anodising of aluminium" Aluminium Verlag, Duesseldorf (1977)

124. Götzelmann, W., Hartinger, L., Gülbas, M.: "Separations using the retardation process" Metalloberflaeche 41 (1987) 208-212, 315-322

125. Schott Glass, Mainz: "Process & etching solutions for the aluminium industry" Techn. Note (1981)

126. as ref. 80 Sheet 1. "Basics for effluent treatment" and Sheet 2 "Electroplating and electroless deposition"

127. Dettner, H.W., Elze, J.: "Handbook of Electroplating", Hanser Verlag, Munich 1963-9

128. private communication

129. Köhler, J.: "Low emission cadmium plating" Metalloberflaeche 37 (1983) 53-4

130. Huss R., Peters, W.: "Pilot plant for low effluent cadmium plating" Galvanotechnik 73 (1982) 1208-1216

131. Anon; "Low discharge cadmium plating" Galvanotechnik 74 (1983) 1197-1200

132. DE 31 38 503 A1 (7.4.83) SEP Ges. "Process for continuous regeneration of chromating solutions"

133. Götzelmann, W., Hartinger, L.: "Reducing cadmium discharges" Galvanotechnik 74 (1983) 140-4

134. as ref. 127. Vol. 2

135. Stallmann, K.: "Electroless nickel" Galvanotechnik 75 (1984) 142-9

136. Frick, W.: "Electroless metal deposition" Galvanotechnik 76 (1985) 548-551

137. Broszeit, E.: "Dispersion coatings" Galvanotechnik 75 (1984) 2-13

138. Ying, Wei-chi, Bonk, R.R.; "Removal of Nickel and Phosphorus from Electroless Nickel Plating Baths." Metal Finishing (1987) Dec. 23-31.

139. Scheve, F.: "Effluent treatment in the computer manufacturing industry". Lecture. TA Esslingen (1980)

140. Hartinger, L.: unpublished work
141. Kunces, D.: "Plate-Out-Pollution Control". Products Finishing (1987) Dec. 85-88.
142. Parker, K.: "Removal of Spent Electroless Nickel Plating Bath". Plating and Surface Finishing (1980) March 48-52.
143. Levy, F.: Doss, S.K.: "Regeneration of Electroless Nickel Solution by Ion Exchange". Plating and Surface Finishing (1987) Sept. 80-81.
144. Läser, L., Huschens, D.: "Criteria for selecting electroless nickel coating" Galvanotechnik 79 (1988) 1141-7
145. "Plating Handbook" Vol. 1 LPW GmbH (1988)
146. Anon; "Installation of a fully automated degreasing, etching & metal colouring plant at Heidelberg Printing Press Manufacturers" Metalloberflaeche 36 (1982) 514-6
147. ZWEZ-Chemie GmbH, "Metal colouring as a closed system" Galvanotechnik 74 (1983) 197 & 1215
148. Cupr, V., Pleva, M.: "Colouring of metals" Metalloberflaeche 30 (1976) 221-6
149. Dammer, H.J.: "Non-cyanide chemical and electrochemical metal stripping" Galvanotechnik 71 (1980) 29-36
150. Hammel, B.: "Metal stripping" Oberflaeche Surf. 23 (1982) 215-220; 238-244
151. Goselwitz, H.: "Air emissions of toxic species" Galvanotechnik 77 (1986) 303-311 and 587-592
152. Anon; "Gas and effluent purification" Chemie Ing. Techn. 54 (1982) 1029-1030
153. Degussa AG, Frankfurt: "Hydrogen peroxide for treating NO_x gases" Umwelt und Degussa (1982)
154. as ref. 80. Sheet 9 "Printed circuit board manufacture"
155. Herrmann, G.: "Printed Circuit Handbook" Leuze Verlag, Saulgau (1982)
156. Saubestre, E.B.: "Stabilizing Electroless Copper Solutions". Plating 59 (1972) 563-566.
157. Hartinger, L.: "Effluent & recycling in the p.c.b industry" Galvanotechnik 79 (1988) 2325-2330
158. Emde, W.: "Perspectives in plant construction – through hole plating and board construction" Galvanotechnik 80 (1989) 2092-2101
159. Hartinger, L., Scheiffelen, B.: unpublished work
160. Regenauer, G.W.: "Disposal of photo-resist containing effluent" Galvanotechnik 79 (1988) 4214-4219
161. Knaff, J.C., Liebsch, W.: "Photo-resists in terms of their environmental and health and safety aspects in the workplace" Galvanotechnik 75 (1984) 115-120
162. Horn, K.: "Treating effluent with dry photo-resist content" Techn. Note, Kalle Co.
163. Süssmuth, W.: "Disposal of alkaline photo-resist stripper and developer from the p.c.b industry" Galvanotechnik 79 (1988) 3834-7
164. Anon; "New approaches to treating and disposal of photo-resist containing effluent" Galvanotechnik 80 (1989) 266-7
165. Mylius, U.v., Leudolph, J.: "Ultrafiltration – a simple and cost-effective process for disposing of photo-resist bearing effluent" Galvanotechnik 77 (1986) 402-6
166. Leudolph, J.: "Membrane methods for treating effluent containing photo-resist" Galvanotechnik 79 (1988) 3827-3833
167. Gülbas, M.: "Effluent and recycling technologies" Metalloberflaeche 42 (1988) (Apr)

168. Horn, R.E.: "Continuous Regeneration of an Electroless Copper Bath". Plating and Surface Finishing (1981) Oct. 50-52.
169. Krulik, G.A., Lipson, M.A., Davison, J.B., Davis, S.C.: "The copper-stat process - a new approach to control and regeneration of electroless copper baths" Galvanotechnik 76 (1985) 1806-1811
170. Hummel, M.: "Cost-savings by recycling, exemplified by the fully-additive process" Galvanotechnik 72 (1981) 1390-2
171. Lieber, H.W.: "Process control and regeneration" Galvanotechnik 70 (1979) 1039-1043
172. Jola, M.: "Automatic concentration measurement and control of electroless copper baths" Galvanotechnik 70 (1979) 1067-1080
173. Schiller, W.: "Control of electroless copper baths" Galvanotechnik 71 (1980) 1165-1171
174. Hartinger, L. (in) ATV "Guide and Handbook of Effluent Treatment" Vol VII "Industrial effluent with inorganic content" (1985) 195-486 Verlag Ernst u. Sohn, Berlin
175. Hartinger, L.: "Chemical processes to remove metals from their complexes in p.c.b processing" Galvanotechnik 71 (1980) 797-805
176. Hoffmann, J., Rochlitzer, B.: "Disposal of electroless copper baths" Werkstoff u. ihre Veredlung 3 (1981) 369-372
177. Hartinger, L.: "Options for regeneration of etch baths" Proc. PCB Seminar (1987)
178. Beyer, G.: "Rinse techniques and their effectiveness in acid and alkaline etching of pcb's. Pt 1." Galvanotechnik 72 (1981) 679-687
179. Bogenschütz, A.F., Jostan, J.L., Marten, A.: "Regeneration of etch baths in the pcb industry" Galvanotechnik 70 (1979) 816-823
180. DP 2008766 (9.3.72) Licentia Patent Agents
181. Bringmann, M.: "Cost-effective and eco-friendly regeneration of spent etch baths by electrolysis" Galvanotechnik 78 (1987) 2985-9
182. OS 2641905 (23.3.78) Kutscherenko et al. (Moscow)
183. OS 2650912 (18.5.78) Hans Hoellmueller Machinery Co, Germany
184. Knipps, K.E.: "Processes to regenerate copper chloride etch baths" Galvanotechnik 75 (1984) 678-9
185. European Patent Specification Nr. 0018848 (26.9.1984). The Electricity Council, London.
186. Hillis, M.R.: "Application of the Capenhurst electrolytic etchant regeneration (CEER) process to cupric chloride and ammoniacal etchants". The Electricity Council Research Centre, Capenhurst, 1983.
187. Filor, U.: "Regeneration & copper recovery from etch baths" Metalloberflaeche 41 (1987) 518-520
188. Beyer, G.: as ref. 178. Pt 2. Galvanotechnik 72 (1981) 793-801
189. Bogenschütz, A.F., Jostan, J.L., Marten, A.: as ref. 179. Pt 2. Galvanotechnik 70 (1979) 940-5
190. Geier, E.: "Effluent free etching using chemicals in salt form" Galvanotechnik 70 (1979) 1165-7
191. Eur. Pat. Appl. Nr. 0158910 Al (30.3.1985). Lancy International Inc.
192. Hans Höllmüller Maschinenbau GmbH, REGASTREM Copper recovery unit for ammoniacal etch baths. Techn. Note (1985)
193. Hoffmann, H.: "Regeneration of ammoniacal etches and copper recovery by liquid-liquid extraction" Galvanotechnik 73 (1982) 311-5

194. Fischer, G.: "Electrolytic metal recovery" Galvanotechnik 74 (1983) 145-150
195. Stupar, M.: "Sulfuric/Peroxide Etchant – A Novel Regeneration Method". Metal Finishing (1988) Febr. 95-99.
196. Schlitter, F.W.: "Environmentally-friendly etching based on hydrogen peroxide-sulphuric acid" Galvanotechnik 79 (1988) 1631-4
197. Faul, W., Kastening, B.: "Electrolytic etching of pcb's in a closed electrolyte circulation loop" Werkstoffe u. ihre Veredlung 1 (1979) (2) 11-13
198. Hiller, F.: "The battery industry" ATV Effluent Treatment Handbook" Vol VII. 487-513. Verlag Ernst u. Sohn, Berlin
199. Hiller, F. *et al.* "Batteries and the Environment" 2nd Ed'n expert Verlag (1990)
200. ATV-Note M 752 "Effluent from manufacture of storage batteries and primary cells" (1986)
201. Hiller, F.: (in) ref. 199, "Introduction"
202. Hiller, F.: (in) ref. 199 "Air cleaning in battery manufacture" pp 38-50
203. Hartinger, L.: "Precipitation of lead from effluent" Metalloberflaeche 27 (1973) 157-9
204. Hartinger, L.: "(in) ref. 199 "Effluent from the battery industry" 74-113
205. Anon; "To what extent can alternatives be found for cadmium ?" Galvanotechnik 81 (1990) 207-8
206. Hiller, F.: (in) ref. 199. "Disposal of primary batteries" 138-145
207. Anon; "Mercury waste from batteries" Galvanotechnik 81 (1990) 206

Index

THE FASTEST DEVELOPING AND MOST VERSATILE,
SURFACE COATING TECHNIQUE IN USE TODAY:–

ELECTROLESS NICKEL PLATING

And the most authoritative book on the subject –

"Electroless Nickel Plating"

by W Riedel

**ISBN 0 904477 12 6   328 pp, 193 figs, 35 tables, 739 refs.
Price £60.00 (US $100.00) post free**

Tells you –

- how to operate electroless nickel baths
- about the physical & chemical properties of the deposits and how to optimise them for your particular requirements (magnetic, corrosion-resistant, high ductility, low porosity)
- how coating a substrate with electroless nickel can transform its properties
- how to test and analyse plating baths for maximum life and use
- how to test coatings, maintain Quality and conform with standards
- how to heat-treat deposits to improve their properties
- how to strip coatings or re-work defective components
- how to process effluent and dispose of spent solutions
- all about plant construction, design, materials instrumentation & process control to minimise supervision costs
- about post-treatments, e.g conversion coatings
- about special electroless nickels and alloys
- about all major international standards & specifications (US, UK, German etc) relating to electroless nickel
- where these coatings are used, whether for specific applications you might have in mind or into what industries they might be introduced
- what it costs to apply electroless nickel coatings or operate your own coating plant
- with over 700 references, where to find just about anything useful ever published on the subject
- how to troubleshoot the process

A BOOK FOR PLATERS AND COATERS, A BOOK FOR ENGINEERS AND
PRODUCT DESIGNERS, A BOOK FOR ANY LIBRARY WANTING TO MAINTAIN
GOOD COVERAGE OF MAIN-LINE ENGINEERING & ELECTRONIC PRACTICE

 **Co-published by Finishing Publications Ltd,
Stevenage UK and the ASM, Materials Park, Ohio**

THE PHOSPHATING OF METALS

By Werner Rausch

in collaboration with the late Heinz Blum, Dieter Funke, Hans Hansen, Friedrich Kaysser, Dieter Oppen, Guenter Siegmund and Wolfgang Ussat

418 pp. 158 figs. ISBN 0 904477-11-8

This comprehensive treatise covers the theoretical principles of phosphating and its practical applications, effluent treatment technology, standards and specifications, plant construction, methods for cold-working of treated surfaces and modern lacquering processes. Methods for testing and quality control, as well as the underlying specifications and standards, have all been improved with variants appropriate to the various applications being devised. These and other advances have all been incorporated into this first English (second German) edition.

CONTENTS (abridged):

1. Historical Development and Technical/ Economic Significance of the Process

2. Phosphoric Acid Pickling Baths
Formation and dissolution of rust and scale; dissolution of iron; phosphoric acid pickling; inhibitors; combined pickling and degreasing.

3. Phosphating Processes
Film-forming processes, solubility equilibria, kinetics of film formation, properties, effects on metal substrate, processes; non-film forming processes, kinetics of film formation, composition of baths, film properties; phosphating in non-aqueous media, systems based on halogenated hydrocarbons, halogen-free solvents, lubricants, wash primers; heat treatment.

4. Technical Applications of Phosphating
Corrosion protection; as a basis for paint, liquid paint application, electrocoating, spray painting, dip painting, high pressure spraying, motor car bodies, powder coating, precoated steel, coil coating; to facilitate cold forming, drawing of wire and sections, lubricants; improvement of sliding properties; electrical insulation.

5. Phosphating Plant
Process units, tanks, heating, mechanical handling, spray equipment, dosing, drying, sludge removal; storage, transport, materials for plant construction, process feed water, rinse tanks and water supply; treatment plants, immersion, spray, conveyor type, for continuous treatment of wire, steam and high pressure treatment, manual and special purpose plant, maintenance.

6. Treatment of Effluents from Phosphating Plant
Legal requirements; effluent materials; management; treatment, neutralisation, chromate removal, nitrite removal, rinsewater recycling, emulsion breaking; treatment plants, batch treatment, continuous treatment; plants with closed loop operation; effluent monitoring and analytical control.

7. Techniques for the Characterisation of Phosphate Films
Determination of coating weight, thickness, porosity, surface roughness, crystal size, phase structure and electrical properties; analysis and identification of films.

8. Analytical Control of Process Chemicals
Determination of free acid, total phosphate, total acid, acidity ratio, point of incipient precipitation; other methods, automatic process control; concentrate testing.

9. Technical Assessment of Phosphate Films
Testing for corrosion protection, adhesion, combined adhesion and corrosion protection, surface friction, electrical insulation.

10. Standards Relating to Phosphating
Federal German Republic; U.K.; Italy; Japan; Sweden; U.S.A.

11. References

FINISHING PUBLICATIONS LTD.,

PO Box 70, 105 Whitney Drive, Stevenage, Herts., SG1 4DF, England. Tel.: 0438 745115 Fax.: 0438 364536

ZINC PLATING

by Herb Geduld *(Columbia Chemical Corp., U.S.)*

ISBN 9 904477 10X

360 + xii pp, 23 figs., 23 tables, 245 refs.

Published jointly by Finishing Publications Ltd. and The American Society for Metals

This book on zinc plating is an effort to fill a void in the ready availability of practical data on zinc plating technology. It is written essentially for the practical zinc plater and plating chemist. Theoretical aspects of zinc deposition are touched upon only in passing. It is based on the author's 37 years of experience in the development of processes for bright zinc plating and the troubleshooting and maintenance of these electrolytes.

CONTENTS (abridged):

1. History of Zinc Plating
Additives: bright cyanide; mid and low cyanide: alkaline non-cyanide; bright acid; other systems; zinc fluoborate; the future.

2. Cleaning and Preplate Treatments
Alkaline soak cleaning; electrolytic cleaning; alkaline descaling or derusting; acid pickling; electrolytic pickling; rinsing.

3. Processing Cycles and Plant Layouts for Zinc Plating
Barrel zinc; equipment; processing cycles and plant layouts; basic processing cycle; horizontal barrels; oblique barrels; rack plating.

4. Cyanide Zinc Plating
Bright cyanide baths: ductile zinc bath; low embrittlement formulation; mid cyanide bath: low cyanide - low metal baths.

5. Zincate or Alkaline Non-Cyanide Zinc Plating
Makeup of baths; operating conditions; preplate and postplate treatments; conversion of cyanide to zincate electrolytes; conversion from one non-cyanide system to another; addition agents.

6. Acid Zinc Plating
Dull acid; ammonium chloride bright; potassium chloride bright; sodium chloride; mixed potassium chloride-ammonium chloride baths.

7. Post-Plate Treatments for Zinc Plate
Nitric acid bright dips; chromate conversion coatings; water soluble resins and lacquers.

8. Practical Problems, Treatment Procedures and Troubleshooting in Bright Zinc Plating
Preventing problems; impurities in baths; treatment procedures; troubleshooting charts; bright zinc plating of high-carbon baths.

9. Mechanical Plating of Zinc
Characteristics, comparison of mechanical plating to bright zinc electroplating; media; post-plate treatments.

10. Continuous Zinc Plating of Steel Strip and Wire
Electrogalvanizing; recent developments in continuous strip plating; plating equipment; plating baths.

11. The Plating of Zinc-Nickel Alloys and Other Zinc Alloy Baths
Nickel-zinc; iron-zinc; cobalt-zinc; tin-zinc; white brass; zinc-cadmium; zinc-graphite composite plating.

12. Analytical Control of Zinc Plating Baths
Cyanide and non-cyanide alkaline baths; acid chloride baths; preparation of solutions for analysis; correcting the chemistry of solutions.

13. Hull Cell Control of Bright Zinc Plating Baths
Use to evaluate and compare brighteners; detecting metallic impurities; determination of covering and throwing power; testing effectiveness of treatment procedures; other tests.

14. Waste Disposal Procedures for Zinc Plating Operations
Destructive; evaporative recovery, electrolytic recovery, plant design.

15. Safety in the Zinc Plating Plant
U.S. federal safety legislation; U.K. legislation; chemical hazards; first aid for poisoning.

Glossary

FINISHING PUBLICATIONS LTD.,
P.O. Box 70, 105 Whitney Drive, Stevenage, Herts. SG1 4DF, England.
Telephone: 0438 745115

WASTE MANAGEMENT CHALLENGES?

SERFILCO has an answer for every stage.

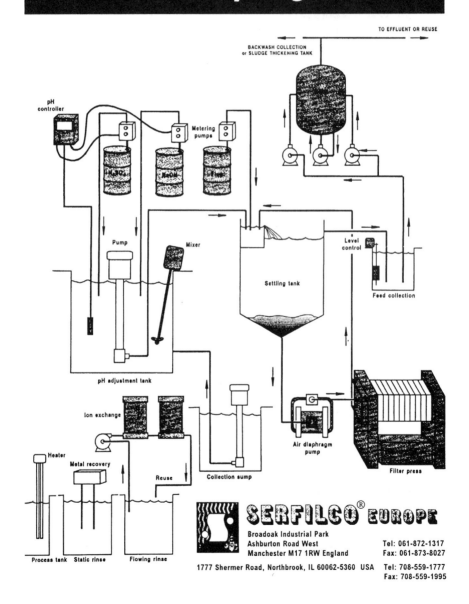

MORE AND MORE PRODUCT DESIGNERS ARE SPECIFYING:

METALLIZED PLASTICS

And here is the book that shows how to do it – from start to finish!

"Metallising of Plastics – A Handbook of Theory & Practice"

Edited by Richard Suchentrunk

ISBN 0-904477-13-4   pp 348   202 Figs. With hundreds of references and Appendix with 200 Patent abstracts.

- ◆ Adhesion of metals to plastics, theory
- ◆ Adhesion, methods of measurement
- ◆ Metallizing of plastics: process engineering
    - aqueous processes (electroless)
    - vacuum processes
- ◆ Processing of plastics for aqueous metallizing
    - type of plastic
    - moulding design and operation
    - moulding tools and feeds
- ◆ Types of plastic for metallizing
    - ABS, ABS-polypropylene blends
    - polypropylene
    - polyamides
    - fluorinated polymers
    - fibre-reinforced plastics
    - polyimides and polyetherimides
    - specialised plastics
- ◆ Quality Assurance
- ◆ Plant for Metallizing
- ◆ Properties of metallized plastics
- ◆ Effluent treatment, environmental, recycling

Target readership: **Product Engineers, Materials Scientists and Technologists, Electroplaters & Vacuum Coaters**

 Co-published by Finishing Publications Ltd, Stevenage UK and the ASM, Materials Park, Ohio

A METAL FINISHING COMPUTER SEARCH FOR YOUR SPECIALIST REQUIREMENTS

Call, Fax or write with details of your needs. Computer searches from $45 upwards. Tell us your requirements and we'll give you a firm quote – so no unpleasant surprises!

If we don't find what you're looking for – then there's No Charge! Searches printed and sent by airmail – or if you prefer, on floppy disk or by Email.

METAL FINISHING INFORMATION SERVICES
PO Box 70, 105 Whitney Drive, Stevenage, Herts.,
SG1 4DF, England. Tel: 01438 745115 Fax: 01438 364536

FOR ULTIMATE AWARENESS – SURFACE TREATMENT TECHNOLOGY ABSTRACTS

Six issues per year, around 8000 abstracts – papers, patents, standards. Covers etching, pickling, mechanical treatments, plating, electroless, anodising, conversion coating, hot-dip galvanising, organic coating, effluent treatment, PVD, CVD Send for free sample copy.

Computer-readable version available.

FINISHING PUBLICATIONS LTD.,
PO Box 70, 105 Whitney Drive, Stevenage, Herts.,
SG1 4DF, England. Tel: 01438 745115 Fax: 01438 364536

NOTES:

NOTES: